Power System Stability and Control

The EPRI Power System Engineering Series
Dr. Neal J. Balu, Editor-in-Chief

POWER SYSTEM STABILITY AND CONTROL

P. KUNDUR

Vice-President, Power Engineering
Powertech Labs Inc., Surrey, British Columbia

Formerly Manager
Analytical Methods and Specialized Studies Department
Power System Planning Division, Ontario Hydro, Toronto, Ontario

and

Adjunct Professor
Department of Electrical and Computer Engineering
University of Toronto, Toronto, Ontario

Edited by

Neal J. Balu
Mark G. Lauby
Power System Planning and Operations Program
Electrical Systems Division
Electric Power Research Institute
3412 Hillview Avenue
Palo Alto, California

McGraw-Hill, Inc.
New York San Francisco Washington, D.C. Auckland Bogotá
Caracas Lisbon London Madrid Mexico City Milan
Montreal New Delhi San Juan Singapore
Sydney Tokyo Toronto

Library of Congress Cataloging-in-Publication Data

Kundur, Prabha.
  Power system stability and control / Prabha Kundur.
      p. cm.
  EPRI Editors, Neal J. Balu and Mark G. Lauby.
  Includes bibliographical references and index.
  ISBN 0-07-035958-X
  1. Electric power system stability. 2. Electric power systems–
+Control. I. Title.
  TK1005.K86 1993
  621.319–dc20   93-21456
                                                             CIP

6 7 8 9 10 11 DOC/DOC 0 9 8 7 6 5 4 3 2 1

ISBN 0-07-035958-X

*The sponsoring editor for this book was Harold B. Crawford, and the
production supervisor was Donald Schmidt.*

Printed and bound by R.R. Donnelley & Sons Company.

Dedicated to My Parents

Contents

PART II EQUIPMENT CHARACTERISTICS AND MODELLING

Contents   xiii

PART III SYSTEM STABILITY: physical aspects, analysis, and improvement

Contents

17 METHODS OF IMPROVING STABILITY   1103

Foreword

To paraphrase the renowned electrical engineer, Charles Steinmetz, the North American interconnected power system is the largest and most complex machine ever devised by man. It is truly amazing that such a system has operated with a high degree of reliability for over a century.

The robustness of a power system is measured by the ability of the system to operate in a state of equilibrium under normal and perturbed conditions. Power system stability deals with the study of the behavior of power systems under conditions such as sudden changes in load or generation or short circuits on transmission lines. A power system is said to be stable if the interconnected generating units remain in synchronism.

The ability of a power system to maintain stability depends to a large extent on the controls available on the system to damp the electromechanical oscillations. Hence, the study and design of controls are very important.

Of all the complex phenomena on power systems, power system stability is the most intricate to understand and challenging to analyze. Electric power systems of the 21st century will present an even more formidable challenge as they are forced to operate closer to their stability limits.

I cannot think of a more qualified person than Dr. Prabha Kundur to write a book on power system stability and control. Dr. Kundur is an internationally recognized authority on power system stability. His expertise and practical experience in developing solutions to stability problems is second to none. Dr. Kundur not only has a thorough grasp of the fundamental concepts but also has worked on solving electric utility system stability problems worldwide. He has taught many courses, made excellent presentations at professional society and industry committee meetings, and has written numerous technical papers on power system stability and control.

It gives me great pleasure to write the Foreword for this timely book, which I am confident will be of great value to practicing engineers and students in the field of power engineering.

Dr. Neal J. Balu
Program Manager
Power System Planning and Operations
Electrical Systems Division
Electric Power Research Institute

Preface

    This book is concerned with understanding, modelling, analyzing, and mitigating power system stability and control problems. Such problems constitute very important considerations in the planning, design, and operation of modern power systems. The complexity of power systems is continually increasing because of the growth in interconnections and use of new technologies. At the same time, financial and regulatory constraints have forced utilities to operate the systems nearly at stability limits. These two factors have created new types of stability problems. Greater reliance is, therefore, being placed on the use of special control aids to enhance system security, facilitate economic design, and provide greater flexibility of system operation. In addition, advances in computer technology, numerical analysis, control theory, and equipment modelling have contributed to the development of improved analytical tools and better system-design procedures. The primary motivation for writing this book has been to describe these new developments and to provide a comprehensive treatment of the subject.

    The text presented in this book draws together material on power system stability and control from many sources: graduate courses I have taught at the University of Toronto since 1979, several EPRI research projects (RP1208, RP2447, RP3040, RP3141, RP4000, RP849, and RP997) with which I have been closely associated, and a vast number of technical papers published by the IEEE, IEE, and CIGRE.

    This book is intended to meet the needs of practicing engineers associated with the electric utility industry as well as those of graduate students and researchers. Books on this subject are at least 15 years old; some well-known books are 30 to 40 years old. In the absence of a comprehensive text, courses on power system stability often tend to address narrow aspects of the subject with emphasis on special analytical techniques. Moreover, both the teaching staff and students do not have ready access to information on the practical aspects. Since the subject requires an understanding of a wide range of areas, practicing engineers just entering this field are faced with the formidable task of gathering the necessary information from widely scattered sources.

This book attempts to fill the gap by providing the necessary fundamentals, explaining the practical aspects, and giving an integrated treatment of the latest developments in modelling techniques and analytical tools. It is divided into three parts. Part I provides general background information in two chapters. Chapter 1 describes the structure of modern power systems and identifies different levels of control. Chapter 2 introduces the stability problem and provides basic concepts, definitions, and classification.

Part II of the book, comprising Chapters 3 to 11, is devoted to equipment characteristics and modelling. System stability is affected by the characteristics of every major element of the power system. A knowledge of the physical characteristics of the individual elements and their capabilities is essential for the understanding of system stability. The representation of these elements by means of appropriate mathematical models is critical to the analysis of stability. Chapters 3 to 10 are devoted to generators, excitation systems, prime movers, ac and dc transmission, and system loads. Chapter 11 describes the principles of active power and reactive power control and develops models for the control equipment.

Part III, comprising Chapters 12 to 17, considers different categories of power system stability. Emphasis is placed on physical understanding of many facets of the stability phenomena. Methods of analysis along with control measures for mitigation of stability problems are described in detail.

The notions of power system stability and power system control are closely related. The overall controls in a power system are highly distributed in a hierarchical structure. System stability is strongly influenced by these controls.

In each chapter, the theory is developed from simple beginnings and is gradually evolved so that it can be applied to complex practical situations. This is supplemented by a large number of illustrative examples. Wherever appropriate, historical perspectives and past experiences are highlighted.

Because this is the first edition, it is likely that some aspects of the subject may not be adequately covered. It is also likely that there may be some errors, typographical or otherwise. I welcome feedback on such errors as well as suggestions for improvements in the event that a second edition should be published.

I am indebted to many people who assisted me in the preparation of this book. Baofu Gao and Sainath Moorty helped me with many of the calculations and computer simulations included in the book. Kip Morison, Solomon Yirga, Meir Klein, Chi Tang, and Deepa Kundur also helped me with some of the results presented.

Atef Morched, Kip Morison, Ernie Neudorf, Graham Rogers, David Wong, Hamid Hamadanizadeh, Behnam Danai, Saeed Arabi, and Lew Rubino reviewed various chapters of the book and provided valuable comments.

David Lee reviewed Chapters 8 and 9 and provided valuable comments and suggestions. I have worked very closely with Mr. Lee for the last 22 years on a number of complex power system stability-related problems; the results of our joint effort are reflected in various parts of the book.

Carson Taylor reviewed the manuscript and provided many helpful suggestions for improving the text. In addition, many stimulating discussions I have had with Mr. Taylor, Dr. Charles Concordia, and with Mr. Yakout Mansour helped me develop a better perspective of current and future needs of power system stability analysis.

Patti Scott and Christine Hebscher edited the first draft of the manuscript. Janet Kibblewhite edited the final draft and suggested many improvements.

I am deeply indebted to Lei Wang and his wife, Xiaolu Meng, for their outstanding work in the preparation of the manuscript, including the illustrations.

I wish to take this opportunity to express my gratitude to Mr. Paul L. Dandeno for the encouragement he gave me and the confidence he showed in me during the early part of my career at Ontario Hydro. It is because of him that I joined the electric utility industry and then ventured into the many areas of power system dynamic performance covered in this book.

I am grateful to the Electric Power Research Institute for sponsoring this book. In particular, I am thankful to Dr. Neal Balu and Mr. Mark Lauby for their inspiration and support. Mark Lauby also reviewed the manuscript and provided many helpful suggestions.

I wish to express my appreciation to Liz Doherty and Patty Jones for helping me with the correspondence and other business matters related to this book.

Finally, I wish to thank my wife, Geetha Kundur, for her unfailing support and patience during the many months I worked on this book.

Prabha Shankar Kundur

PART I

GENERAL
BACKGROUND

Chapter 1

General Characteristics
of Modern Power Systems

The purpose of this introductory chapter is to provide a general description of electric power systems beginning with a historical sketch of their evolution. The basic characteristics and structure of modern power systems are then identified. The performance requirements of a properly designed power system and the various levels of controls used to meet these requirements are also described.

This chapter, together with the next, provides general background information and lays the groundwork for the remainder of the book.

1.1 EVOLUTION OF ELECTRIC POWER SYSTEMS

The commercial use of electricity began in the late 1870s when arc lamps were used for lighthouse illumination and street lighting.

The first complete electric power system (comprising a generator, cable, fuse, meter, and loads) was built by Thomas Edison - the historic Pearl Street Station in New York City which began operation in September 1882. This was a dc system consisting of a steam-engine-driven dc generator supplying power to 59 customers within an area roughly 1.5 km in radius. The load, which consisted entirely of incandescent lamps, was supplied at 110 V through an underground cable system. Within a few years similar systems were in operation in most large cities throughout the world. With the development of motors by Frank Sprague in 1884, motor loads were added to such systems. This was the beginning of what would develop into one of the largest industries in the world.

In spite of the initial widespread use of dc systems, they were almost completely superseded by ac systems. By 1886, the limitations of dc systems were becoming increasingly apparent. They could deliver power only a short distance from the generators. To keep transmission power losses (RI^2) and voltage drops to acceptable levels, voltage levels had to be high for long-distance power transmission. Such high voltages were not acceptable for generation and consumption of power; therefore, a convenient means for voltage transformation became a necessity.

The development of the transformer and ac transmission by L. Gaulard and J.D. Gibbs of Paris, France, led to ac electric power systems. George Westinghouse secured rights to these developments in the United States. In 1886, William Stanley, an associate of Westinghouse, developed and tested a commercially practical transformer and ac distribution system for 150 lamps at Great Barrington, Massachusetts. In 1889, the first ac transmission line in North America was put into operation in Oregon between Willamette Falls and Portland. It was a single-phase line transmitting power at 4,000 V over a distance of 21 km.

With the development of polyphase systems by Nikola Tesla, the ac system became even more attractive. By 1888, Tesla held several patents on ac motors, generators, transformers, and transmission systems. Westinghouse bought the patents to these early inventions, and they formed the basis of the present-day ac systems.

In the 1890s, there was considerable controversy over whether the electric utility industry should be standardized on dc or ac. There were passionate arguments between Edison, who advocated dc, and Westinghouse, who favoured ac. By the turn of the century, the ac system had won out over the dc system for the following reasons:

- Voltage levels can be easily transformed in ac systems, thus providing the flexibility for use of different voltages for generation, transmission, and consumption.

- AC generators are much simpler than dc generators.

- AC motors are much simpler and cheaper than dc motors.

The first three-phase line in North America went into operation in 1893 - a 2,300 V, 12 km line in southern California. Around this time, ac was chosen at Niagara Falls because dc was not practical for transmitting power to Buffalo, about 30 km away. This decision ended the ac versus dc controversy and established victory for the ac system.

In the early period of ac power transmission, frequency was not standardized. Many different frequencies were in use: 25, 50, 60, 125, and 133 Hz. This posed a problem for interconnection. Eventually 60 Hz was adopted as standard in North America, although many other countries use 50 Hz.

The increasing need for transmitting larger amounts of power over longer distances created an incentive to use progressively higher voltage levels. The early ac systems used 12, 44, and 60 kV (RMS line-to-line). This rose to 165 kV in 1922, 220 kV in 1923, 287 kV in 1935, 330 kV in 1953, and 500 kV in 1965. Hydro Quebec energized its first 735 kV in 1966, and 765 kV was introduced in the United States in 1969.

To avoid the proliferation of an unlimited number of voltages, the industry has standardized voltage levels. The standards are 115, 138, 161, and 230 kV for the high voltage (HV) class, and 345, 500 and 765 kV for the extra-high voltage (EHV) class [1,2].

With the development of mercury arc valves in the early 1950s, high voltage dc (HVDC) transmission systems became economical in special situations. The HVDC transmission is attractive for transmission of large blocks of power over long distances. The cross-over point beyond which dc transmission may become a competitive alternative to ac transmission is around 500 km for overhead lines and 50 km for underground or submarine cables. HVDC transmission also provides an asynchronous link between systems where ac interconnection would be impractical because of system stability considerations or because nominal frequencies of the systems are different. The first modern commercial application of HVDC transmission occurred in 1954 when the Swedish mainland and the island of Gotland were interconnected by a 96 km submarine cable.

With the advent of thyristor valve converters, HVDC transmission became even more attractive. The first application of an HVDC system using thyristor valves was at Eel River in 1972 - a back-to-back scheme providing an asynchronous tie between the power systems of Quebec and New Brunswick. With the cost and size of conversion equipment decreasing and its reliability increasing, there has been a steady increase in the use of HVDC transmission.

Interconnection of neighbouring utilities usually leads to improved system security and economy of operation. Improved security results from the mutual emergency assistance that the utilities can provide. Improved economy results from the need for less generating reserve capacity on each system. In addition, the interconnection permits the utilities to make economy transfers and thus take advantage of the most economical sources of power. These benefits have been recognized from the beginning and interconnections continue to grow. Almost all the utilities in the United States and Canada are now part of one interconnected system. The result is a very large system of enormous complexity. The design of such a system and its secure operation are indeed challenging problems.

1.2 STRUCTURE OF THE POWER SYSTEM

Electric power systems vary in size and structural components. However, they all have the same basic characteristics:

- Are comprised of three-phase ac systems operating essentially at constant voltage. Generation and transmission facilities use three-phase equipment.

Industrial loads are invariably three-phase; single-phase residential and commercial loads are distributed equally among the phases so as to effectively form a balanced three-phase system.

- Use synchronous machines for generation of electricity. Prime movers convert the primary sources of energy (fossil, nuclear, and hydraulic) to mechanical energy that is, in turn, converted to electrical energy by synchronous generators.

- Transmit power over significant distances to consumers spread over a wide area. This requires a transmission system comprising subsystems operating at different voltage levels.

Figure 1.1 illustrates the basic elements of a modern power system. Electric power is produced at generating stations (GS) and transmitted to consumers through a complex network of individual components, including transmission lines, transformers, and switching devices.

It is common practice to classify the transmission network into the following subsystems:

1.   Transmission system

2.   Subtransmission system

3.   Distribution system

The *transmission system* interconnects all major generating stations and main load centres in the system. It forms the backbone of the integrated power system and operates at the highest voltage levels (typically, 230 kV and above). The generator voltages are usually in the range of 11 to 35 kV. These are stepped up to the transmission voltage level, and power is transmitted to transmission substations where the voltages are stepped down to the subtransmission level (typically, 69 kV to 138 kV). The generation and transmission subsystems are often referred to as the *bulk power system*.

The *subtransmission system* transmits power in smaller quantities from the transmission substations to the distribution substations. Large industrial customers are commonly supplied directly from the subtransmission system. In some systems, there is no clear demarcation between subtransmission and transmission circuits. As the system expands and higher voltage levels become necessary for transmission, the older transmission lines are often relegated to subtransmission function.

The *distribution system* represents the final stage in the transfer of power to the individual customers. The primary distribution voltage is typically between 4.0 kV and 34.5 kV. Small industrial customers are supplied by primary feeders at this voltage level. The secondary distribution feeders supply residential and commercial customers at 120/240 V.

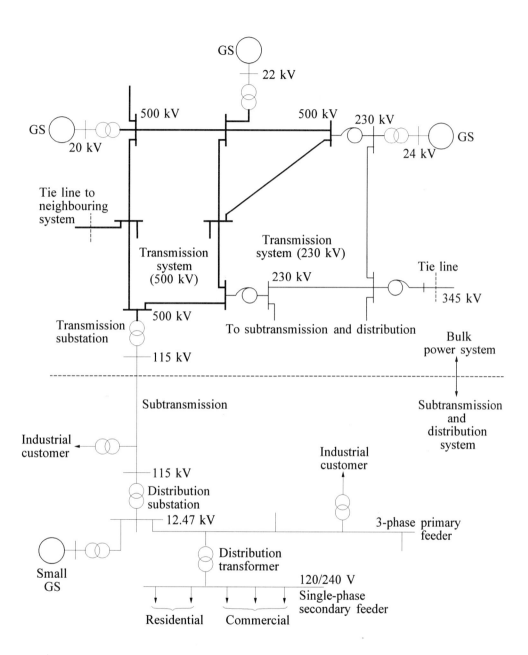

Figure 1.1 Basic elements of a power system

Small generating plants located near the load are often connected to the subtransmission or distribution system directly.

Interconnections to neighbouring power systems are usually formed at the transmission system level.

The overall system thus consists of multiple generating sources and several layers of transmission networks. This provides a high degree of structural redundancy that enables the system to withstand unusual contingencies without service disruption to the consumers.

1.3 POWER SYSTEM CONTROL

The function of an electric power system is to convert energy from one of the naturally available forms to the electrical form and to transport it to the points of consumption. Energy is seldom consumed in the electrical form but is rather converted to other forms such as heat, light, and mechanical energy. The advantage of the electrical form of energy is that it can be transported and controlled with relative ease and with a high degree of efficiency and reliability. A properly designed and operated power system should, therefore, meet the following fundamental requirements:

1. The system must be able to meet the continually changing load demand for active and reactive power. Unlike other types of energy, electricity cannot be conveniently stored in sufficient quantities. Therefore, adequate "spinning" reserve of active and reactive power should be maintained and appropriately controlled at all times.

2. The system should supply energy at minimum cost and with minimum ecological impact.

3. The "quality" of power supply must meet certain minimum standards with regard to the following factors:

    (a) constancy of frequency;

    (b) constancy of voltage; and

    (c) level of reliability.

Several levels of controls involving a complex array of devices are used to meet the above requirements. These are depicted in Figure 1.2 which identifies the various subsystems of a power system and the associated controls. In this overall structure, there are controllers operating directly on individual system elements. In a generating unit these consist of prime mover controls and excitation controls. The prime mover controls are concerned with speed regulation and control of energy supply system variables such as boiler pressures, temperatures, and flows. The function of the

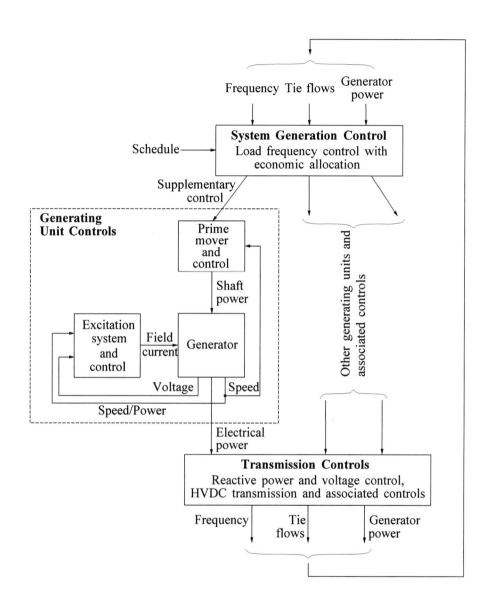

Figure 1.2 Subsystems of a power system and associated controls

excitation control is to regulate generator voltage and reactive power output. The desired MW outputs of the individual generating units are determined by the system-generation control.

The primary purpose of the system-generation control is to balance the total system generation against system load and losses so that the desired frequency and power interchange with neighbouring systems (tie flows) is maintained.

The transmission controls include power and voltage control devices, such as static var compensators, synchronous condensers, switched capacitors and reactors, tap-changing transformers, phase-shifting transformers, and HVDC transmission controls.

The controls described above contribute to the satisfactory operation of the power system by maintaining system voltages and frequency and other system variables within their acceptable limits. They also have a profound effect on the dynamic performance of the power system and on its ability to cope with disturbances.

The control objectives are dependent on the operating state of the power system. Under normal conditions, the control objective is to operate as efficiently as possible with voltages and frequency close to nominal values. When an abnormal condition develops, new objectives must be met to restore the system to normal operation.

Major system failures are rarely the result of a single catastrophic disturbance causing collapse of an apparently secure system. Such failures are usually brought about by a combination of circumstances that stress the network beyond its capability. Severe natural disturbances (such as a tornado, severe storm, or freezing rain), equipment malfunction, human error, and inadequate design combine to weaken the power system and eventually lead to its breakdown. This may result in cascading outages that must be contained within a small part of the system if a major blackout is to be prevented.

Operating states of a power system and control strategies [3,4]

For purposes of analyzing power system security and designing appropriate control systems, it is helpful to conceptually classify the system-operating conditions into five states: normal, alert, emergency, *in extremis*, and restorative. Figure 1.3 depicts these operating states and the ways in which transition can take place from one state to another.

In the *normal state*, all system variables are within the normal range and no equipment is being overloaded. The system operates in a secure manner and is able to withstand a contingency without violating any of the constraints.

The system enters the *alert state* if the security level falls below a certain limit of adequacy, or if the possibility of a disturbance increases because of adverse weather conditions such as the approach of severe storms. In this state, all system variables are still within the acceptable range and all constraints are satisfied. However, the system has been weakened to a level where a contingency may cause

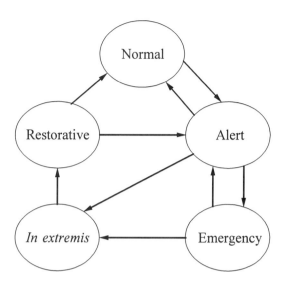

Figure 1.3 Power system operating states

an overloading of equipment that places the system in an emergency state. If the disturbance is very severe, the *in extremis* (or extreme emergency) state may result directly from the alert state.

Preventive action, such as generation shifting (security dispatch) or increased reserve, can be taken to restore the system to the normal state. If the restorative steps do not succeed, the system remains in the alert state.

The system enters the *emergency state* if a sufficiently severe disturbance occurs when the system is in the alert state. In this state, voltages at many buses are low and/or equipment loadings exceed short-term emergency ratings. The system is still intact and may be restored to the alert state by the initiating of emergency control actions: fault clearing, excitation control, fast-valving, generation tripping, generation run-back, HVDC modulation, and load curtailment.

If the above measures are not applied or are ineffective, the system is *in extremis*; the result is cascading outages and possibly a shut-down of a major portion of the system. Control actions, such as load shedding and controlled system separation, are aimed at saving as much of the system as possible from a widespread blackout.

The *restorative state* represents a condition in which control action is being taken to reconnect all the facilities and to restore system load. The system transits from this state to either the alert state or the normal state, depending on the system conditions.

Characterization of the system conditions into the five states as described above provides a framework in which control strategies can be developed and operator actions identified to deal effectively with each state.

For a system that has been disturbed and that has entered a degraded operating state, power system controls assist the operator in returning the system to a normal state. If the disturbance is small, power system controls by themselves may be able to achieve this task. However, if the disturbance is large, it is possible that operator actions such as generation rescheduling or element switching may be required for a return to the normal state.

The philosophy that has evolved to cope with the diverse requirements of system control comprises a hierarchial structure as shown in Figure 1.4. In this structure, there are controllers operating directly on individual system elements such as excitation systems, prime movers, boilers, transformer tap changers, and dc converters. There is usually some form of overall plant controller that coordinates the controls of closely linked elements. The plant controllers are in turn supervised by system controllers at the operating centres. The system-controller actions are coordinated by pool-level master controllers. The overall control system is thus highly distributed, and relies on many different types of telemetering and control signals. Supervisory Control and Data Acquisition (SCADA) systems provide information to indicate the system status. State estimation programs filter monitored data and provide an accurate picture of the system's condition. The human operator is an important link at various levels in this control hierarchy and at key locations on the system. The primary function of the operator is to monitor system performance and manage resources so as to ensure economic operation while maintaining the required quality

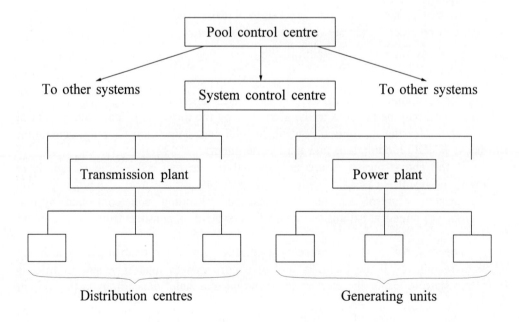

Figure 1.4 Power system control hierarchy

and reliability of power supply. During system emergencies, the operator plays a key role by coordinating related information from diverse sources and developing corrective strategies to restore the system to a more secure state of operation.

1.4 DESIGN AND OPERATING CRITERIA FOR STABILITY

For reliable service, a bulk electricity system must remain intact and be capable of withstanding a wide variety of disturbances. Therefore, it is essential that the system be designed and operated so that the more probable contingencies can be sustained with no loss of load (except that connected to the faulted element) and so that the most adverse possible contingencies do not result in uncontrolled, widespread and cascading power interruptions.

The November 1965 blackout in the northeastern part of the United States and Ontario had a profound impact on the electric utility industry, particularly in North America. Many questions were raised relating to design concepts and planning criteria. These led to the formation of the National Electric Reliability Council in 1968. The name was later changed to the North American Electric Reliability Council (NERC). Its purpose is to augment the reliability and adequacy of bulk power supply in the electricity systems of North America. NERC is composed of nine regional reliability councils and encompasses virtually all the power systems in the United States and Canada. Reliability criteria for system design and operation have been established by each regional council. Since differences exist in geography, load pattern, and power sources, criteria for the various regions differ to some extent [5].

Design and operating criteria play an essential role in preventing major system disturbances following severe contingencies. The use of criteria ensures that, for all frequently occurring contingencies, the system will, at worst, transit from the normal state to the alert state, rather than to a more severe state such as the emergency state or the *in extremis* state. When the alert state is entered following a contingency, operators can take actions to return the system to the normal state.

The following example of design and operating criteria related to system stability is based on those of the Northeast Power Coordinating Council (NPCC) [6]. It does not attempt to provide an exact reproduction of the NPCC criteria but gives an indication of the types of contingencies considered for stability assessment.

Normal design contingencies

The criteria require that the stability of the bulk power system be maintained during and after the most severe of the contingencies specified below, with due regard to reclosing facilities. These contingencies are selected on the basis that they have a significant probability of occurrence given the large number of elements comprising the power system.

The normal design contingencies include the following:

(a)   A permanent three-phase fault on any generator, transmission circuit, transformer or bus section, with normal fault clearing and with due regard to reclosing facilities.

(b)   Simultaneous permanent phase-to-ground faults on different phases of each of two adjacent transmission circuits on a multiple-circuit tower, cleared in normal time.

(c)   A permanent phase-to-ground fault on any transmission circuit, transformer, or bus section with delayed clearing because of malfunction of circuit breakers, relay, or signal channel.

(d)   Loss of any element without a fault.

(e)   A permanent phase-to-ground fault on a circuit breaker, cleared in normal time.

(f)   Simultaneous permanent loss of both poles of a dc bipolar facility.

The criteria require that, following any of the above contingencies, the stability of the system be maintained, and voltages and line and equipment loadings be within applicable limits.

These requirements apply to the following two basic conditions:

(1)   All facilities in service.

(2)   A critical generator, transmission circuit, or transformer out of service, assuming that the area generation and power flows are adjusted between outages by use of ten minute reserve.

Extreme contingency assessment

The extreme contingency assessment recognizes that the interconnected bulk power system can be subjected to events that exceed in severity the normal design contingencies. The objective is to determine the effects of extreme contingencies on system performance in order to obtain an indication of system strength and to determine the extent of a widespread system disturbance even though extreme contingencies do have very low probabilities of occurrence. After an analysis and assessment of extreme contingencies, measures are to be utilized, where appropriate, to reduce the frequency of occurrence of such contingencies or to mitigate the consequences that are indicated as a result of simulating for such contingencies.

The extreme contingencies include the following:

(a)   Loss of the entire capability of a generating station.

(b)   Loss of all lines emanating from a generating station, switching station or substation.

(c)   Loss of all transmission circuits on a common right-of-way.

(d)   A permanent three-phase fault on any generator, transmission circuit, transformer, or bus section, with delayed fault clearing and with due regard to reclosing facilities.

(e)   The sudden dropping of a large-load or major-load centre.

(f)   The effect of severe power swings arising from disturbances outside the NPCC interconnected systems.

(g)   Failure or misoperation of a special protection system, such as a generation rejection, load rejection, or transmission cross-tripping scheme.

System design for stability

The design of a large interconnected system to ensure stable operation at minimum cost is a very complex problem. The economic gains to be realized through the solution to this problem are enormous. From a control theory point of view, the power system is a very high-order multivariable process, operating in a constantly changing environment. Because of the high dimensionality and complexity of the system, it is essential to make simplifying assumptions and to analyze specific problems using the right degree of detail of system representation. This requires a good grasp of the characteristics of the overall system as well as of those of its individual elements.

The power system is a highly nonlinear system whose dynamic performance is influenced by a wide array of devices with different response rates and characteristics. System stability must be viewed not as a single problem, but rather in terms of its different aspects. The next chapter describes the different forms of power system stability problems.

Characteristics of virtually every major element of the power system have an effect on system stability. A knowledge of these characteristics is essential for the understanding and study of power system stability. Therefore, equipment characteristics and modelling will be discussed in Part II. Intricacies of the physical aspects of various categories of the system stability, methods of their analysis, and special measures for enhancing stability performance of the power system will be presented in Part III.

REFERENCES

[1]   H.M. Rustebakke (editor), *Electric Utility Systems and Practices*, John Wiley & Sons, 1983.

[2]   C.A. Gross, *Power System Analysis*, Second Edition, John Wiley & Sons, 1986.

[3]   L.H. Fink and K. Carlsen, "Operating under Stress and Strain," *IEEE Spectrum*, pp. 48-53, March 1978.

[4]   EPRI Report EL 6360-L, "Dynamics of Interconnected Power Systems: A Tutorial for System Dispatchers and Plant Operators," Final Report of Project 2473-15, prepared by Power Technologies Inc., May 1989.

[5]   IEEE Special Publication 77 CH 1221-1-PWR, *Symposium on Reliability Criteria for System Dynamic Performance*, 1977.

[6]   Northeast Power Coordinating Council, "Basic Criteria for Design and Operation of Interconnected Power Systems," October 26, 1990 revision.

Chapter 2

Introduction to the Power System Stability Problem

This chapter presents a general introduction to the power system stability problem including physical concepts, classification, and definition of related terms. Analysis of elementary power system configurations by means of idealized models illustrates some of the fundamental stability properties of power systems. In addition, a historical review of the emergence of different forms of stability problems as power systems evolved and of the developments in the associated methods of analysis is presented. The objective is to provide an overview of the power system stability phenomena and to lay a foundation based on relatively simple physical reasoning. This will help prepare for a detailed treatment of the various aspects of the subject in subsequent chapters.

2.1 BASIC CONCEPTS AND DEFINITIONS

Power system stability may be broadly defined as that property of a power system that enables it to remain in a state of operating equilibrium under normal operating conditions and to regain an acceptable state of equilibrium after being subjected to a disturbance.

Instability in a power system may be manifested in many different ways depending on the system configuration and operating mode. Traditionally, the stability problem has been one of maintaining synchronous operation. Since power systems rely on synchronous machines for generation of electrical power, a necessary condition for satisfactory system operation is that all synchronous machines remain in synchronism or, colloquially, "in step." This aspect of stability is influenced by the dynamics of generator rotor angles and power-angle relationships.

Instability may also be encountered without loss of synchronism. For example, a system consisting of a synchronous generator feeding an induction motor load through a transmission line can become unstable because of the collapse of load voltage. Maintenance of synchronism is not an issue in this instance; instead, the concern is stability and control of voltage. This form of instability can also occur in loads covering an extensive area supplied by a large system.

In the evaluation of stability the concern is the behaviour of the power system when subjected to a transient *disturbance*. The disturbance may be small or large. Small disturbances in the form of load changes take place continually, and the system adjusts itself to the changing conditions. The system must be able to operate satisfactorily under these conditions and successfully supply the maximum amount of load. It must also be capable of surviving numerous disturbances of a severe nature, such as a short-circuit on a transmission line, loss of a large generator or load, or loss of a tie between two subsystems. The system response to a disturbance involves much of the equipment. For example, a short-circuit on a critical element followed by its isolation by protective relays will cause variations in power transfers, machine rotor speeds, and bus voltages; the voltage variations will actuate both generator and transmission system voltage regulators; the speed variations will actuate prime mover governors; the change in tie line loadings may actuate generation controls; the changes in voltage and frequency will affect loads on the system in varying degrees depending on their individual characteristics. In addition, devices used to protect individual equipment may respond to variations in system variables and thus affect the system performance. In any given situation, however, the responses of only a limited amount of equipment may be significant. Therefore, many assumptions are usually made to simplify the problem and to focus on factors influencing the specific type of stability problem. The understanding of stability problems is greatly facilitated by the classification of stability into various categories.

The following sections will explore different forms of power system instability and associated concepts by considering, where appropriate, simple power system configurations. Analysis of such systems using idealized models will help identify fundamental properties of each form of stability problem.

2.1.1 Rotor Angle Stability

Rotor angle stability is the ability of interconnected synchronous machines of a power system to remain in synchronism. The stability problem involves the study of the electromechanical oscillations inherent in power systems. A fundamental factor in this problem is the manner in which the power outputs of synchronous machines vary as their rotors oscillate. A brief discussion of synchronous machine characteristics is helpful as a first step in developing the related basic concepts.

Synchronous machine characteristics

The characteristics and modelling of synchronous machines will be covered in considerable detail in Chapters 3, 4, and 5. Here discussion is limited to the basic characteristics associated with synchronous operation.

A synchronous machine has two essential elements: the field and the armature. Normally, the field is on the rotor and the armature is on the stator. The field winding is excited by direct current. When the rotor is driven by a prime mover (turbine), the rotating magnetic field of the field winding induces alternating voltages in the three-phase armature windings of the stator. The frequency of the induced alternating voltages and of the resulting currents that flow in the stator windings when a load is connected depends on the speed of the rotor. The frequency of the stator electrical quantities is thus synchronized with the rotor mechanical speed: hence the designation "synchronous machine."

When two or more synchronous machines are interconnected, the stator voltages and currents of all the machines must have the same frequency and the rotor mechanical speed of each is synchronized to this frequency. Therefore, the rotors of all interconnected synchronous machines must be in synchronism.

The physical arrangement (spatial distribution) of the stator armature windings is such that the time-varying alternating currents flowing in the three-phase windings produce a rotating magnetic field that, under steady-state operation, rotates at the same speed as the rotor (see Chapter 3, Section 3.1.3). The stator and rotor fields react with each other and an electromagnetic torque results from the tendency of the two fields to align themselves. In a generator, this electromagnetic torque opposes rotation of the rotor, so that mechanical torque must be applied by the prime mover to sustain rotation. The electrical torque (or power) output of the generator is changed only by changing the mechanical torque input by the prime mover. The effect of increasing the mechanical torque input is to advance the rotor to a new position relative to the revolving magnetic field of the stator. Conversely, a reduction of mechanical torque or power input will retard the rotor position. Under steady-state operating conditions, the rotor field and the revolving field of the stator have the same speed. However, there is an angular separation between them depending on the electrical torque (or power) output of the generator.

In a synchronous motor, the roles of electrical and mechanical torques are reversed compared to those in a generator. The electromagnetic torque sustains rotation while mechanical load opposes rotation. The effect of increasing the mechanical load is to retard the rotor position with respect to the revolving field of the stator.

In the above discussion, the terms *torque* and *power* have been used interchangeably. This is common practice in the power system stability literature, since the average rotational velocity of the machines is constant even though there may be small momentary excursions above and below synchronous speed. The per unit values of torque and power are, in fact, very nearly equal.

Power versus angle relationship

An important characteristic that has a bearing on power system stability is the relationship between interchange power and angular positions of the rotors of synchronous machines. This relationship is highly nonlinear. To illustrate this let us consider the simple system shown in Figure 2.1(a). It consists of two synchronous machines connected by a transmission line having an inductive reactance X_L but negligible resistance and capacitance. Let us assume that machine 1 represents a generator feeding power to a synchronous motor represented by machine 2.

The power transferred from the generator to the motor is a function of angular separation (δ) between the rotors of the two machines. This angular separation is due to three components: generator internal angle δ_G (angle by which the generator rotor leads the revolving field of the stator); angular difference between the terminal voltages of the generator and motor (angle by which the stator field of the generator leads that of the motor); and the internal angle of the motor (angle by which the rotor lags the revolving stator field). Figure 2.1(b) shows a model of the system that can be used to determine the power versus angle relationship. A simple model comprising an internal voltage behind an effective reactance is used to represent each synchronous machine. The value of the machine reactance used depends on the purpose of the study. For analysis of steady-state performance, it is appropriate to use the synchronous reactance with the internal voltage equal to the excitation voltage. The basis for such a model and the approximations associated with it are presented in Chapter 3.

A phasor diagram identifying the relationships between generator and motor voltages is shown in Figure 2.1(c). The power transferred from the generator to the motor is given by

$$P = \frac{E_G E_M}{X_T}\sin\delta \tag{2.1}$$

where

$$X_T = X_G + X_L + X_M$$

The corresponding power versus angle relationship is plotted in Figure 2.1(d). With the somewhat idealized models used for representing the synchronous machines, the power varies as a sine of the angle: a highly nonlinear relationship. With more accurate machine models including the effects of automatic voltage regulators, the variation in power with angle would deviate significantly from the sinusoidal relationship; however, the general form would be similar. When the angle is zero, no power is transferred. As the angle is increased, the power transfer increases up to a maximum. After a certain angle, nominally 90°, a further increase in angle results in a decrease in power transferred. There is thus a maximum steady-state power that can be transmitted between the two machines. The magnitude of the maximum power is

Machine 1   Machine 2

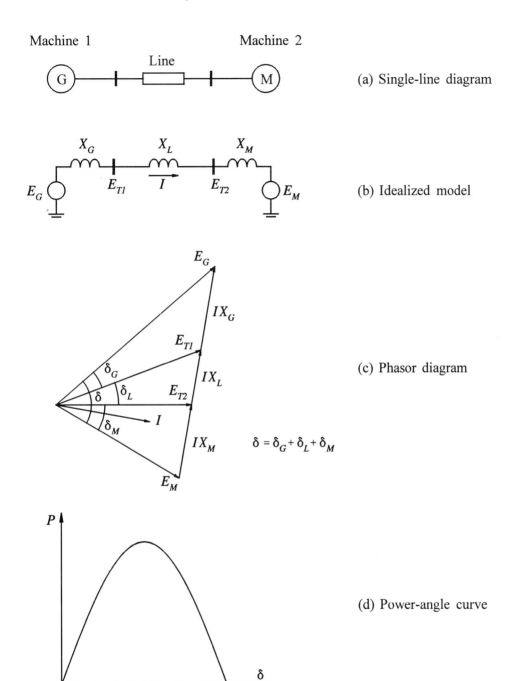

(a) Single-line diagram

(b) Idealized model

(c) Phasor diagram

$$\delta = \delta_G + \delta_L + \delta_M$$

(d) Power-angle curve

Figure 2.1 Power transfer characteristic of
a two-machine system

directly proportional to the machine internal voltages and inversely proportional to the reactance between the voltages, which includes reactance of the transmission line connecting the machines and the reactances of the machines.

When there are more than two machines, their relative angular displacements affect the interchange of power in a similar manner. However, limiting values of power transfers and angular separation are a complex function of generation and load distribution. An angular separation of 90° between any two machines (the nominal limiting value for a two-machine system) in itself has no particular significance.

The stability phenomena

Stability is a condition of equilibrium between opposing forces. The mechanism by which interconnected synchronous machines maintain synchronism with one another is through restoring forces, which act whenever there are forces tending to accelerate or decelerate one or more machines with respect to other machines. Under steady-state conditions, there is equilibrium between the input mechanical torque and the output electrical torque of each machine, and the speed remains constant. If the system is perturbed this equilibrium is upset, resulting in acceleration or deceleration of the rotors of the machines according to the laws of motion of a rotating body. If one generator temporarily runs faster than another, the angular position of its rotor relative to that of the slower machine will advance. The resulting angular difference transfers part of the load from the slow machine to the fast machine, depending on the power-angle relationship. This tends to reduce the speed difference and hence the angular separation. The power-angle relationship, as discussed above, is highly nonlinear. Beyond a certain limit, an increase in angular separation is accompanied by a decrease in power transfer; this increases the angular separation further and leads to instability. For any given situation, the stability of the system depends on whether or not the deviations in angular positions of the rotors result in sufficient restoring torques.

When a synchronous machine loses synchronism or "falls out of step" with the rest of the system, its rotor runs at a higher or lower speed than that required to generate voltages at system frequency. The "slip" between rotating stator field (corresponding to system frequency) and the rotor field results in large fluctuations in the machine power output, current, and voltage; this causes the protection system to isolate the unstable machine from the system.

Loss of synchronism can occur between one machine and the rest of the system or between groups of machines. In the latter case synchronism may be maintained within each group after its separation from the others.

The synchronous operation of interconnected synchronous machines is in some ways analogous to several cars speeding around a circular track while joined to each other by elastic links or rubber bands. The cars represent the synchronous machine rotors and the rubber bands are analogous to transmission lines. When all the cars run side by side, the rubber bands remain intact. If force applied to one of the cars causes it to speed up temporarily, the rubber bands connecting it to the other cars will stretch; this tends to slow down the faster car and speed up the other cars. A chain reaction results until all the cars run at the same speed once again. If the pull on one of the rubber bands exceeds its strength, it will break and one or more cars will pull away from the other cars.

With electric power systems, the change in electrical torque of a synchronous machine following a perturbation can be resolved into two components:

$$\Delta T_e = T_S \Delta\delta + T_D \Delta\omega \qquad (2.2)$$

where

> $T_S\Delta\delta$ is the component of torque change in phase with the rotor angle perturbation $\Delta\delta$ and is referred to as the *synchronizing torque* component; T_S is the synchronizing torque coefficient.

> $T_D\Delta\omega$ is the component of torque in phase with the speed deviation $\Delta\omega$ and is referred to as the *damping torque* component; T_D is the damping torque cocfficicnt.

System stability depends on the existence of both components of torque for each of the synchronous machines. Lack of sufficient synchronizing torque results in instability through an *aperiodic drift* in rotor angle. On the other hand, lack of sufficient damping torque results in *oscillatory instability*.

For convenience in analysis and for gaining useful insight into the nature of stability problems, it is usual to characterize the rotor angle stability phenomena in terms of the following two categories:

(a)   *Small-signal (or small-disturbance) stability* is the ability of the power system to maintain synchronism under small disturbances. Such disturbances occur continually on the system because of small variations in loads and generation. The disturbances are considered sufficiently small for linearization of system equations to be permissible for purposes of analysis. Instability that may result can be of two forms: (i) steady increase in rotor angle due to lack of sufficient synchronizing torque, or (ii) rotor oscillations of increasing amplitude due to lack of sufficient damping torque. The nature of system response to small disturbances depends on a number of factors including the initial operating, the transmission system strength, and the type of generator excitation controls used. For a generator connected radially to a large power system, in the absence of automatic voltage regulators (i.e., with constant field voltage) the instability is due to lack of sufficient synchronizing torque. This results in instability through a non-oscillatory mode, as shown in Figure 2.2(a). With continuously acting voltage regulators, the small-disturbance stability problem is one of ensuring sufficient damping of system oscillations. Instability is normally through oscillations of increasing amplitude. Figure 2.2(b) illustrates

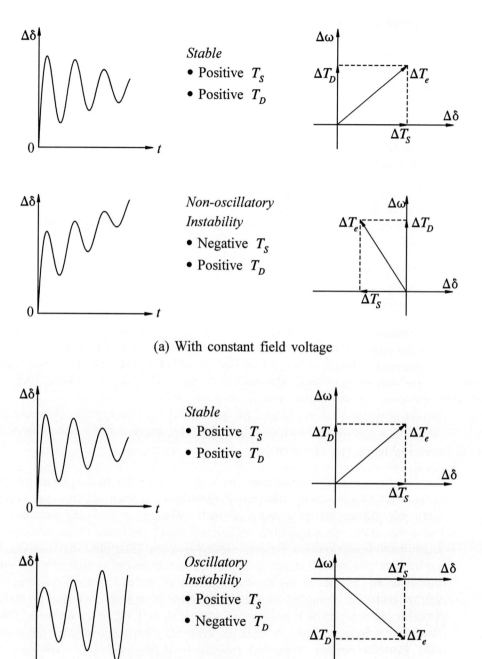

(a) With constant field voltage

(b) With excitation control

Figure 2.2 Nature of small-disturbance response

the nature of generator response with automatic voltage regulators.

In today's practical power systems, small-signal stability is largely a problem of insufficient damping of oscillations. The stability of the following types of oscillations is of concern:

- *Local modes or machine-system modes* are associated with the swinging of units at a generating station with respect to the rest of the power system. The term *local* is used because the oscillations are localized at one station or a small part of the power system.

- *Interarea modes* are associated with the swinging of many machines in one part of the system against machines in other parts. They are caused by two or more groups of closely coupled machines being interconnected by weak ties.

- *Control modes* are associated with generating units and other controls. Poorly tuned exciters, speed governors, HVDC converters and static var compensators are the usual causes of instability of these modes.

- *Torsional modes* are associated with the turbine-generator shaft system rotational components. Instability of torsional modes may be caused by interaction with excitation controls, speed governors, HVDC controls, and series-capacitor-compensated lines.

(b)   *Transient stability* is the ability of the power system to maintain synchronism when subjected to a severe transient disturbance. The resulting system response involves large excursions of generator rotor angles and is influenced by the nonlinear power-angle relationship. Stability depends on both the initial operating state of the system and the severity of the disturbance. Usually, the system is altered so that the post-disturbance steady-state operation differs from that prior to the disturbance.

Disturbances of widely varying degrees of severity and probability of occurrence can occur on the system. The system is, however, designed and operated so as to be stable for a selected set of contingencies. The contingencies usually considered are short-circuits of different types: phase-to-ground, phase-to-phase-to-ground, or three-phase. They are usually assumed to occur on transmission lines, but occasionally bus or transformer faults are also considered. The fault is assumed to be cleared by the opening of appropriate breakers to isolate the faulted element. In some cases, high-speed reclosure may be assumed.

Figure 2.3 illustrates the behaviour of a synchronous machine for stable and unstable situations. It shows the rotor angle responses for a stable case and for two unstable cases. In the stable case (Case 1), the rotor angle increases to a

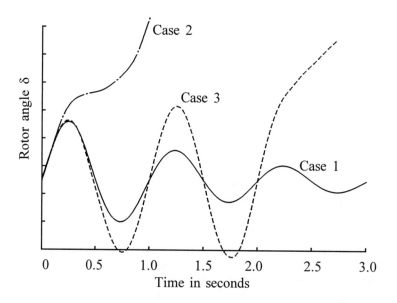

Figure 2.3 Rotor angle response to a transient disturbance

maximum, then decreases and oscillates with decreasing amplitude until it reaches a steady state. In Case 2, the rotor angle continues to increase steadily until synchronism is lost. This form of instability is referred to as *first-swing* instability and is caused by insufficient synchronizing torque. In Case 3, the system is stable in the first swing but becomes unstable as a result of growing oscillations as the end state is approached. This form of instability generally occurs when the postfault steady-state condition itself is "small-signal" unstable, and not necessarily as a result of the transient disturbance.

In large power systems, transient instability may not always occur as first-swing instability; it could be the result of the superposition of several modes of oscillation causing large excursions of rotor angle beyond the first swing.

In transient stability studies the study period of interest is usually limited to 3 to 5 seconds following the disturbance, although it may extend to about ten seconds for very large systems with dominant interarea modes of oscillation.

The term *dynamic stability* has also been widely used in the literature as a class of rotor angle stability. However, it has been used to denote different aspects of the phenomenon by different authors. In North American literature, it has been used mostly to denote small-signal stability in the presence of automatic control devices (primarily generator voltage regulators) as distinct from the classical steady-state stability without automatic controls [1,2]. In the French and German literature, it has been used to denote what we have termed here *transient stability*. Since much confusion has resulted from use of the term *dynamic stability*, both CIGRE and IEEE have recommended that it not be used [3,4].

2.1.2 Voltage Stability and Voltage Collapse

Voltage stability is the ability of a power system to maintain steady acceptable voltages at all buses in the system under normal operating conditions and after being subjected to a disturbance. A system enters a state of voltage instability when a disturbance, increase in load demand, or change in system condition causes a progressive and uncontrollable drop in voltage. The main factor causing instability is the inability of the power system to meet the demand for reactive power. The heart of the problem is usually the voltage drop that occurs when active power and reactive power flow through inductive reactances associated with the transmission network [5-7].

A criterion for voltage stability is that, at a given operating condition for every bus in the system, the bus voltage magnitude increases as the reactive power injection at the same bus is increased. A system is voltage unstable if, for at least one bus in the system, the bus voltage magnitude (V) decreases as the reactive power injection (Q) at the same bus is increased. In other words, a system is voltage stable if V-Q sensitivity is positive for every bus and voltage unstable if V-Q sensitivity is negative for at least one bus.

Progressive drop in bus voltages can also be associated with rotor angles going out of step. For example, the gradual loss of synchronism of machines as rotor angles between two groups of machines approach or exceed 180° would result in very low voltages at intermediate points in the network (see Chapter 13, Section 13.5.3). In contrast, the type of sustained fall of voltage that is related to voltage instability occurs where rotor angle stability is not an issue.

Voltage instability is essentially a local phenomenon; however, its consequences may have a widespread impact. *Voltage collapse* is more complex than simple voltage instability and is usually the result of a sequence of events accompanying voltage instability leading to a low-voltage profile in a significant part of the power system.

Voltage instability may occur in several different ways. In its simple form it can be illustrated by considering the two terminal network of Figure 2.4 [5]. It consists of a constant voltage source (E_S) supplying a load (Z_{LD}) through a series impedance (Z_{LN}). This is representative of a simple radial feed to load or a load area served by a large system through a transmission line.

The expression for current $\tilde{I}$ in Figure 2.4 is

$$\tilde{I} = \frac{\tilde{E}_S}{\tilde{Z}_{LN} + \tilde{Z}_{LD}} \tag{2.3}$$

where $\tilde{I}$ and $\tilde{E}_S$ are phasors, and

$$\tilde{Z}_{LN} = Z_{LN} \angle \theta, \quad \tilde{Z}_{LD} = Z_{LD} \angle \phi$$

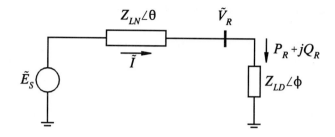

Figure 2.4 A simple radial system for illustration of voltage stability phenomenon

The magnitude of the current is given by

$$I = \frac{E_S}{\sqrt{(Z_{LN}\cos\theta + Z_{LD}\cos\phi)^2 + (Z_{LN}\sin\theta + Z_{LD}\sin\phi)^2}}$$

This may be expressed as

$$I = \frac{1}{\sqrt{F}} \frac{E_S}{Z_{LN}} \tag{2.4}$$

where

$$F = 1 + \left(\frac{Z_{LD}}{Z_{LN}}\right)^2 + 2\left(\frac{Z_{LD}}{Z_{LN}}\right)\cos(\theta - \phi)$$

The magnitude of the receiving end voltage is given by

$$V_R = Z_{LD}I$$

$$= \frac{1}{\sqrt{F}} \frac{Z_{LD}}{Z_{LN}} E_S \tag{2.5}$$

The power supplied to the load is

$$P_R = V_R I \cos\phi$$

$$= \frac{Z_{LD}}{F} \left(\frac{E_S}{Z_{LN}}\right)^2 \cos\phi \tag{2.6}$$

Plots of I, V_R, and P_R are shown in Figure 2.5 as a function of Z_{LN}/Z_{LD}, for the case with $\tan\theta = 10.0$ and $\cos\phi = 0.95$. To make the results applicable to any value of Z_{LN}, the values of I, V_R, and P_R are appropriately normalized.

As the load demand is increased by decreasing Z_{LD}, P_R increases rapidly at first and then slowly before reaching a maximum, after which it decreases. There is thus a maximum value of active power that can be transmitted through an impedance from a constant voltage source.

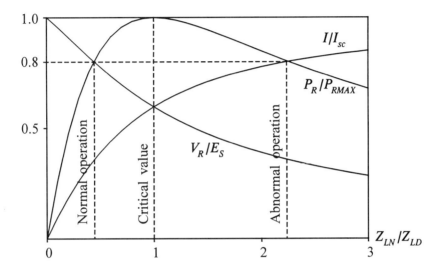

Figure 2.5 Receiving end voltage, current and power as a function
of load demand for the system of Figure 2.4
$(I_{sc} = E_S/Z_{LN};\ \cos\phi = 0.95\ \text{lag};\ \tan\theta = 10.0)$

The power transmitted is maximum when the voltage drop in the line is equal in magnitude to V_R, that is when $Z_{LN}/Z_{LD} = 1$. As Z_{LD} is decreased gradually, I increases and V_R decreases. Initially, at high values of Z_{LD}, the increase in I dominates over the decrease in V_R, and hence P_R increases rapidly with decrease in Z_{LD}. As Z_{LD} approaches Z_{LN}, the effect of the decrease in I is only slightly greater than that of the decrease in V_R. When Z_{LD} is less than Z_{LN}, the decrease in V_R dominates over the increase in I, and the net effect is a decrease in P_R.

The critical operating condition corresponding to maximum power represents the limit of satisfactory operation. For higher load demand, control of power by varying load would be unstable; that is, a decrease in load impedance reduces power. Whether voltage will progressively decrease and the system will become unstable depends on the load characteristics. With a constant-impedance static load characteristic, the system stabilizes at power and voltage levels lower than the desired values. On the other hand, with a constant-power load characteristic, the system becomes unstable through collapse of the load bus voltage. With other characteristics, the voltage is determined by the composite characteristic of the transmission line and load. If the load is supplied by transformers with automatic underload tap-changing (ULTC), the tap-changer action will try to raise the load voltage. This has the effect of reducing the effective Z_{LD} as seen from the system. This in turn lowers V_R still further and leads to a progressive reduction of voltage. This is a simple and pure form of voltage instability.

From the viewpoint of voltage stability, the relationship between P_R and V_R is of interest. This is shown in Figure 2.6 for the system under consideration when the load power factor is equal to 0.95 lag.

From Equations 2.5 and 2.6, we see that the load-power factor has a significant effect on the power-voltage characteristics of the system. This is to be expected since the voltage drop in the transmission line is a function of active as well as reactive power transfer. Voltage stability, in fact, depends on the relationships between P, Q and V. The traditional forms displaying these relationships are shown in Figures 2.7 and 2.8.

Figure 2.7 shows, for the power system of Figure 2.4, curves of the V_R-P_R relationship for different values of load power factor. The locus of critical operating points is shown by the dotted line in the figure. Normally, only the operating points above the critical points represent satisfactory operating conditions. A sudden reduction in power factor (increase in Q_R) can thus cause the system to change from a stable operating condition to an unsatisfactory, and possibly unstable, operating condition represented by the lower part of a V-P curve.

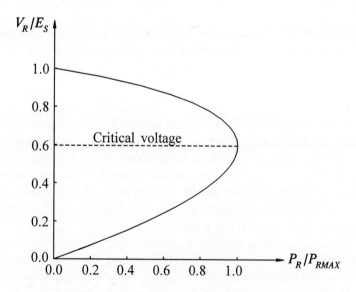

Figure 2.6 Power-voltage characteristics of
the system of Figure 2.4
($\cos\phi = 0.95$ lag; $\tan\theta = 10.0$)

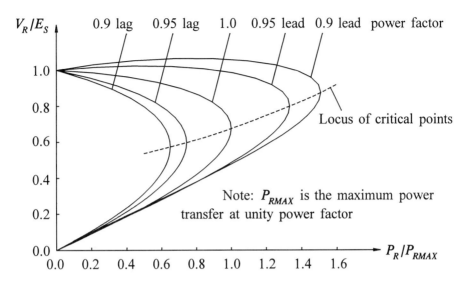

Figure 2.7 V_R-P_R characteristics of the system of
Figure 2.4 with different load-power factors

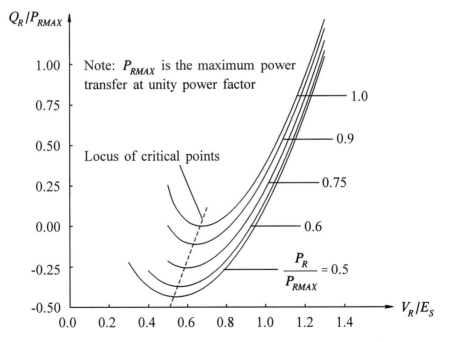

Figure 2.8 V_R-Q_R characteristics of the system of
Figure 2.4 with different P_R/P_{RMAX} ratio

The influence of the reactive power characteristics of the devices at the receiving end (loads and compensating devices) is more apparent in Figure 2.8. It shows a family of curves applicable to the power system of Figure 2.4, each of which represents the relationship between V_R and Q_R for a fixed value of P_R. The system is stable in the region where the derivative dQ_R/dV_R is positive. The voltage stability limit (critical operating point) is reached when the derivative is zero. Thus the parts of the Q-V curves to the right of the minima represent stable operation, and the parts to the left represent unstable operation. Stable operation in the region where dQ_R/dV_R is negative can be achieved only with a regulated reactive power compensation having sufficient control range and a high Q/V gain with a polarity opposite to that of the normal.

The above description of the voltage stability phenomenon is basic and intended to help classification and understanding of different aspects of power system stability. Analysis has been limited to a radial system because it presents a simple, yet clear, picture of the voltage stability problem. In complex practical power systems, many factors contribute to the process of system collapse because of voltage instability: strength of transmission system; power-transfer levels; load characteristics; generator reactive power capability limits; and characteristics of reactive power compensating devices. In some cases, the problem is compounded by uncoordinated action of various controls and protective systems.

For purposes of analysis, it is useful to classify voltage stability into the following two subclasses:

(a)   *Large-disturbance voltage stability* is concerned with a system's ability to control voltages following large disturbances such as system faults, loss of generation, or circuit contingencies. This ability is determined by the system-load characteristics and the interactions of both continuous and discrete controls and protections. Determination of large-disturbance stability requires the examination of the nonlinear dynamic performance of a system over a period of time sufficient to capture the interactions of such devices as ULTCs and generator field-current limiters. The study period of interest may extend from a few seconds to tens of minutes. Therefore, long-term dynamic simulations are required for analysis.

A criterion for large-disturbance voltage stability is that, following a given disturbance and following system-control actions, voltages at all buses reach acceptable steady-state levels.

(b)   *Small-disturbance voltage stability* is concerned with a system's ability to control voltages following small perturbations such as incremental changes in system load. This form of stability is determined by the characteristics of load, continuous controls, and discrete controls *at a given instant of time*. This concept is useful in determining, at any instant, how the system voltage will respond to small system changes.

The basic processes contributing to small-disturbance voltage instability are essentially of a steady-state nature. Therefore, static analysis can be effectively used to determine stability margins, identify factors influencing stability, and examine a wide range of system conditions and a large number of post-contingency scenarios [8].

A criterion for small-disturbance voltage stability is that, at a given operating condition for every bus in the system, the bus voltage magnitude increases as the reactive power injection at the same bus is increased. A system is voltage-unstable if, for at least one bus in the system, the bus voltage magnitude (V) decreases as the reactive power injection (Q) at the same bus is increased. In other words, a system is voltage-stable if V-Q sensitivity is positive for every bus and unstable if V-Q sensitivity is negative for at least one bus.

Voltage instability does not always occur in its pure form. Often the angle and voltage instabilities go hand in hand. One may lead to the other and the distinction may not be clear. However, a distinction between angle stability and voltage stability is important for understanding of the underlying causes of the problems in order to develop appropriate design and operating procedures.

A more detailed discussion of voltage stability, including analytical techniques and methods of preventing voltage collapse, is presented in Chapter 14. A comprehensive treatment of the subject, with an in-depth analysis of the problem, is presented in the companion book *Power System Voltage Stability* by C.W. Taylor.

2.1.3 Mid-Term and Long-Term Stability

The terms *long-term stability* and *mid-term stability* are relatively new to the literature on power system stability. They were introduced as a result of the need to deal with problems associated with the dynamic response of power systems to severe upsets [9-13]. Severe system upsets result in large excursions of voltage, frequency, and power flows that thereby invoke the actions of slow processes, controls, and protections not modelled in conventional transient stability studies. The characteristic times of the processes and devices activated by the large voltage and frequency shifts will range from a matter of seconds (the responses of devices such as generator controls and protections) to several minutes (the responses of devices such as prime mover energy supply systems and load-voltage regulators) [10,14].

Long-term stability analysis assumes that inter-machine synchronizing power oscillations have damped out, the result being uniform system frequency [3,11,15]. The focus is on the slower and longer-duration phenomena that accompany large-scale system upsets and on the resulting large, sustained mismatches between generation and consumption of active and reactive power. These phenomena include: boiler dynamics of thermal units, penstock and conduit dynamics of hydro units, automatic generation control, power plant and transmission system protection/controls, transformer saturation, and off-nominal frequency effects on loads and the network.

The mid-term response represents the transition between short-term and long-term responses. In *mid-term stability* studies, the focus is on synchronizing power oscillations between machines, including the effects of some of the slower phenomena, and possibly large voltage or frequency excursions.

Typical ranges of time periods are as follows:

Short-term or transient: 0 to 10 seconds
Mid-term: 10 seconds to a few minutes
Long-term: a few minutes to 10's of minutes

It should, however, be noted that the distinction between mid-term and long-term stability is primarily based on the phenomena being analyzed and the system representation used, particularly with regard to fast transients and inter-machine oscillations, rather than the time period involved.

Generally, the long-term and mid-term stability problems are associated with inadequacies in equipment responses, poor coordination of control and protection equipment, or insufficient active/reactive power reserves.

Long-term stability is usually concerned with system response to major disturbances that involve contingencies beyond the normal system design criteria. This may entail cascading and splitting of the power system into a number of separate islands with the generators in each island remaining in synchronism. *Stability in this case is a question of whether or not each island will reach an acceptable state of operating equilibrium with minimal loss of load.* It is determined by the overall response of the island as evidenced by its mean frequency, rather than the relative motion of machines. In an extreme case, the system and unit protections may compound the adverse situation and lead to a collapse of the island as a whole or in part.

Other applications of long-term and mid-term stability analysis include dynamic analysis of voltage stability requiring simulation of the effects of transformer tap-changing, generator overexcitation protection and reactive power limits, and thermostatic loads. In this case, inter-machine oscillations are not likely to be important. However, care should be exercised not to neglect some of the fast dynamics.

There is limited experience and literature related to the analysis of long-term and mid-term stability. As more experience is gained and improved analytical techniques for simulation of slow as well as fast dynamics become available, the distinction between mid-term and long-term stability becomes less significant.

2.2 CLASSIFICATION OF STABILITY

Power system stability is a single problem; however, it is impractical to study it as such. As seen in the previous section, instability of a power system can take different forms and can be influenced by a wide range of factors. Analysis of stability problems, identification of essential factors that contribute to instability, and formation of methods of improving stable operation are greatly facilitated by classification of stability into appropriate categories. These are based on the following considerations:

- The physical nature of the resulting instability;

- The size of the disturbance considered;

- The devices, processes, and time span that must be taken into consideration in order to determine stability; and

- The most appropriate method of calculation and prediction of stability.

Figure 2.9 gives an overall picture of the power system stability problem, identifying its classes and subclasses in terms of the categories described in the previous section. As a practical necessity, the classification has been based on a number of diverse considerations, making it difficult to select clearly distinct categories and to provide definitions that are rigorous and yet convenient for practical use. For example, there is some overlap between mid-term/long-term stability and voltage stability. With appropriate models for loads, on-load transformer tap changers and generator reactive power limits, mid-term/long-term stability simulations are ideally suited for dynamic analysis of voltage stability. Similarly, there is overlap between transient, mid-term and long-term stability: all three use similar analytical techniques for simulation of the nonlinear time domain response of the system to large disturbances. Although the three categories are concerned with different aspects of the stability problem, in terms of analysis and simulation they are really extensions of one another without clearly defined boundaries.

While classification of power system stability is an effective and convenient means to deal with the complexities of the problem, the overall stability of the system should always be kept in mind. Solutions to stability problems of one category should not be at the expense of another. It is essential to look at all aspects of the stability phenomena and at each aspect from more than one viewpoint. This requires the development and wise use of different kinds of analytical tools. In this regard, some degree of overlap in the phenomena being analyzed is in fact desirable.

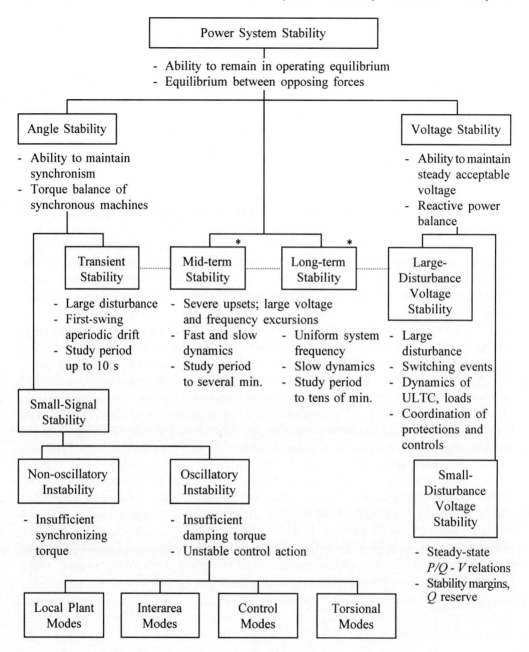

* With availability of improved analytical techniques providing unified approach for analysis of fast and slow dynamics, distinction between mid-term and long-term stability has become less significant.

Figure 2.9 Classification of power system stability

2.3 HISTORICAL REVIEW OF STABILITY PROBLEMS

Power system stability is a complex subject that has challenged power system engineers for many years. A review of the history of the subject is useful for a better understanding of present-day stability problems.

The stability of power systems was first recognized as an important problem in 1920 [16]. Results of the first laboratory tests on miniature systems were reported in 1924 [17]; the first field tests on the stability on a practical power system were conducted in 1925 [18,19].

Early stability problems were associated with remote hydroelectric generating stations feeding into metropolitan load centres over long-distance transmission. For economic reasons, such systems were operated close to their steady- state stability limits. In a few instances, instability occurred during steady-state operation, but it occurred more frequently following short-circuits and other system disturbances [20]. The stability problem was largely influenced by the strength of the transmission system, with instability being the result of insufficient synchronizing torque. The fault-clearing times were slow, being in the order of 0.5 to 2.0 seconds or longer.

The methods of analysis and the models used were dictated by developments in the art of computation and the stability theory of dynamic systems. Slide rules and mechanical calculators were used; hence, the models and methods of analysis had to be simple. In addition, graphical techniques such as the equal-area criterion and circle diagrams were developed. Such techniques were adequate for the analysis of the simple systems that could be treated effectively as two-machine systems. Steady-state and transient stability were treated separately. The former was related to the slope and peak of the power-angle curve; it was taken for granted that damping was positive.

As power systems evolved and interconnections between independent systems were found to be economically attractive, the complexity of the stability problems increased. Systems could no longer be treated as two-machine systems. A significant step towards the improvement of stability calculations was the development in 1930 of the network analyzer (or the ac calculating board). A network analyzer is essentially a scaled model of an ac power system with adjustable resistors, reactors and capacitors to represent transmission network and loads, voltage sources whose magnitude and angle can be adjusted to represent generators, and meters to measure voltages, currents, and power anywhere in the network. This development permitted power-flow analysis of multimachine systems; however, the equation of motion or the swing equation still had to be solved by hand using step-by-step numerical integration.

The theoretical work carried out in the 1920s and early 1930s laid the foundation for the industry's basic understanding of the power system stability phenomena. The principal developments and knowledge of power system stability in this early period came about as a result of the study of long-distance transmission, rather than as an extension of synchronous machine theory. The emphasis was on the network; the generators were viewed as simple voltage sources behind fixed reactances, and loads were considered as constant impedances. This was a practical necessity since the computational tools available during this period were suited for solution of algebraic equations, but not differential equations.

Improvements to system stability came about by way of faster-fault clearing and continuous-acting voltage regulators with no dead band. The benefits of an excitation system with a high degree of response for increasing steady-state stability were in fact recognized in the early 1920s; however, initially this region of "dynamic stability" was not recommended for normal operation but was treated as additional margin in determining operating limits. With the increased realization of the potential benefits of faster-responding excitation systems in limiting first-swing transient instability as well as increasing steady-state power transfer limits, their use became more commonplace. However, the use of high-response exciters in some cases resulted in decreased damping of power swings. Oscillatory instability thus became a cause for concern, while steady-state monotonic instability was virtually eliminated. These trends required better analytical tools. Synchronous machine and excitation system representation had to be more detailed and simulations had to be carried out for longer time periods.

In the early 1950s, electronic analog computers were used for analysis of special problems requiring detailed modelling of the synchronous machine, excitation system, and speed governor. Such simulations were suited for a detailed study of the effects of equipment characteristics rather than the overall behaviour of multimachine systems. The 1950s also saw the development of digital computers: the first digital computer program for power system stability analysis was developed about 1956. The models used in the early stability programs were similar to those of network analyzer studies. It was soon recognized that digital computer programs would allow improvements over network analyzer methods in both the size of the network that could be simulated and the modelling of equipment dynamic characteristics. They would provide the ideal means for the study of stability problems associated with growth in interconnections between formerly separate power systems.

In the 1960s, most of the power systems in the United States and Canada were joined as part of one of two large interconnected systems, one in the east and the other in the west. In 1967, low capacity HVDC ties were also established between the east and west systems. At present, the power systems in the United States and Canada form virtually one large system. While interconnections result in operating economy and increased reliability through mutual assistance, they also contribute to increased complexity of stability problems and increase the consequences of instability. The northeast blackout of November 9, 1965 made this abundantly clear; it brought the problem of stability and the importance of power system reliability beyond the focus of engineers and to the attention of the public and of the regulatory agencies [25].

Much of the industry effort and interest related to system stability since the 1960s has been concentrated on *transient stability*. Power systems are designed and operated to criteria concerning transient stability. As a consequence, the principal tool for stability analysis in power system design and operation has been the transient stability program. Very powerful programs have been developed, with facilities for representing very large systems and detailed equipment models. This has been greatly facilitated by developments in numerical methods and digital computer technology. There have also been significant developments in equipment modelling and testing, particularly for synchronous machines, excitation systems, and loads. In addition, significant improvements in transient stability performance of power systems have been achieved through use of high-speed fault-clearing, high initial-response exciters, series capacitors, and special stability aids.

Accompanying the above trends has been an increased tendency of power systems to exhibit oscillatory instability. Higher-response exciters, while improving transient stability, adversely affect small-signal stability associated with *local plant modes of oscillation* by introducing negative damping. The effects of fast exciters are compounded by the decreasing strength of transmission systems relative to the size of generating stations. Such problems have been solved through use of power system stabilizers (see Chapter 12).

Another source of the oscillatory instability problem has been the formation, as a consequence of growth in interconnections among power systems, of large groups of closely coupled machines connected by weak links. With heavy power transfers, such systems exhibit *interarea modes of oscillation* of low frequency. In many situations, the stability of these modes has become a source of concern.

Present trends in the planning and operation of power systems have resulted in new kinds of stability problems. Financial and regulatory conditions have caused electric utilities to build power systems with less redundancy and operate them closer to transient stability limits. Interconnections are continuing to grow with more use of new technologies such as multiterminal HVDC transmission. More extensive use is being made of shunt capacitors. Composition and characteristics of loads are changing. These trends have contributed to significant changes in the dynamic characteristics of modern power systems. Modes of instability are becoming increasingly more complex and require a comprehensive consideration of the various aspects of system stability. In particular, voltage instability and low-frequency interarea oscillations have become greater sources of concern than in the past. Whereas these problems used to occur in isolated situations, they have now become more commonplace. The need for analyzing the long-term dynamic response following major upsets and ensuring proper coordination of protection and control systems is also being recognized.

Significant research and development work has been undertaken in the last few years to gain a better insight into physical aspects of these new stability problems and to develop analytical tools for their analysis and better system design. Developments in control system theory and numerical methods have had a significant influence on this work. The following chapters describe these new developments and provide a comprehensive treatment of the subject of power system stability.

REFERENCES

[1]   D.N. Ewart and F.P. deMello, "A Digital Computer Program for the Automatic Determination of Dynamic Stability Limits," *IEEE Trans.*, Vol. PAS-86, pp. 867-875, July 1967.

[2]   H.M. Rustebakke (editor), *Electric Utility Systems and Practices*, John Wiley & Sons, 1983.

[3]   CIGRE Working Group 32-03, "Tentative Classification and Terminologies Relating to Stability Problems of Power Systems," *Electra,* No. 56, 1978.

[4]   IEEE Task Force, "Proposed Terms and Definitions for Power System Stability," *IEEE Trans.*, Vol. PAS-101, pp. 1894-1898, July 1982.

[5]   C. Barbier and J.P. Barret, "An Analysis of Phenomena of Voltage Collapse on Transmission System," *Revue Generale d'Electricite,* pp. 672-690, October 1980.

[6]   CIGRE Task Force 38-01-03, "Planning against Voltage Collapse," *Electra,* No. 111, pp. 55-75, March 1987.

[7]   IEEE Special Publication, *Voltage Stability of Power Systems: Concepts, Analytical Tools, and Industry Experience*, 90TH0358-2-PWR, 1990.

[8]   B. Gao, G.K. Morison, and P. Kundur, "Voltage Stability Evaluation Using Modal Analysis," *IEEE Trans.*, Vol. PWRS-7, No. 4, pp. 1529-1542, November 1992.

[9]   D.R. Davidson, D.N. Ewart, and L.K. Kirchmayer, "Long Term Dynamic Response of Power Systems - An Analysis of Major Disturbances," *IEEE Trans.*, Vol. PAS-94, pp. 819-826, May/June 1975.

[10]   C. Concordia, D.R. Davidson, D.N. Ewart, L.K. Kirchmayer, and R.P. Schulz, "Long Term Power System Dynamics - A New Planning Dimension," CIGRE Paper 32-13, 1976.

[11]   EPRI Report EL-596, "Midterm Simulation of Electric Power Systems," Project RP745, June 1979.

[12]   EPRI Report EL-983, "Long Term Power System Dynamics, Phase III," Research Project 764-2, May 1982.

[13] R.P. Schulz, "Capabilities of System Simulation Tools for Analyzing Severe Upsets," *Proceedings of International Symposium on Power System Stability,* Ames, Iowa, pp. 209-215, May 13-15, 1985.

[14] E.G. Cate, K. Hemmaplardh, J.W. Manke, and D.P. Gelopulos, "Time Frame Notion and Time Response of the Methods in Transient, Mid-Term and Long-Term Stability Programs," *IEEE Trans.,* Vol. PAS-103, pp. 143-151, January 1984.

[15] K. Hemmaplardh, J.W. Manke, W.R. Pauly, and J.W. Lamont, "Considerations for a Long Term Dynamics Simulation Program," *IEEE Trans.,* Vol. PWRS-1, pp. 129-135, February 1986.

[16] C.P. Steinmetz, "Power Control and Stability of Electric Generating Stations," *AIEE Trans.,* Vol. XXXIX, Part II, pp. 1215, July-December, 1920.

[17] R.D. Evans and R.C. Bergvall, "Experimental Analysis of Stability and Power Limitations," *AIEE Trans.,* pp. 39-58, 1924.

[18] R. Wilkins, "Practical Aspects of System Stability," *AIEE Trans.,* pp. 41-50, 1926.

[19] R.D. Evans and C.F. Wagner, "Further Studies of Transmission System Stability," *AIEE Trans.,* pp. 51-80, 1926.

[20] AIEE Subcommittee on Interconnections and Stability Factors, "First Report of Power System Stability," *AIEE Trans.,* pp. 261-282, February 1937.

[21] S.B. Crary, *Power System Stability, Vol. I: Steady-State Stability*, John Wiley & Sons, 1945.

[22] E.W. Kimbark, *Power System Stability, Vol. I: Elements of Stability Calculations*, John Wiley & Sons, 1948.

[23] Westinghouse Electric Corporation, *Electric Transmission and Distribution Reference Book*, East Pittsburgh, Pa., 1964.

[24] C. Concordia, "Power System Stability," *Proceedings of the International Symposium on Power System Stability,* Ames, Iowa, pp. 3-5, May 13-15, 1985.

[25] G.S. Vassell, "Northeast Blackout of 1965," *IEEE Power Engineering Review,* pp. 4-8, January 1991.

PART II

EQUIPMENT
CHARACTERISTICS
AND
MODELLING

Chapter 3

Synchronous Machine Theory
and Modelling

Synchronous generators form the principal source of electric energy in power systems. Many large loads are driven by synchronous motors. Synchronous condensers are sometimes used as a means of providing reactive power compensation and controlling voltage. These devices operate on the same principle and are collectively referred to as synchronous machines. As discussed in Chapter 2, the power system stability problem is largely one of keeping interconnected synchronous machines in synchronism. Therefore, an understanding of their characteristics and accurate modelling of their dynamic performance are of fundamental importance to the study of power system stability.

The modelling and analysis of the synchronous machine has always been a challenge. The problem was worked on intensely in the 1920s and 1930s [1,2,3], and has been the subject of several more recent investigations [4-9]. The theory and performance of synchronous machines have also been covered in a number of books [10-14].

In this chapter, we will develop in detail the mathematical model of a synchronous machine and briefly review its steady-state and transient performance characteristics.

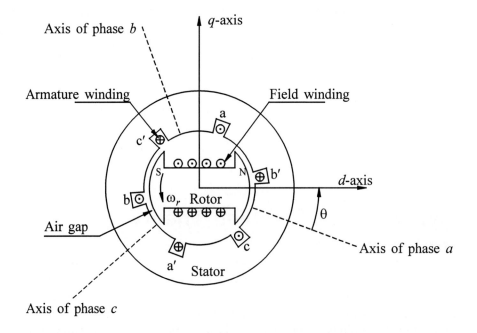

Figure 3.1 Schematic diagram of a three-phase synchronous machine

3.1 PHYSICAL DESCRIPTION

Figure 3.1 shows the schematic of the cross section of a three-phase synchronous machine with one pair of field poles. The machine consists of two essential elements: the field and the armature. The field winding carries direct current and produces a magnetic field which induces alternating voltages in the armature windings.

3.1.1 Armature and Field Structure

The armature windings usually operate at a voltage that is considerably higher than that of the field and thus they require more space for insulation. They are also subject to high transient currents and must have adequate mechanical strength. Therefore, normal practice is to have the armature on the stator. The three-phase windings of the armature are distributed 120° apart in *space* so that, with uniform rotation of the magnetic field, voltages displaced by 120° in *time* phase will be produced in the windings. Because the armature is subjected to a varying magnetic flux, the stator iron is built up of thin laminations to reduce eddy current losses.

When carrying balanced three-phase currents, the armature will produce a magnetic field in the air-gap rotating at synchronous speed (this will be formally shown in Section 3.1.3). The field produced by the direct current in the rotor winding, on the other hand, revolves with the rotor. For production of a steady torque, the fields of stator and rotor must rotate at the same speed. Therefore, the rotor must run at precisely the synchronous speed.

The number of field poles is determined by the mechanical speed of the rotor and electric frequency of stator currents. The synchronous speed is given by

$$n = \frac{120f}{p_f} \tag{3.1}$$

where n is the speed in rev/min, f is the frequency in Hz, and p_f is the number of field poles.

There are two basic rotor structures used, depending on speed. Hydraulic turbines operate at low speeds and hence a relatively large number of poles are required to produce the rated frequency. A rotor with salient or projecting poles and concentrated windings is better suited mechanically to this situation. Such rotors often have damper windings or amortisseurs in the form of copper or brass rods embedded in the pole face. These bars are connected to end rings to form short-circuited windings similar to those of a squirrel cage induction motor, as shown in Figure 3.2(a). They are intended to damp out speed oscillations. The damper windings may also be non-continuous, being wound only about the pole pieces as shown in Figure 3.2(b). The space harmonics of the armature magnetomotive force (mmf) contribute to surface eddy current losses; therefore, pole faces of salient pole machines are usually laminated.

Steam and gas turbines, on the other hand, operate at high speeds. Their generators have round (or cylindrical) rotors made up of solid steel forgings. They have two or four field poles, formed by distributed windings placed in slots milled in the solid rotor and held in place by steel wedges. They often do not have special

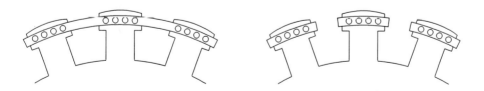

(a) Continuous damper   (b) Non-continuous damper

Figure 3.2 Salient pole rotor construction

damper windings, but the solid steel rotor offers paths for eddy currents which have effects equivalent to amortisseur currents. Some manufacturers provide for additional damping effects and negative-sequence current capability by using metal wedges in the field winding slots as damper bars and interconnecting them to form a damper cage, or by providing separate copper rods underneath the wedges. Figure 3.3 illustrates the rotor structure.

Under steady-state conditions, the only rotor current that exists is the direct current in the field winding. However, under dynamic conditions eddy currents are induced on the rotor surface and slot walls, and in slot wedges or damper windings (if used to produce additional damping). Figure 3.4 shows the rotor current paths of a steam turbine generator.

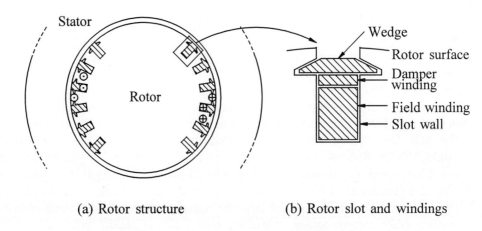

(a) Rotor structure   (b) Rotor slot and windings

Figure 3.3 Solid round rotor construction

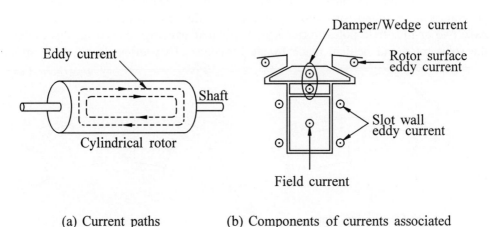

(a) Current paths   (b) Components of currents associated
                                 with an individual rotor slot

Figure 3.4 Current paths in a round rotor

3.1.2 Machines with Multiple Pole Pairs

Machines with more than one pair of field poles will have stator windings made up of a corresponding multiple set of coils. For purposes of analysis, it is convenient to consider only a single pair of poles and recognize that conditions associated with other pole pairs are identical to those for the pair under consideration. Therefore, angles are normally measured in electrical radians or degrees. The angle covered by one pole pair is 2π radians or 360 electrical degrees. The relationship between angle θ in electrical units and the corresponding angle θ_m in mechanical units is

$$\theta = \frac{p_f}{2}\theta_m \tag{3.2}$$

3.1.3 MMF Waveforms

In practice, the armature windings and round rotor machine field windings are distributed in many slots so that the resulting mmf and flux waveforms have nearly sinusoidal space distribution. In the case of salient pole machines, which have field windings concentrated at the poles, shaping of the pole faces is used to minimize harmonics in the flux produced.

First, let us consider the mmf waveform due to the armature windings only. The mmf produced by current flowing in only one coil in phase a is illustrated in Figure 3.5, in which the cross section of the stator has been cut open and rolled out in order to develop a view of the mmf wave.

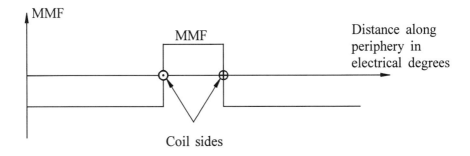

Figure 3.5 MMF waveform due to a single coil

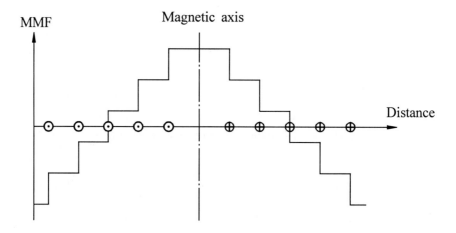

Figure 3.6 MMF waveform due to a number of coils

By adding more coils, the mmf wave distribution shown in Figure 3.6 may be obtained. We see that the mmf waveform is progressing from a square wave toward a sine wave as coils are added. Through use of fractional-pitch windings, the space harmonics can be made small [12]. Machine design aims at minimizing harmonics and, for most analyses of machine performance, it is reasonable to assume that each phase winding produces a sinusoidally distributed mmf wave. The windings are then said to be sinusoidally distributed. The harmonics may be considered as secondary from the viewpoint of machine performance. In addition to causing rotor surface eddy current losses, harmonics contribute to armature leakage reactances.

Rotating magnetic field

Let us now determine the net mmf wave due to the three-phase windings in the stator. Figure 3.7 shows the mmf wave of phase a.

With γ representing the angle along the periphery of the stator with respect to the centre of phase a, the mmf wave due to the three phases may be described as follows:

$$MMF_a = Ki_a \cos\gamma$$

$$MMF_b = Ki_b \cos(\gamma - \frac{2\pi}{3})$$

$$MMF_c = Ki_c \cos(\gamma + \frac{2\pi}{3})$$

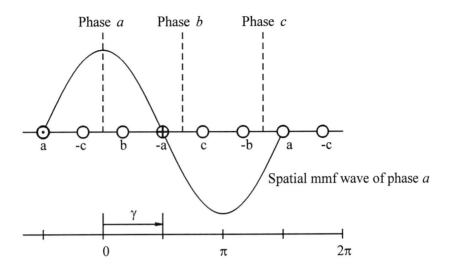

Figure 3.7 Spatial mmf wave of phase a

where i_a, i_b and i_c are the instantaneous values of the phase currents and K is a constant. Each winding produces a stationary mmf wave whose magnitude changes as the instantaneous value of the current through the winding changes. The three mmf waves due to the three phases are displaced 120 electrical degrees apart in space.

With balanced phase currents, and time origin arbitrarily chosen as the instant when i_a is maximum, we have

$$i_a = I_m \cos(\omega_s t)$$

$$i_b = I_m \cos(\omega_s t - \frac{2\pi}{3})$$

$$i_c = I_m \cos(\omega_s t + \frac{2\pi}{3})$$

(3.3)

where $\omega_s = 2\pi f =$ angular frequency of stator currents in electrical rad/s.

The total mmf due to the three phases is given by

$$
\begin{aligned}
MMF_{total} &= MMF_a + MMF_b + MMF_c \\
&= KI_m[\cos(\omega_s t)\cos\gamma + \cos(\omega_s t - \frac{2\pi}{3})\cos(\gamma - \frac{2\pi}{3}) + \\
&\quad\quad \cos(\omega_s t + \frac{2\pi}{3})\cos(\gamma + \frac{2\pi}{3})] \\
&= \frac{3}{2}KI_m\cos(\gamma - \omega_s t)
\end{aligned}
\tag{3.4}
$$

This is the equation of a *travelling wave*. At any instant in time, the total mmf has a sinusoidal spatial distribution. It has a constant amplitude and a space-phase angle $\omega_s t$, which is a function of time. Thus, the entire mmf wave moves at the constant angular velocity of ω_s electrical rad/s. For a machine with p_f field poles, the speed of rotation of the stator field is

$$
\omega_{sm} = \frac{2}{p_f}\omega_s \qquad\qquad \text{mech. rad/s} \tag{3.5a}
$$

or

$$
n_s = \frac{60\omega_{sm}}{2\pi} = \frac{120f}{p_f} \qquad\qquad \text{r/min} \tag{3.5b}
$$

This is the same as the synchronous speed of the rotor given by Equation 3.1. Therefore, for balanced operation the mmf wave due to stator currents is stationary with respect to the rotor.

The stator and rotor mmf waves are shown in Figure 3.8 relative to the rotor structure, again with both stator and rotor cross sections rolled out.

The magnitude of the stator mmf wave and its relative angular position with respect to the rotor mmf wave depend on the synchronous machine load (output). The electromagnetic torque on the rotor acts in a direction so as to bring the magnetic fields into alignment. If the rotor field leads the armature field, the torque acts in opposition to the rotation with the machine acting as a generator. On the other hand, if the rotor field lags the armature field, the torque acts in the direction of rotation with the machine acting as a motor. In other words, for generator action, the rotor field leads the armature field by the forward torque of a prime mover; for motor action, the rotor field lags behind the armature field due to the retarding torque of shaft load (mechanical).

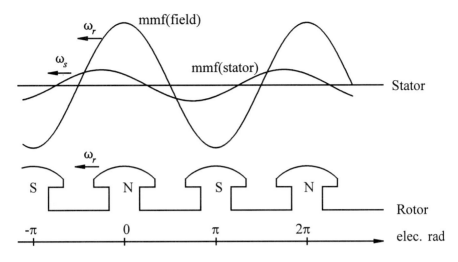

Figure 3.8 Stator and rotor mmf wave shapes

3.1.4 Direct and Quadrature Axes

We see that the magnetic circuits and all rotor windings are symmetrical with respect to both polar axis and the inter-polar axis. Therefore, for the purpose of identifying synchronous machine characteristics, two axes are defined as shown in Figure 3.1:

•   The direct (d) axis, centred magnetically in the centre of the north pole;

•   The quadrature (q) axis, 90 electrical degrees ahead of the d-axis.

The position of the rotor relative to the stator is measured by the angle θ between the d-axis and the magnetic axis of phase a winding.

The selection of the q-axis as leading the d-axis is purely arbitrary. This convention is based on the IEEE standard definition [15], and is widely used. Alternatively, the q-axis could be chosen to lag the d-axis by 90 degrees [16,17].

3.2 MATHEMATICAL DESCRIPTION OF A SYNCHRONOUS MACHINE

In developing equations of a synchronous machine, the following assumptions are made:

(a)   The stator windings are sinusoidally distributed along the air-gap as far as the mutual effects with the rotor are concerned.

(b)   The stator slots cause no appreciable variation of the rotor inductances with rotor position.

(c)   Magnetic hysteresis is negligible.

(d)   Magnetic saturation effects are negligible.

Assumptions (a), (b), and (c) are reasonable. The principal justification comes from the comparison of calculated performances based on these assumptions and actual measured performances. Assumption (d) is made for convenience in analysis. With magnetic saturation neglected, we are required to deal with only linear coupled circuits, making superposition applicable. However, saturation effects are important, and methods of accounting for their effects separately in an approximate manner will be discussed in Section 3.8. The machine equations will be developed first by assuming linear flux-current relationships.

Figure 3.9 shows the circuits involved in the analysis of a synchronous machine. The stator circuits consist of three-phase armature windings carrying alternating currents. The rotor circuits comprise field and amortisseur windings. The field winding is connected to a source of direct current. For purposes of analysis, the currents in the amortisseur (solid rotor and/or damper windings) may be assumed to flow in two sets of closed circuits: one set whose flux is in line with that of the field along the d-axis and the other set whose flux is at right angles to the field axis or along the q-axis. The amortisseur circuits, as discussed previously, take different forms and distinct, electrically independent circuits may not exist. In machine design analysis, a large number of circuits are used to represent amortisseur effects. For system analysis, where the characteristics of the machine as seen from its stator and rotor terminals are of interest, a limited number of circuits may be used. The type of rotor construction and the frequency range over which the model should accurately represent the machine characteristics determine the number of rotor circuits. For system stability studies, it is seldom necessary to represent more than two or three rotor circuits in each axis. In Figure 3.9, for the sake of simplicity only one amortisseur circuit is assumed in each axis, and we will write the machine equations based on this assumption. However, we implicitly consider an arbitrary number of such circuits; the subscript k is used to denote this.

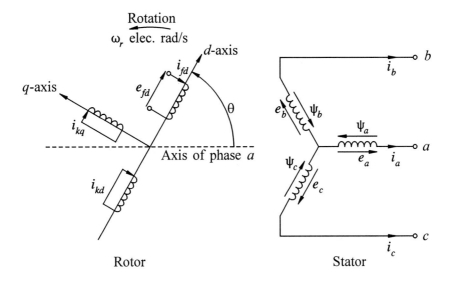

a, b, c :   Stator phase windings
fd :   Field winding
kd :   d-axis amortisseur circuit
kq :   q-axis amortisseur circuit
k = 1, 2, ... n; n = no. of amortisseur circuits
θ = Angle by which d-axis leads the magnetic axis of phase a winding, electrical rad
ω_r = Rotor angular velocity, electrical rad/s

Figure 3.9 Stator and rotor circuits of a synchronous machine

In Figures 3.1 and 3.9, θ is defined as the angle by which the d-axis leads the centreline of phase a winding in the direction of rotation. Since the rotor is rotating with respect to the stator, angle θ is continuously increasing and is related to the rotor angular velocity ω_r and time t as follows:

$$\theta = \omega_r t$$

The electrical performance equations of a synchronous machine can be developed by writing equations of the coupled circuits identified in Figure 3.9. Before we attempt to do this, it is useful to review how the equations of simple circuits may be written.

3.2.1 Review of Magnetic Circuit Equations

Single excited circuit

Consider first the elementary circuit of Figure 3.10, comprising a single exciting coil. The coil has N turns and a resistance of r. It is assumed to have a linear flux-mmf relationship. According to Faraday's law, the induced voltage e_i is

$$e_i = \frac{d\psi}{dt} \tag{3.6}$$

where ψ is the instantaneous value of flux linkage and t is time. The terminal voltage e_1 is given by

$$e_1 = \frac{d\psi}{dt} + ri \tag{3.7}$$

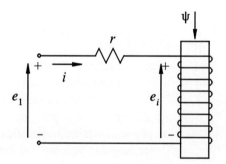

Figure 3.10 Single-excited magnetic circuit

The flux linkage may be expressed in terms of the inductance L of the circuit:

$$\psi = Li \tag{3.8}$$

The inductance, by definition, is equal to flux linkage per unit current. Therefore,

$$L = N\frac{\Phi}{i}$$
$$= N^2 P \tag{3.9}$$

where

$$P = \text{permeance of magnetic path}$$
$$\Phi = \text{flux} = (MMF)P = NiP$$

Coupled circuits

Let us next consider the circuit shown in Figure 3.11, consisting of two magnetically coupled windings. The windings have turns N_1 and N_2, and resistances r_1 and r_2, respectively; the magnetic path is assumed to have a linear flux-mmf relationship. The winding currents i_1 and i_2 are considered positive into the windings, as shown in the figure. The terminal voltages are

$$e_1 = \frac{d\psi_1}{dt} + r_1 i_1 \tag{3.10}$$

$$e_2 = \frac{d\psi_2}{dt} + r_2 i_2 \tag{3.11}$$

The magnetic field is determined by currents in both windings. Therefore, ψ_1 and ψ_2 are the flux linkages with the respective windings produced by the total effect of both currents. Thus

$$\psi_1 = N_1(\Phi_{m1} + \Phi_{l1}) + N_1 \Phi_{m2} \tag{3.12}$$

$$\psi_2 = N_2(\Phi_{m2} + \Phi_{l2}) + N_2 \Phi_{m1} \tag{3.13}$$

where

Φ_{m1} = mutual flux linking both windings due to current in winding 1 acting alone
Φ_{l1} = leakage flux linking winding 1 only
Φ_{m2} = mutual flux linking both windings due to current in winding 2 acting alone
Φ_{l2} = leakage flux linking winding 2 only

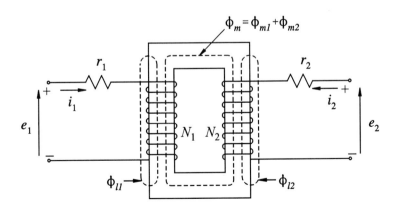

Figure 3.11 Magnetically coupled circuits

The flux linkages can be expressed in terms of self and mutual inductances whose expressions are given below.

Self inductance, by definition, is the flux linkage per unit current in the same winding. Accordingly, the self inductances of windings 1 and 2 are, respectively,

$$L_{11} = N_1(\Phi_{m1} + \Phi_{l1})/i_1 \qquad (3.14)$$

$$L_{22} = N_2(\Phi_{m2} + \Phi_{l2})/i_2 \qquad (3.15)$$

or

$$L_{11} = L_{m1} + L_{l1} \qquad (3.16)$$

$$L_{22} = L_{m2} + L_{l2} \qquad (3.17)$$

where L_{m1} and L_{m2} are the magnetizing inductances, and L_{l1} and L_{l2} the leakage inductances, of the respective windings.

Mutual inductance between two windings, by definition, is the flux linkage with one winding per unit current in the other winding. Therefore, the mutual inductances between windings 1 and 2 are

$$L_{12} = N_1\Phi_{m2}/i_2 \qquad (3.18)$$

and

$$L_{21} = N_2\Phi_{m1}/i_1 \qquad (3.19)$$

If $\mathbf{P}$ is the permeance of the mutual flux path,

$$\Phi_{m1} = N_1 i_1 \mathbf{P} \qquad (3.20)$$

$$\Phi_{m2} = N_2 i_2 \mathbf{P} \qquad (3.21)$$

From Equations 3.18, 3.19, 3.20 and 3.21, we see that

$$L_{12} = L_{21} = N_1 N_2 \mathbf{P} \qquad (3.22)$$

Substitution of Equations 3.16 to 3.19 in Equations 3.12 and 3.13 gives the following expressions for flux linking windings 1 and 2 in terms of self and mutual inductances:

$$\psi_1 = L_{11}i_1 + L_{12}i_2 \qquad (3.23)$$

$$\psi_2 = L_{21}i_1 + L_{22}i_2 \qquad (3.24)$$

In the above equations, it is important to recognize the relative directions of self and mutual flux linkages by the use of an appropriate algebraic sign for the mutual inductance. The mutual inductance is positive if positive currents in the two windings produce self and mutual fluxes in the same direction (i.e., the fluxes add up); otherwise it is negative.

Equations 3.10 and 3.11 for voltage together with Equations 3.23 and 3.24 for flux linkage give the performance equations of the linear static coupled circuits of Figure 3.11. In this form of representation, the self and mutual inductances of the windings are used as parameters. An inductance represents the proportionality between a flux linkage and a current. As seen from Equations 3.9 and 3.22, an inductance is directly proportional to the permeance of the associated flux path.

In developing the equations of magnetic circuits in this section, we have not explicitly specified units of system quantities. These equations are valid in any consistent system of units.

Finally, before we turn to synchronous machine equations, a comment about notation used is appropriate. In circuit analysis, the symbol λ is commonly used to denote flux linkage, whereas in most of the literature on synchronous machines and power system stability the symbol ψ is used. Here we have followed the latter practice, in order to correspond with the published literature and to avoid confusion in later chapters where we use λ to denote eigenvalues.

3.2.2 Basic Equations of a Synchronous Machine

The same general form of the equations derived in the previous section applies to the coupled circuits of Figure 3.9. We will, however, use the generator convention for polarities so that the positive direction of a stator winding current is assumed to be out of the machine. The positive direction of field and amortisseur currents is assumed to be into the machine.

In addition to the large number of circuits involved, the fact that the mutual and self inductances of the stator circuits vary with rotor position complicates the synchronous machine equations. The variations in inductances are caused by the variations in the permeance of the magnetic flux path due to non-uniform air-gap. This is pronounced in a salient pole machine in which the permeances in the two axes are significantly different. Even in a round rotor machine there are differences in the two axes due mostly to the large number of slots associated with the field winding.

The flux produced by a stator winding follows a path through the stator iron, across the air-gap, through the rotor iron, and back across the air-gap. The variations in permeance of this path as a function of the rotor position can be approximated as

$$P = P_0 + P_2 \cos 2\alpha \tag{3.25}$$

In the above equation, α is the angular distance from the d-axis along the periphery as shown in Figure 3.12.

A double frequency variation is produced, since the permeances of the north and south poles are equal. Higher order even harmonics of permeance exist but are small enough to be neglected.

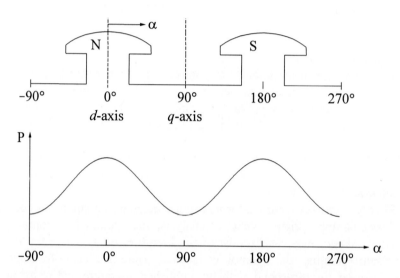

Figure 3.12 Variation of permeance with rotor position

We will use the following notation in writing the equations for the stator and rotor circuits:

e_a, e_b, e_c	= instantaneous stator phase to neutral voltages
i_a, i_b, i_c	= instantaneous stator currents in phases a, b, c
e_{fd}	= field voltage
i_{fd}, i_{kd}, i_{kq}	= field and amortisseur circuit currents
R_{fd}, R_{kd}, R_{kq}	= rotor circuit resistances
l_{aa}, l_{bb}, l_{cc}	= self-inductances of stator windings
l_{ab}, l_{bc}, l_{ca}	= mutual inductances between stator windings
$l_{afd}, l_{akd}, l_{akq}$	= mutual inductances between stator and rotor windings
$l_{ffd}, l_{kkd}, l_{kkq}$	= self-inductances of rotor circuits
R_a	= armature resistance per phase
p	= differential operator d/dt

Stator circuit equations

The voltage equations of the three phases are

$$e_a = \frac{d\psi_a}{dt} - R_a i_a = p\psi_a - R_a i_a \tag{3.26}$$

$$e_b = p\psi_b - R_a i_b \tag{3.27}$$

$$e_c = p\psi_c - R_a i_c \tag{3.28}$$

The flux linkage in the phase *a* winding at any instant is given by

$$\psi_a = -l_{aa}i_a - l_{ab}i_b - l_{ac}i_c + l_{afd}i_{fd} + l_{akd}i_{kd} + l_{akq}i_{kq} \tag{3.29}$$

Similar expressions apply to flux linkages of windings *b* and *c*. The units used are webers, henrys, and amperes. The negative sign associated with the stator winding currents is due to their assumed direction.

As shown below, all the inductances in Equation 3.29 are functions of the rotor position and are thus time-varying.

Stator self-inductances

The self-inductance l_{aa} is equal to the ratio of flux linking phase *a* winding to the current i_a, with currents in all other circuits equal to zero. The inductance is directly proportional to the permeance, which as indicated earlier has a second harmonic variation. The inductance l_{aa} will be a maximum for $\theta = 0°$, a minimum for $\theta = 90°$, a maximum again for $\theta = 180°$, and so on.

Neglecting space harmonics, the mmf of phase *a* has a sinusoidal distribution in space with its peak centred on the phase *a* axis. The peak amplitude of the mmf wave is equal to $N_a i_a$, where N_a is the effective turns per phase. As shown in Figure 3.13, this can be resolved into two other sinusoidally distributed mmf's, one centred on the *d*-axis and the other on the *q*-axis.

The peak values of the two component waves are

$$\text{peak } MMF_{ad} = N_a i_a \cos\theta \tag{3.30}$$

$$\text{peak } MMF_{aq} = N_a i_a \cos(\theta + 90°) = -N_a i_a \sin\theta \tag{3.31}$$

The reason for resolving the mmf into the *d*- and *q*-axis components is that each acts on specific air-gap geometry of defined configuration. Air-gap fluxes per pole along the two axes are

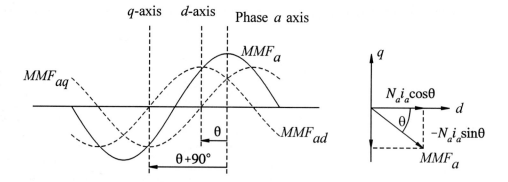

Figure 3.13 Phase a mmf wave and its components

$$\Phi_{gad} = (N_a i_a \cos\theta) P_d \qquad (3.32)$$

$$\Phi_{gaq} = (-N_a i_a \sin\theta) P_q \qquad (3.33)$$

In the above, P_d and P_q are the permeance coefficients of the d- and q-axis, respectively. In addition to the actual permeance, they include factors required to relate flux per pole with peak value of the mmf wave.

The total air-gap flux linking phase a is

$$
\begin{aligned}
\Phi_{gaa} &= \Phi_{gad}\cos\theta - \Phi_{gaq}\sin\theta \\[4pt]
&= N_a i_a (P_d \cos^2\theta + P_q \sin^2\theta) \\[4pt]
&= N_a i_a \left(\frac{P_d + P_q}{2} + \frac{P_d - P_q}{2}\cos2\theta \right)
\end{aligned}
\qquad (3.34)
$$

The self-inductance l_{gaa} of phase a due to air-gap flux is

$$
\begin{aligned}
l_{gaa} &= \frac{N_a \Phi_{gaa}}{i_a} \\[4pt]
&= N_a^2 \left(\frac{P_d + P_q}{2} + \frac{P_d - P_q}{2}\cos2\theta \right) \\[4pt]
&= L_{g0} + L_{aa2}\cos2\theta
\end{aligned}
\qquad (3.35)
$$

The total self-inductance l_{aa} is given by adding to the above the leakage inductance L_{al} which represents the leakage flux not crossing the air-gap:

$$l_{aa} = L_{al} + l_{gaa}$$

$$= L_{al} + L_{g0} + L_{aa2}\cos 2\theta \tag{3.36}$$

$$= L_{aa0} + L_{aa2}\cos 2\theta$$

Since the windings of phases b and c are identical to that of phase a and are displaced from it by 120° and 240° respectively, we have

$$l_{bb} = L_{aa0} + L_{aa2}\cos 2(\theta - \frac{2\pi}{3}) \tag{3.37}$$

$$l_{cc} = L_{aa0} + L_{aa2}\cos 2(\theta + \frac{2\pi}{3}) \tag{3.38}$$

The variation of l_{aa} with θ is shown in Figure 3.14.

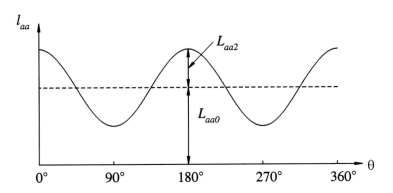

Figure 3.14 Variation of self-inductance of a stator phase

In Equations 3.36, 3.37 and 3.38, the stator self-inductances have a fixed plus second harmonic terms. Higher order harmonic terms have been neglected. In a well designed machine in which the stator and rotor windings produce nearly sinusoidally distributed mmf and flux waves, these higher order harmonic terms are negligible.

Stator mutual inductances

The mutual inductance between any two stator windings also exhibits a second harmonic variation because of the rotor shape. It is always negative, and has the greatest absolute value when the north and south poles are equidistant from the centres of the two windings concerned. For example, l_{ab} has maximum absolute value when $\theta = -30°$ or $\theta = 150°$.

The mutual inductance l_{ab} can be found by evaluating the air-gap flux Φ_{gba} linking phase b when only phase a is excited. As we wish to find the flux linking phase b due to mmf of phase a, θ is replaced by $\theta - 2\pi/3$ in Equation 3.34.

$$
\begin{aligned}
\Phi_{gba} &= \Phi_{gad}\cos(\theta - \frac{2\pi}{3}) - \Phi_{gaq}\sin(\theta - \frac{2\pi}{3}) \\
&= N_a i_a \left[P_d\cos\theta\cos(\theta - \frac{2\pi}{3}) + P_q\sin\theta\sin(\theta - \frac{2\pi}{3}) \right] \\
&= N_a i_a \left[-\frac{P_d + P_q}{4} + \frac{P_d - P_q}{2}\cos(2\theta - \frac{2\pi}{3}) \right]
\end{aligned}
\tag{3.39}
$$

The mutual inductance between phases a and b due to the air-gap flux is

$$
\begin{aligned}
l_{gba} &= \frac{N_a \Phi_{gba}}{i_a} \\
&= -\frac{1}{2}L_{g0} + L_{ab2}\cos(2\theta - \frac{2\pi}{3})
\end{aligned}
\tag{3.40}
$$

where L_{g0} has the same meaning as in the expression for self-inductance l_{gaa} given by Equation 3.35. There is a very small amount of mutual flux around the ends of windings which does not cross the air-gap. With this flux included, the mutual inductance between phases a and b can be written as

$$
\begin{aligned}
l_{ab} = l_{ba} &= -L_{ab0} + L_{ab2}\cos(2\theta - \frac{2\pi}{3}) \\
&= -L_{ab0} - L_{ab2}\cos(2\theta + \frac{\pi}{3})
\end{aligned}
\tag{3.41}
$$

Similarly,

$$
l_{bc} = l_{cb} = -L_{ab0} - L_{ab2}\cos(2\theta - \pi)
\tag{3.42}
$$

$$
l_{ca} = l_{ac} = -L_{ab0} - L_{ab2}\cos(2\theta - \frac{\pi}{3})
\tag{3.43}
$$

From the above equations, it can be readily seen that $L_{ab2}=L_{aa2}$. This is to be expected since the same variation in permeance produces the second harmonic terms in self and mutual inductances. It can also be seen that L_{ab0} is nearly equal to $L_{aa0}/2$.

The variation of mutual inductance between phases a and b as a function of θ is illustrated in Figure 3.15.

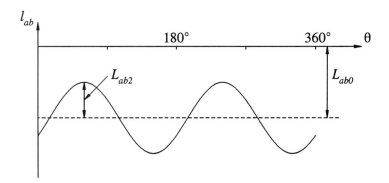

Figure 3.15 Variation of mutual inductance between stator windings

Mutual inductance between stator and rotor windings

With the variations in air-gap due to stator slots neglected, the rotor circuits see a constant permeance. Therefore, the situation in this case is not one of variation of permeance; instead, the variation in the mutual inductance is due to the relative motion between the windings themselves.

When a stator winding is lined up with a rotor winding, the flux linking the two windings is maximum and the mutual inductance is maximum. When the two windings are displaced by 90°, no flux links the two circuits and the mutual inductance is zero.

With a sinusoidal distribution of mmf and flux waves,

$$l_{afd} = L_{afd}\cos\theta \qquad (3.44)$$

$$l_{akd} = L_{akd}\cos\theta \qquad (3.45)$$

$$l_{akq} = L_{akq}\cos(\theta+\frac{\pi}{2})$$
$$= -L_{akq}\sin\theta \qquad (3.46)$$

For considering the mutual inductance between phase b winding and the rotor circuits, θ is replaced by $\theta-2\pi/3$; for phase c winding θ is replaced by $\theta+2\pi/3$.

We now have the expressions for all the inductances that appear in the stator voltage equations. On substituting the expressions for these inductances into Equation 3.29, we obtain

$$\psi_a = -i_a[L_{aa0}+L_{aa2}\cos2\theta]+i_b[L_{ab0}+L_{aa2}\cos(2\theta+\frac{\pi}{3})]$$
$$+i_c[L_{ab0}+L_{aa2}\cos(2\theta-\frac{\pi}{3})]+i_{fd}L_{afd}\cos\theta$$
$$+i_{kd}L_{akd}\cos\theta-i_{kq}L_{akq}\sin\theta \tag{3.47}$$

Similarly,

$$\psi_b = i_a[L_{ab0}+L_{aa2}\cos(2\theta+\frac{\pi}{3})]-i_b[L_{aa0}+L_{aa2}\cos2(\theta-\frac{2\pi}{3})]$$
$$+i_c[L_{ab0}+L_{aa2}\cos(2\theta-\pi)]+i_{fd}L_{afd}\cos(\theta-\frac{2\pi}{3})$$
$$+i_{kd}L_{akd}\cos(\theta-\frac{2\pi}{3})-i_{kq}L_{akq}\sin(\theta-\frac{2\pi}{3}) \tag{3.48}$$

and

$$\psi_c = i_a[L_{ab0}+L_{aa2}\cos(2\theta-\frac{\pi}{3})]+i_b[L_{ab0}+L_{aa2}\cos(2\theta-\pi)]$$
$$-i_c[L_{aa0}+L_{aa2}\cos2(\theta+\frac{2\pi}{3})]+i_{fd}L_{afd}\cos(\theta+\frac{2\pi}{3})$$
$$+i_{kd}L_{akd}\cos(\theta+\frac{2\pi}{3})-i_{kq}L_{akq}\sin(\theta+\frac{2\pi}{3}) \tag{3.49}$$

Rotor circuit equations

The rotor circuit voltage equations are

$$e_{fd} = p\psi_{fd}+R_{fd}i_{fd} \tag{3.50}$$
$$0 = p\psi_{kd}+R_{kd}i_{kd} \tag{3.51}$$
$$0 = p\psi_{kq}+R_{kq}i_{kq} \tag{3.52}$$

The rotor circuits see constant permeance because of the cylindrical structure of the stator. Therefore, the self-inductances of rotor circuits and mutual inductances between each other do not vary with rotor position. Only the rotor to stator mutual inductances vary periodically with θ as given by Equations 3.44, 3.45 and 3.46.

The rotor circuit flux linkages may be expressed as follows:

$$\psi_{fd} = L_{ffd}i_{fd} + L_{fkd}i_{kd} - L_{afd}[i_a\cos\theta + i_b\cos(\theta - \frac{2\pi}{3}) + i_c\cos(\theta + \frac{2\pi}{3})] \qquad (3.53)$$

$$\psi_{kd} = L_{fkd}i_{fd} + L_{kkd}i_{kd} - L_{akd}[i_a\cos\theta + i_b\cos(\theta - \frac{2\pi}{3}) + i_c\cos(\theta + \frac{2\pi}{3})] \qquad (3.54)$$

$$\psi_{kq} = L_{kkq}i_{kq} + L_{akq}[i_a\sin\theta + i_b\sin(\theta - \frac{2\pi}{3}) + i_c\sin(\theta + \frac{2\pi}{3})] \qquad (3.55)$$

3.3 THE *dq0* TRANSFORMATION

Equations 3.26 to 3.28 and Equations 3.47 to 3.49 associated with the stator circuits, together with Equations 3.50 to 3.55 associated with the rotor circuits, completely describe the electrical performance of a synchronous machine. However, these equations contain inductance terms which vary with angle θ which in turn varies with time. This introduces considerable complexity in solving machine and power system problems. A much simpler form leading to a clearer physical picture is obtained by appropriate transformation of stator variables.

We see from Equations 3.53 to 3.55 that stator currents combine into convenient forms in each axis. This suggests the transformation of the stator phase currents into new variables as follows:

$$i_d = k_d\left[i_a\cos\theta + i_b\cos(\theta - \frac{2\pi}{3}) + i_c\cos(\theta + \frac{2\pi}{3})\right] \qquad (3.56)$$

$$i_q = -k_q\left[i_a\sin\theta + i_b\sin(\theta - \frac{2\pi}{3}) + i_c\sin(\theta + \frac{2\pi}{3})\right] \qquad (3.57)$$

The constants k_d and k_q are arbitrary and their values may be chosen to simplify numerical coefficients in performance equations. In most of the literature on synchronous machine theory [3,10,11,12,13,19], k_d and k_q are taken as 2/3, and this choice will be followed here. An alternative transformation with $k_d = k_q = \sqrt{2/3}$ is discussed in Section 3.4.8.

With k_d and k_q equal to 2/3, for balanced sinusoidal conditions, the peak values of i_d and i_q are equal to the peak value of the stator current as shown below.

For the balanced condition,

$$i_a = I_m \sin \omega_s t$$

$$i_b = I_m \sin(\omega_s t - \frac{2\pi}{3})$$

$$i_c = I_m \sin(\omega_s t + \frac{2\pi}{3})$$

Substituting in Equation 3.56 gives

$$i_d = k_d \left[I_m \sin \omega_s t \cos\theta + I_m \sin(\omega_s t - \frac{2\pi}{3})\cos(\theta - \frac{2\pi}{3}) + I_m \sin(\omega_s t + \frac{2\pi}{3})\cos(\theta + \frac{2\pi}{3}) \right]$$

$$= k_d \frac{3}{2} I_m \sin(\omega_s t - \theta)$$

For the peak value of i_d to be equal to I_m, k_d should equal 2/3.
Similarly from Equation 3.57, for the balanced condition

$$i_q = -k_q \frac{3}{2} I_m \cos(\omega_s t - \theta)$$

Again, $k_q = 2/3$ results in the maximum value of i_q being equal to the peak value of stator current.

To give a complete degree of freedom, a third component must be defined so that the three-phase currents are transformed into three variables. Since the two current components i_d and i_q together produce a field identical to that produced by the original set of phase currents, the third component must produce no space field in the air-gap. Therefore, a convenient third variable is the zero sequence current i_0, associated with the symmetrical components:

$$i_0 = \frac{1}{3}(i_a + i_b + i_c) \tag{3.58}$$

Under balanced conditions $i_a + i_b + i_c = 0$ and, therefore, $i_0 = 0$.

The transformation from the *abc* phase variables to the *dq0* variables can be written in the following matrix form:

$$
\begin{bmatrix} i_d \\ i_q \\ i_0 \end{bmatrix} = \frac{2}{3} \begin{bmatrix} \cos\theta & \cos(\theta-\dfrac{2\pi}{3}) & \cos(\theta+\dfrac{2\pi}{3}) \\ -\sin\theta & -\sin(\theta-\dfrac{2\pi}{3}) & -\sin(\theta+\dfrac{2\pi}{3}) \\ \dfrac{1}{2} & \dfrac{1}{2} & \dfrac{1}{2} \end{bmatrix} \begin{bmatrix} i_a \\ i_b \\ i_c \end{bmatrix} \tag{3.59}
$$

The inverse transformation is given by

$$
\begin{bmatrix} i_a \\ i_b \\ i_c \end{bmatrix} = \begin{bmatrix} \cos\theta & -\sin\theta & 1 \\ \cos(\theta-\dfrac{2\pi}{3}) & -\sin(\theta-\dfrac{2\pi}{3}) & 1 \\ \cos(\theta+\dfrac{2\pi}{3}) & -\sin(\theta+\dfrac{2\pi}{3}) & 1 \end{bmatrix} \begin{bmatrix} i_d \\ i_q \\ i_0 \end{bmatrix} \tag{3.60}
$$

The above transformations also apply to stator flux linkages and voltages.

Stator flux linkages in dq0 components

Using the expressions for ψ_a, ψ_b and ψ_c given by Equations 3.47, 3.48 and 3.49, transforming the flux linkages and currents into *dq0* components (Equation 3.59), and with suitable reduction of terms involving trigonometric terms, we obtain the following expressions:

$$
\psi_d = -(L_{aa0}+L_{ab0}+\frac{3}{2}L_{aa2})i_d+L_{afd}i_{fd}+L_{akd}i_{kd}
$$

$$
\psi_q = -(L_{aa0}+L_{ab0}-\frac{3}{2}L_{aa2})i_q+L_{akq}i_{kq}
$$

$$
\psi_0 = -(L_{aa0}-2L_{ab0})i_0
$$

Defining the following new inductances

$$
L_d = L_{aa0}+L_{ab0}+\frac{3}{2}L_{aa2} \tag{3.61}
$$

$$L_q = L_{aa0} + L_{ab0} - \frac{3}{2}L_{aa2} \tag{3.62}$$

$$L_0 = L_{aa0} - 2L_{ab0} \tag{3.63}$$

the flux linkage equations become

$$\psi_d = -L_d i_d + L_{afd} i_{fd} + L_{akd} i_{kd} \tag{3.64}$$

$$\psi_q = -L_q i_q + L_{akq} i_{kq} \tag{3.65}$$

$$\psi_0 = -L_0 i_0 \tag{3.66}$$

The *dq0* components of stator flux linkages are seen to be related to the components of stator and rotor currents through constant inductances.

Rotor flux linkages in dq0 components

Substitution of the expressions for i_d, i_q in Equations 3.53 to 3.55 gives

$$\psi_{fd} = L_{ffd} i_{fd} + L_{fkd} i_{kd} - \frac{3}{2}L_{afd} i_d \tag{3.67}$$

$$\psi_{kd} = L_{fkd} i_{fd} + L_{kkd} i_{kd} - \frac{3}{2}L_{akd} i_d \tag{3.68}$$

$$\psi_{kq} = L_{kkq} i_{kq} - \frac{3}{2}L_{akq} i_q \tag{3.69}$$

Again, all the inductances are seen to be constant, i.e., they are independent of the rotor position. It should, however, be noted that the saturation effects are not considered here. The variations in inductances due to saturation are of a different nature and this will be treated separately.

It is interesting to note that i_0 does not appear in the rotor flux linkage equations. This is because zero sequence components of armature current do not produce net mmf across the air-gap.

While the *dq0* transformation has resulted in constant inductances in Equations 3.64 to 3.69, the mutual inductances between stator and rotor quantities are not reciprocal. For example, the mutual inductance associated with the flux linking the field winding due to current i_d flowing in the *d*-axis stator winding from Equation

3.67 is $(3/2)L_{afd}$, whereas from Equation 3.64 the mutual inductance associated with flux linking the *d*-axis stator winding due to field current is L_{afd}. As discussed in Section 3.4, this problem is overcome by appropriate choice of the per unit system for the rotor quantities.

Stator voltage equations in dq0 components

Equations 3.26 to 3.28 are basic equations for phase voltages in terms of phase flux linkages and currents. By applying the *dq0* transformation of Equation 3.59, the following expressions in terms of transformed components of voltages, flux linkages and currents result:

$$e_d = p\psi_d - \psi_q p\theta - R_a i_d \tag{3.70}$$

$$e_q = p\psi_q + \psi_d p\theta - R_a i_q \tag{3.71}$$

$$e_0 = p\psi_0 - R_a i_0 \tag{3.72}$$

The angle θ, as defined in Figure 3.9, is the angle between the axis of phase *a* and the *d*-axis. The term $p\theta$ in the above equations represents the angular velocity ω_r of the rotor. For a 60 Hz system under steady-state conditions $p\theta = \omega_r = \omega_s = 2\pi60 = 377$ electrical rad/s.

The above equations have a form similar to those of a static coil, except for the $\psi_q p\theta$ and $\psi_d p\theta$ terms. They result from the transformation from a stationary to a rotating reference frame, and represent the fact that a flux wave rotating in synchronism with the rotor will create voltages in the stationary armature coil. The $\psi_q p\theta$ and $\psi_d p\theta$ terms are referred to as *speed voltages* (due to flux change in space) and the terms $p\psi_d$ and $p\psi_q$ as the *transformer voltages* (due to flux change in time).

The speed voltage terms are the dominant components of the stator voltage. Under steady-state conditions, the transformer voltage terms $p\psi_d$ and $p\psi_q$ are in fact equal to zero; there are many transient conditions where the transformer voltage terms can be dropped from the stator voltage equations without causing errors of any significance. However, in other situations they could be important. This will be discussed further in Sections 3.7 and 5.1.

The signs associated with the speed voltage terms in Equations 3.70 and 3.71 are related to the sign conventions assumed for the voltage and flux linkage relationship and to the assumed relative positions of *d*- and *q*- axes. Since we have assumed that the *q*-axis leads the *d*-axis by 90°, the voltage e_q in the *q*-axis is induced by the flux in the *d*-axis. Similarly, the voltage e_d is induced by a flux in an axis lagging the *d*-axis by 90°, i.e., the negative *q*-axis. Therefore, the voltage induced in the *q*-axis due to rotation is $+\omega\psi_d$ and that in the *d*-axis is $-\omega\psi_q$.

Electrical power and torque

The instantaneous three-phase power output of the stator is

$$P_t = e_a i_a + e_b i_b + e_c i_c$$

Eliminating phase voltages and currents in terms of *dq0* components, we have

$$P_t = \frac{3}{2}(e_d i_d + e_q i_q + 2e_0 i_0) \tag{3.73}$$

Under balanced operation, $e_0 = i_0 = 0$ and the expression for power is given by

$$P_t = \frac{3}{2}(e_d i_d + e_q i_q)$$

The electromagnetic torque may be determined from the basic consideration of forces acting on conductors as being the product of currents and the flux. Alternatively, it can be derived by developing an expression for the power transferred across the air-gap.

Using Equations 3.70 to 3.72 to express the voltage components in terms of flux linkages and currents, by recognizing ω_r as the rotor speed $d\theta/dt$, and rearranging, we have

$$\begin{aligned}
P_t = \frac{3}{2}[&(i_d p\psi_d + i_q p\psi_q + 2i_0 p\psi_0) \\
&+ (\psi_d i_q - \psi_q i_d)\omega_r \\
&- (i_d^2 + i_q^2 + 2i_0^2)R_a]
\end{aligned} \tag{3.74}$$

$$\begin{aligned}
= \ &\text{(Rate of change of armature magnetic energy)} \\
&+\text{(power transferred across the air-gap)} \\
&-\text{(armature resistance loss)}
\end{aligned}$$

The air-gap torque T_e is obtained by dividing the power transferred across the air-gap (i.e., power corresponding to the speed voltages) by the rotor speed in mechanical radians per second.

$$\begin{aligned}
T_e &= \frac{3}{2}(\psi_d i_q - \psi_q i_d)\frac{\omega_r}{\omega_{mech}} \\
&= \frac{3}{2}(\psi_d i_q - \psi_q i_d)\frac{p_f}{2}
\end{aligned} \tag{3.75}$$

The flux-linkage equations 3.64 to 3.69 associated with the stator and rotor circuits, together with the voltage equations 3.70 to 3.72 for the stator, the voltage equations 3.50 to 3.52 for the rotor, and the torque equation 3.75, describe the electrical dynamic performance of the machine in terms of the *dq0* components. These equations are usually referred to as Park's equations in honour of R.H. Park who developed the concepts on which the equations are based [3]. The *dq0* transformation given by Equation 3.59 is referred to as Park's transformation. It is based on the two-reaction theory originally developed by Blondel [1] and the further exposition of the concept by Doherty and Nickle [2].

Physical interpretation of dq0 transformation

In Section 3.1.3, we saw that the combined mmf wave due to the currents in the three armature phases travels along the periphery of the stator at a velocity of ω_s rad/s. This is also the velocity of the rotor. Therefore, for balanced synchronous operation, the armature mmf wave appears stationary with respect to the rotor and has a sinusoidal space distribution. Since a sine function can be expressed as a sum of two sine functions, the mmf due to stator windings can be resolved into two sinusoidally distributed mmf waves stationary with respect to the rotor, so that one has its peak over the *d*-axis and the other has its peak over the *q*-axis. Therefore, i_d may be interpreted as the instantaneous current in a fictitious armature winding which rotates at the same speed as the rotor, and remains in such a position that its axis always coincides with the *d*-axis. The value of the current in this winding is such that it results in the same mmf on the *d*-axis as do actual phase currents flowing in the armature windings. A similar interpretation applies to i_q, except that it acts on the *q*-axis instead of the *d*-axis.

The mmfs due to i_d and i_q are stationary with respect to the rotor and act on paths of constant permeance. Therefore, the corresponding inductances L_d and L_q are constant.

For balanced steady-state conditions, the phase currents may be written as follows:

$$i_a = I_m \sin(\omega_s t + \phi) \tag{3.76}$$

$$i_b = I_m \sin(\omega_s t + \phi - \frac{2\pi}{3}) \tag{3.77}$$

$$i_c = I_m \sin(\omega_s t + \phi + \frac{2\pi}{3}) \tag{3.78}$$

where $\omega_s = 2\pi f$ is the angular frequency of stator currents. Using the *dq0* transformation,

$$i_d = I_m \sin(\omega_s t + \phi - \theta) \tag{3.79}$$

$$i_q = -I_m \cos(\omega_s t + \phi - \theta) \tag{3.80}$$

$$i_0 = 0 \tag{3.81}$$

For synchronous operation, the rotor speed ω_r is equal to the angular frequency ω_s of the stator currents. Hence,

$$\theta = \omega_r t = \omega_s t$$

Therefore,

$$i_d = I_m \sin\phi = \text{constant}$$

$$i_q = -I_m \cos\phi = \text{constant}$$

For balanced steady-state operation, i_d and i_q are constant. In other words, alternating phase currents in the *abc* reference frame appear as direct currents in the *dq0* reference frame.

The *dq0* transformation may be viewed as a means of referring the stator quantities to the rotor side. This is analogous to referring secondary side quantities in a transformer to the primary side by means of the turns ratio. The inverse transformation (Equation 3.60) can similarly be viewed as referring the rotor quantities to the stator side.

The analysis of synchronous machine equations in terms of *dq0* variables is considerably simpler than in terms of phase quantities, for the following reasons:

- The dynamic performance equations have constant inductances.

- For balanced conditions, zero sequence quantities disappear.

- For balanced steady-state operation, the stator quantities have constant values. For other modes of operation they vary with time. Stability studies involve slow variations having frequencies below 2 to 3 Hz.

- The parameters associated with *d*- and *q*-axes may be directly measured from terminal tests.

We will show in Section 3.6 that, under balanced steady-state conditions, the *dq0* transformation is equivalent to the use of phasors to represent alternating stator phase quantities. In many ways, the advantages of using *d,q* variables are similar to those of using phasors (instead of dealing directly with time varying sinusoidal quantities) for steady-state analysis of ac circuits.

3.4 PER UNIT REPRESENTATION

In power system analysis, it is usually convenient to use a per unit system to normalize system variables. Compared to the use of physical units (amperes, volts, ohms, webers, henrys, etc.), the per unit system offers computational simplicity by eliminating units and expressing system quantities as dimensionless ratios. Thus,

$$\text{quantity in per unit} = \frac{\text{actual quantity}}{\text{base value of quantity}}$$

A well-chosen per unit system can minimize computational effort, simplify evaluation, and facilitate understanding of system characteristics. Some base quantities may be chosen independently and quite arbitrarily, while others follow automatically depending on fundamental relationships between system variables. Normally, the base values are chosen so that the principal variables will be equal to one per unit under rated condition.

In the case of a synchronous machine, the per unit system may be used to remove arbitrary constants and simplify mathematical equations so that they may be expressed in terms of equivalent circuits. The basis for selection of the per unit system for the stator is straightforward, whereas it requires careful consideration for the rotor. Several alternative per unit systems have been proposed in the literature for the selection of base rotor quantities [18,19]. Only one system will be discussed here as it offers several advantages over others and has found wide acceptance. This system is referred to as *the L_{ad}-base reciprocal per unit system.*

In this section, for the purpose of defining per unit values and showing their relationships to the values in natural units, a superbar will be used to identify per unit quantities. We will, however, drop this convention for subsequent general use to simplify the notation.

3.4.1 Per Unit System for the Stator Quantities

The universal practice is to use the machine ratings as the base values for the stator quantities. In the machine equations developed so far, the stator currents and voltages have been expressed as instantaneous values; where they were sinusoidal quantities, they have been expressed in terms of the peak values and sinusoidal functions of time and frequency.

Let us choose the following base quantities for the stator (denoted by subscript s):

e_{sbase} = peak value of rated line-to-neutral voltage, V

i_{sbase} = peak value of rated line current, A

f_{base} = rated frequency, Hz

The base values of the remaining quantities are automatically set and depend on the above as follows:

$$\omega_{base} = 2\pi f_{base}, \text{ elec. radians/second}$$

$$\omega_{m\,base} = \omega_{base}\left(\frac{2}{p_f}\right), \text{ mech. radians/second}$$

$$Z_{s\,base} = \frac{e_{s\,base}}{i_{s\,base}}, \text{ ohms}$$

$$L_{s\,base} = \frac{Z_{s\,base}}{\omega_{base}}, \text{ henrys}$$

$$\psi_{s\,base} = L_{s\,base}\,i_{s\,base}$$

$$= \frac{e_{s\,base}}{\omega_{base}}, \text{ weber-turns}$$

$$\text{3-phase } VA_{base} = 3E_{RMS\,base}I_{RMS\,base}$$

$$= 3\frac{e_{s\,base}}{\sqrt{2}}\frac{i_{s\,base}}{\sqrt{2}}$$

$$= \frac{3}{2}e_{s\,base}i_{s\,base}, \text{ volt-amperes}$$

$$\text{Torque base} = \frac{\text{3-phase } VA_{base}}{\omega_{m\,base}}$$

$$= \frac{3}{2}\left(\frac{p_f}{2}\right)\psi_{s\,base}i_{s\,base}, \text{ newton-meters}$$

3.4.2 Per Unit Stator Voltage Equations

From Equation 3.70,

$$e_d = p\psi_d - \psi_q\omega_r - R_a i_d$$

Dividing throughout by $e_{s\,base}$, and noting that $e_{s\,base}=i_{s\,base}Z_{s\,base}=\omega_{base}\psi_{s\,base}$, we get

$$\frac{e_d}{e_{s\,base}} = p\left(\frac{1}{\omega_{s\,base}}\frac{\psi_d}{\psi_{s\,base}}\right) - \frac{\psi_q}{\psi_{s\,base}}\frac{\omega_r}{\omega_{base}} - \frac{R_a}{Z_{s\,base}}\frac{i_d}{i_{s\,base}} \qquad (3.82)$$

Expressed in per unit notation,

$$\bar{e}_d = \frac{1}{\omega_{base}} p\bar{\psi}_d - \bar{\psi}_q \bar{\omega}_r - \bar{R}_a \bar{i}_d \qquad (3.83)$$

The unit of time in the above equation is seconds. Time can also be expressed in per unit (or radians) with the base value equal to the time required for the rotor to move one electrical radian at synchronous speed:

$$t_{base} = \frac{1}{\omega_{base}} = \frac{1}{2\pi f_{base}} \qquad (3.84)$$

With time in per unit, Equation 3.83 may be written as

$$\bar{e}_d = \bar{p}\bar{\psi}_d - \bar{\psi}_q \bar{\omega}_r - \bar{R}_a \bar{i}_d \qquad (3.85)$$

Comparing Equation 3.70 and Equation 3.85, we see that the form of the original equation is unchanged, when all quantities involved are expressed in per unit.

Similarly, the per unit forms of Equations 3.71 and 3.72 are

$$\bar{e}_q = \bar{p}\bar{\psi}_q + \bar{\psi}_d \bar{\omega}_r - \bar{R}_a \bar{i}_q \qquad (3.86)$$

$$\bar{e}_0 = \bar{p}\bar{\psi}_0 - \bar{R}_a \bar{i}_0 \qquad (3.87)$$

The per unit time derivative $\bar{p}$ appearing in the above equations is given by

$$\bar{p} = \frac{d}{d\bar{t}} = \frac{1}{\omega_{base}} \frac{d}{dt} = \frac{1}{\omega_{base}} p \qquad (3.88)$$

3.4.3 Per Unit Rotor Voltage Equations

From Equation 3.50, dividing throughout by $e_{fd\,base} = \omega_{base} \psi_{fd\,base} = Z_{fd\,base} i_{fd\,base}$, the per unit field voltage equation may be written as

$$\bar{e}_{fd} = \bar{p}\bar{\psi}_{fd} + \bar{R}_{fd} \bar{i}_{fd} \qquad (3.89)$$

Similarly, the per unit forms of Equations 3.51 and 3.52 are

$$0 = \overline{p\psi}_{kd} + \overline{R}_{kd}\overline{i}_{kd} \tag{3.90}$$

$$0 = \overline{p\psi}_{kq} + \overline{R}_{kq}\overline{i}_{kq} \tag{3.91}$$

The above equations show the form of the rotor circuit voltage equations. However, we have not yet developed a basis for the choice of the rotor base quantities.

3.4.4 Stator Flux Linkage Equations

Using the basic relationship $\psi_{s\,base} = L_{s\,base}\,i_{s\,base}$, the per unit forms of Equations 3.64, 3.65 and 3.66 may be written as

$$\overline{\psi}_d = -\overline{L}_d\overline{i}_d + \overline{L}_{afd}\overline{i}_{fd} + \overline{L}_{akd}\overline{i}_{kd} \tag{3.92}$$

$$\overline{\psi}_q = -\overline{L}_q\overline{i}_q + \overline{L}_{akq}\overline{i}_q \tag{3.93}$$

$$\overline{\psi}_0 = -\overline{L}_0\overline{i}_0 \tag{3.94}$$

where by definition,

$$\overline{L}_{afd} = \frac{L_{afd}}{L_{s\,base}}\frac{i_{fd\,base}}{i_{s\,base}} \tag{3.95}$$

$$\overline{L}_{akd} = \frac{L_{akd}}{L_{s\,base}}\frac{i_{kd\,base}}{i_{s\,base}} \tag{3.96}$$

$$\overline{L}_{akq} = \frac{L_{akq}}{L_{s\,base}}\frac{i_{kq\,base}}{i_{s\,base}} \tag{3.97}$$

3.4.5 Rotor Flux Linkage Equations

Similarly, in per unit form Equations 3.67, 3.68 and 3.69 become

$$\overline{\psi}_{fd} = \overline{L}_{ffd}\overline{i}_{fd} + \overline{L}_{fkd}\overline{i}_{kd} - \overline{L}_{fda}\overline{i}_d \tag{3.98}$$

$$\overline{\psi}_{kd} = \overline{L}_{kdf}\overline{i}_{fd} + \overline{L}_{kkd}\overline{i}_{kd} - \overline{L}_{kda}\overline{i}_d \tag{3.99}$$

$$\overline{\psi}_{kq} = \overline{L}_{kkq}\overline{i}_{kq} - \overline{L}_{kqa}\overline{i}_q \tag{3.100}$$

where by definition,

$$\bar{L}_{fda} = \frac{3}{2} \frac{L_{afd}}{L_{fdbase}} \frac{i_{sbase}}{i_{fdbase}} \tag{3.101}$$

$$\bar{L}_{fkd} = \frac{L_{fkd}}{L_{fdbase}} \frac{i_{kdbase}}{i_{fdbase}} \tag{3.102}$$

$$\bar{L}_{kda} = \frac{3}{2} \frac{L_{akd}}{L_{kdbase}} \frac{i_{sbase}}{i_{kdbase}} \tag{3.103}$$

$$\bar{L}_{kdf} = \frac{L_{fkd}}{L_{kdbase}} \frac{i_{fdbase}}{i_{kdbase}} \tag{3.104}$$

$$\bar{L}_{kqa} = \frac{3}{2} \frac{L_{akq}}{L_{kqbase}} \frac{i_{sbase}}{i_{kqbase}} \tag{3.105}$$

By appropriate choice of per unit system, we have eliminated the factor 3/2 in the rotor flux equations. However, we have not yet tied down the values of the rotor base voltages and currents, which we will proceed to do next.

3.4.6 Per Unit System for the Rotor

The rotor circuit base quantities will be chosen so as to make the flux linkage equations simple by satisfying the following:

(a)   The per unit mutual inductances between different windings are to be reciprocal; for example, $\bar{L}_{afd} = \bar{L}_{fda}$. This will allow the synchronous machine model to be represented by equivalent circuits.

(b)   All per unit mutual inductances between stator and rotor circuits in each axis are to be equal; for example, $\bar{L}_{afd} = \bar{L}_{akd}$.

In order to have $\bar{L}_{fkd}$ equal to $\bar{L}_{kdf}$ so that reciprocity is achieved, from Equations 3.102 and 3.104, it is necessary to have

$$\frac{L_{fkd}}{L_{fdbase}} \frac{i_{kdbase}}{i_{fdbase}} = \frac{L_{fkd}}{L_{kdbase}} \frac{i_{fdbase}}{i_{kdbase}}$$

or

$$L_{kdbase} i^2_{kdbase} = L_{fdbase} i^2_{fdbase} \tag{3.106}$$

Multiplying by ω_{base} gives

$$\omega_{base} L_{kd\,base} i^2_{kd\,base} = \omega_{base} L_{fd\,base} i^2_{fd\,base}$$

Since $\omega_{base} L_{base} i_{base} = e_{base}$,

$$e_{kd\,base} i_{kd\,base} = e_{fd\,base} i_{fd\,base} \qquad (3.107)$$

Therefore, in order for the rotor circuit mutual inductances to be equal, their volt-ampere bases must be equal.

For mutual inductances $\bar{L}_{afd}$ and $\bar{L}_{fda}$ to be equal, from Equations 3.95 and 3.101,

$$\frac{L_{afd}}{L_{s\,base}} \frac{i_{fd\,base}}{i_{s\,base}} = \frac{3}{2} \frac{L_{afd}}{L_{fd\,base}} \frac{i_{s\,base}}{i_{fd\,base}}$$

or

$$L_{fd\,base} i^2_{fd\,base} = \frac{3}{2} L_{s\,base} i^2_{s\,base}$$

Multiplying by ω_{base} and noting that $\omega Li = e$, we get

$$e_{fd\,base} i_{fd\,base} = \frac{3}{2} e_{s\,base} i_{s\,base} \qquad (3.108)$$

$$= 3\text{-phase VA base for stator}$$

Similarly in order for $\bar{L}_{akd} = \bar{L}_{kda}$ and $\bar{L}_{akq} = \bar{L}_{kqa}$,

$$e_{kd\,base} i_{kd\,base} = \frac{3}{2} e_{s\,base} i_{s\,base} \qquad (3.109)$$

and

$$e_{kq\,base} i_{kq\,base} = \frac{3}{2} e_{s\,base} i_{s\,base} \qquad (3.110)$$

These equations imply that in order to satisfy requirement (a) above, the volt-ampere base in all rotor circuits must be the same and equal to the stator three-phase VA base.

So far, we have specified only the product of base voltage and base current for the rotor circuits. The next step is to specify either the base voltage or the base current for these circuits.

The stator self inductances $\bar{L}_d$ and $\bar{L}_q$ are associated with the total flux linkages due to i_d and i_q, respectively. They can be split into two parts: the leakage inductance due to flux that does not link any rotor circuit and the mutual inductance due to flux that links the rotor circuits. As shown in Figure 3.16, the stator leakage flux is made up of the slot leakage, end turn leakage and air-gap leakage. The stator

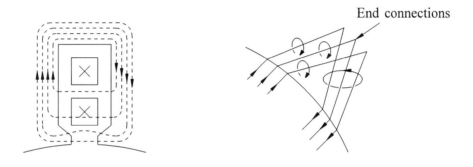

(a) Leakage flux within the slot

(b) Leakage flux around
the end connection

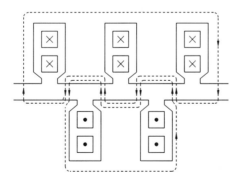

(c) Leakage flux following zigzag path between
stator and rotor tooth faces along the air-gap

Figure 3.16   Stator leakage flux patterns

leakage inductances in the two axes are nearly equal. Denoting the leakage inductance $\bar{L}_l$ and the mutual inductances by $\bar{L}_{ad}$ and $\bar{L}_{aq}$:

$$\bar{L}_d = \bar{L}_l + \bar{L}_{ad} \tag{3.111}$$

and

$$\bar{L}_q = \bar{L}_l + \bar{L}_{aq} \tag{3.112}$$

In order to make all the per unit mutual inductances between the stator and rotor circuits in the *d*-axis equal, from Equations 3.95 and 3.96 it follows that

$$\bar{L}_{ad} = \frac{L_{ad}}{L_{sbase}} = \bar{L}_{afd} = \frac{L_{afd}}{L_{sbase}} \frac{i_{fdbase}}{i_{sbase}}$$

$$= \bar{L}_{akd} = \frac{L_{akd}}{L_{sbase}} \frac{i_{kdbase}}{i_{sbase}}$$

Therefore,

$$i_{fdbase} = \frac{L_{ad}}{L_{afd}} i_{sbase} \tag{3.113}$$

$$i_{kdbase} = \frac{L_{ad}}{L_{akd}} i_{sbase} \tag{3.114}$$

Similarly, for the *q*-axis mutual inductances $\bar{L}_{aq}$ and $\bar{L}_{akq}$ to be equal,

$$i_{kqbase} = \frac{L_{aq}}{L_{akq}} i_{sbase} \tag{3.115}$$

This completes the choice of rotor base quantities.

As stated before, the per unit system used here is referred to as the L_{ad}-base reciprocal per unit system. In this system, the base current in any rotor circuit is defined as that which induces in each phase a per unit voltage equal to per unit $\bar{L}_{ad}$, that is, the same voltage as balanced three-phase unit-peak armature currents.

3.4.7 Per Unit Power and Torque

From Equation 3.73, the instantaneous power at the machine terminal is

$$P_t = \frac{3}{2}(e_d i_d + e_q i_q + 2e_0 i_0)$$

Dividing by the base three-phase $VA = (3/2)e_{s\,base}\,i_{s\,base}$, the expression for per unit may be written as

$$\bar{P}_t = \bar{e}_d \bar{i}_d + \bar{e}_q \bar{i}_q + 2\bar{e}_0 \bar{i}_0 \tag{3.116}$$

Similarly, with base torque $= \dfrac{3}{2}\left(\dfrac{p_f}{2}\right)\psi_{s\,base}\,i_{s\,base}$, the per unit form of Equation 3.75 is

$$\bar{T}_e = \bar{\psi}_d \bar{i}_q - \bar{\psi}_q \bar{i}_d \tag{3.117}$$

3.4.8 Alternative Per Unit Systems and Transformations

Several different alternative per unit systems have been proposed in the literature for the analysis of synchronous machines [4,13,16]. Some analysts, notably Lewis [20], have also suggested the use of an alternative form of transformation from the *abc* reference frame to the *dq0* reference frame, which is similar to that of Equation 3.59, but with factors k_d and k_q equal to $\sqrt{2/3}$ instead of 2/3 and with zero-sequence coefficients equal to $1/\sqrt{2}$. The alternative transformation equations are given by

$$\begin{bmatrix} i_d \\ i_q \\ i_0 \end{bmatrix} = \sqrt{\frac{2}{3}} \begin{bmatrix} \cos\theta & \cos(\theta - \frac{2\pi}{3}) & \cos(\theta + \frac{2\pi}{3}) \\ -\sin\theta & -\sin(\theta - \frac{2\pi}{3}) & -\sin(\theta + \frac{2\pi}{3}) \\ \sqrt{\frac{1}{2}} & \sqrt{\frac{1}{2}} & \sqrt{\frac{1}{2}} \end{bmatrix} \begin{bmatrix} i_a \\ i_b \\ i_c \end{bmatrix} \tag{3.118}$$

and the inverse transformation by

$$
\begin{bmatrix} i_a \\ i_b \\ i_c \end{bmatrix} = \sqrt{\frac{2}{3}} \begin{bmatrix} \cos\theta & -\sin\theta & \sqrt{\frac{1}{2}} \\ \cos(\theta - \frac{2\pi}{3}) & -\sin(\theta - \frac{2\pi}{3}) & \sqrt{\frac{1}{2}} \\ \cos(\theta + \frac{2\pi}{3}) & -\sin(\theta + \frac{2\pi}{3}) & \sqrt{\frac{1}{2}} \end{bmatrix} \begin{bmatrix} i_d \\ i_q \\ i_0 \end{bmatrix}
\tag{3.119}
$$

Such a transformation is orthogonal; i.e., the inverse of the transformation matrix is equal to its transpose. This also means that the transformation is power invariant:

$$
\begin{aligned}
P_t &= e_a i_a + e_b i_b + e_c i_c \\
&= e_d i_d + e_q i_q + e_0 i_0
\end{aligned}
$$

In addition, with this transformation, all mutual inductances would be reciprocal. However, as discussed in reference 19 by Harris, Lawrenson and Stephenson, such a transformation has several fundamental disadvantages which appear to override the advantages. The orthogonal transformation does not correspond to any particular meaningful physical situation. With k_d and k_q equal to $\sqrt{2/3}$, the equivalent d- and q-axis coils would have $\sqrt{3/2}$ times the number of turns as abc coils. This removes the unit-to-unit relationship between abc and $dq0$ variables that exists with the original transformation of Equation 3.59.

Reference 19 provides a thorough and comprehensive analysis of the alternative per unit and transformation systems. It concludes that the transformation of Equation 3.59 together with the L_{ad}-base reciprocal per unit system leads to a system which reflects most closely the physical features of the machine. In addition, the inductances in the resulting equivalent circuits correspond to those normally calculated by machine designers. In view of these advantages, this system is widely used by the electrical utility industry and generator manufacturers.

3.4.9 Summary of Per Unit Equations

Base quantities

Stator base quantities:

> 3-phase VA_{base} = volt-ampere rating of machine, VA
> e_{sbase} = peak phase-to-neutral rated voltage, V

f_{base} = rated frequency, Hz

i_{sbase} = peak line current, A

$\quad = \dfrac{3\text{-phase VA}_{base}}{(3/2)\,e_{sbase}}$

$Z_{sbase} = \dfrac{e_{sbase}}{i_{sbase}}$, Ω

$\omega_{base} = 2\pi f_{base}$, elec. rad/s

$\omega_{mbase} = \omega_{base}\dfrac{2}{p_f}$, mech. rad/s

$L_{sbase} = \dfrac{Z_{sbase}}{\omega_{base}}$, H

$\psi_{sbase} = L_{sbase}i_{sbase}$, Wb·turns

Rotor base quantities:

$i_{fdbase} = \dfrac{L_{ad}}{L_{afd}}i_{sbase}$, A

$i_{kdbase} = \dfrac{L_{ad}}{L_{akd}}i_{sbase}$, A

$i_{kqbase} = \dfrac{L_{aq}}{L_{akq}}i_{sbase}$, A

$e_{fdbase} = \dfrac{3\text{-phase VA}_{base}}{i_{fdbase}}$, V

$Z_{fdbase} = \dfrac{e_{fdbase}}{i_{fdbase}}$, Ω

$\quad = \dfrac{3\text{-phase VA}_{base}}{i_{fdbase}^2}$

$Z_{kdbase} = \dfrac{3\text{-phase VA}_{base}}{i_{kdbase}^2}$, Ω

$Z_{kqbase} = \dfrac{3\text{-phase VA}_{base}}{i_{kqbase}^2}$, Ω

$L_{fdbase} = \dfrac{Z_{fdbase}}{\omega_{base}}$, H

$$L_{kdbase} = \frac{Z_{kdbase}}{\omega_{base}}, \text{H}$$

$$L_{kqbase} = \frac{Z_{kqbase}}{\omega_{base}}, \text{H}$$

$$t_{base} = \frac{1}{\omega_{base}}, \text{s}$$

$$T_{base} = \frac{3\text{-phase VA}_{base}}{\omega_{mbase}}, \text{N·m}$$

Complete set of electrical equations in per unit

In view of the L_{ad}-base per unit system chosen, in per unit

$$L_{afd} = L_{fda} = L_{akd} = L_{kda} = L_{ad}$$

$$L_{akq} = L_{kqa} = L_{aq}$$

$$L_{fkd} = L_{kdf}$$

In the following equations, two q-axis amortisseur circuits are considered, and the subscripts $1q$ and $2q$ are used (in place of kq) to identify them. Only one d-axis amortisseur circuit is considered, and it is identified by the subscript $1d$. **Since all quantities are in per unit, we drop the superbar notation.**

Per unit stator voltage equations:

$$e_d = p\psi_d - \psi_q\omega_r - R_a i_d \tag{3.120}$$

$$e_q = p\psi_q + \psi_d\omega_r - R_a i_q \tag{3.121}$$

$$e_0 = p\psi_0 - R_a i_0 \tag{3.122}$$

Per unit rotor voltage equations:

$$e_{fd} = p\psi_{fd} + R_{fd} i_{fd} \tag{3.123}$$

$$0 = p\psi_{1d} + R_{1d} i_{1d} \tag{3.124}$$

$$0 = p\psi_{1q} + R_{1q} i_{1q} \tag{3.125}$$

$$0 = p\psi_{2q} + R_{2q} i_{2q} \tag{3.126}$$

Per unit stator flux linkage equations:

$$\psi_d = -(L_{ad}+L_l)i_d+L_{ad}i_{fd}+L_{ad}i_{1d} \tag{3.127}$$

$$\psi_q = -(L_{aq}+L_l)i_q+L_{aq}i_{1q}+L_{aq}i_{2q} \tag{3.128}$$

$$\psi_0 = -L_0 i_0 \tag{3.129}$$

Per unit rotor flux linkage equations:

$$\psi_{fd} = L_{ffd}i_{fd}+L_{f1d}i_{1d}-L_{ad}i_d \tag{3.130}$$

$$\psi_{1d} = L_{f1d}i_{fd}+L_{11d}i_{1d}-L_{ad}i_d \tag{3.131}$$

$$\psi_{1q} = L_{11q}i_{1q}+L_{aq}i_{2q}-L_{aq}i_q \tag{3.132}$$

$$\psi_{2q} = L_{aq}i_{1q}+L_{22q}i_{2q}-L_{aq}i_q \tag{3.133}$$

Per unit air-gap torque:

$$T_e = \psi_d i_q - \psi_q i_d \tag{3.134}$$

In writing Equations 3.132 and 3.133, we have assumed that the per unit mutual inductance L_{12q} is equal to L_{aq}. This implies that the stator and rotor circuits in the q-axis all link a single mutual flux represented by L_{aq}. This is acceptable because the rotor circuits represent the overall rotor body effects, and actual windings with physically measurable voltages and currents do not exist.

For power system stability analysis, the machine equations are normally solved with all quantities expressed in per unit, with the exception of time. Usually time t is expressed in seconds, in which case the per unit p in Equations 3.120 to 3.126 is replaced by $(1/\omega_{base})p$.

Per unit reactances

If the frequency of the stator quantities is equal to the base frequency, the per unit reactance of a winding is numerically equal to the per unit inductance. For example,

$$X_d = 2\pi f L_d \quad \Omega$$

Dividing by $Z_{s\,base} = 2\pi f_{base} L_{s\,base}$,

$$\frac{X_d}{Z_{s\,base}} = \frac{2\pi f}{2\pi f_{base}} \frac{L_d}{L_{s\,base}}$$

If $f = f_{base}$, per unit values of X_d and L_d are equal. For this reason, in the literature on synchronous machines, symbols associated with reactances are often used to denote per unit inductances.

3.5 EQUIVALENT CIRCUITS FOR DIRECT AND QUADRATURE AXES

While Equations 3.120 to 3.133 can be used directly to determine synchronous machine performance, it is a common practice to use equivalent circuits to provide a visual description of the machine model.

Before we develop an equivalent circuit to represent complete electrical characteristics of the machine, let us first consider only the d-axis flux linkage. Figure 3.17 shows an equivalent circuit which represents the d-axis stator and rotor flux linkage equations 3.127, 3.130 and 3.131. In this figure, the currents appear as loop currents.

A similar equivalent circuit can be developed for the q-axis flux linkage and current relationships. At this point, it is helpful to introduce the following rotor circuit per unit leakage inductances:

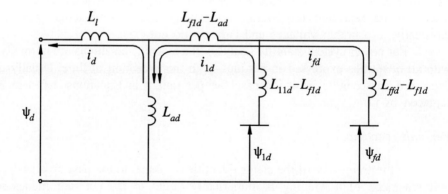

Figure 3.17 The d-axis equivalent circuit illustrating ψ-i relationship

$$L_{fd} = L_{ffd} - L_{fld} \qquad (3.135)$$

$$L_{1d} = L_{11d} - L_{fld} \qquad (3.136)$$

$$L_{1q} = L_{11q} - L_{aq} \qquad (3.137)$$

$$L_{2q} = L_{22q} - L_{aq} \qquad (3.138)$$

Equivalent circuits representing the complete characteristics, including the voltage equations, are shown in Figure 3.18. In these equivalent circuits, voltages as well as flux linkages appear. Therefore, flux linkages are shown in terms of their time derivatives.

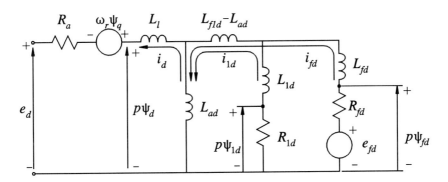

(a) *d*-axis equivalent circuit

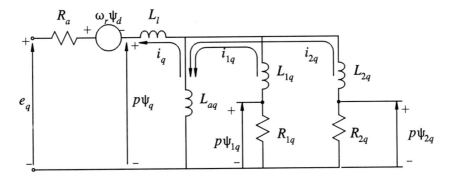

(b) *q*-axis equivalent circuit

Figure 3.18 Complete *d*- and *q*-axis equivalent circuits

In the *d*-axis equivalent circuit, the series inductance L_{fld}-L_{ad} represents the flux linking both the field winding and the amortisseur, but not the armature. It is a very common practice to neglect this series inductance on the grounds that the flux linking the damper circuit is very nearly equal to that linking the armature, because the damper windings are near the air-gap. This would be true in practice if the damper circuits were fully pitched. For short-pitched damper circuits and solid rotor iron paths, this approximation is not strictly valid [19]. In recent years, there has been some emphasis on including the series inductance L_{fld}-L_{ad}, particularly for detailed studies where the identity of the field circuit is to be retained [5,21].

In the case of the *q*-axis, there is no field winding and the amortisseurs represent the overall effects of the damper windings and eddy current paths. Therefore, it is reasonable to assume (as has been done in the development of the *q*-axis equivalent circuit of Figure 3.18 and the related equations) that the armature and damper circuits all link a single ideal mutual flux represented by L_{aq}.

In the literature, it is a widely accepted practice to simplify the *d*- and *q*-axis equivalent circuits as shown in Figure 3.19 which do not show the stator resistance

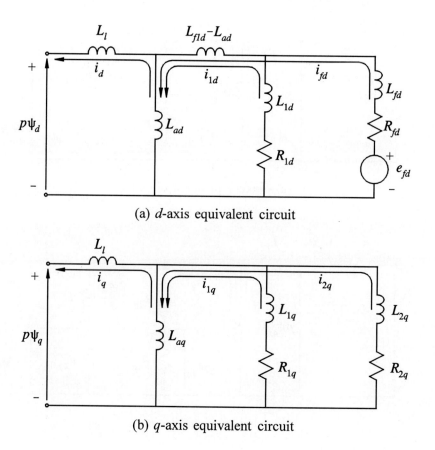

(a) *d*-axis equivalent circuit

(b) *q*-axis equivalent circuit

Figure 3.19 Commonly used simplified equivalent circuits

voltage drops and the speed voltage terms. These equivalent circuits are adequate for determining ψ_d and ψ_q, including their time derivatives.

The equivalent circuits in Figure 3.19 represent rotor flux linkage and voltage equations. So far as the stator is concerned they merely establish ψ_d, ψ_q in terms of i_d, i_q and rotor variables.

Example 3.1

A 555 MVA, 24 kV, 0.9 p.f., 60 Hz, 3 phase, 2 pole synchronous generator has the following inductances and resistances associated with the stator and field windings:

$$l_{aa} = 3.2758 + 0.0458 \cos(2\theta) \text{ mH}$$
$$l_{ab} = -1.6379 - 0.0458 \cos(2\theta + \pi/3) \text{ mH}$$
$$l_{afd} = 40.0 \cos\theta \text{ mH}$$
$$L_{ffd} = 576.92 \text{ mH}$$
$$R_a = 0.0031 \ \Omega$$
$$R_{fd} = 0.0715 \ \Omega$$

a.   Determine L_d and L_q in henrys.

b.   If the stator leakage inductance L_l is 0.4129 mH, determine L_{ad} and L_{aq} in henrys.

c.   Using the machine rated values as the base values for the stator quantities, determine the per unit values of the following in the L_{ad}-base reciprocal per unit system:

$$L_l, \ L_{ad}, \ L_{aq}, \ L_d, \ L_q, \ L_{afd}, \ L_{ffd}, \ L_{fd}, \ R_a, \ R_{fd}$$

Solution

a. From Equations 3.61 and 3.62,

$$L_d = L_{aa0} + L_{ab0} + \frac{3}{2} L_{aa2}$$

$$= 3.2758 + 1.6379 + \frac{3}{2} \times 0.0458$$

$$= 4.9825 \text{ mH}$$

$$L_q = L_{aa0} + L_{ab0} - \frac{3}{2} L_{aa2}$$

$$= 3.2758 + 1.6379 - \frac{3}{2} \times 0.0458$$

$$= 4.8451 \text{ mH}$$

b. $L_{ad} = L_d - L_l$
$$= 4.9825 - 0.4129$$
$$= 4.5696 \text{ mH}$$

$L_{aq} = L_q - L_l$
$$= 4.845 - 0.4129$$
$$= 4.432 \text{ mH}$$

c. The base values of stator and rotor quantities are as follows:

3-phase VA base = 555 MVA

$E_{sbase}(\text{RMS}) = 24/\sqrt{3} = 13.856 \text{ kV}$

$e_{sbase}(\text{peak}) = \sqrt{2} \times 13.856 = 19.596 \text{ kV}$

$I_{sbase}(\text{RMS}) = \dfrac{555 \times 10^6}{3 \times 13.856 \times 10^3} = 13{,}351.2 \text{ A}$

$i_{sbase}(\text{peak}) = \sqrt{2} \times 13{,}351.2 = 18{,}881.5 \text{ A}$

$Z_{sbase} = \dfrac{13.856 \times 10^3}{13{,}351.2} = 1.03784 \ \Omega$

$\omega_{base} = 2\pi 60 = 377 \text{ elec. rad/s}$

$L_{sbase} = \dfrac{1.03784}{377} \times 10^3 = 2.753 \text{ mH}$

$i_{fdbase} = \dfrac{L_{ad}}{L_{afd}} i_{sbase}$
$$= \dfrac{4.5696}{40.0} \times 18{,}881.5$$
$$= 2158.0 \text{ A}$$

$e_{fdbase} = \dfrac{555 \times 10^6}{2158} = 257.183 \text{ kV}$

$Z_{fdbase} = \dfrac{257.183}{2158.0} = 119.18 \ \Omega$

$L_{fdbase} = \dfrac{119.18}{377} \times 10^3 = 316.12 \text{ mH}$

The per unit values are

$L_l = \dfrac{0.4129}{2.753} = 0.15 \text{ pu}$

$L_{ad} = \dfrac{4.5696}{2.753} = 1.66 \text{ pu}$

$L_{aq} = \dfrac{4.432}{2.753} = 1.61 \text{ pu}$

$$L_d = L_{ad} + L_l = 1.66 + 0.15 = 1.81 \text{ pu}$$

$$L_q = L_{aq} + L_l = 1.61 + 0.15 = 1.76 \text{ pu}$$

$$L_{afd} = \frac{L_{afd}\, i_{fdbase}}{L_{sbase}\, i_{sbase}}$$

$$= \frac{40.0}{2.753} \times \frac{2158}{18,881.5} = 1.66 \text{ pu}$$

$$L_{ffd} = \frac{576.92}{316.12} = 1.825 \text{ pu}$$

$$L_{fd} = L_{ffd} - L_{ad}$$

$$= 1.825 - 1.66 = 0.165 \text{ pu}$$

$$R_a = \frac{0.0031}{1.03784} = 0.003 \text{ pu}$$

$$R_{fd} = \frac{0.0715}{119.18} = 0.0006 \text{ pu} \qquad\blacksquare$$

3.6 STEADY-STATE ANALYSIS

The performance of synchronous machines under balanced steady-state conditions may be readily analyzed by applying the per unit equations summarized in Section 3.4.9.

3.6.1 Voltage, Current, and Flux Linkage Relationships

As has been shown in Section 3.3, the *dq0* transformation applied to balanced steady-state armature phase currents results in steady direct currents. This is also true of stator voltages and flux linkages. Since rotor quantities are also constant under steady state, all time derivative terms drop out of machine equations. In addition, zero-sequence components are absent and $\omega_r = \omega_s = 1$ pu.

With $p\psi$ terms set to zero in Equations 3.124, 3.125 and 3.126,

$$R_{1d} i_{1d} = R_{1q} i_{1q} = R_{2q} i_{2q} = 0$$

Therefore, all amortisseur currents are zero. This is to be expected since, under steady state, the rotating magnetic field due to the stator currents is stationary with respect to the rotor. As the amortisseurs are closed circuits with no applied voltage, currents are induced in them only when the magnetic field due to the stator windings or the field winding is changing.

The per unit machine equations (3.120 to 3.134), under balanced steady-state conditions, become

$$e_d = -\omega_r \psi_q - R_a i_d \qquad (3.139)$$

$$e_q = \omega_r \psi_d - R_a i_q \qquad (3.140)$$

$$e_{fd} = R_{fd} i_{fd} \qquad (3.141)$$

$$\psi_d = -L_d i_d + L_{ad} i_{fd} \qquad (3.142)$$

$$\psi_q = -L_q i_q \qquad (3.143)$$

$$\psi_{fd} = L_{ffd} i_{fd} - L_{ad} i_d \qquad (3.144)$$

$$\psi_{1d} = L_{f1d} i_{fd} - L_{ad} i_d \qquad (3.145)$$

$$\psi_{1q} = \psi_{2q} = -L_{aq} i_q \qquad (3.146)$$

Field current

From Equation 3.142,

$$i_{fd} = \frac{\psi_d + L_d i_d}{L_{ad}}$$

Substituting for ψ_d in terms of e_d, i_q from Equation 3.140,

$$i_{fd} = \frac{e_q + R_a i_q + \omega_r L_d i_d}{\omega_r L_{ad}}$$

Replacing the product of synchronous speed and inductance L by the corresponding reactance X,

$$i_{fd} = \frac{e_q + R_a i_q + X_d i_d}{X_{ad}} \qquad (3.147)$$

The above equation is useful in computing the steady-state value of the field current for any specified operating condition. The inductances/reactances appearing in Equations 3.139 to 3.147 are saturated values. This will be discussed in Section 3.8.

3.6.2 Phasor Representation

For balanced steady-state operation, the stator phase voltages may be written as

$$e_a = E_m \cos(\omega_s t + \alpha) \tag{3.148}$$

$$e_b = E_m \cos(\omega_s t - \frac{2\pi}{3} + \alpha) \tag{3.149}$$

$$e_c = E_m \cos(\omega_s t + \frac{2\pi}{3} + \alpha) \tag{3.150}$$

where ω_s is the angular frequency and α is the phase angle of e_a with respect to the time origin.

Applying the *dq* transformation gives

$$e_d = E_m \cos(\omega_s t + \alpha - \theta) \tag{3.151}$$

$$e_q = E_m \sin(\omega_s t + \alpha - \theta) \tag{3.152}$$

The angle θ by which the *d*-axis leads the axis of phase *a* is given by

$$\theta = \omega_r t + \theta_0 \tag{3.153}$$

where θ_0 is the value of θ at *t*=0.

With ω_r equal to ω_s at synchronous speed, substitution for θ in Equations 3.151 and 3.152 yields

$$e_d = E_m \cos(\alpha - \theta_0) \tag{3.154}$$

$$e_q = E_m \sin(\alpha - \theta_0) \tag{3.155}$$

In the above equations, E_m is the peak value of phase voltage. In steady-state analysis, we are interested in RMS values and phase displacements rather than instantaneous or peak values. Using E_t to denote per unit RMS value of armature terminal voltage and noting that in per unit RMS and peak values are equal,

$$e_d = E_t \cos(\alpha - \theta_0) \tag{3.156}$$

$$e_q = E_t \sin(\alpha - \theta_0) \tag{3.157}$$

The *dq* components of armature voltage are scalar quantities. However, in view of the trigonometric relationship between them, they can be expressed as phasors in

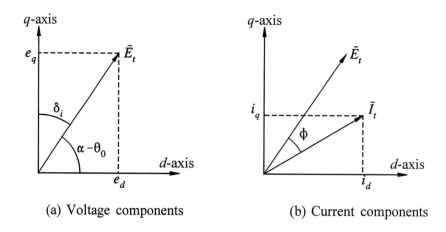

(a) Voltage components (b) Current components

Figure 3.20 Representation of *dq* components of armature
voltage and current as phasors

a complex plane having *d*- and *q*-axes as coordinates. This is illustrated in Figure 3.20
and is conceptually similar to phasor representation of alternating quantities varying
sinusoidally with respect to time. Thus the armature terminal voltage may be
expressed in complex form as

$$\tilde{E}_t = e_d + je_q \qquad (3.158)$$

By denoting δ_i as the angle by which the *q*-axis leads the phasor $\tilde{E}_t$, Equations 3.156
and 3.157 become

$$e_d = E_t \sin\delta_i \qquad (3.159)$$

$$e_q = E_t \cos\delta_i \qquad (3.160)$$

Similarly, the *dq* components of armature terminal current I_t can be expressed as
phasors. If ϕ is the power factor angle, we can write

$$i_d = I_t \sin(\delta_i + \phi) \qquad (3.161)$$

$$i_q = I_t \cos(\delta_i + \phi) \qquad (3.162)$$

and

$$\tilde{I}_t = i_d + ji_q \tag{3.163}$$

From the above analysis, it is clear that in phasor form with dq axes as reference, the RMS armature phase current and voltage can be treated the same way as is done with phasor representation of alternating voltages and currents. This provides the link between the steady-state values of dq components of armature quantities and the phasor representation used in conventional ac circuit analysis.

The relationships between dq components of armature terminal voltage and current are defined by Equations 3.139, 3.140, 3.142 and 3.143. Thus

$$
\begin{aligned}
e_d &= -\omega_r \psi_q - R_a i_d \\
&= \omega_r L_q i_q - R_a i_d \\
&= X_q i_q - R_a i_d
\end{aligned} \tag{3.164}
$$

$$
\begin{aligned}
e_q &= \omega_r \psi_d - R_a i_q \\
&= -X_d i_d + X_{ad} i_{fd} - R_a i_q
\end{aligned} \tag{3.165}
$$

The reactances X_d and X_q are called the direct- and quadrature-axis synchronous reactances, respectively. They represent the inductive effects of the armature mmf wave by separately accounting for its d- and q-axis components. These and other reactances of a synchronous machine will be discussed in detail in a later section.

We have not yet developed a means of identifying the d- and q-axis positions relative to $\tilde{E}_t$. In order to assist us in this regard, let us define a voltage $\tilde{E}_q$ as

$$
\begin{aligned}
\tilde{E}_q &= \tilde{E}_t + (R_a + jX_q)\tilde{I}_t \\
&= (e_d + je_q) + (R_a + jX_q)(i_d + ji_q)
\end{aligned} \tag{3.166}
$$

Substitution of Equations 3.164 and 3.165, followed by reduction of the resulting expression, yields the following expression for $\tilde{E}_q$ in phasor form with d, q axes as reference:

$$\tilde{E}_q = j[X_{ad} i_{fd} - (X_d - X_q) i_d] \tag{3.167}$$

The corresponding phasor diagram is shown in Figure 3.21. We see that the phasor $\tilde{E}_q$ lies along the q-axis. The position of the q-axis with respect to $\tilde{E}_t$ can be identified by computing $\tilde{E}_q$, the voltage behind $R_a + jX_q$.

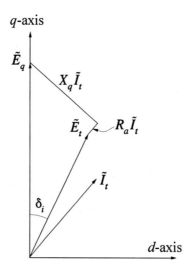

Figure 3.21 Phasor $\tilde{E}_q$ in dq complex plane

3.6.3 Rotor Angle

Under no-load or open-circuit conditions, $i_d=i_q=0$. Substituting in Equations 3.139, 3.140, 3.142 and 3.143 yields

$$\psi_d = L_{ad}i_{fd}$$

$$\psi_q = 0$$

$$e_d = 0$$

$$e_q = X_{ad}i_{fd}$$

Therefore,

$$\tilde{E}_t = e_d + je_q$$

$$= jX_{ad}i_{fd}$$

(3.168)

Under no-load conditions, $\tilde{E}_t$ has only the q-axis component and hence $\delta_i=0$. As the machine is loaded, δ_i increases. Therefore, the angle δ_i is referred to as the *internal rotor angle* or *load angle*. The relationship between power output and the rotor angle is nonlinear and is of fundamental importance in power system stability studies.

The angle δ_i represents the angle by which the q-axis leads the stator terminal voltage phasor $\tilde{E}_t$, and it is given by

$$\delta_i = 90° - (\alpha - \theta_0) \tag{3.169}$$

where α is the phase angle of e_a and θ_0 is the value of θ with respect to the time origin. Therefore, δ_i depends on the angle between the stator and rotor magnetic fields. For any given machine power output, either α or θ_0 may be arbitrarily chosen, but not both.

3.6.4 Steady-State Equivalent Circuit

If saliency is neglected,

$$X_d = X_q = X_s$$

where X_s is the synchronous reactance. Therefore,

$$\tilde{E}_q = \tilde{E}_t + (R_a + jX_s)\tilde{I}_t \tag{3.170}$$

With $X_d = X_q$, from Equation 3.167, the magnitude of $\tilde{E}_q$ is given by

$$E_q = X_{ad}i_{fd} \tag{3.171}$$

The corresponding equivalent circuit is shown in Figure 3.22. The resistance R_a is usually very small and may be neglected.

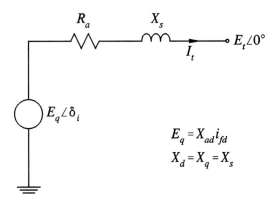

Figure 3.22 Steady-state equivalent circuit with saliency neglected

The voltage E_q may be considered as the effective internal voltage. It is equal in magnitude to $X_{ad}i_{fd}$ and hence represents the excitation voltage due to the field current. The synchronous reactance X_s accounts for the flux produced by the stator currents, i.e., the effect of armature reaction. For a round rotor machine, X_d is nearly equal to X_q and therefore the above equivalent circuit provides a satisfactory representation.

For salient pole machines, X_d is not equal to X_q. The effect of saliency is, however, not very significant so far as the relationships between terminal voltage, armature current, power and excitation over the normal operating range are concerned. The approximate equivalent often provides sufficient insight into the steady-state characteristics. Only at small excitations will the effect of saliency become significant. The approximation also neglects the reluctance torque due to saliency. With modern computing facilities, there is little difficulty in accounting for saliency; therefore, the approximation associated with round rotor theory is not used in detailed calculations.

Active and reactive power

$$S = \tilde{E}_t \tilde{I}_t^*$$

$$= (e_d + je_q)(i_d - ji_q)$$

$$= (e_d i_d + e_q i_q) + j(e_q i_d - e_d i_q)$$

$$P_t = e_d i_d + e_q i_q \tag{3.172}$$

$$Q_t = e_q i_d - e_d i_q \tag{3.173}$$

Steady-state torque is given by

$$T_e = \psi_d i_q - \psi_q i_d$$

$$= (e_d i_d + e_q i_q) + R_a(i_d^2 + i_q^2) \tag{3.174}$$

$$= P_t + R_a I_t^2$$

3.6.5 Procedure for Computing Steady-State Values

For stability analysis, it is necessary to find the initial steady-state values of machine variables as a function of specified terminal quantities. The following steps summarize the procedure for computing these values. It is assumed that all quantities are expressed in per unit.

(a)   Normally, terminal active power P_t, reactive power Q_t, and magnitude of voltage E_t are specified. The corresponding terminal current I_t and power factor angle ϕ are computed as follows:

$$I_t = \frac{\sqrt{P_t^2 + Q_t^2}}{E_t}$$

$$\phi = \cos^{-1}\left(\frac{P_t}{E_t I_t}\right)$$

(b)   The next step is to compute the internal rotor angle δ_i. Since $\tilde{E}_q$ lies along the q-axis, as illustrated in Figure 3.23, the internal angle is given by

$$\delta_i = \tan^{-1}\left(\frac{X_q I_t \cos\phi - R_a I_t \sin\phi}{E_t + R_a I_t \cos\phi + X_q I_t \sin\phi}\right)$$

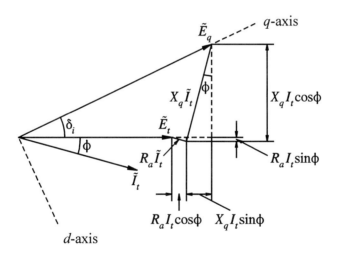

Figure 3.23 Steady-state phasor diagram

(c)   With δ_i known, the *dq* components of stator voltage and current are given by

$$e_d = E_t \sin\delta_i$$

$$e_q = E_t \cos\delta_i$$

$$i_d = I_t \sin(\delta_i + \phi)$$

$$i_q = I_t \cos(\delta_i + \phi)$$

(d)   The remaining machine quantities are computed as follows:

$$\psi_d = e_q + R_a i_q$$

$$\psi_q = -e_d - R_a i_d$$

$$i_{fd} = \frac{e_q + R_a i_q + X_d i_d}{X_{ad}}$$

$$e_{fd} = R_{fd} i_{fd}$$

$$\psi_{fd} = (L_{ad} + L_{fd}) i_{fd} - L_{ad} i_d$$

$$\psi_{1d} = L_{ad}(i_{fd} - i_d)$$

$$\psi_{1q} = \psi_{2q} = -L_{aq} i_q$$

$$T_e = P_t + R_a I_t^2$$

So far we have not discussed the effect of magnetic saturation. It is, however, important to recognize that the inductances L_{ad}, L_{aq}, L_d, L_q and the corresponding reactances vary with saturation and should be accounted for in computing the steady-state machine quantities. The method of accounting for saturation will be discussed in Section 3.8.

Example 3.2

The following are the parameters in per unit on machine rating of a 555 MVA, 24 kV, 0.9 p.f., 60 Hz, 3600 RPM turbine-generator [1]:

$L_{ad} = 1.66$	$L_{aq} = 1.61$	$L_l = 0.15$	$R_a = 0.003$
$L_{fd} = 0.165$	$R_{fd} = 0.0006$	$L_{1d} = 0.1713$	$R_{1d} = 0.0284$
$L_{1q} = 0.7252$	$R_{1q} = 0.00619$	$L_{2q} = 0.125$	$R_{2q} = 0.02368$

L_{fkd} is assumed to be equal to L_{ad}.

(a)   When the generator is delivering rated MVA at 0.9 p.f. (lag) and rated terminal voltage, compute the following:

---

[1] This generator is same as that of Example 3.1, except that amortisseurs are considered here. We will consider this generator in many other examples throughout the book to illustrate different aspects of generator characteristics and stability performance.

(i)   Internal angle δ_i in electrical degrees

(ii) Per unit values of e_d, e_q, i_d, i_q, i_{1d}, i_{1q}, i_{2q}, i_{fd}, e_{fd}, ψ_{fd}, ψ_{1d}, ψ_{1q}, ψ_{2q}

(iii) Air-gap torque T_e in per unit and in newton-meters

Assume that the effect of magnetic saturation at the given operating condition is to reduce L_{ad} and L_{aq} to 83.5% of the values given above.

(b)   Compute the internal angle δ_i and field current i_{fd} for the above operating condition, using the approximate equivalent circuit of Figure 3.22. Neglect R_a.

Solution

(a) With the given operating condition, the per unit values of terminal quantities are

$$P = 0.9, \qquad Q = 0.436, \qquad E_t = 1.0, \qquad I_t = 1.0, \qquad \phi = 25.84°$$

The saturated values of the inductances are

$$L_{ad} = 0.835 \times 1.66 = 1.386$$

$$L_{aq} = 0.835 \times 1.61 = 1.344$$

$$L_d = L_{ad} + L_l = 1.386 + 0.15 = 1.536$$

$$L_q = L_{aq} + L_l = 1.344 + 0.15 = 1.494$$

Following the procedure outlined in Section 3.6.5,

(i) $\delta_i = \tan^{-1}\left(\dfrac{1.494 \times 1.0 \times 0.9 - 0.003 \times 1.0 \times 0.436}{1.0 + 0.003 \times 1.0 \times 0.9 + 1.494 \times 1.0 \times 0.436} \right)$

$\qquad = \tan^{-1}(0.812) = 39.1 \qquad$ electrical degrees

(ii) $e_d = E_t \sin\delta_i = 1.0\sin 39.1 = 0.631 \quad$ pu

$\qquad e_q = E_t \cos\delta_i = 1.0\cos 39.1 = 0.776 \quad$ pu

$\qquad i_d = I_t \sin(\delta_i + \phi) = 1.0\sin(39.1 + 25.84) = 0.906 \quad$ pu

$\qquad i_q = I_t \cos(\delta_i + \phi) = 1.0\cos(39.1 + 25.84) = 0.423 \quad$ pu

$\qquad i_{fd} = \dfrac{e_q + R_a i_q - X_d i_d}{X_{ad}}$

$\qquad\quad = \dfrac{0.776 + 0.003 \times 0.423 + 1.536 \times 0.906}{1.386}$

$\qquad\quad = 1.565 \quad$ pu

$$e_{fd} = R_{fd}i_{fd} = 0.0006 \times 1.565$$

$$= 0.000939 \quad \text{pu}$$

$$\psi_{fd} = (L_{ad}+L_{fd})i_{fd}-L_{ad}i_d$$

$$= (1.386+0.165) \times 1.565 - 1.386 \times 0.907$$

$$= 1.17 \quad \text{pu}$$

$$\psi_{1d} = L_{ad}(i_{fd}-i_d)$$

$$= 1.386 \times (1.565-0.906)$$

$$= 0.913 \quad \text{pu}$$

$$\psi_{1q} = \psi_{2q} = -L_{aq}i_q = -1.344 \times 0.423$$

$$= -0.569 \quad \text{pu}$$

Under steady state,

$$i_{1d} = i_{1q} = i_{2q} = 0$$

(iii) Air-gap torque

$$T_e = P_t+I_t^2 R_a$$

$$= 0.9+1.0^2 \times 0.003$$

$$= 0.903 \quad \text{pu}$$

$$T_{base} = \frac{\text{MVA}_{base} \times 10^6}{\omega_{m\,base}}$$

$$= \frac{555 \times 10^6}{2\pi 60} = 1.472 \times 10^6 \quad \text{N·m}$$

Therefore,

$$T_e = 0.903 \times 1.472 \times 10^6$$

$$= 1.329 \times 10^6 \quad \text{N·m}$$

(b) Using the saturated value of X_{ad},

$$E_q = X_{ad}i_{fd} = 1.386i_{fd}$$

and

$$X_s = X_{ad} + X_l = 1.386 + 0.15 = 1.536$$

From the equivalent circuit of Figure 3.22, with $\tilde{E}_t$ as reference phasor,

$$\tilde{E}_q = \tilde{E}_t + jX_s\tilde{I}_t$$

$$= 1.0 + j1.536(0.9 - j0.436)$$

$$= 1.670 + j1.382$$

$$= 2.17\angle 39.6° \quad \text{pu}$$

$$\delta_i = 39.6°$$

Therefore,

$$i_{fd} = \frac{E_q}{X_{ad}} = \frac{2.17}{1.386} = 1.566 \quad \text{pu}$$

The values of δ_i and i_{fd} computed using the approximate representation are seen to be in good agreement with the accurate calculation. This is to be expected, since X_q is nearly equal to X_d and we are considering rated operating condition.   ■

3.7 ELECTRICAL TRANSIENT PERFORMANCE CHARACTERISTICS

This section examines the fundamental electrical transient characteristics of a synchronous machine by considering the response to a three-phase short-circuit at the terminals. Such an analysis, in addition to providing insight into the machine transient performance, is useful in identifying some of the approximations necessary for its representation in large scale stability studies. We will first consider a simple *RL* circuit, as it helps in understanding of the nature of transient response of a synchronous machine.

3.7.1 Short-Circuit Current in a Simple *RL* Circuit

Consider the *RL* circuit shown in Figure 3.24, with

$$e = E_m\sin(\omega t + \alpha) \tag{3.175}$$

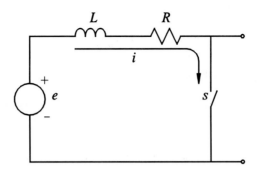

Figure 3.24 *RL* circuit

If the switch *s* is closed at time $t=0$, the current *i* is given by

$$e = L\frac{di}{dt} + Ri \qquad (3.176)$$

Solving for *i*,

$$i = Ke^{-\frac{R}{L}t} + \frac{E_m}{Z}\sin(\omega t + \alpha - \phi) \qquad (3.177)$$

with

$$Z = \sqrt{R^2 + \omega^2 L^2}$$

$$\phi = \tan^{-1}\left(\frac{\omega L}{R}\right)$$

In Equation 3.177, *K* is such that *i* at $t=0^+$ is the same as that at $t=0^-$. If *i* is equal to i_0 at $t=0^-$,

$$K = i_0 - \frac{E_m}{Z}\sin(\alpha - \phi) \qquad (3.178)$$

From Equation 3.177 the short circuit current has two components: a transient unidirectional component and a steady-state alternating component. The presence of the unidirectional component of the short circuit current ensures that the current does not change instantaneously. The unidirectional component decays to zero with a time constant of *L/R*.

3.7.2 Three-Phase Short-Circuit at the Terminals of a Synchronous Machine

If a bolted three-phase fault is suddenly applied to a synchronous machine, the three phase currents are as shown in Figure 3.25.

In general, the fault current in each phase has two distinct components:

(a)   A fundamental frequency component, which decays initially very rapidly (in a few cycles) and then relatively slowly (in several seconds) to a steady-state value.

(b)   A unidirectional component (or a dc offset), which decays exponentially in several cycles.

This is similar to the short-circuit current in the case of the simple *RL* circuit considered in the previous section. However, in the case of a synchronous machine the amplitude of the ac component is not constant because the internal voltage, which is a function of the rotor flux linkages, is not constant. The initial rapid decay of the

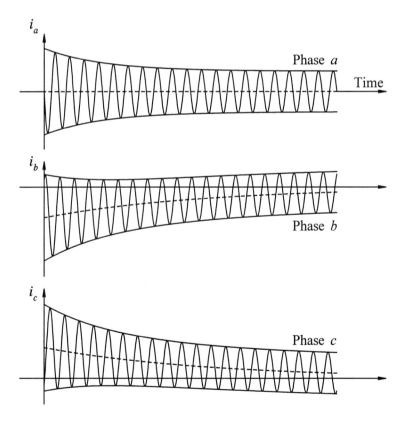

Figure 3.25 Short-circuit currents in the three phases

ac component of the short circuit current is due to the rapid decay of flux linking the subtransient circuits (*1d* and *2q* in Figures 3.18 and 3.19). The slowly decaying part of the ac component is due to the relatively slow decay of flux linking the transient circuits (field and *1q*).

The dc components have different magnitudes in the three phases and decay with a time constant T_a, the armature time constant. This time constant is equal to the ratio of the effective armature inductance (with the unidirectional currents in the armature) to the armature resistance.

In addition to the fundamental frequency and dc components, the short circuit armature currents contain second harmonic components, which depend on subtransient saliency $(X''_q - X''_d)$ [10]. The amplitudes of these components are very small and are usually of little significance.

The currents during short-circuits (either balanced or unbalanced) or any other disturbance can be computed by solving Equations 3.26 to 3.51 in terms of the phase (*abc*) variables, or the corresponding Equations 3.120 to 3.133 in terms of the transformed *dq0* variables. The fundamental frequency components of phase currents are reflected as unidirectional components in the transformed currents i_d and i_q. The dc offset associated with the phase currents is reflected as fundamental frequency components in i_d and i_q.

The field current following the short-circuit is shown in Figure 3.26. It consists of a unidirectional component and an alternating component, corresponding to the ac component and the dc component in the armature phase, respectively.

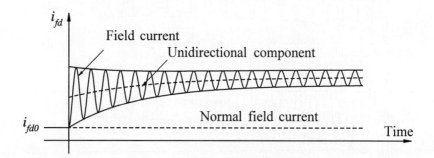

Figure 3.26 Field current response following a stator short-circuit

3.7.3 Elimination of DC Offset in Short-Circuit Current

The analysis of power system performance with the effects of both fundamental frequency and unidirectional components of phase currents included would be complex and computationally very involved. For many classes of power system problems, computation is much simpler if the effects of dc offset in the phase current are either neglected or treated separately. This also makes it easier to distinguish between the important and the unimportant factors influencing the dynamic performance of power systems.

The effects of dc offset in the stator phase currents may be eliminated by neglecting the transformer voltage terms ($p\psi_d$, $p\psi_q$) in the stator voltage equations 3.120 and 3.121:

$$e_d = p\psi_d - \omega_r\psi_q - R_a i_d$$

$$e_q = p\psi_q + \omega_r\psi_d - R_a i_q$$

The transformer voltage terms represent the stator transients and prevent ψ_d and ψ_q from changing instantaneously. It is this fact that produces the dc offset in the phase currents. Omission of $p\psi_d$, $p\psi_q$ terms would therefore eliminate the dc offset and its related effects on the dynamic performance of the machine. By neglecting the $p\psi$ terms, we are not assuming that ψ_d and ψ_q remain constant; on the contrary, we are assuming that they change instantly following a perturbation.

If the stator transients (accounted for by the $p\psi$ terms) are neglected, the resulting armature short circuit current is as shown in Figure 3.27.

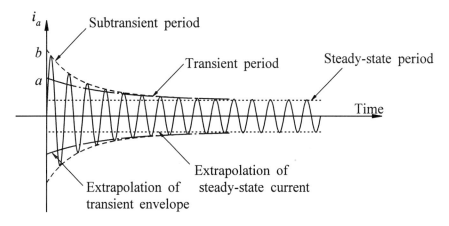

Figure 3.27 Fundamental frequency component of armature current

We see that the resulting current consists of only the fundamental frequency component. The waveform of the current may be divided into three distinct time periods: the *subtransient period*, lasting only for the first few cycles, during which the amplitude decays rapidly; the *transient period*, spanning a longer time, during which the amplitude decays considerably more slowly; and finally the *steady-state period*, during which the amplitude of the current remains constant. The parameters of the synchronous machine that determine the amplitudes of the short-circuit waveform during the three time periods and the rates of decay during the first two periods will be discussed in the next section.

To summarize the results of this section, we have discussed the response of the electrical circuits associated with a synchronous machine by considering a three-phase short circuit. We have identified the significance of neglecting the transformer voltage terms ($p\psi_d$, $p\psi_q$) in the stator voltage equations. This provides valuable guides to the representation of synchronous machines in power system analysis. It is also helpful in understanding the basis for and the significance of parameters widely used to identify the machine parameters.

The need for and the effects of neglecting the transformer voltage terms in power system stability studies will be discussed in Chapter 5.

3.8 MAGNETIC SATURATION

In the development of basic equations of the synchronous machine and the analysis of its characteristics so far, we have ignored the effects of stator and rotor iron saturation. As noted in Section 3.2, this was done to make the analysis simple and manageable. A rigorous treatment of synchronous machine performance including saturation effects is a futile exercise. Any practical method of accounting for saturation effects must be based on semi-heuristic reasoning and judiciously chosen approximations, with due consideration to simplicity of model structure, data availability, and accuracy of results.

Before we discuss the methods of representing saturation in stability studies, it is useful to briefly review the characteristics of synchronous machines with stator terminals open and shorted.

3.8.1 Open-Circuit and Short-Circuit Characteristics

Magnetic circuit data essential to the treatment of saturation are given by the open-circuit characteristic.

Under no-load rated speed conditions, as seen in Section 3.6.3, we have

$$i_d = i_q = \psi_q = e_d = 0$$

and

$$E_t = e_q = \psi_d = L_{ad}i_{fd}$$

Therefore, the open-circuit characteristic (OCC) relating E_t and i_{fd} gives the saturation characteristic of the d-axis.

A typical OCC is shown in Figure 3.28. The straight line tangent to the lower part of the curve is the *air-gap line* indicating the field current (or mmf) required to overcome the reluctance of the air-gap. The departure of the OCC from the air-gap line is an indication of the degree of saturation in the rotor and stator iron.

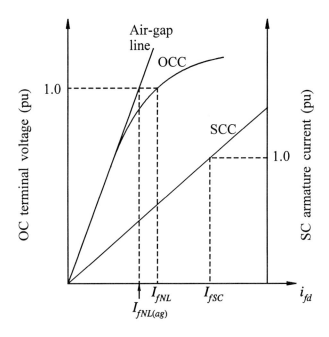

Figure 3.28 Open-circuit and short-circuit characteristics

The short-circuit characteristic (SCC) is also shown in Figure 3.28. This is a plot of the armature current vs. the field current, with the generator operating at rated speed in the steady state and with a three-phase short-circuit placed on the armature terminals. The SCC is linear up to and well beyond the rated armature current, since there is very little or no saturation in the iron under rated short-circuit conditions due to the demagnetizing effect of the armature reaction.

From Figure 3.22, it can be seen that the internally generated voltage is equal to the product of the short-circuit current and synchronous reactance (R_a is much much smaller than X_s and hence can be neglected). When the short-circuit current is one per unit, this voltage is proportional to I_{fSC} as shown in Figure 3.28. Hence,

$$KI_{fSC} = 1.0X_{s(unsat)}$$

Corresponding to the one per unit voltage on the air-gap line,

$$KI_{fNL(ag)} = 1.0$$

Therefore, the unsaturated value of X_s is given by

$$X_{s(unsat)} = \frac{I_{fSC}}{I_{fNL(ag)}} \tag{3.179}$$

The saturated value of X_s corresponding to rated voltage is given by

$$X_{s(sat)} = \frac{I_{fSC}}{I_{fNL}} \tag{3.180}$$

where I_{fNL} and I_{fSC} are the values of field current required to give rated terminal voltage on the OCC and rated armature current on the SCC, respectively.

The *short-circuit ratio* (SCR) is defined as the ratio of the field current required to produce rated voltage at rated speed and no load to the field current required to produce rated armature current under a steady three-phase short-circuit condition. That is,

$$SCR = \frac{I_{fNL}}{I_{fSC}} = \frac{1}{X_{s(sat)}} \tag{3.181}$$

If there were no saturation, the SCR would be equal to the reciprocal of the unsaturated value of synchronous reactance. The SCR reflects the degree of saturation and therefore has significance with respect to both the performance of the machine and its cost. A lower SCR is indicative of a larger change in field current required to maintain constant terminal voltage for a given change in load. Therefore, a machine with a low SCR requires an excitation system that is able to provide large changes of field current to maintain system stability.

On the other hand, when the SCR is lower, the size, weight and cost of the machine are lower. With improvements in excitation systems and the associated controls, there has been a trend toward the use of generators of lower SCR and, consequently, lower cost.

3.8.2 Representation of Saturation in Stability Studies

In the representation of magnetic saturation for stability studies, the following assumptions are usually made:

(a)   The leakage inductances are independent of saturation. The leakage fluxes are in air for a considerable portion of their paths so that they are not significantly affected by saturation of the iron portion. As a result, the only elements of the equivalent circuits of Figure 3.18 that saturate are the mutual inductances L_{ad} and L_{aq}.

(b)   The leakage fluxes do not contribute to the iron saturation. The leakage fluxes
        are usually small and their paths coincide with that of the main flux for only
        a small part of its path. By this assumption, saturation is determined by the air-
        gap flux linkage.

(c)   The saturation relationship between the resultant air-gap flux and the mmf
        under loaded conditions is the same as under no-load conditions. This allows
        the saturation characteristics to be represented by the open-circuit saturation
        curve, which is usually the only saturation data readily available.

(d)   There is no magnetic coupling between the d- and q-axes as a result of
        nonlinearities introduced by saturation; i.e., currents in the windings of one
        axis do not produce flux that link with the windings of the other axis.

With the above assumptions, the effects of saturation may be represented as

$$L_{ad} = K_{sd}L_{adu} \qquad (3.182)$$

$$L_{aq} = K_{sq}L_{aqu} \qquad (3.183)$$

where L_{adu} and L_{aqu} are the unsaturated values of L_{ad} and L_{aq}. The saturation factors
K_{sd} and K_{sq} identify the degrees of saturation in the d- and q-axis, respectively. We
will first discuss how d-axis saturation is represented and then consider saturation of
the q-axis.

        According to assumption (c) above, the degree of d-axis saturation is
determined from the OCC. Referring to Figure 3.29, for an operating point defined
by point "a" on the OCC, the saturation factor K_{sd} is given by

$$K_{sd} = \frac{\psi_{at}}{\psi_{at0}} \qquad (3.184)$$

It can be shown by simple proportion that K_{sd} is also given by

$$K_{sd} = \frac{I_0}{I} \qquad (3.185)$$

        In our representation of saturation, we will use the expression given by
Equation 3.184. This gives the degree of saturation for any specified value of air-gap
flux linkage or voltage.
        Our next step is to identify a convenient way of mathematically representing
the deviation of OCC from the air-gap line.

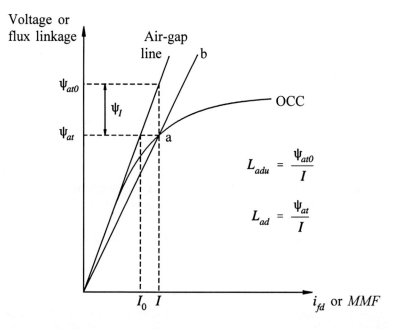

Figure 3.29 Open-circuit characteristic showing effects of saturation

Defining

$$\psi_I = \psi_{at0} - \psi_{at} \tag{3.186}$$

the expression for the saturation factor becomes

$$K_{sd} = \frac{\psi_{at}}{\psi_{at} + \psi_I} \tag{3.187}$$

The saturation curve may be divided into three segments: unsaturated segment I, nonlinear segment II, and fully saturated linear segment III. The threshold values ψ_{T1} and ψ_{T2} define the boundaries of the three segments as shown in Figure 3.30.

For segment I defined by $\psi_{at} \leq \psi_{T1}$

$$\psi_I = 0 \tag{3.188}$$

For segment II defined by $\psi_{T1} < \psi_{at} \leq \psi_{T2}$, ψ_I can be expressed by a suitable mathematical function. Here we will use an exponential function:

$$\psi_I = A_{sat} e^{B_{sat}(\psi_{at} - \psi_{T1})} \tag{3.189}$$

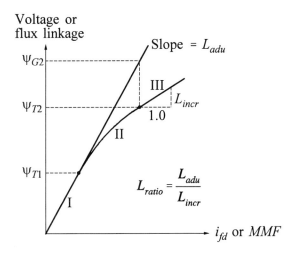

Figure 3.30 Representation of saturation characteristic

where A_{sat} and B_{sat} are constants depending on the saturation characteristic in the segment II portion.

When $\psi_{at}=\psi_{T1}$, from Equation 3.189, $\psi_I=A_{sat}$. Hence, this representation results in a small discontinuity at the junction of segments I and II. However, A_{sat} is normally very small and the discontinuity is inconsequential.

For segment III defined by $\psi_{at}>\psi_{T2}$,

$$\psi_I = \psi_{G2}+L_{ratio}(\psi_{at}-\psi_{T2})-\psi_{at} \tag{3.190}$$

where L_{ratio}, as defined in Figure 3.30, is the ratio of the slope of the air-gap line to the incremental slope of segment III of the OCC.

With the above method of representation, the saturation characteristic for any given machine is completely specified by ψ_{T1}, ψ_{T2}, ψ_{G2}, A_{sat}, B_{sat} and L_{ratio}.

The value of K_{sd}, for any given operating condition, is computed as a function of the corresponding air-gap flux linkage given by

$$\psi_{at} = \sqrt{\psi_{ad}^2+\psi_{aq}^2} \tag{3.191}$$

where ψ_{ad} and ψ_{aq} are the *d*- and *q*-axis components of air-gap or mutual flux linkages, identified in Figure 3.31.

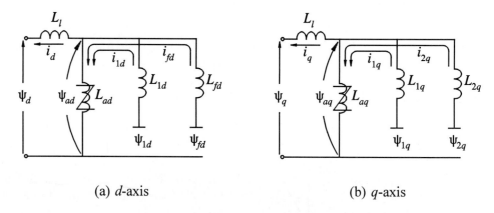

(a) *d*-axis   (b) *q*-axis

Figure 3.31 Equivalent circuits identifying nonlinear
elements and air-gap flux linkages

The *d*- and *q*-axis air-gap flux linkages are given by

$$\psi_{ad} \;=\; \psi_d + L_l i_d \;=\; (e_q + R_a i_q) + L_l i_d \tag{3.192}$$

$$\psi_{aq} \;=\; \psi_q + L_l i_q \;=\; (-e_d - R_a i_d) + L_l i_q \tag{3.193}$$

Therefore, ψ_{at} in per unit is equal to the air-gap voltage

$$\tilde{E}_a \;=\; \tilde{E}_t + (R_a + jX_l)\tilde{I}_t \tag{3.194}$$

The saturation factor K_{sd} can thus be determined, for any given values of terminal voltage and current, by first computing $\tilde{E}_a$ and then using Equations 3.187 and 3.188, 3.189 or 3.190.

For salient pole machines, because the path for *q*-axis flux is largely in air, L_{aq} does not vary significantly with saturation of the iron portion of the path. Therefore, K_{sq} is assumed to be equal to 1.0 for all loading conditions.

In the case of round rotor machines, there is magnetic saturation in both axes. The saturation factor K_{sq} can be determined from the no-load saturation characteristic of the *q*-axis. However, *q*-axis saturation data is usually not available; hence, K_{sq} is assumed to be equal to K_{sd}. This is equivalent to assuming that the reluctance of the magnetic path is homogeneous around the rotor periphery. Improved saturation modelling using *q*-axis saturation characteristics derived from finite-element analysis or tests on loaded machines is discussed in the following section.

3.8.3 Improved Modelling of Saturation

The method of representing saturation described in the previous section is based on a number of idealizing assumptions. For accurate representation under all loading conditions these assumptions are not strictly valid. Several improvements in saturation modelling have been proposed in the literature to overcome these limitations. These include consideration of the following effects:

(a)   Power and load angle dependency of the direct and quadrature axis saturation [4,7,22]. This variation occurs because of higher saturation of rotor iron under load and variation in permeance in the air-gap with load angle. The mutual fluxes in the stator at no-load and loaded conditions are similar because the stator voltage is close to rated value under both conditions. On the other hand, due to large excitation under loaded conditions, particularly when overexcited, the rotor flux is considerably more dense and hence saturation is considerably higher than under no-load. For a given stator terminal voltage and active power output, L_{ad} is smaller when overexcited and larger when underexcited. In contrast, L_{aq} is smaller when underexcited since the corresponding load angle is higher. Accurate representation of saturation effects including power and load angle dependency requires a significant amount of computational effort, which cannot be justified for stability studies.

(b)   Cross coupling between d- and q-axes [6,9,23]. Due to the nonlinearity introduced by saturation, the permeability pattern is not symmetric around the d-axis. This results in dissymmetry in flux linkages; that is, d-axis currents produce q-axis flux linkages and vice versa. The cross coupling phenomenon in fact invalidates the fundamental assumption on which Park's $dq0$ transformation is based. However, the use of a rigorous mathematical model of a synchronous machine including nonlinear effects is neither practical nor justified. The cross-coupling effects are of a secondary nature and may be represented approximately or neglected altogether.

(c)   Quadrature-axis saturation [5,6,9,24]. For round rotor machines, experience has shown that the q-axis saturates appreciably more than the d-axis. This is due to the presence of rotor teeth in the magnetic path of the q-axis. For accurate prediction of steady-state rotor angle and field excitation, it would be necessary to account for the q-axis saturation characteristic as distinct from the d-axis characteristic. The errors introduced by assuming the q-axis saturation to be same as the d-axis open-circuit curve are shown in Figure 3.32. The figure shows the differences between measured and computed values of rotor angle and field current as a function of reactive power output of a 500 MW generator at the Lambton coal-fired generating station in Ontario [5]. The error in field current is greater in the overexcited region with the highest error being on the order of 4%; the error in rotor angle is higher in the underexcited region with the highest value being as high as 10°.

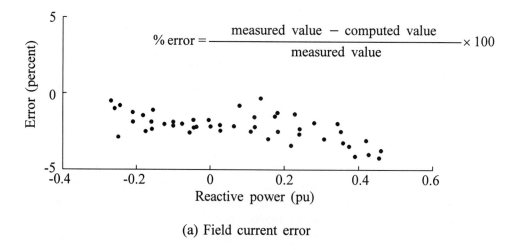

(a) Field current error

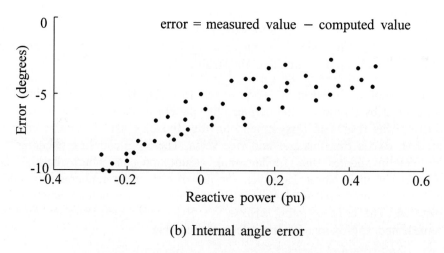

(b) Internal angle error

Figure 3.32 Field current and internal angle errors
with conventional saturation representation

The *q*-axis saturation characteristic is not readily available from the generator manufacturers. It can, however, be easily determined from steady-state measurements of field current and rotor angle at different values of terminal voltage, active and reactive power output. These measurements also provide the *d*-axis saturation characteristics under load, which should be more accurate than using the open-circuit saturation curves. Figure 3.33 shows the *d*- and *q*-axis saturation characteristics derived from steady-state measurements on the Lambton unit [5].

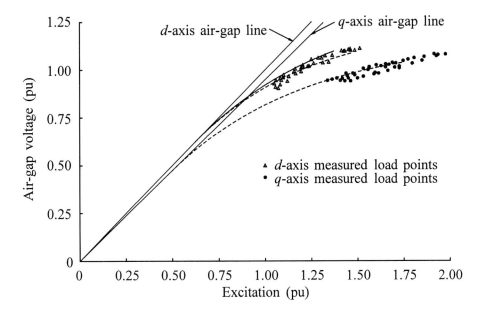

Figure 3.33 Lambton saturation curves derived from steady-state
field current and rotor angle measurements

An accurate method of saturation representation including the effects of cross coupling and q-axis saturation, based on finite element analysis, is presented in references 6 and 9. The overall effect of this improved representation on system stability is assessed in reference 8. The beneficial effect of using such a representation appears to be only marginal for the case investigated.

In general, very complex saturation models based on specially developed data are not justified for stability studies. The approach described in Section 3.8.2 using readily available data should be adequate in most situations. In critical cases, the best approach is to use d- and q-axis saturation characteristics based on simple measurement under steady-state on-load conditions, such as those shown in Figure 3.33.

Use of Potier reactance[1]

The voltage E_p behind the Potier reactance (X_p) is often used in place of E_a to identify the saturation level. The use of X_p, instead of X_l, is believed to make an empirical allowance for the difference between the load saturation and open-circuit

---

[1] For a description of the Potier reactance and its physical significance, see references 12 and 14.

characteristic. An important factor causing the discrepancy between the load and no-load saturation is the difference in the field leakage flux under the two conditions. The Potier reactance accounts for this difference and is higher than the stator leakage reactance due to the appreciably higher field leakage under loaded conditions.

The use of E_p, because of the manner in which X_p is computed, gives accurate values of field current under steady-state loaded conditions. However, its use to represent saturation under transient conditions, when the amortisseur effects are included, leads to unrealistic results in some situations. Its use in stability studies is, therefore, not recommended.

Example 3.3

The open-circuit saturation curve of the 555 MVA generator considered in Examples 3.1 and 3.2 is shown in Figure E3.1. The per unit resistance and inductances associated with the stator circuits are as follows:

$$R_a = 0.003 \qquad L_l = 0.15 \qquad L_{adu} = 1.66 \qquad L_{aqu} = 1.61$$

(a)   If the field current required to generate rated E_t on the air-gap line is 1300 amperes, determine the base values of the field current and field voltage. Compare with the values computed in Example 3.1.

(b)   If segment II of the saturation curve is to be represented by the function defined by Equation 3.189, determine the constants A_{sat} and B_{sat}.

(c)   With the armature terminal voltage at rated value, for each of the following generator output conditions expressed in per unit of the MVA rating:

(i) $P_t = 0$, $Q_t = 0$
(ii) $P_t = 0.4$, $Q_t = 0.2$ (overexcited)
(iii) $P_t = 0.9$, $Q_t = 0.436$
(iv) $P_t = 0.9$, $Q_t = 0$
(v) $P_t = 0.9$, $Q_t = -0.2$ (underexcited)

compute the air-gap voltage E_a, saturation factor K_{sd}, internal rotor angle δ_i, and field current i_{fd}. Assume that the open-circuit curve represents the saturation characteristics of both d- and q-axis. Comment on the effect of reactive power output on δ_i and i_{fd}.

(d)   How would the results of (c) change if the q-axis saturation characteristic differs from that of the d-axis as shown in Figure E3.2?

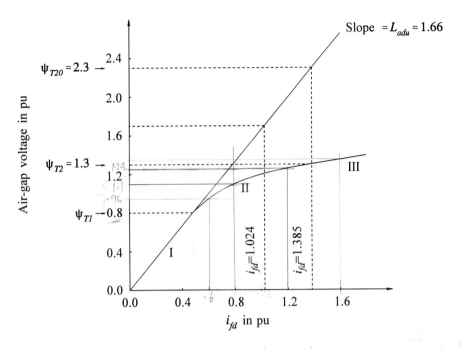

Figure E3.1 Open-circuit saturation curve

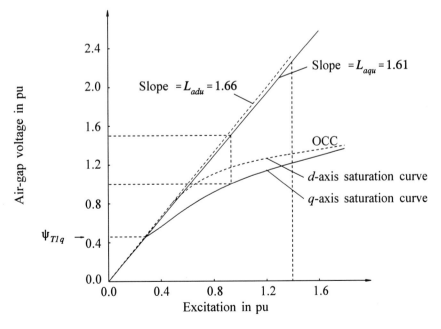

Figure E3.2 The *d*- and *q*-axis saturation characteristics

Solution

(a) As discussed in Section 3.8.1, under no-load rated speed conditions,

$$E_t = e_q = L_{ad}i_{fd}$$

The slope of the air-gap line represents the unsaturated value of L_{ad}. Therefore, the field current required to produce 1.0 per unit E_t on the air-gap line is

$$i_{fd} = \frac{1.0}{L_{adu}} \quad \text{per unit}$$

$$= \frac{1.0}{1.66} \quad \text{per unit}$$

Since this is equal to 1300 amperes, the base field current is

$$i_{fd\,base} = 1300 \times 1.66 = 2158 \quad \text{A}$$

$$e_{fd\,base} = \frac{\text{VA rating}}{i_{fd\,base}} = \frac{555 \times 10^6}{2158} \quad \text{V}$$

$$= 257.183 \quad \text{kV}$$

As is to be expected, this agrees with the value computed in Example 3.1.

(b) From Equation 3.189,

$$\psi_I = A_{sat}e^{B_{sat}(\psi_{at}-\psi_{TI})}$$

We will compute A_{sat} and B_{sat} by considering two points on the saturation curve corresponding to $\psi_{at}=1.2$ and 1.3 in Figure E3.1. With $\psi_{at}=1.2$, $\psi_I=1.7-1.2=0.5$, and with $\psi_{at}=1.3$, $\psi_I=2.3-1.3=1.0$. Therefore,

$$0.5 = A_{sat}e^{B_{sat}(1.2-0.8)}$$

and

$$1.0 = A_{sat}e^{B_{sat}(1.3-0.8)}$$

Solving gives

$$A_{sat} = 0.03125$$
$$B_{sat} = 6.931$$

(c)   From the steady-state equations summarized in Section 3.6.5,

$$I_t = \frac{\sqrt{P_t^2 + Q_t^2}}{E_t}$$

$$\phi = \cos^{-1}\left(\frac{P_t}{I_t E_t}\right)$$

$$\tilde{E}_a = \tilde{E}_t + (R_a + jX_l)\tilde{I}_t$$

$$\psi_{at} = E_a$$

From Figure E3.1, ψ_{T1}=0.8 and ψ_{T2}=1.3. For $0.8 < \psi_{at} < 1.3$,

$$\psi_I = 0.03125e^{6.931(\psi_{at} - 0.8)}$$

$$K_{sd} = K_{sq} = \frac{\psi_{at}}{\psi_{at} + \psi_I}$$

$$X_{ad} = K_{sd}X_{adu}; \quad X_d = X_{ad} + X_l$$

$$X_{aq} = K_{sq}X_{aqu}; \quad X_q = X_{aq} + X_l$$

$$\delta_i = \tan^{-1}\left(\frac{X_q I_t \cos\phi - R_a I_t \sin\phi}{E_t + R_a I_t \cos\phi + X_q I_t \sin\phi}\right)$$

$$e_q = E_t \cos\delta_i$$

$$i_d = I_t \sin(\delta_i + \phi)$$

$$i_q = I_t \cos(\delta_i + \phi)$$

$$i_{fd} = \frac{e_q + R_a i_q + X_d i_d}{X_{ad}}$$

Table E3.1 summarizes the results for the given operating conditions.

Table E3.1

P_t	Q_t	E_a (pu)	K_{sd}	δ_i (deg)	i_{fd} (pu)
0	0	1.0	0.889	0	0.678
0.4	0.2	1.033	0.868	25.3	1.016
0.9	0.436	1.076	0.835	39.1	1.565
0.9	0	1.012	0.882	54.6	1.206
0.9	-0.2	0.982	0.899	64.6	1.089

From the results, the reactive power output is seen to have a significant effect on δ_i as well as i_{fd}. The reason for this is readily apparent from the phasor diagrams shown in Figure E3.3. These diagrams are based on the simplified model of Figure 3.22, which neglects saliency.

(d)   Constants approximating the nonlinear portion (segment II) of the q-axis saturation curve of Figure E3.2, determined by considering two points on the curve corresponding $\psi_{at}=1.0$ and 1.2, are

$$A_{satq} = 0.077$$
$$B_{satq} = 3.465$$

The threshold value defining the beginning of segment II is

$$\psi_{T1q} = 0.46$$

Corresponding to any air-gap voltage $E_a=\psi_{at}$, the q-axis saturation factor is

$$K_{sq} = \frac{\psi_{at}}{\psi_{at}+\psi_{Iq}}$$

where

$$\psi_{Iq} = 0.077e^{3.465(\psi_{at}-0.46)}$$

The d-axis saturation representation is same as in (c).

Table E3.2 summarizes the results obtained with distinct d- and q-axis saturation representations.

Table E3.2

P_t	Q_t	E_a (pu)	K_{sd}	K_{sq}	δ_i (deg)	i_{fd} (pu)
0	0	1.0	0.889	0.667	0	0.678
0.4	0.2	1.033	0.868	0.648	21.0	1.013
0.9	0.436	1.076	0.835	0.623	34.6	1.559
0.9	0	1.012	0.882	0.660	47.5	1.194
0.9	-0.2	0.982	0.899	0.676	55.9	1.074

By comparing with the results of Table E3.1, the effect of increased q-axis saturation is seen to be quite significant, particularly in the underexcited operating condition.

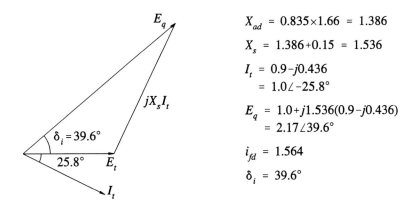

$$X_{ad} = K_{sd}X_{adu}$$

$$E_q = X_{ad}i_{fd}$$

(a) Simplified steady-state model

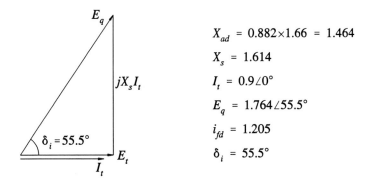

$X_{ad} = 0.835 \times 1.66 = 1.386$

$X_s = 1.386 + 0.15 = 1.536$

$I_t = 0.9 - j0.436$
$\quad = 1.0\angle -25.8°$

$E_q = 1.0 + j1.536(0.9 - j0.436)$
$\quad = 2.17\angle 39.6°$

$i_{fd} = 1.564$

$\delta_i = 39.6°$

(b) Phasor diagram with $P = 0.9$, $Q = 0.436$, $E_t = 1.0$

$X_{ad} = 0.882 \times 1.66 = 1.464$

$X_s = 1.614$

$I_t = 0.9\angle 0°$

$E_q = 1.764\angle 55.5°$

$i_{fd} = 1.205$

$\delta_i = 55.5°$

(c) Phasor diagram with $P = 0.9$, $Q = 0$, $E_t = 1.0$

Figure E3.3 Phasor diagrams showing the effect
of varying Q on i_{fd} and δ_i
(Continued on next page)

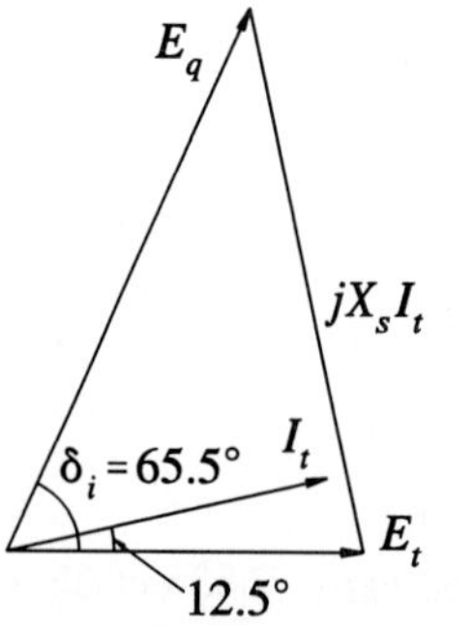

$$X_{ad} = 0.899 \times 1.66 = 1.492$$

$$X_s = 1.64$$

$$I_t = 0.92\angle 12.5°$$

$$E_q = 1.62\angle 65.5°$$

$$i_{fd} = 1.086$$

$$\delta_i = 65.5°$$

(d) Phasor diagram with $P = 0.9$, $Q = -0.2$, $E_t = 1.0$

Figure E3.3 (*Continued*) Phasor diagrams showing
the effect of varying Q on i_{fd} and δ_i

Example 3.4

For the generator of Example 3.3, determine the steady-state relationship between armature current and field current, with the terminal voltage maintained at rated value and with:

(a)   Active power constant at 0, 0.25 pu, 0.5 pu and 0.9 pu

(b)   Power factor constant at 0.6 lead, 0.8 lead, 1.0, 0.8 lag, and 0.6 lag

Assume that the d- and q-axis saturation characteristics are as shown in Figure E3.2. Discuss the characteristics determined.

Solution

(a) With P equal to the specified value, Q is varied between -0.6 and $+0.6$. For each value of Q, the I_t, E_a, K_{sd}, K_{sq}, δ_i, e_q, i_d, i_q, and i_{fd} values are computed as described in Example 3.3. Curves showing the relationship between armature current and field current for $P=0$, 0.25, 0.5 and 0.9 pu are in Figure E3.4. These are known as *V curves* because of their characteristic shape.

(b) With power factor at specified value, I_t is varied between 0.0 and 1.5. The corresponding values of Q, P and i_{fd} are computed using the same approach as in (a). The loci of constant power factor are shown as dashed lines in Figure E3.4. These are commonly referred to as *compounding curves*.

For a given P, armature current I_t is a minimum at unity power factor (UPF). Under such a condition the field provides all the magnetizing current required and the armature current is associated with only active power. The generator neither supplies nor absorbs reactive power.

If the field current is increased beyond the value corresponding to UPF, not all of the field current is required to magnetize the generator. The stator current, which now has a reactive component in addition to the active component, increases and lags the stator terminal voltage. The reactive power output Q is positive and the power factor is lagging. The generator under such a condition is said to be overexcited.

If, on the other hand, the field current is decreased beyond the value corresponding to the UPF condition, the field current is insufficient to magnetize the machine, and the balance of magnetizing current is drawn from the system to which the machine is connected. This inductive component of stator current absorbed by the generator increases as the field current is reduced. The magnitude of the total stator current therefore increases. The machine absorbs Q; i.e., the Q output is negative. The stator current leads the stator voltage and hence the power factor is considered to be leading. The generator under these conditions is said to be underexcited.

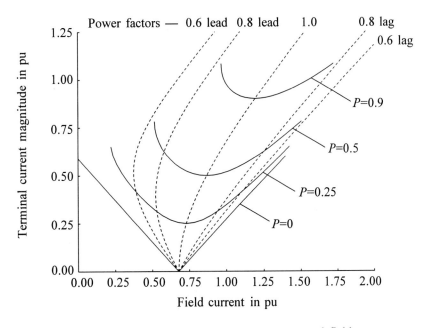

Figure E3.4 Relationship between armature and field current
for varying values of P and power factors

3.9 EQUATIONS OF MOTION

The equations of central importance in power system stability analysis are the rotational inertia equations describing the effect of unbalance between the electromagnetic torque and the mechanical torque of the individual machines. In this section, we will develop these equations in *per unit form* and define parameters that are used to represent mechanical characteristics of synchronous machines in stability studies.

3.9.1 Review of Mechanics of Motion

Before we develop the equations of motion of a synchronous machine, it is useful to review the quantities and relationships associated with the mechanics of motion. These are summarized in Table 3.1. Since it is easier to visualize quantities associated with rotation by analogy with those associated with the more familiar linear motion, the latter are also included in the table.

3.9.2 Swing Equation

As we are introducing new per unit equations and parameters, once again we temporarily resort to the use of superbars to identify per unit quantities.

When there is an unbalance between the torques acting on the rotor, the net torque causing acceleration (or deceleration) is

$$T_a = T_m - T_e \qquad (3.195)$$

where

T_a = accelerating torque in N·m

T_m = mechanical torque in N·m

T_e = electromagnetic torque in N·m

In the above equation, T_m and T_e are positive for a generator and negative for a motor.

The combined inertia of the generator and prime mover is accelerated by the unbalance in the applied torques. Hence, the equation of motion is

$$J\frac{d\omega_m}{dt} = T_a = T_m - T_e \qquad (3.196)$$

where

J = combined moment of inertia of generator and turbine, kg·m^2

ω_m = angular velocity of the rotor, mech. rad/s

t = time, s

Table 3.1

Linear Motion			Rotation		
Quantity	Symbol/ Equation	MKS unit	Quantity	Symbol/ Equation	MKS unit
Length	s	meter (m)	Angular displacement	θ	radian (rad)
Mass	M	kilogram (kg)	Moment of inertia	$J=\int r^2 dm$	kg·m^2
Velocity	$v=ds/dt$	meter/second (m/s)	Angular velocity	$\omega=d\theta/dt$	rad/s
Acceleration	$a=dv/dt$	m/s^2	Angular acceleration	$\alpha=d\omega/dt$	rad/s^2
Force	$F=Ma$	newton (N)	Torque	$T=J\alpha$	newton-meter (N·m) or J/rad
Work	$W=\int F ds$	joule (J)	Work	$W=\int T d\theta$	J, or W·s
Power	$p=dW/dt$ $=Fv$	watt (W)	Power	$p=dW/dt$ $=T\omega$	W

The above equation can be normalized in terms of per unit *inertia constant H*, defined as the kinetic energy in watt-seconds at rated speed divided by the VA base. Using ω_{0m} to denote rated angular velocity in mechanical radians per second, the inertia constant is

$$H = \frac{1}{2} \frac{J\omega_{0m}^2}{VA_{base}} \qquad (3.197)$$

The moment of inertia J in terms of H is

$$J = \frac{2H}{\omega_{0m}^2} VA_{base}$$

Substituting the above in Equation 3.196 gives

$$\frac{2H}{\omega_{0m}^2}VA_{base}\frac{d\omega_m}{dt} = T_m - T_e$$

Rearranging yields

$$2H\frac{d}{dt}\left(\frac{\omega_m}{\omega_{0m}}\right) = \frac{T_m - T_e}{VA_{base}/\omega_{0m}}$$

Noting that $T_{base} = VA_{base}/\omega_{0m}$, the equation of motion in per unit form is

$$2H\frac{d\bar{\omega}_r}{dt} = \bar{T}_m - \bar{T}_e \tag{3.198}$$

In the above equation,

$$\bar{\omega}_r = \frac{\omega_m}{\omega_{0m}} = \frac{\omega_r/p_f}{\omega_0/p_f} = \frac{\omega_r}{\omega_0}$$

where ω_r is angular velocity of the rotor in electrical rad/s, ω_0 is its rated value, and p_f is number of field poles.

If δ is the angular position of the rotor in electrical radians with respect to a synchronously rotating reference and δ_0 is its value at $t=0$,

$$\delta = \omega_r t - \omega_0 t + \delta_0 \tag{3.199}$$

Taking the time derivative, we have

$$\frac{d\delta}{dt} = \omega_r - \omega_0 = \Delta\omega_r \tag{3.200}$$

and

$$\frac{d^2\delta}{dt^2} = \frac{d\omega_r}{dt} = \frac{d(\Delta\omega_r)}{dt}$$

$$= \omega_0\frac{d\bar{\omega}_r}{dt} = \omega_0\frac{d(\Delta\bar{\omega}_r)}{dt} \tag{3.201}$$

Substituting for $d\bar{\omega}_r/dt$ given by the above equation in Equation 3.198, we get

$$\frac{2H}{\omega_0}\frac{d^2\delta}{dt^2} = \bar{T}_m - \bar{T}_e \qquad\qquad (3.202)$$

It is often desirable to include a component of damping torque, not accounted for in the calculation of T_e, separately. This is accomplished by adding a term proportional to speed deviation in the above equation as follows:

$$\frac{2H}{\omega_0}\frac{d^2\delta}{dt^2} = \bar{T}_m - \bar{T}_e - K_D\Delta\bar{\omega}_r \qquad\qquad (3.203)$$

From Equation 3.200,

$$\Delta\bar{\omega}_r = \frac{\Delta\omega_r}{\omega_0} = \frac{1}{\omega_0}\frac{d\delta}{dt}$$

Equation 3.203 represents the equation of motion of a synchronous machine. It is commonly referred to as the *swing equation* because it represents swings in rotor angle δ during disturbances.

Per unit moment of inertia

Substituting in Equation 3.203 gives

$$\frac{2H}{\omega_0}\frac{d^2\delta}{dt^2} = \bar{T}_m - \bar{T}_e - \frac{K_D}{\omega_0}\frac{d\delta}{dt} \qquad\qquad (3.204)$$

In Equations 3.203 and 3.204, K_D is the damping factor or coefficient in pu torque/pu speed deviation.

If it is desired to use per unit value of time $\bar{t}$, Equation 3.204 becomes

$$2H\omega_0\frac{d^2\delta}{d\bar{t}^2} = \bar{T}_m - \bar{T}_e - K_D\frac{d\delta}{d\bar{t}} \qquad\qquad (3.205)$$

Some authors (for example, reference 19) refer to $2H\omega_0$ as the per unit moment of inertia $\bar{J}$.

3.9.3 Mechanical Starting Time

From Equation 3.198,

$$\frac{d\bar{\omega}_r}{dt} = \frac{1}{2H}\bar{T}_a$$

Integrating with respect to time gives

$$\bar{\omega}_r = \frac{1}{2H}\int_0^t \bar{T}_a\, dt \tag{3.206}$$

Let T_M be the time required for rated torque to accelerate the rotor from standstill to rated speed. From Equation 3.206, with $\bar{\omega}_r = 1.0$, $\bar{T}_a = 1.0$ and with the starting value of $\bar{\omega}_r = 0$, we have

$$1.0 = \frac{1}{2H}\int_0^{T_M} 1.0\, dt = \frac{T_M}{2H}$$

Therefore,

$$T_M = 2H \qquad \text{s}$$

and T_M is called the *mechanical starting time.* The symbol M is also used in the literature to denote this time.

3.9.4 Calculation of Inertia Constant

As defined in Section 3.9.2, the inertia constant is given by

$$H = \frac{\text{stored energy at rated speed in MW·s}}{\text{MVA rating}}$$

Calculation of H from moment of inertia in MKS units

Stored energy = kinetic energy

$$= \frac{1}{2}J\omega_{0m}^2 \quad \text{W·s}$$

$$= \frac{1}{2}J\omega_{0m}^2 \times 10^{-6} \quad \text{MW·s}$$

where

$$J \quad = \text{ moment of inertia in kg·m}^2$$
$$\omega_{0m} = \text{ rated speed in mech. rad/s}$$
$$\quad = 2\pi \frac{RPM}{60}$$

Therefore,

$$
\begin{aligned}
H &= \frac{1}{2} \frac{J\omega_{0m}^2 \times 10^{-6}}{\text{MVA rating}} \\
&= \frac{1}{2} \frac{J(2\pi RPM/60)^2 \times 10^{-6}}{\text{MVA rating}} \\
&= 5.48 \times 10^{-9} \frac{J(RPM)^2}{\text{MVA rating}}
\end{aligned}
\tag{3.207}
$$

Calculation of H from WR² in English units

Sometimes the moment of inertia of the rotor is given in terms of WR^2, which is equal to the weight of rotating parts multiplied by the square of radians of gyration in lb·ft². Then, moment of inertia in slug·ft²=WR^2/32.2.

The following relationship between MKS units and English units is useful in converting WR^2 to J:

$$1 \text{ m} \quad = 3.281 \text{ ft}$$
$$1 \text{ kg} \quad = 2.205 \text{ lb (mass)} = 0.0685 \text{ slug}$$
$$1 \text{ slug·ft}^2 = \frac{1}{0.0685 \times 3.281^2} = 1.356 \text{ kg·m}^2$$

The moment of inertia J in kg·m² is related to WR^2 as follows:

$$J = \frac{WR^2}{32.2} \times 1.356$$

Substituting the above expression for J in Equation 3.207 gives

$$H = \frac{5.48\times10^{-9}\times1.356(WR^2)(RPM)^2}{MVA\ rating\times32.2}$$

$$= \frac{2.31\times10^{-10}(WR^2)(RPM)^2}{MVA\ rating}\quad \text{in MW·s/MVA}$$

(3.208)

Typical values of H

Table 3.2 gives the normal range within which the inertia constant H lies, for thermal and hydraulic generating units. The values of H given are in MW·s per MVA rating of the generator, and represent the combined inertia of the generator and the turbine.

Table 3.2

Type of generating unit	H
Thermal unit (a) 3600 r/min (2-pole) (b) 1800 r/min (4-pole)	2.5 to 6.0 4.0 to 10.0
Hydraulic unit	2.0 to 4.0

Example 3.5

If the WR^2 of the rotor (including the turbine rotor) of the 555 MVA generating unit of Examples 3.1 and 3.2 is 654,158 lb·ft^2, compute the following:

(a) Moment of inertia J, kg·m^2
(b) Inertia constant H, MW·s/MVA rating
(c) Stored energy, MW·s at rated speed
(d) The mechanical starting time, s

Solution

(a) From the relationships developed in Section 3.9.4,

$$J = \frac{WR^2}{32.2}\times1.356 = \frac{654,158\times1.356}{32.2}$$

$$= 27,547.8\quad \text{kg·m}^2$$

(b) Inertia constant

$$H = 5.48 \times 10^{-9} \frac{J(RPM)^2}{MVA \ \text{rating}}$$

$$= \frac{5.48 \times 10^{-9} \times 27{,}547.8 \times 3600^2}{555}$$

$$= 3.525 \quad \text{MW·s/MVA}$$

(c) Stored energy at rated speed

$$E = H \times MVA \ \text{rating} = 3.525 \times 555$$

$$= 1956.4 \quad \text{MW·s}$$

(d) Mechanical starting time

$$T_M = 2H = 2 \times 3.525$$

$$= 7.05 \quad \text{s} \qquad\qquad\qquad ■$$

3.9.5 Representation in System Studies

For analysis of power system dynamic performance, the component models are expressed in the state-space form (see Chapter 12, Section 12.1) or the block diagram form.

The state-space form requires the component models to be expressed as a set of first order differential equations. The swing equation 3.203, expressed as two first order differential equations, becomes

$$\frac{d\Delta\bar{\omega}_r}{dt} = \frac{1}{2H}(\bar{T}_m - \bar{T}_e - K_D \Delta\bar{\omega}_r) \qquad (3.209)$$

$$\frac{d\delta}{dt} = \omega_0 \Delta\bar{\omega}_r \qquad (3.210)$$

In the above equations, time t is in seconds, rotor angle δ is in electrical radians, and ω_0 is equal to $2\pi f$. In later chapters, when we use the above equations we will not use superbars to identify per unit quantities. We will assume the variables $\Delta\omega_r$, T_m and T_e to be in per unit. However, t will be expressed in seconds and ω_0 in electrical radians per second.

The block diagram form representation of Equations 3.209 and 3.210 is shown in Figure 3.34.

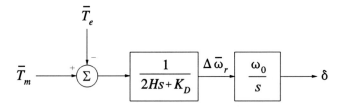

Figure 3.34 Block diagram representation of swing equations

In the block diagram, s is the Laplace operator; it replaces d/dt of Equations 3.209 and 3.210. As noted earlier, symbols T_M and M are often used in place of $2H$.

REFERENCES

[1]   A. Blondel, "The Two-Reaction Method for Study of Oscillatory Phenomena in Coupled Alternators," *Revue Générale de L'Electricité*, Vol. 13, pp. 235-251, February 1923; pp. 515-531, March 1923.

[2]   R.E. Doherty and C.A. Nickle, "Synchronous Machine I and II," *AIEE Trans.*, Vol. 45, pp. 912-942, 1926.

[3]   R.H. Park, "Two-Reaction Theory of Synchronous Machines - Generalized Method of Analysis - Part I," *AIEE Trans.*, Vol. 48, pp. 716-727, 1929; Part II, Vol. 52, pp. 352-355, 1933.

[4]   G. Shackshaft and P.B. Henser, "Model of Generator Saturation for Use in Power System Studies," *Proc. IEE* (London), Vol. 126, No. 8, pp. 759-763, 1979.

[5]   EPRI Report EL-1424, "Determination of Synchronous Machine Stability Constants," Vol. 2, prepared by Ontario Hydro, December 1980.

[6]   EPRI Report EL-3359, "Improvement in Accuracy of Prediction of Electrical Machine Constants, and Generator Model for Subsynchronous Resonance Conditions," Final Report of EPRI Projects RP 1288-1 and RP, Vols. 1, 2 and 3, prepared by General Electric Company, 1984.

[7]   D.C. Macdonald, A.B.J. Reece, and P.J. Turner, "Turbine-Generator Steady-State Reactances," *Proc. IEE* (London), Vol. 132, No. 3, pp. 101-108, 1985.

[8]   R.P. Schulz, et al., "Benefit Assessment of Finite-Element Based Generator

Saturation Model," *IEEE Trans.*, Vol. PWRS-2, pp. 1027-1033, November 1987.

[9]   S.H. Minnich et al., "Saturation Functions for Synchronous Generators from Finite Elements," *IEEE Trans.*, Vol. EC-2, pp. 680-692, December 1987.

[10]   C. Concordia, *Synchronous Machines*, John Wiley & Sons, 1951.

[11]   E.W. Kimbark, *Power System Stability, Vol. III: Synchronous Machines*, John Wiley & Sons, 1956.

[12]   A.E. Fitzgerald and C. Kingsley, *Electric Machinery*, Second Edition, McGraw-Hill, 1961.

[13]   B. Adkins, *The General Theory of Electrical Machines*, Chapman and Hall, 1964.

[14]   G.R. Slemon, *Magnetoelectric Devices*, John Wiley & Sons, 1966.

[15]   ANSI/IEEE Standard 100-1977, *IEEE Standard Dictionary of Electrical and Electronic Terms*.

[16]   IEEE Committee Report, "Recommended Phasor Diagram for Synchronous Machines," *IEEE Trans.*, Vol. PAS-88, pp. 1593-1610, 1969.

[17]   P.M. Anderson and A.A Fouad, *Power System Control and Stability*, Iowa State University Press, Ames, Iowa, 1977.

[18]   A.W. Rankin, "Per-Unit Impedances of Synchronous Machines," *AIEE Trans.*, Vol. 64, Part I, pp. 569-573, August 1945; Part II, pp. 839-845, December 1945.

[19]   M.R. Harris, P.J. Lawrenson, and J.M. Stephenson, "Per-Unit Systems with Special Reference to Electric Machines," IEE Monograph, Cambridge University Press, 1970.

[20]   W.A. Lewis, "A Basic Analysis of Synchronous Machine - Part I," *AIEE Trans.*, Vol. 77, pp. 436-456, 1958.

[21]   I.M. Canay, "Causes of Discrepancies on Calculation of Rotor Quantities and Exact Equivalent Diagram of Synchronous Machines," *IEEE Trans.*, Vol. PAS-88, pp. 1114-1120, July 1969.

[22]   E.F. Fuchs and E.A. Erdelyi, "Non-linear Theory of Turbine-Alternators,"

IEEE Trans., Vol. PAS-91, pp. 583-599, 1972.

[23] El-Serafi et al., "Experimental Study of the Saturation and the Cross-Magnetizing Phenomenon in Saturated Synchronous Machines," *IEEE Trans.*, Vol. EC-3, pp. 815-823, 1988.

[24] I.M. Canay, "Extended Synchronous Machine Model for the Calculation of Transient Processes and Stability," *Electrical Machines & Electromechanics*, Vol. 1, pp. 137-150, 1977.

Chapter 4

Synchronous Machine Parameters

The synchronous machine equations developed in Chapter 3 have the inductances and resistances of the stator and rotor circuits as parameters. These are referred to as *fundamental* or *basic parameters* and are identified by the elements of the *d*- and *q*-axis equivalent circuits shown in Figure 3.18. While the fundamental parameters completely specify the machine electrical characteristics, they cannot be directly determined from measured responses of the machine. Therefore, the traditional approach to assigning machine data has been to express them in terms of derived parameters that are related to observed behaviour as viewed from the terminals under suitable test conditions. In this chapter, we will define these derived parameters and develop their relationships to the fundamental parameters.

4.1 OPERATIONAL PARAMETERS

A convenient method of identifying the machine electrical characteristics is in terms of operational parameters relating the armature and field terminal quantities. Referring to Figure 4.1, the relationship between the incremental values of terminal quantities may be expressed in the operational form as follows:

$$\Delta \psi_d(s) = G(s)\Delta e_{fd}(s) - L_d(s)\Delta i_d(s) \tag{4.1}$$

$$\Delta \psi_q(s) = -L_q(s)\Delta i_q(s) \tag{4.2}$$

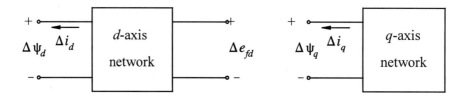

Figure 4.1 The *d*- and *q*-axis networks identifying terminal quantities

where

> $G(s)$ is the stator to field transfer function
> $L_d(s)$ is the *d*-axis operational inductance
> $L_q(s)$ is the *q*-axis operational inductance

In the above equations, s is the familiar Laplace operator and the prefix Δ denotes incremental or perturbed values.

Equations 4.1 and 4.2 are true for any number of rotor circuits. In fact, R.H. Park in his original paper [1] expressed the stator flux equations without specifying the number of rotor circuits. With the equations in operational form, the rotor can be considered as a distributed parameter system. The operational parameters may be determined either from design calculations or more readily from frequency response measurements.

When a finite number of rotor circuits are assumed, the operational parameters can be expressed as a ratio of polynomials in s. The orders of the numerator and denominator polynomials of $L_d(s)$ and $L_q(s)$ are equal to the number of rotor circuits assumed in the respective axes, and $G(s)$ has the same denominator as $L_d(s)$, but a different numerator of order one less than the denominator.

We will develop here the expressions for the operational parameters of the model represented by the equivalent circuits of Figure 4.2. This model structure is generally considered adequate for stability studies and is widely used in large scale stability programs. The rotor characteristics are represented by the field winding and a damper winding in the *d*-axis and two damper windings in the *q*-axis. The mutual inductances L_{fld} and L_{ad} are assumed to be equal; this makes all mutual inductances in the *d*-axis equal. In the next section we will consider the effect of not making this simplifying assumption.

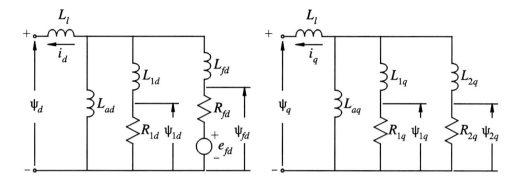

Figure 4.2 Structure of commonly used model

With equal mutual inductances, Equations 3.127, 3.130 and 3.131 for d-axis flux linkages in the operational form become

$$\psi_d(s) = -L_d i_d(s) + L_{ad} i_{fd}(s) + L_{ad} i_{1d}(s) \tag{4.3}$$

$$\psi_{fd}(s) = -L_{ad} i_d(s) + L_{ffd} i_{fd}(s) + L_{ad} i_{1d}(s) \tag{4.4}$$

$$\psi_{1d}(s) = -L_{ad} i_d(s) + L_{ad} i_{fd}(s) + L_{11d} i_{1d}(s) \tag{4.5}$$

The operational forms of Equations 3.123 and 3.124 for rotor voltages are

$$e_{fd}(s) = s\psi_{fd}(s) - \psi_{fd}(0) + R_{fd} i_{fd}(s) \tag{4.6}$$

$$0 = s\psi_{1d}(s) - \psi_{1d}(0) + R_{1d} i_{1d}(s) \tag{4.7}$$

where $\psi_d(0)$, $\psi_{fd}(0)$ and $\psi_{1d}(0)$ denote initial values of the flux linkages. It is preferable to express the above equations in terms of incremental values about the initial operating condition so that the initial values drop out; this makes it more convenient to manipulate the operational equations. Substituting for the flux linkages in terms of the currents, the rotor voltage equations in incremental form become

$$\begin{aligned}
\Delta e_{fd}(s) &= s\Delta\psi_{fd}(s) + R_{fd}\Delta i_{fd}(s) \\
&= -sL_{ad}\Delta i_d(s) + (R_{fd} + sL_{ffd})\Delta i_{fd}(s) + sL_{ad}\Delta i_{1d}(s)
\end{aligned} \tag{4.8}$$

$$0 = s\Delta\psi_{1d}(s)+R_{1d}\Delta i_{fd}(s)$$

$$= -sL_{ad}\Delta i_d(s)+sL_{ad}\Delta i_{fd}(s)+(R_{1d}+sL_{11d})\Delta i_{1d}(s) \qquad (4.9)$$

Our objective is to express the d-axis equations in the form of Equation 4.1, and this can be achieved by eliminating the rotor currents in terms of the terminal quantities e_{fd} and i_d. Accordingly, solution of Equations 4.8 and 4.9 gives

$$\Delta i_{fd}(s) = \frac{1}{D(s)}[(R_{1d}+sL_{11d})\Delta e_{fd}(s)+sL_{ad}(R_{1d}+sL_{1d})\Delta i_d(s)] \qquad (4.10)$$

$$\Delta i_{1d}(s) = \frac{1}{D(s)}[-sL_{ad}\Delta e_{fd}(s)+sL_{ad}(R_{fd}+sL_{fd})\Delta i_d(s)] \qquad (4.11)$$

where

$$D(s) = s^2(L_{11d}L_{ffd}-L_{ad}^2)+s(L_{11d}R_{fd}+L_{ffd}R_{1d})+R_{1d}R_{fd} \qquad (4.12)$$

Given that

$$L_d = L_{ad}+L_l$$

$$L_{ffd} = L_{ad}+L_{fd}$$

$$L_{11d} = L_{ad}+L_{1d}$$

substitution of Equations 4.10 and 4.11 in the incremental form of Equation 4.3 then gives the relationship between the d-axis quantities in the desired form:

$$\Delta\psi_d(s) = G(s)\Delta e_{fd}(s)-L_d(s)\Delta i_d(s)$$

The expressions for the d-axis operational parameters are given by

$$L_d(s) = L_d\frac{1+(T_4+T_5)s+T_4T_6s^2}{1+(T_1+T_2)s+T_1T_3s^2} \qquad (4.13)$$

$$G(s) = G_0 \frac{(1+sT_{kd})}{1+(T_1+T_2)s+T_1T_3s^2} \tag{4.14}$$

where

$$G_0 = \frac{L_{ad}}{R_{fd}} \qquad\qquad T_{kd} = \frac{L_{1d}}{R_{1d}}$$

$$T_1 = \frac{L_{ad}+L_{fd}}{R_{fd}} \qquad\qquad T_2 = \frac{L_{ad}+L_{1d}}{R_{1d}} \tag{4.15}$$

$$T_3 = \frac{1}{R_{1d}}\left(L_{1d}+\frac{L_{ad}L_{fd}}{L_{ad}+L_{fd}}\right) \qquad T_4 = \frac{1}{R_{fd}}\left(L_{fd}+\frac{L_{ad}L_l}{L_{ad}+L_l}\right)$$

$$T_5 = \frac{1}{R_{1d}}\left(L_{1d}+\frac{L_{ad}L_l}{L_{ad}+L_l}\right) \qquad T_6 = \frac{1}{R_{1d}}\left(L_{1d}+\frac{L_{ad}L_{fd}L_l}{L_{ad}L_l+L_{ad}L_{fd}+L_{fd}L_l}\right)$$

Equations 4.13 and 4.14 can be expressed in the factored form:

$$L_d(s) = L_d \frac{(1+sT_d')(1+sT_d'')}{(1+sT_{d0}')(1+sT_{d0}'')} \tag{4.16}$$

$$G(s) = G_0 \frac{(1+sT_{kd})}{(1+sT_{d0}')(1+sT_{d0}'')} \tag{4.17}$$

The expression for the q-axis operational inductance may be written by inspection and recognizing the similarities between d- and q-axis equivalent circuits. In the factored form, it is given by

$$L_q(s) = L_q \frac{(1+sT_q')(1+sT_q'')}{(1+sT_{q0}')(1+sT_{q0}'')} \tag{4.18}$$

The time constants associated with the expressions for $L_d(s)$, $L_q(s)$ and $G(s)$ in the factored form as seen in the following section represent important machine parameters.

4.2 STANDARD PARAMETERS

Following a disturbance, currents are induced in the machine rotor circuits. As shown in Section 3.7 for a short-circuit, some of these induced rotor currents decay more rapidly than others. Machine parameters that influence rapidly decaying components are called the *subtransient* parameters, while those influencing the slowly decaying components are called the *transient* parameters and those influencing sustained components are the *synchronous* parameters.

The synchronous machine characteristics of interest are the effective inductances (or reactances) as seen from the terminals of the machine and associated with the fundamental frequency currents during sustained, transient and subtransient conditions. In addition to these inductances, the corresponding time constants which determine the rate of decay of currents and voltages form the *standard parameters* used in specifying synchronous machine electrical characteristics. These standard parameters, as discussed below, can be determined from the expressions for the operational parameters $L_d(s)$, $G(s)$ and $L_q(s)$.

The constants T'_{d0}, T''_{d0}, T'_d and T''_d are the four principal d-axis time constants of the machine. Their relationships to fundamental parameters are determined by equating the respective numerators and denominators of Equations 4.13 and 4.16. Thus,

$$(1+sT'_{d0})(1+sT''_{d0}) = 1+s(T_1+T_2)+s^2(T_1T_3) \qquad (4.19)$$

$$(1+sT'_d)(1+sT''_d) = 1+s(T_4+T_5)+s^2(T_4T_6) \qquad (4.20)$$

The expressions for the four time constants can be determined accurately by solving the above equations. However, such expressions would be very complex. Simpler expressions can be developed by making some reasonable approximations.

Parameters based on classical definitions

The solution of Equations 4.19 and 4.20 is considerably simplified by recognizing that the value of R_{1d} is very much larger than R_{fd}. This makes T_2 and T_3 very much smaller than T_1, and T_5 and T_6 very much smaller than T_4. Hence

$$(1+sT'_{d0})(1+sT''_{d0}) \approx (1+sT_1)(1+sT_3) \qquad (4.21)$$

$$(1+sT'_d)(1+sT''_d) \approx (1+sT_4)(1+sT_6) \qquad (4.22)$$

We thus have the following approximate relationships

$$
\begin{aligned}
T'_{d0} &\approx T_1 \\
T''_{d0} &\approx T_3 \\
T'_d &\approx T_4 \\
T''_d &\approx T_6
\end{aligned}
\tag{4.23}
$$

The expressions for T_1 to T_6 in terms of the fundamental parameters are given by Equation 4.15. These time constants are in per unit (radians). They have to be divided by rated angular frequency ($\omega_0 = 2\pi f$) to be converted to seconds. For a 60 Hz system, $\omega_0 = 376.991$ (usually approximated to be 377) radians per second.

With the stator terminals open ($\Delta i_d = 0$), from Equations 4.1 and 4.17, we have

$$
\Delta \psi_d(s) = G_0 \frac{1 + sT_{kd}}{(1 + sT'_{d0})(1 + sT''_{d0})} \Delta e_{fd}
\tag{4.24}
$$

The above indicates that, for open-circuit conditions, the d-axis stator flux and hence the terminal voltage respond to a change in field voltage with time constants T'_{d0} and T''_{d0}. Since R_{1d} is much larger than R_{fd}, T''_{d0} is much smaller than T'_{d0}. Thus T''_{d0} is associated with the initial change and is referred to as the d-axis open-circuit *subtransient time constant*. The time constant T'_{d0} represents a slow change corresponding to the transient period and is referred to as the d-axis open-circuit *transient time constant*. The time constants T'_d and T''_d represent the transient and subtransient short-circuit time constants, respectively.

Let us now examine the effective values of $L_d(s)$ under steady-state, transient, and subtransient conditions.

Under steady-state conditions, with $s = 0$ Equation 4.16 becomes

$$
L_d(0) = L_d
\tag{4.25}
$$

This represents the d-axis *synchronous inductance*.

During a rapid transient, as s tends to infinity, the limiting value of $L_d(s)$ is given by

$$
\begin{aligned}
L''_d &= L_d(\infty) \\
&= L_d \left(\frac{T'_d T''_d}{T'_{d0} T''_{d0}} \right)
\end{aligned}
\tag{4.26}
$$

This represents the effective inductance $\Delta\psi_d/\Delta i_d$ immediately following a sudden change and is called the d-axis *subtransient inductance*.

In the absence of the damper winding, the limiting value of inductance is

$$L_d' = L_d(\infty)$$

$$= L_d\left(\frac{T_d'}{T_{d0}'}\right) \tag{4.27}$$

This is referred to as the d-axis *transient inductance*.

Substitution of the expressions of the time constants from Equations 4.15 and 4.23 in Equations 4.26 and 4.27 gives the following alternative expressions for L_d'' and L_d' in terms of the mutual and leakage inductances.

$$L_d'' = L_l + \frac{L_{ad}L_{fd}L_{1d}}{L_{ad}L_{fd}+L_{ad}L_{1d}+L_{fd}L_{1d}} \tag{4.28}$$

and

$$L_d' = L_l + \frac{L_{ad}L_{fd}}{L_{ad}+L_{fd}} \tag{4.29}$$

The above expressions for the subtransient and transient inductances can also be derived from the principle of constant flux linkages, which states that the flux linking an inductive circuit with a finite resistance and emf cannot change instantly. Following a disturbance, the rotor flux linkages do not therefore change instantly. For conditions immediately following a disturbance, the equivalent circuit of Figure 3.17, with incremental rotor flux linkages ($\Delta\psi_{fd}$, $\Delta\psi_{1d}$) set to zero, reduces to that shown in Figure 4.3.

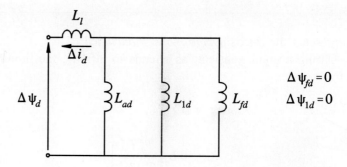

Figure 4.3 Equivalent circuit for incremental values, immediately following a disturbance

From Figure 4.3, the effective inductance $\Delta\psi_d/\Delta i_d$ representing L_d'' is seen to be the same as that given by Equation 4.28. With the damper winding absent ($L_{1d}=\infty$), the effective inductance corresponds to L_d' given by Equation 4.29.

The expression for the q-axis parameters may be readily written by recognizing the similarities in the structure of d- and q-axis equivalent circuits. Thus, the q-axis open-circuit transient and subtransient time constants are given by

$$T_{q0}' = \frac{L_{aq}+L_{1q}}{R_{1q}} \tag{4.30}$$

$$T_{q0}'' = \frac{1}{R_{2q}}\left(L_{2q}+\frac{L_{aq}L_{1q}}{L_{aq}+L_{1q}}\right) \tag{4.31}$$

and the subtransient and transient inductances are given by

$$L_q'' = L_l+\frac{L_{aq}L_{1q}L_{2q}}{L_{aq}L_{1q}+L_{aq}L_{2q}+L_{1q}L_{2q}} \tag{4.32}$$

$$L_q' = L_l+\frac{L_{aq}L_{1q}}{L_{aq}+L_{1q}} \tag{4.33}$$

The q-axis synchronous inductance is given by the steady-state value of $L_q(s)$, which is equal to L_q.

The expressions derived above for the machine standard parameters are based on the assumptions that during the subtransient period $R_{fd}=R_{1q}=0$ and that during the transient period $R_{1d}=R_{2q}=\infty$. These assumptions have been used in the classical theory on synchronous machines [2,3]. However, in recent years there has been some concern [4,5] that significant errors could occur between the values of parameters calculated using the above assumptions and those derived from test measurements such as those described in IEEE Standard 115-1983. Expressions which more closely reflect the definition of the standard parameters are derived below.

Accurate expressions for standard parameters

The exact values of T_{d0}' and T_{d0}'' are given by the poles of $L_d(s)$ and those of T_d' and T_d'' by the zeros of $L_d(s)$. In other words, we need to use the exact solutions of Equations 4.19 and 4.20. From Equation 4.19, the poles of $L_d(s)$ are given by

$$s^2+\frac{T_1+T_2}{T_1T_3}s+\frac{1}{T_1T_3} = 0 \tag{4.34}$$

However, the exact expressions for the poles of $L_d(s)$ become cumbersome and difficult to handle. The expressions can be simplified considerably without much loss of accuracy if it is recognized that $4T_1T_3$ is much less than $(T_1+T_2)^2$. With this simplification, the roots of Equation 4.34 reduce to

$$s_1 = -\frac{1}{T_1+T_2}$$

$$s_2 = -\frac{T_1+T_2}{T_1T_3}$$

The open-circuit time constants are equal to the negatives of the reciprocals of the roots.

$$T_{d0}' = T_1+T_2 \qquad (4.35)$$

$$T_{d0}'' = \frac{T_1T_3}{T_1+T_2} \qquad (4.36)$$

Similarly, by solving for the roots of the numerator of $L_d(s)$, we have

$$T_d' = T_4+T_5 \qquad (4.37)$$

$$T_d'' = \frac{T_4T_6}{T_4+T_5} \qquad (4.38)$$

The transient and subtransient inductances are given by substituting for the above time constants in Equations 4.26 and 4.27. Thus,

$$L_d' = L_d\frac{T_4+T_5}{T_1+T_2} \qquad (4.39)$$

$$L_d'' = L_d\frac{T_4T_6}{T_1T_3} \qquad (4.40)$$

The accurate and approximate (classical) expressions for the standard parameters are summarized in Table 4.1. These expressions apply to a synchronous machine model represented by the equivalent circuit of Figure 4.2 which considers two rotor circuits in each axis with equal mutual inductances.

It should be noted that most stability programs assume that the input data in terms of transient and subtransient parameters are based on the simplifying assumptions from classical theory. However the data provided by some generator manufacturers and those obtained by standard test procedures correspond to the exact values of these parameters. There could thus be some inconsistency associated with the definition and use of standard parameters. As illustrated in Example 4.1, the discrepancies between the two definitions are likely to be significant primarily for the q-axis parameters.

In this book, unless otherwise specified, we will assume the classical definition.

Parameters including unequal mutual effects

In deriving the expressions for the standard parameters above, it is assumed that all mutual inductances in the d-axis are equal. The reciprocal per unit system chosen makes the mutual inductances between the armature and field and between the armature and damper equal. The mutual inductance between field and damper, however, could be different from these mutual inductances.

Although the assumption of all d-axis mutual inductances being equal gives good results in calculating armature quantities, it could lead to significant errors in calculating the field current during transient conditions [6].

The d-axis equivalent circuit including unequal mutual effects and the various mutual and leakage fluxes involved are shown in Figure 4.4.

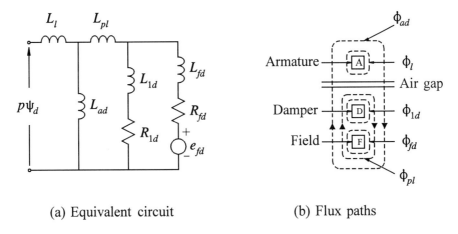

(a) Equivalent circuit   (b) Flux paths

Figure 4.4 Unequal mutual effects in d-axis

Table 4.1

Expressions for Standard Parameters of Synchronous Machine

Parameter	Classical Expression	Accurate Expression
T'_{d0}	T_1	T_1+T_2
T'_d	T_4	T_4+T_5
T''_{d0}	T_3	$T_3[T_1/(T_1+T_2)]$
T''_d	T_6	$T_6[T_4/(T_4+T_5)]$
L'_d	$L_d(T_4/T_1)$	$L_d(T_4+T_5)/(T_1+T_2)$
L''_d	$L_d(T_4T_6)/(T_1T_3)$	$L_d(T_4T_6)/(T_1T_3)$

with

$$T_1 = \frac{L_{ad}+L_{fd}}{R_{fd}} \qquad\qquad T_2 = \frac{L_{ad}+L_{1d}}{R_{1d}}$$

$$T_3 = \frac{1}{R_{1d}}\left(L_{1d}+\frac{L_{ad}L_{fd}}{L_{ad}+L_{fd}}\right) \qquad T_4 = \frac{1}{R_{fd}}\left(L_{fd}+\frac{L_{ad}L_l}{L_{ad}+L_l}\right)$$

$$T_5 = \frac{1}{R_{1d}}\left(L_{1d}+\frac{L_{ad}L_l}{L_{ad}+L_l}\right) \qquad T_6 = \frac{1}{R_{1d}}\left(L_{1d}+\frac{L_{ad}L_lL_{fd}}{L_{ad}L_l+L_{ad}L_{fd}+L_{fd}L_l}\right)$$

Notes: 1. Similar expressions apply to q-axis parameters.
2. All parameters are in per unit.
3. Time constants in seconds are obtained by dividing the per unit values given in the table by $\omega_0 = 2\pi f$.
4. All mutual inductances in d-axis assumed equal.

The series inductance, $L_{pl}=L_{fld}-L_{ad}$, corresponds to the peripheral leakage flux (ϕ_{pl}) which links the field and damper, but not the armature.

Assuming that $R_{fd}=0$ during the subtransient period and that $R_{1d}=\infty$ during the transient period, following the approach used previously, the expressions for the standard parameters may be written by inspection of the equivalent circuit as follows:

$$L_d = L_{ad}+L_l$$

$$L_d' = L_l+\cfrac{1}{\cfrac{1}{L_{ad}}+\cfrac{1}{L_{fd}+L_{pl}}} = L_l+\frac{L_{ad}(L_{fd}+L_{pl})}{L_{ad}+L_{fd}+L_{pl}}$$

$$L_d'' = L_l+\frac{L_{1d}L_{fd}L_{ad}+L_{1d}L_{pl}L_{ad}+L_{ad}L_{fd}L_{pl}}{L_{ad}L_{fd}+L_{ad}L_{1d}+L_{1d}L_{fd}+L_{1d}L_{pl}+L_{fd}L_{pl}}$$

$$T_{d0}' = \frac{L_{ad}+L_{fd}+L_{pl}}{R_{fd}} \tag{4.41}$$

$$T_{d0}'' = \frac{1}{R_{1d}}\left(L_{1d}+\frac{L_{fd}(L_{ad}+L_{pl})}{L_{pl}+L_{fd}+L_{ad}}\right)$$

$$T_d' = \frac{1}{R_{fd}}\left(L_{fd}+L_{pl}+\frac{L_{ad}L_l}{L_{ad}+L_l}\right)$$

$$T_d'' = \frac{1}{R_{1d}}\left(L_{1d}+\frac{L_{ad}L_{pl}L_{fd}+L_lL_{fd}L_{ad}+L_lL_{fd}L_{pl}}{L_{fd}L_{ad}+L_{fd}L_l+L_{pl}L_{ad}+L_{pl}L_l+L_{ad}L_l}\right)$$

The above expressions are based on the approximations associated with the classical definition of the parameters. Accurate expressions applicable to a more complex model structure consisting of these rotor circuits in each axis and unequal d-axis mutual inductances are given in reference 7.

Parameters of salient pole machines:

In the discussion of standard parameters so far, we have considered a model structure with two rotor circuits in each axis. This is applicable to a round rotor machine. However, for a laminated salient pole machine, the damper winding is the only rotor circuit in the q-axis; therefore, only one q-axis rotor circuit (denoted by the subscript $1q$) is applicable. The parameters of this rotor circuit are such that it represents rapidly decaying subtransient effects. The second rotor circuit (denoted by subscript $2q$) is ignored and no distinction is made between transient and synchronous (steady-state) conditions. Hence, the expressions for the q-axis parameters of a salient pole machine are as follows:

$$L_q = L_l + L_{aq}$$

$$L_q'' = L_l + \frac{L_{aq}L_{1q}}{L_{aq}+L_{1q}} \qquad (4.42)$$

$$T_{q0}'' = \frac{L_{aq}+L_{1q}}{R_{1q}}$$

The transient parameters L_q' and T_{q0}' are not applicable in this case.

In the d-axis, it is appropriate to consider two rotor circuits (field and damper) and the expressions previously derived are applicable to salient pole machines.

Reactances:

In per unit, the subtransient, transient and synchronous reactances are equal to the corresponding inductances. Hence, common practice is to identify synchronous machine parameters in terms of the reactances, instead of the inductances.

Typical values of standard parameters:

Table 4.2 gives ranges within which generator parameters normally lie.

From the expressions for machine parameters summarized in Table 4.1, it is readily apparent that

$$X_d \geq X_q > X_q' \geq X_d' > X_q'' \geq X_d'' \qquad (4.43)$$

$$T_{d0}' > T_d' > T_{d0}'' > T_d'' > T_{kd} \qquad (4.44)$$

$$T_{q0}' > T_q' > T_{q0}'' > T_q'' \qquad (4.45)$$

Table 4.2

Parameter		Hydraulic Units	Thermal Units
Synchronous Reactance	X_d	0.6 - 1.5	1.0 - 2.3
	X_q	0.4 - 1.0	1.0 - 2.3
Transient Reactance	X_d'	0.2 - 0.5	0.15 - 0.4
	X_q'	-	0.3 - 1.0
Subtransient Reactance	X_d''	0.15 - 0.35	0.12 - 0.25
	X_q''	0.2 - 0.45	0.12 - 0.25
Transient OC Time Constant	T_{d0}'	1.5 - 9.0 s	3.0 - 10.0 s
	T_{q0}'	-	0.5 - 2.0 s
Subtransient OC Time Constant	T_{d0}''	0.01 - 0.05 s	0.02 - 0.05 s
	T_{q0}''	0.01 - 0.09 s	0.02 - 0.05 s
Stator Leakage Inductance	X_l	0.1 - 0.2	0.1 - 0.2
Stator Resistance	R_a	0.002 - 0.02	0.0015 - 0.005

Notes: 1. Reactance values are in per unit with stator base values equal to the corresponding machine rated values.
2. Time constants are in seconds.

Example 4.1

The following are the per unit values of the standard parameters of the 555 MVA, 0.9 p.f., 60 Hz turbine generator considered in the examples of Chapter 3:

$$L_d = 1.81 \qquad L_q = 1.76 \qquad L_l = 0.15 \qquad R_a = 0.003$$
$$L_d' = 0.3 \qquad L_q' = 0.65 \qquad L_d'' = 0.23 \qquad L_q'' = 0.25$$
$$T_{d0}' = 8.0 \ s \qquad T_{q0}' = 1.0 \ s \qquad T_{d0}'' = 0.03 \ s \qquad T_{q0}'' = 0.07 \ s$$

The transient and subtransient parameters are based on the *classical definitions*. The parameters given correspond to unsaturated values of L_{ad} and L_{aq}.

(a)   Assuming that $L_{fkd}=L_{ad}$, determine the per unit values of the fundamental parameters, i.e., elements of the *d*- and *q*-axis equivalent circuits.

(b)   Using the fundamental parameters computed in (a) above, and without making the simplifying assumptions of the classical definitions, calculate the accurate values of the transient and subtransient parameters. How do they compare with the values based on the classical definitions?

(c)   If, at the rated output conditions, the effect of magnetic saturation is to reduce L_{ad} and L_{aq} to 83.5% of their unsaturated values, find the corresponding values of the standard parameters based on the classical definitions.

Solution

(a) We first compute the unsaturated mutual inductances

$$L_{ad} = L_d - L_l = 1.81 - 0.15 = 1.66 \quad \text{pu}$$
$$L_{aq} = L_q - L_l = 1.76 - 0.15 = 1.61 \quad \text{pu}$$

We then compute the rotor leakage inductances from the expressions for transient and subtransient inductances. From Equation 4.29, the expression for L'_d based on the classical definition is

$$L'_d = L_l + \frac{L_{ad} L_{fd}}{L_{ad} + L_{fd}}$$

Substituting the respective numerical values gives

$$0.3 = 0.15 + \frac{1.66 \times L_{fd}}{1.66 + L_{fd}}$$

Solving for L_{fd} yields

$$L_{fd} = 0.165 \quad \text{pu}$$

Similarly, from the expression for L'_q given by Equation 4.33, we obtain

$$0.65 = 0.15 + \frac{1.61 \times L_{1q}}{1.61 + L_{1q}}$$

Solving for L_{1q} gives

$$L_{1q} = 0.7252 \quad \text{pu}$$

From Equation 4.28

$$L_d'' = L_l + \frac{L_{ad}L_{fd}L_{1d}}{L_{ad}L_{fd} + L_{ad}L_{1d} + L_{fd}L_{1d}}$$

Substituting the respective numerical values gives

$$0.23 = 0.15 + \frac{1.66 \times 0.165 \times L_{1d}}{1.66 \times 0.165 + 1.66 \times L_{1d} + 0.165 \times L_{1d}}$$

Solving for L_{1d}, we have

$$L_{1d} = 0.1713 \quad \text{pu}$$

Similarly, from the expression for L_q'' given by Equation 4.32, we obtain

$$0.25 = 0.15 + \frac{1.61 \times 0.7252 \times L_{2q}}{1.61 \times 0.7252 + (1.61 + 0.7252) L_{2q}}$$

and solving for L_{2q} yields

$$L_{2q} = 0.125 \quad \text{pu}$$

Next, we compute the rotor resistances from the expressions for the open-circuit time constants. From Equations 4.15 and 4.23, the expressions for T_{d0}' and T_{d0}'', based on the classical definitions, are

$$T_{d0}' = \frac{L_{ad} + L_{fd}}{R_{fd}} \quad \text{pu}$$

and

$$T_{d0}'' = \frac{1}{R_{fd}} \left(L_{1d} + \frac{L_{ad}L_{fd}}{L_{ad} + L_{fd}} \right) \quad \text{pu}$$

Substituting the numerical values and noting that the time constant in per unit is equal to 377 times the time constant in seconds, we obtain

$$R_{fd} = \frac{1.66 + 0.165}{8.0 \times 377}$$

$$= 0.000605 \quad \text{pu}$$

and

$$R_{1d} = \frac{1}{0.03 \times 377}\left(0.1713 + \frac{1.66 \times 0.165}{1.66 + 0.165}\right)$$

$$= 0.0284 \quad \text{pu}$$

Similarly, using Equations 4.30 and 4.31 to compute the q-axis rotor circuit resistances, we obtain

$$R_{1q} = 0.0062 \quad \text{pu}$$

and

$$R_{2q} = 0.0237 \quad \text{pu}$$

The following is a summary of the per unit values of the fundamental parameters:

$R_a = 0.003$	$L_l = 0.15$	$L_{ad} = 1.66$	$L_{aq} = 1.61$
$L_{fd} = 0.165$	$R_{fd} = 0.0006$	$L_{1d} = 0.1713$	$R_{1d} = 0.0284$
$L_{1q} = 0.7252$	$R_{1q} = 0.0062$	$L_{2q} = 0.125$	$R_{2q} = 0.0237$

(b) The accurate expressions for the d-axis transient and subtransient constants are summarized in Table 4.1.

Substituting the fundamental parameters computed above in the expressions for the time constants T_1 to T_6 and dividing by 377 to convert to seconds, we obtain

$$
\begin{aligned}
T_1 &= 8.0 \ \text{s} & T_2 &= 0.171 \ \text{s} & T_3 &= 0.03 \ \text{s} \\
T_4 &= 1.326 \ \text{s} & T_5 &= 0.0288 \ \text{s} & T_6 &= 0.023 \ \text{s}
\end{aligned}
$$

The accurate value of the d-axis transient inductance is

$$L_d' = L_d \frac{T_4 + T_5}{T_1 + T_2}$$

$$= 1.81\left(\frac{1.326 + 0.0288}{8.0 + 0.171}\right)$$

$$= 0.3 \quad \text{pu}$$

The accurate values of transient and subtransient open-circuit time constants are

$$T'_{d0} = T_1 + T_2 = 8.0 + 0.171$$

$$= 8.171 \text{ s}$$

$$T''_{d0} = \frac{T_1 T_3}{T_1 + T_2}$$

$$= \frac{8.0 \times 0.03}{8.0 + 0.03}$$

$$= 0.0294 \text{ s}$$

Similarly, for the q-axis, we obtain

$$T_1 = 1.0 \text{ s} \qquad T_2 = 0.1943 \text{ s} \qquad T_3 = 0.07 \text{ s}$$
$$T_4 = 0.3693 \text{ s} \qquad T_5 = 0.0294 \text{ s} \qquad T_6 = 0.0269 \text{ s}$$

The accurate value of the q-axis transient inductance is

$$L'_q = L_q \frac{T_4 + T_5}{T_1 + T_2}$$

$$= 1.76 \frac{0.3693 + 0.0294}{1.0 + 0.1943}$$

$$= 0.5875 \text{ pu}$$

The accurate values of the q-axis open-circuit time constants are

$$T'_{q0} = T_1 + T_2 = 1.0 + 0.1943$$

$$= 1.1943 \text{ s}$$

$$T''_{q0} = T_3 \frac{T_1}{T_1 + T_2}$$

$$= 0.07 \frac{1.0}{1.1943}$$

$$= 0.0586 \text{ s}$$

There are no approximations associated with the classical definitions of the subtransient inductances; hence, accurate values of L''_d and L''_q are the same as the given values.

The following is a comparison of the accurate and approximate values of the parameters.

Parameter	Value based on Classical Definition	Accurate Value
L_d'	0.3	0.3
T_{d0}'	8.0	8.171
T_{d0}''	0.03	0.0294
L_q'	0.65	0.5875
T_{q0}'	1.0	1.1943
T_{q0}''	0.07	0.0586

The differences are significant only for the q-axis.

(c) At the rated output condition, the saturated values of mutual inductances are

$$L_{ad} = 0.835 \times L_{adu} = 0.835 \times 1.66$$

$$= 1.386 \quad \text{pu}$$

$$L_{aq} = 0.835 \times L_{aqu} = 0.835 \times 1.61$$

$$= 1.344 \quad \text{pu}$$

The corresponding values of the standard parameters, based on the classical definitions, are

$$L_d = 1.386 + 0.15 = 1.536 \quad \text{pu}$$

$$L_q = 1.344 + 0.15 = 1.494 \quad \text{pu}$$

$$L_d' = 0.15 + \frac{1.386 \times 0.165}{1.386 + 0.165}$$

$$= 0.2974 \quad \text{pu}$$

$$L_d'' = 0.15 + \frac{1.386 \times 0.165 \times 0.1713}{1.386 \times 0.165 + 1.386 \times 0.1713 + 0.165 \times 0.1713}$$

$$= 0.2292 \quad \text{pu}$$

$$L_q' = 0.15 + \frac{1.344 \times 0.7252}{1.344 + 0.7252}$$

$$= 0.621 \quad \text{pu}$$

$$L_q'' = 0.15 + \frac{1.344 \times 0.7252 \times 0.125}{1.344 \times 0.7252 + 1.344 \times 0.125 + 0.7252 \times 0.125}$$

$$= 0.2488 \quad \text{pu}$$

$$T_{d0}' = \frac{1.386 + 0.165}{0.0006 \times 377} = 6.86 \quad \text{s}$$

$$T_{d0}'' = \frac{1}{0.0284 \times 377} \left(0.1713 + \frac{1.386 \times 0.165}{1.386 + 0.165} \right)$$

$$= 0.0298 \quad \text{s}$$

$$T_{q0}' = \frac{1.344 + 0.7252}{0.0062 \times 377} = 0.885 \quad \text{s}$$

$$T_{q0}'' = \frac{1}{0.0237 \times 377} \left(0.125 + \frac{1.344 \times 0.7252}{1.344 + 0.7252} \right)$$

$$= 0.0667 \quad \text{s} \qquad\qquad\qquad ■$$

4.3 FREQUENCY-RESPONSE CHARACTERISTICS

It is of interest to examine the frequency-response characteristics of the operational parameters and relate them to standard parameters. Such characteristics provide useful insight into the dynamic characteristics of the machine and may be readily sketched using the asymptotic approximation.

With the rotor effects represented by two circuits in the d-axis, $L_d(s)$ and $G(s)$ are given by Equations 4.16 and 4.17, respectively. Figure 4.5 shows the magnitude of $L_d(s)$ as a function of frequency. The transient and subtransient time constants and inductances have been used to identify the corner points for the asymptotic approximation. The plot shown is for the generator considered in Example 4.1. The general shape of the frequency response characteristic, however, is applicable to any synchronous machine. The effective inductance is equal to the synchronous inductance L_d at frequencies below 0.02 Hz, the transient inductance L_d' in the range 0.2 Hz to 2 Hz, and the subtransient inductance L_d'' beyond 10 Hz. In stability studies, the frequency range of interest is that corresponding to L_d'.

Figure 4.6 shows the magnitude of $G(s)$ as a function of frequency for the same machine, with G_0 normalized so that it is equal to 1.0. From the plot we see that the effective gain drops off considerably at high frequencies. This indicates that high-frequency variations in field voltage are not reflected in the stator flux linkage and hence other stator quantities.

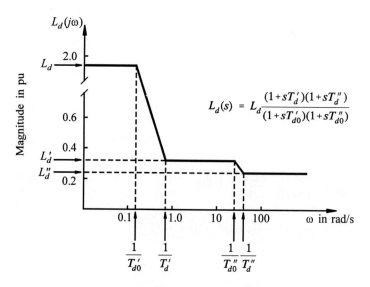

Figure 4.5 Variation of magnitude of $L_d(s)$

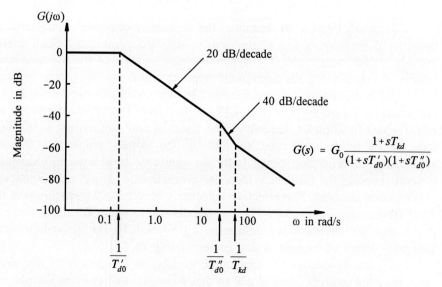

Figure 4.6 Variation of magnitude of $G(s)$ with frequency

Armature time constant

As discussed in Section 3.7.2, the armature time constant, T_a, gives the rate of decay of the unidirectional component (dc offset) of armature phase currents following a three-phase short-circuit at the terminals.

The unidirectional components of currents, which have unequal magnitudes in the three phases, produce a stationary mmf wave. This induces fundamental frequency currents in the rotor circuits, keeping the flux linking with these circuits constant. The flux path due to the direct current in the armature is similar to those corresponding to subtransient inductances L_d'' and L_q''. Since the rotor moves at synchronous speed with respect to the mmf wave produced by the direct currents in the stator, the mmf wave alternately sees conditions corresponding to L_d'' and L_q''. Hence, the effective inductance seen by the direct currents in the armature lies between L_d'' and L_q''. It will be shown later that this is also the inductance L_2, seen by the negative sequence currents applied to the stator windings. Therefore,

$$T_a = \frac{1}{R_a}\left(\frac{L_d'' + L_q''}{2}\right) \qquad \text{per unit} \tag{4.46}$$

The value of T_a lies between 0.03 and 0.35 s.

4.4 DETERMINATION OF SYNCHRONOUS MACHINE PARAMETERS

The conventional method of determining synchronous machine parameters is from short-circuit tests on unloaded machines. The test procedures are specified in IEEE Standard 115-1983 [8]. These tests provide X_d, X_q, X_d', X_d'', T_{d0}', T_d', T_{d0}'' and T_d''. They do not, however, provide q-axis transient and subtransient constants. In addition, they do not include measurement of the field circuit during the short-circuit tests, and consequently the field circuit is not specifically identified. The limitations of these procedures for providing data suitable for stability studies have been recognized for some time.

Several alternative testing and analytical methods have been proposed and used to obtain better models:

- Enhanced sudden short-circuit tests;

- Stator decrement test;

- Frequency-response tests

  — Standstill frequency response,
  — Open-circuit frequency response,

— On-line frequency response;

• Analysis of design data; and

• Quadrature axis saturation measurements.

The special features of theses methods are briefly discussed below. A detailed description of these methods is, however, beyond the scope of this book. Readers may refer to the references provided for additional information.

Enhanced short-circuit tests

Improved methods of utilizing results from sudden short-circuit tests to determine more accurate d-axis parameters are described in references 9 and 10. The most important feature of these methods is the utilization of rotor current measurements during the short-circuit tests to identify the field-circuit characteristics more accurately.

Among the disadvantages of the short-circuit approach are the inability to provide q-axis parameters accurately and the necessity of exposing the machine to a severe shock.

Decrement tests

These tests are similar in approach to the sudden short-circuit tests in that the time responses of machine variables following a sudden disturbance are measured to identify machine characteristics. References 10, 11 and 12 use this approach.

In this approach, with operating conditions arranged such that current is flowing only in the direct axis (i_q=0), the unit is tripped and the resulting terminal voltage and field current decay are used to extract parameters for a model in much the same way as with sudden short-circuit data. A similar test is performed with current flowing only in the quadrature axis (i_d=0) to obtain q-axis data. To maintain unsaturated conditions, these tests are conducted at partial load and reduced voltage.

These tests provide both d- and q-axis data. However, they are somewhat difficult and expensive to conduct. For most machines, it is difficult or impossible to attain unsaturated conditions which unfortunately complicates the testing and the analysis of results. The tests require rescheduling of generation and several unit trips, which is often not practical.

Frequency-response tests

a. Standstill frequency response (SSFR):

In the SSFR technique, all tests are conducted with the unit at rest (rotor stationary) and disconnected from the system. The rotor must be aligned to two particular positions with respect to the stator during the tests. With the stator excited by a low level (±60 A, ±20 V) source over the range of frequencies from 1 mHz to 1 kHz, the following responses are measured:

With the stator winding excited and field shorted

$$sG(s) = \frac{\Delta i_{fd}(s)}{\Delta i_d(s)} \tag{4.47}$$

and

$$Z_d(s) = \frac{\Delta e_d(s)}{\Delta i_d(s)} \tag{4.48}$$

With the stator winding excited (field condition immaterial)

$$Z_q(s) = -\frac{\Delta e_q(s)}{\Delta i_q(s)} \tag{4.49}$$

The latter two, $Z_d(s)$ and $Z_q(s)$, are the d- and q-axis operational impedances as viewed from the armature terminals. The operational inductances are computed by subtracting the armature resistance from these impedances:

$$L_d(s) = \frac{Z_d(s) - R_a}{s} \tag{4.50}$$

and

$$L_q(s) = \frac{Z_q(s) - R_a}{s} \tag{4.51}$$

where R_a is the dc resistance of one armature phase and $s = j\omega$.

The parameters of the d-axis equivalent circuit are obtained using transfer function approximations for the functions $L_d(s)$ and $sG(s)$. The function $sG(s)$ rather than $G(s)$ is used because the former can be measured at the same time as $Z_d(s)$, with the field shorted. The q-axis parameters are obtained using the transfer function for $L_q(s)$. The order of the transfer function depends on the number of rotor circuits assumed in the respective axes.

Because the tests are conducted at very low flux levels, the results must be corrected to bring them from the "toe" of the saturation curve to normal unsaturated levels. This is done by minor adjustment of the mutual reactance in each axis.

This technique has been used in references 7, 13, 14, and 15. In addition, it is now a "trial use" standard of the IEEE [16]. SSFR testing is relatively easy to perform either in the factory or during a maintenance outage on a unit.

Besides the required adjustment for unsaturated conditions, there are two other limitations to this type of testing, both of which involve rotational effects. Where damper windings are employed, they are often just overlapped and may not form a good connection at standstill. In addition, the extent to which rotation causes the slot wedges to form a low impedance path to the rotor is largely unknown.

At low and high frequencies, the data obtained are expected to be good. In the mid-frequency range it is expected to be good for most machines with no specific damper windings or the equivalent, short dampers and non-magnetic retaining rings [17].

Details of models developed from SSFR tests on three large generators are given in references 7 and 14. The measured $L_d(s)$, $sG(s)$ and $L_q(s)$ characteristics for one of these generators (500 MW, 3800 RPM generator at Lambton GS in Ontario) are shown in Figures 4.7, 4.8 and 4.9. Second- and third-order transfer function approximations to the measured characteristics are also shown in the figures. For this generator, second-order transfer functions are seen to be good approximations for both d- and q-axis. However, this may not always be the case; depending on the rotor construction, third-order transfer functions may be more appropriate in some cases [7,14].

b. Open-circuit frequency response (OCFR):

Open-circuit frequency-response testing allows confirmation of some of the SSFR data in the middle of the frequency range for the d-axis only [7]. For this test, the unit is operated on open-circuit at reduced voltage. The field is excited at various frequencies and the field-to-stator frequency-response measured. The difference between this response and the equivalent one from the standstill tests gives some indication of rotational effects.

The test is normally done at more than one voltage to examine saturation effects. By conducting the test with various signal amplitudes, slot wedge conduction effects can be assessed.

c. On-line frequency response (OLFR):

In many respects, on-line frequency response (OLFR) testing is the "proof in the pudding" as far as small-signal verification of machine models is concerned. Here the machine is tested under the same conditions as those under which the model is expected to perform, although over a restricted operating range.

For this test, the machine is operated near rated (or at reduced) load preferably over a substantial impedance to the system. The excitation is modulated by either sinusoidal or random noise. Components are resolved on the two axes and data similar to those of the SSFR tests are used to derive a model. The frequency range of usable data in this case is more limited than that of the SSFR tests, but the SSFR data are

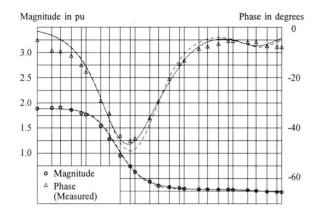

$$L_d(s) = 1.89 \frac{(1+s1.03)(1+s0.32)(1+s0.0042)}{(1+s5.35)(1+s0.41)(1+s0.0053)}$$

$$L_d(s) = 1.89 \frac{(1+s0.77)(1+s0.0038)}{(1+s5.06)(1+s0.0049)}$$

Figure 4.7 Lambton GS - Variation of $L_d(s)$ with frequency at standstill with field closed

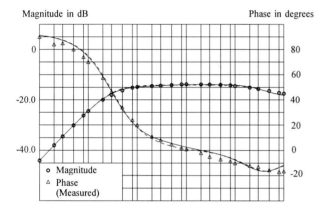

$$sG(s) = s4.25 \frac{(1+s0.446)(1+s0.003)}{(1+s5.35)(1+s0.41)(1+s0.0053)}$$

$$sG(s) = s4.25 \frac{(1+s0.0028)}{(1+s5.06)(1+s0.0049)}$$

Figure 4.8 Lambton GS - Variation of $sG(s)$ with frequency at standstill with field closed

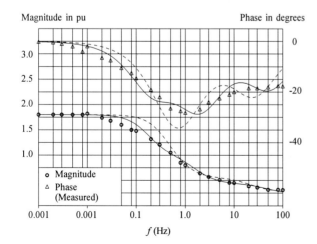

$$L_q(s) = 1.8 \frac{(1+s0.562)(1+s0.052)(1+s0.0022)}{(1+s1.020)(1+s0.118)(1+s0.0040)}$$

$$L_q(s) = 1.8 \frac{(1+s0.128)(1+s0.0043)}{(1+s0.417)(1+s0.0086)}$$

Figure 4.9 Lambton GS - Variation of $L_q(s)$ with frequency at standstill

expected to be good at the frequency extremes. The OLFR testing allows the middle of the frequency range to be filled in with data for both axes which include rotational effects. The disadvantage of this test is that it requires testing on an operating unit connected to the system, possibly under special system or unit conditions. In addition, it does not provide large signal response information.

Details of OLFR measurement techniques and model identification procedures are given in reference 18. Results of OLFR tests on two large units and improved models developed from the tests are given in references 7 and 19. In reference 20, models for three large generators based on SSFR and OLFR tests are validated by comparing the results of simulations with measured responses involving line switching. For one generator, models derived from short-circuit tests and decrement tests are also validated.

Calculation of machine parameters from design data

Improved generator models developed from design information are described in reference 21. Reference 22 presents more recent work using finite element analysis.

Work done on two and three dimensional finite element models in the steady state has shown good agreement with measured results. In addition, two dimensional finite element models have been developed which provide frequency response data that compare well with measured results. The degree to which conduction occurs across slot wedges appears to have a significant effect on the model at higher frequencies. The significance of rotational effects and the effect of disturbance amplitude on the model require additional investigation.

REFERENCES

[1]   R.H. Park, "Two-Reaction Theory of Synchronous Machines," *AIEE Trans.*, Part I, Vol. 48, pp. 716-730, 1929; Part II, Vol. 52, pp. 352-355, 1933.

[2]   C. Concordia, *Synchronous Machines - Theory and Performance*, John Wiley & Sons, 1952.

[3]   B. Adkins, *The General Theory of Electric Machines*, Chapman and Hall, 1964.

[4]   G. Shackshaft, "New Approach to the Determination of Synchronous Machine Parameters from Tests," *Proc. IEE* (London), Vol. 121, No. 11, pp. 1385-1392, 1974.

[5]   IEEE Committee Report, "Supplementary Definitions and Associated Test Methods for Obtaining Parameters for Synchronous Machine Stability Simulations," *IEEE Trans.*, Vol. PAS-99, pp. 1625-1633, July/August 1980.

[6]   I.M. Canay, "Causes of Discrepancies on Calculation of Rotor Quantities and Exact Equivalent Diagram of Synchronous Machines," *IEEE Trans.*, Vol. PAS-88, pp. 1114-1120, July 1969.

[7]   EPRI Report EL-1424, "Determination of Synchronous Machine Stability Constants," Vol. 2, prepared by Ontario Hydro, December 1980.

[8]   ANSI/IEEE Standard 115-1983, *IEEE Guide: Test Procedures for Synchronous Machines.*

[9]   Y. Takeda and B. Adkins, "Determination of Synchronous Machine Parameters Allowing for Unequal Mutual Inductances," *Proc. IEE* (London), Vol. 121, No. 12, pp. 1501-1504, December 1974.

[10]   G. Shackshaft and A.T. Poray, "Implementation of New Approach to Determination of Synchronous Machine Parameters from Tests," *Proc. IEE* (London), Vol. 124, No. 12, pp. 1170-1178, 1977.

[11]   F.P. deMello and J.R. Ribeiro, "Derivation of Synchronous Machine Parameters from Tests," *IEEE Trans.*, Vol. PAS-96, pp. 1211-1218, July/August 1977.

[12]   F.P. deMello and L.N. Hannett, "Validation of Synchronous Machine Models and Determination of Model Parameters from Tests," *IEEE Trans.*, Vol. PAS-100, pp. 662-672, February 1981.

[13]   M.E. Coultes and W. Watson, "Synchronous Machine Models by Standstill Frequency Response Tests," *IEEE Trans.*, Vol. PAS-100, pp. 1480-1489, April 1981.

[14]   P.L. Dandeno and A.T. Poray, "Development of Detailed Turbogenerator Equivalent Circuits from Standstill Frequency Response Measurements," *IEEE Trans.*, Vol. PAS-100, pp. 1646-1653, April 1981.

[15]   M.E. Coultes, "Standstill Frequency Response Tests," Presented at the IEEE Symposium on Synchronous Machine Modelling for Power System Studies, IEEE Publication 83TH0101-6-PWR, pp. 26-30.

[16]   IEEE Standard 115A-1984, *IEEE Trial Use Standard Procedures for Obtaining Synchronous Machine Parameters by Standstill Frequency Response Testing.*

[17]   A.G. Jack and T.J. Bedford, "A Study of the Frequency Response of Turbogenerators with Special Reference to Nanticoke GS," *IEEE Trans.*, Vol.

EC-2, pp. 496-505, September 1987.

[18]   M.E. Coultes, P. Kundur and G.J. Rogers, "On-Line Frequency Response Tests and Identification of Generator Models," *IEEE Trans.*, Vol. EC-2, pp. 38-42, September 1987.

[19]   P.L. Dandeno, P. Kundur, A.T. Poray, and H.M. Zein El-Din, "Adaptation and Validation of Turbogenerator Model Parameters through On-Line Frequency Response Measurements," *IEEE Trans.*, Vol. EC-2, pp. 1656-1661, September 1987.

[20]   P.L. Dandeno, P. Kundur, A.T. Poray, and M.E. Coultes, "Validation of Turbogenerator Stability Models by Comparison with Power System Tests," *IEEE Trans.*, Vol. EC-2, pp. 1637-1645, September 1987.

[21]   R.P. Schulz, W.D. Jones, and D.N. Ewart, "Dynamic Models of Turbine Generators Derived from Solid Rotor Equivalent Circuits," *IEEE Trans.*, Vol. PAS-92, pp. 926-933, May/June 1973.

[22]   J.W. Dougherty and S.M. Minnich, "Finite Element Modelling of Large Turbine Generators: Calculations versus Load Test Data," *IEEE Trans.*, Vol. PAS-100, pp. 3921-3929, August 1981.

Chapter 5

Synchronous Machine Representation in Stability Studies

The per unit equations summarized in Section 3.4.9 describe completely the electrical dynamic performance of a synchronous machine. However, except for the analysis of very small systems, these equations cannot be used directly for system stability studies. In this chapter the simplifications required for the representation of synchronous machines in stability studies are discussed. Also considered are various degrees of approximations that can be made to simplify the machine model, minimizing data requirements and computational effort.

5.1 SIMPLIFICATIONS ESSENTIAL FOR LARGE-SCALE STUDIES

For stability analysis of large systems, it is necessary to neglect the following from Equations 3.120 and 3.121 for stator voltage:

- The transformer voltage terms, $p\psi_d$ and $p\psi_q$.

- The effect of speed variations.

The reasons for and the effects of these simplifications are discussed below.

5.1.1 Neglect of Stator $p\psi$ Terms

As discussed in Section 3.7, the $p\psi_d$ and $p\psi_q$ terms represent the stator transients. With these terms neglected, the stator quantities contain only fundamental frequency components and the stator voltage equations appear as algebraic equations. This allows the use of steady-state relationships for representing the interconnecting transmission network.

The transients associated with the network decay very rapidly and there is little justification for modelling their effects in stability studies. The network transients cannot be neglected unless machine stator transients are also neglected; otherwise we would have an inconsistent set of equations representing the various elements of the power system. Inclusion of the network transients increases the order of the overall system model considerably, and hence limits the size of the system that can be simulated. In addition, a system representation with machine stator and network transients contains high frequency transients. This requires small time steps for numerical integration, resulting in an enormous increase in computational costs. Also, time responses of system variables containing high frequency components are difficult to analyze and interpret from the viewpoint of system stability. For these reasons, stability analysis of practical power systems consisting of thousands of buses and hundreds of generators would be impossible without the simplification resulting from the neglect of machine stator transients. While we have identified the need for omitting the stator $p\psi$ terms from computational considerations, we have not yet established that the computed machine response with this simplification is acceptable for the study of system stability. This has been investigated in references 1 to 3, and we will discuss the results of one of the cases considered in reference 1.

Figure 5.1 shows the system studied. It consists of a salient-pole generator connected to an infinite bus through two transmission lines. The disturbance applied is a three-phase short-circuit at the sending end of one of the lines, cleared in 0.09 s by isolating the faulted circuit. The responses of generator variables computed with and without inclusion of the stator $p\psi$ terms are compared in Figures 5.2, 5.3 and 5.4.

When the stator $p\psi$ terms are omitted, we see from Figure 5.2 that i_d and i_q have only unidirectional components; these correspond to the fundamental frequency component of phase currents. The resulting air-gap torque is unidirectional and small in magnitude; it is due to the stator resistance losses.

On the other hand, when the $p\psi$ terms are included, i_d and i_q contain fundamental frequency (60 Hz) components, which correspond to the dc offset in the phase currents (discussed in Section 3.7). These in turn result in the following components of air-gap torque:

- A fundamental frequency oscillatory component, due to interaction with the rotor field.

- A unidirectional component, due to rotor resistance losses caused by the fundamental frequency currents induced in the rotor.

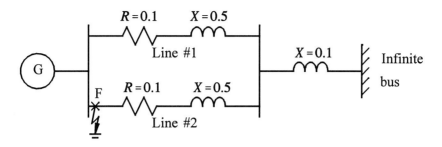

Disturbance:

   3-phase fault at F; cleared in 0.09 s by opening line #2

Generator parameters in per unit:

   $L_{ad} = 1.0$ $L_{aq} = 0.6$ $L_l = 0.18$ $L_{fd} = 0.13$

   $L_{1d} = 0.11$ $L_{1q} = 0.13$ $R_a = 0.005$ $R_{fd} = 0.00075$

   $R_{1d} = 0.02$ $R_{1q} = 0.04$ $H = 3.5$

Figure 5.1 System configuration and parameters

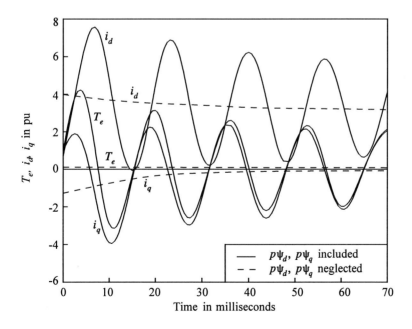

Figure 5.2 Effect of neglecting stator transients on air-gap
torque and *d-q* components of stator currents

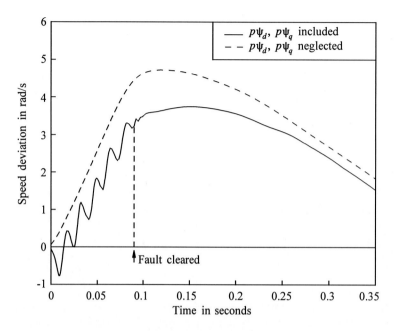

Figure 5.3 Effect of neglecting stator transients
on speed deviation

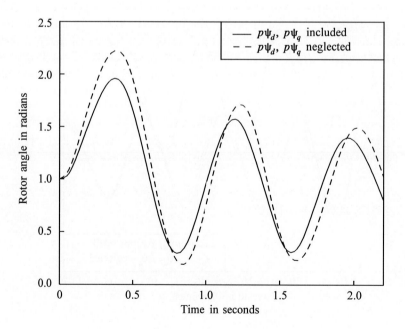

Figure 5.4 Effect of neglecting stator transients
on rotor angle swings

The unidirectional component of torque, due to rotor resistive losses, can be quite high and has a braking effect. Therefore, it is referred to as the *dc braking torque*; its effect is to reduce the acceleration of the rotor following the disturbance. The effect of the oscillatory component is to decelerate the rotor during the first half cycle and to accelerate it to near its initial speed during the second half cycle, and so on during subsequent cycles. The net effect of the oscillatory torque is therefore a reduction of the mean speed of the rotor [4]. The overall effect of these two components for a close-up simultaneous three-phase fault could be large enough to initially cause retardation of the rotor or a *back swing*. This could have a significant beneficial effect on system stability as seen from plots of speed deviation and rotor angle in Figures 5.3 and 5.4.

Since, for large-scale stability studies, it is necessary to neglect the $p\psi$ terms, the effects of the unidirectional (dc) braking torque and the oscillatory torque may be estimated and included in the calculations. References 1, 3, 5 and 6 give methods of estimating the unidirectional braking torque, and reference 4 gives a method to account for the oscillatory torque. These would be applicable only to simultaneous three-phase faults and need to be considered only for machines very close to the fault.

However, in practical studies of power system stability, the effects of dc braking and oscillatory torque are not normally included for the following reasons:

(a)   Multi-phase faults are mostly sequentially developed; that is, they start as single-phase faults and develop into two-phase or three-phase faults. If each phase were to become involved as its voltage reached the peak of the voltage wave, the short circuit current would lag by 90° and would therefore start as a sinusoidal wave beginning at zero on the sine wave. In such a case, there is no requirement for dc offset in the fault current. Records of oscillograms of actual short circuit currents appear to show very little dc offset in the phase currents. Therefore, one cannot depend on the existence of dc braking to assist in stability.

(b)   Even if some dc offset exists, ignoring its effect introduces a slight degree of conservatism.

With the stator transients neglected, the per unit stator voltage Equations 3.120 and 3.121 appear as algebraic equations:

$$e_d = -\psi_q \omega_r - R_a i_d \qquad (5.1)$$

$$e_q = \psi_d \omega_r - R_a i_q \qquad (5.2)$$

5.1.2 Neglecting the Effect of Speed Variations on Stator Voltages

Another simplifying assumption normally made is that the per unit value of ω_r is equal to 1.0 in the stator voltage equations. This is not the same as saying that speed is constant; it assumes that speed changes are small and do not have a significant effect on the voltage.

The assumption of per unit $\omega_r = 1.0$ (i.e., $\omega_r = \omega_0$ rad/s) in the stator voltage equations does not contribute to computational simplicity in itself. The primary reason for making this assumption is that it counterbalances the effect of neglecting $p\psi_d, p\psi_q$ terms so far as the low-frequency rotor oscillations are concerned [7,8]. We will demonstrate this in Example 5.1.

With per unit $\omega_r = 1.0$, the stator voltage equations reduce to

$$e_d = -\psi_q - R_a i_d \tag{5.3}$$

$$e_q = \psi_d - R_a i_q \tag{5.4}$$

All other equations (3.123 to 3.134) summarized in Section 3.4.9 remain the same. Organization of the machine equations in a form suitable for small-signal and transient stability studies is described in Chapters 12 and 13.

Relationship between per unit P_e and T_e

The terminal electric power in per unit is given by

$$P_t = e_d i_d + e_q i_q$$

Substituting for e_d and e_q from Equations 5.3 and 5.4 gives

$$P_t = (-\psi_q - R_a i_d) i_d + (\psi_d - R_a i_q) i_q$$

$$= (\psi_d i_q - \psi_q i_d) - R_a (i_d^2 + i_q^2) \tag{5.5}$$

$$= T_e - R_a I_t^2$$

The air-gap power, measured behind R_a, is given by

$$P_e = P_t + R_a I_t^2$$

$$= T_e \tag{5.6}$$

The per unit air-gap power P_e so computed is in fact the power at synchronous speed and is equal to the per unit air-gap torque T_e.

Normally, $P_e = \omega_r T_e$. However, the assumption of $\omega_r = 1.0$ pu in the stator voltage equation is also reflected in the torque equation, making $P_e = T_e$. This fact is often overlooked.

Example 5.1

With zero armature resistance, show that for small perturbations the effect of neglecting the speed variations (i.e., assuming $\omega_r = \omega_0$) in the stator voltage equations is to counterbalance the effect of neglecting $p\psi_d$, $p\psi_q$ terms.

Solution

With $R_a = 0$ and time t in seconds, the stator voltage Equations 3.120 and 3.121 become

$$e_d = \frac{1}{\omega_0}p\psi_d - \psi_q\frac{\omega_r}{\omega_0} \tag{E5.1}$$

$$e_q = \frac{1}{\omega_0}p\psi_q + \psi_d\frac{\omega_r}{\omega_0} \tag{E5.2}$$

where ω_r and ω_0 are expressed in rad/s.

For small perturbation,

$$\Delta e_d = \frac{1}{\omega_0}p(\Delta\psi_d) - \Delta\psi_q - \psi_{q0}\frac{\Delta\omega_r}{\omega_0} \tag{E5.3}$$

$$\Delta e_q = \frac{1}{\omega_0}p(\Delta\psi_q) + \Delta\psi_d + \psi_{d0}\frac{\Delta\omega_r}{\omega_0} \tag{E5.4}$$

Since

$$e_d = E_t\sin\delta \tag{E5.5}$$

$$e_q = E_t\cos\delta \tag{E5.6}$$

from Equations E5.1, E5.2, at steady state (derivative terms absent) we have

$$e_{d0} = E_t\sin\delta_0 = -\psi_{q0}$$

$$e_{q0} = E_t\cos\delta_0 = \psi_{d0}$$

Hence, for small perturbations, Equations E5.5 and E5.6 become

$$\Delta e_d = (E_t \cos\delta_0)\Delta\delta = \psi_{d0}\Delta\delta \qquad (E5.7)$$

$$\Delta e_q = -(E_t \sin\delta_0)\Delta\delta = \psi_{q0}\Delta\delta \qquad (E5.8)$$

From Equations E5.3, E5.4, E5.7, and E5.8, with $\Delta\omega_r = p(\Delta\delta)$, we may write

$$\psi_{d0}\Delta\delta = \frac{1}{\omega_0}p(\Delta\psi_d) - \Delta\psi_q - \psi_{q0}\frac{1}{\omega_0}p(\Delta\delta) \qquad (E5.9)$$

$$\psi_{q0}\Delta\delta = \frac{1}{\omega_0}p(\Delta\psi_q) + \Delta\psi_d + \psi_{d0}\frac{1}{\omega_0}p(\Delta\delta) \qquad (E5.10)$$

We will compare the expressions for Δe_d and Δe_q, as given by Equations E5.7 and E5.8, with and without the effects of $p\Delta\psi$ and $p\Delta\delta$ terms.

(a) With both $p(\Delta\psi)$ and $p(\Delta\delta)$ terms included:

Rearranging Equation E5.9, we have

$$\Delta\delta = \frac{\dfrac{1}{\omega_0}p(\Delta\psi_d) - \Delta\psi_q}{\psi_{d0} + \dfrac{\psi_{q0}}{\omega_0}p} \qquad (E5.11)$$

From Equation E5.10,

$$\Delta\psi_d = (\psi_{q0} - \frac{\psi_{d0}}{\omega_0}p)\Delta\delta - \frac{1}{\omega_0}p(\Delta\psi_q) \qquad (E5.12)$$

Substituting Equation E5.12 in E5.11, rearranging and simplifying, we get

$$\boxed{\psi_{d0}\Delta\delta = -\Delta\psi_q} \qquad (E5.13)$$

Similarly, from Equation E5.10,

$$\Delta\delta = \frac{\dfrac{1}{\omega_0}p(\Delta\psi_q) + \Delta\psi_d}{\psi_{q0} - \dfrac{\psi_{d0}}{\omega_0}p} \qquad (E5.14)$$

and from Equation E5.9,

$$\Delta\psi_q = -(\psi_{d0} + \frac{\psi_{q0}}{\omega_0}p)\Delta\delta + \frac{1}{\omega_0}p(\Delta\psi_d) \tag{E5.15}$$

Substituting Equation E5.15 in E5.14, rearranging and simplifying, we find

$$\boxed{\psi_{q0}\Delta\delta = \Delta\psi_d} \tag{E5.16}$$

(b) With both p(Δψ) and p(Δδ) terms neglected:

Equations E5.9 and E5.10, with $p\Delta\psi_d = p\Delta\psi_q = p\Delta\delta = 0$, simplify to

$$\boxed{\psi_{d0}\Delta\delta = -\Delta\psi_q} \tag{E5.17}$$

$$\boxed{\psi_{q0}\Delta\delta = \Delta\psi_d} \tag{E5.18}$$

These are same as Equations E5.13 and E5.16.

(c) With only p(Δψ) terms neglected:

From Equation E5.9, with the $p\psi_d$ term neglected, we have

$$\psi_{d0}\Delta\delta = -\Delta\psi_q - \frac{\psi_{q0}}{\omega_0}p(\Delta\delta)$$

Hence,

$$\Delta\delta = -\frac{\Delta\psi_q}{\psi_{d0} + \frac{\psi_{q0}}{\omega_0}p}$$

Multiplying both sides by ψ_{d0} gives

$$\boxed{\psi_{d0}\Delta\delta = -\Delta\psi_q\left[\frac{\psi_{d0}}{\psi_{d0} + \frac{\psi_{q0}}{\omega_0}p}\right]} \tag{E5.19}$$

Similarly from Equation E5.10, with the $p\psi_q$ term neglected, we can show that

$$\psi_{q0}\Delta\delta = \Delta\psi_d\left[\cfrac{\psi_{q0}}{\psi_{q0}+\cfrac{\psi_{d0}}{\omega_0}p}\right] \qquad (E5.20)$$

The above two equations differ from Equations E5.13 and E5.16.

(d) With only p(Δδ) terms neglected:

With $p\Delta\delta=0$, from Equation E5.9, we have

$$\psi_{d0}\Delta\delta = \frac{1}{\omega_0}p(\Delta\psi_d)-\Delta\psi_q \qquad (E5.21)$$

and from Equation E5.10, we have

$$\psi_{q0}\Delta\delta = \frac{1}{\omega_0}p(\Delta\psi_q)+\Delta\psi_d \qquad (E5.22)$$

or

$$\Delta\psi_d = \psi_{q0}\Delta\delta-\frac{1}{\omega_0}p(\Delta\psi_q) \qquad (E5.23)$$

Substituting Equation E5.23 into Equation E5.21 yields

$$\psi_{d0}\Delta\delta = \frac{1}{\omega_0}p(\psi_{q0}\Delta\delta-\frac{p}{\omega_0}\Delta\psi_q)-\Delta\psi_q$$

Grouping terms involving Δδ and rearranging, we have

$$\psi_{d0}\Delta\delta = -\Delta\psi_q\left[\psi_{d0}\cfrac{1+p^2/\omega_0^2}{\psi_{d0}-\psi_{q0}\cfrac{p}{\omega_0}}\right] \qquad (E5.24)$$

Similarly,

$$\psi_{q0}\Delta\delta = \Delta\psi_d\left[\psi_{q0}\cfrac{1+p^2/\omega_0^2}{\psi_{q0}+\psi_{d0}\cfrac{p}{\omega_0}}\right] \qquad (E5.25)$$

Again, the above equations differ from Equations E5.13 and E5.16.

We see that the expressions for Δe_d (i.e., $\psi_{d0}\Delta\delta$) and Δe_q (i.e., $\psi_{q0}\Delta\delta$) with both $p(\Delta\psi)$ and $p(\Delta\delta)$ terms included equal those with both terms neglected, but not those with only one of these terms neglected. Therefore, the effect of neglecting $p(\Delta\delta)$ terms is to counterbalance the effect of neglecting $p(\Delta\psi)$ terms.   ∎

5.2 SIMPLIFIED MODEL WITH AMORTISSEURS NEGLECTED

The first order of simplification to the synchronous machine model is to neglect the amortisseur effects. This minimizes data requirements since the machine parameters related to the amortisseurs are often not readily available. In addition, it may contribute to reduction in computational effort by reducing the order of the model and allowing larger integration steps in time-domain simulations.

With the amortisseurs neglected, the stator voltage Equations 5.3 and 5.4 are unchanged. The remaining equations (3.123 to 3.133) simplify as follows.

Flux linkages:

$$\psi_d = -L_d i_d + L_{ad} i_{fd} \tag{5.7}$$

$$\psi_q = -L_q i_q \tag{5.8}$$

$$\psi_{fd} = -L_{ad} i_d + L_{ffd} i_{fd} \tag{5.9}$$

Rotor voltage:

$$e_{fd} = p\psi_{fd} + R_{fd} i_{fd}$$

or

$$p\psi_{fd} = e_{fd} - R_{fd} i_{fd} \tag{5.10}$$

Equation 5.10 is now the only differential equation associated with the electrical characteristics of the machine. In the above equations all quantities, including time, are in per unit.

Alternative form of machine equations

In the literature on synchronous machines, Equations 5.7 to 5.10 are often written in terms of the following variables:

$$E_I = L_{ad}i_{fd} = \text{voltage proportional to } i_{fd}$$

$$E_q' = \frac{L_{ad}}{L_{ffd}}\psi_{fd} = \text{voltage proportional to } \psi_{fd}$$

$$E_{fd} = \frac{L_{ad}}{R_{fd}}e_{fd} = \text{voltage proportional to } e_{fd}$$

In terms of the new variables, Equation 5.7 becomes

$$\psi_d = -L_d i_d + E_I \tag{5.11}$$

Multiplying Equation 5.9 by L_{ad}/L_{ffd} throughout and expressing in terms of the new variables, we get

$$E_q' = -\frac{L_{ad}^2}{L_{ffd}}i_d + E_I \tag{5.12}$$

Since, from Equation 4.29,

$$L_d' = L_l + \frac{L_{ad}L_{fd}}{L_{ad}+L_{fd}}$$

$$= (L_d - L_{ad}) + \frac{L_{ad}L_{fd}}{L_{ad}+L_{fd}}$$

$$= L_d - \frac{L_{ad}^2}{L_{ffd}}$$

then

$$L_d - L_d' = \frac{L_{ad}^2}{L_{ffd}}$$

Substituting in Equation 5.12 gives

$$E_q' = E_I - (L_d - L_d')i_d \tag{5.13}$$

Multiplying Equation 5.10 by L_{ad}/L_{ffd} throughout, we have

$$p\left(\frac{L_{ad}}{L_{ffd}}\psi_{fd}\right) = \frac{L_{ad}}{R_{fd}}\frac{R_{fd}}{L_{ffd}}e_{fd} - \frac{R_{fd}}{L_{ffd}}L_{ad}i_{fd}$$

Expressing this in terms of the new variables gives

$$pE_q' = \frac{1}{T_{d0}'}(E_{fd}-E_I) \qquad\qquad (5.14)$$

where T_{d0}' is the open-circuit transient time constant defined in Section 4.2.

The following is a summary of the alternative form of machine equations:

$$\psi_d = -L_d i_d + E_I$$
$$\psi_q = -L_q i_q$$
$$E_q' = E_I - (L_d - L_d')i_d$$
$$pE_q' = \frac{1}{T_{d0}'}(E_{fd}-E_I)$$

The above form is used to represent synchronous machines in many stability programs, particularly the older programs. In this formulation, E_q' is the state variable instead of ψ_{fd}. In a way, this is a carryover from the techniques used in the network analyzer days. As seen in the following development of the phasor diagram, E_q' is the q-axis component of the voltage behind transient reactance X_d'. Therefore, the use of E_q' as the state variable allows a simple enhancement to the classical model to account for the field circuit dynamics. It should, however, be recognized that the parameters L_d, L_q, L_d' and T_{d0}' which appear in this formulation are functions of magnetic saturation.

Phasor diagram for transient conditions:

As we account for only the fundamental frequency components of stator quantities, we can use phasor representation to illustrate the transient conditions. In order to do this it is first necessary to express E_q', E_I and E_q in terms of d- and q-axis components of terminal voltage and current.

Since in per unit $X_d = L_d$, from Equations 5.4 and 5.7,

$$e_q = \psi_d - R_a i_q$$
$$= -X_d i_d + X_{ad} i_{fd} - R_a i_q$$
$$= -X_d i_d + E_I - R_a i_q$$

Therefore

$$E_I = e_q + X_d i_d + R_a i_q$$

Multiplying by j, we have

$$jE_I = je_q + jX_d i_d + jR_a i_q$$

In terms of phasor notation,

$$\tilde{E}_I = \tilde{e}_q + jX_d \tilde{i}_d + R_a \tilde{i}_q \tag{5.15}$$

From Equation 5.13, with $X_d' = L_d'$,

$$\begin{aligned} E_q' &= e_q + X_d i_d + R_a i_q - X_d i_d + X_d' i_d \\ &= e_q + X_d' i_d + R_a i_q \end{aligned}$$

Multiplying by j gives

$$jE_q' = je_q + jX_d' i_d + jR_a i_q$$

In phasor notation,

$$\tilde{E}_q' = \tilde{e}_q + jX_d' \tilde{i}_d + R_a \tilde{i}_q \tag{5.16}$$

We see that the phasors $\tilde{E}_I$ and $\tilde{E}_q'$ both lie along the q-axis. In Section 3.6.2, we saw that $\tilde{E}_q$ also lies along the q-axis.

Rearranging Equation 3.167 and substituting E_I for $X_{ad} i_{fd}$, we get

$$\tilde{E}_I = \tilde{E}_q + j(X_d - X_q)\tilde{i}_d \tag{5.17}$$

Figure 5.5 shows the phasor diagram representing $\tilde{E}_q'$, $\tilde{E}_q$ and $\tilde{E}_I$ given by Equations 5.15, 5.16 and 5.17.

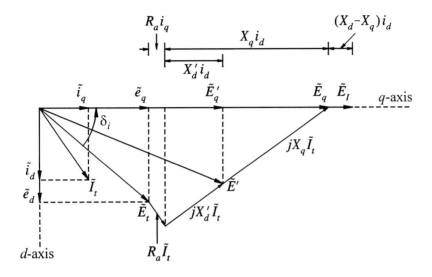

$$\tilde{E}_t = e_d + je_q = \tilde{e}_d + \tilde{e}_q$$

$$\tilde{E}' = \tilde{E}_t + (R_a + jX_d')\tilde{I}_t$$

$$\tilde{E}_q' = q\text{-axis component of } \tilde{E}'$$

$$= \tilde{e}_q + R_a\tilde{i}_q + jX_d'\tilde{i}_d$$

$$\tilde{E}_q = \text{Voltage behind } R_a + jX_q$$

$$= \tilde{E}_t + (R_a + jX_q)\tilde{I}_t$$

$$= \tilde{e}_q + R_a\tilde{i}_q + jX_q\tilde{i}_d$$

$$\tilde{E}_I = \tilde{E}_q + j(X_d - X_q)\tilde{i}_d$$

Figure 5.5 Synchronous machine phasor diagram
in terms of E_q, E_I, and E_q'

5.3 CONSTANT FLUX LINKAGE MODEL

5.3.1 Classical Model

For studies in which the period of analysis is small in comparison to T'_{do}, the machine model of Section 5.2 is often simplified by assuming E'_q (or ψ_{fd}) constant throughout the study period. This assumption eliminates the only differential equation associated with the electrical characteristics of the machine.

A further approximation, which simplifies the machine model significantly, is to ignore transient saliency by assuming $X'_d = X'_q$, and to assume that the flux linkage ψ_{1q} (associated with the q-axis rotor circuit corresponding to X'_q) also remains constant. With these assumptions, as shown below, the voltage behind the transient impendence $R_a + jX'_d$ has a constant magnitude.

The d- and q-axis equivalent circuits with only one circuit in each axis are shown in Figure 5.6.

The per unit flux linkages identified in the d-axis are given by

$$\psi_{ad} = -L_{ad}i_d + L_{ad}i_{fd} \tag{5.18}$$

$$\psi_d = \psi_{ad} - L_l i_d \tag{5.19}$$

$$\psi_{fd} = \psi_{ad} + L_{fd}i_{fd} \tag{5.20}$$

From Equation 5.20

$$i_{fd} = \frac{\psi_{fd} - \psi_{ad}}{L_{fd}} \tag{5.21}$$

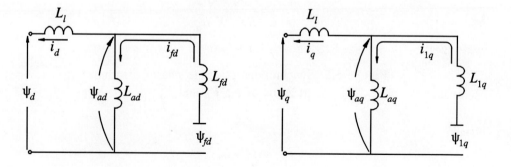

Figure 5.6 The d- and q-axis equivalent circuits with one rotor circuit in each axis

Substituting in Equation 5.18 gives

$$\psi_{ad} = -L_{ad}i_d + \frac{L_{ad}}{L_{fd}}(\psi_{fd} - \psi_{ad})$$

Rearranging to express ψ_{ad} in terms of ψ_{fd}, we find

$$\psi_{ad} = L'_{ad}\left(-i_d + \frac{\psi_{fd}}{L_{fd}}\right) \tag{5.22}$$

where

$$L'_{ad} = \frac{1}{\dfrac{1}{L_{ad}} + \dfrac{1}{L_{fd}}} = L'_d - L_l \tag{5.23}$$

Similarly, for the q-axis

$$\psi_{aq} = L'_{aq}\left(-i_q + \frac{\psi_{1q}}{L_{1q}}\right) \tag{5.24}$$

where

$$L_{aq} = L'_d - L_l$$

From Equation 5.1, the d-axis stator voltage is given by

$$e_d = -R_a i_d - \omega\psi_q$$

$$= -R_a i_d + \omega(L_l i_q - \psi_{aq})$$

where $\omega = \omega_r = \omega_0 = 1.0$ pu. Substituting for ψ_{aq} from Equation 5.24 gives

$$e_d = -R_a i_d + \omega L_l i_q - \omega L'_{aq}\left(-i_q + \frac{\psi_{1q}}{L_{1q}}\right)$$

$$= -R_a i_d + \omega(L_l + L'_{aq})i_q - \omega L'_{aq}\left(\frac{\psi_{1q}}{L_{1q}}\right) \tag{5.25}$$

$$= -R_a i_d + X'_q i_q + E'_d$$

where

$$E_d' = -\omega L_{aq}'\left(\frac{\psi_{1q}}{L_{1q}}\right) \tag{5.26}$$

Similarly, the q-axis stator voltage is given by

$$e_q = -R_a i_q - X_d' i_d + E_q' \tag{5.27}$$

where

$$E_q' = \omega L_{ad}'\left(\frac{\psi_{fd}}{L_{fd}}\right) \tag{5.28}$$

With transient saliency neglected ($X_d'=X_q'$), the stator terminal voltage is

$$e_d + je_q = (E_d' + jE_q') - R_a(i_d + ji_q) + X_d'(i_q - ji_d)$$

$$= (E_d' + jE_q') - R_a(i_d + ji_q) - jX_d'(i_d + ji_q)$$

Using phasor notation, we have

$$\tilde{E}_t = \tilde{E}' - (R_a + jX_d')\tilde{I}_t \tag{5.29}$$

where

$$\tilde{E}' = E_d' + jE_q'$$

$$= L_{ad}'\left(-\frac{\psi_{1q}}{L_{1q}} + j\frac{\psi_{fd}}{L_{fd}}\right)$$

The corresponding equivalent is shown in Figure 5.7.

   With rotor flux linkages (ψ_{fd} and ψ_{1q}) constant, E_d' and E_q' are constant. Therefore, the magnitude of E' is constant. As the rotor speed changes, the d- and q-axes move with respect to any general reference coordinate system whose R-I axes rotate at synchronous speed, as shown in Figure 5.8. Hence, the components E_R' and E_I' change.

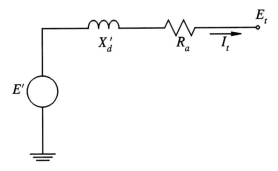

Figure 5.7 Simplified transient model

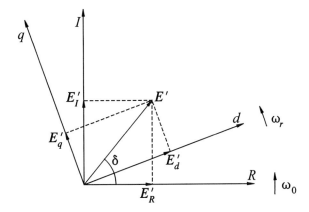

Figure 5.8 The *R-I* and *d-q* coordinate systems

The magnitude of E' can be determined by computing its pre-disturbance value.

$$\tilde{E}' = \tilde{E}_{t0} + (R_a + jX_d')\tilde{I}_{t0}$$

Its magnitude is then assumed to remain constant throughout the study period. Since R_a is small, it is usual to neglect it.

With the components E_d' and E_q' each having a constant magnitude, E' will have constant orientation with respect to d- and q-axes, as the rotor speed changes. Therefore, the angle of E' with respect to synchronously rotating reference axes $(R\text{-}I)$ can be used as a measure of the rotor angle.

This model offers considerable computational simplicity; it allows the transient electrical performance of the machine to be represented by a simple voltage source of fixed magnitude behind an effective reactance. It is commonly referred to as the *classical model*, since it was used extensively in early stability studies.

5.3.2 Constant Flux Linkage Model Including the Effects of Subtransient Circuits

With the subtransient circuits included (see Figure 5.9), the expression developed for the direct axis air-gap flux linkage in the previous section changes as follows:

$$\psi_{ad} = -L_{ad}i_d + L_{ad}i_{fd} + L_{ad}i_{1d}$$

$$= -L_{ad}i_d + \frac{L_{ad}}{L_{fd}}(\psi_{fd} - \psi_{ad}) + \frac{L_{ad}}{L_{1d}}(\psi_{1d} - \psi_{ad})$$

$$= L_{ad}''\left(-i_d + \frac{\psi_{fd}}{L_{fd}} + \frac{\psi_{1d}}{L_{1d}}\right) \tag{5.30}$$

where

$$L_{ad}'' = \frac{1}{\dfrac{1}{L_{ad}} + \dfrac{1}{L_{fd}} + \dfrac{1}{L_{1d}}} = L_d'' - L_l \tag{5.31}$$

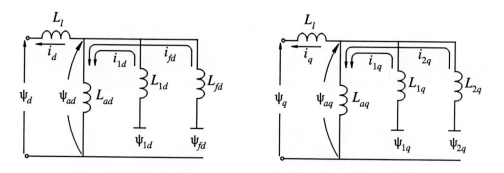

Figure 5.9 Equivalent circuits with subtransient circuits

Similarly, for the q-axis

$$\psi_{aq} = L_{aq}''\left(-i_q + \frac{\psi_{1q}}{L_{1q}} + \frac{\psi_{2q}}{L_{2q}}\right) \tag{5.32}$$

where

$$L_{aq}'' = L_q'' - L_l$$

The d-axis stator voltage is

$$e_d = -R_a i_d + \omega(L_l i_q - \psi_{aq})$$

$$= -R_a i_d + \omega(L_l + L''_{aq})i_q - \omega L''_{aq}(\frac{\psi_{1q}}{L_{1q}} + \frac{\psi_{2q}}{L_{2q}}) \quad (5.33)$$

$$= -R_a i_d + X''_q i_q + E''_d$$

Similarly, the q-axis stator voltage is

$$e_q = -R_a i_q - X''_d i_d + E''_q \quad (5.34)$$

where

$$E''_d = -\omega L''_{aq}\left(\frac{\psi_{1q}}{L_{1q}} + \frac{\psi_{2q}}{L_{2q}}\right) \quad (5.35)$$

$$E''_q = \omega L''_{ad}\left(\frac{\psi_{fd}}{L_{fd}} + \frac{\psi_{1d}}{L_{1d}}\right) \quad (5.36)$$

With subtransient saliency neglected, $X''_d = X''_q$. We then have

$$\tilde{E}_t = e_d + je_q$$

$$= (E''_d + jE''_q) - (R_a + jX''_d)(i_d + ji_q) \quad (5.37)$$

$$= \tilde{E}'' - (R_a + jX''_d)\tilde{I}_t$$

The corresponding equivalent circuit is shown in Figure 5.10. With constant rotor flux linkages, E''_d and E''_q are constant.

This model is used in short-circuit programs for computing the initial value of the fundamental frequency component of short-circuit currents. As the rotor flux linkages cannot change instantaneously, the value E'' is equal to its prefault value.

Such a constant flux linkage model would not be generally acceptable for stability studies, since the subtransient time constants associated with the decay of ψ_{1d} and ψ_{2q} are very small compared to the study period, and hence the assumption of constant rotor flux linkage of all rotor circuits, including the subtransient circuits, is not reasonable. The model, however, is valid if rotor flux linkage variations are accounted for, in which case the magnitude of E'' varies as determined by Equations 3.123 to 3.126.

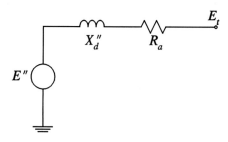

Figure 5.10 Simplified model for subtransient period

5.3.3 Summary of Simple Models for Different Time Frames

Figure 5.11 summarizes the simple models of the synchronous machine applicable to the three time frames: subtransient, transient, and steady state. The model applicable to steady-state conditions was developed in Chapter 3 (Section 3.6.4) and is included here since it has the same structure as the models developed in this chapter for transient and subtransient conditions. The subtransient and transient models assume constant rotor flux linkages, and the steady-state model assumes constant field current. These models neglect saliency effects and stator resistance and offer considerable structural and computational simplicity.

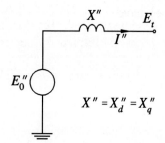

E_0'' is the predisturbance value of internal voltage given by

$$\tilde{E}_0'' = \tilde{E}_{t0} + jX''\tilde{I}_{t0}$$

(a) Subtransient model

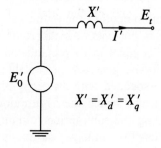

E_0' is the internal voltage

$$\tilde{E}_0' = \tilde{E}_{t0} + jX'\tilde{I}_{t0}$$

(b) Transient model

Figure 5.11 Simple synchronous machine models (*Continued on next page*)

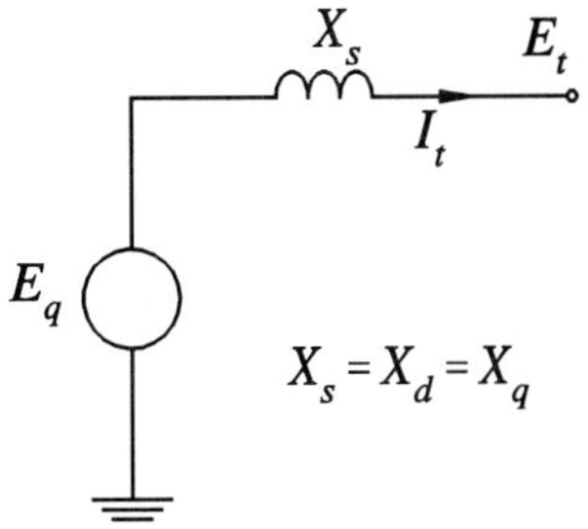

$$\tilde{E}_q = \tilde{E}_{t0} + jX_s\tilde{I}_{t0}$$
$$|E_q| = X_{ad}i_{fd} = E_I$$

(c) Steady-state model

Figure 5.11 (*Continued*) Simple synchronous machine models

5.4 REACTIVE CAPABILITY LIMITS

In voltage stability and long-term stability studies, it is important to consider the reactive capability limits of synchronous machines. In this section, we will develop capability curves which identify these limits.

5.4.1 Reactive Capability Curves

Synchronous generators are rated in terms of the maximum MVA output at a specified voltage and power factor (usually 0.85 or 0.9 lagging) which they can carry continuously without overheating. The active power output is limited by the prime mover capability to a value within the MVA rating. The continuous reactive power output capability is limited by three considerations: armature current limit, field current limit, and end region heating limit.

Armature current limit

The armature current results in an RI^2 power loss, and the energy associated with this loss must be removed so as to limit the increase in temperature of the conductor and its immediate environment. Therefore, one of the limitations on generator rating is the maximum current that can be carried by the armature without exceeding the heating limitations.

The per unit complex output power is

$$S = P + jQ = \tilde{E}_t\tilde{I}_t^*$$

$$= |E_t||I_t|(\cos\phi + j\sin\phi)$$

where ϕ is the power factor angle.

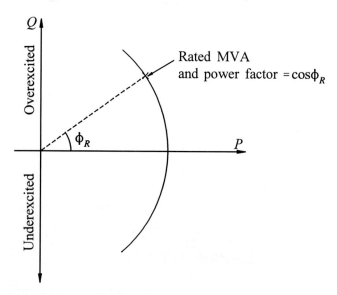

Figure 5.12   Armature current heating limit

Therefore, in the P-Q plane the armature current limit, as shown in Figure 5.12, appears as a circle with centre at the origin and radius equal to the MVA rating.

Field current limit

Because of the heat resulting from the $R_{fd}i_{fd}^2$ power loss, the field current imposes a second limit on the operation of the generator.

The constant field current locus may be developed by the steady-state equivalent circuit developed in Section 3.6.4 and reproduced in Section 5.3.3. With $X_d = X_q = X_s$, the equivalent circuit of Figure 5.11(c) gives the relationship between E_t, I_t and E_q (equal to $X_{ad}i_{fd}$). The corresponding phasor diagram, with R_a neglected, is shown in Figure 5.13.

Equating the components along and perpendicular to the phasor $\tilde{E}_t$, we get

$$(X_{ad}i_{fd})\sin\delta_i = X_s I_t \cos\phi \tag{5.38}$$

$$(X_{ad}i_{fd})\cos\delta_i = E_t + X_s I_t \sin\phi \tag{5.39}$$

Rearranging yields

$$I_t\cos\phi = \frac{X_{ad}i_{fd}\sin\delta_i}{X_s}$$

$$I_t\sin\phi = \frac{X_{ad}i_{fd}\cos\delta_i - E_t}{X_s}$$

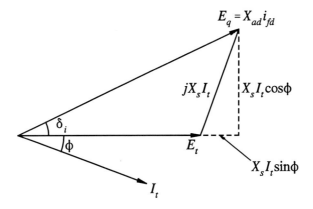

Figure 5.13 Steady-state phasor diagram

Therefore,

$$P = E_t I_t \cos\phi = \frac{X_{ad}}{X_s} E_t i_{fd} \sin\delta_i \tag{5.40}$$

$$Q = E_t I_t \sin\phi = \frac{X_{ad}}{X_s} E_t i_{fd} \cos\delta_i - \frac{E_t^2}{X_s} \tag{5.41}$$

The relationship between the active and reactive powers for a given field current is a circle centred at $-E_t^2/X_s$ on the Q-axis and with $(X_{ad}/X_s)E_t i_{fd}$ as the radius. Therefore, the effect of the maximum field current rating on the capability of the machine may be illustrated on the P-Q plane as shown in Figure 5.14.

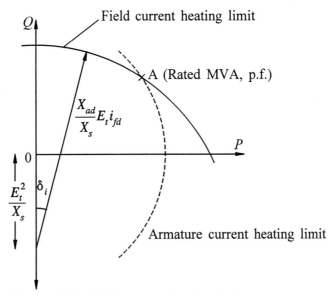

Figure 5.14 Field current heating limit

In any balanced design, the thermal limits for the field and armature intersect at a point A, which represents the machine nameplate MVA and power factor rating.

End region heating limit

The localized heating in the end region of the armature imposes a third limit on the operation of a synchronous machine. As explained below, this limit affects the capability of the machine in the underexcited condition.

Figure 5.15 is a schematic of the end-turn region of a generator. The end-turn leakage flux, as shown in the figure, enters and leaves in a direction perpendicular (axial) to the stator laminations. This causes eddy currents in the laminations, resulting in localized heating in the end region. The high field currents corresponding to the overexcited condition keep the retaining ring saturated, so that end leakage flux is small. However, in the underexcited region the field current is low and the retaining ring is not saturated; this permits an increase in armature end leakage flux [9]. Also, in the underexcited condition, the flux produced by the armature currents adds to the flux produced by the field current; therefore, the end-turn flux enhances the axial flux in the end region and the resulting heating effect may severely limit the generator output, particularly in the case of a round rotor machine. This is illustrated in Figure 5.16, which also includes the limit imposed by the armature current heating effects.

The field current and armature current heating limits when plotted on a P-Q plane depend on the armature voltage. Figure 5.17 shows these limits for a 588 MVA, 22 kV, 0.85 power factor machine at two values of armature voltage, 1.0 pu and 0.95 pu.

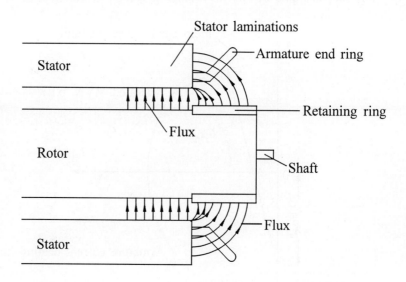

Figure 5.15 Sectional view of end region of a generator

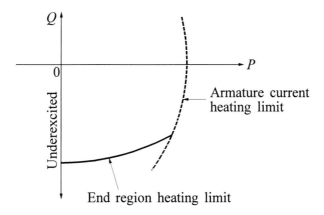

Figure 5.16 End region heating limit

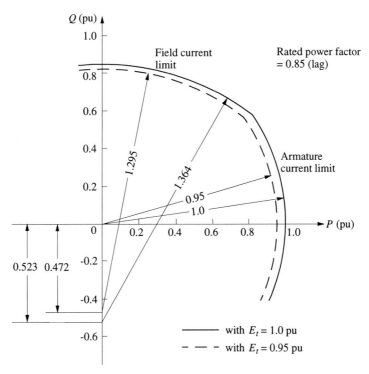

Figure 5.17 Effect of reducing the armature voltage
on the generator capability curve

The above limits on the operation of the generator are those imposed by the capabilities of the machine itself and are determined by the design of the machine. Additional limits may be imposed by power system stability limits.

Figure 5.18 shows the reactive capability curves of a 400 MVA hydrogen-cooled steam turbine-driven generator at rated armature voltage. The effectiveness of cooling and hence the allowable machine loading depend on the hydrogen pressure. The base MVA in this case is the rated MVA at 45 PSIG (pounds/square inch gauge) hydrogen pressure. For each pressure, segment AB represents the field heating limit, segment BC the armature heating limit, and segment CD the end region heating limit. Also shown in the figure are the loci of constant power factor.

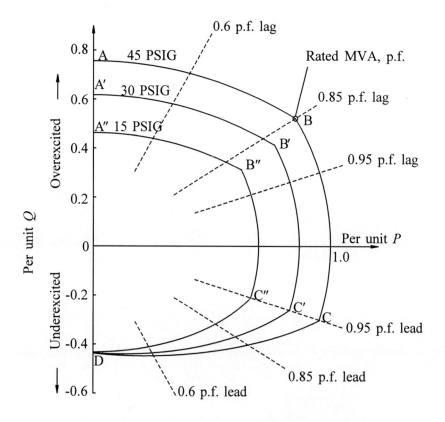

Figure 5.18 Reactive capability curves of a hydrogen-cooled generator at rated voltage

5.4.2 V Curves and Compounding Curves

As discussed in Example 3.4 the curve showing the relation between armature current and field current at a constant terminal voltage and with constant active power output is known as a *V curve*. The V curves for the 400 MVA generator considered in Section 5.4.1 are shown in Figure 5.19. These are solid curves shown for three values of P (0.5, 0.7, 0.85 pu). The dashed lines are loci of constant power factor and are known as *compounding curves*. Each of these curves shows how field current has to vary in order to maintain a constant power factor.

As illustrated in Example 3.4, V curves and compounding curves can be readily computed using the steady-state equations summarized in Section 3.6.5. They can also be determined approximately by using the equivalent circuit of Figure 5.11(c).

Also shown in Figure 5.19 are the reactive capability limits for one value of hydrogen pressure (45 PSIG). The three segments AB, BC and CD correspond to the field current limit, armature current limit, and end region heating limit, respectively. Since the characteristics shown in Figure 5.19 apply at rated stator terminal voltage, the per unit values of armature current and apparent power output are equal, and hence both are shown along the ordinate. The field current plotted along the abscissa is the normalized value, with 1.0 pu representing the field current corresponding to rated MVA output and power factor.

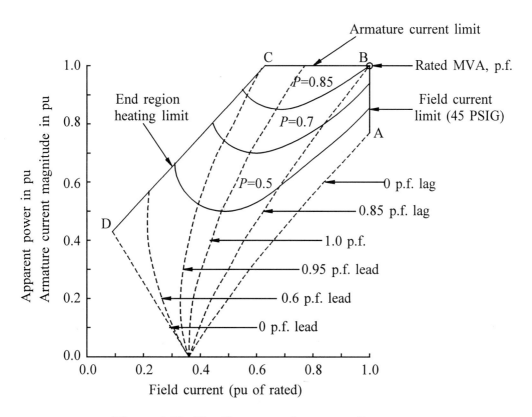

Figure 5.19 The V curves and compounding curves
for a generator at rated armature voltage

REFERENCES

[1]   Kundur Prabhashankar, "Digital Simulation and Analysis of Power System Dynamic Performance," Ph.D. thesis, University of Toronto, May 1967.

[2]   Kundur Prabhashankar and W. Janischewskyj, "Digital Simulation and Analysis of Power System Dynamic Performance - Effects of Synchronous Machine Amortisseur Circuits and Control Mechanisms," Paper CP 94-PWR, presented at the IEEE Winter Power Meeting, January/February 1968.

[3]   R.G. Harley and B. Adkins, "Calculation of the Angular Back Swing Following a Short Circuit of a Loaded Alternator," *Proc. IEE*, Vol. 117, No. 2, pp. 377-386, 1970.

[4]   G. Shackshaft, "Effect of Oscillatory Torque on the Movement of Generator Rotors," *Proc. IEE*, Vol. 117, No. 10, pp. 1969-1974, 1970.

[5]   E.W. Kimbark, *Power System Stability, Vol. III: Synchronous Machines*, John Wiley & Sons, 1956.

[6]   D.B. Mehta and B. Adkins, "Transient Torque and Load Angle of a Synchronous Generator Following Several Types of System Disturbance," *Proc. IEE*, Vol. 107A, pp. 61-74, 1960.

[7]   G. Kron, Discussion of paper by R.H. Park, "Two-Reaction Theory of Synchronous Machines - Part II," *AIEE Trans.*, Vol. 52, pp. 352-355, June 1933.

[8]   D.N. Ewart and R.P. Schulz, "Face Multi-machine Power System Simulator Program," *Proc. PICA Conference*, pp. 133-153, May 1969.

[9]   S.B. Farnham and R.W. Swartnout, "Field Excitation in Relation to Machine and System Operation," *AIEE Trans.*, pp. 1215-1223, December 1953.

Chapter 6

AC Transmission

This chapter will review the characteristics and modelling of ac transmission elements and develop methods of power flow analysis in transmission systems.

The focus is on those aspects of the transmission system characteristics that affect system stability and voltage control. Specifically, the objectives are as follows:

(a)   To develop performance equations and models for transmission lines;

(b)   To examine the power transfer capabilities of transmission lines as influenced by voltage, reactive power, thermal, and system stability considerations;

(c)   To develop models for representation of two-winding, three-winding, and phase-shifting transformers;

(d)   To examine factors influencing the flow of active power and reactive power through transmission networks; and

(e)   To describe analytical techniques for the analysis of power flow in transmission systems.

The above require consideration of balanced steady-state operation of the transmission system. As such, we will model and analyze the performance of transmission elements in terms of their single-phase equivalents.

6.1 TRANSMISSION LINES

Electrical power is transferred from generating stations to consumers through overhead lines and cables.

Overhead lines are used for long distances in open country and rural areas, whereas cables are used for underground transmission in urban areas and for underwater crossings. For the same rating, cables are 10 to 15 times more expensive than overhead lines and they are therefore only used in special situations where overhead lines cannot be used; the distances in such applications are short.

6.1.1 Electrical Characteristics

(a) Overhead lines

A transmission line is characterized by four parameters: series resistance R due to the conductor resistivity, shunt conductance G due to leakage currents between the phases and ground, series inductance L due to magnetic field surrounding the conductors, and shunt capacitance C due to the electric field between conductors.

Detailed derivations from first principles for the line parameters can be found in standard books on power systems [1-7]. References 8 and 9 provide data related to transmission line configurations used in practice. Here, we will briefly summarize salient points relating to line parameters.

Series Resistance (R). The resistances of lines accounting for stranding and skin effect are determined from manufacturers' tables.

Shunt Conductance (G). The shunt conductance represents losses due to leakage currents along insulator strings and corona. In power lines, its effect is small and usually neglected.

Series Inductance (L). The line inductance depends on the partial flux linkages within the conductor cross section and external flux linkages. For overhead lines, the inductances of the three phases are different from each other unless the conductors have equilateral spacing, a geometry not usually adopted in practice. The inductances of the three phases with non-equilateral spacing can be equalized by transposing the lines in such a way that each phase occupies successively all three possible positions.

For a transposed three-phase line, the inductance per phase is [1]

$$L = 2\times10^{-7}\ln\frac{D_{eq}}{D_s} \qquad \text{H/m} \qquad (6.1)$$

In the above equation, D_s is the self geometric mean distance, taking into account the conductor composition, stranding, and bundling; it is also called the geometric mean radius. And D_{eq} is the geometric mean of the distances between the conductors of the three phases a, b, and c:

$$D_{eq} = (d_{ab}d_{bc}d_{ca})^{1/3} \tag{6.2}$$

Shunt Capacitance (C). The potential difference between the conductors of a transmission line causes the conductors to be charged; the charge per unit of potential difference is the capacitance between conductors. When alternating voltages are applied to the conductors, a charging current flows due to alternate charging and discharging of the capacitances. For a three-phase transposed line, the capacitance of each phase to neutral is [1]

$$C = \frac{2\pi k}{\ln(D_{eq}/r)} \qquad \text{F/m} \tag{6.3}$$

where r is the conductor radius, D_{eq} is given by Equation 6.2, and k is the permittivity of the dielectric medium. For parallel-circuit lines, the "modified geometric mean distance" of the conductors of the same phase replaces r in Equation 6.3 [1].

The earth presents an equipotential surface and will hence influence the capacitance per phase. This can be accounted for by using the concept of "images" [1-5].

(b) Underground cables [3,4,6,7]

Underground cables have the same basic parameters as overhead lines: series resistance and inductance; shunt capacitance and conductance.

However, the values of the parameters and hence the characteristic of cables differ significantly from those of overhead lines for the following reasons:

1.   The conductors in a cable are much closer to each other than are the conductors of overhead lines.

2.   The conductors in a cable are surrounded by metallic bodies such as shields, lead or aluminium sheets, and steel pipes.

3.   The insulating material between conductors in a cable is usually impregnated paper, low-viscosity oil, or an inert gas.

6.1.2 Performance Equations

In the previous section, we identified the parameters of a transmission line per unit length. These are distributed parameters; that is, the effects represented by the parameters are distributed throughout the length of the line.

If the line is assumed transposed, we can analyze the line performance on a per-phase basis. Figure 6.1 shows the relationship between current and voltage along one phase of the line in terms of the distributed parameters, with

$$z = R + j\omega L = \text{series impedance per unit length/phase}$$
$$y = G + j\omega C = \text{shunt admittance per unit length/phase}$$
$$l = \text{length of the line}$$

The voltages and currents shown are phasors representing sinusoidal time-varying quantities.

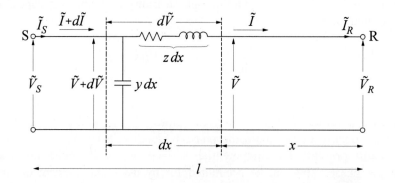

Figure 6.1 Voltage and current relationship of a distributed parameter line

Consider a differential section of the line of length dx at a distance x from the receiving end. The differential voltage across the elemental length is given by

$$d\tilde{V} = \tilde{I}(z\,dx)$$

Hence,

$$\frac{d\tilde{V}}{dx} = \tilde{I}z \qquad\qquad (6.4)$$

The differential current flowing into the shunt admittance is

$$d\tilde{I} = \tilde{V}(y\,dx)$$

Hence,

$$\frac{d\tilde{I}}{dx} = \tilde{V}y \qquad\qquad (6.5)$$

Differentiating Equations 6.4 and 6.5 with respect to x, we obtain

$$\frac{d^2\tilde{V}}{dx^2} = z\frac{d\tilde{I}}{dx} = yz\tilde{V} \tag{6.6}$$

and

$$\frac{d^2\tilde{I}}{dx^2} = y\frac{d\tilde{V}}{dx} = yz\tilde{I} \tag{6.7}$$

We will establish the boundary conditions by assuming that voltage V_R and current I_R are known at the receiving end ($x=0$). The general solution of Equations 6.6 and 6.7 for voltage and current at a distance x from the receiving end is

$$\tilde{V} = \frac{\tilde{V}_R+Z_C\tilde{I}_R}{2}e^{\gamma x} + \frac{\tilde{V}_R-Z_C\tilde{I}_R}{2}e^{-\gamma x} \tag{6.8}$$

$$\tilde{I} = \frac{\tilde{V}_R/Z_C+\tilde{I}_R}{2}e^{\gamma x} - \frac{\tilde{V}_R/Z_C-\tilde{I}_R}{2}e^{-\gamma x} \tag{6.9}$$

where

$$Z_C = \sqrt{z/y} \tag{6.10}$$

$$\gamma = \sqrt{yz} = \alpha+j\beta \tag{6.11}$$

The constant Z_C is called the *characteristic impedance* and γ is called the *propagation constant*.

The constants γ and Z_C are complex quantities. The real part of the propagation constant γ is called the *attenuation constant* α, and the imaginary part the *phase constant* β. Thus the exponential term $e^{\gamma x}$ may be expressed as follows:

$$e^{\gamma x} = e^{(\alpha+j\beta)x} = e^{\alpha x}(\cos\beta x+j\sin\beta x) \tag{6.12}$$

Therefore, the first term in Equation 6.8 increases in magnitude and advances in phase as the distance from the receiving end increases. This term is called the *incident voltage*.

The expanded form of the second exponential term is

$$e^{-\gamma x} = e^{-\alpha x}(\cos\beta x-j\sin\beta x) \tag{6.13}$$

As a result, the second term in Equation 6.8 decreases in magnitude and is retarded in phase from the receiving end toward the sending end. It is called the *reflected voltage*.

At any point along the line, the voltage is equal to the sum of the incident and reflected components at that point. Since Equation 6.9 is similar to Equation 6.8, the current is also composed of incident and reflected components.

If a line is terminated in its characteristic impedance Z_C, V_R is equal to $Z_C I_R$, and there is no reflected wave. Such a line is called a *flat line* or an *infinite line* (since a line of infinite length cannot have a reflected wave). Power lines, unlike communication lines, are not usually terminated in their characteristic impedances.

From Equations 6.12 and 6.13, we see that the incident and reflected components of voltage and current at any instant in time appear as sinusoidal curves or waves along the length of the line. In addition to this variation, the voltage and current components at any point along the line vary in time, since V_R and I_R in Equations 6.8 and 6.9 are phasors representing sinusoidal time-varying quantities. Thus, the incident and reflected components of voltage and current represent travelling waves. They are similar to the travelling waves in water. The total instantaneous voltage and current along the transmission line are not travelling, but they can each be interpreted as the sum of two such travelling waves.

For typical power lines, G is practically zero and $R \ll \omega L$. Therefore,

$$Z_C = \sqrt{\frac{R+j\omega L}{j\omega C}} \approx \sqrt{\frac{L}{C}\left(1-j\frac{R}{2\omega L}\right)} \tag{6.14}$$

$$\gamma = \sqrt{(R+j\omega L)j\omega C} \approx j\omega\sqrt{LC}\left(1-j\frac{R}{2\omega L}\right) \tag{6.15}$$

If losses are completely neglected, Z_C is a real number (i.e., a pure resistance), and γ is an imaginary number.

For a *lossless line*, Equations 6.8 and 6.9 simplify to

$$\tilde{V} = \tilde{V}_R\cos\beta x + jZ_C\tilde{I}_R\sin\beta x \tag{6.16}$$

and

$$\tilde{I} = \tilde{I}_R\cos\beta x + j(\tilde{V}_R/Z_C)\sin\beta x \tag{6.17}$$

Thus, the voltage and current vary harmonically along the line length. A full cycle of voltage and current in space along the line length corresponds to 2π radians. The length corresponding to one full cycle is called the *wavelength* λ. If β is the phase shift in radians per meter, the wavelength in meters is

$$\lambda = \frac{2\pi}{\beta} \qquad (6.18)$$

6.1.3 Natural or Surge Impedance Loading

Since G is negligible and R is small, high-voltage lines are assumed to be lossless when we are dealing with lightning and switching surges. Hence, the characteristic impedance Z_C with losses neglected is commonly referred to as the *surge impedance*. It is equal to $\sqrt{L/C}$ and has the dimension of a pure resistance.

The power delivered by a transmission line when it is terminated by its surge impedance is known as the *natural load* or *surge impedance load* (SIL):

$$SIL = \frac{V_0^2}{Z_C} \quad W \qquad (6.19)$$

where V_0 is the rated voltage of the line. If V_0 is the line-to-neutral voltage, SIL given by the above equation is the per-phase value; if V_0 is the line-to-line value, then SIL is the three-phase value.

From Equations 6.16 and 6.17, the voltage and current along the length of a lossless line at SIL are given by

$$\tilde{V} = \tilde{V}_R e^{\gamma x} \qquad (6.20)$$

and

$$\tilde{I} = \tilde{I}_R e^{\gamma x} \qquad (6.21)$$

where $\gamma = j\beta = j\omega\sqrt{LC}$.

At SIL, transmission lines (lossless) exhibit the following special characteristics:

- $\tilde{V}$ and $\tilde{I}$ have constant amplitude along the line.

- $\tilde{V}$ and $\tilde{I}$ are in phase throughout the length of the line.

- The phase angle between the sending end and receiving end voltages (currents) is equal to βl (see Figure 6.2).

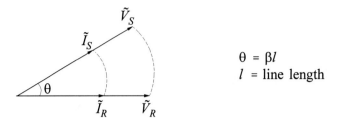

$\theta = \beta l$

l = line length

Figure 6.2 Sending end and receiving end voltage and
current relationships of a lossless line at SIL

At the natural load, the reactive power generated by C is equal to the reactive
power absorbed by L, for each incremental length of the line. Hence, no reactive
power is absorbed or generated at either end of the line, and the voltage and current
profiles are flat. This is an optimum condition with respect to control of voltage and
reactive power.

As we will see in subsequent sections of this chapter, the natural or surge
impedance loading of a line serves as a convenient *reference quantity* for evaluating
and expressing its capability.

6.1.4 Equivalent Circuit of a Transmission Line

Equations 6.8 and 6.9 provide a complete description of the performance of
transmission lines. However, for purposes of analysis involving interconnection with
other elements of the system, it is more convenient to use equivalent circuits which
represent the performance of the lines only as seen from their terminals.

By letting $x=l$ in Equation 6.8 and rearranging, we have

$$\tilde{V}_S = \tilde{V}_R \frac{e^{\gamma l}+e^{-\gamma l}}{2} + Z_C \tilde{I}_R \frac{e^{\gamma l}-e^{-\gamma l}}{2}$$

$$= \tilde{V}_R \cosh(\gamma l) + Z_C \tilde{I}_R \sinh(\gamma l) \tag{6.22}$$

Similarly from Equation 6.9, we have

$$\tilde{I}_S = \tilde{I}_R \cosh(\gamma l) + \frac{\tilde{V}_R}{Z_C} \sinh(\gamma l) \tag{6.23}$$

A π circuit with *lumped* parameters, as shown in Figure 6.3, can be used to represent
the above relationships.

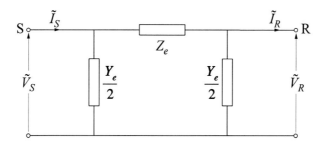

Figure 6.3 Equivalent π circuit of a transmission line

From the equivalent circuit, the sending end voltage is

$$\tilde{V}_s = Z_e\left(\tilde{I}_R + \frac{Y_e}{2}\tilde{V}_R\right) + \tilde{V}_R$$

$$= \left(\frac{Z_e Y_e}{2} + 1\right)\tilde{V}_R + Z_e\tilde{I}_R \qquad (6.24)$$

Comparing Equation 6.24 with 6.22, we have

$$Z_e = Z_C \sinh(\gamma l) \qquad (6.25)$$

and

$$\frac{Z_e Y_e}{2} + 1 = \cosh(\gamma l)$$

Therefore,

$$\frac{Y_e}{2} = \frac{1}{Z_C}\frac{\cosh(\gamma l) - 1}{\sinh(\gamma l)}$$

$$= \frac{1}{Z_C}\tanh\left(\frac{\gamma l}{2}\right) \qquad (6.26)$$

Equations 6.25 and 6.26 give the elements of the equivalent circuit. These elements reflect the voltage and current relationships given by Equations 6.22 and 6.23 exactly.

Nominal π equivalent circuit

If $\gamma l \ll 1$, the expressions for Z_e and Y_e may be approximated as follows:

$$Z_e = Z_C \sinh(\gamma l)$$

$$\approx Z_C(\gamma l) \tag{6.27}$$

$$\approx zl = Z$$

and

$$\frac{Y_e}{2} = \frac{1}{Z_C}\tanh\left(\frac{\gamma l}{2}\right)$$

$$\approx \frac{1}{Z_C}\frac{\gamma l}{2} \tag{6.28}$$

$$\approx \frac{yl}{2} = \frac{Y}{2}$$

In Equations 6.27 and 6.28, Z and Y represent the total series impedance (zl) and total shunt admittance (yl), respectively. The resultant circuit model is called the *nominal π* equivalent circuit. Generally, the approximation is good if

$l<10,000/f$ km (170 km at 60 Hz) for overhead lines;
$l<3,000/f$ km (50 km at 60 Hz) for underground cables.

Classification of line length

Overhead lines may be classified according to length, based on the approximations justified in their modelling:

(a)   Short lines: lines shorter than about 80 km (50 mi). They have negligible shunt capacitance, and may be represented by their series impedance.

(b)   Medium-length lines: lines with lengths in the range of 80 km to about 200 km (125 mi). They may be represented by the nominal π equivalent circuit.

(c)   Long lines: lines longer than about 200 km. For such lines the distributed effects of the parameters are significant. They need to be represented by the equivalent π circuit. Alternatively, they may be represented by cascaded sections of shorter lengths, with each section represented by a nominal π equivalent.

6.1.5 Typical Parameters

(a) Overhead lines

Table 6.1 gives typical parameters of overhead lines of nominal voltage ranging from 230 kV to 1,100 kV.

Table 6.1 Typical overhead transmission line parameters

Nominal Voltage	230 kV	345 kV	500 kV	765 kV	1,100 kV
R (Ω/km)	0.050	0.037	0.028	0.012	0.005
$x_L = \omega L$ (Ω/km)	0.488	0.367	0.325	0.329	0.292
$b_C = \omega C$ (μs/km)	3.371	4.518	5.200	4.978	5.544
α (nepers/km)	0.000067	0.000066	0.000057	0.000025	0.000012
β (rad/km)	0.00128	0.00129	0.00130	0.00128	0.00127
Z_C (Ω)	380	285	250	257	230
SIL (MW)	140	420	1000	2280	5260
Charging MVA/km $= V_0^2 b_C$	0.18	0.54	1.30	2.92	6.71

Notes: 1. Rated frequency is assumed to be 60 Hz.
2. Bundled conductors used for all lines listed, except for the 230 kV line.
3. R, x_L, and b_C are per-phase values.
4. SIL and charging MVA are three-phase values.

We see that the surge impedance lies within the range of 230 to 290 Ω for EHV and UHV overhead lines. For a 230 kV line, it is about 380 Ω. The value of β is practically the same for all lines. This is to be expected since $\sqrt{LC}$ is the propagation velocity of electromagnetic waves. For overhead lines it is slightly slower than the velocity of light (300,000 km/s). At 60 Hz, β is nearly equal to 1.27×10^{-3} rad/km. The corresponding wavelength ($\lambda = 2\pi/\beta$) is approximately 5,000 km.

An easy-to-remember approximate equivalent circuit applicable to an overhead line of 160 km (100 mi) length and of any voltage rating is shown in Figure 6.4.

For example, for the 500 kV line whose parameters are listed in Table 6.1, with a line length of 160 km, we have

$$X_L = 160 \times 0.325 = 52.0 \ \Omega$$

and

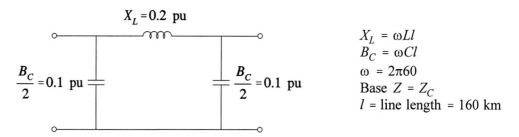

Figure 6.4 Approximate equivalent circuit for an overhead line of any
voltage rating, with parameters in per unit of surge impedance

$$B_C = 160 \times 5.20 \times 10^{-6} = 8.32 \times 10^{-4} \text{ siemens}$$

Expressed in per unit of Z_C (250 Ω),

$$X_L = 52.0/250 = 0.208 \text{ pu}$$
$$B_C = 8.32 \times 10^{-4} \times 250 = 0.208 \text{ pu}$$

(b) Underground cables

Table 6.2 gives typical parameters of cables. Two types of cables are included:
direct-buried paper-insulated lead-covered (PILC) and high-pressure pipe type (PIPE),
with nominal voltages of 115, 230, and 500 kV.

Table 6.2 Typical cable parameters

Nominal Voltage	115 kV	115 kV	230 kV	230 kV	500 kV
Cable Type	PILC	PIPE	PILC	PIPE	PILC
R (Ω/km)	0.0590	0.0379	0.0277	0.0434	0.0128
$x_L = \omega L$ (Ω/km)	0.3026	0.1312	0.3388	0.2052	0.2454
$b_C = \omega C$ (μs/km)	230.4	160.8	245.6	298.8	96.5
α (nepers/km)	0.00081	0.000656	0.000372	0.000824	0.000127
β (rad/km)	0.00839	0.00464	0.00913	0.00787	0.00487
Z_C (Ω)	36.2	28.5	37.1	26.2	50.4
SIL (MW)	365	464	1426	2019	4960
Charging MVA/km $= V_0^2 b_C$	3.05	2.13	13.0	15.8	24.1

From the table we see that underground cables have very high shunt capacitance. The characteristic impedance Z_C of a cable is about one-tenth to one-fifth of that for an overhead line of the same voltage rating.

6.1.6 Performance Requirements of Power Transmission Lines

In Section 6.1.2, we developed the performance equations of transmission lines. The basic equations apply to communication lines as well as power lines. However, the performance requirements of power and communication lines are significantly different.

Communication lines transmit signals of many relatively high frequencies and are very long compared to the wavelengths involved. Fidelity and strength of the signals at the receiving end are the primary considerations. Consequently, termination at the characteristic impedance of the line is the only practical way of operation to avoid distortion on the line. The energy associated with communication lines is small; consequently, efficiency is of minor interest.

In contrast, efficiency, economy and reliability of supply are factors of prime importance in the case of power transmission. There is only one frequency, and distortion is not a problem in the same sense as it is in communication lines. The lengths of most power lines are a fraction of the normal wavelength; hence, the lines can be terminated on equivalent load impedances which are much lower than their characteristic impedances.

If the power line is very long (greater than 500 km), terminating close to the characteristic impedance becomes imperative. To increase power levels that can be transmitted, either the characteristic impedance has to be reduced (by adding compensation) or the transmission voltage has to be increased.

Voltage regulation, *thermal limits*, and *system stability* are the factors that determine the power transmission capability of power lines. In what follows, we will discuss these aspects of power transmission line performance. Wherever appropriate, we will consider a lossless line, as it offers considerable simplicity and a better insight into the performance characteristics of transmission lines.

6.1.7 Voltage and Current Profile under No-Load [3,10]

(a) Receiving end open-circuited

When the receiving end is open, $I_R = 0$. Equations 6.8 and 6.9 then reduce to

$$\tilde{V} = \frac{\tilde{V}_R}{2} e^{\gamma x} + \frac{\tilde{V}_R}{2} e^{-\gamma x} \tag{6.29}$$

$$\tilde{I} = \frac{\tilde{V}_R}{2Z_C} e^{\gamma x} - \frac{\tilde{V}_R}{2Z_C} e^{-\gamma x} \tag{6.30}$$

For a lossless line, $\gamma=j\beta$, and the above two equations simplify to

$$\tilde{V} = \tilde{V}_R\cos(\beta x) \tag{6.31}$$

$$\tilde{I} = j(\tilde{V}_R/Z_C)\sin(\beta x) \tag{6.32}$$

The voltage[1] and current at the sending end are obtained by substituting line length l for x:

$$\tilde{E}_S = \tilde{V}_R\cos\beta l$$
$$= \tilde{V}_R\cos\theta \tag{6.33}$$

and

$$\tilde{I}_S = j(\tilde{V}_R/Z_C)\sin\theta$$
$$= j(\tilde{E}_S/Z_C)\tan\theta \tag{6.34}$$

where $\theta=\beta l$. The angle θ is referred to as the *electrical length* or the *line angle*, and is expressed in radians.

Based on Equations 6.31, 6.32 and 6.33, the line voltage and current can be expressed in terms of sending end voltage E_S as follows:

$$\tilde{V} = \tilde{E}_S\frac{\cos\beta x}{\cos\theta} \tag{6.35}$$

$$\tilde{I} = j\frac{\tilde{E}_S}{Z_C}\frac{\sin\beta x}{\cos\theta} \tag{6.36}$$

As an example, let us consider the voltage and current profiles for a 300 km, 500 kV line with the sending voltage at rated value and the receiving end open-circuited. The line is assumed to be lossless with $\beta=0.0013$ rad/km and $Z_C=250$ Ω.

The electrical length of the line is

$$\theta = 300\times0.0013 = 0.39 \text{ rad}$$
$$= 22.3°$$

---

[1] Since the sending end voltage in this case is a controlled voltage, it is denoted by the symbol E, instead of V.

From Equation 6.33,

$$V_R = \frac{1}{\cos 22.3°} = 1.081 \text{ pu}$$

and from Equation 6.34

$$\tilde{I}_S = j(\tilde{E}_S/Z_C)\tan\theta$$

Expressing I_S in per unit with base current equal to that corresponding to the natural load, its magnitude is

$$I_S = E_S \tan\theta \quad \text{pu}$$

$$= 1.0\tan 22.3°$$

$$= 0.411 \quad \text{pu}$$

From Equations 6.35 and 6.36, the voltage and current magnitudes as functions of distance x from the receiving end are given by

$$V = \frac{1.0\cos(0.0013x)}{\cos 22.3°}$$

$$= 1.0812\cos(0.0013x) \quad \text{pu}$$

and

$$I = \frac{1.0\sin(0.0013x)}{\cos 22.3°}$$

$$= 1.0812\sin(0.0013x) \quad \text{pu}$$

The voltage and current profiles are shown in Figure 6.5. The current represents the capacitive charging current of the line, expressed in per unit of current at SIL.

The only line parameter, other than line length, that affects the results of Figure 6.5 is β. Since β is practically the same for overhead lines of all voltage levels (see Table 6.1), the results are universally applicable, not just for a 500 kV line.

The receiving end voltage V_R for the 300 km line is 1.081 pu, that is, 8.1% higher than the sending end voltage. For a 600 km line, the receiving-end open-circuit voltage would be 1.407 pu. A line of about 1,200 km (one quarter wavelength) would have infinitely high receiving end voltage on open-circuit.

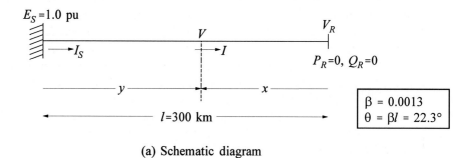

(a) Schematic diagram

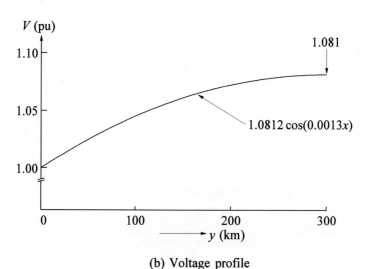

(b) Voltage profile

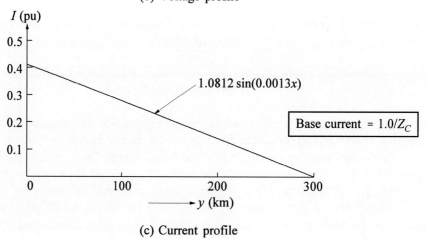

(c) Current profile

Figure 6.5 Voltage and current profiles for a 300 km lossless
line with receiving end open-circuited

The rise in voltage at the receiving end on open-circuit is due to the flow of line charging (capacitive) current through line inductance. This phenomenon was first noticed by Ferranti on overhead lines supplying a lightly loaded (and hence highly capacitive) cable network; it is therefore referred to as the *Ferranti effect*.

In the above calculations, the sending end voltage has been assumed to be constant. In practice, following a sudden opening of the line at the receiving end, the sending end voltage will rise due to the capacitive current of the line flowing through the source impedance (mostly inductive reactance). Appropriate forms of reactive power compensation should be provided on long lines to keep the rise in voltage to acceptable levels. This is discussed in Chapter 11.

(b) Line connected to sources at both ends

For simplicity, let us assume that the line is symmetrical; i.e., it is connected to identical sources at the two ends. Let E_S and E_R denote the voltages at the sending end and receiving end, respectively. From Equations 6.8 and 6.9, with $x = l$ and $\theta = \beta l$, we have

$$\tilde{E}_S = \frac{\tilde{E}_R + Z_C \tilde{I}_R}{2} e^{\gamma l} + \frac{\tilde{E}_R - Z_C \tilde{I}_R}{2} e^{-\gamma l} \tag{6.37}$$

Hence,

$$\tilde{I}_R = \frac{2\tilde{E}_S - \tilde{E}_R(e^{\gamma l} + e^{-\gamma l})}{Z_C(e^{\gamma l} - e^{-\gamma l})} \tag{6.38}$$

Substituting the above expression for $\tilde{I}_R$ in Equations 6.8 and 6.9, we have

$$\tilde{V} = \frac{\tilde{E}_S - \tilde{E}_R e^{-\gamma l}}{e^{\gamma l} - e^{-\gamma l}} e^{\gamma x} + \frac{\tilde{E}_R e^{\gamma l} - \tilde{E}_S}{e^{\gamma l} - e^{-\gamma l}} e^{-\gamma x} \tag{6.39}$$

$$\tilde{I} = \frac{\tilde{E}_S - \tilde{E}_R e^{-\gamma l}}{Z_C(e^{\gamma l} - e^{-\gamma l})} e^{\gamma x} - \frac{\tilde{E}_R e^{\gamma l} - \tilde{E}_S}{Z_C(e^{\gamma l} - e^{-\gamma l})} e^{-\gamma x} \tag{6.40}$$

Since $\tilde{E}_S = \tilde{E}_R$, we have

$$\tilde{V} = \frac{\tilde{E}_S - \tilde{E}_S e^{-\gamma l}}{e^{\gamma l} - e^{-\gamma l}} e^{\gamma x} + \frac{\tilde{E}_S e^{\gamma l} - \tilde{E}_S}{e^{\gamma l} - e^{-\gamma l}} e^{-\gamma x} \tag{6.41}$$

$$\tilde{I} = \frac{\tilde{E}_S - \tilde{E}_S e^{-\gamma l}}{Z_C(e^{\gamma l} - e^{-\gamma l})} e^{\gamma x} - \frac{\tilde{E}_S e^{\gamma l} - \tilde{E}_S}{Z_C(e^{\gamma l} - e^{-\gamma l})} e^{-\gamma x} \tag{6.42}$$

For a lossless line, $\gamma = j\beta$. With $\theta = \beta l$, we have

$$\tilde{V} = \tilde{E}_S \frac{\cos\beta(l/2 - x)}{\cos(\theta/2)} \tag{6.43}$$

$$\tilde{I} = -j\frac{\tilde{E}_S}{Z_C} \frac{\sin\beta(l/2 - x)}{\cos(\theta/2)} \tag{6.44}$$

The voltage and current profiles are shown in Figure 6.6 for a 400 km line with $E_S = E_R = 1.0$ pu. The generators at the sending end and receiving end should be capable of absorbing the reactive power due to line charging. If this exceeds the underexcited reactive power capability of the connected generators, compensation may have to be provided.

If E_S and E_R are not equal, the voltage and current profiles are not symmetrical and the highest voltage is not at midpoint, but is nearer to the end with higher voltage.

6.1.8 Voltage-Power Characteristics [4,10]

(a) Radial line with fixed sending end voltage

Corresponding to a load of $P_R + jQ_R$ at the receiving end, we have

$$\tilde{I}_R = \frac{P_R - jQ_R}{\tilde{V}_R^*}$$

From Equation 6.8, with $x = l$, we have

$$\tilde{E}_S = \frac{\tilde{V}_R + Z_C(P_R - jQ_R)/\tilde{V}_R^*}{2} e^{\gamma l} + \frac{\tilde{V}_R - Z_C(P_R - jQ_R)/\tilde{V}_R^*}{2} e^{-\gamma l} \tag{6.45}$$

With $\gamma = j\beta$ (lossless line) and $\theta = \beta l$,

$$\tilde{E}_S = \tilde{V}_R \cos\theta + jZ_C \sin\theta \left(\frac{P_R - jQ_R}{\tilde{V}_R^*}\right) \tag{6.46}$$

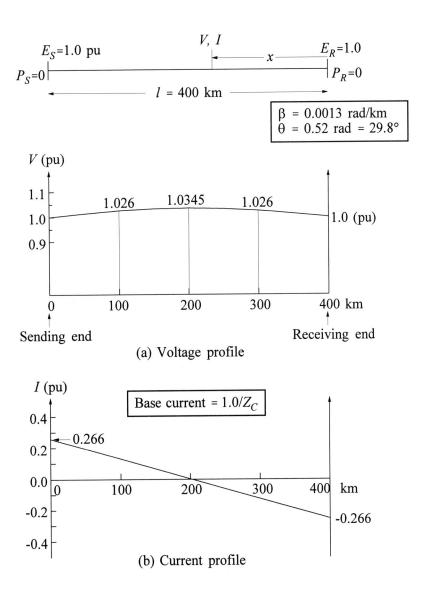

Figure 6.6 Voltage and current profile of a 400 km
lossless line under no-load

The above equation can be solved for V_R for any given load and sending end voltage. Figure 6.7 shows a typical relationship between the receiving end voltage and load, for a fixed sending end voltage. The results shown are for a 300 km line with $E_S=1.0$ pu. The constant β for the line is assumed to be 0.0013 rad/km. The load is *normalized* by dividing P_R by P_o, the natural load (SIL), *so that the results are applicable to overhead lines of all voltage levels*. From Figure 6.7, several fundamental properties of ac transmission are evident:

- There is an inherent maximum limit of power that can be transmitted at any load power factor. Obviously, there has to be such a limit since, with E_S constant, the only way to increase power is by lowering the load impedance. This will result in increased current, but decreased V_R and large line losses. Up

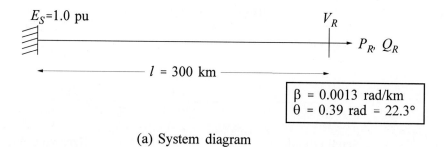

(a) System diagram

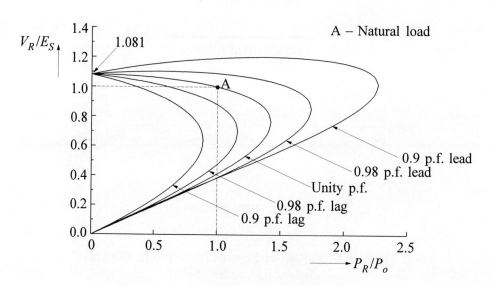

(b) Receiving end voltage versus power characteristic

Figure 6.7 Voltage-power characteristics of
a 300 km lossless radial line

to a certain point the increase of current dominates the decrease of V_R, thereby resulting in an increased P_R. Finally, the decrease in V_R is such that the trend reverses.

- Any value of power below the maximum can be transmitted at two different values of V_R. The normal operation is at the upper value, within narrow limits around 1.0 pu. At the lower voltage, the current is higher and may exceed thermal limits. The feasibility of operation at the lower voltage also depends on load characteristics and may lead to voltage instability. This will be discussed in Chapter 14.

- The load power factor has a significant influence on V_R and the maximum power that can be transmitted. The power limit and V_R are lower with lagging power factors (inductive load, Q_R positive). With leading power factors (capacitive load, Q_R negative), the top portion of the voltage profile tends to be flatter and maximum power is higher. This means that the receiving end voltage can be regulated by the addition of shunt capacitive compensation.

The effect of line length is depicted in Figure 6.8, which shows the performance of lines with lengths of 200, 300, 400, 600 and 800 km with *unity power factor* load. The results show that, for longer lines, V_R is extremely sensitive to variations in P_R. For lines longer than 600 km ($\theta > 45°$), V_R at natural load is the lower of the two voltages which satisfy the Equation 6.46. Such operation is likely to be voltage unstable (see Chapter 14).

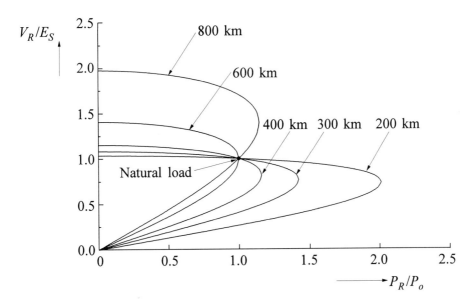

Figure 6.8 Relationship between receiving end voltage,
line length, and load of a lossless radial line

(b) Line connected to sources at both ends

As in the no-load case (Section 6.1.7), we will assume the magnitudes of the source voltages at the two ends to be equal. Under load, E_S leads E_R in phase. Because the magnitudes of E_S and E_R are equal, the following conditions exist:

• The midpoint voltage is midway in phase between E_S and E_R.

• The power factor at midpoint is unity.

• With $P_R > P_o$, both ends supply reactive power to the line; with $P_R < P_o$, both ends absorb reactive power from the line.

The phasor diagram for $P_R < P_o$ is shown in Figure 6.9.

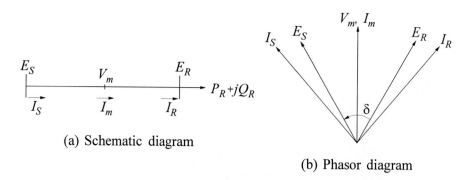

(a) Schematic diagram

(b) Phasor diagram

Figure 6.9 Voltage and current phase relationships
with E_S equal to E_R, and P_R less than P_o

Figure 6.8 may be used to analyze how V_m varies with the power transmitted. With the magnitudes of E_S and E_R equal to 1.0 pu and the length equal to half that of the actual line, plots of V_R shown in Figure 6.8 give the values of V_m. For example, the variations in V_m with load of a 400 km line connected to sources at both ends are the same as the variations in V_R of a 200 km radial line.

At 400 km, the performance of the line is significantly improved by having sources at both ends. However, an 800 km symmetrical line would have unacceptably large voltage variations at midpoint.

Although we have considered a line connected to identical sources at the two ends, the observations made here are sufficiently general and provide a physical understanding helpful in dealing with more complex cases.

6.1.9 Power Transfer and Stability Considerations

Assuming a lossless line, from Equation 6.16, with $x=l$, and $\theta=\beta l$, we have

$$\tilde{E}_S = \tilde{E}_R\cos\theta + jZ_C\tilde{I}_R\sin\theta$$

Expressing I_R in terms of P_R and Q_R, we have

$$\tilde{E}_S = \tilde{E}_R\cos\theta + jZ_C\sin\theta\left(\frac{P_R - jQ_R}{\tilde{E}_R^*}\right) \tag{6.47}$$

As shown in Figure 6.9(b), let δ be the angle by which E_S leads E_R, i.e., the *load angle* or the *transmission angle*.

With E_R as reference phasor, E_S may be written as

$$\tilde{E}_S = E_S e^{j\delta} = E_S(\cos\delta + j\sin\delta) \tag{6.48}$$

Equating real and imaginary parts of Equations 6.47 and 6.48, we have

$$E_S\cos\delta = E_R\cos\theta + Z_C(Q_R/E_R)\sin\theta \tag{6.49}$$

$$E_S\sin\delta = Z_C(P_R/E_R)\sin\theta \tag{6.50}$$

Rearranging Equation 6.50 yields

$$P_R = \frac{E_S E_R}{Z_C\sin\theta}\sin\delta \tag{6.51}$$

The above equation gives a very important expression for power transferred across a line. It is valid for a synchronous as well as an asynchronous load at the receiving end. The only approximation is that line losses are neglected.

For a *short line*, $\sin\theta$ can be replaced by θ in radians. Hence,

$$Z_C\sin\theta = Z_C\theta = \sqrt{L/C}\,\omega\sqrt{LC}\,l = \omega Ll$$
$$= X_L \quad \text{the series inductive reactance}$$

Therefore, the expression for power transferred reduces to the more familiar form:

$$P_R \approx \frac{E_S E_R}{X_L} \sin\delta \qquad (6.52)$$

If $E_S = E_R = V_o$, the rated voltage, then the natural load is

$$P_o = \frac{E_S E_R}{Z_C}$$

and Equation 6.51 becomes

$$P_R = \frac{P_o}{\sin\theta} \sin\delta \qquad (6.53)$$

With the voltage magnitudes fixed, the power transmitted is a function of only the transmission angle δ. When P_R is equal to the natural load (P_o), $\delta = \theta$.

Figure 6.10(a) shows this relationship for a 400 km line, for which $\theta = 0.52$ rads and $\sin\theta = 0.497$. It is interesting to compare this with the voltage-power characteristic of Figure 6.8. The characteristic corresponding to the 400 km symmetrical line (equivalent to the 200 km radial line in Figure 6.8) is reproduced in Figure 6.10(b).

From Figures 6.10(a) and (b), we see that there is a maximum power that can be transmitted. As the load angle is increased (i.e., as the sending end synchronous system is advanced with respect to the receiving end synchronous system), the transmitted power increases according to Figure 6.10(a) and Equation 6.53. This is accompanied by a reduction in the midpoint voltage V_m [Figure 6.10(b)] and an increase in the midpoint current I_m so that there is an increase in power. Up to a certain point the increase in I_m dominates over the decrease of V_m. When the load angle reaches 90°, the transmitted power reaches its maximum value. Beyond this, the decrease in V_m is greater than the accompanying increase in I_m; hence, their product decreases with any further increase in transmission angle. The system, as explained below, is unstable when this condition is reached.

The sending end and receiving end systems may be considered in terms of equivalent synchronous machines. The load angle δ is then a measure of the relative position of the rotors of those two machines. Beyond the point of maximum power, an increase in the torque of the sending end machine results in an increase of δ, but the transmitted power decreases. This causes the sending end machine to accelerate and the receiving end machine to decelerate, resulting in a further increase of δ. This is a runaway situation, and the two machines (or the systems they represent) lose synchronism.

The maximum power that can be transmitted represents the small-signal or steady-state stability limit. For the 400 km line considered in Figures 6.10(a) and (b), this limit is equal to $P_o/0.497$ or 2.012 times the natural load.

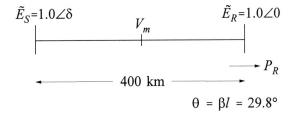

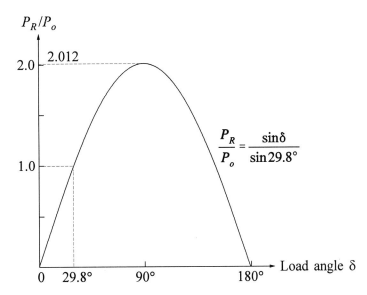

(a) Power/angle characteristics

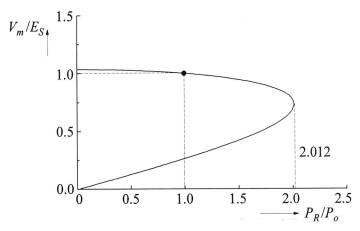

(b) Midpoint voltage as a function of power transmitted

Figure 6.10 The P_R-δ and V_m-P_R characteristics of 400 km lossless line transmitting power between two large systems

The stability analysis considered above represents a highly idealized situation. In particular, the assumption that E_S and E_R have constant magnitude is not realistic; the dynamic characteristics of the sending and receiving end systems need to be considered for accurate analysis. However, the analysis presented is useful for understanding the phenomenon and the performance characteristics of transmission lines. Chapter 12 provides a comprehensive description of the small-signal stability problem.

If the receiving end system is a nonsynchronous load, there is still a maximum value of power that can be transmitted, as illustrated in Figure 6.10(b), but maintenance of synchronism would not be an issue.

Reactive power requirements

The relationship between the receiving end reactive power and the voltages at the two ends is given by Equation 6.49 which is rewritten here for convenience.

$$E_S \cos\delta = E_R \cos\theta + Z_C (Q_R/E_R) \sin\theta$$

Rearranging, we get

$$Q_R = \frac{E_R(E_S \cos\delta - E_R \cos\theta)}{Z_C \sin\theta}$$

Similarly, the sending end reactive power is given by

$$Q_S = \frac{-E_S(E_R \cos\delta - E_S \cos\theta)}{Z_C \sin\theta}$$

If the magnitudes of E_S and E_R are equal, then

$$Q_R = -Q_S$$

$$= \frac{E_S^2(\cos\delta - \cos\theta)}{Z_C \sin\theta}$$

Figure 6.11 shows the terminal reactive power requirements of lines of different lengths as a function of active power transmitted. Both active power and reactive power have been normalized by dividing by the natural load P_o. When $P_R<P_o$, there is an excess of line charging; Q_S is negative and Q_R is positive, indicating reactive power absorption by the systems at both ends. When $P_R>P_o$, reactive power is supplied to the line from both ends.

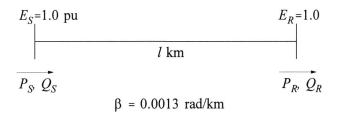

$E_S = 1.0$ pu $E_R = 1.0$

l km

P_S, Q_S P_R, Q_R

$\beta = 0.0013$ rad/km

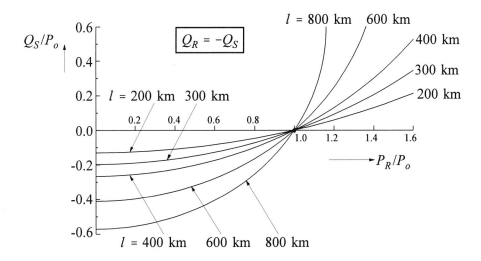

Figure 6.11   Terminal reactive power as a function of
power transmitted for different line lengths

Transmission lines can be operated with varying load and nearly constant voltage at both ends, if adequate sources of reactive power are available at the two ends.

6.1.10 Effect of Line Loss on *V-P* and *Q-P* Characteristics

In the analysis of transmission line performance presented so far, we have neglected line losses. We will now examine the effect of line resistance on V_R-P_R and Q_S-P_R characteristics by considering a 300 km, 500 kV line having the following parameters:

$R = 0.028$ Ω/km $x_L = 0.325$ Ω/km $b_C = 5.20$ μs/km
$\alpha = 0.000057$ nepers/km $\beta = 0.0013$ rad/km

The line is assumed to supply a radial load of *unity power factor*, with the sending end voltage E_S maintained constant at 1.0 pu.

The relationships between V_R and P_R and between Q_S and P_R are shown in Figure 6.12, with and without the line resistance included. The values of P_R and Q_S plotted are normalized values with 1.0 pu equal to P_o (1,008 MW).

We see from Figure 6.12(a) that the effect of the line resistance is to reduce the maximum power that can be transmitted by about 8.5%.

The lower portion of the Q_S-P_R curves shown in Figure 6.12(b) corresponds to the upper portion of the V_R-P_R characteristics, where the receiving end voltage V_R is closer to rated value. The high values of Q_S in the upper portion of the Q_S-P_R curve are due to the high values of line current (hence high XI^2 line loss) corresponding to the lower portion of the V_R-P_R curve. We see that the effect of line resistance on the computed value of Q_S in the normal lower portion is significant only when P_R exceeds P_o.

6.1.11 Thermal Limits

The heat produced by current flow in transmission lines has two undesirable effects:

• Annealing and gradual loss of mechanical strength of the aluminium conductor caused by continued exposure to temperature extremes

• Increased sag and decreased clearance to ground due to conductor expansion at higher temperatures

The second of the above two effects is generally the limiting factor in setting the maximum permissible operating temperature. At this limit, the resulting line sag approaches the statutory minimum ground clearance.

The maximum allowable conductor temperatures based on annealing considerations are 127°C for conductors with high aluminum content and 150°C for other conductors.

The allowable maximum current (i.e., the ampacity) depends on the ambient temperature and the wind velocity. The thermal time constant is on the order of 10 to 20 minutes. Therefore, distinction is usually made between continuous rating and limited time rating. Depending on the pre-contingency current, temperature and wind velocity, the limited time rating may be used during emergencies. As an example, the 230 kV line whose parameters are given in Table 6.1 has summer and winter emergency ratings of 1,880 A and 2,040 A, respectively. These are design values based on extreme values of ambient temperature, wind velocity, and solar radiation.

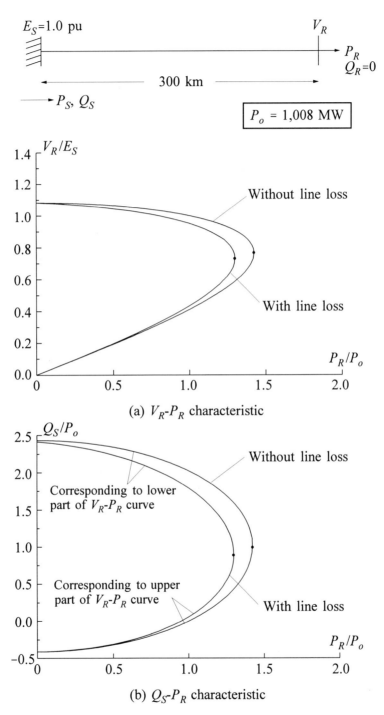

(a) V_R-P_R characteristic

(b) Q_S-P_R characteristic

Figure 6.12 The V_R-P_R and Q_S-P_R characteristics of
a 300 km, 500 kV line supplying a radial load

6.1.12 Loadability Characteristics

The concept of "line loadability" is useful in developing a fuller understanding of power transfer capability as influenced by voltage level and line length. Line loadability is defined as the degree of line loading (expressed in percent of SIL) permissible given the thermal, voltage drop, and stability limits. This concept was first introduced by H.P. St. Clair in 1953 [11]. Based on practical considerations and experience, St. Clair developed transmission line power-transfer capability curves covering voltage levels between 34.5 kV and 330 kV and line lengths up to 400 mi (approximately 645 km). These curves, known as *St. Clair curves*, have been a valuable tool for transmission planning engineers for quickly estimating the maximum line loading limits. This work was later extended in reference 12 by presenting an analytical basis for the St. Clair curves so as to be able to cover higher voltage levels (up to 1,500 kV) and longer line lengths (600 mi or 960 km).

Figure 6.13 shows the universal loadability curve for overhead *uncompensated* transmission lines applicable to all voltage levels. The curve, which is based on the results presented in reference 12, shows the limiting values of power that can be transmitted as a function of line length. Three factors influence the limiting values of power: thermal limit, voltage drop limit, and the small-signal or steady-state stability limit. In determining the loadability curve, it is assumed that the maximum allowable voltage drop along the line is 5% and that the minimum allowable steady-state stability margin is 30%. Referring to Figure 6.14, the percent steady-state stability margin is defined as

$$\text{Percent stability margin} = \frac{P_{max} - P_{limit}}{P_{max}} \times 100$$

As shown in Figure 6.14, for a 30% stability margin, the load angle δ is 44°. The calculation of stability limit includes the effects of the equivalent system reactances at the two ends of the line. In reference 12, the system strength at each end is taken to be that corresponding to 50 kA fault duty which represents a well-developed system.

Since the resistances of extra-high voltage (EHV) and ultra-high voltage (UHV) lines are very much smaller than their reactances, such lines closely approximate a lossless line. Since the parameter β is practically the same for all overhead lines, the loadabilities expressed in per unit of SIL are universally applicable to lines of all voltage classes.

As identified in Figure 6.13, the limits to line loading are governed by the following considerations:

- Thermal limits for lines up to 80 km (50 mi)

- Voltage drop limits for lines between 80 km and 320 km (200 mi) long

- Stability limits for lines longer than 320 km

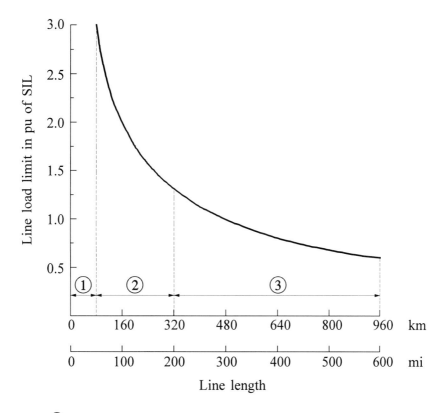

Figure 6.13 Transmission line loadability curve

① 0–80 km: Region of thermal limitation
② 80–320 km: Region of voltage drop limitation
③ 320–960 km: Region of small-signal (steady-state) stability limitation

For lines longer than 480 km (300 mi), the loadability is less than SIL. The loadability limits can be increased by "compensating" the lines.

Alternative forms of line compensation and consideration that influence their selection are discussed in Chapter 11.

The universal loadability curve discussed here provides a simple means of visualizing power-transfer capabilities of transmission lines. It is useful for developing conceptual guides to line loadability and preliminary planning of transmission systems. However, it must be used with some caution. Large complex power systems require detailed assessment of their performance and consideration of additional factors that influence their performance.

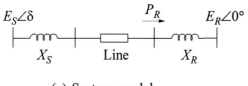

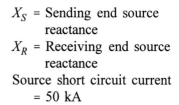

X_S = Sending end source reactance

X_R = Receiving end source reactance

Source short circuit current = 50 kA

(a) System model

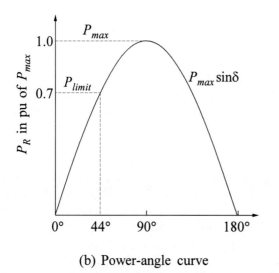

(b) Power-angle curve

For 30% stability margin:

$$P_{limit} = 0.7P_{max}$$
$$\delta = \sin^{-1}0.7$$
$$= 44°$$

Figure 6.14 Steady-state stability margin calculation

Effect of using bundled conductors

Bundled conductors are commonly used on EHV lines to control the voltage gradient at the surface of the conductors and thus avoid unacceptably high radio noise, audible noise, and corona loss.

The use of bundled conductors decreases the self geometric mean distance. Hence, it has an added advantage of reducing the characteristic impedance Z_C by decreasing the series inductance and increasing the shunt capacitance of the line. The reduction in Z_C is on the order of 10 to 20%. Consequently, SIL or natural load increases, thus contributing to the increase in loadability.

6.2 TRANSFORMERS

Transformers enable utilization of different voltage levels across the system. From the viewpoints of efficiency and power-transfer capability, the transmission voltages have to be high, but it is not practically feasible to generate and consume power at these voltages. In modern electric power systems, the transmitted power undergoes four to five voltage transformations between the generators and the ultimate consumers. Consequently, the total MVA rating of all the transformers in a power system is about five times the total MVA rating of all the generators.

In addition to voltage transformation, transformers are often used for control of voltage and reactive power flow. Therefore, practically all transformers used for bulk power transmission and many distribution transformers have taps in one or more windings for changing the turns ratio. From the power system viewpoint, changing the ratio of transformation is required to compensate for variations in system voltages. Two types of tap-changing facilities are provided: off-load tap changing and under-load tap changing (ULTC).[1] The off-load tap-changing facilities require the transformer to be de-energized for tap changing; they are used when the ratio will need to be changed only to meet long-term variations due to load growth, system expansion, or seasonal changes. The ULTC is used when the changes in ratio need to be frequent; for example, to take care of daily variations in system conditions. The taps normally allow the ratio to vary in the range of ±10% to ±15%.

Transformers may be either three-phase units or three single-phase units. The latter type of construction is normally used for large EHV transformers and for distribution transformers. Large EHV transformers are of single-phase design due to the cost of spare, insulation requirements, and shipping considerations. The distribution systems serve single-phase loads and are supplied by single-phase transformers.

When the voltage transformation ratio is small, *autotransformers* are normally used. The primary and secondary windings of autotransformers are interconnected so that the power to be transformed by magnetic coupling is only a portion of the total power transmitted through the transformer. There is thus inherent metallic connection between the primary side and secondary side circuits; this is unlike the conventional two-winding transformer which isolates the two circuits.

Autotransformers are usually Y connected, with neutrals solidly grounded to minimize the propagation of disturbances occurring on one side into the other side. It is a common practice to add a low-capacity delta-connected tertiary winding. The tertiary winding provides a path for third harmonic currents, thereby reducing their flow on the network. It also assists in stabilizing the neutral. Reactive compensation is often provided through use of switched reactors and capacitors on a tertiary bus

---

[1] Under-load tap changing is also referred to by other names such as on-load tap changing (OLTC) and load tap changing (LTC).

(see Chapter 11).

As compared to the *conventional two-winding transformer*, the autotransformer has advantages of lower cost, higher efficiency, and better regulation. These advantages become less significant as the transformation ratio increases; hence, autotransformers are used for low transformation ratios (for example, 500/230 kV).

In interconnected systems, it sometimes becomes necessary to make electrical connections that form loop circuits through one or more power systems. To control the circulation of power and prevent overloading certain lines, it is usually necessary in such situations to use *phase-angle transformers*. Often it is necessary to vary the extent of phase shift to suit changing system conditions; this requires provision of on-load phase-shifting capability. Voltage transformation may also be required in addition to phase shift.

The transformer is a well-known device. The basic principle of its operation is covered in standard textbooks [2,5,7]. References 2 and 8 provide information related to physical realization of various types of transformers and their performance characteristics. Here, we will focus on representation of transformers in stability and power-flow studies.

6.2.1 Representation of Two-Winding Transformers

Basic equivalent circuit in physical units:

The basic equivalent circuit of a two-winding transformer with all quantities in physical units is shown in Figure 6.15. The subscripts p and s refer to primary and secondary quantities, respectively.

The magnetizing reactance X_{mp} is very large and is usually neglected. For special studies requiring representation of transformer saturation, the magnetizing reactance representation may be approximated by moving it to the primary or the secondary terminals and treating it as a voltage-dependent variable shunt reactance.

Per unit equivalent circuit:

With appropriate choice of primary and secondary side base quantities, the equivalent circuit can be simplified by eliminating the ideal transformer. However, this is not always possible and the base quantities often have to be chosen independent of the actual turns ratio. It is therefore necessary to consider an off-nominal turns ratio.

From the equivalent circuit of Figure 6.15, with X_{mp} neglected, we have

$$\tilde{v}_p = Z_p \tilde{i}_p + \frac{n_p}{n_s} \tilde{v}_s - \frac{n_p}{n_s} Z_s \tilde{i}_s \tag{6.54}$$

$$\tilde{v}_s = \frac{n_s}{n_p} \tilde{v}_p - \frac{n_s}{n_p} Z_p \tilde{i}_p + Z_s \tilde{i}_s \tag{6.55}$$

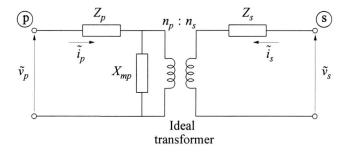

$Z_p = R_p + jX_p \; ; \; Z_s = R_s + jX_s$

R_p, R_s = primary and secondary winding resistances

X_p, X_s = primary and secondary winding leakage reactances

n_p, n_s = number of turns of primary and secondary winding

X_{mp} = magnetizing reactance referred to the primary side

Figure 6.15 Basic equivalent circuit of a two-winding transformer

Let

$Z_{p0} = Z_p$ at nominal primary side tap position

$Z_{s0} = Z_s$ at nominal secondary side tap position

n_{p0} = primary side nominal number of turns

n_{s0} = secondary side nominal number of turns

Expressing Equations 6.54 and 6.55 in terms of the above nominal values,

$$\tilde{v}_p = \left(\frac{n_p}{n_{p0}} \right)^2 Z_{p0} \tilde{i}_p + \frac{n_p}{n_s} \tilde{v}_s - \frac{n_p}{n_s} \left(\frac{n_s}{n_{s0}} \right)^2 Z_{s0} \tilde{i}_s \qquad (6.56)$$

$$\tilde{v}_s = \frac{n_s}{n_p} \tilde{v}_p - \frac{n_s}{n_p} \left(\frac{n_p}{n_{p0}} \right)^2 Z_{p0} \tilde{i}_p + \left(\frac{n_s}{n_{s0}} \right)^2 Z_{s0} \tilde{i}_s \qquad (6.57)$$

Here, we have assumed that both leakage reactance and resistance of a transformer winding are proportional to the square of the number of turns. This assumption is generally valid for the leakage reactance, but not for the resistance. Since the resistance is much smaller than the leakage reactance and since the deviation of the actual turns ratio from the nominal turns ratio is not very large, the resulting approximation is acceptable. For convenience, we will assume that both primary and secondary windings are connected so as to form a Y-Y connected three-phase bank.

With the nominal number of turns related to the base voltages as follows:

$$\frac{n_{p0}}{n_{s0}} = \frac{v_{pbase}}{v_{sbase}}$$

and

$$v_{pbase} = Z_{pbase}\,i_{pbase}, \qquad v_{sbase} = Z_{sbase}\,i_{sbase}$$

Equations 6.56 and 6.57 in per unit form become

$$\bar{v}_p = \bar{n}_p^2 \bar{Z}_{p0}\bar{i}_p + \frac{\bar{n}_p}{\bar{n}_s}\bar{v}_s - \bar{n}_s^2 \frac{\bar{n}_p}{\bar{n}_s}\bar{Z}_{s0}\bar{i}_s \tag{6.58}$$

$$\bar{v}_s = \frac{\bar{n}_s}{\bar{n}_p}\bar{v}_p - \bar{n}_p^2 \frac{\bar{n}_s}{\bar{n}_p}\bar{Z}_{p0}\bar{i}_p + \bar{n}_s^2 \bar{Z}_{s0}\bar{i}_s \tag{6.59}$$

where the superbars denote per unit values, with $\bar{v}_p$, $\bar{v}_s$, $\bar{i}_p$, $\bar{i}_s$ equal to per unit values of *phasor* voltages and currents, and

$$\bar{n}_p = \frac{n_p}{n_{p0}} \tag{6.60}$$

$$\bar{n}_s = \frac{n_s}{n_{s0}} \tag{6.61}$$

The per unit equivalent circuit representing Equations 6.58 and 6.59 is shown in Figure 6.16.

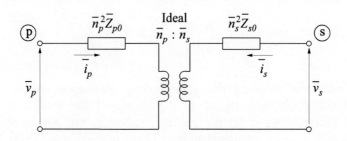

Figure 6.16 Per unit equivalent circuit

Standard equivalent circuit:

The equivalent circuit of Figure 6.16 can be reduced to the standard form shown in Figure 6.17, where $\bar{n}$ is the per unit turns ratio:

$$\bar{n} = \frac{\bar{n}_p}{\bar{n}_s} = \frac{n_p n_{s0}}{n_{p0} n_s} \tag{6.62}$$

and

$$\begin{aligned}
\bar{Z}_e &= \bar{n}_s^2 (\bar{Z}_{p0} + \bar{Z}_{s0}) \\
&= \left(\frac{n_s}{n_{s0}}\right)^2 (\bar{Z}_{p0} + \bar{Z}_{s0})
\end{aligned} \tag{6.63}$$

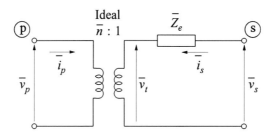

Figure 6.17 Standard equivalent circuit for a transformer

The equivalent circuit of Figure 6.17 is widely used for representation of two-winding transformers in power flow and stability studies. The IEEE common format for exchange of solved power flow cases uses this representation [13].

We see from Equation 6.63 that Z_e does not change with $\bar{n}_p$. Therefore, if the tap is on the primary side, only $\bar{n}$ changes.

If the actual turns ratio is equal to n_{p0}/n_{s0}, then $\bar{n}=1.0$, and the ideal transformer vanishes. When the actual turns ratio is not equal to the nominal turns ratio, $\bar{n}$ represents the off-nominal ratio (ONR).

The equivalent circuit of Figure 6.17 can be used to represent a transformer with a fixed (or off-load) tap on one side and an under-load tap changer (ULTC) on the other side. The off-nominal turns ratio is assigned to the side with ULTC and Z_e has a value corresponding to the fixed-tap position of the other side, as given by Equation 6.63.

Equivalent π circuit representation [14]:

In digital computer analysis of power flow, it is not convenient to represent an ideal transformer. We will therefore reduce the equivalent circuit of Figure 6.17 to the form of a π network of Figure 6.18(a).

From Figure 6.17, the terminal current at bus p is

$$\bar{i}_p = (\bar{v}_t - \bar{v}_s)\frac{\bar{Y}_e}{\bar{n}}$$

$$= (\frac{\bar{v}_P}{\bar{n}} - \bar{v}_s)\frac{\bar{Y}_e}{\bar{n}} \qquad (6.64)$$

$$= (\bar{v}_p - \bar{n}\bar{v}_s)\frac{\bar{Y}_e}{\bar{n}^2}$$

where $\bar{Y}_e = 1/\bar{Z}_e$. Similarly, the terminal current at bus s is

$$\bar{i}_s = (\bar{n}\bar{v}_s - \bar{v}_p)\frac{\bar{Y}_e}{\bar{n}} \qquad (6.65)$$

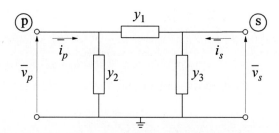

(a) General π network

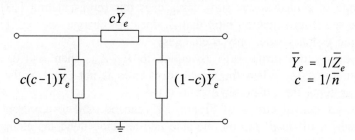

$$\bar{Y}_e = 1/\bar{Z}_e$$
$$c = 1/\bar{n}$$

(b) Equivalent π circuit

Figure 6.18 Transformer representation with ONR

The corresponding terminal currents for the π network shown in Figure 6.18(a) are

$$\bar{i}_p = y_1(\bar{v}_p - \bar{v}_s) + y_2 \bar{v}_p \qquad (6.66)$$

$$\bar{i}_s = y_1(\bar{v}_s - \bar{v}_p) + y_3 \bar{v}_s \qquad (6.67)$$

Equating the corresponding admittance terms in Equations 6.64 and 6.66, we have

$$y_1 = \frac{1}{n}\bar{Y}_e = c\bar{Y}_e \qquad (6.68)$$

and

$$y_2 = \left(\frac{1}{n^2} - \frac{1}{n}\right)\bar{Y}_e = (c^2 - c)\bar{Y}_e \qquad (6.69)$$

where $c = \dfrac{1}{n}$. Similarly, from Equations 6.65 and 6.67,

$$y_3 = (1 - c)\bar{Y}_e \qquad (6.70)$$

The equivalent π circuit with parameters expressed in terms of the ONR and transformer leakage impedance is shown in Figure 6.18(b).

Consideration of three-phase transformer connections

The standard equivalent circuit of Figure 6.17 represents the single-phase equivalent of a three-phase transformer. In establishing the ONR, the nominal turns ratio (n_{p0}/n_{s0}) is taken to be equal to the ratio of line-to-line base voltages on both sides of the transformer irrespective of the winding connections (Y-Y, Δ-Δ, or Y-Δ). For Y-Y and Δ-Δ connected transformers, this makes the ratios of the base voltages equal to the ratios of the nominal turns of the primary and secondary windings of each transformer phase. For a Y-Δ connected transformer, this in addition accounts for the factor $\sqrt{3}$ due to the winding connection.

In the case of a Y-Δ connected transformer, a 30° phase shift is introduced between line-to-line voltages on the two sides of the transformer. The line-to-neutral voltages and line currents are similarly shifted in phase due to the winding connections. As we will illustrate in Section 6.4, it is usually not necessary to take this phase shift into consideration in system studies. Thus, the single-phase equivalent circuit of a Y-Δ transformer does not account for the phase shift, except in so far as the phase shift of voltages due to the impedance of the transformer.

Example of modelling two-winding transformers

As an example, let us consider a 60 Hz, two-winding, three-phase transformer with the following data:

MVA rating	: 42.00 MVA
Primary (HV) nominal voltage	: 110.00 kV
Secondary (LV) nominal voltage	: 28.40 kV
Winding connections (HV/LV)	: Y/Δ
Resistance	: $\bar{R}_{p0} + \bar{R}_{s0} = 0.00411$ pu on rating/phase
Leakage reactance	: $\bar{X}_{p0} + \bar{X}_{s0} = 0.1153$ pu on rating/phase
Off-load tap changer on HV side	: 4 steps, 2.75 kV/step
Under-load tap changer on LV side	: ±2.84 kV in 16 steps

Let us examine the condition when the LV winding is initially at its nominal position, and the HV winding is manually set two steps above its nominal position, i.e., at 115.5 kV. The parameters of the standard equivalent circuit (Figure 6.17) with the ONR on the LV (ULTC) side and values expressed in per unit of the transformer rated values are as follows:

Initial off-nominal turns ratio:

$$\bar{n} = \frac{28.4}{28.4}\frac{110}{115.5} = 0.95238$$

Per unit equivalent impedance:

$$\bar{Z}_e = \left(\frac{115.5}{110}\right)^2 (0.00411 + j0.1153)$$

$$= 0.00453 + j0.12712 \quad \text{pu}$$

Maximum per unit turns ratio:

$$\bar{n}_{max} = \frac{31.24}{28.4}\frac{110}{115.5} = 1.04762$$

Minimum per unit turns ratio:

$$\bar{n}_{min} = \frac{25.56}{28.4}\frac{110}{115.5} = 0.85714$$

Per unit turns ratio step:

$$\Delta\bar{n} = \frac{2.84}{16\times28.4}\frac{110}{115.5} = 0.0059524$$

Now, if the common system voltage and MVA base values are

      Primary system voltage base   : 115.0 kV
      Secondary system voltage base   : 28.4 kV
      System MVA base   : 100 MVA

the corresponding per unit parameters of the equivalent circuit are as follows:

Initial off-nominal turns ratio:

$$\bar{n} = 0.95238\frac{28.4}{28.4}\frac{115}{110} = 0.99567$$

Per unit equivalent impedance:

$$\bar{Z}_e = (0.00453+j0.12712)\left(\frac{110}{115}\right)^2\frac{100}{42}$$

$$= 0.009868+j0.27692$$

Maximum per unit turns ratio:

$$\bar{n}_{max} = 1.04762\frac{28.4}{28.4}\frac{115}{110} = 1.09524$$

Minimum per unit turns ratio:

$$\bar{n}_{min} = 0.85714\frac{28.4}{28.4}\frac{115}{110} = 0.89610$$

Per unit turns ratio step:

$$\Delta\bar{n} = 0.005924\frac{28.4}{28.4}\frac{115}{110} = 0.006193$$

    The equivalent π circuit (Figure 6.18) parameters representing the initial tap position are as follows:

$$y_1 = \frac{1}{\bar{n}\bar{Z}_e} = \frac{1}{0.99567(0.009868+j0.27692)}$$

$$= 0.12908-j3.62226$$

$$y_2 = \left(\frac{1}{\bar{n}}-1\right)y_1 = \left(\frac{1}{0.99567}-1\right)\frac{1}{0.99567(0.009868+j0.27692)}$$

$$= 0.00056-j0.01575$$

$$y_3 = \left(1 - \frac{1}{n}\right)\frac{1}{Z_e} = \left(1 - \frac{1}{0.99567}\right)\frac{1}{0.009868 + j0.27692}$$

$$= -0.00056 + j0.01568$$

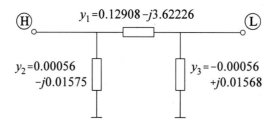

$$y_1 = 0.12908 - j3.62226$$

$y_2 = 0.00056$
$\quad -j0.01575$

$y_3 = -0.00056$
$\quad +j0.01568$

6.2.2 Representation of Three-Winding Transformers

Figure 6.19 shows the single-phase equivalent of a three-winding transformer under balanced conditions. The effect of the magnetizing reactance has been neglected, and the transformer is represented by three impedances connected to form a star. The common star point is fictitious and unrelated to the system neutral.

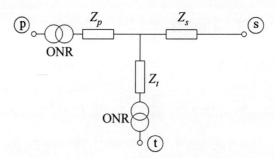

Figure 6.19 Equivalent circuit of a three-winding transformer

The three windings of the transformer may have different MVA ratings. However, the per unit impedances must be expressed on the same MVA base. As in the case of the two-winding transformer equivalent circuit developed in the previous section, off-nominal turns ratios are used to account for the differences between the ratios of actual turns and the base voltages. The values of the equivalent impedances Z_p, Z_s and Z_t may be obtained by standard short-circuit tests as follows [1]:

Z_{ps} = leakage impedance measured in primary with secondary shorted and tertiary open

Z_{pt} = leakage impedance measured in primary with tertiary shorted and secondary open

Z_{st} = leakage impedance measured in secondary with tertiary shorted and primary open

With the above impedances in ohms referred to the same voltage base, we have

$$Z_{ps} = Z_p + Z_s$$
$$Z_{pt} = Z_p + Z_t \qquad (6.71)$$
$$Z_{st} = Z_s + Z_t$$

Hence,

$$Z_p = \frac{1}{2}(Z_{ps} + Z_{pt} - Z_{st})$$

$$Z_s = \frac{1}{2}(Z_{ps} + Z_{st} - Z_{pt}) \qquad (6.72)$$

$$Z_t = \frac{1}{2}(Z_{pt} + Z_{st} - Z_{ps})$$

In large transformers, Z_s is small and may even be negative.

Example of modelling three-winding transformers

We will consider a 60 Hz, three-winding, three-phase transformer with the following data:

MVA rating	: 750 MVA
High/low/tertiary nominal voltages	: 500/240/28 kV
Winding connections (H/L/T)	: Y/Y/Δ

Measured positive-sequence impedances in pu on transformer MVA rating and nominal voltages at nominal tap position:

$Z_{H-L} = 0.0015 + j0.1339$
$Z_{L-T} = 0 + j0.1895$
$Z_{T-H} = 0 + j0.3335$

ULTC at high voltage side: 500±50 kV in 20 steps.

Neglecting the magnetizing reactance, the equivalent star circuit with ULTC at nominal tap position is

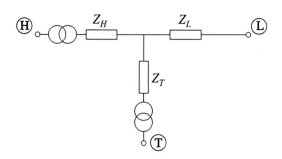

$$Z_H = \frac{Z_{H-L} + Z_{T-H} - Z_{L-T}}{2} = 0.00075 + j0.13895$$

$$Z_L = \frac{Z_{H-L} + Z_{L-T} - Z_{T-H}}{2} = 0.00075 - j0.00505$$

$$Z_T = \frac{Z_{L-T} + Z_{T-H} - Z_{H-L}}{2} = -0.00075 + j0.19455$$

Equivalent delta circuit with parameters in pu on transformer MVA rating and nominal voltages:

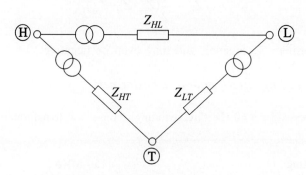

$$\Sigma = Z_H Z_L + Z_H Z_T + Z_L Z_T$$

$$= -0.02535 + j0.0002914$$

$$Z_{HL} = \frac{\Sigma}{Z_T} = 0.0020 + j0.1303$$

$$Z_{LT} = \frac{\Sigma}{Z_H} = 0.0011 + j0.1824$$

$$Z_{HT} = \frac{\Sigma}{Z_L} = -0.7859 - j4.9029$$

Equivalent delta circuit with parameters in pu on system MVA base of 100 MVA and voltage bases (H/L/T) of 500/220/27.6 kV:

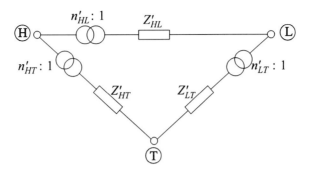

$$Z'_{HL} = Z_{HL}\frac{100}{750}\left(\frac{240}{220}\right)^2 = 0.00032 + j0.02067$$

$$Z'_{LT} = Z_{LT}\frac{100}{750}\left(\frac{28.0}{27.6}\right)^2 = 0.00015 + j0.02504$$

$$Z'_{TH} = Z_{TH}\frac{100}{750}\left(\frac{28.0}{27.6}\right)^2 = -0.10784 - j0.6728$$

$$n'_{HL} = \frac{500}{500}\frac{220}{240} = 0.91667$$

$$n'_{LT} = \frac{240}{220}\frac{27.6}{28.0} = 1.07532$$

$$n'_{HT} = \frac{500}{500}\frac{27.6}{28.0} = 0.98571$$

ULTC data:

$$n'_{HLmax} = \frac{550}{500}\frac{500}{500}\frac{220}{240} = 1.00833$$

$$n'_{HLmin} = \frac{450}{500}\frac{500}{500}\frac{220}{240} = 0.8250$$

$$\Delta n'_{HL} = \frac{1.00833 - 0.825}{20} = 0.00917$$

$$n'_{HTmax} = \frac{550}{500}\frac{500}{500}\frac{27.6}{28.0} = 1.08429$$

$$n'_{HTmin} = \frac{450}{500}\frac{500}{500}\frac{27.6}{28.0} = 0.88714$$

$$\Delta n'_{HT} = \frac{1.08429 - 0.8814}{20} = 0.01014$$

We should recognize that the ULTC action at the high-voltage side changes the ONRs n'_{HL} and n'_{HT}; these two ONRs cannot be adjusted independently.

The three branches of the delta equivalent circuit can each be represented by an equivalent circuit as shown in Figure 6.18.

The equivalent π circuits representing the initial ULTC tap position are as follows.

H-L branch:

$$y_1 = \frac{1}{n'_{HL} Z'_{HL}} = \frac{1}{0.91667(0.00032 + j0.02067)}$$

$$= 0.81687 - j52.76457$$

$$y_2 = \left(\frac{1}{n'_{HL}} - 1 \right) y_1 = \left(\frac{1}{0.91667} - 1 \right)(0.81687 - j52.26457)$$

$$= 0.074258 - j4.79657$$

$$y_3 = \left(1 - \frac{1}{n'_{HL}} \right) \frac{1}{Z'_{HL}} = \left(1 - \frac{1}{0.91667} \right) \frac{1}{0.00032 + j0.02067}$$

$$= -0.06807 + j4.39687$$

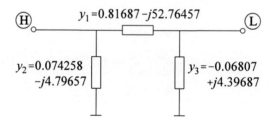

L-T branch:

$$y_1 = \frac{1}{n'_{LT} Z'_{LT}} = \frac{1}{1.07532(0.00015 + j0.02504)}$$

$$= 0.22247 - j37.13748$$

$$y_2 = \left(\frac{1}{n'_{LT}} - 1 \right) y_1 = \left(\frac{1}{1.07532} - 1 \right)(0.22247 - j37.13748)$$

$$= -0.01558 + j2.60127$$

$$y_3 = \left(1 - \frac{1}{n'_{LT}} \right) \frac{1}{Z'_{LT}} = \left(1 - \frac{1}{1.07532} \right) \frac{1}{0.00015 + j0.02504}$$

$$= 0.01676 - j2.79719$$

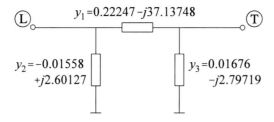

H-T branch:

$$y_1 = \frac{1}{n'_{HT}Z'_{HT}} = \frac{1}{0.98571(-0.10784-j0.67280)}$$

$$= -0.23564+j1.47010$$

$$y_2 = \left(\frac{1}{n'_{HT}}-1\right)y_1 = \left(\frac{1}{0.98571}-1\right)(-0.23564+j1.47010)$$

$$= -0.00342+j0.02131$$

$$y_3 = \left(1-\frac{1}{n'_{HT}}\right)\frac{1}{Z'_{HT}} = \left(1-\frac{1}{0.98571}\right)\frac{1}{-0.10784-j0.67280}$$

$$= 0.00337-j0.02101$$

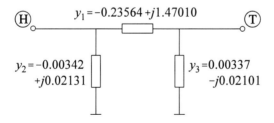

6.2.3 Phase-Shifting Transformers [14]

A phase-shifting transformer can be represented by the equivalent circuit shown in Figure 6.20. It consists of an admittance in series with an ideal transformer having a *complex turns ratio*, $\tilde{n}=n\angle\alpha$. The phase angle step size may not be equal at different tap positions. However, equal step size is usually used in power flow and transient stability programs.

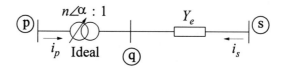

Figure 6.20 Phase-shifting transformer representation

By definition:

$$\frac{\tilde{v}_p}{\tilde{v}_q} = n\angle\alpha = n(\cos\alpha + j\sin\alpha)$$

$$= a_s + jb_s \tag{6.73}$$

where α is the phase shift from bus p to bus q; it is positive when $\tilde{v}_p$ leads $\tilde{v}_q$. Since there is no power loss in an ideal transformer,

$$\tilde{v}_p\tilde{i}_p^* = -\tilde{v}_q\tilde{i}_s^* \tag{6.74}$$

Therefore, the transformer current at bus p is

$$\tilde{i}_p = -\frac{1}{a_s - jb_s}\tilde{i}_s$$

$$= \frac{Y_e}{a_s - jb_s}(\tilde{v}_q - \tilde{v}_s) \tag{6.75}$$

Substituting for $\tilde{v}_q$ from Equation 6.73, we get

$$\tilde{i}_p = \frac{Y_e}{a_s - jb_s}\left[\frac{1}{a_s + jb_s}\tilde{v}_p - \tilde{v}_s\right]$$

$$= \frac{Y_e}{a_s^2 + b_s^2}[\tilde{v}_p - (a_s + jb_s)\tilde{v}_s] \tag{6.76}$$

From Equation 6.75,

$$\tilde{i}_s = -(a_s - jb_s)\tilde{i}_p$$

Substituting for $\tilde{i}_p$ from Equation 6.76 gives

$$\tilde{i}_s = \frac{Y_e}{a_s + j b_s}[(a_s + j b_s)\tilde{v}_s - \tilde{v}_p] \qquad (6.77)$$

Combining Equations 6.76 and 6.77, we obtain the following matrix equation relating the phase-shifter terminal voltages and currents

$$\begin{bmatrix} \tilde{i}_p \\ \tilde{i}_s \end{bmatrix} = \begin{bmatrix} \dfrac{Y_e}{a_s^2 + b_s^2} & \dfrac{-Y_e}{a_s - j b_s} \\ \dfrac{-Y_e}{a_s + j b_s} & Y_e \end{bmatrix} \begin{bmatrix} \tilde{v}_p \\ \tilde{v}_s \end{bmatrix} \qquad (6.78)$$

We see that the admittance matrix in the above equation is *not symmetrical*, that is, the transfer admittance from p to s is not equal to the transfer admittance from s to p. Therefore, a π equivalent circuit is not possible.

     If the turns ratio is real (i.e., $a_s = n$ and $b_s = 0$), the model reduces to the equivalent π circuit shown in Figure 6.18(b).

Example of modelling a phase-shifting transformer

     Let us consider a three-phase, two-winding phase shifter with the following data:

MVA rating	: 300 MVA
Primary/secondary base voltages	: 240/240 kV
Resistance per phase	: 0
Leakage reactance per phase	: 0.145 pu
Phase-shift range and steps	: ±40°, 36 steps
System voltage base (primary/secondary)	: 220/230 kV
System MVA base	: 100 MVA

Leakage reactance in pu on system voltage and MVA base:

$$X_e = 0.145 \times \frac{100}{300} \times \left(\frac{240}{230}\right)^2$$

$$= 0.05263 \quad \text{pu}$$

Off-nominal turns ratio:

$$n = \frac{240}{220} \times \frac{230}{240} = 1.04545$$

Phase-shift angle limits:

$$\alpha_{max} = 40°$$

$$\alpha_{min} = -40°$$

The impedance of the transformer changes with the phase-shift angle. The following table (provided by the manufacturer) gives values of the impedance multiplier as a function of the angle.

Angle in degrees	±40	±29.5	±25.1	±20.6	0
Impedance multiplier	1.660	1.331	1.228	1.144	1.0

The admittance matrix of Equation 6.78 representing the phase shifter is

$$\mathbf{Y}_s = \begin{bmatrix} \dfrac{Y_e}{a_s^2+b_s^2} & \dfrac{-Y_e}{a_s-jb_s} \\ \dfrac{-Y_e}{a_s+jb_s} & Y_e \end{bmatrix}$$

As an illustration, we will determine the elements of the admittance matrix for two values of α.

(a) $\alpha = 0$:

$$Y_e = \frac{1}{jX_e} = \frac{1}{j0.05263} = -j19.0006 \quad \text{pu}$$

The turns ratio of the ideal phase shifter is

$$a_s + jb_s = n(\cos\alpha + j\sin\alpha)$$

$$= 1.0455(\cos 0 + j\sin 0) = 1.0455 + j0$$

The corresponding admittance matrix $\mathbf{Y}_s$ is

$$\mathbf{Y}_s = \begin{bmatrix} -j17.3844 & j18.1745 \\ j18.1745 & -j19.0006 \end{bmatrix}$$

(b) α corresponding to the 10th step:

$$\alpha = \frac{40}{36} \times 10 = 11.11°$$

The turns ratio is

$$a_s + jb_s = n(\cos 11.11° + j\sin 11.11°)$$

$$= 1.02585 + j0.20147$$

The phase-shifter leakage reactance at this value of α by interpolation is

$$X_e = \left[1.0 + \frac{11.11(1.144 - 1.0)}{20.6}\right] \times 0.05263$$

$$= 0.05672$$

Hence,

$$Y_e = \frac{1}{jX_e} = j17.6305$$

The admittance matrix $\mathbf{Y}_s$, with $a_s + jb_s = 1.02585 + j0.20147$ and $Y_e = j17.6305$, is

$$\mathbf{Y}_s = \begin{bmatrix} -j16.1310 & (-3.2499 + j16.5479) \\ (3.2499 + j16.5479) & -j17.6305 \end{bmatrix}$$

6.3 TRANSFER OF POWER BETWEEN ACTIVE SOURCES

We will now examine factors influencing transfer of active and reactive power between two sources connected by an inductive reactance as shown in Figure 6.21. Such a system is representative of two sections of a power system interconnected by a transmission system, with power transfer from one section to the other.

We have considered a purely inductive reactance interconnecting the two sources. This is because impedances representing transmission lines, transformers, and generators are predominantly inductive. When the full network is represented by an appropriate model for each of its elements and then reduced to a two-bus system, the resulting impedance will be essentially an inductive reactance. The shunt capacitances of transmission lines do not explicitly appear in the model shown in Figure 6.21; their effects are implicitly represented by the net reactive power transmitted. Analysis of transmission of active and reactive power through an inductive reactance thus gives useful insight into the characteristics of ac transmission systems.

Referring to Figure 6.21, the complex power at the receiving end is

$$\tilde{S}_R = P_R + jQ_R = \tilde{E}_R \tilde{I}^* = \tilde{E}_R \left[\frac{\tilde{E}_S - \tilde{E}_R}{jX} \right]^*$$

$$= E_R \left[\frac{E_s \cos\delta + jE_s \sin\delta - E_R}{jX} \right]^*$$

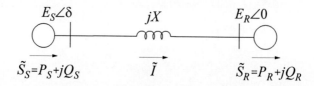

$$\tilde{S}_S = P_S + jQ_S \qquad \tilde{I} \qquad \tilde{S}_R = P_R + jQ_R$$

(a) Equivalent system diagram

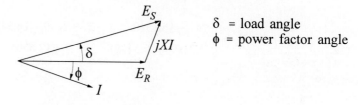

δ = load angle
ϕ = power factor angle

(b) Phasor diagram

Figure 6.21 Power transfer between two sources

Hence,

$$P_R = \frac{E_S E_R}{X} \sin\delta \qquad (6.79)$$

$$Q_R = \frac{E_S E_R \cos\delta - E_R^2}{X} \qquad (6.80)$$

Similarly,

$$P_S = \frac{E_S E_R}{X} \sin\delta \qquad (6.81)$$

$$Q_S = \frac{E_S^2 - E_S E_R \cos\delta}{X} \qquad (6.82)$$

Equations 6.79 to 6.82 describe the way in which active power and reactive power are transferred between active parts of a power network. Let us examine the dependence of active power and reactive power transfer on the source voltages by considering separately the effects of differences in voltage magnitudes and angles.

(a)   We will look first at the condition with $\delta = 0$. Equations 6.79 to 6.82 become

$$P_R = P_S = 0$$

and

$$Q_R = \frac{E_R(E_S - E_R)}{X}$$

$$Q_S = \frac{E_S(E_S - E_R)}{X}$$

The active power transfer is now zero. With $E_S > E_R$, Q_S and Q_R are positive, that is, reactive power is transferred from the sending end to the receiving end. The corresponding phasor diagram is shown in Figure 6.22(a). With $E_S < E_R$, Q_S and Q_R are negative, indicating that reactive power flows from the receiving end to the sending end. The phasor diagram is shown in Figure 6.22(b).

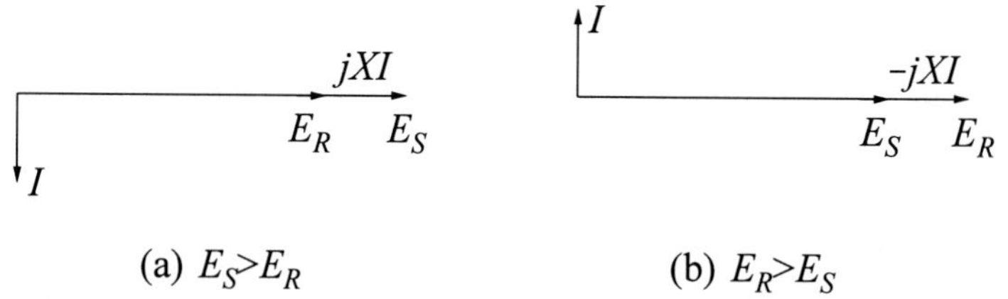

(a) $E_S > E_R$ (b) $E_R > E_S$

Figure 6.22 Phasor diagrams with $\delta = 0$

An alternative way of interpreting the above results is as follows:

- Transmission of lagging current through an inductive reactance causes a drop in receiving end voltage.

- Transmission of leading current through an inductive reactance causes a rise in receiving end voltage.

In each case,

$$Q_S - Q_R = \frac{(E_S - E_R)^2}{X} = XI^2$$

Therefore, the reactive power consumed by X is XI^2.

(b)   We will next consider the condition with $E_S = E_R$, but with $\delta \neq 0$. From Equations 6.79 to 6.82, we now have

$$P_R = P_S = \frac{E^2}{X} \sin\delta$$

$$Q_S = -Q_R = \frac{E^2}{X}(1 - \cos\delta)$$

$$= \frac{1}{2}XI^2$$

With δ positive, P_S and P_R are positive, that is, active power flows from the sending end to the receiving end. With δ negative, the direction of active power flow reverses. In each case there is no reactive power transferred from one end to the other; instead, each end supplies half of the XI^2 consumed by X. The corresponding phasor diagrams are shown in Figure 6.23.

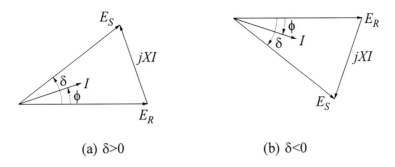

(a) δ>0   (b) δ<0

Figure 6.23 Phasor diagram with $E_S = E_R$

If the current I is in phase with E_R (i.e., the receiving end power factor is unity), the phasor diagram is as shown in Figure 6.24. In this case, the magnitude of E_S is only slightly larger than E_R. The sending end supplies all of the XI^2 consumed by X.

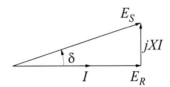

Figure 6.24 Phasor diagram with I in phase with E_R

We see that the active power transferred (P_R) is a function of voltage magnitudes and δ. However, for satisfactory operation of the power system, the voltage magnitude at any bus cannot deviate significantly from the nominal value. Therefore, control of active power transfer is achieved primarily through variations in angle δ.

(c)   Finally, let us consider a general case applicable to any values of δ, E_S and E_R. The current I is

$$I = \frac{E_S \cos\delta + jE_S \sin\delta - E_R}{jX} \tag{6.83}$$

From Equations 6.80, 6.82, and 6.83, we have

$$Q_S - Q_R = \frac{E_S^2 + E_R^2 - 2E_S E_R \cos\delta}{X}$$

$$= \frac{(XI)^2}{X} = XI^2 \tag{6.84}$$

If, in addition to inductive reactance X, we consider the series resistance R of the network, then

$$Q_{loss} = XI^2 = X\frac{P_R^2 + Q_R^2}{E_R^2} \tag{6.85}$$

$$P_{loss} = RI^2 = R\frac{P_R^2 + Q_R^2}{E_R^2} \tag{6.86}$$

We see from Equation 6.84 that the reactive power absorbed by X for all conditions is XI^2. This leads us to the concept of "reactive power loss," a companion term to active power loss RI^2 associated with resistive elements.

As seen from Equations 6.85 and 6.86, an increase of reactive power transmitted increases active as well as reactive power losses. This has an impact on efficiency of power transmission and voltage regulation.

From the above analysis, we can draw the following conclusions:

• Active power transfer depends mainly on the angle by which the sending end voltage leads the receiving end voltage.

• Reactive power transfer depends mainly on voltage magnitudes. It is transmitted from the side with higher voltage magnitude to the side with lower voltage magnitude.

• Reactive power cannot be transmitted over long distances since it would require a large voltage gradient to do so.

• An increase in reactive power transfer causes an increase in active as well as reactive power losses.

Although we have considered a simple system, the general conclusions are applicable to any practical system. In fact, the basic characteristics of ac transmission reflected in these conclusions have a dominant effect on the way in which we operate and control the power system.

6.4 POWER-FLOW ANALYSIS

So far in this chapter, we have considered simple system configurations and idealizing assumptions to gain an understanding of basic characteristics of ac transmission. In this section, we will describe analytical techniques for detailed analysis of power flow in large complex networks.

The power-flow (load-flow) analysis involves the calculation of power flows and voltages of a transmission network for specified terminal or bus conditions. Such calculations are required for the analysis of steady-state as well as dynamic performance of power systems.

The system is assumed to be balanced; this allows a single-phase representation of the system. For bulk power system studies, common practice is to represent the composite loads as seen from bulk power delivery points (see Chapter 7, Section 7.1). Therefore, the effects of distribution system voltage control devices on loads are represented implicitly.

In this section we will describe the power-flow analysis as it applies to the steady-state performance of the power system. The basic network equations presented here also apply to their representation in the analysis of system stability; however, as we will see in later chapters, some of the constraints vary depending on the type of stability problem being solved.

Bus classification

Associated with each bus are four quantities: active power P, reactive power Q, voltage magnitude V, and voltage angle θ.

The following types of buses (nodes) are represented, and at each bus two of the above four quantities are specified:

- Voltage-controlled (PV) bus: Active power and voltage magnitude are specified. In addition, limits to the reactive power are specified depending on the characteristics of the individual devices. Examples are buses with generators, synchronous condensers, and static var compensators.

- Load (PQ) bus: Active and reactive power are specified. Normally loads are assumed to have constant power. If the effect of distribution transformer ULTC operation is neglected, load P and Q are assumed to vary as a function of bus voltage.

- Device bus: Special boundary conditions associated with devices such as HVDC converters are recognized.

- Slack (swing) bus: Voltage magnitude and phase angle are specified. Because the power losses in the system are not known a priori, at least one bus must have unspecified P and Q. Thus the slack bus is the only bus with known voltage.

In some applications, it is desirable to keep the Q associated with the slack bus within reasonable limits; otherwise, the power-flow solution may become unrealistic. With Q at a limiting value, only the angle of the slack bus voltage is known.

Representation of network elements

Transmission lines are represented by equivalent π circuits with lumped parameters as described in Section 6.1.4. Shunt capacitors and reactors are represented as simple admittance elements connected to ground.

Transformers with off-nominal turns ratio are represented by equivalent π circuits as described in Section 6.2.2. Any phase shifts introduced due to transformer connections (such as Δ-Y connections) are not usually represented. In radial networks, such phase shifts do not affect power-flow analysis since currents and voltages are shifted by the same angle. In closed-loop networks, utilities take special care to connect transformer windings so that there is no net phase shift introduced in a common direction round a loop; otherwise, circulating current will flow, which is normally unacceptable. Figure 6.25 shows a scheme for connecting transformer windings with due regard to the resulting phase shifts.

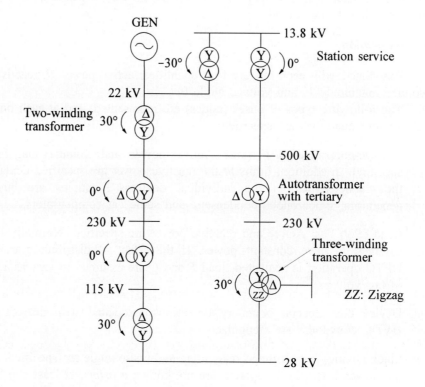

Figure 6.25 Illustration of a scheme for transformer-winding connections

Phase-shifting transformers, which are provided specifically for controlling power flow, are represented as illustrated in Section 6.2.3. This representation may also be used to account for phase shift introduced by the transformer-winding connection (Y-Δ, Y-Zigzag) in special situations where this is desired; the phase-shift angle in such cases, however, remains fixed.

As we are considering only the balanced operation of the power system, each element is modelled in terms of its single-phase equivalent (positive sequence).

6.4.1 Network Equations

The relationships between network bus (node) voltages and currents may be represented by either loop equations or node equations [1]. Node equations are normally preferred because the number of independent node equations is smaller than the number of independent loop equations.

The network equations in terms of the *node admittance matrix* can be written as follows:

$$
\begin{bmatrix} \tilde{I}_1 \\ \tilde{I}_2 \\ \dots \\ \tilde{I}_n \end{bmatrix} = \begin{bmatrix} Y_{11} & Y_{12} & \cdots & Y_{1n} \\ Y_{21} & Y_{22} & \cdots & Y_{2n} \\ \dots & \dots & \dots & \dots \\ Y_{n1} & Y_{n2} & \cdots & Y_{nn} \end{bmatrix} \begin{bmatrix} \tilde{V}_1 \\ \tilde{V}_2 \\ \dots \\ \tilde{V}_n \end{bmatrix}
\tag{6.87}
$$

where

n is the total number of nodes
Y_{ii} is the self admittance of node i
     = sum of all the admittances terminating at node i
Y_{ij} is mutual admittance between nodes i and j
     = negative of the sum of all admittances between nodes i and j
$\tilde{V}_i$ is the phasor voltage to ground at node i
$\tilde{I}_i$ is the phasor current flowing into the network at node i

The effects of generators, nonlinear loads, and other devices (for example, dynamic reactive compensators, HVDC converters) connected to the network nodes are reflected in the node current. Constant impedance (linear) loads are, however, included in the node admittance matrix.

We will illustrate the formulation of the node equation by considering the simple three-bus system depicted in Figure 6.26.

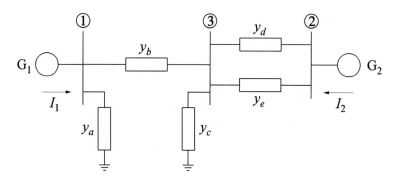

Figure 6.26 Single-line diagram of a three-bus system

The elements of the node admittance matrix are

$$
\begin{array}{lll}
Y_{11} = y_a + y_b & Y_{12} = 0 & Y_{13} = -y_b \\
Y_{21} = 0 & Y_{22} = y_d + y_e & Y_{23} = -(y_d + y_e) \\
Y_{31} = -y_b & Y_{32} = -(y_d + y_e) & Y_{33} = y_b + y_c + y_d + y_e
\end{array}
$$

The node currents are

I_1 = current into node 1 from generator G_1
I_2 = current into node 2 from generator G_2
I_3 = 0

The node equation for the network of Figure 6.22 is

$$
\begin{bmatrix} \tilde{I}_1 \\ \tilde{I}_2 \\ 0 \end{bmatrix} = \begin{bmatrix} Y_{11} & 0 & Y_{13} \\ 0 & Y_{22} & Y_{23} \\ Y_{31} & Y_{32} & Y_{33} \end{bmatrix} \begin{bmatrix} \tilde{V}_1 \\ \tilde{V}_2 \\ \tilde{V}_3 \end{bmatrix}
$$

We can make the following general observations regarding the node admittance matrix:

(a)   It is sparse with the degree of sparsity increasing with the network size.

(b)   It is singular if floating (i.e., if there are no shunt branches to ground).

(c)   It has weak diagonal dominance, i.e.,

$$
|Y_{ii}| \geq \sum_{j \neq i} |Y_{ij}|
$$

(d)   It is symmetrical, if there are no phase-shifting transformers.

Nonlinear power-flow equations

Equations 6.87 would be *linear* if the current injections $\tilde{I}$ were known. However, in practice, the current injections are not known for most nodes. The current at any node k is related to P, Q and $\tilde{V}$ as follows:

$$\tilde{I}_k = \frac{P_k - jQ_k}{\tilde{V}_k^*} \tag{6.88}$$

For the PQ nodes, P and Q are specified; and for the PV nodes, P and the magnitude of $\tilde{V}$ are specified. For other types of nodes, the relationships between P, Q, $\tilde{V}$ and $\tilde{I}$ are defined by the characteristics of the devices connected to the nodes. Clearly the boundary conditions imposed by the different types of nodes make the problem *nonlinear* and therefore power-flow equations are solved iteratively using techniques such as the Gauss-Seidel or Newton-Raphson method. The principles of application of these methods are briefly described below. References 15 and 16 provide comprehensive reviews of numerical methods for power flow analysis.

6.4.2 Gauss-Seidel Method

This method is based on the iterative approach proposed by Seidel in 1874 (Academy of Science, Munich). For application to the power-flow problem, from Equations 6.87 and 6.88, for the kth node we can write

$$\frac{P_k - jQ_k}{\tilde{V}_k^*} = Y_{kk}\tilde{V}_k + \sum_{\substack{i=1 \\ i \neq k}}^{n} Y_{ki}\tilde{V}_i \tag{6.89}$$

from which the voltage $\tilde{V}_k$ may be expressed as

$$\tilde{V}_k = \frac{P_k - jQ_k}{Y_{kk}\tilde{V}_k^*} - \frac{1}{Y_{kk}}\sum_{\substack{i=1 \\ i \neq k}}^{n} Y_{ki}\tilde{V}_i \tag{6.90}$$

Equation 6.89 is the heart of the iterative algorithm. The iterations begin with an informed guess of the magnitude and angle of the voltages at all load buses, and of the voltage angle at all generator buses.

For a load bus, P and Q are known, and Equation 6.90 is used to compute the voltage $\tilde{V}_k$ by using the best available voltages for all the buses. In other words, the upgraded values of bus voltages are used as soon as they are available. For example, for the pth iteration, the best values of bus voltages for computing the voltage V_k at bus k are $V_1^p, V_2^p, \cdots, V_{k-1}^p, V_k^{p-1}, V_{k+1}^{p-1}, \cdots, V_n^{p-1}$.

If the kth bus is a generator bus, the following procedure is used:

(a)   Rearranging Equation 6.89, we have

$$Q_k = -\mathrm{Im}\left[\tilde{V}_k^* \sum_{i=1}^{n} Y_{ki} \tilde{V}_i\right] \qquad (6.91)$$

where Q_k is calculated by using the best available values of bus voltages. If Q_k is within the limits $Q_{k\,max}$ and $Q_{k\,min}$, it is used in Equation 6.90 to compute the updated value of $\tilde{V}_k$. Its real and imaginary components are multiplied by the ratio of the *specified* value of the magnitude of the generator voltage to the magnitude of its updated value, thus complying with the magnitude constraint. In other words, the magnitude of the voltage is forced to be the specified value and Equations 6.90 and 6.91 are solved to compute the angle.

(b)   If Q_k computed by Equation 6.91 exceeds either the maximum or minimum limit, it is set equal to the limit. The updated value of $\tilde{V}_k$ is computed by treating the generator bus as a *PQ* node.

The iterations are continued until the real and imaginary components of voltages at each bus computed by successive iterations converge to a specified tolerance.

The Gauss-Seidel method has slow convergence because of weak diagonal dominance of the node admittance matrix. Acceleration factors are often used to speed up the convergence:

$$\text{Accelerated } \tilde{V}_k^{new} = \tilde{V}_k^{old} + c(\tilde{V}_k^{new} - \tilde{V}_k^{old}) \qquad (6.92)$$

where c is the acceleration factor, which is typically on the order of 1.4 to 1.7.

6.4.3 Newton-Raphson (N-R) Method

This is an iterative technique for solving a set of nonlinear equations. Let the following represent n such equations in n unknowns:

$$f_1(x_1, x_2, \cdots, x_n) = b_1$$
$$f_2(x_1, x_2, \cdots, x_n) = b_2$$
$$\cdots \quad \cdots \quad \cdots \quad \cdots \quad \cdots \quad \cdots \qquad (6.93)$$
$$f_n(x_1, x_2, \cdots, x_n) = b_n$$

If the iterations start with an initial estimate of $x_1^0, x_2^0, \cdots, x_n^0$ for the n unknowns and if $\Delta x_1, \Delta x_2, \cdots, \Delta x_n$ are the corrections necessary to the estimates so that the equations are exactly satisfied, we have

$$f_1(x_1^0+\Delta x_1, \; x_2^0+\Delta x_2, \; \cdots, \; x_n^0+\Delta x_n) = b_1$$

$$f_2(x_1^0+\Delta x_1, \; x_2^0+\Delta x_2, \; \cdots, \; x_n^0+\Delta x_n) = b_2 \qquad (6.94)$$

$$\cdots \quad \cdots \quad \cdots \quad \cdots \quad \cdots \quad \cdots \quad \cdots \quad \cdots \quad \cdots \quad \cdots \quad \cdots$$

$$f_n(x_1^0+\Delta x_1, \; x_2^0+\Delta x_2, \; \cdots, \; x_n^0+\Delta x_n) = b_n$$

Each of the above equations can be expanded using Taylor's theorem. The expanded form of the ith equation is

$$f_i(x_1^0+\Delta x_1, x_2^0+\Delta x_2, \cdots, x_n^0+\Delta x_n)$$

$$= f_i(x_1^0, x_2^0, \cdots, x_n^0) + \left(\frac{\partial f_i}{\partial x_1}\right)_0 \Delta x_1 + \left(\frac{\partial f_i}{\partial x_2}\right)_0 \Delta x_2 + \cdots + \left(\frac{\partial f_i}{\partial x_n}\right)_0 \Delta x_n$$

$$+ \text{ terms with higher powers of } \Delta x_1, \Delta x_2, \cdots, \Delta x_n$$

$$= b_i$$

The terms of higher powers can be neglected, if our initial estimate is close to the true solution.

The resulting linear set of equations in matrix form is

$$
\begin{bmatrix}
b_1 - f_1(x_1^0, x_2^0, \cdots, x_n^0) \\
b_2 - f_2(x_1^0, x_2^0, \cdots, x_n^0) \\
\cdots \cdots \cdots \cdots \cdots \cdots \cdots \cdots \\
b_n - f_n(x_1^0, x_2^0, \cdots, x_n^0)
\end{bmatrix}
=
\begin{bmatrix}
\left(\dfrac{\partial f_1}{\partial x_1}\right)_0 & \left(\dfrac{\partial f_1}{\partial x_2}\right)_0 & \cdots & \left(\dfrac{\partial f_1}{\partial x_n}\right)_0 \\
\left(\dfrac{\partial f_2}{\partial x_1}\right)_0 & \left(\dfrac{\partial f_2}{\partial x_2}\right)_0 & \cdots & \left(\dfrac{\partial f_2}{\partial x_n}\right)_0 \\
\cdots & \cdots & \cdots & \cdots \\
\left(\dfrac{\partial f_n}{\partial x_1}\right)_0 & \left(\dfrac{\partial f_n}{\partial x_2}\right)_0 & \cdots & \left(\dfrac{\partial f_n}{\partial x_n}\right)_0
\end{bmatrix}
\begin{bmatrix}
\Delta x_1 \\
\Delta x_2 \\
\cdots \\
\Delta x_n
\end{bmatrix}
\qquad (6.95)
$$

or

$$\Delta \mathbf{f} = \mathbf{J}\Delta\mathbf{x} \qquad (6.96)$$

where $\mathbf{J}$ is referred to as the *Jacobian*. If the estimates $x_1^0, \cdots, x_n^0$ were exact, then $\Delta\mathbf{f}$ and $\Delta\mathbf{x}$ would be zero. However, as $x_1^0, \cdots, x_n^0$ are only estimates, the errors $\Delta\mathbf{f}$ are finite. Equation 6.95 provides a linearized relationship between the errors $\Delta\mathbf{f}$ and the corrections $\Delta\mathbf{x}$ through the Jacobian of the simultaneous equations. A solution for $\Delta\mathbf{x}$ can be obtained by applying any suitable method for the solution of a set of linear equations. Updated values of x are calculated from

$$x_i^1 = x_i^0 + \Delta x_i$$

The process is repeated until the errors Δf_i are lower than a specified tolerance. The iterations have quadratic convergence. The Jacobian has to be recalculated at each step.

This method is sometimes referred to as *Newton's method*. However, it is more commonly called the *Newton-Raphson method* after J. Raphson (1648-1715) who wrote the iteration method in the form now commonly used.

Application of the N-R method to power-flow solution

To apply the Newton-Raphson method, each complex equation represented by Equation 6.89 has to be rewritten as two real equations in terms of two real variables instead of one complex variable. This is because Equation 6.89 is not an analytic function of the complex voltages due to the conjugate term $\tilde{V}_k^*$, and as a consequence the complex derivatives do not exist.

Most production-type power-flow programs use the power equation form with polar coordinates which we will use here. For any node k, we have

$$\tilde{S}_k = P_k + jQ_k = \tilde{V}_k \tilde{I}_k^* \tag{6.97}$$

From Equation 6.87,

$$\tilde{I}_k = \sum_{m=1}^{n} \tilde{Y}_{km} \tilde{V}_m \tag{6.98}$$

Substitution of $\tilde{I}_k$ given by Equation 6.98 in Equation 6.97 yields

$$P_k + jQ_k = \tilde{V}_k \sum_{m=1}^{n} (G_{km} - jB_{km}) \tilde{V}_m^* \tag{6.99}$$

The product of phasors $\tilde{V}_k$ and $\tilde{V}_m^*$ may be expressed as

$$\tilde{V}_k \tilde{V}_m^* = (V_k e^{j\theta_k})(V_m e^{-j\theta_m}) = V_k V_m e^{j(\theta_k - \theta_m)}$$

$$= V_k V_m (\cos\theta_{km} + j\sin\theta_{km}) \qquad (\theta_{km} = \theta_k - \theta_m) \tag{6.100}$$

Therefore, the expressions for P_k and Q_k may be written in real form as follows:

$$P_k = V_k \sum_{m=1}^{n} (G_{km}V_m\cos\theta_{km} + B_{km}V_m\sin\theta_{km}) \tag{6.101}$$

$$Q_k = V_k \sum_{m=1}^{n} (G_{km}V_m\sin\theta_{km} - B_{km}V_m\cos\theta_{km})$$

Thus, P and Q at each bus are functions of voltage magnitude V and angle θ of *all* buses.

If the active power and reactive power at each bus are specified, using superscript *sp* to denote specified values, we may write the LF equation:

$$P_1(\theta_1, \cdots, \theta_n, V_1, \cdots, V_n) = P_1^{sp}$$

$$\cdots \quad \cdots \quad \cdots \quad \cdots \quad \cdots \quad \cdots \quad \cdots \quad \cdots$$

$$P_n(\theta_1, \cdots, \theta_n, V_1, \cdots, V_n) = P_n^{sp} \tag{6.102}$$

$$Q_1(\theta_1, \cdots, \theta_n, V_1, \cdots, V_n) = Q_1^{sp}$$

$$\cdots \quad \cdots \quad \cdots \quad \cdots \quad \cdots \quad \cdots \quad \cdots \quad \cdots$$

$$Q_n(\theta_1, \cdots, \theta_n, V_1, \cdots, V_n) = Q_n^{sp}$$

Following the general procedure described earlier for the application of the N-R method (Equation 6.95), we have

$$
\begin{bmatrix}
P_1^{sp} - P_1(\theta_1^0, \cdots, \theta_n^0, V_1^0, \cdots, V_n^0) \\
\cdots \quad \cdots \quad \cdots \quad \cdots \quad \cdots \quad \cdots \quad \cdots \\
P_n^{sp} - P_n(\theta_1^0, \cdots, \theta_n^0, V_1^0, \cdots, V_n^0) \\
Q_1^{sp} - Q_1(\theta_1^0, \cdots, \theta_n^0, V_1^0, \cdots, V_n^0) \\
\cdots \quad \cdots \quad \cdots \quad \cdots \quad \cdots \quad \cdots \quad \cdots \\
Q_n^{sp} - Q_n(\theta_1^0, \cdots, \theta_n^0, V_1^0, \cdots, V_n^0)
\end{bmatrix}
=
\begin{bmatrix}
\frac{\partial P_1}{\partial \theta_1} & \cdots & \frac{\partial P_1}{\partial \theta_n} & \frac{\partial P_1}{\partial V_1} & \cdots & \frac{\partial P_1}{\partial V_n} \\
\cdots & \cdots & \cdots & \cdots & \cdots & \cdots \\
\frac{\partial P_n}{\partial \theta_1} & \cdots & \frac{\partial P_n}{\partial \theta_n} & \frac{\partial P_n}{\partial V_1} & \cdots & \frac{\partial P_n}{\partial V_n} \\
\frac{\partial Q_1}{\partial \theta_1} & \cdots & \frac{\partial Q_1}{\partial \theta_n} & \frac{\partial Q_1}{\partial V_1} & \cdots & \frac{\partial Q_1}{\partial V_n} \\
\cdots & \cdots & \cdots & \cdots & \cdots & \cdots \\
\frac{\partial Q_n}{\partial \theta_1} & \cdots & \frac{\partial Q_n}{\partial \theta_n} & \frac{\partial Q_n}{\partial V_1} & \cdots & \frac{\partial Q_n}{\partial V_n}
\end{bmatrix}
\begin{bmatrix}
\Delta\theta_1 \\
\cdots \\
\Delta\theta_n \\
\Delta V_1 \\
\cdots \\
\Delta V_n
\end{bmatrix}
$$

or

$$
\begin{bmatrix} \Delta \mathbf{P} \\ \Delta \mathbf{Q} \end{bmatrix} = \underbrace{\begin{bmatrix} \dfrac{\partial \mathbf{P}}{\partial \boldsymbol{\theta}} & \dfrac{\partial \mathbf{P}}{\partial \mathbf{V}} \\[2ex] \dfrac{\partial \mathbf{Q}}{\partial \boldsymbol{\theta}} & \dfrac{\partial \mathbf{Q}}{\partial \mathbf{V}} \end{bmatrix}}_{\text{Jacobian}} \begin{bmatrix} \Delta \boldsymbol{\theta} \\ \Delta \mathbf{V} \end{bmatrix} \qquad (6.103)
$$

The sparsity of each submatrix of the Jacobian is the same as that of the node admittance matrix. For efficient solution of the above equation, a suitable method, such as sparsity-oriented triangular factorization method (discussed in Section 6.4.6), must be used.

In formulating Equation 6.103, we have assumed that all buses are *PQ* buses. For a *PV* bus, only *P* is specified and the magnitude of *V* is fixed. Therefore, terms corresponding to ΔQ and ΔV would be absent for each of the *PV* buses. Thus the Jacobian would have only one row and one column for each *PV* bus.

Sensitivity analysis using the Jacobian

When we reach a solution with the N-R method, we have a linearized model around the given operating point:

$$
\begin{bmatrix} \mathbf{J} \end{bmatrix} \begin{bmatrix} \Delta \boldsymbol{\theta} \\ \Delta \mathbf{V} \end{bmatrix} = \begin{bmatrix} \Delta \mathbf{P} \\ \Delta \mathbf{Q} \end{bmatrix} \qquad (6.104)
$$

We can, therefore, easily compute the expected small changes in θ and V for small changes in P and Q. This type of sensitivity information is useful for estimating expected voltage changes which result from the installation of reactive compensation.

As we will see in Chapter 14, the Jacobian also provides very useful information regarding voltage stability.

6.4.4 Fast Decoupled Load-Flow (FDLF) Methods [17,18]

These techniques take advantage of the physical weak coupling between P and V, and between Q and θ (see Section 6.3). They also make a number of approximations which simplify the power-flow problem.

The basic algorithm for FDLF may be derived as follows from Equation 6.103:

$$
\begin{bmatrix} \Delta \mathbf{P} \\ \Delta \mathbf{Q} \end{bmatrix} = \begin{bmatrix} \dfrac{\partial \mathbf{P}}{\partial \boldsymbol{\theta}} & \dfrac{\partial \mathbf{P}}{\partial \mathbf{V}} \\[2ex] \dfrac{\partial \mathbf{Q}}{\partial \boldsymbol{\theta}} & \dfrac{\partial \mathbf{Q}}{\partial \mathbf{V}} \end{bmatrix} \begin{bmatrix} \Delta \boldsymbol{\theta} \\ \Delta \mathbf{V} \end{bmatrix}
$$

The first step in applying the P-θ/Q-V decoupling is to neglect the coupling submatrices $\partial\mathbf{P}/\partial\mathbf{V}$ and $\partial\mathbf{Q}/\partial\boldsymbol\theta$ in Equation 6.103, giving two separate equations:

$$\Delta\mathbf{P} = \frac{\partial\mathbf{P}}{\partial\boldsymbol\theta}\Delta\boldsymbol\theta$$
$$= \mathbf{H}\Delta\boldsymbol\theta \qquad (6.105)$$

and

$$\Delta\mathbf{Q} = \frac{\partial\mathbf{Q}}{\partial\mathbf{V}}\Delta\mathbf{V}$$
$$= \mathbf{L}\,\Delta\mathbf{V} \qquad (6.106)$$

The elements of matrices $\mathbf{H}$ and $\mathbf{L}$ are derived from Equation 6.101 as follows [17]:

$$H_{km} = \frac{\partial P_k}{\partial\theta_m} = V_k V_m (G_{km}\sin\theta_{km} - B_{km}\cos\theta_{km}) \qquad \text{for } m \neq k$$

and

$$H_{kk} = \frac{\partial P_k}{\partial\theta_k} = -B_{kk}V_k^2 - Q_k$$

Similarly,

$$L_{km} = \frac{\partial Q_k}{\partial V_m} = V_k(G_{km}\sin\theta_{km} - B_{km}\cos\theta_{km})$$
$$= H_{km}/V_m \qquad \text{for } m \neq k$$

and

$$L_{kk} = -B_{kk}V_k + Q_k/V_k$$

Equations 6.105 and 6.106 may be solved alternately by the decoupled N-R method, re-evaluating and re-triangularizing matrices $\mathbf{H}$ and $\mathbf{L}$ at each iteration. However, further simplifications can be made by recognizing that in practical power systems the following approximations are valid:

$$\cos\theta_{km} \approx 1; \qquad G_{km}\sin\theta_{km} << B_{km}; \qquad Q_k << B_{kk}V_k^2$$

Therefore, Equations 6.105 and 6.106 simplify to

$$\Delta P = (VB'V)\Delta\theta \qquad\qquad (6.107)$$

$$\Delta Q = (VB'')\Delta V \qquad\qquad (6.108)$$

At this stage, matrices B' and B'' are identical and equal to $-B$, where B is the network susceptance matrix.

The following simplifications further improve the convergence rate of the iterative process [17]:

(a)   The network elements that predominantly affect reactive power flows (i.e., shunt reactances and off-nominal ratio transformer taps) are omitted from B'. Similarly, phase-shifter effects are omitted from B''.

(b)   The left-hand V terms in Equations 6.107 and 6.108 are moved to the left-hand sides of the equations, and the influence of reactive power flows on the calculation of $\Delta\theta$ is removed by setting the right-hand V terms to 1.0 pu in Equation 6.107.

Equations 6.107 and 6.108 now simplify to

$$\Delta P/V = B'\Delta\theta \qquad\qquad (6.109)$$

$$\Delta Q/V = B''\Delta V \qquad\qquad (6.110)$$

The method originally proposed in reference 17, in addition to the approximations stated above, neglects the effects of series resistances in B'; it is called the *XB scheme*. The method proposed later in reference 18 neglects the effects of series resistance in B'' and is called the *BX scheme*.

The matrices B' and B'' are both real and sparse. They contain only those network admittances that are constant. Therefore, they have to be triangularized only once at the beginning. Matrix B'' is symmetrical so that only the upper triangular factor needs to be stored. If phase shifters are absent, B' is also symmetrical.

The advantage of using Equations 6.109 and 6.110 is that very fast repeat solutions of $\Delta\theta$ and ΔV can be obtained using constant triangular factors B' and B''. These solutions may be iterated with each other in some defined manner toward the exact solution. The power-flow solution is reached when $\Delta P/V$ and $\Delta Q/V$ become less than the specified solution tolerance. At the ith iteration ΔP and ΔQ are computed as follows by using Equations 6.101 and 6.102:

$$\Delta P = P^{sp} - P(\theta^{i-1}, V^{i-1}) \qquad\qquad (6.111)$$

$$\Delta Q = Q^{sp} - Q(\theta^{i-1}, V^{i-1}) \qquad\qquad (6.112)$$

This ensures that the full system equations are satisfied in the final solution. Equations 6.109 and 6.110 merely establish the corrections for ΔV and $\Delta\theta$ at each iteration step.

6.4.5 Comparison of the Power-Flow Solution Methods

The Gauss-Seidel method is the oldest of the power-flow solution methods. It is simple, reliable, and usually tolerant of poor voltage and reactive power conditions. In addition, it has low computer memory requirements. However, the computation time increases rapidly with system size. This method has a slow convergence rate and exhibits convergence problems when the system is stressed due to high levels of active power transfer.

The Newton-Raphson method has a very good convergence rate (quadratic). The computation time increases only linearly with system size. This method has convergence problems when the initial voltages are significantly different from their true values; it is therefore not suited for a "flat" voltage start.[1] Once the voltage solution is near the true solution, however, the convergence is very rapid. The Newton-Raphson method is therefore particularly suited for applications involving large systems requiring very accurate solutions.

The convergence properties of the Newton-Raphson method complement those of the Gauss-Seidel method. Therefore, many power-flow programs provide both solution techniques. The solution may be started with the Gauss-Seidel method and then switched to the Newton-Raphson method to obtain a rapid well-converged solution.

Fast decoupled load-flow (FDLF) methods are basically approximations to the Newton-Raphson (N-R) method. In the N-R method the Jacobian is required for computing $\Delta\theta$ and ΔV. Therefore, the Jacobian has an impact on the convergence of the iterative solution but does not directly affect the final solution. The approximations made in the FDLF methods generally result in a small increase in the number of iterations. However, the computation effort is significantly reduced since the Jacobian does not have to be recalculated and refactorized in each iteration. In addition, the computer memory requirements are reduced. The convergence rate of the FDLF methods is linear as compared to the quadratic rate of the N-R method. The FDLF methods are less sensitive to the initial voltage and reactive power conditions than the N-R method. The fast decoupled *XB* method is not suited for systems with high *R/X* ratio; the *BX* method is better-suited for such systems. For most system conditions the FDLF methods provide rapid solution with good accuracy. However, for system conditions with very large angles across lines and with special control devices that strongly influence active and reactive power flows, full N-R formulation may be required.

---

[1] If a previously solved case for the same network with generally similar operating conditions is not available, a common practice is to start the solution with all load bus voltages at one per unit magnitude and zero angle, and all generator bus voltage magnitudes at specified magnitude and zero angle. This is known as a *flat* voltage start.

6.4.6 Sparsity-Oriented Triangular Factorization

Analysis of the power-flow problem using methods such as the N-R method and the FDLF method requires the solution of sparse linear matrix equations. Sparsity-oriented triangular factorization is commonly used for solving these equations.

A sparse linear matrix equation has the form

$$\mathbf{Ax} = \mathbf{b} \tag{6.113}$$

For any given $\mathbf{b}$, the above equation can be solved for $\mathbf{x}$, by triangular factorization of $\mathbf{A}$ as follows:

$$(\mathbf{LDU})\mathbf{x} = \mathbf{b} \tag{6.114}$$

where

      $\mathbf{L}$ is lower triangular matrix
      $\mathbf{U}$ is upper triangular matrix
      $\mathbf{D}$ is diagonal matrix

The matrices $\mathbf{L}$ and $\mathbf{U}$ are also sparse. If $\mathbf{A}$ is symmetrical, $\mathbf{L}$ is the transpose of $\mathbf{U}$ and need not be computed or stored.

Equation 6.114 is solved for $\mathbf{x}$ in terms of $\mathbf{b}$ by forward and backward substitution. Forward substitution reduces Equation 6.114 to the form

$$\left[\begin{array}{c} \diagdown \\ \mathbf{U} \end{array}\right]\left[\,\mathbf{x}\,\right] = \left[\,\mathbf{b'}\,\right] \tag{6.115}$$

This in effect is triangular factorization due to Gauss elimination. Solution of $\mathbf{x}$ is found by back substitution: the last equation gives x_n, inserted into $(n-1)^{\text{th}}$ equation gives x_{n-1}, and so on.

Sparsity techniques using optimal ordering are essential to this approach for the solution of large network equations [19]. The efficiency of sparse matrix methods can be enhanced by using sparse vector methods [20]. Reference 21 provides a detailed discussion of sparse matrix concepts and methods.

6.4.7 Network Reduction

The size of the network can be reduced by elimination of the passive nodes. If $I_k=0$, node k can be eliminated by replacing the elements of the remaining $n-1$ rows and columns with

$$y'_{ij} = y_{ij} - \frac{y_{ik}y_{kj}}{y_{kk}} \qquad (6.116)$$

for $i=1, 2, \cdots, k-1, k+1, \cdots, n$ and $j=1, 2, \cdots, k-1, k+1, \cdots, n$.

By successive application of Equation 6.116, any desired number of passive nodes can be eliminated. Equation 6.116 is called *Kron's reduction formula*.

If sparsity techniques are used, it is not advisable to apply node elimination indiscriminately. In general, reduced systems become denser; therefore, the best compromise is to eliminate only those nodes that do not contribute to an increase in the number of branches when eliminated.

Reference 22 describes an efficient network reduction technique that takes advantage of certain special properties of sparse matrix factorization.

REFERENCES

[1]   W.D. Stevenson, Jr., *Elements of Power System Analysis*, Third Edition, McGraw-Hill, 1975.

[2]   O. Elgerd, *Electric Energy Systems Theory: An Introduction*, McGraw-Hill, 1971.

[3]   J. Zaborszky and J.W. Rittenhouse, *Electric Power Transmission*, Vols. 1 and 2, 3rd reprint, The Rensselaer Book Store, Troy, N.Y., 1977.

[4]   B.M. Weedy, *Electric Power Systems*, John Wiley & Sons, Third Edition, 1979.

[5]   C.A. Gross, *Power System Analysis*, Second Edition, John Wiley & Sons, 1986.

[6]   C.F. Wagner and R.D. Evans, *Symmetrical Components*, McGraw-Hill, 1933.

[7]   M.E. El-Hawary, *Electrical Power Systems: Design and Analysis*, Reston Publishing Company, 1983.

[8]   *Electrical Transmission and Distribution Reference Book*, Westinghouse Electric Corporation, East Pittsburgh, Pa., 1964.

[9]   J.J. Leforest (editor), *Transmission Line Reference Book-345 kV and Above*, Second Edition, EPRI, 1982.

[10]   T.J.E. Miller (editor), *Reactive Power Control in Electric Systems*, Wiley-Interscience, 1982.

[11]   H.P. St. Clair, "Practical Concepts in Capability and Performance of Transmission Lines," *AIEE Trans.*, Vol. 72, pp. 1152-1157, December 1953.

[12]   R.D. Dunlop, R. Gutman, and R.P. Marchenko, "Analytical Development of Loadability Characteristics for EHV and UHV Transmission Lines," *IEEE Trans.*, Vol. PAS-98, pp. 606-617, March/April 1979.

[13]   IEEE Committee Report, "Common Format for the Exchange of Solved Load Flow Data," *IEEE Trans.*, Vol. PAS-92, pp. 1916-1925, November/December 1973.

[14]   G.W. Stagg and A.H. El-Abiad, *Computer Methods in Power System Analysis*, McGraw-Hill, 1968.

[15]   M.A. Laughton and M.W.H. Davies, "Numerical Techniques in Solution of Power System Load-Flow Problems," *Proceedings of IEE*, Vol. III, No. 9, 1964.

[16]   B. Stott, "Review of Load-Flow Calculation Methods," *Proceedings of IEEE*, Vol. 62, pp. 916-929, July 1974.

[17]   B. Stott and D. Alsac, "Fast Decoupled Load Flow," *IEEE Trans.*, Vol. PAS-93, pp. 859-869, May/June 1974.

[18]   R.A.M. Van Amerongen, "A General-Purpose Version of the Fast Decoupled Load Flow," *IEEE Trans.*, Vol. PWRS-4, pp. 760-770, May 1989.

[19]   W.F. Tinney and J.W. Walker, "Direct Solutions of Sparse Network Equations by Optimally Ordered Triangular Factorization," *Proceedings of IEEE*, Vol. 55, pp. 1801-1809, November 1967.

[20]   W.F. Tinney, V. Brandwajn, and S.M. Chan, "Sparse Vector Methods," *IEEE Trans.*, Vol. PAS-104, pp. 295-301, February 1985.

[21]   F.L. Alvarado, W.F. Tinney, and M.K. Enns, "Sparsity in Large-Scale Network Computation," *Control and Dynamic Systems*, Vol. 41: *Analysis and Control System Techniques for Electric Power Systems*, Part 1, pp. 207-271, Academic Press, Inc., 1991.

[22]   W.F. Tinney and J.M. Bright, "Adaptive Reductions for Power Flow Equivalents," *IEEE Trans.*, Vol. PWRS-2, No. 2, pp. 351-360, May 1987.

Chapter 7

Power System Loads

Stable operation of a power system depends on the ability to continuously match the electrical output of generating units to the electrical load on the system. Consequently, load characteristics have an important influence on system stability.

The modelling of loads is complicated because a typical load bus represented in stability studies is composed of a large number of devices such as fluorescent and incandescent lamps, refrigerators, heaters, compressors, motors, furnaces, and so on. The exact composition of load is difficult to estimate. Also, the composition changes depending on many factors including time (hour, day, season), weather conditions, and state of the economy.

Even if the load composition were known exactly, it would be impractical to represent each individual component as there are usually millions of such components in the total load supplied by a power system. Therefore, load representation in system studies is based on a considerable amount of simplification.

This chapter will discuss basic load-modelling concepts, load composition and component characteristics, and acquisition of load model parameters. As induction motors constitute a major portion of the system loads, the characteristics and modelling of induction motors are discussed in detail.

7.1 BASIC LOAD-MODELLING CONCEPTS

In power system stability and power flow studies, the common practice is to represent the composite load characteristics as seen from bulk power delivery points. As illustrated in Figure 7.1, the aggregated load represented at a transmission

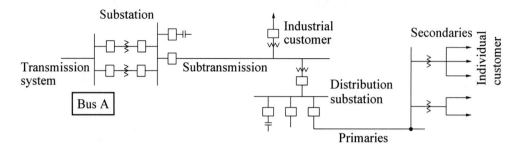

Figure 7.1 Power system configuration identifying parts of the system represented as load at a bulk power delivery point (bus A)

substation (bus A) usually includes, in addition to the connected load devices, the effects of substation step-down transformers, subtransmission feeders, distribution feeders, distribution transformers, voltage regulators, and reactive power compensation devices.

The load models are traditionally classified into two broad categories: static models and dynamic models.

7.1.1 Static Load Models

A static load model expresses the characteristics of the load at any instant of time as *algebraic functions* of the bus voltage magnitude and frequency at that instant [1]. The active power component P and the reactive power component Q are considered separately.

Traditionally, the voltage dependency of load characteristics has been represented by the *exponential model*:

$$P = P_0(\bar{V})^a$$

$$Q = Q_0(\bar{V})^b$$

(7.1)

In this and other load models described in this section,

$$\bar{V} = \frac{V}{V_0}$$

where P and Q are active and reactive components of the load when the bus voltage magnitude is V. The subscript 0 identifies the values of the respective variables at the initial operating condition.

The parameters of this model are the exponents a and b. With these exponents equal to 0, 1, or 2, the model represents constant power, constant current, or constant impedance characteristics, respectively. For composite loads, their values depend on the aggregate characteristics of load components.

The exponent a (or b) is nearly equal to the slope dP/dV (or dQ/dV) at $V=V_0$. For composite system loads, the exponent a usually ranges between 0.5 and 1.8; the exponent b is typically between 1.5 and 6. A significant characteristic of the exponent b is that it varies as a nonlinear function of voltage. This is caused by magnetic saturation in distribution transformers and motors. At higher voltages, Q tends to be significantly higher.

Reference 2 gives a summary of data available in the literature on voltage dependency of load. References 3 and 4 give results of measurements which provide information about the variation of the load characteristics with time of day, season, and/or temperature. In the absence of specific information, the most commonly accepted static load model is to represent active power as constant current (i.e., $a=1$) and reactive power as constant impedance (i.e., $b=2$) [2].

An alternative model which has been widely used to represent the voltage dependency of loads is the *polynomial model*:

$$P = P_0\left[p_1\bar{V}^2 + p_2\bar{V} + p_3\right]$$

$$Q = Q_0\left[q_1\bar{V}^2 + q_2\bar{V} + q_3\right]$$

(7.2)

This model is commonly referred to as the *ZIP* model, as it is composed of constant impedance (*Z*), constant current (*I*), and constant power (*P*) components [1]. The parameters of the model are the coefficients p_1 to p_3 and q_1 to q_3, which define the proportion of each component.

The frequency dependency of load characteristics is usually represented by multiplying the exponential model or the polynomial model by a factor as follows:

$$P = P_0(\bar{V})^a\,(1+K_{pf}\Delta f)$$

$$Q = Q_0(\bar{V})^b\,(1+K_{qf}\Delta f)$$

(7.3)

or

$$P = P_0\left[p_1\bar{V}^2 + p_2\bar{V} + p_3\right](1+K_{pf}\Delta f)$$

$$Q = Q_0\left[q_1\bar{V}^2 + q_2\bar{V} + q_3\right](1+K_{qf}\Delta f)$$

(7.4)

where Δf is the frequency deviation $(f-f_0)$. Typically, K_{pf} ranges from 0 to 3.0, and K_{qf} ranges from -2.0 to 0 [2]. The bus frequency f is usually not a state variable in the system model used for stability analysis. Therefore, it is evaluated by computing the time derivative of the bus voltage angle.

A comprehensive static model which offers the flexibility of accommodating several forms of load representation is as follows [5]:

$$P = P_0[P_{ZIP} + P_{EX1} + P_{EX2}] \qquad (7.5)$$

where

$$P_{ZIP} = p_1 \bar{V}^2 + p_2 \bar{V} + p_3$$

$$P_{EX1} = p_4 (\bar{V})^{a1} (1 + K_{pf1} \Delta f) \qquad (7.6)$$

$$P_{EX2} = p_5 (\bar{V})^{a2} (1 + K_{pf2} \Delta f)$$

The expression for the reactive component of the load has a similar structure. The reactive power compensation associated with the load is represented separately.

The static models given by Equations 7.1 to 7.6 are not realistic at low voltages, and may lead to computational problems. Therefore, stability programs usually make provisions for switching the load characteristic to the constant impedance model when the bus voltage falls below a specified value. In the load model used in the EPRI Extended Transient/Midterm Stability Program (ETMSP) [5], the exponents a_1, a_2, b_1, and b_2 are varied as a function of voltage below a threshold value of bus voltage, and the constant power and constant current components are switched to constant impedance representation.

7.1.2 Dynamic Load Models

The response of most composite loads to voltage and frequency changes is fast, and the steady state of the response is reached very quickly. This is true at least for modest amplitudes of voltage/frequency change. The use of static models described in the previous sections is justified in such cases.

There are, however, many cases where it is necessary to account for the dynamics of load components. Studies of interarea oscillations, voltage stability, and long-term stability often require load dynamics to be modelled. Study of systems with large concentrations of motors also requires representation of load dynamics.

Typically, motors consume 60 to 70% of the total energy supplied by a power system. Therefore, the dynamics attributable to motors are usually the most significant aspects of dynamic characteristics of system loads. Modelling of motors is discussed in Section 7.2.

Other dynamic aspects of load components that require consideration in stability studies include the following [1,2,6]:

(a)   Extinction of discharge lamps below a certain voltage and their restart when the voltage recovers. Discharge lamps include mercury vapour, sodium vapour, and fluorescent lamps. These extinguish at voltages in the range of 0.7 to 0.8 pu. When the voltage recovers, they restart after 1 or 2 seconds delay.

(b)   Operation of protective relays, such as thermal and overcurrent relays. Many industrial motors have starters with electromagnetically held contactors. These drop open at voltages in the range of 0.55 to 0.75 pu; the dropout time is on the order of a few cycles. Small motors on refrigerators and air conditioners have only thermal overload protections, which typically trip in about 10 to 30 seconds.

(c)   Thermostatic control of loads, such as space heaters/coolers, water heaters, and refrigerators. Such loads operate longer during low-voltage conditions. As a result, the total number of these devices connected to the system will increase in a few minutes after a drop in voltage. Air conditioners and refrigerators also exhibit such characteristics under sustained low-frequency conditions.

(d)   Response of ULTCs on distribution transformers, voltage regulators, and voltage-controlled capacitor banks. These devices are not explicitly modelled in many studies. In such cases, their effects must be implicitly included in the equivalent load that is represented at the bulk power delivery point.

   As these devices restore distribution voltages following a disturbance, the power supplied to voltage sensitive loads is restored to the pre-disturbance levels. The control action begins about 1 minute after the change in voltage, and the voltage restoration within the capability of these devices is completed in a total time of 2 to 3 minutes.

A composite load model which allows the representation of the wide range of characteristics exhibited by the various load components is shown in Figure 7.2. It is similar to the complex or aggregate load models described in references 6 and 7. The model has provision for representing aggregations of small induction motors, large induction motors, static load characteristics (Equations 7.5 and 7.6), discharge lighting, thermostatically controlled loads, transformer saturation effects, and shunt capacitors.

Thermostatically controlled loads:

   Figure 7.3 shows a simple model of the aggregated thermostatically controlled (that is, constant energy) loads suggested in references 8 and 9. In this model, G is the load conductance, G_0 is the initial value of G, $K_L G_0$ is the maximum value of G which represents a condition with all such loads on, and T is a time constant.
   A more realistic model of the thermostatically controlled loads used in the ETMSP [5] is shown in Figure 7.4. The basis for the model is as follows:

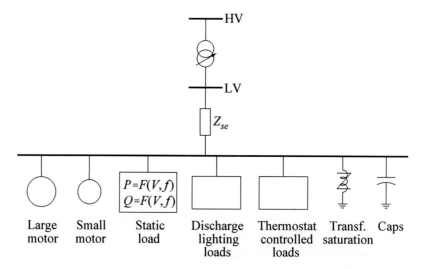

Figure 7.2 Composite static and dynamic load model

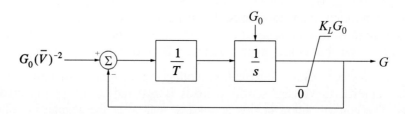

Figure 7.3 A simple model for thermostatically controlled loads

The dynamic equation of a heating device may be written as

$$K\frac{d\tau_H}{dt} = P_H - P_L \tag{7.7}$$

where

τ_H = temperature of heated area
τ_A = ambient temperature
P_H = power from the heater
　　 = $K_H GV^2$
P_L = heat loss by escape to ambient area
　　 = $K_A(\tau_H - \tau_A)$
G = load conductance

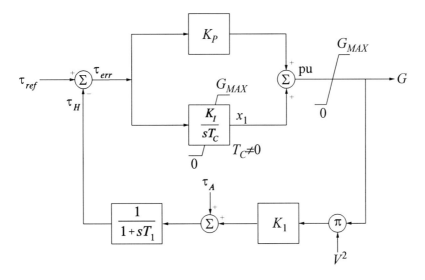

K_P = gain of proportional controller
K_I = gain of integral controller
T_C = time constant of integral controller, s
τ_{ref} = reference temperature
τ_A = ambient temperature
T_1 = load time constant, s
K_1 = gain associated with load model
G_0 = initial value of G
G_{MAX} = maximum value of G

Figure 7.4 A realistic model for thermostatically controlled loads

Substituting the expressions for P_H and P_L in Equation 7.7 gives

$$K\frac{d\tau_H}{dt} = K_H G V^2 - K_A(\tau_H - \tau_A) \tag{7.8}$$

Rearranging, we have

$$\frac{d\tau_H}{dt} = \frac{K_H}{K}GV^2 + \frac{K_A}{K}\tau_A - \frac{K_A}{K}\tau_H$$

or

$$\frac{d\tau_H}{dt} = \frac{K_1}{T_1}GV^2 + \frac{1}{T_1}\tau_A - \frac{1}{T_1}\tau_H \tag{7.9}$$

where

$$T_1 = \frac{K}{K_A}$$

and

$$K_1 = \frac{K_H}{K_A}$$

The temperature τ_H is compared with the reference temperature, and the error signal controls the load conductance through a proportional plus integral controller. When all the thermostatically controlled loads supplied by the load bus are on, G reaches its maximum value G_{MAX}.

From Figure 7.4, under pre-disturbance conditions τ_H is equal to τ_{ref}. Hence,

$$\tau_{ref} = K_1 V_0^2 G_0 + \tau_A$$

or

$$K_1 = \frac{\tau_{ref} - \tau_A}{V_0^2 G_0} \qquad (7.10)$$

Discharge lighting loads:

Figure 7.5 shows a model suitable for representing the characteristics of discharge lighting loads in stability studies. At bus voltages less than V_1, the lamps extinguish. For voltages greater than V_1, P and Q vary as nonlinear functions of V.

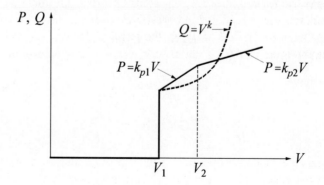

Figure 7.5 Discharge lighting characteristics

7.2 MODELLING OF INDUCTION MOTORS

As indicated in the previous section, motors form a major portion of the system loads. Induction motors in particular form the workhorse of the electric power industry; hence modelling of motors is important in system stability studies. In this section, we will develop in detail the mathematical model for an induction machine.

7.2.1 Equations of an Induction Machine

An induction machine carries alternating currents in both the stator and rotor windings. In a three-phase induction machine, the stator windings are connected to a balanced three-phase supply. The rotor windings are either short-circuited internally or connected through slip rings to a passive external circuit. The distinctive feature of the induction machine is that the rotor currents are induced by electromagnetic induction from the stator. This is the reason for the designation "induction machine."

The stator windings of a three-phase induction machine are similar to those of a synchronous machine (see Chapter 3, Section 3.1). When balanced three-phase currents of frequency f_s Hz are applied, the stator windings produce a field rotating at synchronous speed given by

$$n_s = \frac{120 f_s}{p_f} \tag{7.11}$$

where n_s is in r/min and p_f is the number of poles (2 per three-phase winding set).

When there is a relative motion between the stator field and the rotor, voltages are induced in the rotor windings. The frequency (f_r Hz) of the induced rotor voltages depends on the relative speeds of the stator field and the rotor. The current in each rotor winding is equal to the induced voltage divided by the rotor circuit impedance at the rotor frequency f_r. The rotor current reacting with the stator field produces a torque which accelerates the rotor in the direction of the stator field rotation. As the rotor speed n_r approaches the speed n_s of the stator field, the induced rotor voltages and currents approach zero. To develop a positive torque, n_r must be less than n_s. The rotor thus travels at a speed $n_s - n_r$ r/min in the backward direction with respect to the stator field. The *slip* speed of the rotor in per unit of the synchronous speed is

$$s = \frac{n_s - n_r}{n_s} \quad \text{per unit} \tag{7.12}$$

The frequency f_r of the induced rotor voltages is equal to the slip frequency $s f_s$.

At no load, the machine operates with negligible slip. If a mechanical load is applied, the slip increases (i.e., rotor speed decreases) such that the induced voltage and current produce the torque required by the load. The machine thus operates as a *motor*.

If the rotor is driven by a prime mover at a speed greater than that of the stator field, the slip is negative (i.e., rotor speed is greater than n_s). The polarities of the induced voltages are reversed so that the resulting torque is opposite in direction to that of rotation. The machine now operates as an induction *generator*.

We now develop a mathematical representation of induction machines appropriate for use in system studies. In the model, the effect of slots on the motor performance is ignored. In well-designed motors, their effect on motor performance is negligible. While our interest in this chapter is in the performance of induction machines as motors, the models we develop are sufficiently general to be applicable for generator as well as motor modes of operation. The general procedure followed here is similar to the approach used for modelling synchronous machines in Chapters 3 to 5. First we write the basic machine equations in terms of phase (a, b, c) variables; then, we transform the equations into the d-q reference frame; and finally, we make appropriate simplifications necessary to conform to the system equations in large-scale stability studies. In developing the model of an induction machine, it is worth noting the following aspects of its characteristics which differ from those of a synchronous machine:

- The rotor has a symmetrical structure. This makes the d- and q-axis equivalent circuits identical.

- The rotor speed is not fixed, but varies with load. This has an impact on the selection of the d-q reference frame.

- There is no excitation source applied to the rotor windings. Consequently, the dynamics of the rotor circuits are determined by slip, rather than by excitation control.

- The currents induced in the shorted rotor windings produce a field with the same number of poles as that produced by the stator winding. Rotor windings may therefore be modelled by an equivalent three-phase winding.

Basic equations of an induction machine

Figure 7.6 shows the circuits applicable to the analysis of an induction machine. The stator circuits consist of three-phase windings a, b, and c, distributed 120° apart in space. The rotor circuits have three distributed windings A, B, and C. The stator and rotor phases may be Y or Δ connected. In Figure 7.6, they are shown to be Y-connected.

There are two types of rotor structure. The *wound rotor* type has conventional three-phase windings brought out through three slip rings on the shaft so that they may be connected to an external circuit. The *squirrel-cage rotor* consists of a number of uninsulated bars in slots short-circuited by end rings at both ends. In the case of a squirrel-cage rotor, the rotor voltages v_A, v_B, and v_C are zero.

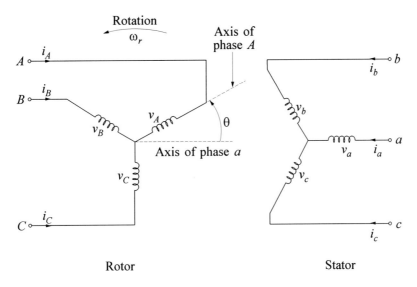

Figure 7.6 Stator and rotor circuits of an induction machine

For convenience in analysis, we consider only one pair of poles and measure all angles in electrical radians or degrees (see Section 3.1.2 of Chapter 3).

In Figure 7.6, θ is defined as the angle by which the axis of phase A rotor winding leads the axis of phase a stator winding in the direction of rotation. With a constant rotor angular velocity of ω_r in electrical rad/s,

$$\theta = \omega_r t \tag{7.13}$$

With a constant slip s,

$$\theta = (1-s)\omega_s t \tag{7.14}$$

where ω_s is the angular velocity of the stator field in electrical rad/s.

Neglecting saturation, hysteresis, and eddy currents, and assuming purely sinusoidal distribution of flux waves, the machine equations may be written as follows.

Stator voltage equations:

$$v_a = p\psi_a + R_s i_a$$

$$v_b = p\psi_b + R_s i_b \tag{7.15}$$

$$v_c = p\psi_c + R_s i_c$$

Rotor voltage equations:

$$v_A = p\psi_A + R_r i_A$$

$$v_B = p\psi_B + R_r i_B \qquad (7.16)$$

$$v_C = p\psi_C + R_r i_C$$

In the above equations, ψ represents the flux linking the winding denoted by the subscript, R_s the stator phase resistance, R_r the rotor phase resistance, and p the differential operator d/dt. The positive directions of currents are into the windings.

For power system studies, the slot effects may be neglected and the rotor may be considered to have a symmetrical structure. Hence, only the mutual inductances between stator and rotor windings are functions of rotor position defined by angle θ.

The flux linkage in the stator phase a winding at any instant is given by

$$\psi_a = L_{aa} i_a + L_{ab}(i_b + i_c) + L_{aA}[i_A \cos\theta + i_B \cos(\theta + 120°) + i_C \cos(\theta - 120°)] \quad (7.17)$$

where L_{aa} is the self-inductance of the stator windings, L_{ab} the mutual inductance between stator windings, and L_{aA} is the maximum value of mutual inductance between stator and rotor windings. Similar expressions apply to flux linkages of windings b and c of the stator.

The flux linkage in the rotor phase A winding is

$$\psi_A = L_{AA} i_A + L_{AB}(i_B + i_C) + L_{aA}[i_a \cos\theta + i_b \cos(\theta - 120°) + i_c \cos(\theta + 120°)] \quad (7.18)$$

Similar expressions apply to ψ_B and ψ_C.

With no neutral currents due to winding connections or balanced conditions,

$$i_a + i_b + i_c = 0$$

$$i_A + i_B + i_C = 0 \qquad (7.19)$$

Let

$$L_{ss} = L_{aa} - L_{ab}$$

$$L_{rr} = L_{AA} - L_{AB} \qquad (7.20)$$

Then the expressions fc. ψ_a and ψ_A may be written as

$$\psi_a = L_{ss}i_a + L_{aA}[i_A\cos\theta + i_B\cos(\theta + 120°) + i_C\cos(\theta - 120°)] \tag{7.21}$$

$$\psi_A = L_{rr}i_A + L_{aA}[i_a\cos\theta + i_b\cos(\theta - 120°) + i_c\cos(\theta + 120°)] \tag{7.22}$$

The d-q transformation

As for synchronous machine equations (see Section 3.3 of Chapter 3), the above equations for the induction machine can be simplified by appropriate transformation of phase variables into components along rotating axes. In the case of the synchronous machine, we chose axes rotating with the rotor. For an induction machine the preferred reference frame is one with the *axes rotating at synchronous speed* [10,11]. The q-axis is assumed to be 90° ahead of the d-axis in the direction of rotation. If the d-axis is so chosen that it coincides with the phase a axis at time $t=0$, its displacement from phase a axis at any time t is $\omega_s t$.

The transformation of the stator phase currents into d and q variables is as follows:

$$i_{ds} = \frac{2}{3}[i_a\cos\omega_s t + i_b\cos(\omega_s t - 120°) + i_c\cos(\omega_s t + 120°)]$$
$$i_{qs} = -\frac{2}{3}[i_a\sin\omega_s t + i_b\sin(\omega_s t - 120°) + i_c\sin(\omega_s t + 120°)] \tag{7.23}$$

The inverse transformation is

$$i_a = i_{ds}\cos\omega_s t - i_{qs}\sin\omega_s t$$
$$i_b = i_{ds}\cos(\omega_s t - 120°) - i_{qs}\sin(\omega_s t - 120°) \tag{7.24}$$
$$i_c = i_{ds}\cos(\omega_s t + 120°) - i_{qs}\sin(\omega_s t + 120°)$$

Similar transformations apply to stator flux linkages and to stator voltages.

Let us now identify the corresponding transformations for rotor quantities in relation to the synchronously rotating d and q axes. Let θ_r be the angle by which d-axis leads phase A axis of the rotor. If the rotor slip is s, the d-axis is advancing with respect to a point on the rotor at the rate

$$\frac{d\theta_r}{dt} = p\theta_r = s\omega_s \tag{7.25}$$

The transformation of rotor currents into d and q components is as follows:

$$i_{dr} = \frac{2}{3}[i_A\cos\theta_r + i_B\cos(\theta_r - 120°) + i_C\cos(\theta_r + 120°)]$$

$$i_{qr} = -\frac{2}{3}[i_A\sin\theta_r + i_B\sin(\theta_r - 120°) + i_C\sin(\theta_r + 120°)]$$

$$(7.26)$$

The inverse transformation is

$$i_A = i_{dr}\cos\theta_r - i_{qr}\sin\theta_r$$

$$i_B = i_{dr}\cos(\theta_r - 120°) - i_{qr}\sin(\theta_r - 120°)$$

$$i_C = i_{dr}\cos(\theta_r + 120°) - i_{qr}\sin(\theta_r + 120°)$$

$$(7.27)$$

Similar transformations apply to rotor flux linkages and voltages.

Basic machine equations in d-q reference frame

From Equations 7.14 and 7.25,

$$\theta = \omega_s t - \theta_r \tag{7.28}$$

From Equations 7.21 and 7.22, we can show that the stator and rotor flux linkages can be expressed in terms of the d and q components as follows.

Stator flux linkages:

$$\psi_{ds} = L_{ss}i_{ds} + L_m i_{dr}$$

$$\psi_{qs} = L_{ss}i_{qs} + L_m i_{qr}$$

$$(7.29)$$

Rotor flux linkages:

$$\psi_{dr} = L_{rr}i_{dr} + L_m i_{ds}$$

$$\psi_{qr} = L_{rr}i_{qr} + L_m i_{qs}$$

$$(7.30)$$

with $L_m = 3/2 L_{aA}$.

The stator voltages in terms of the d and q components are

$$v_{ds} = R_s i_{ds} - \omega_s \psi_{qs} + p\psi_{ds}$$

$$v_{qs} = R_s i_{qs} + \omega_s \psi_{ds} + p\psi_{qs}$$

$$(7.31)$$

and the rotor voltages are

$$v_{dr} = R_r i_{dr} - (p\theta_r)\psi_{qr} + p\psi_{dr}$$

$$v_{qr} = R_r i_{qr} + (p\theta_r)\psi_{dr} + p\psi_{qr}$$

(7.32)

The term $p\theta_r$ in Equation 7.32 is the slip angular velocity given by

$$p\theta_r = s\omega_s$$

(7.33)

This represents the relative angular velocity between the rotor and the reference d-q axes.

We have expressed the machine equations in a *synchronously rotating d-q reference frame*. For balanced synchronous operation, both stator and rotor mmf waves rotate at synchronous speed. The mmf waves due to stator windings are resolved into two sinusoidally distributed mmf waves rotating at synchronous speed so that one has its peak over the d-axis and the other has its peak over the q-axis. Therefore, i_{ds} and i_{qs} represent currents in fictitious windings rotating at synchronous speed and remaining in such positions that their axes always coincide with the d- and q-axis, respectively. A similar interpretation applies to the rotor currents i_{dr} and i_{qr}.

From Equations 7.31 and 7.32 we see that the expression for each voltage consists of three terms: The Ri drop term, the transient $p\psi$ term, and the speed voltage term. The first two are familiar terms associated with the voltage of any coil. The speed voltage terms, however, are peculiar to the specific situation at hand. The terms $\omega_s\psi_{qs}$ and $\omega_s\psi_{ds}$ in the stator voltage equations represent voltages created in the stationary windings by the synchronously rotating flux wave. Similarly, the terms $(p\theta_r)\psi_{qr}$ and $(p\theta_r)\psi_{dr}$ in the rotor voltage equations represent voltages created in the rotor windings which move at the slip speed $(p\theta_r = s\omega_s)$ with respect to the synchronously rotating flux waves. For motor action, s and $p\theta_r$ are positive. Conversely, for generator action, s and $p\theta_r$ are negative.

Electrical power and torque:

The instantaneous power input to the stator is

$$P_s = v_a i_a + v_b i_b + v_c i_c$$

In terms of the d and q components, the above expression becomes

$$P_s = \frac{3}{2}(v_{ds} i_{ds} + v_{qs} i_{qs})$$

(7.34)

Similarly, the instantaneous power input to the rotor is

$$P_r = \frac{3}{2}(v_{dr}i_{dr} + v_{qr}i_{qr})$$
(7.35)

The electromagnetic torque developed is obtained as the power associated with the speed voltages divided by the shaft speed in mechanical radians per second. From Equation 7.32, the speed voltage terms associated with v_{dr} and v_{qr} are $-\psi_{qr}(p\theta_r)$ and $\psi_{dr}(p\theta_r)$, respectively. Substituting in Equation 7.35, the power associated with the speed voltage is

$$\frac{3}{2}(\psi_{dr}i_{qr} - \psi_{qr}i_{dr})(p\theta_r)$$

The rotor speed with respect to the d, q axes is $-(p\theta_r)(2/p_f)$. Hence, the electromagnetic torque is

$$T_e = \frac{3}{2}(\psi_{qr}i_{dr} - \psi_{dr}i_{qr})\frac{p_f}{2}$$
(7.36)

Acceleration equation:

The electromagnetic torque developed by the motor drives the mechanical load. If there is a mismatch between the electromagnetic torque and the mechanical load torque (T_m), the differential torque accelerates the rotor mass. Consequently,

$$T_e - T_m = J\frac{d\omega_m}{dt} = J\frac{d^2\theta}{dt^2}$$
(7.37)

where ω_m is the angular velocity of the rotor in mechanical radians per second, and J is the polar moment of inertia of the rotor and the connected load. The load torque varies with speed. A commonly used expression for the load torque is

$$T_m = T_0(\bar{\omega}_r)^m$$
(7.38)

where $\bar{\omega}_r$ is the rotor speed expressed in per unit of synchronous speed. An alternative expression often used for the load torque is

$$T_m = T_0[A\bar{\omega}_r^2 + B\bar{\omega}_r + C]$$
(7.39)

7.2.2 Steady-State Characteristics

For balanced steady-state operation, the stator currents may be written as

$$i_a = I_m \cos(\omega_s t + \alpha)$$

$$i_b = I_m \cos(\omega_s t + \alpha - 120°) \qquad (7.40)$$

$$i_c = I_m \cos(\omega_s t + \alpha + 120°)$$

where α is the phase angle of i_a with respect to the time origin. Applying the d-q transformation (Equation 7.23), we have

$$i_{ds} = I_m \cos\alpha$$
$$\qquad\qquad\qquad (7.41)$$
$$i_{qs} = I_m \sin\alpha$$

Thus, for balanced steady-state operation the stator currents appear as direct currents in the d-q reference frame. Similar expressions apply to stator voltages and rotor currents.

From Equation 7.24, the stator current may be written as

$$i_s = i_a = i_{ds}\cos\omega_s t - i_{qs}\sin\omega_s t$$
$$\qquad\qquad\qquad (7.42)$$
$$= i_{ds}\cos\omega_s t + i_{qs}\cos(\omega_s t + 90°)$$

Using I_s to denote per unit RMS stator current, Equation 7.42 may be written in the phasor form:

$$\tilde{I}_s = I_{ds} + jI_{qs} \qquad (7.43)$$

where $I_{ds} = i_{ds}/\sqrt{2}$ and $I_{qs} = i_{qs}/\sqrt{2}$. In a similar manner, the stator phase voltages and rotor phase currents can be expressed in the phasor form:

$$\tilde{V}_s = (v_{ds} + jv_{qs})/\sqrt{2} \qquad (7.44)$$

$$\tilde{I}_r = (i_{dr} + ji_{qr})/\sqrt{2} \qquad (7.45)$$

Under steady-state conditions, $p\psi$ terms in Equations 7.31 and 7.32 disappear. Substituting in Equation 7.31 the expressions for flux linkages given by Equation 7.29, we may write

$$v_{ds} = R_s i_{ds} - \omega_s L_{ss} i_{qs} - \omega_s L_m i_{qr}$$

$$v_{qs} = R_s i_{qs} + \omega_s L_{ss} i_{ds} + \omega_s L_m i_{dr} \tag{7.46}$$

From Equations 7.43 to 7.46, we have

$$\tilde{V}_s = R_s \tilde{I}_s + j\omega_s L_{ss} \tilde{I}_s + j\omega_s L_m \tilde{I}_r$$

$$= R_s \tilde{I}_s + j\omega_s (L_{ss} - L_m)\tilde{I}_s + j\omega_s L_m (\tilde{I}_s + \tilde{I}_r) \tag{7.47}$$

$$= R_s \tilde{I}_s + jX_s \tilde{I}_s + jX_m (\tilde{I}_s + \tilde{I}_r)$$

where

$$X_s = \omega_s(L_{ss} - L_m) = \text{stator leakage reactance}$$

$$X_m = \omega_s L_m = \text{magnetizing reactance}$$

With the rotor circuits shorted, $v_{dr} = v_{qr} = 0$. Hence, from Equations 7.30, 7.32, and 7.33, we have

$$v_{dr} = 0 = R_r i_{dr} - s\omega_s(L_{rr} i_{qr} + L_m i_{qs})$$

$$v_{qr} = 0 = R_r i_{qr} + s\omega_s(L_{rr} i_{dr} + L_m i_{ds}) \tag{7.48}$$

From Equations 7.43, 7.45, and 7.48, we may write

$$\tilde{V}_r = 0 = \frac{R_r}{s}\tilde{I}_r + j\omega_s L_{rr}\tilde{I}_r + j\omega_s L_m \tilde{I}_s$$

$$= \frac{R_r}{s}\tilde{I}_r + jX_r\tilde{I}_r + jX_m(\tilde{I}_s + \tilde{I}_r) \tag{7.49}$$

where

$$X_r = \omega_s(L_{rr} - L_m) = \text{rotor leakage reactance}$$

Equivalent circuit

Equations 7.47 and 7.49 represent the steady-state performance of the induction machine. These equations can be represented by the equivalent circuit shown in Figure 7.7, which accounts for quantities in one phase.

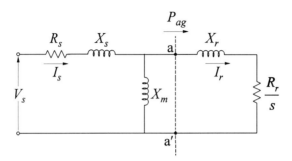

Figure 7.7 Equivalent circuit of a three-phase induction machine

In the equivalent circuit *all quantities have been referred to the stator side.* The directions of currents shown are positive when operating as a motor, in which case the slip *s* is positive.

The power transferred across the air-gap to the rotor is

$$P_{ag} = \frac{R_r}{s} I_r^2 \tag{7.50}$$

The rotor resistance loss is

$$P_{lr} = R_r I_r^2 \tag{7.51}$$

Therefore, the mechanical power transferred to the shaft is

$$\begin{aligned}
P_{sh} &= P_{ag} - P_{lr} \\
&= \frac{R_r}{s} I_r^2 - R_r I_r^2 \\
&= R_r \frac{1-s}{s} I_r^2
\end{aligned} \tag{7.52}$$

This represents the mechanical power transferred to the shaft. An alternative form of induction machine equivalent circuit is shown in Figure 7.8 in which the rotor power is separated into resistance loss and shaft power.

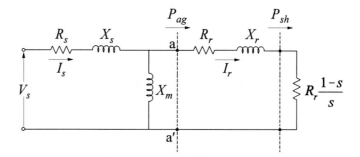

Figure 7.8 Alternative form of induction machine equivalent circuit

The above represent per phase values. For a three-phase motor, the electromagnetic torque developed by the motor is

$$T_e = \frac{3P_{sh}}{\omega_m}$$

where ω_m is the angular velocity of the rotor in mechanical rad/s given by

$$\omega_m = \omega_r \frac{2}{p_f}$$

$$= \omega_s (1-s) \frac{2}{p_f}$$

Hence,

$$T_e = 3\frac{p_f}{2} \frac{R_r}{s\omega_s} I_r^2 \qquad (7.53)$$

where $\omega_s = 2\pi f$ and p_f is the number of poles.

Torque-slip characteristic

The torque is slip dependent. For analysis of torque-slip relationships, the equivalent circuit of Figure 7.7 may be simplified by replacing the part of the equivalent to the left of nodes a and a' by its Thevenin's equivalent. Figure 7.9 shows the resulting simplified equivalent circuit.

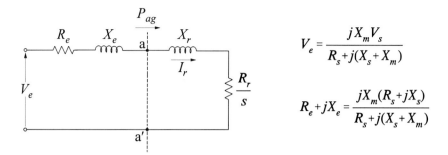

$$V_e = \frac{jX_m V_s}{R_s + j(X_s + X_m)}$$

$$R_e + jX_e = \frac{jX_m(R_s + jX_s)}{R_s + j(X_s + X_m)}$$

Figure 7.9 Equivalent circuit suitable for evaluating torque-slip relationships

From Figure 7.9, the rotor current is

$$\tilde{I}_r = \frac{\tilde{V}_e}{(R_e + R_r/s) + j(X_e + X_r)} \tag{7.54}$$

From Equation 7.53, the torque is

$$T_e = 3\frac{p_f}{2}\left(\frac{R_r}{s\omega_s}\right)\frac{V_e^2}{(R_e + R_r/s)^2 + (X_e + X_r)^2} \tag{7.55}$$

Figure 7.10 shows a typical relationship between torque and slip/speed. At standstill (i.e., at starting), speed is zero and slip s is equal to 1.0 pu. Between zero and synchronous speed, the machine performs as a motor. Beyond synchronous speed, slip is negative, representing generator operation.

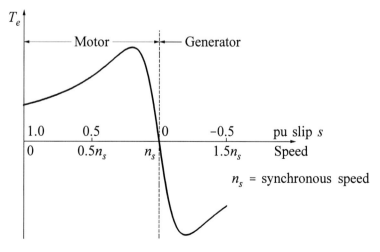

Figure 7.10 Typical torque-speed characteristic of an induction machine

The maximum torque occurs when

$$\frac{R_r}{s} = Z'$$ (7.56)

where

$$Z' = \sqrt{R_e^2 + (X_e + X_r)^2}$$ (7.57)

Therefore, per unit slip at maximum torque is

$$s_{Tmax} = \frac{R_r}{Z'}$$ (7.58)

From Equation 7.55, the maximum torque is

$$T_{max} = 3\frac{p_f}{2\omega_s}\frac{0.5V_e^2}{R_e + Z'}$$
$$= 3\frac{1}{\omega_{ms}}\frac{0.5V_e^2}{R_e + Z'}$$ (7.59)

where $\omega_s = 2\pi f$ and ω_{ms} is the synchronous speed in mechanical radians per second.

From Equation 7.58, we see that the slip at maximum torque is directly proportional to the rotor resistance R_r. However, as seen from Equation 7.59, the value of maximum torque is independent of R_r and is affected principally by the stator and rotor leakage reactances.

Effect of rotor resistance on efficiency

The efficiency of an induction motor is highly dependent on the operating slip. If we neglect all losses except that due to rotor resistance loss, then the efficiency is given by the ratio of mechanical power transferred to the shaft to the air-gap power. Therefore, from Equations 7.50 and 7.52, the ideal efficiency is

$$\eta = \frac{P_{sh}}{P_{ag}} \times 100 = 1 - s \quad \text{per cent}$$ (7.60)

It is evident from the above relationship that, to achieve high efficiency, an induction motor must operate at low values of slip. Therefore, most three-phase induction motors are designed so that the per unit slip at normal full-load operation is less than 0.05. For large three-phase induction motors the actual maximum efficiency may be higher than 95%.

7.2.3 Alternative Rotor Constructions

The selection of an appropriate value of rotor resistance of an induction motor involves design compromises. High efficiency at normal operating conditions requires a low rotor resistance. On the other hand, a high rotor resistance is required to produce a high starting torque and to keep the magnitude of the starting current low and the power factor high.

The use of a *wound rotor* is one way of meeting the need for varying the rotor resistance at different operating conditions. At starting, resistors are connected in series with the rotor windings through slip rings. As the rotor speed picks up, the external resistance is reduced. For normal running, external resistance is made zero so that the full load slip is small. Figure 7.11 shows the effect of varying the rotor resistance on the shaft torque. Wound rotor motors are, however, more expensive than squirrel-cage motors.

Special squirrel-cage arrangements can also be used to obtain a high value of effective resistance at starting and a low value of the resistance at full-load operation. One such arrangement of rotor bars frequently used is the *double squirrel-cage* shown in Figure 7.12. It consists of two layers of bars, both short-circuited by end rings. The

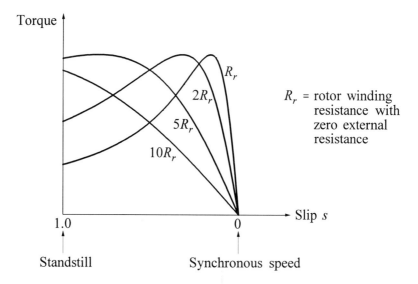

Figure 7.11 Torque-slip curves showing the effect
of rotor circuit resistance

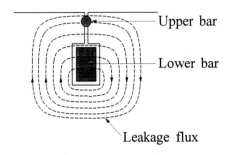

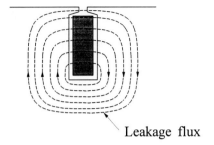

Figure 7.12 Double squirrel-cage rotor bars

Figure 7.13 Deep-bar rotor construction

upper bars are small in cross section and have a high resistance. They are placed near the rotor surface so that the leakage flux sees a path of high reluctance; consequently, they have a low leakage inductance. The lower bars have a large cross section, a lower resistance, and a higher leakage inductance. At starting, rotor frequency is high and very little current flows through the lower bars; the effective resistance of the rotor then is that of the high-resistance upper bars. At normal low slip operation, leakage reactances are negligible, and the rotor current flows largely through the low-resistance lower bars; the effective rotor resistance is equal to that of the two sets of bars in parallel.

The use of *deep, narrow rotor bars* as shown in Figure 7.13 produces torque-speed characteristics similar to those of a double-cage rotor. As is evident from the flux paths shown in the figure, the leakage inductance of the top cross section of the rotor bar is relatively low; the lower cross sections have progressively higher leakage inductance. At starting, due to the high rotor frequency, the current is concentrated towards the top layers of the rotor bar. As the rotor accelerates and slip decreases, the current distribution becomes more uniform. At normal full-load operation, the current distribution is nearly uniform and the effective resistance is low.

The equivalent circuit of a single-cage induction motor (i.e., with one rotor winding) is shown in Figure 7.7. This is extended in Figure 7.14 to represent a motor with double-cage rotor. It may be represented by an equivalent single rotor circuit as shown in Figure 7.15, with slip-dependent rotor parameters:

$$R_r(s) = R_{r0} \frac{m^2 + ms^2 R_1/R_{r0}}{m^2 + s^2} \tag{7.61}$$

$$X_r(s) = X_1 + \frac{R_{r0}(mR_1/R_2)}{m^2 + s^2} \tag{7.62}$$

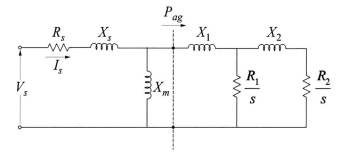

Figure 7.14 Equivalent circuit of an induction
motor with a double-cage rotor

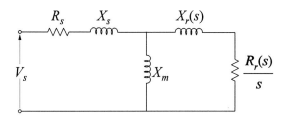

Figure 7.15 Equivalent single rotor circuit representation of
a motor with a double-cage rotor or a deep-bar rotor

where

$$R_{r0} = \frac{R_1 R_2}{R_1 + R_2} \tag{7.63}$$

$$m = \frac{R_1 + R_2}{X_2} \tag{7.64}$$

The deep-bar effects may be represented by an equivalent in the form of a ladder network. For system studies, however, it is preferable to use a quasi steady-state approach, representing the rotor by a single rotor circuit whose parameters vary as a function of slip as shown in Figure 7.15.

The rotor parameters can be expressed as a function of slip based on the eddy current distribution within a rectangular bar [22]:

$$R_r(s) = \frac{R_{r0}}{2} \frac{\beta(\sinh\beta + \sin\beta)}{(\cosh\beta - \cos\beta)} \tag{7.65}$$

$$X_r(s) = X_{r0} + \frac{R_{r0}}{2} \frac{\beta(\sinh\beta - \sin\beta)}{(\cosh\beta - \cos\beta)} \qquad (7.66)$$

$$\beta = \sqrt{|s|}\,B \qquad (7.67)$$

where B is a deep-bar factor which determines the motor starting torque, X_{r0} is the rotor leakage reactance not associated with the rotor bar (for example, zigzag leakage reactance), and R_{r0} is the rotor bar running resistance.

The deep-bar factor is a function of the base frequency, the resistivity (ρ) and the depth (d) of the rotor bar, and the permeability of free space (μ_0):

$$B = 2d\sqrt{\mu_0\omega_0/(2\rho)}$$

In the limit, as s tends to zero, the deep-bar rotor resistance is equal to R_{r0}.

7.2.4 Representation of Saturation

The magnetizing reactance X_m varies with magnetic saturation. This is represented as a function of the air-gap voltage V_{ag}, in a manner similar to that used for synchronous machines.

Induction motors are normally designed so that their leakage reactances saturate at high-current levels. This assists in meeting the requirement for a high starting torque. For accurate analysis of conditions involving high-current levels, saturation representation should include variation of leakage reactance as shown in Figure 7.16.

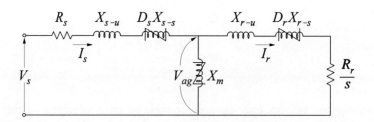

Figure 7.16 Induction motor equivalent circuit including effects of magnetic saturation

The stator and rotor leakage reactances are separated into saturating components (X_{s-s}, X_{r-s}) and unsaturated components (X_{s-u}, X_{r-u}). The saturating components of the reactances represent leakage flux that concentrate in the slot tooth tips which saturate and limit the magnetic field at the slot mouth [12]. The unsaturating portions of the reactances represent end leakage and peripheral leakage.

The leakage reactance saturation depends on the current. The limiting value of current (I_{sat}) at which saturation begins is typically in the range of 1.3 to 3.0 pu. The saturation of the leakage reactances due to current I can be represented by the describing function D [5,12] as follows:

For $I \leq I_{sat}$, $D = 1.0$

For $I > I_{sat}$, $D = \dfrac{2}{\pi}\left[\tan^{-1}\left(\dfrac{\gamma}{\sqrt{1-\gamma}}\right) + \gamma\sqrt{1-\gamma^2}\right]$ (7.68)

where

$$\gamma = \frac{I_{sat}}{I} = \frac{\text{current at which saturation begins}}{\text{current through the leakage inductance}}$$

7.2.5 Per Unit Representation

As in the case of a synchronous machine (see Section 3.4 of Chapter 3), we choose the following base quantities for the stator:

$v_{s\,base}$ = peak value of rated phase voltage, V
$i_{s\,base}$ = peak value of rated phase current, A
f_{base} = rated frequency, Hz

The base values of the remaining quantities are automatically set:

ω_{base} = $2\pi f_{base}$, elec. rad/s
$\omega_{m\,base}$ = $\omega_{base}(2/p_f)$, mech. rad/s
$Z_{s\,base}$ = $v_{s\,base}/i_{s\,base}$, Ω
$L_{s\,base}$ = $v_{s\,base}/(i_{s\,base}\omega_{base})$, H
$\psi_{s\,base}$ = $v_{s\,base}/\omega_{base}$, Wb·turns
3-phase VA_{base} = $3/2\,(v_{s\,base}\,i_{s\,base})$, VA
Torque base   = $3/2\,(p_f/2)\psi_{s\,base}\,i_{s\,base}$, N·m

With the rotor quantities referred to the stator side, the above base values also apply to the rotor.
From Equation 7.31,

$$v_{ds} = R_s i_{ds} - \omega_s \psi_{qs} + p\psi_{ds}$$

Dividing by $v_{s\,base}$, and noting that $v_{s\,base} = Z_{s\,base}\,i_{s\,base} = \omega_{base}\,\psi_{s\,base}$, we get

$$\frac{v_{ds}}{v_{sbase}} = \frac{R_s}{Z_{sbase}}\frac{i_{ds}}{i_{sbase}} - \frac{\omega_s}{\omega_{base}}\frac{\psi_{qs}}{\psi_{sbase}} + p\left(\frac{1}{\omega_{base}}\frac{\psi_{ds}}{\psi_{sbase}}\right)$$

In per unit notation,

$$\bar{v}_{ds} = \bar{R}_s\bar{i}_{ds} - \bar{\omega}_s\bar{\psi}_{qs} + \bar{p}\bar{\psi}_{ds} \tag{7.69}$$

Similarly,

$$\bar{v}_{qs} = \bar{R}_s\bar{i}_{qs} + \bar{\omega}_s\bar{\psi}_{ds} + \bar{p}\bar{\psi}_{qs} \tag{7.70}$$

Using the same approach to express Equation 7.32 in per unit form, we find

$$\bar{v}_{dr} = \bar{R}_r\bar{i}_{dr} - (\bar{p}\theta_r)\bar{\psi}_{qr} + \bar{p}\bar{\psi}_{dr} \tag{7.71}$$

$$\bar{v}_{qr} = \bar{R}_r\bar{i}_{qr} + (\bar{p}\theta_r)\bar{\psi}_{dr} + \bar{p}\bar{\psi}_{qr} \tag{7.72}$$

where

$$\bar{p}\theta_r = \frac{1}{\omega_{base}}(p\theta_r) = s\bar{\omega}_s = \frac{\omega_s - \omega_r}{\omega_s} \tag{7.73}$$

From Equations 7.29 and 7.30, by dividing throughout by $\psi_{sbase} = L_{sbase}i_{sbase}$, the per unit flux linkage equations may be written as

$$\bar{\psi}_{ds} = \bar{L}_{ss}\bar{i}_{ds} + \bar{L}_m\bar{i}_{dr} \tag{7.74}$$

$$\bar{\psi}_{qs} = \bar{L}_{ss}\bar{i}_{qs} + \bar{L}_m\bar{i}_{qr} \tag{7.75}$$

$$\bar{\psi}_{dr} = \bar{L}_{rr}\bar{i}_{dr} + \bar{L}_m\bar{i}_{ds} \tag{7.76}$$

$$\bar{\psi}_{qr} = \bar{L}_{rr}\bar{i}_{qr} + \bar{L}_m\bar{i}_{qs} \tag{7.77}$$

From Equation 7.36, dividing by $T_{base} = 3/2(\psi_{sbase}i_{sbase})(p_f/2)$, we have

$$\frac{T_e}{T_{base}} = \frac{3/2(\psi_{qr}i_{dr} - \psi_{dr}i_{qr})(p_f/2)}{3/2(\psi_{sbase}i_{sbase})(p_f/2)}$$

or

$$\bar{T}_e = \bar{\Psi}_{qr}\bar{i}_{dr} - \bar{\Psi}_{dr}\bar{i}_{qr} \tag{7.78}$$

Dividing Equation 7.37 by $T_{base} = VA_{base}/\omega_{m\,base}$ gives

$$\frac{T_e}{T_{base}} - \frac{T_m}{T_{base}} = J\left(\frac{\omega_{m\,base}}{VA_{base}}\right)\omega_{m\,base}\, p\left(\frac{\omega_m}{\omega_{m\,base}}\right)$$

or

$$p(\bar{\omega}_r) = \frac{1}{2H}(\bar{T}_e - \bar{T}_m)$$

or

$$\bar{p}(\omega_r) = \frac{1}{2H\omega_{base}}(\bar{T}_e - \bar{T}_m) \tag{7.79}$$

where

$$H = \frac{1}{2}\frac{J\omega_{m\,base}^2}{VA_{base}}$$

$$\bar{\omega}_r = \frac{\omega_m}{\omega_{m\,base}} = \frac{\omega_r/p_f}{\omega_{base}/p_f} = \frac{\omega_r}{\omega_{base}}$$

The parameter H is the combined inertia constant of the motor and the mechanical load.

Dividing Equation 7.38 by T_{base} to obtain an expression for the load torque in per unit, we have

$$\frac{T_m}{T_{base}} = \frac{T_0}{T_{base}}(\bar{\omega}_r)^m$$

or

$$\bar{T}_m = \bar{T}_0(\bar{\omega}_r)^m \tag{7.80}$$

Similarly, the per unit form of Equation 7.39, giving the alternative expression for load torque, is

$$\bar{T}_m = \bar{T}_0[A\bar{\omega}_r^2 + B\bar{\omega}_r + C]$$

(7.81)

Equations 7.69 to 7.81 represent the per unit dynamic equations of an induction motor.

The per unit time derivative $\bar{p}$ appearing in these equations is related to the derivative with time in seconds as follows:

$$\bar{p} = \frac{d}{d\bar{t}} = \frac{1}{\omega_{base}}\frac{d}{dt} = \frac{1}{\omega_{base}}p$$

For convenience in introducing the per unit equations, we have used superbars to denote per unit quantities. Further analysis of induction machines in this book will be solely in per unit form, and we will drop the superbar notation.

7.2.6 Representation in Stability Studies

For representation in power system stability studies, $p\psi_{ds}$ and $p\psi_{qs}$ are neglected in the stator voltage relations (Equations 7.69 and 7.70). These terms represent stator transients; their neglect corresponds to ignoring the dc component in the stator transient currents, permitting representation of only fundamental frequency components. As with synchronous machines, this simplification is essential to ensure compatibility with the models used for representing other system components, particularly the transmission network (see Chapter 5, Section 5.1.1).

With the stator transients neglected and the rotor windings shorted, the *per unit* induction motor electrical equations may be summarized as follows.

Stator voltages:

$$v_{ds} = R_s i_{ds} - \omega_s \psi_{qs}$$

(7.82)

$$v_{qs} = R_s i_{qs} + \omega_s \psi_{ds}$$

(7.83)

Rotor voltages:

$$v_{dr} = 0 = R_r i_{dr} - p\theta_r \psi_{qr} + p\psi_{dr}$$

(7.84)

$$v_{qr} = 0 = R_r i_{qr} + p\theta_r \psi_{dr} + p\psi_{qr}$$

(7.85)

Flux linkages:

$$\psi_{ds} = L_{ss}i_{ds} + L_m i_{dr} \tag{7.86}$$

$$\psi_{qs} = L_{ss}i_{qs} + L_m i_{qr} \tag{7.87}$$

$$\psi_{dr} = L_m i_{ds} + L_{rr}i_{dr} \tag{7.88}$$

$$\psi_{qr} = L_m i_{qs} + L_{rr}i_{qr} \tag{7.89}$$

with

$$L_{ss} = L_s + L_m \quad \text{and} \quad L_{rr} = L_r + L_m$$

where L_s and L_r are per unit stator and rotor leakage inductances.

The above equations assume a single rotor winding. As discussed earlier, the model for a machine with a *double-cage* rotor or a *deep-bar* rotor can be reduced to an equivalent single rotor circuit machine model whose parameters vary with slip.

To reduce Equations 7.82 to 7.89 to a form suitable for implementation in a stability program, we eliminate the rotor currents and express the relationship between stator current and voltage in terms of a voltage behind the transient reactance. Thus, from Equation 7.88,

$$i_{dr} = \frac{\psi_{dr} - L_m i_{ds}}{L_{rr}} \tag{7.90}$$

and, upon substitution in Equation 7.86,

$$\begin{aligned}
\psi_{ds} &= L_{ss}i_{ds} + \frac{L_m(\psi_{dr} - L_m i_{ds})}{L_{rr}} \\[2mm]
&= \frac{L_m}{L_{rr}}\psi_{dr} + \left(L_{ss} - \frac{L_m^2}{L_{rr}}\right)i_{ds}
\end{aligned} \tag{7.91}$$

Similarly from Equations 7.87 and 7.89, we have

$$\psi_{qs} = \frac{L_m}{L_{rr}}\psi_{qr} + \left(L_{ss} - \frac{L_m^2}{L_{rr}}\right)i_{qs} \tag{7.92}$$

Substituting the above expression for ψ_{qs} in Equation 7.82, we may write

$$v_{ds} = R_s i_{ds} - X'_s i_{qs} + v'_d \tag{7.93}$$

and, similarly substituting the expression for ψ_{ds} in Equation 7.83,

$$v_{qs} = R_s i_{qs} + X_s' i_{ds} + v_q'$$ (7.94)

where

$$v_d' = -\frac{\omega_s L_m}{L_{rr}} \psi_{qr}$$ (7.95)

$$v_q' = \frac{\omega_s L_m}{L_{rr}} \psi_{dr}$$ (7.96)

$$X_s' = \omega_s \left(L_{ss} - \frac{L_m^2}{L_{rr}} \right)$$ (7.97)

The reactance X_s' is the *transient reactance* of the induction machine.

The stator voltage equations may be combined and expressed in the phasor form:

$$v_{ds} + j v_{qs} = (R_s + j X_s')(i_{ds} + j i_{qs}) + (v_d' + j v_q')$$

Noting that in per unit RMS and peak values are equal, we see that the above relationship may be expressed as

$$\tilde{V}_s = (R_s + j X_s')\tilde{I}_s + \tilde{V}'$$ (7.98)

From the above equation, it is evident that the induction machine can be represented by the simple transient equivalent circuit of Figure 7.17.

By eliminating the rotor currents and expressing the rotor flux linkages in terms of v_d' and v_q', Equations 7.84 and 7.85 can be written in the following form:

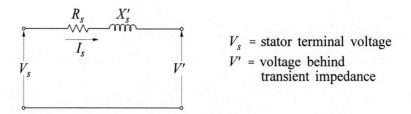

V_s = stator terminal voltage
V' = voltage behind
      transient impedance

Figure 7.17 Induction machine transient-equivalent circuit

$$p(v_d') = -\frac{1}{T_0'}[v_d' + (X_s - X_s')i_{qs}] + p\theta_r v_q' \qquad (7.99)$$

$$p(v_q') = -\frac{1}{T_0'}[v_q' - (X_s - X_s')i_{ds}] - p\theta_r v_d' \qquad (7.100)$$

where

$$T_0' = \frac{L_r + L_m}{R_r} = \frac{L_{rr}}{R_r} \qquad (7.101)$$

$$X_s = \omega_s(L_s + L_m) = \omega_s L_{ss} \qquad (7.102)$$

$$p\theta_r = \frac{\omega_s - \omega_r}{\omega_s} \qquad (7.103)$$

The constant T_0' is the *transient open-circuit time constant* (expressed in radians) of the induction machine; it characterizes the decay of the rotor transients when the stator is open-circuited. In the above equations, p is the per unit time derivative; that is, time t is in radians.

In system studies, we normally prefer to express time in seconds. *The above equations are also applicable with time t and the time constant T_0' expressed in seconds. In this case, $p\theta_r$ is the slip speed in radians per second.*

Equations 7.99 and 7.100 describe the rotor circuit dynamics. The rotor acceleration equation, with time expressed in seconds, is

$$p(\bar{\omega}_r) = \frac{1}{2H}(T_e - T_m) \qquad (7.104)$$

From Equation 7.78, the per unit electromagnetic torque is

$$T_e = \psi_{qr} i_{dr} - \psi_{dr} i_{qr}$$

Eliminating the rotor currents by expressing them in terms of the stator currents and rotor flux linkages (see Equation 7.90), we find

$$T_e = \psi_{qr}\left(\frac{\psi_{dr}-L_m i_{ds}}{L_{rr}}\right) - \psi_{dr}\left(\frac{\psi_{qr}-L_m i_{qs}}{L_{rr}}\right)$$

$$= -\psi_{qr}\left(\frac{L_m}{L_{rr}}\right)i_{ds} + \psi_{dr}\left(\frac{L_m}{L_{rr}}\right)i_{qs}$$

$$= \frac{v'_d i_{ds} + v'_q i_{qs}}{\omega_s}$$

With $\omega_s = 1.0$ pu,

$$T_e = v'_d i_{ds} + v'_q i_{qs} \tag{7.105}$$

The load torque T_m (required for the solution of Equation 7.104) is given by Equation 7.80 or 7.81.

In Equations 7.99 and 7.100, the term $s\omega_s = p\theta_r$ represents the angular velocity between the rotor and the reference d-q axes rotating at synchronous speed ω_s. As ω_r changes, $s\omega_s$ is computed as being equal to $\omega_s - \omega_r$.

In the above induction machine equations, rotor slip with respect to the stator field does not appear explicitly. During system swings (electromechanical oscillations), the frequency of stator voltages and currents deviates from the synchronous frequency ω_s. This is reflected in the machine equations as oscillations in v_{ds}, v_{qs}, i_{ds} and i_{qs}. If the transient variation in slip were required, it would be necessary to compute the rate of change of stator voltage angle to establish the angular velocity of the stator field.

Simplified induction machine model

For many applications, particularly those involving small motors, it is not necessary to account for the dynamics of the rotor electrical circuits. In such cases, the rotor-circuit dynamics may be assumed to be very fast (i.e., T'_0 very small), and $p(v'_d)$ and $p(v'_q)$ are set to zero in Equations 7.99 and 7.100. With this simplification, the induction motor representation can be based on steady-state theory, including the equivalent circuit of Figure 7.9 and torque-slip relationship given by Equation 7.55.

Induction motor parameters

The equivalent-circuit parameters of an induction motor are given by design data. If the design data are not readily available, the parameters may be estimated from standard specifications as described in reference 13.

As an example of induction motor parameters, Figure 7.18 shows the data for a 150 HP double-cage induction motor and the corresponding torque and current characteristics.

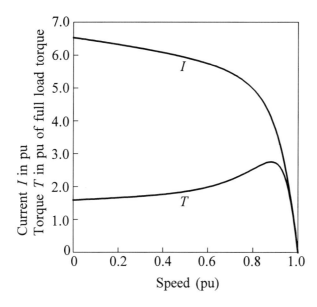

Motor specification data

Full-load efficiency	=	0.9151
Full-load power factor	=	0.8895
Full-load slip	=	0.0166
No-load loss (kW)	=	2.594
Starting current (pu)	=	6.4961
Starting torque ratio	=	1.5921
Pull-out torque	=	2.650
Pull-out slip	=	0.0914
kVA rating	=	137.0
Rated voltage (V)	=	400
Rated current (A)	=	198

Parameters of equivalent circuit in pu of 137 kVA and 400 V

$R_s = 0.0425$ $X_{s-u} = 0.0435$ $X_{s-s} = 0.0435$ $X_m = 2.9745$
$R_1 = 0.0739$ $X_{r-u} = 0.0329$ $X_{r-s} = 0.0329$ $R_2 = 0.0249$
$X_2 = 0.0739$ $H = 0.6$ $I_{sat\,s} = 3.0$ $I_{sat\,r} = 3.0$
Load torque exponent $m = 2.0$
(See Figures 7.14 and 7.16 for definition of parameters.)

Figure 7.18 Typical 150 HP double-cage induction motor data

7.3 SYNCHRONOUS MOTOR MODEL

A synchronous motor is modelled in the same manner as a synchronous generator (see Chapters 3 to 5). The only difference is that, instead of a prime mover providing mechanical torque input to the generator, the motor drives a mechanical load. As in the case of an induction motor, a commonly used expression for the load torque is

$$T_m = T_0 \omega_r^m \qquad (7.106)$$

The rotor acceleration equation is given by

$$\frac{d\omega_r}{dt} = \frac{1}{2H}(T_e - T_m) \qquad (7.107)$$

where H is the combined inertia constant of the motor and load.

The data to model the synchronous motor are identical to those of a synchronous generator, except for the exponent m associated with the load characteristic.

7.4 ACQUISITION OF LOAD-MODEL PARAMETERS

There are two basic approaches to the determination of system-load characteristics:

•   Measurement-based approach

•   Component-based approach

7.4.1 Measurement-Based Approach

In this approach, the load characteristics are measured at representative substations and feeders at selected times of the day and season. These are used to extrapolate the parameters of loads throughout the system. References 14 and 15 describe load models derived from staged tests, and reference 16 presents models derived from actual system transients. Reference 3 describes an alternative approach in which load characteristics are monitored continuously from naturally occurring system variations.

Steady-state load-voltage characteristics

Composite load characteristics are normally measured at the highest voltage level for which the voltage of radially connected load can be adjusted by transformer tap changers. This is usually the highest distribution voltage level. The steady-state characteristics can be determined by adjusting the load voltage via the transformer tap changers over a range of voltage above and below the nominal value. Distribution tap changers and switched capacitors must be blocked to obtain meaningful results. The measured responses of voltage, active power, and reactive power are fitted to the polynomial and/or exponential expressions (Equation 7.1, 7.2, or 7.5).

Steady-state load-frequency characteristics

It is usually much more difficult to measure system load-frequency characteristics. To measure composite load-frequency characteristics, an isolated system must be formed and the frequency varied over the desired range. To obtain valid data, care must be taken to separate the effects of voltage changes and frequency changes. For many of the results reported in the literature this is not done, and the dP/df characteristic determined is a composite of the effects due to frequency change and the resulting voltage change.

Dynamic load-voltage characteristics

The small-signal dynamic characteristics of composite loads can be determined relatively easily from simple system tests. Figure 7.19 shows the testing configuration that may be employed when loads are supplied by two tap-changing transformers.

Initially, one tap changer is adjusted upward and the other downward by a few taps, keeping the load voltage constant. One of the transformers is then tripped; this produces not only a voltage magnitude change but also an instantaneous angle change at the load bus. By varying the initial tap positions, it is possible to obtain a range of voltage changes in both positive and negative directions. By selecting the tap positions appropriately, it is also possible to produce an angle change with only a very small

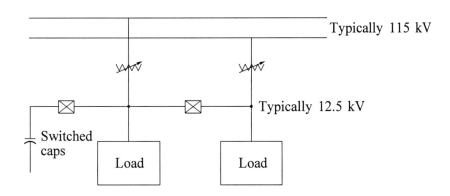

Figure 7.19 Typical station configuration for
testing load characteristics

voltage change. This is useful in separating the effects of voltage magnitude and angle change.

If there is a switched capacitor bank at the load bus, it can be switched in and out to produce a voltage magnitude change at the load without an angle change.

The responses obtained in this fashion are essentially *small-signal*. Reference 14 describes load models derived from such tests at Ontario Hydro. Measured time responses of load-voltage magnitude, angle, active power, and reactive power are used to determine the parameters of a model with a structure as shown in Figure 7.20. The model structure assumes that the composite load appears as an induction motor and shunt static load.

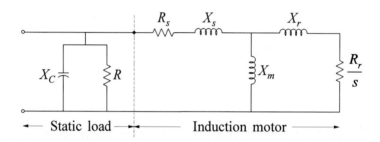

Figure 7.20 Composite induction motor/static load equivalent for representing an industrial load

A least square difference technique is used to optimize the match between the measured and model responses as the model parameters are adjusted. The result is a model whose parameters are chosen to match the small-signal response. However, because of the realistic model structure chosen, the model is likely to give reasonable results under large-disturbance conditions.

Reference 14 also describes *large-signal* models developed for industrial loads from responses monitored by transient monitoring systems during naturally occurring disturbances.

Reference 15 describes models derived from staged tests on generating station auxiliary systems and at bulk supply points supplying both industrial and residential loads in England. The resulting model consists of a constant impedance static load in parallel with a single rotor circuit induction motor. In addition, a nonlinear shunt reactance is included to account for magnetic saturation characteristics of motors and transformers.

7.4.2 Component-Based Approach

This approach was developed by EPRI under several research projects beginning in 1976 [16-19]. It involves building up the load model from information on its constituent parts as illustrated in Figure 7.21. The load supplied at a bulk power

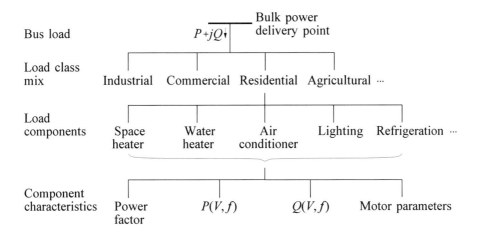

Figure 7.21 Component-based modelling approach

delivery point is categorized into *load classes* such as residential, commercial, industrial, agricultural, and mining. Each category of load class is represented in terms of *load components* such as lighting, air conditioning, space heating, water heating, and refrigeration.

The characteristics of individual appliances have been studied in detail and techniques have been developed to aggregate individual loads to produce a composite load model [16,17].

The EPRI LOADSYN program [18] converts data on the load class mix, components, and their characteristics into the form required for power flow and stability programs. Typical default data have been developed for load composition and characteristics for each class of load. A similar approach is used in reference 20, in which end-user energy consumption data are used to identify load mix data.

In reference 17, load characteristics determined by the component-based load-modelling approach are compared with measured load characteristics on the Long Island Lighting Company, Rochester Gas and Electric Company, and Montana-Dakota Utility systems. The load models correctly represented the steady-state active power versus voltage responses, but could not represent accurately the steady-state reactive power versus voltage responses. Also, the models could not represent accurately the transient responses of active and reactive power. These discrepancies between measured and model responses were subsequently resolved, as reported in reference 21, by using physically based models as opposed to the polynomial representation. Single-phase induction motor models were used to represent compressors and other rotating loads. Separate models were developed to represent compressor and non-compressor loads.

7.4.3 Sample Load Characteristics [2,14,16,20]

(a) Component static characteristics

Table 7.1 summarizes typical voltage and frequency-dependent characteristics of a
number of load components.

Table 7.1 Static characteristics of load components

Component	Power factor	$\partial P/\partial V$	$\partial Q/\partial V$	$\partial P/\partial f$	$\partial Q/\partial f$
Air conditioner					
3-phase central	0.90	0.088	2.5	0.98	-1.3
1-phase central	0.96	0.202	2.3	0.90	-2.7
Window type	0.82	0.468	2.5	0.56	-2.8
Water heaters, Range top, oven, Deep fryer	1.0	2.0	0	0	0
Dishwasher	0.99	1.8	3.6	0	-1.4
Clothes washer	0.65	0.08	1.6	3.0	1.8
Clothes dryer	0.99	2.0	3.2	0	-2.5
Refrigerator	0.8	0.77	2.5	0.53	-1.5
Television	0.8	2.0	5.1	0	-4.5
Incandescent lights	1.0	1.55	0	0	0
Fluorescent lights	0.9	0.96	7.4	1.0	-2.8
Industrial motors	0.88	0.07	0.5	2.5	1.2
Fan motors	0.87	0.08	1.6	2.9	1.7
Agricultural pumps	0.85	1.4	1.4	5.0	4.0
Arc furnace	0.70	2.3	1.6	-1.0	-1.0
Transformer (unloaded)	0.64	3.4	11.5	0	-11.8

(b) Load class static characteristics

Table 7.2 summarizes the sample characteristics of different load classes.

Table 7.2

Load class	Power factor	$\partial P/\partial V$	$\partial Q/\partial V$	$\partial P/\partial f$	$\partial Q/\partial f$
Residential					
Summer	0.9	1.2	2.9	0.8	-2.2
Winter	0.99	1.5	3.2	1.0	-1.5
Commercial					
Summer	0.85	0.99	3.5	1.2	-1.6
Winter	0.9	1.3	3.1	1.5	-1.1
Industrial	0.85	0.18	6.0	2.6	1.6
Power plant auxiliaries	0.8	0.1	1.6	2.9	1.8

(c) Dynamic characteristics

The following are sample data for induction motor equivalents representing three different types of load (see Figure 7.7 for definition of parameters).

(i)   The composite dynamic characteristics of a feeder supplying predominantly a commercial load:

$R_s = 0.001$ $X_s = 0.23$ $X_r = 0.23$
$X_m = 3.0$ $R_r = 0.02$ $H = 0.663$ $m = 5.0$

(ii)   A large industrial motor:

$R_s = 0.007$ $X_{s-u} = 0.0409$ $X_{s-s} = 0.0409$ $X_{r-u} = 0.0267$
$X_{r-s} = 0.0267$ $X_m = 3.62$ $R_r = 0.0062$ $H = 1.6$
$m = 2.0$ $I_{sat\,s} = 3.0$ $I_{sat\,r} = 3.0$
Deep-bar factor $= 6.59$

(iii)   A small industrial motor:

$R_s = 0.078$ $X_s = 0.065$ $X_r = 0.049$
$X_m = 2.67$ $R_r = 0.044$ $H = 0.5$ $m = 2.0$

REFERENCES

[1]   IEEE Task Force Report, "Load Representation for Dynamic Performance Analysis," Paper 92WM126-3 PWRD, presented at the IEEE PES Winter Meeting, New York, January 26-30, 1992.

[2]   C. Concordia and S. Ihara, "Load Representation in Power System Stability Studies," *IEEE Trans.*, Vol. PAS-101, pp. 969-977, April 1982.

[3]   K. Srinivasan, C.T. Nguyen, Y. Robichaud, A. St. Jacques, and G.J. Rogers, "Load Response Coefficients Monitoring System: Theory and Field Experience," *IEEE Trans.*, Vol. PAS-100, pp. 3818-3827, August 1981.

[4]   T. Ohyama, A. Watanabe, K. Nishimura, and S. Tsuruta, "Voltage Dependence of Composite Loads in Power Systems," *IEEE Trans.*, Vol. PAS-104, pp. 3064-3073, November 1985.

[5]   EPRI Report of Project 1208-9, "Extended Transient-Midterm Stability Program," Final report, prepared by Ontario Hydro, December 1992.

[6]   H.K. Clark, "Experience with Load Models in Simulation of Dynamic Phenomena," Paper presented at the Panel Discussion on Load Modelling Impact on System Dynamic Performance, IEEE PES Winter Meeting, New York, February 1, 1989.

[7]   H.K. Clark and T.F. Laskowski, "Transient Stability Sensitivity to Detailed Load Models: A Parametric Study," Paper A78 559-7, IEEE PES Summer Meeting, Los Angeles, July 16-21, 1978.

[8]   Klaus-Martin Graf, "Dynamic Simulation of Voltage Collapse Processes in EHV Power Systems," *Proceeding: Bulk Power System Voltage Phenomena-Voltage Stability and Security*, EPRI EL-6183, Section 6.3, January 1989.

[9]   C.W. Taylor, *Power System Voltage Stability*, McGraw-Hill, 1993.

[10]   D.S. Brereton, D.G. Lewis, and C.C. Young, "Representation of Induction-Motor Loads during Power System Stability Studies," *AIEE Trans.*, Vol. 76, Part III, pp. 451-460, August 1957.

[11]   A.E. Fitzgerald and C. Kingsley, *Electric Machinery*, Second Edition, McGraw-Hill, 1961.

[12]   G.J. Rogers and D.S. Benaragama, "An Induction Motor Model with Deep-Bar

Effect and Leakage Inductance Saturation," *Archiv Für Elektrotechnik*, Vol. 60, pp. 193-201, 1978.

[13]   G.J. Rogers, J. Di Manno, and R.T.H. Alden, "An Aggregate Induction Motor Model for Industrial Plants," *IEEE Trans.*, Vol. PAS-103, pp. 683-690, 1984.

[14]   S.A.Y. Sabir and D.C. Lee, "Dynamic Load Models Derived from Data Acquired during System Transients," *IEEE Trans.*, Vol. PAS-101, pp. 3365-3372, September 1982.

[15]   G. Shackshaft, O.C. Symons, and J.G. Hadwick, "General Purpose Model of Power System Loads," *Proc IEE*, Vol. 124, No. 8, pp. 715-723, August 1977.

[16]   EPRI Report of Project RP849-3, "Determining Load Characteristics for Transient Performance," EPRI EL-840, Vols. 1 to 3, prepared by the University of Texas at Arlington, May 1979.

[17]   EPRI Report of Project RP849-1, "Determining Load Characteristics for Transient Performance," EPRI EL-850, prepared by General Electric Company, March 1981.

[18]   EPRI Report of Project RP849-7, "Load Modelling for Power Flow and Transient Stability Studies," EPRI EL-5003, Prepared by General Electric Company, January 1987.

[19]   T. Frantz, T. Gentile, S. Ihara, N. Simons, and M. Waldron, "Load Behaviour Observed in LILCO and RG&E Systems," *IEEE Trans.*, Vol. PAS-103, No. 4, pp. 819-831, April 1984.

[20]   E. Vaahedi, M.A. El-Kady, J.A. Libaque-Esaine, and V.F. Carvalho, "Load Models for Large-Scale Stability Studies from End-User Consumption," *IEEE Trans.*, Vol. PWRS-2, pp. 864-872, November 1987.

[21]   R.J. Frowd, R. Podmore, and M. Waldron, "Synthesis of Dynamic Load Models for Stability Studies," *IEEE Trans.*, Vol. PAS-101, pp. 127-135, January 1982.

[22]   B. Adkins and R.G. Harley, *The General Theory of Alternating Current Machine*, Chapman and Hall, 1975.

Chapter 8

Excitation Systems

The basic function of an excitation system is to provide direct current to the synchronous machine field winding. In addition, the excitation system performs control and protective functions essential to the satisfactory performance of the power system by controlling the field voltage and thereby the field current.

The control functions include the control of voltage and reactive power flow, and the enhancement of system stability. The protective functions ensure that the capability limits of the synchronous machine, excitation system, and other equipment are not exceeded.

This chapter describes the characteristics and modelling of different types of synchronous generator excitation systems. In addition, it discusses dynamic performance criteria and provides definitions of related terms useful in the identification and specification of excitation system requirements. This subject has been covered in several IEEE reports [1-8]. These serve as useful references to utilities, manufacturers, and system analysts by establishing a common nomenclature, by standardizing models, and by providing guides for specifications and testing. Models and terminologies used in this chapter largely conform to these publications.

8.1 EXCITATION SYSTEM REQUIREMENTS

The performance requirements of the excitation system are determined by considerations of the synchronous generator as well as the power system [6,9].

315

Generator considerations

The basic requirement is that the excitation system supply and automatically adjust the field current of the synchronous generator to maintain the terminal voltage as the output varies within the *continuous capability* of the generator. This requirement can be visualized from the generator *V*-curves, such as those shown in Figure 5.19 of Chapter 5. Margins for temperature variations, component failures, emergency overrating, etc., must be factored in when the steady-state power rating is determined. Normally, the exciter rating varies from 2.0 to 3.5 kW/MVA generator rating.

In addition, the excitation system must be able to respond to transient disturbances with field forcing consistent with the generator instantaneous and short-term capabilities. The generator capabilities in this regard are limited by several factors: rotor insulation failure due to high field voltage, rotor heating due to high field current, stator heating due to high armature current loading, core end heating during underexcited operation, and heating due to excess flux (volts/Hz). The thermal limits have time-dependent characteristics, and the short-term overload capability of the generators may extend from 15 to 60 seconds. To ensure the best utilization of the excitation system, it should be capable of meeting the system needs by taking full advantage of the generator's short-term capabilities without exceeding their limits.

Power system considerations

From the power system viewpoint, the excitation system should contribute to effective control of voltage and enhancement of system stability. It should be capable of responding rapidly to a disturbance so as to enhance transient stability, and of modulating the generator field so as to enhance small-signal stability.

Historically, the role of the excitation system in enhancing power system performance has been growing continually. Early excitation systems were controlled manually to maintain the desired generator terminal voltage and reactive power loading. When the voltage control was first automated, it was very slow, basically filling the role of an alert operator. In the early 1920s, the potential for enhancing small-signal and transient stability through use of continuous and fast-acting regulators was recognized. Greater interest in the design of excitation systems developed, and exciters and voltage regulators with faster response were soon introduced to the industry. Excitation systems have since undergone continuous evolution. In the early 1960s, the role of the excitation system was expanded by using auxiliary stabilizing signals, in addition to the terminal voltage error signal, to control the field voltage to damp system oscillations. This part of excitation control is referred to as the *power system stabilizer*. Modern excitation systems are capable of providing practically instantaneous response with high ceiling voltages. The combination of high field-forcing capability and the use of auxiliary stabilizing signals contributes to substantial enhancement of the overall system dynamic performance. This will be discussed in detail in Chapters 12, 13, and 17.

To fulfill the above roles satisfactorily, the excitation system must satisfy the following requirements:

- Meet specified response criteria.

- Provide limiting and protective functions as required to prevent damage to itself, the generator, and other equipment.

- Meet specified requirements for operating flexibility.

- Meet the desired reliability and availability, by incorporating the necessary level of redundancy and internal fault detection and isolation capability.

8.2 ELEMENTS OF AN EXCITATION SYSTEM

Figure 8.1 shows the functional block diagram of a typical excitation control system for a large synchronous generator. The following is a brief description of the various subsystems identified in the figure.

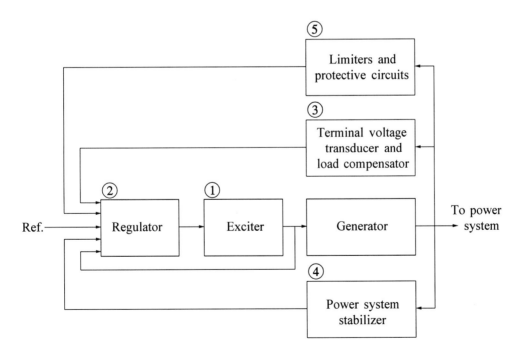

Figure 8.1 Functional block diagram of a synchronous generator excitation control system

(1)   *Exciter.* Provides dc power to the synchronous machine field winding, constituting the power stage of the excitation system.

(2)   *Regulator.* Processes and amplifies input control signals to a level and form appropriate for control of the exciter. This includes both regulating and excitation system stabilizing functions (rate feedback or lead-lag compensation).

(3)   *Terminal voltage transducer and load compensator.* Senses generator terminal voltage, rectifies and filters it to dc quantity, and compares it with a reference which represents the desired terminal voltage. In addition, load (or line-drop, or reactive) compensation may be provided, if it is desired to hold constant voltage at some point electrically remote from the generator terminal (for example, partway through the step-up transformer).

(4)   *Power system stabilizer.* Provides an additional input signal to the regulator to damp power system oscillations. Some commonly used input signals are rotor speed deviation, accelerating power, and frequency deviation.

(5)   *Limiters and protective circuits.* These include a wide array of control and protective functions which ensure that the capability limits of the exciter and synchronous generator are not exceeded. Some of the commonly used functions are the field-current limiter, maximum excitation limiter, terminal voltage limiter, volts-per-Hertz regulator and protection, and underexcitation limiter. These are normally distinct circuits and their output signals may be applied to the excitation system at various locations as a summing input or a gated input. For convenience, they have been grouped and shown in Figure 8.1 as a single block.

8.3 TYPES OF EXCITATION SYSTEMS

Excitation systems have taken many forms over the years of their evolution. They may be classified into the following three broad categories based on the excitation power source used [4,8]:

- DC excitation systems
- AC excitation systems
- Static excitation systems

This section provides a description of the above classes of excitation systems, the different forms they take, and their general structure. Details regarding various regulating and protective functions often included with the excitation systems will be covered in Section 8.5.

8.3.1 DC Excitation Systems

The excitation systems of this category utilize dc generators as sources of excitation power and provide current to the rotor of the synchronous machine through slip rings. The exciter may be driven by a motor or the shaft of the generator. It may be either self-excited or separately excited. When separately excited, the exciter field is supplied by a pilot exciter comprising a permanent magnet generator.

DC excitation systems represent early systems, spanning the years from the 1920s to the 1960s. They lost favour in the mid-1960s and were superseded by ac exciters.

The voltage regulators for such systems range all the way from the early non-continuously acting rheostatic type to the later systems utilizing many stages of magnetic amplifiers and rotating amplifiers [10,11].

DC excitation systems are gradually disappearing, as many older systems are being replaced by ac or static type systems. In some cases, the voltage regulators alone have been replaced by modern solid-state electronic regulators. As many of the dc excitation systems are still in service, they still require modelling in stability studies.

Figure 8.2 shows a simplified schematic representation of a typical dc excitation system with an amplidyne voltage regulator. It consists of a dc commutator exciter which supplies direct current to the main generator field through slip rings. The exciter field is controlled by an amplidyne.

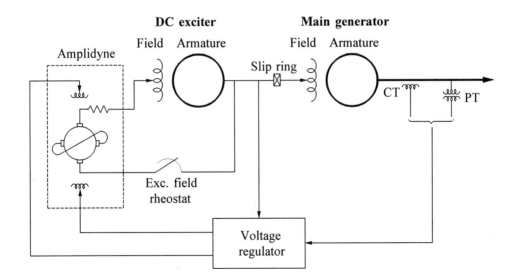

Figure 8.2 DC excitation system with amplidyne voltage regulator

An amplidyne is a special type of the general class of rotating amplifiers known as metadynes [11,12]. It is a dc machine of special construction having two sets of brushes 90 electrical degrees apart, one set on its direct (*d*) axis and the other set on its quadrature (*q*) axis. The control-field windings are located on the *d*-axis. A compensating winding in series with the *d*-axis load produces flux equal and opposite to the *d*-axis armature current, thereby cancelling negative feedback of the armature reaction. The brushes on the *q*-axis are shorted, and very little control-field power is required to produce a large current in the *q*-axis armature. The *q*-axis current produces the principal magnetic field, and the power required to sustain the *q*-axis current is supplied mechanically by the motor driving the amplidyne. The result is a device with power amplification on the order of 10,000 to 100,000 and a time constant in the range from 0.02 to 0.25 seconds.

In the excitation system of Figure 8.2, the amplidyne provides incremental changes to the exciter field in a "buck-boost" scheme. The exciter output provides the rest of its own field by self-excitation. If the amplidyne regulator is out of service, the exciter field is on "manual control" and is changed through adjustment of the field rheostat.

8.3.2 AC Excitation Systems

The excitation systems of this category utilize alternators (ac machines) as sources of the main generator excitation power. Usually, the exciter is on the same shaft as the turbine generator. The ac output of the exciter is rectified by either controlled or non-controlled rectifiers to produce the direct current needed for the generator field. The rectifiers may be stationary or rotating.

The early ac excitation systems used a combination of magnetic and rotating amplifiers as regulators [11]. Most new systems use electronic amplifier regulators.

AC excitation systems can thus take many forms depending on the rectifier arrangement, method of exciter output control, and source of excitation for the exciter [13-17]. The following is a description of different forms of ac excitation systems in use.

(a) Stationary rectifier systems

With stationary rectifiers, the dc output is fed to the field winding of the main generator through slip rings.

When non-controlled rectifiers are used, the regulator controls the field of the ac exciter, which in turn controls the exciter output voltage. A simplified one-line diagram of such a *field controlled alternator rectifier excitation system* is shown in Figure 8.3. In the system shown, which is representative of the General Electric ALTERREX[1] excitation system [14], the alternator exciter is driven from the main

---

[1] ALTERREX is a trademark of General Electric Co.

generator rotor. The exciter is self-excited with its field power derived through thyristor rectifiers. The voltage regulator derives its power from the exciter output voltage.

An alternative form of field-controlled alternator rectifier system uses a pilot exciter as the source of exciter field power.

When controlled rectifiers (thyristors) are used, the regulator directly controls the dc output voltage of the exciter. Figure 8.4 shows the schematic diagram of such an *alternator supplied controlled-rectifier system*, representative of the General Electric ALTHYREX[1] System [17]. The voltage regulator controls the firing of the thyristors. The exciter alternator is self-excited and uses an independent static voltage regulator to maintain its output voltage. Since the thyristors directly control the exciter output, this system inherently provides high initial response (small response time).

As shown in Figures 8.3 and 8.4, two independent modes of regulation are provided: (1) ac regulator to automatically maintain the main generator stator terminal voltage at a desired value corresponding to the ac reference; and (2) dc regulator to maintain constant generator field voltage as determined by the dc reference. The dc regulator or manual control mode caters to situations where the ac regulator is faulty or needs to be disabled. The input signals to the ac regulator include auxiliary inputs which provide additional control and protective functions which will be described in Section 8.5.

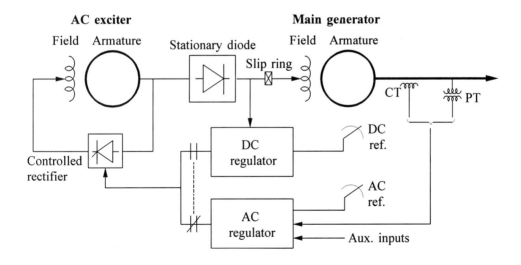

Figure 8.3 Field-controlled alternator rectifier excitation system

---

[1] ALTHYREX is a trademark of General Electric Co.

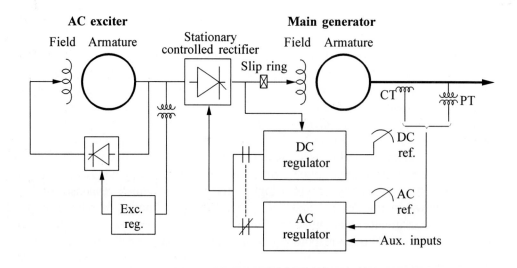

Figure 8.4 Alternator-supplied controlled-rectifier excitation system

(b) Rotating rectifier systems

With rotating rectifiers, the need for slip rings and brushes is eliminated, and the dc output is directly fed to the main generator field. As shown in Figure 8.5, the armature of the ac exciter and the diode rectifiers rotate with the main generator field. A small ac pilot exciter, with a permanent magnet rotor (shown as N S in the figure), rotates with the exciter armature and the diode rectifiers. The rectified output of the pilot exciter stator energizes the stationary field of the ac exciter. The voltage regulator controls the ac exciter field, which in turn controls the field of the main generator.

Such a system is referred to as a *brushless excitation system*. It was developed to avoid problems with the use of brushes that were perceived to exist when supplying the high field currents of very large generators; for example, the power supplied to the field of a 600 MW generator is on the order of 1 MW. However, with well-maintained brushes and slip rings, these perceived problems did not actually develop. AC excitation systems with and without brushes have performed equally well.

High initial-response performance of brushless excitation can be achieved by special design of the ac exciter and high voltage forcing of the exciter stationary field winding. An example of such a system is the Westinghouse high initial response brushless excitation system [13].

Brushless excitation systems do not allow direct measurement of generator field current or voltage. Manual control of main generator voltage is provided by an adjustable dc input setting to the thyristor gating circuits. For the sake of simplicity, the functions of the control circuitry are not shown in detail in Figure 8.5.

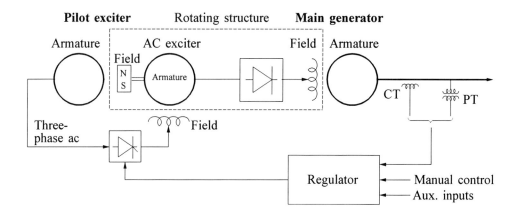

Figure 8.5 Brushless excitation system

8.3.3 Static Excitation Systems

All components in these systems are static or stationary. Static rectifiers, controlled or uncontrolled, supply the excitation current directly to the field of the main synchronous generator through slip rings. The supply of power to the rectifiers is from the main generator (or the station auxiliary bus) through a transformer to step down the voltage to an appropriate level, or in some cases from auxiliary windings in the generator.

The following is a description of three forms of static excitation systems that have been widely used.

(a) Potential-source controlled-rectifier systems

In this system, the excitation power is supplied through a transformer from the generator terminals or the station auxiliary bus, and is regulated by a controlled rectifier (see Figure 8.6). This type of excitation system is also commonly known as a *bus-fed* or *transformer-fed* static system.

This system has a very small inherent time constant. The maximum exciter output voltage (ceiling voltage) is, however, dependent on the input ac voltage. Hence, during system-fault conditions causing depressed generator terminal voltage, the available exciter ceiling voltage is reduced. This limitation of the excitation system is, to a large extent, offset by its virtually instantaneous response and high post-fault field-forcing capability [18,19]. In addition, it is inexpensive and easily maintainable. For generators connected to large power systems such excitation systems perform satisfactorily [18]. Compounding ancillaries, such as those described below, are not normally justified; they are likely important for generators feeding power directly into small industrial networks with slow fault-clearing.

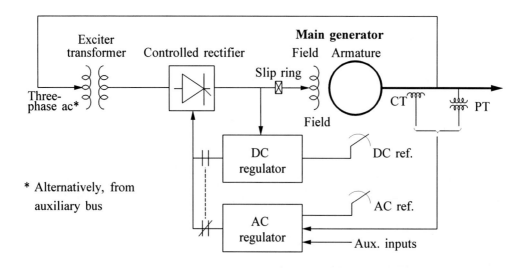

Figure 8.6 Potential-source controlled-rectifier excitation system

Examples of this type of excitation system are: Canadian General Electric Silicomatic excitation system, Westinghouse type PS excitation system, General Electric potential source static excitation system, and ABB, Reyrolle-Parsons, GEC-Eliott, Toshiba, Mitsubishi, and Hitachi static excitation systems.

(b) Compound-source rectifier systems

The power to the excitation system in this case is formed by utilizing the current as well as the voltage of the main generator. This may be achieved by means of a power potential transformer (PPT) and a saturable-current transformer (SCT) as illustrated in Figure 8.7. Alternatively, the voltage and current sources may be combined by utilizing a single excitation transformer, referred to as a saturable-current potential transformer (SCPT).

The regulator controls the exciter output through controlled saturation of the excitation transformer. When the generator is not supplying a load, the armature current is zero and the potential source supplies the entire excitation power. Under loaded conditions, part of the excitation power is derived from the generator current. During a system-fault condition, with severely depressed generator terminal voltage, the current input enables the exciter to provide high field-forcing capability.

Examples of this type of excitation system are General Electric SCT-PPT and SCPT static excitation systems.

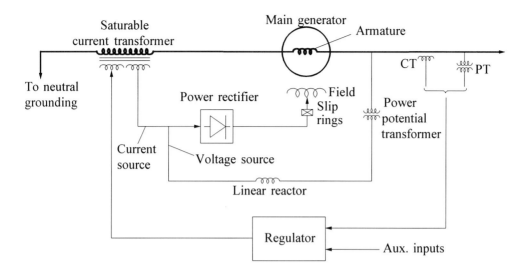

Figure 8.7 Compound-source rectifier excitation system

(c) Compound-controlled rectifier excitation systems

This system utilizes controlled rectifiers in the exciter output circuits and the compounding of voltage and current-derived sources within the generator stator to provide excitation power. The result is a high initial-response static excitation system with full "fault-on" forcing capability.

An example of this type of system is the compound power-source GENERREX[1] excitation system [15,16]. Figure 8.8 shows an elementary single-line diagram of the system. The voltage source is formed by a set of three-phase windings placed in three slots in the generator stator and a series linear reactor. The current source is obtained from current transformers mounted in the neutral end of the stator windings. These sources are combined through transformer action and the resultant ac output is rectified by stationary power semiconductors. The means of control is provided by a combination of diodes and thyristors connected to form a shunt bridge. A static ac voltage regulator controls the firing circuits of the thyristors and thus regulates the excitation to the generator field.

The excitation transformer consists of three single-phase units with three windings: current (C) and potential (P) primary windings, and a secondary output winding (F). Under fault conditions, the fault current flowing through the excitation transformer "C" windings provides the field-forcing capability when the generator voltage is depressed.

---

[1] GENERREX is a trademark of General Electric Co.

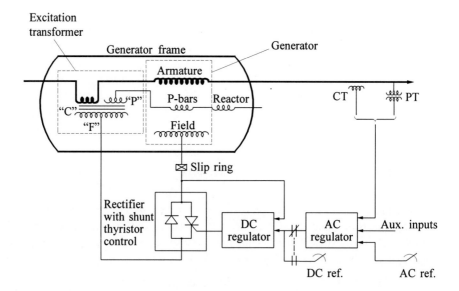

Figure 8.8 GENERREX compound-controlled rectifier
excitation system ©IEEE1976 [16]

The reactor serves two functions: contribution to the desired compounding characteristic of the excitation system and reduction of fault currents for faults in the excitation system or the generator.

The excitation transformers and reactors are contained in an excitation dome that is bolted to the top of the generator frame, forming an integral part of the frame.

Field flashing for static exciters:

Since the source of power to a static excitation system is the main generator, it is in effect self-excited. The generator cannot produce any voltage until there is some field current. It is therefore necessary to have another source of power for a few seconds to initially provide the field current and energize the generator. This process of build-up of generator field flux is called *field flashing*. The usual field-flashing source is a station battery.

8.3.4 Recent Developments and Future Trends

The advances in excitation control systems over the last 20 years have been influenced by developments in solid-state electronics. Developments in analog-integrated circuitry have made it possible to easily implement complex control strategies.

The latest development in excitation systems has been the introduction of digital technology. Thyristors continue to be used for the power stage. The control, protection, and logic functions have been implemented digitally, essentially duplicating the functions previously provided by analog circuitry.

The digital controls are likely to be used extensively in the future as they provide a cheaper and possibly more reliable alternative to analog circuitry. They have the added advantage of being more flexible, allowing easy implementation of more complex control strategies, and interfacing with other generator control and protective functions.

8.4 DYNAMIC PERFORMANCE MEASURES

The effectiveness of an excitation system in enhancing power system stability is determined by some of its key characteristics. In this section, we identify and define performance measures which determine these characteristics and serve as a basis for evaluating and specifying the performance of the excitation control system. Figure 8.9 shows the representation of the overall excitation control system in the classical form used for describing feedback control systems.

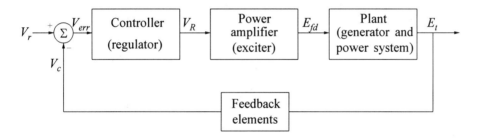

Figure 8.9 Excitation control system in the classical feedback control form

The performance of the excitation control system depends on the characteristics of excitation system, the generator, and the power system. Since the system is nonlinear, it is convenient to classify its dynamic performance into large-signal performance and small-signal performance. For large-signal performance, the nonlinearities are significant; for small-signal performance, the response is effectively linear.

8.4.1 Large-Signal Performance Measures [7]

Large-signal performance measures provide a means of assessing the excitation system performance for severe transients such as those encountered in the consideration of transient, mid-term and long-term stability of the power system. Such measures are based on the quantities defined below. To permit maximum flexibility in the design, manufacture, and application of excitation equipment, some of the performance measures are defined "under specified conditions"; these conditions may be specified as appropriate for the specific situation.

(a) Excitation system ceiling voltage: The maximum direct voltage that the excitation system is able to supply from its terminals under specified conditions [7,20].

Ceiling voltage is indicative of the field-forcing capability of the excitation system; higher ceiling voltages tend to improve transient stability.

For potential source and compound source static excitation systems, whose supply depends on the generator voltage and current, the ceiling voltage is defined at specified supply voltage and current. For excitation systems with rotating exciters, the ceiling voltage is determined at rated speed.

(b) Excitation system ceiling current: The maximum direct current that the excitation system is able to supply from its terminals for a specified time [7,20].

When prolonged disturbances are a concern, the ceiling current may be based on the excitation system thermal duty.

(c) Excitation system voltage time response: The excitation system output voltage expressed as a function of time under specified conditions [7,20].

(d) Excitation system voltage response time: The time in seconds for the excitation voltage to attain 95% of the difference between the ceiling voltage and rated load-field voltage under specified conditions [7,20].

The *rated load field voltage* is the generator field voltage under rated continuous load conditions with the field winding at (i) 75°C for windings designed to operate at rating with a temperature rise of 60°C or less; or (ii) 100°C for windings designed to operate at rating with a temperature rise greater than 60°C.

(e) High initial-response excitation system: An excitation system having a voltage response time of 0.1 seconds or less [7]. It represents a high response and fast-acting system.

(f) Excitation system nominal response[1]: The rate of increase of the excitation system output voltage determined from the excitation system voltage response curve, divided by the rated field voltage. This rate, if maintained constant, would develop the same voltage-time area as obtained from the actual curve over the first half-second interval (unless a different time interval is specified) [7,20].

---

[1] Historically, the excitation system nominal response has been referred to as the *excitation system response ratio* (see 1978 version of [7], and [20]).

The nominal response is determined by initially operating the excitation system at the rated load field voltage (and field current) and then suddenly creating the three-phase terminal voltage input signal conditions necessary to drive the excitation system voltage to ceiling. It should include any delay time that may be present before the excitation system responds to the initiating disturbance.

Referring to Figure 8.10, the excitation response is illustrated by line ac. This line is determined by establishing area acd equal to area abd:

$$\text{Nominal response} = \frac{cd}{(ao)(oe)}$$

where

$$oe = 0.5 \text{ s}$$
$$ao = \text{rated load field voltage}$$

The basis for considering a nominal time span of 0.5 s in the above definition is that, following a severe disturbance, the generator rotor angle swing normally peaks between 0.4 s and 0.75 s. The excitation system must act within this time period to be effective in enhancing transient stability. Accordingly, 0.5 s was chosen for the definition time period of nominal response.

In the past, the nominal response has been a well-established and useful criterion for evaluating the large-signal performance of excitation systems. With older and slower excitation systems, this was an acceptable performance measure, but it is

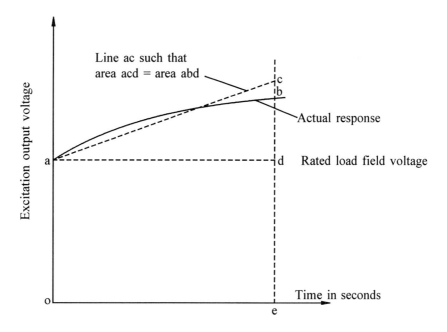

Figure 8.10 Excitation system nominal response

not adequate to cover many of the modern excitation systems. In particular, it is not a good figure of merit for excitation systems supplied from the generator or the power system, due to the reduced capability of such systems during a system fault.

For high initial-response excitation systems, the nominal response merely establishes the required ceiling voltage. The ceiling voltage and voltage response time are more meaningful parameters for such systems.

8.4.2 Small-Signal Performance Measures [3,7]

Small-signal performance measures provide a means of evaluating the response of the closed-loop excitation control systems to incremental changes in system conditions. In addition, small-signal performance characteristics provide a convenient means for determining or verifying excitation system model parameters for system studies.

Small-signal performance may be expressed in terms of performance indices used in feedback control system theory:

*   Indices associated with time response; and
*   Indices associated with frequency response

The typical time response of a feedback control system to a step change in input is shown in Figure 8.11. The associated indices are rise time, overshoot, and settling time.

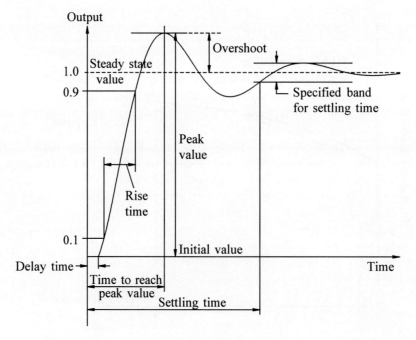

Figure 8.11 Typical time response to step input. © IEEE 1990 [7]

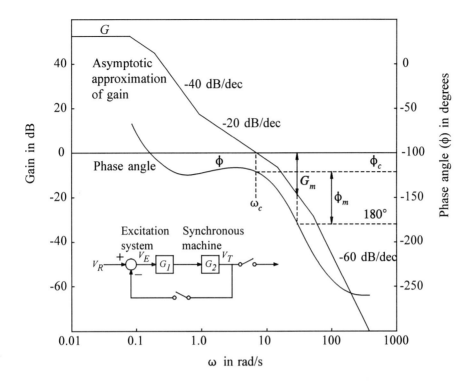

Figure 8.12 Typical open-loop frequency response of an excitation control system with generator open-circuited. © IEEE 1990 [7]

A typical open-loop frequency response characteristic of an excitation control system with the generator open-circuited is shown in Figure 8.12.

The performance indices associated with the open-loop frequency response are the low frequency gain G, crossover frequency ω_c, phase margin ϕ_m, and gain margin G_m. Larger values of G provide better steady-state voltage regulation, and larger crossover frequency ω_c indicates faster response. Larger values of phase margin ϕ_m and gain margin G_m provide a more stable excitation control loop. (The reference here is to excitation control system stability and not power system synchronous stability.)

In tuning the voltage regulator, an improvement to one index will most likely be to the detriment of other indices. For example, an increase in regulator gain will shift the gain curve in Figure 8.12 upward. This has the beneficial effect of increasing the low-frequency gain and crossover frequency, but has the undesirable effect of decreasing the gain and phase margins. In general, a phase margin of 40° or more and a gain margin of 6 dB or more are considered a good design practice for obtaining a stable, non-oscillatory voltage regulator system.

Figure 8.13 shows the corresponding closed-loop frequency response with the generator open-circuited.

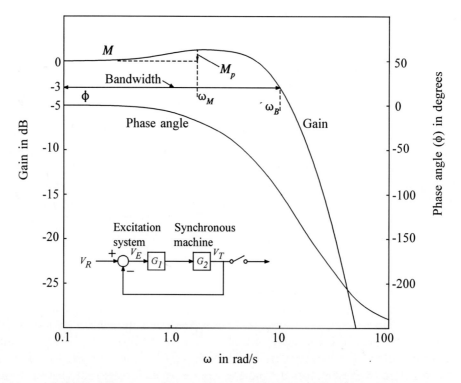

Figure 8.13 Typical closed-loop frequency response with generator open-circuited. © IEEE 1990 [7]

The indices of interest associated with the closed-loop frequency response are the bandwidth ω_B and peak value M_p.

A high value of M_p (>1.6) is indicative of an oscillatory system exhibiting large overshoot in its transient response. In general, a value of M_p between 1.1 and 1.5 is considered a good design practice.

Bandwidth is an important closed-loop frequency response index. Larger values indicate faster response. It approximately describes filtering or noise-rejection characteristics of the system.

Generally accepted values of performance indices characterizing good feedback control system performance are:

Gain margin $\geq$ 6 dB

Phase margin $\geq$ 40°

Overshoot = 5-15%

M_p = 1.1-1.6

It is not possible to define such generally acceptable ranges of values for other small-signal performance indices: rise time, settling time, and bandwidth. These indices are a measure of the relative speed of the control action. They are primarily determined by the synchronous machine dynamic characteristics.

The performance indices given above are applicable to any feedback control system having a single major feedback loop, i.e., a single controlled-output variable. Therefore, they are applicable to an excitation control system with the synchronous machine on open circuit or feeding an isolated load. Stable operation of the excitation control system with the generator off-line is ensured based on these performance indices and associated analytical techniques [3]. On the other hand, synchronous machines connected to a power system form a complex multiloop, multivariable, high-order control system. For such a system, the performance indices identified above are not applicable. The state-space approach using eigenvalue techniques is an effective method of assessing the performance of such complex systems. This is covered in detail in Chapter 12.

8.5 CONTROL AND PROTECTIVE FUNCTIONS

A modern excitation control system is much more than a simple voltage regulator. It includes a number of control, limiting, and protective functions which assist in fulfilling the performance requirements identified in Section 8.1. The extensive nature of these functions and the manner in which they interface with each other are illustrated in Figure 8.14. Any given excitation system may include only some or all of these functions, depending on the requirements of the specific application and on the type of exciter. The philosophy is to have the control functions regulate specific quantities at the desired level, and the limiting functions prevent certain quantities from exceeding set limits. If any of the limiters fail, then the protective functions remove appropriate components or the unit from service.

The following is a brief description of the various control and protective functions, and the associated elements identified in Figure 8.14.

8.5.1 AC and DC Regulators

The basic function of the ac regulator is to maintain the generator stator voltage. In addition, other auxiliary control and protection functions act through the ac regulator to control the generator field voltage as shown in Figure 8.14.

The dc regulator holds constant generator field voltage and is commonly referred to as *manual control*. It is used primarily for testing and start-up, and to cater to situations where the ac regulator is faulty. In this mode of operation, it is the field voltage that is regulated; only operator intervention by adjusting the setpoint will modify the field voltage. In some excitation systems, facilities for automatic setpoint tracking are provided. This will cause the manual setpoint to continually track the generator excitation variation due to the ac regulator and thus minimize the voltage

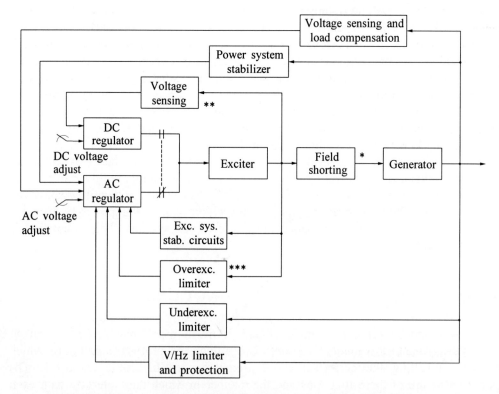

*   Field-shorting circuits are applicable to ac and static exciters only.
**   Some systems have open-loop dc regulator.
***   Overexcitation limiter may also be used with dc regulator

Figure 8.14 Excitation system control and protective circuits

and reactive power excursions in the event the ac regulator is removed from service abruptly. Care must be taken to ensure that a trip of the unit operating on manual control does not leave the generator in an overexcited condition.

8.5.2 Excitation System Stabilizing Circuits

Excitation systems comprised of elements with significant time delays have poor inherent dynamic performance. This is particularly true of dc and ac type excitation systems. Unless a very low steady-state regulator gain is used, the excitation control (through feedback of generator stator voltage) is unstable when the generator is on open circuit. Therefore, excitation control system stabilization, comprising either series or feedback compensation, is used to improve the dynamic performance of the control system. The most commonly used form of compensation is a derivative feedback as shown in Figure 8.15. The effect of the compensation is to minimize the phase shift introduced by the time delays over a selected frequency range [3]. This results in a stable off-line performance of the generator, such as that existing just prior to synchronization or following a load rejection. The feedback parameters can also be adjusted to improve the on-line performance of the generating unit. Depending on the type of excitation system, there may be many levels of excitation control system stabilization involving the major outer loop and minor inner loops. Static excitation systems have negligible inherent time delays and do not require excitation control-system stabilization to ensure stable operation with the generator off-line (see closure of reference 21).

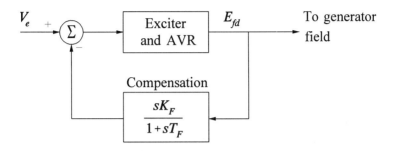

Figure 8.15 Derivative feedback excitation control system stabilization

8.5.3 Power System Stabilizer (PSS)

The power system stabilizer uses auxiliary stabilizing signals to control the excitation system so as to improve power system dynamic performance. Commonly used input signals to the power system stabilizer are shaft speed, terminal frequency and power. Power system dynamic performance is improved by the damping of system oscillations. This is a very effective method of enhancing small-signal stability performance.

The principle of operation of power system stabilizers and their structure and tuning are discussed in detail in Chapters 12 and 17.

8.5.4 Load Compensation

The automatic voltage regulator (AVR) normally controls the generator stator terminal voltage. Sometimes, load compensation is used to control a voltage which is representative of the voltage at a point either within or external to the generator. This is achieved by building additional circuitry into the AVR loop as shown in Figure 8.16. The compensator has adjustable resistance (R_c) and inductive reactance (X_c) that simulate the impedance between the generator terminals and the point at which the voltage is being effectively controlled. Using this impedance and the measured armature current, a voltage drop is computed and added to or subtracted

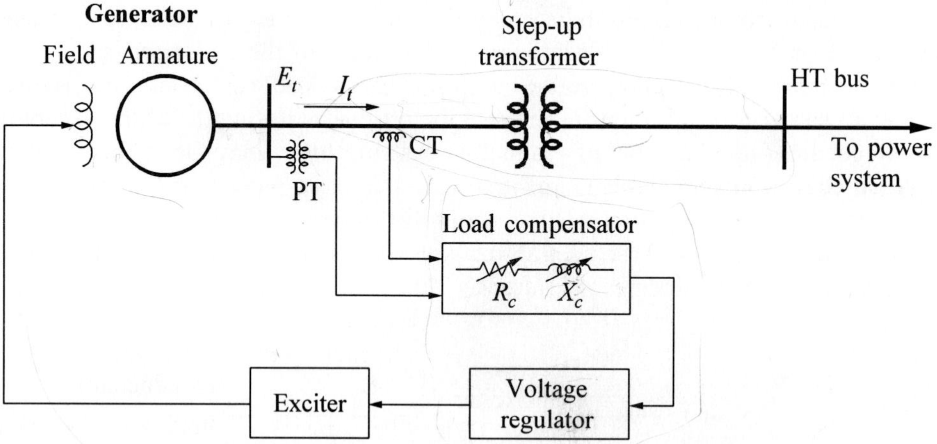

Figure 8.16 Schematic diagram of a load compensator

from the terminal voltage. The magnitude of the resulting compensated voltage (V_c), which is fed to the AVR, is given by

$$V_c = |\tilde{E}_t + (R_c + jX_c)\tilde{I}_t| \tag{8.1}$$

With R_c and X_c positive in Equation 8.1, the voltage drop across the compensator is added to the terminal voltage. The compensator regulates the voltage at a point within the generator and thus provides voltage droop. This is used to ensure proper sharing of reactive power between generators bussed together at their terminals, sharing a common step-up transformer. Such an arrangement is commonly used with hydro electric generating units and cross-compound thermal units. The compensator functions as a *reactive-current compensator* by creating an artificial coupling between the generators. Without this provision, one of the generators would try to control the terminal voltage slightly higher than the other; hence, one generator would tend to supply all of the required reactive power while the other would absorb reactive power to the extent allowed by underexcited limits.

With R_c and X_c negative, the compensator regulates the voltage at a point beyond the machine terminals. This form of compensation is used to compensate for the voltage drop across the step-up transformer, when two or more units are connected through individual transformers. Typically, 50% to 80% of the transformer impedance is compensated, ensuring voltage droop at the paralleling point so that generators can operate in parallel satisfactorily. This device is commonly referred to as a *line-drop compensator* although it is practically always used to compensate only for transformer drop. The nomenclature appears to have been derived from a similar compensator used on distribution system voltage regulators (see Chapter 11, Section 11.2).

In most cases, the resistance component of the impedance to be compensated is negligible and R_c may be set to zero.

Alternative forms of reactive-current and line-drop compensators are described in references 8 and 22.

8.5.5 Underexcitation Limiter [23-26]

The underexcitation limiter (UEL) is intended to prevent reduction of generator excitation to a level where the small-signal (steady-state) stability limit or the stator core end-region heating limit (see Chapter 5, Figure 5.16) is exceeded. This limiter is also referred to by other names such as underexcitation reactive-ampere limiter (URAL) and minimum excitation limiter (MEL).

The control signal of the UEL is derived from a combination of either voltage and current or active and reactive power of the generator. The limits are determined by the signal exceeding a reference level. There are a wide variety of implementations of the UEL function. Some UEL applications act on the voltage error signal of the AVR; when the UEL set limit is reached, a nonlinear element (such as a diode) begins to conduct and the limiter output signal is combined with other signals controlling the excitation system. In a more widely used form of UEL application, the limiter output signal is fed into an auctioneering circuit (high-value gate) which gives control to the larger of the voltage regulator and UEL signals; when the UEL set limit is reached, the limiter is given full control of the excitation system until the limiter signal is below the set limit.

Methods of setting the UEL characteristics are described in references 23 to 25. The settings should be based on the needed protection, i.e., system instability or stator core heating. In addition, the limiter performance should be coordinated with the generator loss-of-excitation protection (see Chapter 13). Figure 8.17 indicates the way in which the UEL characteristic (represented on a P-Q plane) is usually coordinated with the calculated small-signal stability limit and the loss-of-excitation (LOE) relay characteristic [28]. If the UEL is used to protect against stator end-region heating, the coordination is done in a similar manner, except that the stability limit is replaced by the heating limit.

If the input signals to the UEL are the generator stator voltage and current, the limiting characteristic appears circular on a P-Q plane as shown in Figure 8.17. With active and reactive power as input signals, the limiting characteristic would be a straight line.

Care should be taken to ensure that the UEL performance during a transient disturbance is not to the detriment of the power system performance [26,27].

8.5.6 Overexcitation Limiter

The purpose of the overexcitation limiter (OXL) is to protect the generator from overheating due to prolonged field overcurrent. This limiter is also commonly referred to as the maximum excitation limiter (MXL).

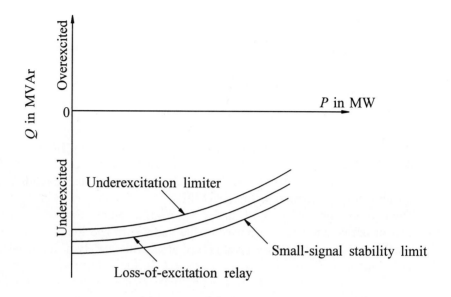

Figure 8.17 Coordination between UEL, LOE relay and stability limit

The generator field winding is designed to operate continuously at a value corresponding to rated load conditions. The permissible thermal overload of the field winding of round rotor generators, as specified by ANSI Standard C50.13-1977, is given by the solid curve of Figure 8.18. The curve passes through the following points:

Time (seconds)	10	30	60	120
Field voltage/current	208	146	125	112
(Percent of rated)				

The actual implementation of overexcitation limiting function varies depending on the manufacturer and vintage of the unit. Limiters supplied by two manufacturers are described in references 28 and 29.

The overexcitation limiting function typically detects the high field current condition and, after a time delay, acts through the ac regulator to ramp down the excitation to a preset value (typically 100% to 110% of rated field current). If this is unsuccessful, it trips the ac regulator, transfers control to the dc regulator, and repositions the setpoint to a value corresponding to the rated value. If this also does not reduce the excitation to a safe value, the limiter will initiate an exciter field breaker trip and a unit trip.

Two types of time delays are used: (a) fixed time and (b) inverse time. The fixed time limiters operate when the field current exceeds the pickup value for a fixed set time, irrespective of the degree of overexcitation. The inverse time limiters operate with the time delay matching the field thermal capability, as shown in Figure 8.18.

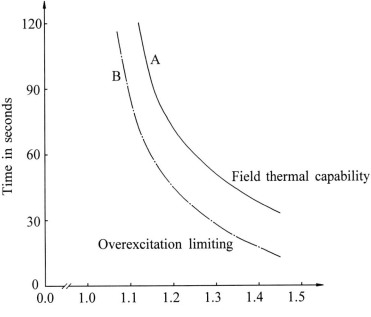

Figure 8.18 Coordination of overexcitation limiting
with field thermal capability

Exciters with very high ceiling voltages may be provided with an additional
field current limiter, which acts instantaneously through the ac regulator and limits the
field current to the short time limit (typically 160% of rated value).

8.5.7 Volts-per-Hertz Limiter and Protection

These are used to protect the generator and step-up transformer from damage
due to excessive magnetic flux resulting from low frequency and/or overvoltage.
Excessive magnetic flux, if sustained, can cause serious overheating and may result
in damage to the unit transformer and to the generator core.

The ratio of per unit voltage to per unit frequency, referred to as volts per
hertz (V/Hz), is a readily measurable quantity that is proportional to magnetic flux.
Typical V/Hz limitations for generators (GEN) and step-up transformers (XFMR) are
shown in the following table.

V/Hz (pu)		1.25	1.2	1.15	1.10	1.05
Damage Time in Minutes	GEN	0.2	1.0	6.0	20.0	∞
	XFMR	1.0	5.0	20.0	∞	

The unit step-up transformer low voltage rating is frequently 5% below the generator voltage rating; therefore, V/Hz limiting and protection requirements are usually determined by the transformer limitation. If, however, the generator and transformer voltage ratings are the same, the generator limitation would be more restrictive.

The *V/Hz limiter* (or regulator, as it is sometimes called) controls the field voltage so as to limit the generator voltage when the V/Hz value exceeds a preset value.

The *V/Hz protection* trips the generator, when the V/Hz value exceeds a preset value for a specified time. Usually, a dual-level protection is provided, one with a higher V/Hz setting and a shorter time setting, and the other with a lower V/Hz setting and a longer time setting. When used in conjunction with a V/Hz limiter, it serves as a backup.

For many units, the V/Hz protection becomes overvoltage protection above 60 Hz.

8.5.8 Field-Shorting Circuits

Since rectifiers cannot conduct in the reverse direction, the exciter current cannot be negative in the case of ac and static exciters. Under conditions of pole slipping and system short circuits, the induced current in the generator field winding may be negative. If a path is not provided for this negative current to flow, very high voltages may result across the field circuit. Therefore, special circuitry is usually provided to bypass the exciter to allow negative field current to flow. This takes the form of either a field-shorting circuit, commonly referred to as *crowbar*, or a varistor [8,30].

A crowbar consists of a thyristor and a field discharge resistor (FDR) connected across the generator field as shown in Figure 8.19. The thyristor is gated in response to an overvoltage condition that is created by the induced current not initially having a path in which to flow. The thyristor so gated conducts induced field current through the field discharge resistor.

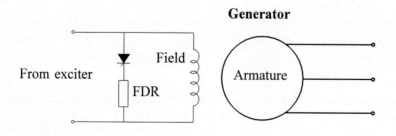

Figure 8.19 Field bypass circuit using a crowbar

A varistor is a nonlinear resistor. When connected across the generator field winding, as shown in Figure 8.20, it provides an effective means of bypassing the exciter under conditions of high induced voltage. With normal exciter voltage across the varistor, it has very high resistance and hence carries negligible current. As the voltage across the varistor increases beyond a threshold value, its resistance decreases and current through it increases very rapidly. Thus the varistor provides a low resistance path to induced negative field current and limits the voltage across the field and exciter.

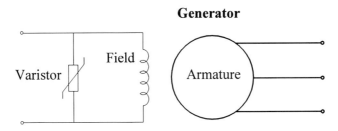

Figure 8.20 Field bypass circuit using a varistor

In some cases, no special field shorting circuits are provided. The amortisseurs associated with the solid rotor iron provide paths for the induced rotor currents. This is sufficient to limit the induced voltage to a level that is below the withstand capabilities of the generator field and the exciter. Since the field is not shorted, it carries no current in the negative direction.

8.6 MODELLING OF EXCITATION SYSTEMS

Mathematical models of excitation systems are essential for the assessment of desired performance requirements, for the design and coordination of supplementary control and protective circuits, and for system stability studies related to the planning and operation of power systems. The detail of the model required depends on the purpose of the study. Referring to Figure 8.14, the control and protective features that impact on transient and small-signal stability studies are the voltage regulator, power system stabilizer, and excitation control stabilization. The limiter and protective circuits identified in the figure normally need to be considered only for mid-term, long-term, and voltage stability studies. Some excitation systems are provided with fast-acting terminal voltage limiters in conjunction with power system stabilizers; these have to be modelled in transient stability simulations.

In this section, modelling of excitation systems is described. We begin with consideration of an appropriate per unit system, then describe models for the various components, and finally present complete models for selected types of excitation systems.

The material presented in this section conforms to the IEEE committee reports on excitation system modelling. The IEEE work on standardization of these models began in the 1960s, and the first set of models was published in 1968 [1]. This work was extended, and improved models to reflect advances in equipment and better modelling practices were published in 1981 [4]. These were updated and refined in 1992 [8].

8.6.1 Per Unit System

In choosing the per unit system for exciter output voltage and current, there are several options.

First, the per unit system used for the main synchronous machine field circuit would appear to be the obvious choice. While this system was chosen to simplify the synchronous machine equations (see Chapter 3, Section 3.4), it is not considered suitable for expressing the exciter output quantities. This is because, for normal operating conditions, the per unit exciter output voltage would be very small, being on the order of 0.001.

Second, for excitation system specification purposes it has become standard practice to use the rated-load field voltage as 1.0 per unit. However, this is not convenient for use in the formulation of synchronous machine and excitation system equations for system studies.

The third choice is to have 1.0 per unit exciter output voltage equal to the field voltage required to produce rated synchronous machine armature terminal voltage on the air-gap line; 1.0 per unit exciter output current is the corresponding synchronous machine field current. This per unit system is universally used in power system stability studies as it offers considerable simplicity. Here, we refer to this system as the *non-reciprocal per unit system* to distinguish it from the reciprocal per unit system used for modelling synchronous machines.

Excitation system models must interface with the synchronous machine model at both the field terminals and armature terminals. The input control signals to the excitation system are the synchronous machine stator quantities and rotor speed. The per unit systems used for expressing these input variables are the same as those used for modelling the synchronous machine. Thus, a change of per unit system is required only for those related to the field circuit.

We now develop the relationship between the per unit values of the exciter output voltage/current expressed in the non-reciprocal system and the synchronous machine field voltage/current expressed in the L_{ad}-base reciprocal system (see Section 3.4). For the synchronous machine, under open-circuit conditions, $i_d=i_q=0$. Substituting in Equations 3.139, 3.140, 3.142 and 3.143 yields

$$e_d = -\psi_q = -L_q i_q = 0 \tag{8.2}$$

$$e_q = \psi_d = L_{ad} i_{fd} \tag{8.3}$$

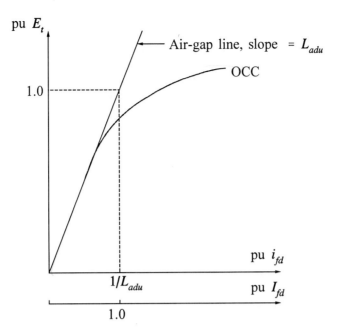

Figure 8.21 Synchronous machine open circuit characteristics

Referring to Figure 8.21, the field current required to produce 1.0 per unit stator terminal voltage on the air-gap line (slope = L_{adu}) is determined by

$$E_t = e_q = L_{adu}i_{fd} = 1.0 \quad \text{pu} \tag{8.4}$$

Therefore, in the reciprocal per unit system, the field current i_{fd} required to generate rated stator terminal voltage on the air-gap line is given by

$$i_{fd} = \frac{1}{L_{adu}} \quad \text{pu}$$

The corresponding field voltage is

$$e_{fd} = R_{fd}i_{fd} = \frac{R_{fd}}{L_{adu}} \quad \text{pu}$$

By definition, the corresponding value of exciter output current I_{fd} is equal to 1.0 per unit. Therefore,

$$I_{fd} = L_{adu} i_{fd} \tag{8.5}$$

and the corresponding exciter output voltage is

$$E_{fd} = \frac{L_{adu}}{R_{fd}} e_{fd} \tag{8.6}$$

Physically, exciter output voltage/current and generator field voltage/current are the same; distinction is made only in their per unit values to allow independent selection of the per unit systems for modelling excitation systems and synchronous machines. This is illustrated in Figure 8.22.

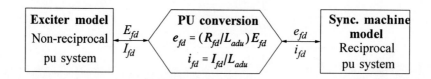

Figure 8.22 Per unit conversion at the interface between excitation system and synchronous machine field circuit

Under steady-state conditions, the per unit values of E_{fd} and I_{fd} are equal. During a transient condition, however, E_{fd} and I_{fd} differ; E_{fd} is determined by the excitation system and I_{fd} is determined by the dynamics of the field circuits.

A few interesting observations regarding the above per unit conversion are appropriate at this time:

1. The factor L_{adu}/R_{fd} in Equation 8.6 is the steady-state value of $G(s)$ (see Chapter 4, Equations 4.14 and 4.17), in the absence of saturation.

2. Equation 4.24 in terms of ΔE_{fd} becomes

$$\Delta \psi_d(s) = \frac{1+sT_{kd}}{(1+sT'_{d0})(1+sT''_{d0})} \Delta E_{fd} \tag{8.7A}$$
$$= \bar{G}(s)\Delta E_{fd}$$

The steady-state value of $\bar{G}(s)$ is equal to 1.0. There is thus a one to one relationship between ψ_d and E_{fd}.

With amortisseurs neglected, Equation 8.7A becomes

$$\Delta \psi_d(s) = \frac{1}{1+sT'_{do}} \Delta E_{fd} \qquad (8.7B)$$

3.   From the steady equations developed in Section 3.6.3, with the generator on open circuit, we have

$$\Delta \psi_d = \Delta e_q = \Delta E_t$$

Substitution in Equation 8.7B yields the following *open-circuit transfer function* of the generator:

$$\frac{\Delta E_t(s)}{\Delta E_{fd}(s)} = \frac{1}{1+sT'_{do}} \qquad (8.7C)$$

Example 8.1

The following are the parameters in per unit on machine rating of the 555 MVA, 0.9 p.f., 24 kV turbine generator considered in Examples 3.1, 3.2 and 3.3 of Chapter 3:

$L_{adu} = 1.66$	$L_{aqu} = 1.61$	$L_l = 0.15$
$L_{fd} = 0.165$	$R_{fd} = 0.0006$	$R_a = 0.003$

(a)   The field current required to generate rated stator voltage E_t on the air-gap line is 1300 A and the corresponding field voltage is 92.95 V. Determine the base values of E_{fd} and I_{fd} in the non-reciprocal per unit system and the base values of e_{fd} and i_{fd} in the reciprocal per unit system.

(b)   Compute the per unit values of E_{fd} and I_{fd}, when the generator is delivering rated MVA at rated power factor and terminal voltage. Assume that the corresponding values of the saturation factors K_{sd} and K_{sq} are equal to 0.835.

Solution

(a) By definition, the base values of E_{fd} and I_{fd} are respectively equal to the field voltage and field current required to produce rated air-gap line voltage. Hence,

$$E_{fd\ base} = 92.95 \text{ V}$$
$$I_{fd\ base} = 1300 \text{ A}$$

From Equations 8.5 and 8.6, the base values of e_{fd} and i_{fd} are

$$\begin{aligned} i_{fd\ base} &= L_{adu}I_{fd\ base} \\ &= 1.66 \times 1300 = 2158 \text{ A} \end{aligned}$$

$$e_{fd\ base} = (L_{adu}/R_{fd})E_{fd\ base}$$
$$= (1.66/0.0006)92.95 = 257.2 \text{ kV}$$

The above base values of e_{fd} and i_{fd}, as expected, agree with the values computed in Example 3.1.

(b) From the results of Example 3.2, at the rated output conditions,

$$e_{fd} = 0.000939 \text{ pu}$$
$$i_{fd} = 1.565 \text{ pu}$$

The corresponding per unit values of E_{fd} and I_{fd} are

$$E_{fd} = (L_{adu}/R_{fd})e_{fd}$$
$$= (1.66/0.0006)0.000939 = 2.598 \text{ pu}$$
$$I_{fd} = L_{adu}i_{fd}$$
$$= 1.66 \times 1.565 = 2.598 \text{ pu} \qquad \blacksquare$$

Specification of temperature

The base exciter output voltage depends on the synchronous machine field resistance, which in turn depends on the field temperature. The standard temperatures used for calculating the base exciter output voltage are 100°C for thermal units (operating temperature rise greater than 60°C) and 75°C for hydraulic units (operating temperature rise 60°C or less) [5]. However, care should be exercised in using these temperatures for modelling of excitation systems.

The value of the field resistance used should correspond to the resistance under the actual operating conditions being simulated, as closely as possible. The value of T'_{do} should be consistent with this value of field resistance.

The field resistance corrected to a specified operating temperature may be calculated as follows [31]:

$$R_s = R_t \left(\frac{t_s + k}{t_t + k} \right) \qquad (8.8)$$

where

t_s = specified operating temperature, °C
t_t = temperature corresponding to known or measured value of winding resistance, °C
R_s = winding resistance at temperature t_s
R_t = winding resistance at temperature t_t

The characteristic constant k depends on winding material. It is equal to 234.5 for pure copper and 225 for aluminium based on a volume conductivity of 62% pure copper [31].

Example 8.2

The per unit R_{fd} of a steam-turbine-driven generator is 0.00063 at 75°C and the corresponding value of T'_{do} is 7.6125 s. The base exciter output voltage at the standard temperature of 100°C is 105.575 V. If the generator is operating at a temperature of 60°C, find consistent values of R_{fd}, T'_{do}, and base E_{fd} at this temperature. Assume that the constant k of the field winding material is 234.5.

Solution

The field resistance at 60°C is

$$R_{fd\ 60} = 0.00063\frac{234.5+60}{234.5+75}$$
$$= 0.0006\ \text{pu}$$

The time constant T'_{do} is inversely proportional to R_{fd}. Hence at 60°C it is

$$T'_{do} = T'_{do}\frac{R_{fd\ 75}}{R_{fd\ 60}}$$

$$= 7.6125\times\frac{0.00063}{0.0006} = 8.0\ \ \text{s}$$

The base E_{fd} is directly proportional to R_{fd}. Hence at 60°C,

$$\text{Base } E_{fd\ 60} = \text{base } E_{fd\ 100}\left(\frac{234.5+60}{234.5+100}\right)$$

$$= 105.575\times0.8804 = 92.95\ \text{V} \qquad\blacksquare$$

8.6.2 Modelling of Excitation System Components

The basic elements which form different types of excitation systems are the dc exciters (self or separately excited); ac exciters; rectifiers (controlled or non-controlled); magnetic, rotating, or electronic amplifiers; excitation system stabilizing feedback circuits; signal sensing and processing circuits. We describe here models for these individual elements. In the next section we will consider modelling of complete excitation systems.

(a) Separately excited dc exciter

The circuit model of the exciter is shown in Figure 8.23.

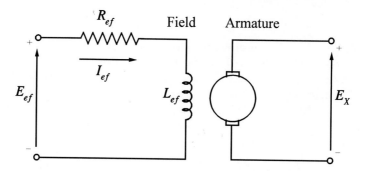

Figure 8.23 Separately excited dc exciter

For the exciter field circuit, we write

$$E_{ef} = R_{ef}I_{ef} + \frac{d\psi}{dt} \tag{8.9}$$

with

$$\psi = L_{ef}I_{ef}$$

Neglecting field leakage, the exciter output voltage E_X is given by

$$E_X = K_X\psi \tag{8.10}$$

where K_X depends on the speed and winding configuration of the exciter armature.

The output voltage E_X is a nonlinear function of the exciter field current I_{ef} due to magnetic saturation. The voltage E_X is also affected by the load on the exciter. The common practice [1,4] in dc exciter modelling is to account for saturation and load regulation approximately by combining the two effects and using the constant-resistance load-saturation curve, as shown in Figure 8.24.

The air-gap line is tangent to the lower linear portion of the open circuit saturation curve. Let R_g be the slope of the air-gap line and ΔI_{ef} denote the departure of the load saturation curve from the air-gap line. From Figure 8.24, we write

$$I_{ef} = \frac{E_X}{R_g} + \Delta I_{ef} \tag{8.11}$$

where ΔI_{ef} is a nonlinear function of E_X and may be expressed as

$$\Delta I_{ef} = E_X S_e(E_X) \tag{8.12}$$

where $S_e(E_X)$ is the *saturation function* dependent on E_X.

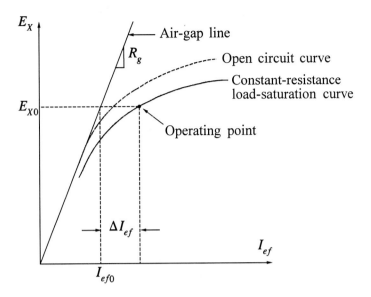

Figure 8.24 Exciter load-saturation curve

From Equations 8.9 to 8.12, we have

$$E_{ef} = \frac{R_{ef}}{R_g}E_X + R_{ef}S_e(E_X)E_X + \frac{1}{K_X}\frac{dE_X}{dt} \tag{8.13}$$

The above equation gives the relationship between the output E_X and the input voltage E_{ef}. A convenient per unit system for this equation is one with base values of E_X and I_{ef} chosen to be equal to those values required to give rated synchronous machine voltage on the air-gap line. Thus,

$$
\begin{aligned}
E_{Xbase} &= E_{fdbase} \\
I_{efbase} &= E_{fdbase}/R_g \\
R_{gbase} &= R_g
\end{aligned}
\tag{8.14}
$$

Dividing Equation 8.13 by E_{Xbase}, we have

$$\frac{E_{ef}}{E_{Xbase}} = \frac{R_{ef}}{R_g}\frac{E_X}{E_{Xbase}} + R_{ef}S_e(E_X)\frac{E_X}{E_{Xbase}} + \frac{1}{K_X}\frac{d}{dt}\left(\frac{E_X}{E_{Xbase}}\right)$$

In per unit form, we have

$$\bar{E}_{ef} = \frac{R_{ef}}{R_g}\bar{E}_X[1+\bar{S}_e(\bar{E}_X)]+\frac{1}{K_X}\frac{d\bar{E}_X}{dt} \tag{8.15}$$

In the above equation, $\bar{S}_e(\bar{E}_X)$ is the per unit saturation function defined as follows:

$$\bar{S}_e(\bar{E}_X) = \frac{\Delta\bar{I}_{ef}}{\bar{E}_X} = R_g S_e(E_X) \tag{8.16}$$

From Figure 8.25, with E_X and I_{ef} expressed in per unit, the per unit saturation function is given by [4]

$$\bar{S}_e(\bar{E}_X) = \frac{A-B}{B} \tag{8.17}$$

The parameter K_X defined by Equation 8.10 may be written as

$$K_X = \frac{E_X}{\psi} = \frac{E_X}{L_{ef}I_{ef}} = \frac{R_g\bar{E}_X}{L_{ef}\bar{I}_{ef}}$$

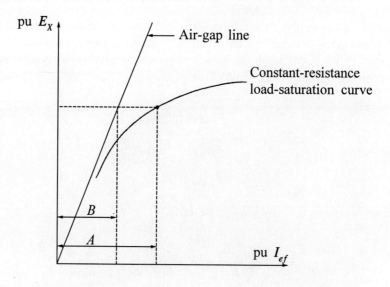

Figure 8.25 Exciter saturation characteristic

Corresponding to any given operating point (I_{ef0}, E_{X0}), let

$$L_{fu} = L_{ef}\frac{\overline{I}_{ef0}}{\overline{E}_{X0}} \tag{8.18}$$

Therefore, $K_X = R_g/L_{fu}$. Substituting in Equation 8.15, we get

$$\overline{E}_{ef} = K_E\overline{E}_X + S_E(\overline{E}_X)\overline{E}_X + T_E\frac{d\overline{E}_X}{dt} \tag{8.19}$$

where

$$K_E = \frac{R_{ef}}{R_g}$$

$$T_E = \frac{L_{fu}}{R_g} \tag{8.20}$$

$$S_E(\overline{E}_X) = \overline{S}_e(\overline{E}_X)\frac{R_{ef}}{R_g}$$

Equation 8.19 represents the input-output relationship of the exciter. For a separately excited exciter, the input voltage E_{ef} is the regulator output V_R. The output voltage E_X of a dc exciter is directly applied to the field of the synchronous machine. Therefore, the exciter may be represented in block diagram form as shown in Figure 8.26. *In the diagram, all variables are in per unit; however, the superbar notation denoting this has been dropped.*

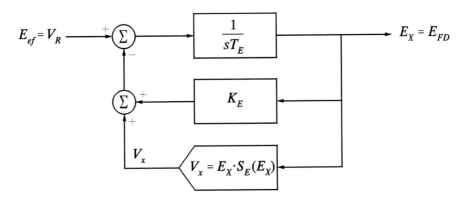

Commonly used representation: $V_X = A_{EX}e^{B_{EX}E_X}$

Figure 8.26 Block diagram of a dc exciter

The adjustment of field-circuit resistance R_{ef} affects K_E as well as the saturation function $S_E(E_X)$, but not the integration time T_E of the forward loop.

There are several convenient mathematical expressions that may be used to approximate the effect of exciter saturation. A commonly used expression is the exponential function

$$V_X = E_X S_E(E_X) = A_{EX} e^{B_{EX} E_X} \tag{8.21}$$

The block diagram of Figure 8.26 provides a convenient means of representing the dc exciter in stability studies. However, the effective gain and time constant of the exciter are not readily apparent from it. These are more evident when the block diagram is reduced to the standard form by considering small-signal response:

$$\Delta E_{ef} = \Delta V_R \longrightarrow \boxed{\dfrac{K}{1+sT}} \longrightarrow \Delta E_X = \Delta E_{FD}$$

For any operating point with $E_X = E_{FD} = E_{FD0}$, the effective gain K and time constant of the exciter for small perturbations are

$$K = \frac{1}{B_{EX} S_E(E_{FD0}) + K_E} \tag{8.22}$$

$$T = \frac{T_E}{B_{EX} S_E(E_{FD0}) + K_E} \tag{8.23}$$

where $S_E(E_{X0}) = A_{EX} e^{B_{EX} E_{FD0}}$.

(b) Self-excited dc exciter

Figure 8.27 shows a circuit model of a self-excited dc exciter.

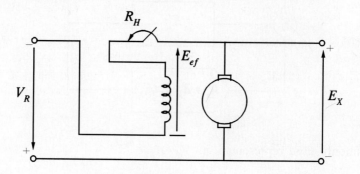

Figure 8.27 Self-excited dc exciter

For this exciter, the regulator output V_R is in series with the exciter shunt field. Therefore, the per unit voltage across the exciter field (without the explicit per unit notation) is

$$E_{ef} = V_R + E_X \tag{8.24}$$

The relationship between the per unit values of E_{ef} and E_X developed for the separately excited exciter also applies in this case. Substituting for E_{ef} given by Equation 8.24 in Equation 8.15, we have

$$V_R + E_X = \frac{R_{ef}}{R_g} E_X [1 + S_e(E_X)] + \frac{1}{K_X} \frac{dE_X}{dt}$$

This reduces to

$$V_R = K_E E_X + S_E(E_X) E_X + T_E \frac{dE_X}{dt} \tag{8.25}$$

where

$$K_E = \frac{R_{ef}}{R_g} - 1$$

$$T_E = \frac{L_{fu}}{R_g} \tag{8.26}$$

$$S_E = S_e(E_X) \frac{R_{ef}}{R_g}$$

The block diagram of Figure 8.26 also applies to the self-excited dc exciter. The value of K_E, however, is now equal to $R_{ef}/R_g - 1$ as compared to R_{ef}/R_g for the separately excited case.

The station operators usually track the voltage regulator by periodically adjusting the rheostat setpoint so as to make the voltage regulator output zero. This is accounted for by selecting the value of K_E so that the initial value of V_R is equal to zero. *The parameter K_E is therefore not fixed, but varies with the operating condition.*

(c) AC exciters and rectifiers

The ac exciter representation (excluding rectification) recommended in reference 8 for use in large-scale stability studies is shown in Figure 8.28.

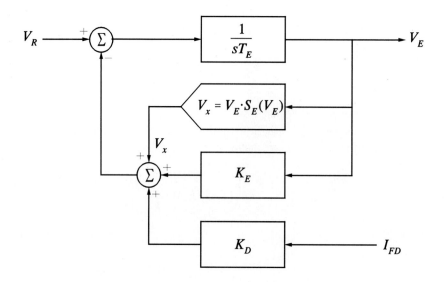

Figure 8.28 Block diagram of an ac exciter

The general structure of the model is similar to that of the dc exciter. However, in this case the load regulation due to the armature reaction effect is accounted for distinctly, and the no-load saturation curve is used to define the saturation function S_E. The exciter internal voltage V_E is the no-load voltage as determined by the saturation function. The main generator field current I_{FD} represents the exciter load current, and the negative feedback of $K_D I_{FD}$ accounts for the armature reaction demagnetizing effect. The constant K_D depends on the ac exciter synchronous and transient reactances [32]. Figure 8.29 illustrates the calculation of the saturation function S_E for a specified value of V_E.

The per unit saturation function is

$$S_E(V_E) = \frac{A-B}{B} \tag{8.27}$$

Any convenient mathematical expression can be used to represent the saturation function. As in the case of dc exciters, a commonly used expression for $V_X = V_E S_E(V_E)$ is the exponential function given by Equation 8.21.

Three-phase full-wave bridge rectifier circuits are commonly used to rectify the ac exciter output voltage. The effective ac source impedance seen by the rectifier is predominantly an inductive reactance. As described in Chapter 10, the effect of this inductive reactance (referred to as the *commutating reactance*) is to delay the process of commutation, i.e., transfer of current from one valve to another. This produces a

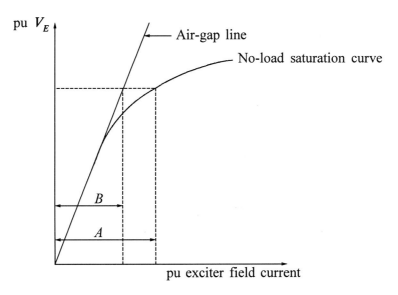

Figure 8.29 AC exciter saturation characteristic

decrease in the average output voltage of the rectifier as its load current increases. Reference 33 shows that a three-phase full-wave bridge rectifier circuit operates in one of three distinct modes as the rectifier load current varies from no load to the short circuit level. The mode of operation depends on the commutating voltage drop (equal to the product of commutating reactance and load current). ✓

The equations defining the rectifier regulation as a function of commutation voltage drop may be expressed as follows [8,34]:

$$E_{FD} = F_{EX}V_E \qquad (8.28)$$

where

$$F_{EX} = f(I_N) \qquad (8.29)$$

and

$$I_N = \frac{K_C I_{FD}}{V_E} \qquad (8.30)$$

The constant K_C depends on the commutating reactance. The expressions for the function $f(I_N)$ characterizing the three modes of rectifier circuit operation are

Mode 1: $\qquad f(I_N) = 1.0 - 0.577 I_N,\qquad$ if $I_N \leq 0.433$

Mode 2: $\qquad f(I_N) = \sqrt{0.75 - I_N^2},\qquad$ if $0.433 < I_N < 0.75$ $\qquad$ (8.31)

Mode 3: $\qquad f(I_N) = 1.732(1.0 - I_N),\qquad$ if $0.75 \leq I_N \leq 1.0$

Now I_N should not be greater than 1.0, but if for some reason it is, F_{EX} should be set to zero [8].

The rectifier regulation effects identified above may be depicted in the block diagram form as shown in Figure 8.30.

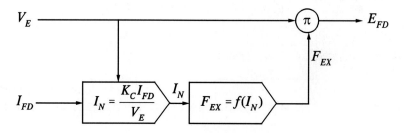

Figure 8.30 Rectifier regulation model

Referring to Figures 8.28 and 8.29, the exciter output voltage E_{FD} is simulated as the ac exciter internal voltage V_E reduced by the armature reaction ($I_{FD}K_D$) and rectifier regulation (F_{EX}).

(d) Amplifiers

Amplifiers may be the magnetic, rotating, or electronic type. Magnetic and electronic amplifiers are characterized by a gain and may also include a time constant. As such they may be represented by the block diagram of Figure 8.31.

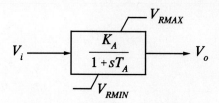

Figure 8.31 Amplifier model

The amplifier output is limited by saturation or power-supply limitations; this is represented by "non-windup" limits V_{RMAX} and V_{RMIN} in Figure 8.31. A description of such limits is provided later in this section.

The output limits of some amplifiers having power supplies from generator or auxiliary bus voltage vary with generator terminal voltage. In such cases, V_{RMAX} and V_{RMIN} vary directly with generator terminal voltage E_t.

The transfer function of an amplidyne is derived in reference 12 and has the general form shown in Figure 8.32.

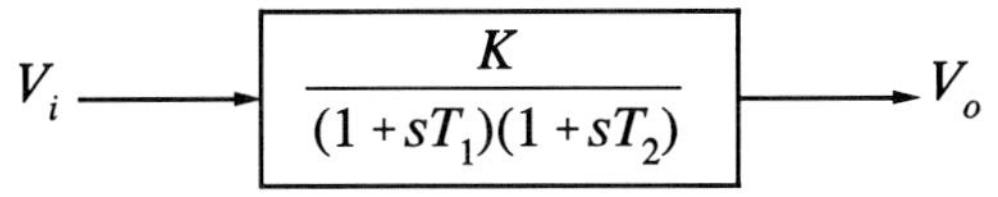

$$V_i \longrightarrow \boxed{\dfrac{K}{(1+sT_1)(1+sT_2)}} \longrightarrow V_o$$

Figure 8.32 Amplidyne model

(e) Excitation system stabilizing circuit

There are several ways of physically realizing the stabilizing function identified in Figure 8.15. Some excitation systems use series transformers as shown in Figure 8.33.

The transformer equations in Laplace notation are

$$V_1 = R_1 i_1 + sL_1 i_1 + sMi_2$$
$$V_2 = R_2 i_2 + sL_2 i_2 + sMi_1 \tag{8.32}$$

where subscripts 1 and 2 denote primary and secondary quantities; R, L, and M denote resistance, leakage inductance and mutual inductance, respectively.

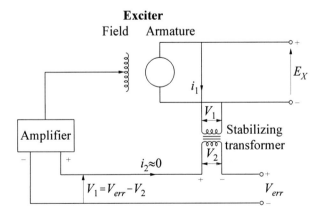

Figure 8.33 Excitation system stabilizing transformer

The secondary of the transformer is connected to a high impedance circuit. Therefore, neglecting i_2, we have

$$V_1 = (R_1 + sL_1)i_1$$

$$V_2 = sMi_1 \tag{8.33}$$

Thus

$$\frac{V_1}{V_2} = \frac{sM}{R_1 + sL_1}$$

$$= \frac{sK_F}{1 + sT_F} \tag{8.34}$$

where $K_F = M/R$ and $T_F = L_1/R$.

(f) Windup and non-windup limits

In the modelling of excitation systems, it is necessary to distinguish between windup and non-windup limits. Such limits are encountered with integrator blocks, single time constant blocks, and lead-lag blocks.

Figures 8.34(a) and (b) show the differences between the two types of limits when applied to an integrator block.

Representation:

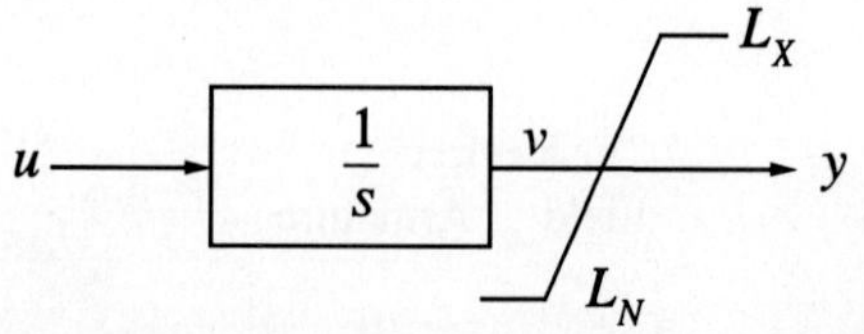

System equation: $\dfrac{dv}{dt} = u$

Limiting action:

If $L_N < v < L_X$, then $y = v$

If $v \geq L_X$,   then $y = L_X$

If $v \leq L_N$,   then $y = L_N$

Figure 8.34 (a) Integrator with windup limits

Representation:

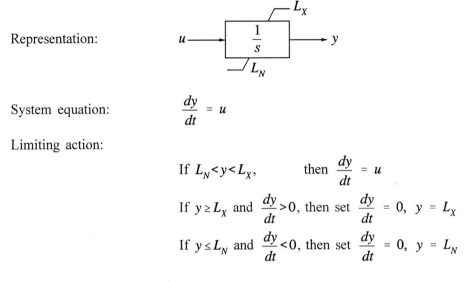

System equation:

$$\frac{dy}{dt} = u$$

Limiting action:

If $L_N < y < L_X$,   then $\frac{dy}{dt} = u$

If $y \geq L_X$ and $\frac{dy}{dt} > 0$, then set $\frac{dy}{dt} = 0,\ y = L_X$

If $y \leq L_N$ and $\frac{dy}{dt} < 0$, then set $\frac{dy}{dt} = 0,\ y = L_N$

Figure 8.34 (b) Integrator with non-windup limits

With windup limits the variable v is not limited. Therefore, the output variable y cannot come off a limit until v comes within the limit. With non-windup limits, the output variable y is limited; it comes off the limit as soon as the input u changes sign.

Figures 8.35(a) and (b) show the difference between the two types of limits when applied to a single time constant block. The significance of the two types of limits is similar to that for an integrator. With a windup limit, the output y cannot come off a limit until v comes within the limit. With a non-windup limit, however, the output y comes off the limit as soon as the input u re-enters the range within limits.

Representation:

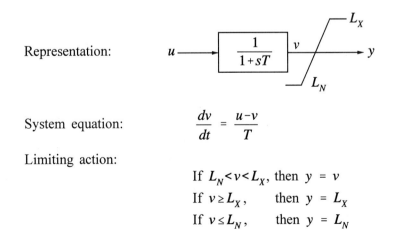

System equation:

$$\frac{dv}{dt} = \frac{u-v}{T}$$

Limiting action:

If $L_N < v < L_X$, then $y = v$

If $v \geq L_X$,   then $y = L_X$

If $v \leq L_N$,   then $y = L_N$

Figure 8.35 (a) Single time constant block with windup limits

Representation:

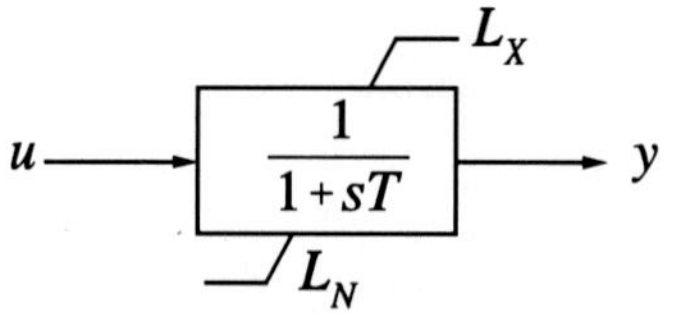

$$\frac{1}{1+sT}$$

$u \longrightarrow \boxed{\dfrac{1}{1+sT}} \longrightarrow y$ with limits L_X and L_N

System equation:

$$f = \frac{u-y}{T}$$

Limiting action:

If $L_N < y < L_X$, then $\dfrac{dy}{dt} = f$

If $y \geq L_X$ and $f > 0$, then set $\dfrac{dy}{dt} = 0,\ y = L_X$

If $y \leq L_N$ and $f < 0$, then set $\dfrac{dy}{dt} = 0,\ y = L_N$

Figure 8.35 (b) Single time constant block with non-windup limits

Representation:

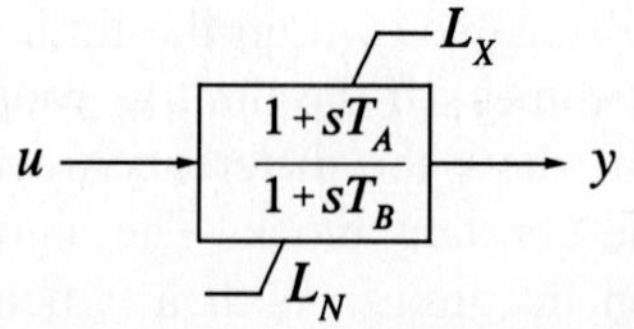

$u \longrightarrow \boxed{\dfrac{1+sT_A}{1+sT_B}} \longrightarrow y$ with limits L_X and L_N $T_A < T_B$

Physical realization:

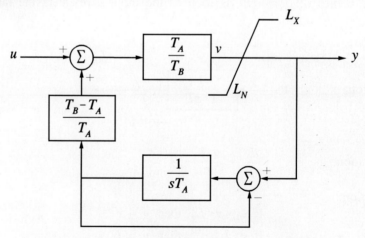

Figure 8.36 Lead-lag function with non-windup limits
(Continued on next page)

Limiting action:

$$\text{If } L_N \leq v \leq L_X, \text{ then } y = v$$
$$\text{If } v > L_X, \qquad \text{then } y = L_X$$
$$\text{If } v < L_N, \qquad \text{then } y = L_N$$

Figure 8.36 (*Continued*) Lead-lag function with non-windup limits

With a lead-lag block, the interpretation of the action of a windup limit is straightforward and is similar to that of a single time constant block. However, the way in which a non-windup limit can be realized is not unique; the interpretation of the limiting action should therefore be based on the physical device represented by the block. Figure 8.36 illustrates such limiting action associated with electronic implementation of lead-lag functions.

(g) Gating functions

Gating or auctioneering circuits are used when it is required to give control to one of two input signals, depending on their relative size with respect to each other. Figure 8.37 illustrates the functions of a low value (LV) gate and a high value (HV) gate, and the symbols used to represent them in block diagrams.

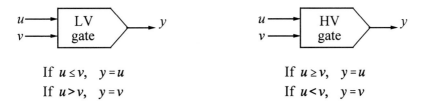

$$\text{If } u \leq v, \quad y = u \qquad\qquad\qquad \text{If } u \geq v, \quad y = u$$
$$\text{If } u > v, \quad y = v \qquad\qquad\qquad \text{If } u < v, \quad y = v$$

Figure 8.37 Low- and high-value gating functions

(h) Terminal voltage transducer and load compensator [8]

The block diagram representation of these elements is shown in Figure 8.38. The time constant T_R represents rectification and filtering of the synchronous machine terminal voltage. The parameters of the load compensator (described in Section 8.5.4) are R_C and X_C. The input variables E_t and I_t are in phasor form. When load compensation is not used, R_C and X_C are set to zero.

The voltage transducer output V_C forms the principal control signal to the excitation system. If a load compensator is not used and T_R is negligible, $V_C = E_t$.

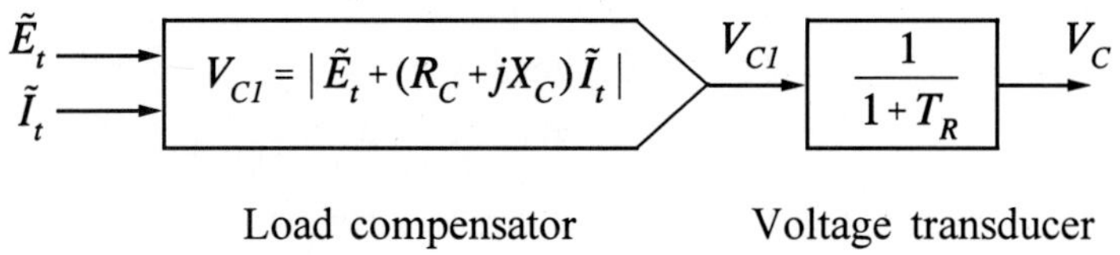

$$\tilde{E}_t \longrightarrow \boxed{V_{CI} = |\tilde{E}_t + (R_C + jX_C)\tilde{I}_t|} \xrightarrow{V_{CI}} \boxed{\frac{1}{1+T_R}} \xrightarrow{V_C}$$
$$\tilde{I}_t \longrightarrow$$

Load compensator   Voltage transducer

Figure 8.38 Terminal voltage transducer and load compensator model

8.6.3 Modelling of Complete Excitation Systems

Figure 8.39 depicts the general structure of a detailed excitation system model having a one-to-one correspondence with the physical equipment. While this model structure has the advantage of retaining a direct relationship between model parameters and physical parameters, such detail is considered too great for general system studies. Therefore, model reduction techniques are used to simplify and obtain a practical model appropriate for the type of study for which it is intended.

The parameters of the reduced model are selected such that the gain and phase characteristics of the reduced model match those of the detailed model over the frequency range of 0 to 3 Hz. In addition, all significant nonlinearities that impact on system stability are accounted for. With a reduced model, however, direct correspondence between the model parameters and the actual system parameters is generally lost.

The appropriate structure for the reduced model depends on the type of excitation system. The IEEE has standardized 12 model structures in block diagram form for representing the wide variety of excitation systems currently in use [8].

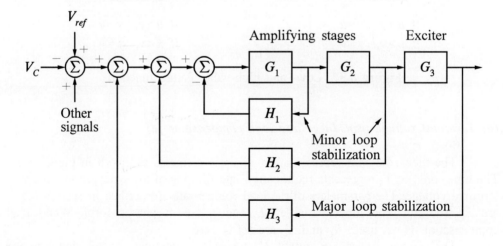

Figure 8.39 Structure of a detailed excitation system model

These are intended for use in *transient stability and small-signal stability studies*. For purposes of illustration, we will consider five of these models. These are shown in Figures 8.40 to 8.44, with slight modification of the block diagram conventions to conform to the recommendations made in reference 35. The figures include a brief description of the key features and sample data. The suffix "A" accompanying the designations is for the purpose of differentiating these models developed in 1992 from similar models developed previously in 1981 [4].

The principal input signal to each of the excitation systems is the output V_C of the voltage transducer shown in Figure 8.38. At the first summing point, the signal V_C is subtracted from the voltage regulator reference V_{ref} and the output V_S of the power system stabilizer, if used, is added to produce the actuating signal which controls the excitation system. Additional signals, such as the underexcitation limiter output (V_{UEL}), come into play only during extreme or unusual conditions. Under steady state, $V_S=0$ and V_{ref} takes on a value unique to the synchronous machine loading condition so that the error signal results in the required field voltage E_{fd}. This is illustrated in Example 8.3.

1. Type DC1A exciter model

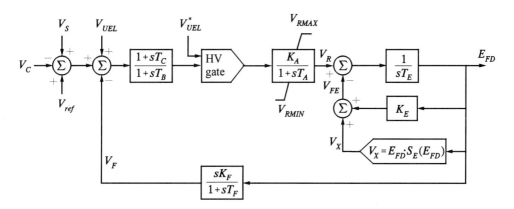

* Alternate input point

Figure 8.40 IEEE type DC1A excitation system model. © IEEE 1992 [8]

The type DC1A exciter model represents field-controlled dc commutator exciters, with continuously acting voltage regulators. The exciter may be separately excited or self-excited, the latter type being more common. When self-excited, K_E is selected so that initially $V_R=0$, representing operator action of tracking the voltage regulator by periodically trimming the shunt field rheostat setpoint.

Sample data

Self-excited dc exciter:

K_A=187 T_A=0.89 T_E=1.15 A_{EX}=0.014 B_{EX}=1.55
K_F=0.058 T_F=0.62 T_B=0.06 T_C=0.173 T_R=0.05
V_{RMAX}=1.7 V_{RMIN}=-1.7

K_E is computed so that initially V_R=0, and the load compensator is not used.

2. Type AC1A exciter model

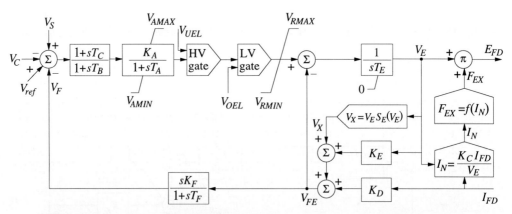

Figure 8.41 IEEE type AC1A excitation system model. © IEEE 1992 [8]

The type AC1A exciter model represents a field-controlled alternator excitation system with non-controlled rectifiers, and is applicable to brushless excitation systems. The diode rectifier characteristic imposes a lower limit of zero on the exciter output voltage. The exciter field is supplied by a pilot exciter, and the voltage regulator power supply is not affected by external transients.

Sample data

Exciter and regulator:

K_A=400.0 T_A=0.02 T_B=0 T_C=0 K_F=0.03
T_F=1.0 K_E=1.0 T_E=0.8 K_D=0.38 K_C=0.2
V_{RMAX}=7.3 V_{RMIN}=-6.6 V_{AMAX}=15.0 V_{AMIN}=-15.0 A_{EX}=0.1
B_{EX}=0.03

The load compensator is not used, and T_R is negligible.

3. Type AC4A exciter model

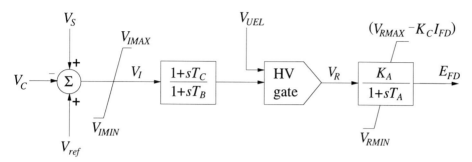

Figure 8.42 IEEE type AC4A excitation system model. © IEEE 1992 [8]

The type AC4A exciter model represents an alternator-supplied controlled-rectifier excitation system – a high initial-response excitation system utilizing a full-wave thyristor bridge circuit. Excitation system stabilization is usually provided in the form of a series lag-lead network (transient gain reduction). The time constant associated with the regulator and firing of thyristors is represented by T_A. The overall gain is represented by K_A. The rectifier operation is confined to mode 1 region. Rectifier regulation effects on exciter output limits are accounted for by constant K_C.

Sample data

Exciter and regulator:

$$K_A=200.0 \quad T_A=0.04 \quad T_C=1.0 \quad T_B=12.0 \quad V_{RMAX}=5.64$$
$$V_{RMIN}=-4.53 \quad K_C=0 \quad V_{IMAX}=1.0 \quad V_{IMIN}=-1.0$$

4. Type ST1A exciter model

The type ST1A exciter model represents a potential-source controlled-rectifier system. The excitation power is supplied through a transformer from generator terminals; therefore, the exciter ceiling voltage is directly proportional to the generator terminal voltage. The effect of rectifier regulation on ceiling voltage is represented by K_C. The model provides flexibility to represent series lag-lead or rate feedback stabilization. Because of the very high field-forcing capability of the system, a field-current limiter is sometimes employed; the limit is defined by I_{LR} and the gain by K_{LR}.

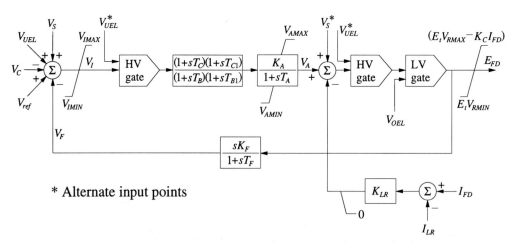

Figure 8.43 IEEE type ST1A excitation system model. © IEEE 1992 [8]

Sample data

Exciter and regulator:

K_A=200.0 T_A=0 V_{RMAX}=7.0 V_{RMIN}=−6.4 K_C=0.04

K_{LR}=4.54 I_{LR}=4.4 T_B, T_C, T_{B1}, T_{C1}, K_F, T_F not used

V_{IMAX}, V_{IMIN}, V_{AMAX}, V_{AMIN} are not represented.

Voltage transducer and load compensator:

T_R=0.015 R_C=0 X_C=0

5. Type ST2A exciter model

The type ST2A exciter model represents a compound-source rectifier excitation system. The exciter power source is formed by phasor combination of main generator armature voltage and current. The regulator controls the exciter output through controlled saturation of the power transformer. The parameter T_E represents the integration rate associated with the control windings; E_{FDMAX} represents the limit on exciter output due to magnetic saturation.

Sample data

K_A=120.0 T_A=0.15 K_E=1.0 T_E=0.5 K_C=0.65

K_F=0.02 T_F=0.56 V_{RMAX}=1.2 V_{RMIN}=−1.2 E_{FDMAX}=3.55

K_P=1.19 K_I=1.62

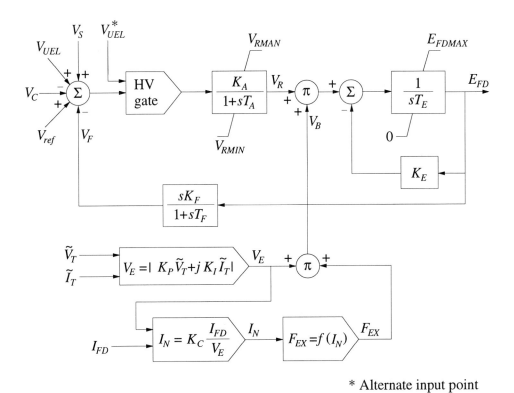

* Alternate input point

Figure 8.44 IEEE type ST2A excitation system model. © IEEE 1992 [8]

Example 8.3

A generator is operating under steady-state conditions with an E_{fd} of 2.598 pu and E_t=1.0 pu.

(a)   If it is equipped with a type AC4A excitation system represented by the block diagram of Figure 8.42, determine the value of V_{ref}.

(b)   If it is equipped with a self-excited dc excitation system represented by the block diagram of Figure 8.40, determine the values of K_E and V_{ref}.

The parameters of the excitation systems are the same as for the sample data provided with the figures.

Solution

(a) *Type AC4A excitation system of Figure 8.42*

When E_{fd} is 2.598 pu, we have

$$V_R = E_{fd}/K_A$$

$$= \frac{2.598}{200.0} = 0.013 \quad \text{pu}$$

Since a load compensator is not used,

$$V_C = E_t = 1.0 \text{ pu}$$

Under steady-state operation, $V_I = V_R$ and $V_S = 0$. Since the generator is operating under normal conditions, $V_{UEL} = 0$. Therefore, from Figure 8.42 we see that

$$V_{ref} = V_C + V_I$$

$$= 1.0 + 0.013 = 1.013 \quad \text{pu}$$

(b) Self-excited dc exciter of Figure 8.40

In this case, K_E takes a value such that $V_R = 0$. With E_{fd} at a steady-state value and $V_R = 0$, $V_{FE} = 0$. Hence,

$$K_E E_{fd} = -V_X$$

$$= -A_{EX} e^{B_{EX} E_{fd}} = -0.014 e^{1.55 \times 2.598}$$

$$= -0.7852$$

Therefore,

$$K_E = -\frac{0.7852}{2.598}$$

$$= -0.3022$$

Under steady-state normal operation, $V_F = 0$ and $V_S = 0$. With $V_R = 0$,

$$V_{ref} = V_C = E_t = 1.0 \text{ pu} \qquad \blacksquare$$

Modelling of limiters

The standard models shown in Figures 8.40 to 8.44 do not include representation of limiting circuits, namely, the underexcitation limiter, V/Hz limiter, and maximum excitation limiter. These circuits do not come into play under normal conditions and are not usually modelled in transient and small-signal stability studies. They may, however, be important for long-term stability and voltage stability studies.

The actual implementation of these limiting functions varies widely depending on the manufacturer, the vintage of the equipment, and the requirements specified by the utility. Therefore, models for these circuits have to be established on a case by case basis. Here we will illustrate how such devices are modelled by considering specific examples.

(a) Underexcitation limiter:

Figure 8.45 shows the model of an underexcitation limiter used in conjunction with a type ST1A (static) exciter. The parameters K_C and K_R determine the characteristics of the limiter on the *P-Q* plane. The output V_{UEL} of the limiter is applied to the HV gate of the exciter model of Figure 8.43.

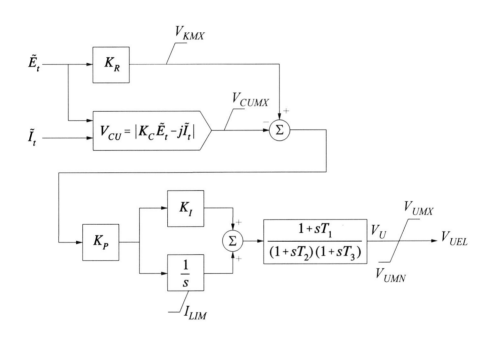

Sample data:

K_P = 0.015 K_I = 10.0 T_1 = 6.4 s T_2 = 0.8 s
T_3 = 0.64 s V_{KMX} = 4.0 V_{CUMX} = 4.0
V_{UMX} = 0.2 V_{UMN} = -0.2 I_{LIM} = -0.012
K_R = Radius of UEL characteristic
K_C = Centre of UEL characteristic

Figure 8.45 An example of UEL model [26]

(b) V/Hz limiter:

An example of the V/Hz limiter model is shown in Figure 8.46. The operation of the limiter is quite straightforward. When the per unit V/Hz value exceeds the limiting value of V_{ZLM}, a strong negative signal drives the excitation down. The V_{ZLM} limit is set typically at 1.07 to 1.09 pu.

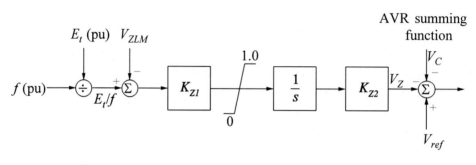

Sample data:

$$V_{ZLM} = 1.07 \qquad K_{ZI} = 1000 \qquad K_{Z2} = 0.007$$

Figure 8.46 V/Hz limiter model

(c) Field-current or overexcitation limiter:

Figure 8.47 shows the model of a field-current limiter circuit. It is designed to have a limiting action as shown in Figure 8.47(b). A high setting provides almost instantaneous limiting at 1.6 times full-load current (FLC). A low setting of 1.05×FLC in conjunction with a ramp timing function provides a limiting action with time delay dependent on the level of field current. For example, a field current level of 1.325×FLC will be allowed for 15 s, followed by a reduction in current level to 1.05×FLC over the next 15 s.

Referring to the block diagram of Figure 8.47(a), when I_{fd} exceeds the high setting I_{FLM1}, signal V_{F1} of control loop ① acts to reduce excitation instantaneously. When the field current is below I_{FLM1}, the limiting action is through the control loop ②. The magnitude of the control signal V_{F2} and the value of gain K_2 determine the time delay and ramping action. Once the field current reaches the low setting I_{FLM2}, the select switch changes to the low select position; this ensures that the field current, in the event of a second disturbance, does not exceed the low setting for a minimum period to allow cooling of the machine. When the field current is below I_{FLM2}, the signal V_{F4} helps to reset rapidly the integrator output to zero.

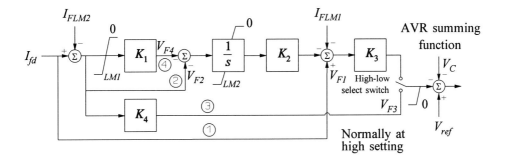

(a) Block diagram representation

Sample data:

I_{FLM1} = 1.6×full load I_{fd} I_{FLM2} = 1.05×full load I_{fd}
K_1 = 150 K_2 = 0.248 K_3 = 12.6
K_4 = 140 $LM1$ = −0.085 $LM2$ = −3.85

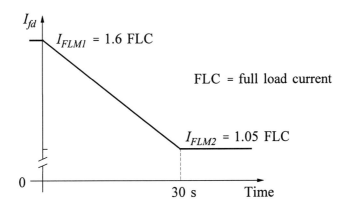

(b) Limiting characteristics

Figure 8.47 Field-current limiter model

8.6.4 Field Testing for Model Development and Verification

Although most of the data related to excitation system models can be obtained from factory tests, such data can only be considered as typical. The actual settings are usually determined on site during installation and commissioning of the equipment. It is therefore desirable to determine the model parameters by performing tests on the actual equipment on site.

The nature of the tests required will depend on the type of excitation system being tested. A general procedure for model verification and development is as follows:

1.   Obtain circuit diagrams, block diagrams, nominal settings and setting ranges. Construct a detailed block diagram of the complete excitation system, identifying gains, time constants, and non-linearities.

2.   With the generator (synchronous machine) shut down, perform frequency response tests and/or transient response tests on the individual elements comprising the excitation system. Identify their transfer functions, non-linearities, saturation characteristics, and ceiling limits. Using these data, validate as much of the detailed block diagram of the excitation system as possible and modify the diagram as necessary.

3.   Perform frequency response and time-response tests with the generator running at rated speed and producing rated voltage on open-circuit. Measure the overall linear response and responses at various points of the system to a step change in terminal voltage. Perform additional tests with the generator operating near rated load. Validate the detailed model of the complete system by comparison with the measured responses.

4.   Reduce the detailed model to fit the standard model applicable to the specific type of excitation system using techniques of classical control theory. Validate the responses of the reduced model against the measured responses.

The above procedure is quite involved and time-consuming for the older slow-response exciter. For a high-response excitation system, frequency response tests are usually not required as the number of time constants within the major feedback loop is small. For static excitation systems, model parameters can usually be obtained from design data; field tests are required only for verification.

Techniques for field testing, performance verification and model development of excitation systems are described in references 3, 7, and 36.

REFERENCES

[1] IEEE Committee Report, "Computer Representation of Excitation Systems," *IEEE Trans.*, Vol. PAS-87, pp. 1460-1464, June 1968.

[2] IEEE Committee Report, "Proposed Excitation System Definitions for Synchronous Machines," *IEEE Trans.*, Vol. PAS-88, pp. 1248-1258, August 1969.

[3] IEEE Committee Report, "Excitation System Dynamic Characteristics," *IEEE Trans.*, Vol. PAS-92, pp. 64-75, January/February 1973.

[4] IEEE Committee Report, "Excitation System Models for Power System Stability Studies," *IEEE Trans.*, Vol. PAS-100, pp. 494-509, February 1981.

[5] *IEEE Standard Definitions for Excitation Systems for Synchronous Machines*, IEEE Standard 421.1-1986.

[6] *IEEE Guide for the Preparation of Excitation System Specifications*, IEEE Standard 421.4-1987.

[7] *IEEE Guide for Identification, Testing and Evaluation of the Dynamic Performance of Excitation Control Systems*, IEEE Standard 421.2-1990 (revision to IEEE Standard 421A-1978).

[8] *IEEE Recommended Practice for Excitation System Models for Power System Stability Studies*, IEEE Standard 421.5-1992.

[9] *Guide to the Characteristics, Performance, and Hardware Requirements in the Specification of Excitation Systems*, Report of the Governor and Excitation Control System Committee of the Canadian Electrical Association, 1982.

[10] Westinghouse Electric Corporation, *Electric Transmission and Distribution Reference Book*, East Pittsburgh, Pa., 1964.

[11] E.W. Kimbark, *Power System Stability, Vol. III: Synchronous Machines*, John Wiley & Sons, 1956.

[12] A.E. Fitzgerald and C. Kingsley, *Electric Machinery*, Second Edition, McGraw-Hill, 1961.

[13] T.L. Dillman, J.W. Skooglund, F.W. Keay, W.H. South, and C. Raczkowski, "A High Initial Response Brushless Excitation System," *IEEE Trans.*, Vol.

PAS-90, pp. 2089-2094, September/October 1971.

[14]   J.S. Bishop, D.H. Miller, and A.C. Shartrand, "Experience with ALTERREX Excitation for Large Turbine-Generators," Paper presented at the Joint IEEE/ASME Power Conference, Miami Beach, September 15-19, 1974.

[15]   R.K. Gerlitz, R.E. Gorman, and M. Temoshok, "The GENERREX Excitation System for Large Steam Turbine-Generators," Paper GE3-3003, Pacific Coast Electric Association Engineering and Operation Conference, Culver City, Calif., March 20-21, 1975.

[16]   P.H. Beagles, K. Carlsen, M.L. Crenshaw, and M. Temoshok, "Generator and Power System Performance with the GENERREX Excitation System," *IEEE Trans.,* Vol. PAS-95, pp. 489-493, March/April 1976.

[17]   H.M. Rustebakke (editor), *Electric Utility Systems and Practices,* John Wiley & Sons, 1983.

[18]   F. Peneder and R. Bertschi, "Static Excitation Systems with and without a Compounding Ancillary," *Brown Boveri Review* 7-85, pp. 343-348.

[19]   D.C. Lee and P. Kundur, "Advanced Excitation Controls for Power System Stability Enhancement," CIGRE Paper 38-01, Paris, France, 1986.

[20]   *IEEE Standard Dictionary of Electrical and Electronics Terms* (ANSI), IEEE Standard 100-1988.

[21]   P. Kundur, M. Klein, G.J. Rogers, and M.S. Zywno, "Application of Power System Stabilizers for Enhancement of Overall System Stability," *IEEE Trans.,* Vol. PWRS-4, No. 2, pp. 614-626, May 1989.

[22]   A.S. Rubenstein and W.W. Walkley, "Control of Reactive kVA with Modern Amplidyne Voltage Regulators," *AIEE Trans.,* Part III, pp. 961-970, December 1957.

[23]   A.S. Rubenstein and M. Temoshok, "Underexcited Reactive Ampere Limit for Modern Amplidyne Voltage Regulator," *AIEE Trans.,* Vol. PAS-73, pp. 1433-1438, December 1954.

[24]   J.T. Carleton, P.O. Bobo, and D.A. Burt, "Minimum Excitation Limit for Magnetic Amplifier Regulating System," *AIEE Trans.,* Vol. PAS-73, pp. 869-874, August 1954.

[25]   I. Nagy, "Analysis of Minimum Excitation Limits of Synchronous Machines,"

IEEE Trans., Vol. PAS-89, No. 6, pp. 1001-1008, July/August 1970.

[26]   J.R. Ribeiro, "Minimum Excitation Limiter Effects on Generator Response to System Disturbances," *IEEE Trans. on Energy Conversion,* Vol. 6, No. 1, pp. 29-38, March 1991.

[27]   Discussion of reference 26 by D.C. Lee, R.E. Beaulieu, P. Kundur, and G.J. Rogers.

[28]   M.S. Baldwin and D.P. McFadden, "Power Systems Performance as Affected by Turbine-Generator Controls Response during Frequency Disturbances," *IEEE Trans.,* Vol. PAS-100, pp. 2486-2494, May 1981.

[29]   General Electric Instructions GEK-15014C, *Inverse Time Maximum Excitation Limit.*

[30]   P. Kundur and P.L. Dandeno, "Implementation of Synchronous Machine Models into Power System Stability Programs," *IEEE Trans.,* Vol. PAS-102, pp. 2047-2054, July 1983.

[31]   *IEEE Guide: Test Procedures for Synchronous Machines,* IEEE Standard 115-1983.

[32]   R.W. Ferguson, H. Herbst, and R.W. Miller, "Analytical Studies of the Brushless Excitation System," *AIEE Trans.,* Part III, pp. 961-970, 1957.

[33]   R.L. Witzke, J.V. Kresser, and J.K. Dillard, "Influence of AC Reactance on Voltage Regulation of 6-Phase Rectifiers," *AIEE Trans.,* Vol. 72, pp. 244-253, July 1953.

[34]   L.L. Freris, "Analysis of a Hybrid Bridge Rectifier," *Direct Current,* pp. 22-23, February 1966.

[35]   IEEE Task Force, "Conventions for Block Diagram Representations," *IEEE Trans.,* Vol. PWRS-1, No. 3, pp. 95-100, August 1986.

[36]   IEEE Tutorial Course Text, "Power System Stabilization via Excitation Control - Chapter IV: Field Testing Techniques," Publication 81 EHO 175-0 PWR.

Chapter 9

Prime Movers and
Energy Supply Systems

The prime sources of electrical energy supplied by utilities are the kinetic energy of water and the thermal energy derived from fossil fuels and nuclear fission. The prime movers convert these sources of energy into mechanical energy that is, in turn, converted to electrical energy by synchronous generators. The prime mover governing systems provide a means of controlling power and frequency, a function commonly referred to as load-frequency control or automatic generation control (AGC). Figure 9.1 portrays the functional relationship between the basic elements associated with power generation and control.

This chapter examines the characteristics of prime movers and energy supply systems and develops appropriate models suitable for their representation in power system dynamic studies. The principles and implementation of automatic generation control are described in Chapter 11.

The focus here is on those characteristics of the power plants that impact on the overall performance of the power system, and not on a detailed study of the associated processes.

9.1 HYDRAULIC TURBINES AND GOVERNING SYSTEMS

Hydraulic turbines are of two basic types: impulse turbines and reaction turbines. They are briefly described here; the reader may refer to reference 1 for a more detailed description.

The *impulse-type turbine* (also known as Pelton wheel) is used for high heads - 300 metres or more. The runner is at atmospheric pressure, and the whole of the

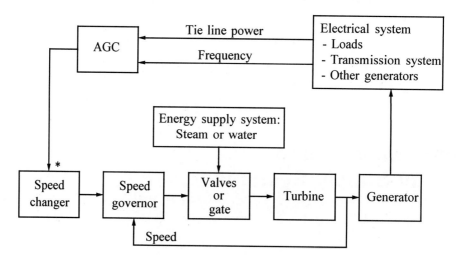

* AGC applied only to selected units

Figure 9.1 Functional block diagram of power generation and control system

pressure drop takes place in stationary nozzles that convert potential energy to kinetic energy. The high-velocity jets of water impinge on spoon-shaped buckets on the runner, which deflect the water axially through about 160°; the change in momentum provides the torque to drive the runner, the energy supplied being entirely kinetic.

In a *reaction turbine* the pressure within the turbine is above atmospheric; the energy is supplied by the water in both kinetic and potential (pressure head) forms. The water first passes from a spiral casing through stationary radial guide vanes and gates around its entire periphery. The gates control water flow. There are two subcategories of reaction turbines: Francis and propeller.

The *Francis turbine* is used for heads up to 360 metres. In this type of turbine, water flows through guide vanes impacting on the runner tangentially and exiting axially.

The *propeller turbine*, as the name implies, uses propeller-type wheels. It is for use on low heads - up to 45 metres. Either fixed blades or variable-pitch blades may be used. The variable-pitch blade propeller turbine, commonly known as the *Kaplan wheel*, has high efficiency at all loads.

The performance of a hydraulic turbine is influenced by the characteristics of the water column feeding the turbine; these include the effects of water inertia, water compressibility, and pipe wall elasticity in the penstock. The effect of water inertia is to cause changes in turbine flow to lag behind changes in turbine gate opening. The effect of elasticity is to cause travelling waves of pressure and flow in the pipe; this phenomenon is commonly referred to as "water hammer." For a detailed analysis of hydraulic transients the reader may refer to references 2 and 3.

Precise modelling of hydraulic turbines requires inclusion of transmission-line-like reflections which occur in the elastic-walled pipe carrying compressible fluid. Typically, the speed of propagation of such travelling waves is about 1200 metres per second. Therefore, travelling wave models may be required only if penstocks are long.

In what follows, we will first develop models of the hydraulic turbine and penstock system without travelling wave effects and assuming that there is no surge tank. We will then identify special governing requirements of hydraulic turbines. Finally, we will extend the model to include the effects of water hammer and surge tank.

9.1.1 Hydraulic Turbine Transfer Function [4-9]

The representation of the hydraulic turbine and water column in stability studies is usually based on the following assumptions:

1.   The hydraulic resistance is negligible.

2.   The penstock pipe is inelastic and the water is incompressible.

3.   The velocity of the water varies directly with the gate opening and with the square root of the net head.

4.   The turbine output power is proportional to the product of head and volume flow.

The essential elements of the hydraulic plant are depicted in Figure 9.2.

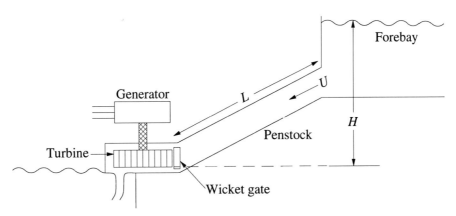

Figure 9.2 Schematic of a hydroelectric plant

The turbine and penstock characteristics are determined by three basic equations relating to the following:

(a)   Velocity of water in the penstock
(b)   Turbine mechanical power
(c)   Acceleration of water column

The *velocity of the water* in the penstock is given by

$$U = K_u G \sqrt{H} \qquad (9.1)$$

where

U = water velocity
G = gate position
H = hydraulic head at gate
K_u = a constant of proportionality

For small displacements about an operating point,

$$\Delta U = \frac{\partial U}{\partial H} \Delta H + \frac{\partial U}{\partial G} \Delta G$$

Substituting the appropriate expressions for the partial derivatives and dividing through by $U_0 = K_u G_0 \sqrt{H_0}$ yields

$$\frac{\Delta U}{U_0} = \frac{\Delta H}{2H_0} + \frac{\Delta G}{G_0}$$

or

$$\Delta \bar{U} = \frac{1}{2} \Delta \bar{H} + \Delta \bar{G} \qquad (9.2)$$

where the subscript 0 denotes initial steady-state values, the prefix Δ denotes small deviations, and the superbar "−" indicates normalized values based on *steady-state operating values*.

The *turbine mechanical power* is proportional to the product of pressure and flow; hence,

$$P_m = K_p H U \qquad (9.3)$$

Linearizing by considering small displacements, and normalizing by dividing both sides by $P_{m0} = K_p H_0 U_0$, we have

$$\frac{\Delta P_m}{P_{m0}} = \frac{\Delta H}{H_0} + \frac{\Delta U}{U_0}$$

or

$$\Delta \bar{P}_m = \Delta \bar{H} + \Delta \bar{U} \qquad (9.4)$$

Substituting for $\Delta \bar{U}$ from Equation 9.2 yields

$$\Delta \bar{P}_m = 1.5 \Delta \bar{H} + \Delta \bar{G} \qquad (9.5A)$$

Alternatively, by substituting for ΔH from Equation 9.2, we may write

$$\Delta \bar{P}_m = 3 \Delta \bar{U} - 2 \Delta \bar{G} \qquad (9.5B)$$

The *acceleration of water column* due to a change in head at the turbine, characterized by Newton's second law of motion, may be expressed as

$$(\rho L A) \frac{d \Delta U}{dt} = -A (\rho a_g) \Delta H \qquad (9.6)$$

where

L = length of conduit
A = pipe area
ρ = mass density
a_g = acceleration due to gravity
$\rho L A$ = mass of water in the conduit
$\rho a_g \Delta H$ = incremental change in pressure at turbine gate
t = time in seconds

By dividing both sides by $A \rho a_g H_0 U_0$, the acceleration equation in normalized form becomes

$$\frac{LU_0}{a_g H_0} \frac{d}{dt}\left(\frac{\Delta U}{U_0}\right) = -\frac{\Delta H}{H_0}$$

or

$$T_w \frac{d\Delta \bar{U}}{dt} = -\Delta \bar{H} \tag{9.7}$$

where by definition,

$$T_w = \frac{LU_0}{a_g H_0} \tag{9.8}$$

Here T_w is referred to as the *water starting time*. It represents the time required for a head H_0 to accelerate the water in the penstock from standstill to the velocity U_0. It should be noted that T_w varies with load. Typically, T_w at full load lies between 0.5 s and 4.0 s.

Equation 9.7 represents an important characteristic of the hydraulic plant. A descriptive explanation of the equation is that if back pressure is applied at the end of the penstock by closing the gate, then the water in the penstock will decelerate. That is, if there is a positive pressure change, there will be a negative acceleration change.

From Equations 9.2 and 9.7, we can express the relationship between change in velocity and change in gate position as

$$T_w \frac{d\Delta \bar{U}}{dt} = 2(\Delta \bar{G} - \Delta \bar{U}) \tag{9.9}$$

Replacing d/dt with the Laplace operator s, we may write

$$T_w s \Delta \bar{U} = 2(\Delta \bar{G} - \Delta \bar{U})$$

or

$$\Delta \bar{U} = \frac{1}{1 + \frac{1}{2}T_w s} \Delta \bar{G} \tag{9.10}$$

Substituting for $\Delta \bar{U}$ from Equation 9.5B and rearranging, we obtain

$$\frac{\Delta \bar{P}_m}{\Delta \bar{G}} = \frac{1 - T_w s}{1 + \frac{1}{2} T_w s} \tag{9.11}$$

Equation 9.11 represents the "classical" transfer function of a hydraulic turbine. It shows how the turbine power output changes in response to a change in gate opening for an *ideal lossless* turbine.

Non-ideal turbine

The transfer function of a non-ideal turbine may be obtained by considering the following general expression for perturbed values of water velocity (flow) and turbine power:

$$\Delta \bar{U} = a_{11} \Delta \bar{H} + a_{12} \Delta \bar{\omega} + a_{13} \Delta \bar{G} \tag{9.12}$$

$$\Delta \bar{P}_m = a_{21} \Delta \bar{H} + a_{22} \Delta \bar{\omega} + a_{23} \Delta \bar{G} \tag{9.13}$$

where $\Delta \bar{\omega}$ is the per unit speed deviation. The speed deviations are small, especially when the unit is synchronized to a large system; therefore, the terms related to $\Delta \bar{\omega}$ may be neglected. Consequently,

$$\Delta \bar{U} = a_{11} \Delta \bar{H} + a_{13} \Delta \bar{G} \tag{9.14}$$

$$\Delta \bar{P}_m = a_{21} \Delta \bar{H} + a_{23} \Delta \bar{G} \tag{9.15}$$

The coefficients a_{11} and a_{13} are partial derivatives of flow with respect to head and gate opening, and the coefficients a_{21} and a_{23} are partial derivatives of turbine power output with respect to head and gate opening. These a coefficients depend on machine loading and may be evaluated from the turbine characteristics at the operating point.

With Equations 9.14 and 9.15 replacing Equations 9.2 and 9.5A, the transfer function between $\Delta \bar{P}_m$ and $\Delta \bar{G}$ becomes

$$\frac{\Delta \bar{P}_m}{\Delta \bar{G}} = a_{23} \frac{1 + (a_{11} - a_{13} a_{21} / a_{23}) T_w s}{1 + a_{11} T_w s} \tag{9.16}$$

The a coefficients vary considerably from one turbine type to another. For an ideal lossless Francis type turbine:

$$a_{11} = 0.5, \qquad a_{13} = 1.0, \qquad a_{21} = 1.5, \qquad a_{23} = 1.0$$

Typical measured values of the a coefficients for a 40 MW unit with Francis turbine are as follows [6]:

Load level	a_{11}	a_{13}	a_{21}	a_{23}
100% of rated	0.58	1.1	1.40	1.5
No load	0.57	1.1	1.18	1.5

Special characteristics of hydraulic turbine

The transfer function given by Equation 9.11 or 9.16 represents a "non-minimum phase" system.[1] The special characteristic of the transfer function may be illustrated by considering the response to a step change in gate position.

For a step change in $\bar{G}$, for the ideal turbine, the initial value theorem gives

$$\Delta \bar{P}_m(0) = \lim_{s \to \infty} s \frac{1}{s} \frac{1 - T_w s}{1 + 0.5 T_w s}$$

$$= -2.0$$

and the final value theorem gives

$$\Delta \bar{P}_m(\infty) = \lim_{s \to 0} s \frac{1}{s} \frac{1 - T_w s}{1 + 0.5 T_w s}$$

$$= 1.0$$

---

[1] Systems with poles or zeros in the right half of the s-plane are referred to as non-minimum phase systems; they do not have the minimum amount of phase shift for a given magnitude plot. Such systems cannot be uniquely identified by a knowledge of magnitude versus frequency plot alone.

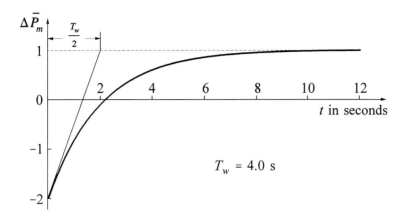

Figure 9.3 Change in turbine mechanical power following
a unit step change in gate position

The complete time response is given by

$$\Delta \bar{P}_m(t) = \left[1 - 3e^{-(2/T_w)t}\right]\Delta \bar{G}$$

Figure 9.3 shows a plot of the response of an ideal turbine model with $T_w = 4.0$.
Immediately following a unit increase in gate position, the mechanical power actually
decreases by 2.0 per unit. It then increases exponentially with a time constant of $T_w/2$
to a steady-state value of 1.0 per unit above the initial steady-state value.

   The initial power surge is opposite to that of the direction of change in gate
position. This is because, when the gate is suddenly opened, the flow does not change
immediately due to water inertia; however, the pressure across the turbine is reduced,
causing the power to reduce. With a response determined by T_w, the water accelerates
until the flow reaches the new steady value which establishes the new steady-power
output.

   Figure 9.4 shows the responses of power, head, and water velocity of a
hydraulic turbine with $T_w = 1.0$ s when the gate opening is reduced by 0.1 pu by

(a)   A step change

(b)   A 1-second ramp.

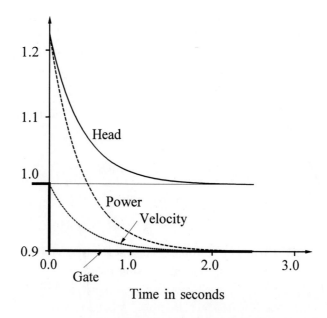

(a) Step reduction in gate opening

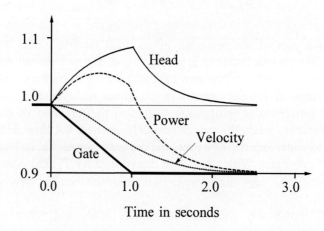

(b) 1-second ramp reduction in gate opening

Figure 9.4 Hydraulic turbine response to a step change and a ramp
change in gate position, with initial values of head power,
velocity, and gate position equal to 1.0 pu

Electrical analog

In understanding the performance of a hydraulic turbine system, it is useful to visualize a lumped-parameter electrical analog as shown in Figure 9.5. The hydraulic and electrical systems are nearly equivalent, with the water velocity u, gate opening G, and head H, corresponding to the current I, load conductance G, and voltage V, respectively.

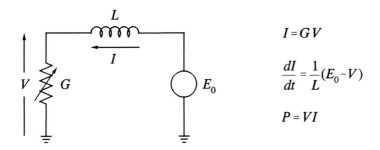

$$I = GV$$

$$\frac{dI}{dt} = \frac{1}{L}(E_0 - V)$$

$$P = VI$$

Figure 9.5   Electrical analog of a hydraulic turbine

When the load is suddenly decreased by a step reduction in conductance G, the current I does not change instantly; however, the voltage across the load suddenly increases because of the reduction in conductance (or increase in resistance). This causes the output power to suddenly increase initially. With a rate determined by the inductance L, the current I decreases exponentially until a new steady value is reached establishing the new steady output power. The responses of I, V, and P are very similar to those of velocity, head, and power shown in Figure 9.4(a) for a step reduction in gate position.

9.1.2 Nonlinear Turbine Model Assuming Inelastic Water Column

The linear model given by Equation 9.16 represents the small-signal performance of the turbine. It is useful for control system tuning using linear analysis techniques (frequency response, root locus, etc.). Because of the simplicity of its structure, this model provides insight into the basic characteristics of the hydraulic system.

In the past, the hydraulic turbine representation in system stability studies was largely based on the transfer function of Equation 9.11 or Equation 9.16 [10]. However, such a model is inadequate for studies involving large variations in power output and frequency [11]. In this section, we describe a nonlinear model which is more appropriate for large-signal time-domain simulations.

Once again we consider a simple hydraulic system configuration with unrestricted head and tail race, and with either a very large or no surge tank.

Assuming a rigid conduit and incompressible fluid, the basic hydrodynamic equations are

$$U = K_u G \sqrt{H} \tag{9.17}$$

$$P = K_p HU \tag{9.18}$$

$$\frac{dU}{dt} = -\frac{a_g}{L}(H - H_0) \tag{9.19}$$

$$Q = AU \tag{9.20}$$

where

U = water velocity
G = ideal gate opening
H = hydraulic head at gate
H_0 = initial steady-state value of H
P = turbine power
Q = water-flow rate
A = pipe area
L = length of conduit
a_g = acceleration due to gravity
t = time in seconds

Since we are interested in the large-signal performance, we normalize the above equations based on *rated values*. Equations 9.17 and 9.18 in normalized form become

$$\frac{U}{U_r} = \frac{G}{G_r}\left(\frac{H}{H_r}\right)^{\frac{1}{2}} \tag{9.21}$$

$$\frac{P}{P_r} = \frac{U}{U_r}\frac{H}{H_r} \tag{9.22}$$

where the subscript r denotes rated values. In per unit notation, the above equations may be written as

$$\bar{U} = \bar{G}(\bar{H})^{1/2} \tag{9.23}$$

$$\bar{P} = \bar{U}\bar{H} \tag{9.24}$$

From Equation 9.23,

$$\bar{H} = \left(\frac{\bar{U}}{\bar{G}}\right)^2 \tag{9.25}$$

Similarly, the per unit form of Equation 9.19 is

$$\frac{d}{dt}\left(\frac{U}{U_r}\right) = -\frac{a_g}{L}\frac{H_r}{U_r}\left(\frac{H}{H_r}-\frac{H_0}{H_r}\right)$$

or

$$\frac{d\bar{U}}{dt} = -\frac{1}{T_W}(\bar{H}-\bar{H}_0) \tag{9.26A}$$

or in Laplace notation

$$\frac{\bar{U}}{\bar{H}-\bar{H}_0} = \frac{-1}{T_W s} \tag{9.26B}$$

where T_W is the *water starting time at rated load*.[1] It has a fixed value for a given turbine-penstock unit and is given by

$$T_W = \frac{LU_r}{a_g H_r} = \frac{LQ_r}{a_g A H_r} \tag{9.27}$$

The mechanical power output P_m is

$$P_m = P-P_L \tag{9.28}$$

where P_L represents the fixed power loss of the turbine given by

$$P_L = U_{NL}H \tag{9.29}$$

---

[1] From Equations 9.8 and 9.27, the water starting time T_w at any load is related to its value T_W at rated load by

$$T_w = \frac{U_0}{U_r}\frac{H_r}{H_0}T_W$$

with U_{NL} representing the no-load water velocity. In normalized form, we have

$$\frac{P_m}{P_r} = \frac{P}{P_r} - \frac{P_L}{P_r} = \left(\frac{U}{U_r} - \frac{U_{NL}}{U_r}\right)\frac{H}{H_r}$$

or

$$\bar{P}_m = (\bar{U} - \bar{U}_{NL})\bar{H} \qquad (9.30)$$

The above equation gives the per unit value of the turbine power output on a base equal to the turbine MW rating. In system stability studies, solution of the machine swing equation requires turbine mechanical torque on a base equal to either the generator MVA rating or a common MVA base. Hence,

$$\bar{T}_m = \left(\frac{\omega_0}{\omega}\right)\bar{P}_m\left(\frac{P_r}{MVA_{base}}\right) = \frac{1}{\bar{\omega}}(\bar{U} - \bar{U}_{NL})\bar{H}\bar{P}_r \qquad (9.31)$$

where

$\bar{\omega}$ = per unit speed
MVA_{base} = base MVA on which turbine torque is to be made per unit
$\bar{P}_r$ = per unit turbine rating = $\dfrac{\text{turbine MW rating}}{MVA_{base}}$

In the above equations, G is the *ideal gate opening* based on the change from no load to full load being equal to 1 per unit. This is related to the real gate opening g as shown in Figure 9.6. The *real gate opening* is based on the change from the fully closed to the fully open position being equal to 1 per unit [12].

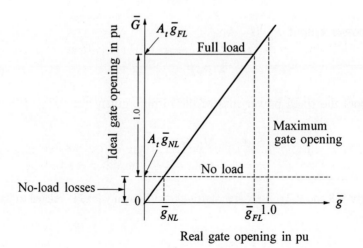

Figure 9.6 Relationship between ideal and real gate openings

The ideal gate opening is related to real gate opening as follows

$$\bar{G} = A_t \bar{g} \tag{9.32}$$

where A_t is the turbine gain given by

$$A_t = \frac{1}{\bar{g}_{FL} - \bar{g}_{NL}} \tag{9.33}$$

Equations 9.25 to 9.27 and 9.31 to 9.33 completely describe the water column and turbine characteristics. These may be represented in block diagram form as shown in Figure 9.7.

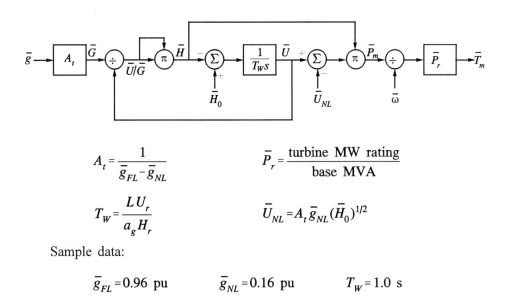

$$A_t = \frac{1}{\bar{g}_{FL} - \bar{g}_{NL}} \qquad \bar{P}_r = \frac{\text{turbine MW rating}}{\text{base MVA}}$$

$$T_W = \frac{L U_r}{a_g H_r} \qquad \bar{U}_{NL} = A_t \bar{g}_{NL} (\bar{H}_0)^{1/2}$$

Sample data:

$$\bar{g}_{FL} = 0.96 \text{ pu} \qquad \bar{g}_{NL} = 0.16 \text{ pu} \qquad T_W = 1.0 \text{ s}$$

Figure 9.7 Hydraulic turbine block diagram assuming inelastic water column

The turbine and penstock model may be rearranged and expressed in terms of two equations, one representing the water column and the other the turbine.

Water column equation:

$$\frac{d\bar{U}}{dt} = -\frac{1}{T_W}(\bar{H} - \bar{H}_0) = -\frac{1}{T_W}\left[\left(\frac{\bar{U}}{A_t \bar{g}}\right)^2 - \bar{H}_0\right] \tag{9.34}$$

Turbine equation:

$$\bar{T}_m = \frac{\bar{U}-\bar{U}_{NL}}{\bar{\omega}}\left(\frac{\bar{U}}{A_t\bar{g}}\right)^2 \bar{P}_r \qquad (9.35)$$

From Equation 9.34, by considering the steady-state condition corresponding to no load, we have

$$\bar{U}_{NL} = A_t\bar{g}_{NL}(\bar{H}_0)^{1/2} \qquad (9.36)$$

Usually $\bar{H}_0 = 1.0$.

Example 9.1

The data related to the turbine, penstock, and generator of a hydraulic power plant are as follows:

Generator rating = 140 MVA   Turbine rating = 127.4 MW
Penstock length = 300 m   Piping area = 11.15 m^2
Rated hydraulic head = 165 m   Water-flow rate at rated load = 85 m^3/s
Gate opening at rated load = 0.94 pu   Gate opening at no load = 0.06 pu

(a)   Calculate (i) the velocity of water in the penstock, and (ii) water starting time, at full load.

(b)   Determine the classical transfer function of the turbine relating the change in power output to change in gate position at rated load.

(c)   Determine the nonlinear model of the turbine, assuming an inelastic water column. Identify the values of the parameters and variables of the model at rated output. The turbine mechanical power/torque is to be expressed on a common MVA base of 100.

Solution

(a) (i) Velocity of water in the penstock at rated load is

U_r = flow rate at rated load/piping area

$$= \frac{85 \text{ m}^3/\text{s}}{11.15 \text{ m}^2}$$

$$= 7.62 \text{ m/s}$$

(ii) Water starting time T_W at full load, from Equation 9.27, is

$$T_W = \frac{L U_r}{a_g H_r}$$

$$= \frac{300 \times 7.62}{9.81 \times 165} = 1.41 \text{ s}$$

(b) From Equation 9.11, the classical transfer function of the turbine at rated load is

$$\frac{\text{pu } \Delta P_m}{\text{pu } \Delta G} = \frac{1 - T_W s}{1 + 0.5 T_W s}$$

$$= \frac{1 - 1.41 s}{1 + 0.705 s}$$

(c) Referring to Figure 9.7, the parameters of the nonlinear turbine model are as follows:

Turbine gain

$$A_t = \frac{1}{\bar{g}_{FL} - \bar{g}_{NL}}$$

$$= \frac{1}{0.94 - 0.06} = 1.136$$

No-load velocity

$$\bar{U}_{NL} = A_t \bar{g}_{NL} \sqrt{\bar{H}_0}$$

$$= 1.136 \times 0.06 \times 1.0$$

$$= 0.068 \text{ pu}$$

The per unit conversion factor is

$$\bar{P}_r = \frac{\text{turbine MW rating}}{\text{MVA}_{base}}$$

$$= \frac{127.4}{100} = 1.274$$

Figure E9.1 shows the turbine model. The values of the variables corresponding to rated output are identified on the model.

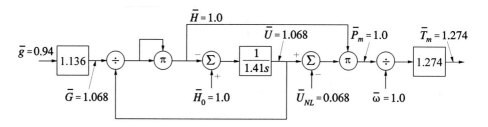

Figure E9.1 ∎

9.1.3 Governors for Hydraulic Turbines

The basic function of a governor is to control speed and/or load. The general principles of load/frequency control will be described in Chapter 11. Here, we discuss the special requirements of governing hydraulic turbines, their physical realization and modelling in system studies.

The primary speed/load control function involves feeding back speed error to control the gate position. In order to ensure satisfactory and stable parallel operation of multiple units, the speed governor is provided with a droop characteristic. The purpose of the droop is to ensure equitable load sharing between generating units (see Chapter 11, Section 11.1.1). Typically, the steady-state droop is set at about 5%, such that a speed deviation of 5% causes 100% change in gate position or power output; this corresponds to a gain of 20. For a hydro turbine, however, such a governor with a simple steady-state droop characteristic would be unsatisfactory. This is illustrated in the following example.

Example 9.2

A simplified block diagram representation of the speed control of a hydraulic generating unit feeding an isolated load is shown in Figure E9.2. The turbine is represented by the classical model and the speed governor by pure gain $K_G = 1/R$. The generator is represented in terms of the combined inertia of the generator and turbine rotors. (For derivation of the transfer function used see Chapter 3, Section 3.9.)

If $T_W = 2.0$ s, $T_M = 10.0$ s, and $K_D = 0.0$, determine (i) the lowest value of the droop R for which the speed governing is stable, and (ii) the value of R for which the speed control action is critically damped.

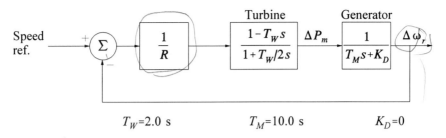

$$T_W = 2.0 \text{ s} \qquad\qquad T_M = 10.0 \text{ s} \qquad\qquad K_D = 0$$

Figure E9.2

Solution

The characteristic equation (of the form $1+GH=0$) of the closed-loop system is

$$1 + \frac{1-2s}{1+s} \frac{1}{10s} \frac{1}{R} = 0$$

or

$$10Rs^2 + (10R-2)s + 1 = 0$$

(i) For stability, the roots of the characteristic equation have to be in the left side of the complex s-plane. In case of a quadratic, a sufficient and necessary condition is that all quadratic coefficients are positive. Hence,

$$10R > 0, \qquad \text{i.e.,} \qquad R > 0$$

and

$$10R - 2 > 0, \qquad \text{i.e.,} \qquad R > 0.2$$

The smallest value of R resulting in stable response is thus 0.2 or 20%. In other words, the speed governor gain K_G has to be less than 5. With the standard 5% droop (or gain of 20), the speed control would be unstable.

(ii) For critical damping

$$(10R - 2)^2 - 4(10R) = 0$$

Solving yields

$$R_1 = 0.746 \qquad R_2 = 0.0536$$

with $R_1 = 0.746$ corresponding to the critical damping (damping ratio $\zeta = 1$) and stable response. And $R_2 = 0.0536$ is less than the limiting value of 0.2; it corresponds to $\zeta = -1.0$ and represents unstable operation. Therefore, $R = 0.746$ or gain $K_G = 1.34$ is required for critical damping. ∎

Requirement for a transient droop

As discussed in Section 9.1.1, hydro turbines have a peculiar response due to water inertia: a change in gate position produces an initial turbine power change which is opposite to that sought. For stable control performance, a large transient (temporary) droop with a long resetting time is therefore required. This is accomplished by the provision of a rate feedback or transient gain reduction compensation as shown in Figure 9.8. The rate feedback retards or limits the gate movement until the water flow and power output have time to catch up. The result is a governor which exhibits a high droop (low gain) for fast speed deviations, and the normal low droop (high gain) in the steady state. Example 9.3 illustrates the effect of the transient droop compensation on the stability characteristics of the governing system.

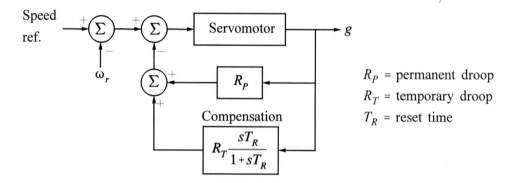

Figure 9.8 Governor with transient droop compensation

Mechanical-hydraulic governor

On older units the governing function is realized using mechanical and hydraulic components. Figure 9.9 shows a simplified schematic of a mechanical-hydraulic governor. Speed sensing, permanent droop feedback, and computing functions are achieved through mechanical components; functions involving higher power are achieved through hydraulic components. A dashpot is used to provide transient droop compensation. A bypass arrangement is usually provided to disable the dashpot if so desired.

The transfer function of the relay valve and gate servomotor is

$$\frac{g}{a} = \frac{K_1}{s} \tag{9.37}$$

The transfer function of the pilot valve and pilot servo is

$$\frac{a}{b} = \frac{K_2}{1+sT_P} \tag{9.38}$$

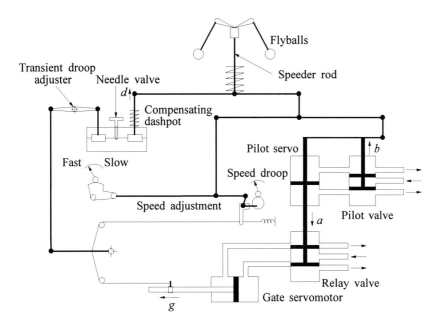

Figure 9.9 Schematic of a mechanical-hydraulic governor for a hydro turbine

where K_2 is determined by the feedback lever ratio, and T_P by port areas of the pilot valve and K_2 [9]. Combining Equations 9.37 and 9.38 yields

$$\frac{g}{b} = \frac{K_1 K_2}{s(1+sT_P)} = \frac{K_s}{s(1+sT_P)} \qquad (9.39)$$

where K_s is the servo gain and T_P is the pilot valve/servomotor time constant. The servo gain K_s is determined by the pilot valve feedback lever ratio.

Assuming that the dashpot fluid flow through the needle valve is proportional to the dashpot pressure, the dashpot transfer function is

$$\frac{d}{g} = R_T \frac{sT_R}{1+sT_R} \qquad (9.40)$$

The temporary droop R_T is determined by the lever ratio, and the reset or washout time T_R is determined by the needle valve setting.

Water is not a very compressible fluid; if the gate is closed too rapidly, the resulting pressure could burst the penstock. Consequently, the gate movement is rate limited. Often, the rate of gate movement is limited even further in the buffer region near full closure to provide cushioning.

A block diagram representation of the governing system suitable for system stability studies is shown in Figure 9.10. This diagram together with the diagram of

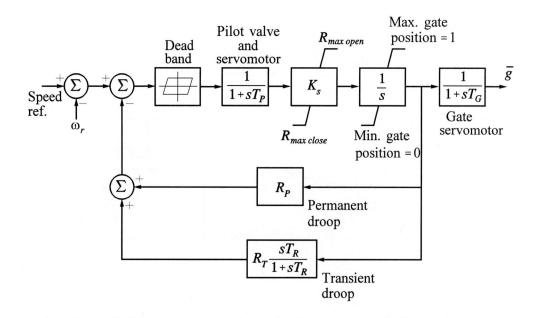

Parameters: Sample data

T_P = pilot valve and   0.05 s
    servomotor time constant
K_s = servo gain   5.0
T_G = main servo time constant   0.2 s
R_P = permanent droop   0.04
R_T = temporary droop   0.4
T_R = reset time   5.0 s

Constraints:

Maximum gate position limit = 1.0
Minimum gate position limit = 0
$R_{max\ open}$ = maximum gate opening rate   0.16 pu/s
$R_{max\ close}$ = maximum gate closing rate   0.16 pu/s
$R_{max\ buff}$ = maximum gate closing rate   0.04 pu/s
    in buffered region
g_{buff} = buffered region in pu of   0.08 pu
    servomotor stroke

Figure 9.10 Model of governors for hydraulic turbines

Figure 9.7 provides a complete model of the hydraulic turbine and speed-governing system.

The governor model shown in Figure 9.10 has provision for representing the effects of dead bands. However, it is usually difficult to get data that identify their magnitudes and locations. Consequently, dead-band effects are not usually modelled in system studies.

Electrohydraulic governor

Modern speed governors for hydraulic turbines use electrohydraulic systems. Functionally, their operation is very similar to that of mechanical-hydraulic governors. Speed sensing, permanent droop, temporary droop, and other measuring and computing functions are performed electrically. The electric components provide greater flexibility and improved performance with regard to dead bands and time lags. The dynamic characteristics of electric governors are usually adjusted to be essentially similar to those of the mechanical-hydraulic governors.

Tuning of speed-governing systems

The basis for selection of hydraulic turbine governor settings is covered in references 7, 13, and 14. We will review this briefly here and discuss it in detail in Chapter 16.

There are two important considerations in the selection of governor settings:

1.   Stable operation during system-islanding conditions or isolated operation; and

2.   Acceptable speed of response for loading and unloading under normal synchronous operation.

For stable operation under islanding conditions, the optimum choice of the temporary droop R_T and reset time T_R is related to the water starting time T_W and mechanical starting time $T_M = 2H$ (see Chapter 3, Section 3.9.3) as follows [14]:

$$R_T = [2.3 - (T_W - 1.0)0.15]\frac{T_W}{T_M} \qquad (9.41)$$

$$T_R = [5.0 - (T_W - 1.0)0.5]T_W \qquad (9.42)$$

In addition, the servosystem gain K_s should be set as high as is practically possible. The above settings ensure good stable performance when the unit is at full load supplying an isolated load. This represents the most severe requirement and ensures stable operation for all situations involving system islanding.

For loading and unloading during normal interconnected system operation, the above settings result in too slow a response. For satisfactory loading rates, the reset time T_R should be less than 1.0 s, preferably close to 0.5 s.

The above conflicting requirements may be met by using the dashpot bypass arrangement as follows:

- With the dashpot not bypassed, the settings satisfy the requirements for system-islanding conditions or isolated operation.

- With the dashpot bypassed, the reset time T_R has a reduced value, resulting in acceptable loading rates.

The dashpot is normally not bypassed, so that in the event of a disturbance leading to an islanding operation, the speed control would be stable. The dashpot is bypassed for brief periods during loading and unloading.

Example 9.3

Figure E9.3 shows the block diagram of the speed-governing system of a hydraulic unit supplying an isolated load. The speed governor representation includes a transient droop compensation $G_c(s)$ and a governor time constant T_G of 0.5 s. (Such a representation may be readily shown to be a linear approximation to the governor model shown in Figure 9.10, with $T_P=0$.) The generator is represented by its equation of motion with a mechanical starting time T_M of 10.0 s and a system-damping coefficient of 1.0 per unit.

Examine the stability of the speed-governing system, by considering its open-loop frequency response with

(a)   No transient droop compensation, i.e., $G_c(s)=1$
(b)   A transient droop compensation having the transfer function

$$G_c(s) = \frac{1+T_R s}{1+(R_T/R_P)T_R s}$$

where T_R and R_T are determined by Equations 9.41 and 9.42.

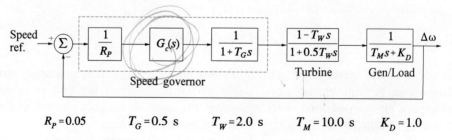

$$R_P=0.05 \qquad T_G=0.5 \text{ s} \qquad T_W=2.0 \text{ s} \qquad T_M=10.0 \text{ s} \qquad K_D=1.0$$

Figure E9.3 Block diagram of speed-governing system

Solution

(a) *Without transient droop compensation [$G_c(s) =1.0$]*

The feedback control system of Figure E9.3 has the standard form

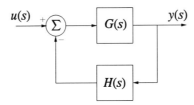

The open-loop transfer function of the system is

$$GH(s) = \frac{20(1-2s)}{(1+0.5s)(1+s)(1+10s)}$$ (E9.1)

This system has a zero corresponding to

$$s = 0.5$$

and three poles corresponding to

$$s = -0.1, -1.0, -2.0$$

The frequency response function for the system is obtained by setting $s=j\omega$ so that

$$GH(j\omega) = \frac{20(1-j2\omega)}{(1+j0.5\omega)(1+j\omega)(1+j10\omega)}$$ (E9.2)

The frequency response may be computed using the linear approximation (Bode plot). The corresponding corner frequencies are

$$\omega = 0.1, 0.5, 1.0, 2.0 \quad \text{rad/s}$$

The linear approximation, particularly for the phase plot, tends to be very approximate in this case. We will therefore compute the frequency response accurately, with the magnitude characteristic given by

$$M(\omega) = \frac{20\sqrt{1+4\omega^2}}{\sqrt{1+0.25\omega^2}\sqrt{1+\omega^2}\sqrt{1+100\omega^2}}$$

and the phase characteristic given by

$$\phi(\omega) = -\tan^{-1}2\omega -\tan^{-1}0.5\omega -\tan^{-1}\omega -\tan^{-1}10\omega$$

The magnitude and phase as a function of frequency are plotted in Figure E9.4. We see from the figure that the crossover frequency ω_{c1}, which is the value of ω for which the magnitude is unity (0 dB), is 2.5 rad/s (0.4 Hz). The phase angle at crossover is $-290°$. The gain and phase margins are

$$G_{m1} = -12 \text{ dB}$$

$$\phi_{m1} = -110°$$

The uncompensated system is hence unstable.

(b) *With transient droop compensation*

From Equations 9.41 and 9.42

$$T_R = [5.0-(2.0-1.0)0.5]2.0 = 9.0 \text{ s}$$

and

$$R_T = [2.3-(2.0-1.0)0.15](2.0/10.0) = 0.43$$

Hence the transfer function of the compensation is

$$G_c(s) = \frac{1+9.0s}{1+(0.43/0.05)9.0s}$$

$$= \frac{1+9.0s}{1+77.4s}$$

This adds a pole at $s=-0.0129$ and a zero at $s=-0.111$. The open-loop frequency response function of the overall system is

$$GH(j\omega) = \frac{20(1-j2\omega)(1+j9\omega)}{(1+j0.5\omega)(1+j\omega)(1+j10\omega)(1+j77.4\omega)}$$

The corresponding magnitude and phase plots are shown in Figure E9.4. The crossover frequency ω_{c2} is 0.25 rad/s and the phase at crossover is $-138°$. The gain and phase margins are

$$G_{m2} = 6 \text{ dB}$$

$$\phi_{m2} = 42°$$

The compensated system is thus stable and the above values of gain and phase margins are considered good values from the viewpoint of compensator design. A lower value of reset time T_R would increase the crossover frequency, and decrease the gain and phase margins. The net effect would be a more oscillatory but faster control system.

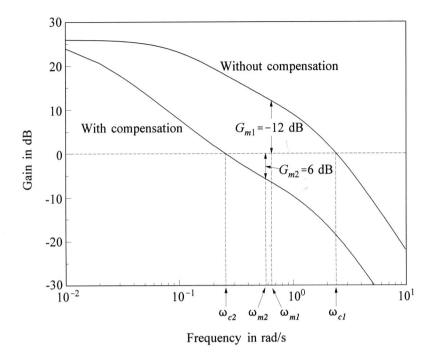

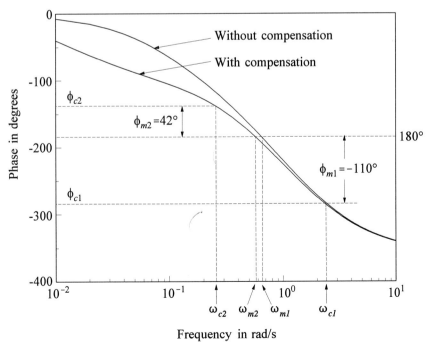

Figure E9.4 Open-loop frequency response characteristics
with and without transient droop compensation ∎

PID governor [11,15]

Some electrohydraulic governors are provided with three-term controllers with proportional-integral-derivative (PID) action, as shown in Figure 9.11. These allow the possibility of higher response speeds by providing both transient gain reduction and transient gain increase. The derivative action is beneficial for isolated operation, particularly for plants with large water starting time (T_W=3 s or more). Typical values are K_P=3.0, K_I=0.7, and K_D=0.5. However, the use of a high derivative gain or transient gain increase will result in excessive oscillations and possibly instability when the generating unit is strongly connected to an interconnected system. Therefore, the derivative gain is usually set to zero; without the derivative action, the transfer function of a PID (now PI) governor is equivalent to that of the mechanical-hydraulic governor. The proportional and integral gains may be selected to result in the desired temporary droop and reset time.

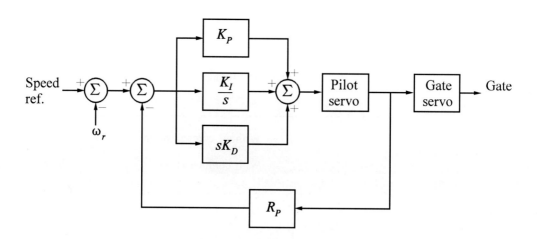

Figure 9.11 PID governor

9.1.4 Detailed Hydraulic System Model

The conventional hydraulic system models developed in Sections 9.1.1 and 9.1.2 neglect the effects of water compressibility and pipe elasticity. In addition, the surge tank, if present, was assumed to be ideal and to isolate the tunnel from the penstock. While these assumptions are valid for a wide range of system studies, there are many applications where a detailed hydraulic system model is necessary to accurately account for the dynamic interaction between the hydraulic system and the power system.

We will first develop the pressure-flow wave equations in a closed conduit and then apply them to derive a detailed model of a hydraulic plant.

Wave equation of flow in a conduit [2,6]

Consider the flow of water through the conduit shown in Figure 9.12(a). When the gate is partially closed suddenly, a pressure wave is set up which moves upstream. A short section of the conduit illustrating the effect of the pressure wave in stretching the conduit walls is shown in Figure 9.12(b). The relationships identified in the figure correspond to the instant when the wave front is at a section of the conduit distance x from the reservoir.

Let the pressure p in the slice Δx increase by Δp. The equation of motion (Newton's second law) of the water in the pipe section is

$$(A\Delta x\,\rho)\frac{dU}{dt} = -A\Delta p \tag{9.43}$$

where ρ = mass density. The change in pressure in terms of change in head is given by

$$\Delta p = \rho a_g \Delta H$$

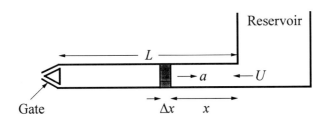

Reservoir

L

$\rightarrow a$ $\leftarrow U$

Gate Δx x

(a) Hydraulic system configuration

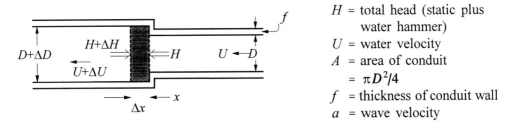

$D+\Delta D$ $H+\Delta H$ $\Leftarrow H$ $U \leftarrow D$ f

$U+\Delta U$

$\rightarrow x$

Δx

H = total head (static plus water hammer)
U = water velocity
A = area of conduit
$\quad = \pi D^2/4$
f = thickness of conduit wall
a = wave velocity

(b) Stretching of the conduit wall at the wave front

Figure 9.12   Water flow through an elastic conduit

where a_g =acceleration due to gravity. By taking infinitesimally small values of Δx, ΔU, and Δt, Equation 9.43 may be written as

$$\boxed{\frac{\partial U}{\partial t} = -a_g \frac{\partial H}{\partial x}} \qquad (9.44)$$

The increase in volume due to stretching of the conduit walls is

$$\Delta V_E = \frac{AD\Delta x}{Ef}\Delta p \qquad (9.45)$$

where

E = Young's modulus of elasticity of pipe material
f = thickness of pipe wall

The change in volume of water in the section due to compressibility of water is

$$\Delta V_C = \frac{A\Delta x}{K}\Delta p \qquad (9.46)$$

where

K = bulk modulus of compression of water

The increase in mass of water in the pipe section due to the combined effects of pipe elasticity and water compressibility is

$$\Delta m = \rho(\Delta V_E + \Delta V_C)$$

$$= \rho A \Delta x \left[\frac{1}{K} + \frac{D}{Ef}\right]\Delta p \qquad (9.47)$$

This should be equal to the change in mass of water within the pipe section during the period Δt given by the difference in flow into the section and out of the section as follows:

$$\Delta m = \rho A U \Delta t - \rho(U + \Delta U)A\Delta t$$

$$= -\rho A \Delta U \Delta t \qquad (9.48)$$

Equating the expressions for Δm given by Equations 9.47 and 9.48, we have

$$\frac{\Delta U}{\Delta x} = -\left(\frac{1}{K} + \frac{D}{Ef}\right)\frac{\Delta p}{\Delta t}$$

$$= -\left(\frac{1}{K} + \frac{D}{Ef}\right)\rho a_g \frac{\Delta H}{\Delta t} \tag{9.49}$$

By taking infinitesimally small incremental values, the above equation may be written as

$$\boxed{\frac{\partial U}{\partial x} = -\alpha \frac{\partial H}{\partial t}} \tag{9.50}$$

where

$$\alpha = \rho a_g \left(\frac{1}{K} + \frac{D}{Ef}\right) \tag{9.51}$$

Equations 9.44 and 9.50 are the basic hydraulic equations which determine the flow of a compressible fluid through a uniform elastic pipe, with friction neglected. These equations are similar to the electrical transmission line equations; the fluid velocity U and head H are analogous to the transmission line current and voltage, respectively. These equations may be conveniently solved using Laplace transforms. As shown in reference 6, the solution is given by

$$H_2 = H_1 \operatorname{sech}(T_e s) - Z_0 Q_2 \tanh(T_e s) \tag{9.52}$$

$$Q_1 = Q_2 \cosh(T_e s) + \frac{1}{Z_0} H_2 \sinh(T_e s) \tag{9.53}$$

where

T_e = elastic time

$$= \frac{\text{conduit length } L}{\text{wave velocity } a} = \frac{L}{\sqrt{a_g/\alpha}} \tag{9.54}$$

Z_0 = hydraulic surge impedance of the conduit

$$= \frac{\text{wave velocity } a}{A a_g} = \frac{1}{A\sqrt{a_g \alpha}} \tag{9.55}$$

Q = water flow = AU

and the subscripts 1 and 2 refer to the conditions at the upstream and downstream ends of the conduit, respectively.

Typical values of wave velocity ($a = \sqrt{a_g / \alpha}$) for water are 1220 m/s for steel conduit and 1420 m/s for rock tunnels.

Expressing Equations 9.52 and 9.53 in per unit form with rated head H_r and rated flow Q_r as base values, we obtain

$$\bar{H}_2 = \bar{H}_1 \text{sech}(T_e s) - Z_n \bar{Q}_2 \tanh(T_e s) \tag{9.56}$$

$$\bar{Q}_1 = \bar{Q}_2 \cosh(T_e s) + \frac{1}{Z_n} \bar{H}_2 \sinh(T_e s) \tag{9.57}$$

where Z_n is the normalized value of the hydraulic surge impedance of the conduit given by

$$Z_n = Z_0 \left(\frac{Q_r}{H_r} \right) \tag{9.58}$$

In per unit, water flow is equal to velocity, since

$$\frac{Q}{Q_r} = \frac{AU}{AU_r}$$

Therefore, Equation 9.57 is valid with the per unit flow $\bar{Q}$ replaced by per unit velocity $\bar{U}$.

In the above formulation, friction has been assumed to be negligible. Following the approach used in reference 6, the effect of head loss due to friction may be approximated by modifying Equation 9.56 as follows:

$$\bar{H}_2 = \bar{H}_1 \text{sech}(T_e s) - Z_n \bar{U}_2 \tanh(T_e s) - k_f \bar{U}_2 |\bar{U}_2| \tag{9.59}$$

where k_f is the head loss constant due to friction.

Writing Equations 9.57 and 9.59 in terms of *deviations* of head and velocity from steady-state values, we obtain

$$h_2 = h_1 \text{sech}(T_e s) - Z_n u_2 \tanh(T_e s) - \phi u_2 \tag{9.60}$$

$$u_1 = u_2 \cosh(T_e s) + \frac{1}{Z_n} h_2 \sinh(T_e s) \tag{9.61}$$

where

h = deviation of head ($H - H_0$) in pu

u = deviation of velocity $(U-U_0)$ in pu
ϕ = friction coefficient $=2k_f|U_{20}|$
U_{20} = initial steady-state value of downstream velocity U_2

Model of hydraulic plant with no surge tank

Referring to Figure 9.13, for a large reservoir, the deviation in head at the upstream end of the penstock is zero; that is, $h_w = 0$. Therefore, based on Equation 9.60, the expression for turbine head deviation is

$$h_t = -Z_p u_t \tanh(T_{ep}s) - \phi_p u_t \qquad (9.62)$$

Consequently the transfer function relating head and flow at the turbine end of the penstock may be written as

$$F(s) = \frac{u_t}{h_t} = \frac{-1}{\phi_p + Z_p \tanh(T_{ep}s)} \qquad (9.63)$$

where

Z_p = normalized hydraulic impedance of penstock
ϕ_p = friction coefficient of the penstock
    = $2k_p|U_{t0}|$
T_{ep} = elastic time of penstock

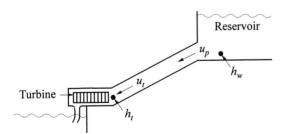

h_t = turbine head deviation
h_w = reservoir head deviation
u_t = turbine velocity deviation
u_p = upper penstock velocity
       deviation

Figure 9.13 Hydraulic plant with no surge tank

The transfer function $F(s)$ of Equation 9.63 represents a *distributed-parameter system*, with

$$\tanh(T_{ep}s) = \frac{1-e^{-2T_{ep}s}}{1+e^{-2T_{ep}s}} = \frac{sT_{ep}\prod\limits_{n=1}^{n=\infty}\left[1+\left(\dfrac{sT_{ep}}{n\pi}\right)^2\right]}{\prod\limits_{n=1}^{n=\infty}\left[1+\left(\dfrac{2sT_{ep}}{(2n-1)\pi}\right)^2\right]}$$

(9.64)

The infinite product expansions of Equation 9.64 are required to preserve all the characteristic roots of $F(s)$. However, the transfer function may be approximated by a *lumped-parameter equivalent* by retaining an appropriate number of terms of the expansions, depending on the purpose of the study and the accuracy required [16].

Lumped-parameter approximations

With $n=0$, $\tanh(T_{ep}s) = T_{ep}s$ and

$$F(s) = \frac{-1}{\phi_p + Z_p T_{ep}s}$$

(9.65)

From Equations 9.54, 9.55, 9.58, and 9.27, we see that

$$Z_p T_{ep} = \frac{Q_r}{H_r}\frac{L}{Aa_g} = T_{Wp}$$

(9.66)

Thus, $Z_p T_{ep}$ is equal to the water starting time T_{Wp} of the penstock at rated load. With friction neglected, $F(s)$ given by Equation 9.65 is the same as Equation 9.26B for u/h derived in Section 9.1.2. The approximation $n=0$ is therefore equivalent to assuming the water column to be inelastic.

With $n=1$, the first pole and zero of the tanh function (representing the fundamental harmonic of the water column) are preserved. The corresponding expression for the transfer function $F(s)$ is

$$F(s) = \frac{-1}{\phi_p + F_2(s)}$$

(9.67)

where

$$F_2(s) = Z_p \tanh(T_{ep}s) \approx \frac{sT_{Wp}\left[1 + s^2\left(\dfrac{T_{ep}}{\pi}\right)^2\right]}{\left[1 + s^2\left(\dfrac{2T_{ep}}{\pi}\right)^2\right]} \qquad (9.68)$$

For most power system stability studies the above approximation should be adequate.

Example 9.4

The parameters of a penstock are as follows:

> Water starting time T_{Wp} = 1.0 s (at full load)
> Elastic time T_{ep} = 0.5 s

The head loss due to friction may be neglected. Examine the accuracy of lumped-parameter approximations to the water hammer effects by considering the frequency response characteristics of the $\Delta P_m/\Delta G$ transfer function at full load.

Solution

The normalized hydraulic impedance of the penstock is

$$Z_p = \frac{T_{Wp}}{T_{ep}} = \frac{1.0}{0.5} = 2.0$$

From Equation 9.63, with $\phi_p = 0$, the water column transfer function is

$$\frac{h_t(s)}{u_t(s)} = -Z_p \tanh(T_{ep}s) \qquad (E9.3)$$

$$= -2.0\tanh(0.5s)$$

Combining the above with Equations 9.2 and 9.5B gives

$$\frac{\Delta P_m(s)}{\Delta G(s)} = \frac{1 - Z_p \tanh(T_{ep}s)}{1 + \dfrac{1}{2}Z_p \tanh(T_{ep}s)} \qquad (E9.4)$$

$$= \frac{1.0 - 2.0\tanh(0.5s)}{1 + \tanh(0.5s)}$$

where

$$\tanh(0.5s) = \frac{1-e^{-s}}{1+e^{-s}} = \frac{0.5s \prod_{n=1}^{n=\infty} \left[1+\left(\frac{0.5s}{n\pi}\right)^2\right]}{\prod_{n=1}^{n=\infty} \left[1+\left(\frac{s}{(2n-1)\pi}\right)^2\right]}$$

With $n=0$ (inelastic water column)

$$\frac{\Delta P_m(s)}{\Delta G(s)} = \frac{1.0-s}{1.0+0.5s}$$

This represents the classical transfer function.

Figure E9.5(a) shows the magnitude versus frequency plots of the transfer function $\Delta P_m/\Delta G$ given by Equation E9.4, when $\tanh(0.5s)$ is represented exactly and by approximations corresponding to $n=0$, $n=1$, and $n=2$. The corresponding phase characteristics are shown in Figure E9.5(b). It is seen that the classical model is valid up to about 0.1 Hz. With $n=1$ (i.e., with the fundamental component of the water column represented), the lumped-parameter approximation is valid up to about 1.0 Hz.

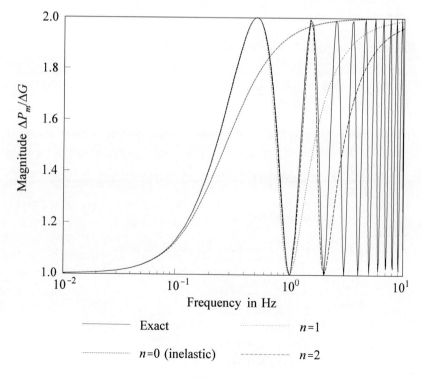

Figure E9.5(a)

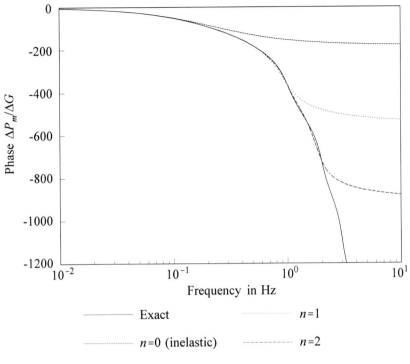

Phase $\Delta P_m/\Delta G$

Frequency in Hz

——— Exact   ·············· $n=1$

·············· $n=0$ (inelastic)   - - - - - $n=2$

Figure E9.5(b)   ■

Model of hydraulic plant with surge tank

A surge tank is sometimes installed near the turbine to reduce the pressure increase which accompanies rapid closure of turbine gates. The kinetic energy of the flowing water in the penstock is converted to potential energy in the surge tank, thus relieving the pressure. A differential surge tank, shown in Figure 9.14, has two compartments: a small area riser connected by orifices to a large area tank. The riser helps suppress water hammer and the tank provides the water storage/supply function.

The following are the equations of the various components of the hydraulic system: tunnel, surge tank, penstock, and the reservoir. All equations are in per unit form. Water velocities and heads are in terms of deviations from steady-state values.

Equation 9.60 applied to the tunnel yields

$$h_r = h_w \operatorname{sech}(T_{ec}s) - Z_c u_c \tanh(T_{ec}s) - \phi_c u_c \tag{9.69}$$

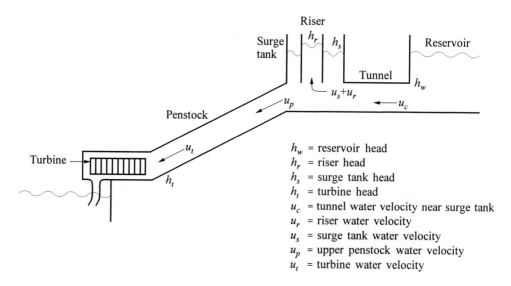

Figure 9.14 Hydraulic plant with surge tank

For a large reservoir, the deviation in head at the upstream end of the tunnel is zero; hence, $h_w=0$. In addition, the wave effects in the tunnel are insignificant and the $T_{ec}s$ term may be approximated by setting $n=0$ in the infinite product expansion form of Equation 9.64. Therefore, Equation 9.69 simplifies to

$$h_r = -sT_{Wc}u_c - \phi_c u_c \tag{9.70}$$

where

$$T_{Wc} = \text{water starting time of the tunnel} = Z_c T_{ec}$$
$$\phi_c = \text{friction coefficient of the tunnel} = 2k_c|U_{c0}|$$

The flow rates and velocities at the surge tank are related by the continuity equation

$$u_c = (u_s+u_r)+u_p \tag{9.71}$$

The water velocities in the surge tank and the riser are related to the riser head by the equation

$$u_s+u_r = sT_s h_r \tag{9.72}$$

where T_s is the surge tank riser time.

The surge tank and riser heads are related by the equation

$$h_r = h_s + \phi_s(u_s + u_r) \tag{9.73}$$

where

$$\phi_s = \text{surge tank friction coefficient} = 2k_f |U_{s0} + U_{r0}|$$

Under steady-state conditions, the time derivatives of all variables are zero. Hence, from Equation 9.72, using total values rather than incremental values, we have

$$U_{s0} + U_{r0} = T_s \frac{dH_r}{dt} = 0$$

Therefore, $\phi_s = 0$ and Equation 9.73 reduces to

$$h_r = h_s \tag{9.74}$$

Application of Equations 9.60 and 9.61 to the penstock yields

$$h_t = h_r \text{sech}(T_{ep}s) - Z_p u_t \tanh(T_{ep}s) - \phi_p u_t \tag{9.75}$$

$$u_p = u_t \cosh(T_{ep}s) + \frac{1}{Z_p} h_t \sinh(T_{ep}s) \tag{9.76}$$

where

$$Z_p = \text{hydraulic impedance of penstock}$$
$$\phi_p = \text{friction coefficient of penstock} = 2k_p |U_{t0}|$$
$$T_{ep} = \text{penstock elastic time}$$

From Equations 9.70 to 9.74, the tunnel and surge tank transfer function may be written as

$$F_1(s) = -\frac{h_s}{u_p} = \frac{\phi_c + sT_{Wc}}{1 + sT_s\phi_c + s^2 T_{Wc}T_s} \tag{9.77}$$

The overall transfer function relating the water velocity to head at the turbine, formed by using Equations 9.74 to 9.77, is

$$F(s) = \frac{u_t}{h_t} = -\frac{1+\left[F_1(s)/Z_p\right]\tanh(T_{ep}s)}{\phi_p+F_1(s)+Z_p\tanh(T_{ep}s)}$$

$$= -\frac{1+\left[F_1(s)/Z_p^2\right]F_2(s)}{\phi_p+F_1(s)+F_2(s)}$$

(9.78)

where

$$F_2(s) = Z_p\tanh(T_{ep}s)$$

(9.79)

As noted earlier, $F(s)$ represents a distributed-parameter system and an infinite product expansion is required to accurately represent the function $F_2(s)$. A lumped-parameter approximation to $F_2(s)$, with $n=1$ in the expression for $\tanh(T_{ep}s)$, is given by Equation 9.68.

The complete hydraulic system is represented by the transfer function $F(s)$ relating turbine velocity (flow) to the turbine head. Figure 9.15 shows the block diagram of the hydraulic turbine with $F(s)$ representing the hydraulic system. The diagram is based on Figure 9.7 and has been modified to allow representation of water hammer and surge tank effects.

The following are sample values of parameters associated with the representation of $F(s)$ [6]:

$T_{ec} = 13$ s $Z_c = 4$ $\phi_c = 0.009$ $T_s = 900$ s

$T_{ep} = 0.25$ s $Z_p = 4$ $\phi_p = 0.001$

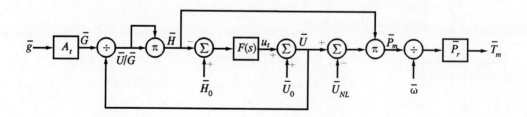

Figure 9.15 Hydraulic turbine block diagram including water hammer and surge tank effects

If there is no surge tank, T_{Wc} and ϕ_c are zero, and the transfer function $F_1(s)$ vanishes in Equation 9.78. The transfer function $F(s)$ reduces to that given by Equation 9.67. If in addition the water hammer and friction effects in the penstock are neglected, $F(s)$ simplifies to the conventional form

$$F(s) = -\frac{1}{sT_{Wp}} \tag{9.80}$$

If water hammer effects in the penstock are to be neglected and surge tank is to be modelled, $F_2(s)$ in Equation 9.78 is set equal to sT_{Wp}.

9.1.5 Guidelines for Modelling Hydraulic Turbines

The modelling detail required for any given study depends on the scope of the study and the system characteristics. The following are general guidelines for the selection of appropriate models.

Governor tuning studies

From the results of Example 9.4, we see that there could be significant differences between frequency response characteristics of the exact elastic model and the simplified inelastic model of the water column, beyond 0.1 Hz. These differences have negligible effect on the conventional governor with transient droop compensation or the equivalent PI controller. The larger bandwidth of PID controllers requires a more accurate representation of the penstock water column. Usually a lumped-parameter approximation (with $n=1$ or 2) should be acceptable.

The surge tank natural period is on the order of several minutes. Its representation in governor tuning studies is usually not necessary.

Transient stability studies

The hydraulic turbine governors have a very slow response from the viewpoint of transient stability. Their effects are likely to be more significant in studies of small isolated systems. A nonlinear turbine model assuming inelastic water column (Figure 9.7) would be adequate for such studies. The speed governor model should account for gate position and rate limits (Figure 9.10).

Small-signal stability studies

The speed governors have negligible effect on local plant mode oscillations of frequencies on the order of 1.0 Hz. However, the effect on low frequency interarea oscillations of frequencies below 0.5 Hz may be significant. These effects can be modelled adequately by linearizing the nonlinear turbine-penstock model of Figure 9.7

and the governor model of Figure 9.10. For plants with long penstocks, it may be prudent to consider the water hammer effects.

Small-signal studies are also ideally suited for investigating interactions between the hydraulic system dynamics and the network power oscillations.

Long-term dynamic studies

Depending on the nature of the problem, a very detailed nonlinear representation of the turbine and water column dynamics may be required. This includes travelling wave effects and surge tank dynamics. Such studies are invaluable for studying problems associated with special plant layouts and establishing appropriate operating procedures.

9.2 STEAM TURBINES AND GOVERNING SYSTEMS

A steam turbine converts stored energy of high pressure and high temperature steam into rotating energy, which is in turn converted into electrical energy by the generator. The heat source for the boiler supplying the steam may be a nuclear reactor or a furnace fired by fossil fuel (coal, oil, or gas).

Steam turbines with a variety of configurations have been built depending on unit size and steam conditions. They normally consist of two or more turbine sections or cylinders coupled in series. Each turbine section consists of a set of moving blades attached to rotor and a set of stationary vanes. The moving blades are called buckets. The stationary vanes, referred to as nozzle sections, form nozzles or passages in which steam is accelerated to high velocity. The kinetic energy of this high velocity steam is converted into shaft torque by the buckets.

A turbine with multiple sections may be either *tandem-compound or cross-compound*. In a tandem-compound turbine, the sections are all on one shaft, with a single generator. In contrast, a cross-compound turbine consists of two shafts, each connected to a generator and driven by one or more turbine sections; however, it is designed and operated as a single unit with one set of controls. The cross-compounding results in greater capacity and improved efficiency, but is more expensive. It is seldom used now; most new units placed in service in recent years have been of the tandem-compound design.

Fossil-fuelled units can be of tandem-compound or cross-compound design. Typical configurations of tandem-compound turbines and cross-compound turbines for fossil-fuelled units are shown in Figures 9.16 and 9.17, respectively. Tandem-compound units run at 3600 r/min. Cross-compound units may have both shafts rotating at 3600 r/min, or more commonly, one shaft at 3600 r/min and the other at 1800 r/min.[1]

---

[1] Turbine speeds given here correspond to 60 Hz systems. For 50 Hz systems, the corresponding speeds are 3000 r/min and 1500 r/min.

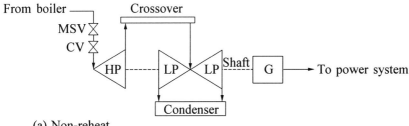

(a) Non-reheat

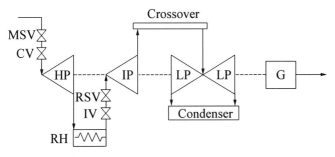

(b) Single reheat

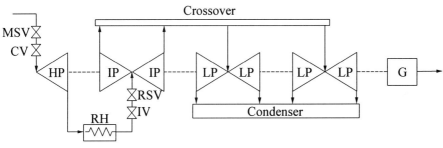

(c) Single reheat

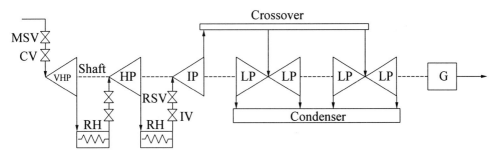

(d) Double reheat

Figure 9.16 Common configurations of tandem-compound
steam turbine of fossil-fuelled units

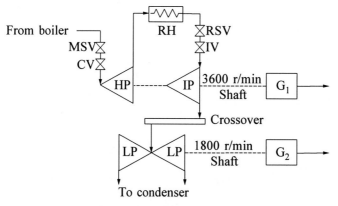

(a) Single reheat, 3600/1800 r/min shaft speeds

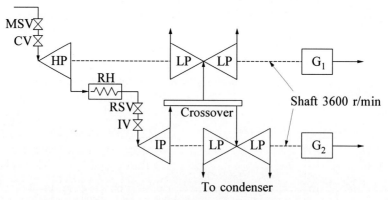

(b) Single reheat, 3600/3600 r/min shaft speeds

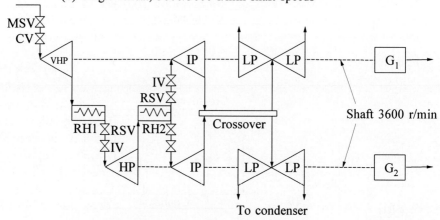

(c) Double reheat, 3600/3600 r/min shaft speeds

Figure 9.17 Examples of cross-compound steam turbine configurations

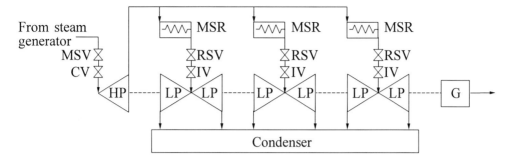

Figure 9.18 An example of nuclear unit turbine configuration

Depending on the turbine configuration, fossil-fuelled units consist of high pressure (HP), intermediate pressure (IP), and low pressure (LP) turbine sections. They may be of either *reheat* type or *non-reheat* type. In a reheat type turbine, the steam upon leaving the HP section returns to the boiler, where it is passed through a reheater (RH) before returning to the IP section. Reheating improves efficiency.

Some units have neither an IP turbine section nor a reheater, in which case the steam passes directly to the LP section. On the other hand, some units have two reheater sections.

The steam exhausted from the turbine is expanded to subatmospheric pressure and condensed in a condenser before returning to the boiler to repeat the cycle.

Nuclear units usually have tandem-compound turbines and run at 1800 r/min. A typical configuration of the turbine is shown in Figure 9.18. It consists of one HP section and three LP sections. The HP exhaust steam passes through the moisture-separator-reheater (MSR) before entering the LP turbines. The moisture separator reduces moisture content of the steam entering the LP section, thereby reducing moisture losses and erosion rates. High-pressure steam is used to reheat the HP exhaust (not shown in figure).

As shown in Figures 9.16, 9.17 and 9.18, large steam turbines for fossil-fuelled and nuclear units are equipped with four sets of valves: main inlet stop valves (MSVs) and control valves (CVs), and the reheater stop valves (RSVs) and intercept valves (IVs). Normally, there are at least two of each of these valves in parallel. Many turbines actually have four or more control valves, operating either in parallel or sequentially. The stop valves are primarily emergency trip valves and are not normally used for control of speed and load. The main inlet control (governor) valves modulate the steam flow through the turbine during normal operation. Control valves as well as intercept valves are responsive to overspeed following a sudden loss of electrical load.

The control valves are usually of the plug diffuser type whereas the intercept valves may be either plug type or butterfly type (used normally for nuclear units). The steam flow area versus valve position characteristics of the two types of valves are shown in Figure 9.19.

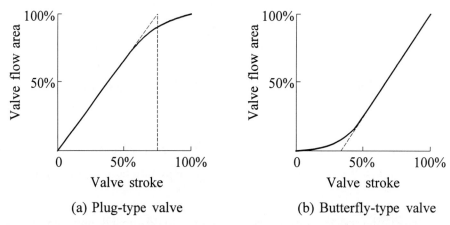

(a) Plug-type valve   (b) Butterfly-type valve

Figure 9.19 Steam valve flow area versus position characteristics

In the following sections we describe the characteristics and modelling of steam turbines and the associated governing systems. In addition, we discuss the requirements for protecting steam turbines during abnormal frequency conditions.

Torsional characteristics of the turbine-generator shaft system and their impact on the performance of the power system are considered in Chapter 15.

9.2.1 Modelling of Steam Turbines

Before we develop the model of a complete turbine, let us first derive the transfer function of a steam vessel and the expression for power developed by a turbine stage.

Time constant of a steam vessel

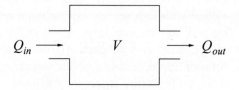

Figure 9.20 Steam vessel

The continuity equation for the vessel is

$$\frac{dW}{dt} = V\frac{d\rho}{dt} = Q_{in} - Q_{out} \tag{9.81}$$

where

W = weight of steam in the vessel (kg)
 = $V\rho$
V = volume of vessel (m^3)
ρ = density of steam (kg/m^3)
Q = steam mass flow rate (kg/s)
t = time (s)

Assuming the flow out of the vessel to be proportional to pressure in the vessel,

$$Q_{out} = \frac{Q_0}{P_0}P \qquad (9.82)$$

where

P = pressure of steam in the vessel (kPa)
P_0 = rated pressure
Q_0 = rated flow out of vessel

With *constant temperature* in the vessel,

$$\frac{d\rho}{dt} = \frac{dP}{dt}\frac{\partial\rho}{\partial P} \qquad (9.83)$$

The change in density of steam with respect to pressure ($\partial\rho/\partial P$) at a given temperature may be determined from steam tables.
From Equations 9.81, 9.82, and 9.83, we have

$$Q_{in} - Q_{out} = V\frac{\partial\rho}{\partial P}\frac{dP}{dt}$$

$$= V\frac{\partial\rho}{\partial P}\frac{P_0}{Q_0}\frac{dQ_{out}}{dt} \qquad (9.84)$$

$$= T_V\frac{dQ_{out}}{dt}$$

where

$$T_V = \frac{P_0}{Q_0}V\frac{\partial\rho}{\partial P} \qquad (9.85)$$

In Laplace form, Equation 9.84 may be written as

$$Q_{in} - Q_{out} = T_V s Q_{out}$$

or

$$\frac{Q_{out}}{Q_{in}} = \frac{1}{1 + T_V s} \qquad (9.86)$$

Equation 9.86 represents the transfer function of the steam vessel, and T_V is its time constant.

Torque developed by a steam turbine stage

In modern steam turbines the force on each rotor blade, and hence the turbine torque, is proportional to the steam flow rate. Thus

$$T_m = kQ \qquad (9.87)$$

where k is a proportional constant.

Example 9.5

The following data relate to a reheater steam volume:

$$Q_0 = 230 \text{ kg/s} \qquad V = 115 \text{ m}^3$$
$$P_0 = 4140 \text{ kPa} \qquad \frac{\partial \rho}{\partial P} = 0.0035$$

Calculate the reheater time constant.

Solution

From Equation 9.85, the reheater time constant is

$$T_R = \frac{P_0}{Q_0} V \frac{\partial \rho}{\partial P}$$

$$= \frac{4140}{230} \times 115 \times 0.0035 = 7.25 \text{ s}$$

∎

Complete turbine model

To illustrate the modelling of steam turbines, let us consider a fossil-fuelled single reheat tandem-compound turbine, a type in common use. The basic configuration identifying the turbine elements that need to be considered for purposes of modelling is shown in Figure 9.21(a).

Steam enters the HP section through the control valve and the inlet piping. The housing for the control valves is called the steam chest. A substantial amount of steam is stored in the chest and the inlet piping to the HP section. The HP exhaust steam is passed through the reheater. The reheat steam flows into the IP turbine section through the reheat intercept valve (IV) and the inlet piping. The crossover piping provides a path for the steam from IP section exhaust to the LP inlet. Since the stop valves merely provide a backup means of stopping steam flow, they need not be modelled in system studies and are not shown in Figure 9.21(a).

As indicated earlier, the control valves modulate the steam flow through the turbine for load/frequency control during normal operation. The response of steam flow to a change in control valve opening exhibits a time constant T_{CH} due to the charging time of the steam chest and the inlet piping to the HP section. This time constant is on the order of 0.2 s to 0.3 s.

The intercept valve is normally used only for rapid control of turbine mechanical power in the event of an overspeed. It is very effective for this purpose, since it is ahead of the reheater and controls steam flow to IP and LP sections, which generate nearly 70% of the total turbine power. The steam flow in the IP and LP sections can change only with the buildup of pressure in the reheater volume. The reheater holds a substantial amount of steam and the time constant T_{RH} associated with it is in the range of 5 s to 10 s. The steam flow into the LP sections experiences an additional time constant T_{CO} associated with the crossover piping; this is on the order of 0.5 s.

Figure 9.21(b) shows the block diagram representation of the tandem-compound reheat turbine. The model accounts for the effects of inlet steam chest, reheater, and the nonlinear characteristics of the control and intercept valves. The representation of the reheater differs from those of steam chest and LP inlet crossover piping. This is to allow computation of reheater pressure to account for the effects of intercept valve actuation. Care must be taken in selecting a per unit system for specifying the turbine parameters and variables. A convenient per unit system is one with base power equal to the *maximum turbine power* at rated main steam pressure with the control valves fully open. In this system, CV position is 1.0 pu when fully open. The sum of the power fractions of the various turbine sections $(F_{HP}+F_{IP}+F_{LP})$ is equal to 1.0.

The per unit turbine power P_{mech} so computed is multiplied by the ratio of P_{MAX} in MW to MVA base to obtain the turbine mechanical power in per unit of a common MVA base used for representing the complete power system. In this system per unit torque is defined as

$$\bar{T}_m = \frac{\omega_{base}}{P_{base}} T_m \tag{9.88}$$

Thus, at steady state, per unit mechanical torque is equal to per unit mechanical power.

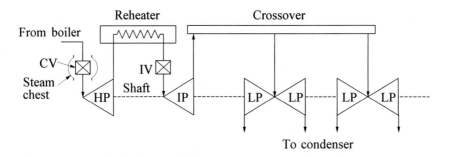

(a) Turbine configuration

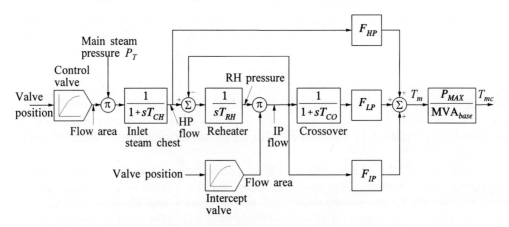

(b) Block diagram representation

Parameters

T_{CH} = time constant of main inlet volumes and steam chest
T_{RH} = time constant of reheater
T_{CO} = time constant of crossover piping and LP inlet volumes
T_m = total turbine torque in per unit of maximum turbine power
T_{mc} = total turbine mechanical torque in per unit of common MVA base
P_{MAX} = maximum turbine power in MW
F_{HP}, F_{IP}, F_{LP} = fraction of total turbine power generated by HP, IP,
                LP sections, respectively
MVA_{base} = common MVA base

Figure 9.21 Single reheat tandem-compound steam turbine model

Typical values of parameters of the model shown in Figure 9.21 applicable to a tandem-compound single reheat turbine of fossil-fuelled units are

$$F_{HP} = 0.3 \qquad F_{IP} = 0.3 \qquad F_{LP} = 0.4$$
$$T_{CH} = 0.3 \text{ s} \qquad T_{RH} = 7.0 \text{ s} \qquad T_{CO} = 0.5 \text{ s}$$

The model of Figure 9.21 can also be used to represent a nuclear unit turbine having the configuration shown in Figure 9.19. In this case, there is no IP section and $F_{IP}=0$. Typical values of the parameters are

$$F_{HP} = 0.3 \qquad F_{IP} = 0.0 \qquad F_{LP} = 0.7$$
$$T_{CH} = 0.3 \text{ s} \qquad T_{RH} = 5.0 \text{ s} \qquad T_{CO} = 0.2 \text{ s}$$

This model neglects main steam flow used for reheating the HP exhaust.

The most significant time constant encountered in controlling steam flow and the turbine power is that associated with the reheater. Therefore, the responses of the reheat turbines are significantly slower than those of the non-reheat turbine.

Simplified transfer function

From Figure 9.21(b), a simplified transfer function of the turbine relating perturbed values of the turbine torque (ΔT_m) and control valve position (ΔV_{CV}) may be written as follows:

$$
\frac{\Delta T_m}{\Delta V_{CV}} = \frac{F_{HP}}{1+sT_{CH}} + \frac{1-F_{HP}}{(1+sT_{CH})(1+sT_{RH})}
$$

$$
= \frac{1+sF_{HP}T_{RH}}{(1+sT_{CH})(1+sT_{RH})}
$$

(9.89)

In writing the above transfer function, it is assumed that T_{CO} is negligible in comparison with T_{RH}. In addition, the control valve characteristic is assumed to be linear.

Figure 9.22 shows the response of a tandem-compound turbine with $T_{RH}=7$ s, for a ramp down of control valve opening by 0.1 pu in 1 second. It is interesting to compare these results with those of the hydraulic turbine shown in Figure 9.4. It is clear that the response of a steam turbine has no peculiarity such as that exhibited by a hydraulic turbine due to water inertia. The governing requirements of steam turbines, in this respect, are more straightforward. The control action, as shown in Example 9.6, is stable with the normal speed regulation of about 5%, and there is no need for transient droop compensation.

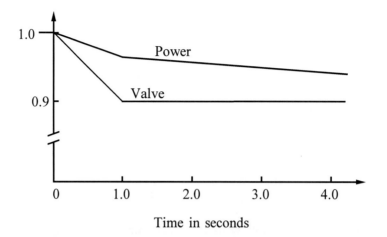

Figure 9.22 Steam turbine response to a 1-second ramp change in CV opening $T_{RH} = 7.0$ s, $F_{HP} = 0.3$; T_{CH} and T_{CO} negligible

Example 9.6

Figure E9.6 shows a simplified block diagram of speed control of a thermal-generating unit feeding an isolated load. The steam turbine is of the tandem-compound single reheat type whose transfer function may be approximated by

$$\frac{\Delta T_m}{\Delta V_{cv}} = \frac{1 + F_{HP} T_{RH} s}{1 + T_{RH} s}$$

where $T_{RH} = 6.0$ s and $F_{HP} = 0.333$. The mechanical starting time T_M of the generator is 12.0 s.

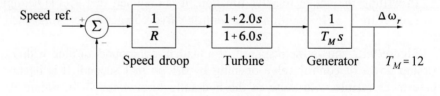

Figure E9.6

Determine (i) the lowest value of R for which the speed control is stable, and (ii) the value of R for which the speed control action is critically damped.

Solution

(i) The characteristic equation $(1+GH=0)$ of the closed-loop system is

$$1+\left(\frac{1+2.0s}{1+6.0s}\frac{1}{12s}\frac{1}{R}\right) = 0$$

This simplifies to

$$s^2+\frac{12R+2}{72R}s+\frac{1}{72R} = 0$$

This is in the standard form

$$s^2+2\zeta\omega_n s+\omega_n^2 = 0$$

For stability the following conditions must be satisfied

$$\frac{1}{72R}>0 \quad \text{or} \quad R>0$$

and

$$\frac{12R+2}{72R}>0 \quad \text{or} \quad R>-\frac{1}{6}$$

Therefore, any positive value of R will result in a stable response.

With the standard 4 to 5% droop, the speed control is quite stable. Unlike a hydraulic turbine, the steam turbine does not require transient droop compensation.

(ii) For critical damping, $\zeta=1.0$. Therefore,

$$2\zeta\omega_n = 2\times1.0\times\frac{1}{\sqrt{72R}} = \frac{12R+2}{72R}$$

Hence,

$$288R = (12R+2)^2$$

Simplifying, we find

$$R^2-1.667R+0.0278 = 0$$

Solving for R, we have

$$R = 0.017 \quad \text{or} \quad 1.65$$

From practical considerations, the lower of the above two values is of significance. Thus, critical damping can be obtained with $R=1.7\%$, i.e., a gain of 59. ∎

Detailed generic model

A generic model structure applicable to all commonly encountered steam turbine configurations is shown in Figure 9.23. By neglecting appropriate time constants and setting some of the power fractions to zero, this model can be used to represent any of the turbine configurations shown in Figures 9.16 and 9.17.

The time constant T_1 represents the main inlet volume and steam chest time constant. The time constants T_2 and T_3 represent reheater time constants. For a single reheat turbine, T_2 is neglected and K_2 and K_3 are set to zero. For example, a single reheat cross-compound turbine of Figure 9.17(b) may be represented by setting the parameters as follows:

$$
\begin{array}{lllll}
K_1 = F_{HP} & K_2 = 0 & K_3 = 0 & K_4 = 0 & K_5 = F_{IP} \\
K_6 = F_{LP1} & K_7 = 0 & K_8 = 0 & K_9 = F_{LP2} & \\
T_1 = T_{CH} & T_3 = T_{RH} & T_4 = T_{CO1} & T_5 = T_{CO2} & \\
T_2 \text{ is neglected} & & & &
\end{array}
$$

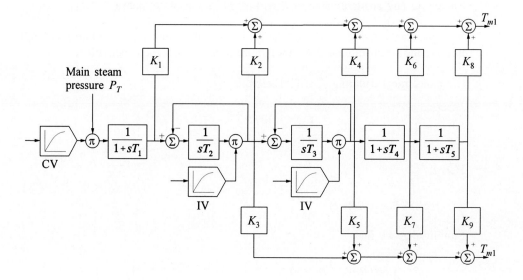

Figure 9.23 Generic model for steam turbines

Enhanced model for significant IV control

For simulations involving closing of IVs for significant periods causing high reheater pressure (for example, fast valving for enhancement of transient stability) a more rigorous representation of the turbine may be necessary [17,18]. An enhancement of the model of Figure 9.21 to account for the effects of reheater pressure is shown in Figure 9.24. The negative feedback of reheat pressure to the HP section accounts for the reduction in pressure difference across the section when the reheater pressure is high. The non-windup limit associated with the reheat pressure accounts for safety valve action.

The multiplier A accounts for the increase in flow through the intercept valves due to the pressure difference across the valve. As the valve opens, after being closed for some time, the flow is limited by the critical pressure ratio. Based on the equation for compressible flow of an ideal gas, assuming an isentropic expansion [19], the expression for the multiplier A is given by

$$A = A_k \sqrt{r^{2/k} - r^{(k+1)/k}} \tag{9.90}$$

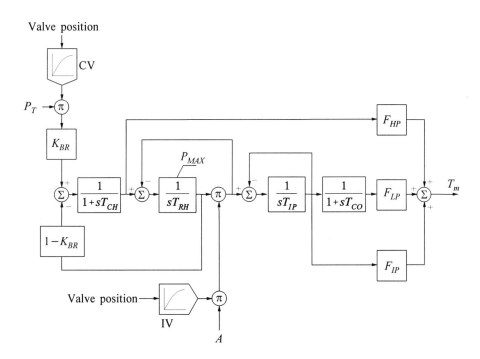

Figure 9.24 Enhanced steam turbine model

where

$$r = \frac{\text{downstream pressure}}{\text{upstream pressure}}$$

$$= \frac{\text{IP turbine pressure}}{\text{RH pressure}}(1-P_{drop}) \quad \text{with } r \ge r_{critical}$$

$$P_{drop} = \text{pressure drop across the valve}$$

For superheated steam (fossil-fuelled units),

$$k = 1.3 \quad \text{and} \quad r_{critical} = 0.547$$

and for wet or saturated steam (nuclear unit),

$$k = 1.135 \quad \text{and} \quad r_{critical} = 0.577$$

Typically P_{drop} is 2% (i.e., 0.02 pu) at full load with valves wide open. In the model, a block has been added, accounting for time constant T_{IP} of IP inlet piping and allowing for computation of IV downstream pressure. IP turbine pressure is required for the calculation of the multiplier A. The critical flow condition that exists as the IVs open is in fact largely caused by the low pressure in IP turbine.

Sample data for the parameters of the model of Figure 9.24 are

$$T_{CH} = 0.42 \text{ s} \qquad T_{RH} = 4.2 \text{ s} \qquad T_{IP} = 0.1 \text{ s} \qquad T_{CO} = 0.7 \text{ s}$$
$$K_{BR} = 1.073 \qquad F_{HP} = 0.25 \qquad F_{IP} = 0.25 \qquad F_{LP} = 0.5$$

For a nuclear unit turbine, as noted earlier, there is no IP section; T_{IP} represents the time constant of the inlet piping.

9.2.2 Steam Turbine Controls

The governing systems for steam turbines have three basic functions: normal speed/load control, overspeed control, and overspeed trip. In addition, the turbine controls include a number of other functions such as start-up/shutdown controls and auxiliary pressure control.

The *speed/load control* function is similar to that for hydraulic units. It is a fundamental requirement for any generating unit. In the case of steam turbines, it is achieved through control of the CVs. The speed control function provides the governor with a 4 to 5% speed droop. This enables the generating unit to operate satisfactorily in parallel with other units with proper division of load. The load control function is achieved by adjusting the speed/load reference. The principles of speed/load control are discussed in detail in Chapter 11. The net effect of this control is to adjust the position of the CVs to control admission of steam to the turbine.

The overspeed control and protection requirements are peculiar to steam turbines, and are of critical importance for their safe operation. The integrity of the turbine depends on the ability of the turbine controls to limit the speed of the rotor, following a reduction of electrical load, to well below the typical design maximum speed of 120%. Steam turbines of the reheat type have two separate valving systems that can be used for rapidly controlling the steam energy supplied to the turbine, one system involving the CVs and the other involving the IVs.

The *overspeed control* is the first line of defense against excessive speed. Its function is to limit the overspeed that occurs on partial or full load rejection and to return the turbine to a steady-state condition such that the turbine is ready for reloading. The objective is to prevent overspeed tripping following a load rejection. Typically, the overspeed trip is set at 110 to 115% of rated speed; the overspeed control attempts to limit overspeed to about 0.5 to 1% below the overspeed trip level. The overspeed control involves fast control of the CVs as well as the IVs. The use of the IVs is very effective in this regard, since they control steam flow to IP and LP turbine sections, which together develop 60 to 80% of the total turbine power. Because of the large amount of stored steam in the reheater, the rapid closure of the CVs alone would not be effective in limiting overspeed.

The *overspeed* or *emergency* trip is a backup protection in the event of failure of normal and overspeed control to limit the rotor speed to a safe level. The overspeed trip is designed to be independent of the overspeed control. The trip function, in addition to fast closing the main and reheat stop valves, trips the boiler.

The characteristics of steam valves, as shown in Figure 9.19, are highly nonlinear. Therefore, compensation is often used to linearize the steam flow response with respect to the control signal. The following are the alternative forms of compensation used:

- A forward loop series compensation comprising a function generator having a characteristic reciprocal to that of the steam valve

- A minor loop feedback compensation comprising a function generator with a characteristic similar to that of the valve

- A major loop feedback compensation in the form of a proportional feedback around the steam valve

Typically, four or more parallel operated CVs are used. Each of these valves admits steam through a nozzle section. The different nozzle sections are distributed around the periphery of the first stage turbine section. The CVs may be controlled in unison (full arc) or in sequence (partial arc).

Under normal operation, the CVs are opened sequentially; with only some of the CVs open, the steam admission is along only a partial arc, rather than uniformly along the circumference. This is referred to as "partial-arc admission" mode of operation [21]. The advantage of this mode of operation is that it ensures higher efficiency at partial load.

During start-up, however, the steam is admitted symmetrically through all nozzle sections to reduce thermal stresses. The CVs are maintained fully open, and stop valves are used to control steam flow. This mode of operation, referred to as "full-arc admission" operation, is used until the load reaches a specified level. Depending on the type of turbine governing system, the control of the stop valves is achieved through either auxiliary speed control equipment or the main speed control. Additional measures to minimize thermal stresses in the turbine include limiting the rate of speed increase and the rate of load increase. The speed ramping during start-up is handled by an acceleration control.

Systems used for the above turbine control functions and other auxiliary functions have continually evolved over the years. Older turbine governor designs used mechanical-hydraulic control. Electrohydraulic control was introduced in the 1960s, and its use has been gradually growing. Most governors supplied today are electrohydraulic or digital electrohydraulic. References 21, 22, and 23 provide descriptions of the elements of the control systems. The following are brief descriptions of the functional characteristics of these elements and their models.

Mechanical-hydraulic control (MHC)

The functional block diagram of an MHC system used for governing a steam turbine is shown in Figure 9.25. The basic elements of the governing system are a speed governor, speed relays, and hydraulic servomotors.

The *speed governor* is essentially a mechanical transducer which transforms the shaft speed into a position output, as shown schematically in Figure 9.26. While mechanical speed governors come in different forms, they all operate on the same basic principle of the classical flyball governor: rotor speed signal (ω_r) is converted

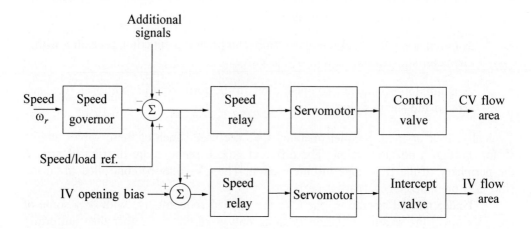

Figure 9.25 Functional block diagram of MHC turbine-governing system

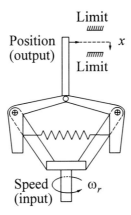

Figure 9.26 Mechanical speed governor [22]

to linear displacement by means of centrifugal forces opposed by a spring. Their performance equation is

$$x = K_G \omega_r \qquad (9.91)$$

The speed governor output is compared with a speed/load reference determined by the speed changer position. The resulting error signal is used to control the CVs as well as the IVs. However, for normal speed/load control only the CVs respond. The IVs are held fully open by a bias (IV opening bias) signal. On overspeed, due to the resulting large speed error signal, the bias is overcome and IVs are closed rapidly. When the control signal is restored to a value less than the bias, IVs are again fully opened. Additional signals may be used, as described later, to provide special control and protection functions. The speed governor speeder rod cannot develop the forces necessary to control the steam valves. Therefore, a pilot valve and a spring-loaded servomotor (commonly referred to as *speed relay*) are used to amplify the speed governor signal. Figure 9.27 shows the basics of speed relay-pilot valve combination. A downward displacement of the speeder rod from the speed governor lowers the pilot valve. The oil supply system then forces the servomotor piston upward, allowing increased steam flow to the turbine. The transfer function of the speed relay has the forms shown in Figure 9.28.

On very large turbines, additional amplification to the energy levels necessary to move the steam valves is obtained using *hydraulic servomotors*. They use a high-pressure fire resistant fluid for auxiliary power. A schematic diagram of the servomotor is shown in Figure 9.29(a) and the block diagram representation is shown in Figure 9.29(b). The position limits may correspond to either the fully open valve position or the setting of a load limiter.

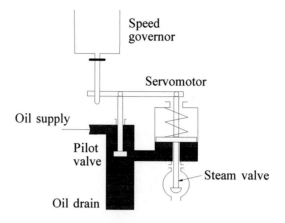

Figure 9.27 Speed relay pilot valve

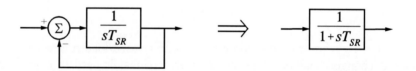

Figure 9.28 Alternative forms of speed relay representation

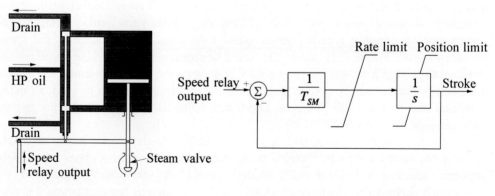

(a) Schematic diagram   (b) Block diagram model

Figure 9.29 Hydraulic servomotor

The basic elements and the general principle of operation of all MHC governing systems are similar. However, the details of the control logic used vary widely depending on the make and vintage of the turbine.

Modelling of MHC governors

For studies involving small deviations in speed (frequency), only normal speed regulation or primary speed control needs to be considered. Such studies include transient and small-signal stability studies. For normal-speed regulation, it is standard practice to use the CVs with proportional control on speed error. A generic model shown in Figure 9.30 can be used to model this turbine control function.

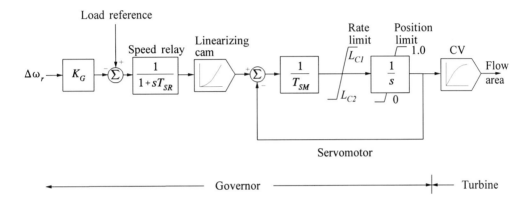

Figure 9.30 Generic speed-governing system model representing normal speed/load control function

As an approximation, the valve characteristic may be assumed to be perfectly compensated, and the valve and cam nonlinearities may be neglected.

Sample values of the parameters are:

$$K_G = 20 \ (5\% \ \text{droop}) \qquad T_{SR} = 0.1 \ \text{s} \quad T_{SM} = 0.3 \ \text{s}$$

$$\text{Rate limits:} \quad L_{C1} = 0.1 \ \text{pu/s (opening)}$$
$$L_{C2} = -1.0 \ \text{pu/s (closing)}$$

For studies involving significant turbine control action, the performance of overspeed controls and other related auxiliary devices needs to be modelled [18,24]. These controls have nonlinear and discrete characteristics, and the control logic is very specific to the make and vintage of the turbine. Their input signals could be rotor speed, acceleration, electrical power, generator current, or circuit-breaker opening. Although there are many variations of control logic used, the general principles of these controls are similar. We will illustrate this by providing several examples.

Figure 9.31 shows the block diagram representation of an MHC speed-governing system, including the overspeed control function, applicable to one make [24]. The model shown accounts for CV and IV controls, valve nonlinearities, linearizing cams, and an "auxiliary governor" for limiting overspeed. Under steady-state conditions and during speed deviations, the IVs are kept fully open by the opening bias (IVOB); only the CVs provide speed regulation. The auxiliary governor, when speed exceeds its setting V_1 (ranging from 1% to 3% over the rated speed), acts in parallel with the main governor so as to effectively increase the gain of the speed control loop by a factor of about 8. This causes the CVs as well as the IVs to close rapidly and thus limit overspeed.

Figure 9.32 shows another example of the MHC speed-governing system described in references 18 and 20. The CV position is determined by the speed relay output signal S_1. The IVs respond to the lower of the two signals applied to the *low-*

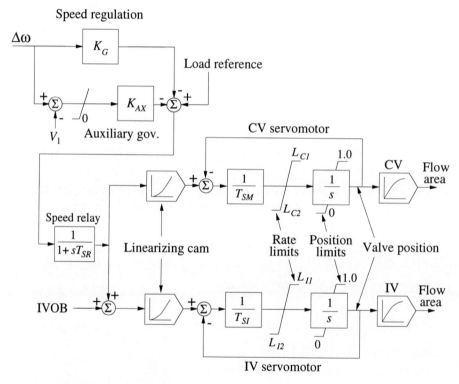

Typical values of parameters:

$K_G = 20$ $V_1 = 0.02$ $K_{AX} = 149$ $T_{SR} = 0.7$ IVOB = 1.17
$T_{SM} = 0.23$ $L_{CI} = 1.0$ $L_{C2} = -3.0$
$T_{SI} = 0.23$ $L_{II} = 1.0$ $L_{I2} = -2.5$

Figure 9.31 MHC turbine-governing system with auxiliary governor

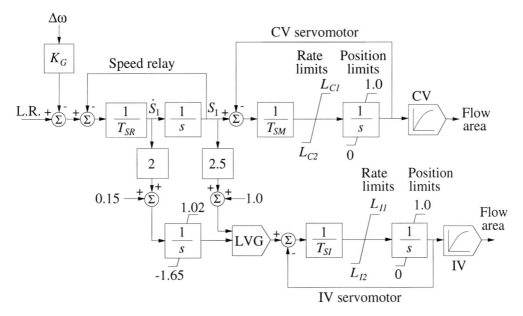

Typical values of parameters:

$K_G = 20$ $T_{SR} = 0.1$ s $T_{SM} = 0.2$ s $T_{SI} = 0.2$ s
$L_{CI} = 0.2$ $L_{C2} = -0.5$ $L_{II} = 0.2$ $L_{I2} = -0.5$

Figure 9.32 MHC turbine governing system. ©IEEE 1990 [18]

value gate (LVG); these signals are derived from S_1 and the derivative of S_1 (proportional to negative of rotor acceleration). For small speed deviations, the CVs respond to the normal speed/load control command which determines S_1, and the IVs remain fully open due to the opening bias (LVG input is at 1.02). During an overspeed condition, the IVs transiently respond by closing rapidly, driven by the lower of the two signals $\dot{S}_1$ and S_1 which depend on rotor acceleration and speed respectively. The control of IV through signal S_1 has a gain of 2.5 and a bias of 1.0. With K_G=20 (5% droop) and load reference at 100%, the signal S_1 becomes effective in controlling IV when $\Delta\omega \geq 5\%$ and the effective speed control gain is 50 (2% droop).

Electrohydraulic control (EHC)

The EHC systems use electronic circuits in place of mechanical components associated with the MHC in the low-power portions. EHC systems offer more flexibility and adaptability, allowing incorporation of a number of features that cannot be obtained on mechanical systems. This contributes to faster response and improved linearity. However, the overall functional requirements of the EHC systems are essentially very similar to those of the MHC system.

Figures 9.33 and 9.34 show two examples of EHC speed-governing systems.

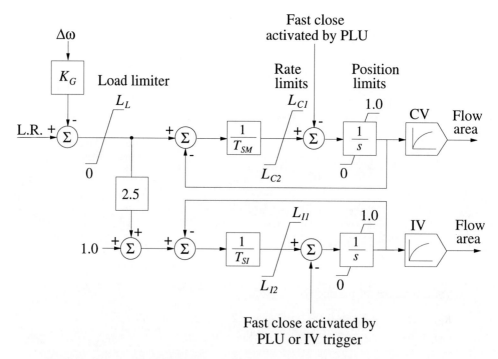

Typical values of parameters:

$K_G = 20$ $T_{SM} = 0.1$ $T_{SI} = 0.1$
$L_{CI} = 0.1$ $L_{C2} = -0.2$ $L_{II} = 0.1$ $L_{I2} = -0.2$

(a) Block diagram of governing system

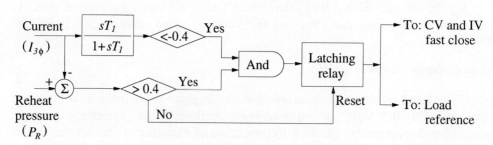

(b) Power load unbalance relay logic

Figure 9.33 EHC governing system with PLU relay and IV trigger

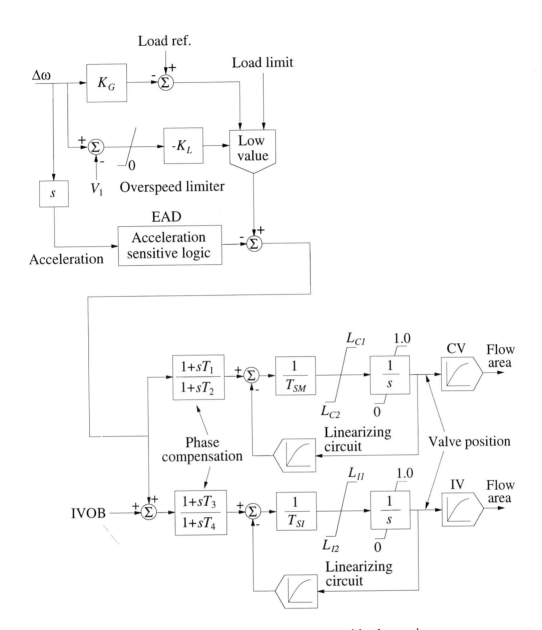

Figure 9.34 EHC governing system with electronic overspeed limiter and acceleration detection

The speed governor of Figure 9.33 has a similar steady-state IV speed versus flow characteristic to the MHC governor of Figure 9.32. In addition, it has two special features for limiting overspeed: IV trigger and power load unbalance relay [18,20].

The *IV trigger* is armed when the load (measured by reheat pressure) is greater than 0.1 per unit. It fast closes the IVs when the control error signal to the IV servovalve is less than −0.1 per unit. This represents a condition where per unit speed deviation $\Delta\omega$ is greater than $0.05LR+0.002$. Following an IV trip, the IV control is blocked for 1 second to permit hydraulic transients to decay, after which it is free to respond to speed control. The opening rate of the IV is limited to 0.1 pu/s by limiting the input signal to the servovalve.

The *power load unbalance* (PLU) *relay* is designed to fast close CVs and IVs under load rejection conditions that might lead to excessive overspeed. It is provided with selectivity to distinguish such conditions from stable system fault conditions. The PLU relay logic is shown in Figure 9.33(b). The relay circuit trips when the difference between turbine power and generator load exceeds a preset amount (0.4 pu) and the load decreases faster than a preset rate (equivalent to going from rated to zero load in about 35 ms). The turbine power is measured by a signal from the cold reheat pressure and the generator load by a signal derived from the three-phase currents. The use of current rather than electrical power helps discriminate between load loss and temporary system faults; under fault conditions current increases whereas power decreases. The tripping of the PLU relay causes the following actions:

- The CVs and IVs close completely.

- The load reference signal is removed from the CV and IV control signals.

- The load reference runs back to a minimum value at a rate of about 2.5% per second.

About 1 second after tripping, the CVs and IVs respond to their control signals. However, the load reference positions of these signals are not restored until the PLU relay resets, at which point the load reference runback stops. The load reference is not restored to the value prior to the disturbance, but remains at the value attained at the time the PLU relay resets.

The EHC speed-governing system shown in Figure 9.34 represents a fast-acting and highly responsive system [24]. Two forms of overspeed controls are provided: an electronic acceleration detection (EAD) function and an electronic overspeed limiting function. The EAD uses the governor end-of-shaft speed (with shaft torsional signals suitably filtered out) which it differentiates electronically to obtain acceleration. If acceleration exceeds the preset rate for a specified time, a prescribed valve opening and closing sequence is initiated, thereby limiting overspeed to an acceptable level. The overspeed limiter consists of a circuit which increases the governor gain by a factor of about 4 if speed exceeds a preset level of about 2%. Phase compensation circuits are provided as shown so that the phase characteristics of the control loops may be adjusted to ensure stable control performance.

Digital electrohydraulic (DEH) control system

A DEH system uses a digital controller, which is interfaced with the turbine valve actuators through an analog hybrid section [25]. The digital control system offers considerable flexibility since the control functions can be implemented through software.

The normal speed control function is set to provide the standard 4 to 5% regulation. The response of the governor is very fast, the time constant being on the order of 0.03 s. Filtering is normally provided to the speed signal to eliminate small variations in measured frequency. Additionally, programmable dead band may be provided to prevent governor response to small frequency variations.

In the DEH system described in reference 25, overspeed protection has two components: load overspeed anticipator (LOA) and overspeed sensing. The LOA function closes the CVs and the IVs rapidly when the main generator breaker opens, if the initial turbine power (determined by measuring IP section steam pressure) is greater than 30% of rated. The overspeed sensing function closes the CVs and IVs when speed exceeds 103% of rated; it is a discontinuous control causing the valves to close rapidly and then reopen. The LOA function does not affect the unit performance when connected to the power system. The overspeed sensing, however, could be triggered during severe system swings and should be designed so that it does not have an adverse effect on system performance. The modelling of the overall DEH turbine control is similar to that of Figures 9.33 and 9.34.

Initial pressure regulator

Often an auxiliary pressure control referred to as the *initial pressure regulator* (also known as *throttle pressure control*) is used to maintain the main steam pressure above a minimum value (usually 90% of rated pressure) [18,25]. When the pressure drops below the setpoint, the initial pressure regulator will introduce a closing signal, as shown in Figure 9.35. This effectively reduces the load reference at a rate of about 100% load/min, until the throttle pressure is restored above the setpoint or a minimum valve position (typically, corresponding to 20% flow) is reached.

The purpose of the initial pressure regulator is to protect the turbine against water induction from the boiler due to uncontrolled throttle pressure reduction during a boiler disturbance. However, the regulator may be triggered during a system transient, causing a rapid decrease in frequency. This in turn may result in a rapid load-pickup by the generators with the possibility of the throttle pressure dropping below the setpoint [25].

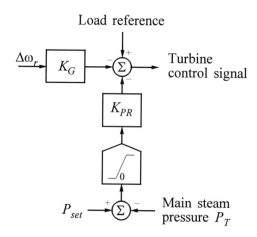

Figure 9.35 Initial pressure regulator

Generic model of steam turbine controls

A general model which can accommodate essentially all significant features of the different forms of steam turbine controls is shown in Figure 9.36. It allows representation of special control logic such as the power load unbalance (PLU) relay, IV trigger, fast valving (see Chapter 17, Section 17.1.8), acceleration and speed sensitive overspeed limit controls. It also includes facilities for representing feedback as well as series linearity compensation for valve nonlinearities, and phase compensation. The CV and IV controls have facilities for representing time constants associated with additional power amplification stages, and the speed governor model can account for transducer and filter time constant. Dead-band effects can be represented at five different locations.

9.2.3 Steam Turbine Off-Frequency Capability

Both the turbine and generator can tolerate only limited off-frequency operation; the turbine, however, is the more restrictive.

The principal risk in off-frequency operation of a steam turbine is the vibration and resonance of long low-pressure turbine blades. The blade vibration stresses are dependent on the excitation forces and the natural vibratory response characteristics of the blade's structural system. The predominant source of excitation is the stimulus produced by natural variations in steam flow. The magnitude of the excitation increases with increased steam flow.

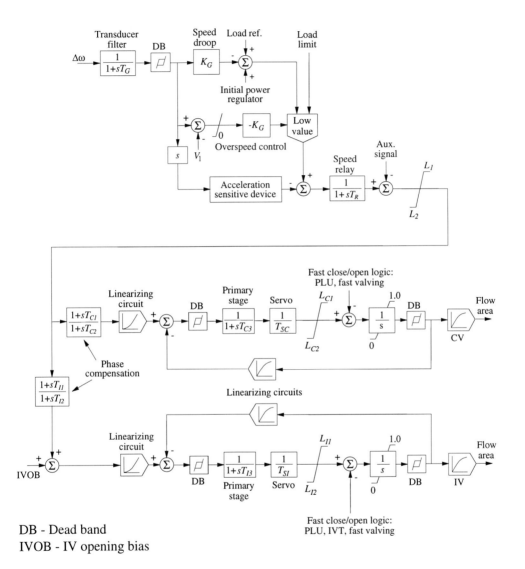

DB - Dead band
IVOB - IV opening bias

Figure 9.36 Generic steam turbine controls model

The vibratory response of a blade depends on the nearness of the excitation frequency to the blade natural frequency and on blade damping. For a resonant condition, the vibratory stress can be several hundred times greater than the stress during nonresonant operating conditions. The most critical blades are the ones in the last three rows in the LP turbine and, in some cases, the last row in the IP turbine. These blades have their natural frequencies controlled or tuned so that their vibration modes will not resonate at normal frequency operation.

There are three principal natural vibration modes associated with steam turbine blades [26]. The first is a tangential mode with blades vibrating in phase in the plane of maximum blade flexibility, perpendicular to the axis of the unit. The second mode is also an in-phase vibration, but with deflection essentially in an axial direction. The third is a torsional mode with the vibration of the blade group in approximately an axial direction.

The diagram shown in Figure 9.37, known as the Campbell diagram, illustrates how a change in turbine speed can provide steam excitation frequencies that coincide with the natural frequencies of a blade. The three nearly horizontal bands represent the characteristics of the three natural vibration modes of a long slender blade of an LP turbine section. The width of these bands indicates the scatter band and variations in natural frequencies due to manufacturing tolerances. The diagonal lines drawn at integral multiples (1 through 7) of turbine operating speed represent the stimulus frequencies inherent in the steam flow. An intersection of a blade natural vibration mode band and a steam stimulus frequency line represents a condition of resonance. Turbines are designed so that such resonant conditions are avoided at rated speed. However, departure from rated speed will bring the stimulus frequencies closer to one or more of the blade natural frequencies with the resulting higher vibrational stresses.

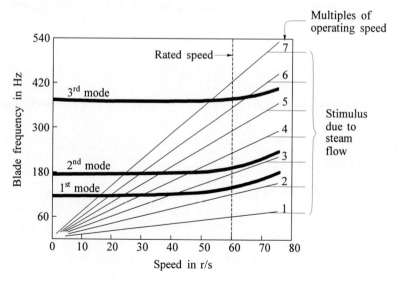

Figure 9.37 Typical Campbell diagram showing blade vibration characteristics

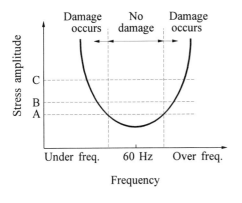

Figure 9.38 Increase in vibration amplitude with off-frequency operation

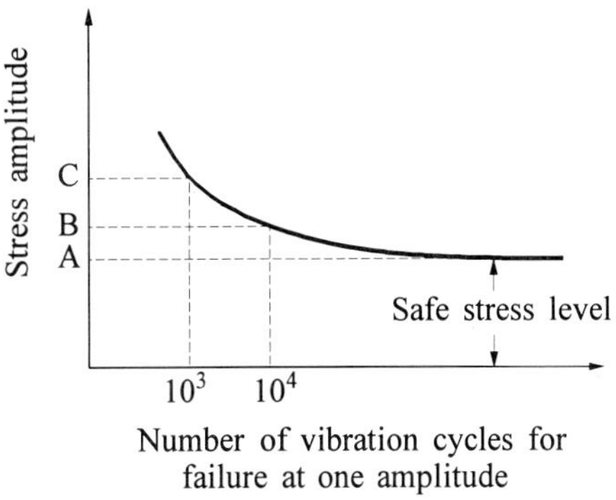

Figure 9.39 Stress versus number of cycles to failure

Figures 9.38 and 9.39 illustrate the relationship between blade failure due to fatigue and off-frequency operation. Below stress level A, the vibration stress amplitude is low enough that no damage results. Operation at stress level B would produce a failure at 10,000 cycles, and at level C failure would occur at 1000 cycles.

Operation at off-nominal frequencies is time-restricted depending on specific blade designs. Figure 9.40 is a composite representation of steam turbine off-nominal-frequency limitations of a large sample of turbines built by five different manufacturers [27]. The figure shows that sustained operation within the band 59.5 Hz to 60.5 Hz would not have any effect on blade life, while the dotted areas above 60.5 Hz and below 59.5 Hz are areas of restricted time operating frequency limits. Operation outside these areas is not recommended.

The characteristics shown in Figure 9.40 represent the composite operating limits of a large number of units and are useful for evaluating the requirements for protective relaying schemes. The applicable limit for a specific turbine may be less restrictive.

The effect of off-frequency operation in a given frequency band is cumulative, but independent of the time accumulated in any other band. The abnormal frequency capability curves are applicable whenever the unit is connected to the system. These curves also apply when the unit is not connected to the system, if it is operated at abnormal frequency while supplying its auxiliary load. During the period when the unit is being brought up to speed, is being tested at no-load for operation of overspeed trip device, or is being shut down, blade life will not be significantly affected if proper procedures are followed.

The most frequently encountered *overfrequency* condition is that of a sudden reduction of generator power resulting from a generator breaker trip. Under these conditions, the governor droop characteristic may result in a steady-state speed as high as 105% following the transient speed rise, and therefore speed reduction to near rated speed should be initiated promptly. For such sudden power reduction, units equipped with modern electrohydraulic turbine-governing systems provide this speed reduction automatically.

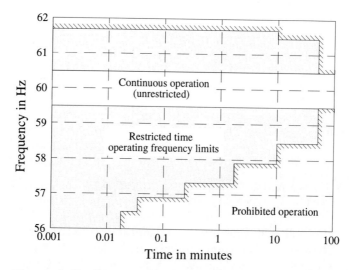

Figure 9.40 Steam turbine partial or full-load operating limitations during abnormal frequency, representing composite worst-case limitations of five manufacturers. ©ANSI/IEEE - 1987 [27]

For partial load reduction or system-islanding conditions, the system frequency will depend on the composite droop characteristic. Assuming a 5% droop characteristic, a generator-load mismatch of 50% would cause approximately a 2.5% (1.5 Hz) rise in frequency. Referring to Figure 9.40, the operating time-limit at this frequency is 35 minutes which is within the practical range of operator action to reduce speed settings. For higher frequencies, automatic control action such as use of overfrequency relay to initiate runback of the governor load reference is often employed.

Underfrequency operation of a steam turbine poses a more critical problem since generation cannot be increased more than full output. Underfrequency protective relays are normally employed to protect the unit. As an example, such relays may be set to trip the unit in 10 s if frequency remains below 57.5 Hz, or instantaneously if frequency drops below 56.0 Hz.

In order to prevent extended operation at lower than normal frequencies during system disturbances, load-shedding schemes are commonly employed to reduce the connected load to a level that can be supplied by the available generation. This is discussed in Chapter 11 (Section 11.1.7). If system load shedding is provided, then it is considered as the primary turbine underfrequency protection. Appropriate load shedding can cause the system frequency to return to normal before the turbine abnormal frequency limit is reached. Turbine underfrequency tripping should be considered as the last line of defense, as it may cause an area blackout.

9.3 THERMAL ENERGY SYSTEMS

9.3.1 Fossil-Fuelled Energy Systems [28]

In a fossil-fuelled power plant, fossil fuels such as coal, oil, or gas are used as the primary energy source to produce heat by combustion which, in turn, is transferred to the cold feed-water to generate (superheated) steam. Turbines convert the steam energy to mechanical energy which eventually produces electrical energy via the generators. The steam exhausting from the high pressure turbine is cycled back through the furnace for reheating (superheating) before it is admitted back to the lower pressure turbines. The exhaust steam, at close to saturation condition, is cooled in condensers (which are maintained at a very high vacuum - about 5 kPa) and converted back into liquid form. Large cooling water pumps provide the necessary heat removal capacity in the condensers. The condensate is then cycled back to the furnace as feed-water at high pressure, after various stages of pumping, feed-heating (to improve the energy conversion efficiency), and de-aerating.

Figure 9.41 illustrates the interrelationships among the various subsystems of a fossil-fuelled power plant.

Primary fuel system

The primary fuel system converts the primary fossil fuel into thermal energy. The raw fuel is transformed into thermal energy in the *furnace*. Fuel enters the furnace as controlled flow of oil, gas, or coal in the form of fine particles suspended in an airstream. A controlled amount of air is also injected to ensure complete combustion. This results in extremely high temperatures within the flame volume and the surrounding furnace. Energy is transferred to waterwall tubes carrying colder liquid from the drum or to the superheaters and reheaters carrying steam. The heat transfer is both radiative and convective.

The by-products of the combustion process are in both gaseous (flue gas) and solid (ash) forms (in coal fired plants). The solid by-products are removed in the furnace and only the gaseous by-products (flue gas) leave the furnace. Energy is extracted from the flue gases by various heat exchange systems. After passage through these heat exchanger banks (superheaters, reheaters, economizer, air heater, etc.) and ductworks, these gases are exhausted into the atmosphere via the induced draft (ID) fans through the chimneys or stacks.

The fuel flow enters the furnace through burner nozzles which, in some furnace designs, can be tilted upward or downward to position the flame in the furnace. The position of the flame affects the radiative heat transfer and is utilized in controlling superheat/reheat temperatures.

Both fuel and air entering the furnace are controlled to provide desired heat generation, while the ID fans are controlled to maintain the furnace at a suitable sub-atmospheric pressure. If a furnace (or pulverizer) trip occurs, it must be purged by air

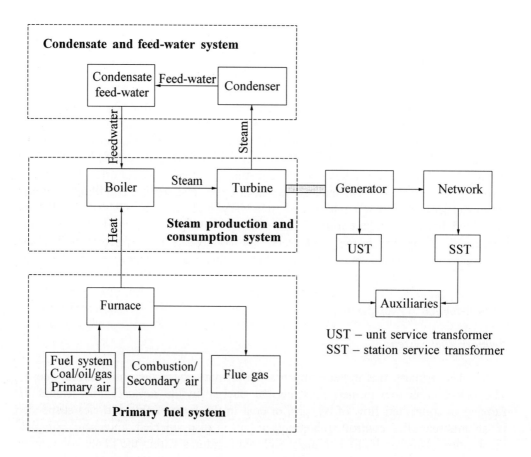

Figure 9.41 Subsystems of a fossil-fuelled power plant

for a preset minimum time before being placed in service again, to prevent duct explosions.

The *fuel system* supplies the fuel (coal, oil or gas) to the furnace. Oil- and gas-fired units can respond to load demand changes more quickly than coal-fired units. This is because solid fuels such as coal require additional processing, such as pulverizing and drying, before they can be used for combustion. Furthermore, pulverized coal systems make use of several pulverizers in parallel, each with a limited capacity and requiring an additional warm-up period when started. In the case of a rapid decrease in firing demand, the burner management controls may automatically trip a number of pulverizers. When pulverizers are tripped, the firing rate is limited to a level corresponding to the number of pulverizers remaining in service. Therefore, the re-loading rate is dictated by the time required for bringing the tripped or shutdown pulverizers back into service.

The *secondary air system* provides the supply of combustion air to the furnace windbox, whence it enters the furnace via fuel air and auxiliary air dampers. The source of the air is the forced draft (FD) fan. The demand for the flow rate is received from the fuel and air controller in the form of a desired flow rate to maintain adequate and complete combustion in the furnace. The air controller logic monitors the actual (measured) flow rate and manipulates inlet vanes of the FD fans to control the air flow rate. The air flow passes through secondary air heaters where it is preheated by extracting heat from exhaust flue gases before it enters the windbox. Usually, the windbox pressure is controlled by manipulation of a number of fuel and auxiliary air dampers so that a required differential pressure exists between the windbox and the furnace.

The *flue gas system* includes the flue gas path from furnace exit to the chimney stacks via the induced draft (ID) fans. The flow rate of the flue gas is controlled by manipulating the variable inlet vanes for the ID fans by the furnace pressure controller. The heating zones in the flue gas path include the primary and secondary superheaters, reheater, economizer, and primary and secondary air heaters. Each section in the flue gas path acts as a heat exchanger, transferring heat from hot gases to colder steam, feed-water, or air. The first two levels of heat exchangers (secondary superheater and reheater sections) receive radiative heat directly from the furnace in addition to convective heat transfer from the gases.

Steam production and utilization system

The steam is produced in the *boiler* and superheated in the superheater and reheater. The thermal energy of the steam is converted to mechanical energy in the various sections of the turbine. Boiler designs fall into two categories: (a) drum type boilers, and (b) once-through boilers. Figure 9.42 shows schematics of the two types of boilers.

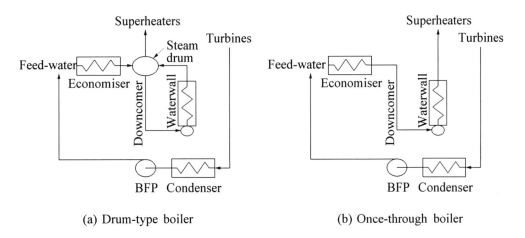

(a) Drum-type boiler   (b) Once-through boiler

Figure 9.42 Fossil-fuelled unit boiler types

Drum-type boilers:

Drum-type or recirculation boilers rely on natural or forced circulation of drum liquid to absorb energy from the hot furnace walls, called waterwalls, for generating steam. The boiler receives feed-water which has been preheated in the economiser and provides saturated steam outflow. The circulation loop consists of almost saturated water being carried from the drum via downcomer tube banks which is then passed through waterwall tubes (which, in effect, form the walls of the furnace - hence the nomenclature) where they absorb radiant and convective heat from the flame and hot flue gases. This system operates at subcritical pressures as long as there is some finite density difference between the steam and water phases. Low pressure subcritical boilers depend totally on the natural recirculation phenomena resulting from the static thermal head difference between the water in the downcomers and the steam-water mixture in the heated steam generating tubes. Higher pressure (but still subcritical) boilers usually make use of the circulating pump to supplement the thermal head. Recirculation boilers make use of a drum to separate steam from the recirculation water so that it can proceed to the superheaters as a heatable vapour; hence, recirculation boilers are referred to as drum type boilers.

The steaming rate is predominantly a function of heat absorbed in the furnace waterwalls. Any discrepancy from the steam utilization rate (in turbines) results in boiler pressure variation. To protect against overpressurization, relief valves are used.

Once-through boilers:

In a once-through boiler, there is no recirculation of water within the furnace. Instead, feed-water, at high pressure, flows straight through the waterwalls where it receives furnace heat and is converted to steam which then passes through the superheaters and eventually enters the HP turbine. The operation of the feed-water system, including the *boiler feed pump* (BFP), is needed to produce the through flow, and a suitable means must be provided to dispose of the circulated flow without incurring loss of heat or working fluid. This is normally done with a steam-turbine bypass system.

The once-through design has been used for both the high subcritical and the supercritical pressure ranges. In the supercritical range (i.e., above 22,120 kPa), the fluid's physical properties change continuously from those of a liquid (water) to those of a vapour (steam). There is no saturation temperature or two-phase mixture. Instead, the temperature rises steadily, and the specific heat and its rate of rise vary considerably during the process. Therefore, the nature of supercritical steam generation rules out the use of a boiler drum to separate steam from water.

Unlike in a drum type boiler, there is no drum level to be controlled in a once-through boiler. There are three major control loops involving megawatt, throttle pressure, and steam temperature. These are controlled by manipulating the governor valve, feed-water valve, and firing rates either individually or in a coordinated mode by manipulating all three loops simultaneously.

A once-through boiler has less stored energy than a drum-type boiler. Therefore, a generating unit with a once-through boiler is more responsive to changes in boiler firing rate. On the other hand, a drum-type boiler unit can deliver power without any fuel flow for a longer time. Neither type of boiler can supply full load steam flow for more than 1 or 2 minutes; however, both types can supply sufficient steam to support unit service load at normal throttle pressure for at least 30 minutes [29].

Control systems

The control systems associated with fossil-fuelled plants vary with the manufacturer and vintage of the unit. Nevertheless, some generalizations about the plant controls are possible. Most boilers have several controlled outputs and several manipulated variables which affect these outputs. Usually the firing rate, pumping rate, and throttle valve settings are the controlled variables. The controlled output variables of interest are temperature, pressure, electrical power, and speed. If, in the process, the plant variables exceed safe limits, the protection systems are designed to reduce the plant output or trip the unit.

The control systems associated with a thermal plant can be classified into two broad categories:

- Overall unit control

- Process parameter controls

There are four *overall unit control* strategies or modes of operation: boiler following (turbine leading), turbine following (boiler leading), integrated or coordinated boiler-turbine control, and sliding pressure control.

Under the *boiler-following* or *turbine-leading* mode of control, changes in generation are initiated by turbine control valves. The boiler controls respond to the resulting changes in steam flow and pressure by changing steam production. A difference between steam production and steam demand results in a change in boiler pressure. The throttle pressure deviation from its setpoint value is used as an error signal by the combustion controls to regulate fuel and air input to the furnace. Usually, a proportional-plus-integral (reset) controller is used to maintain appropriate firing rate at different load levels. In some cases, the turbine first-stage pressure is incorporated as a feedforward signal to the combustion controls to improve initial response.

Under the *turbine-following* or *boiler-leading* mode of control, changes in generation are initiated by changing input to the boiler. The MW demand signal is applied to the combustion controls. The turbine control valves regulate boiler pressure; fast action of the valves maintains essentially constant pressure.

Figure 9.43 shows the relative responses of the boiler-following and the turbine-following modes of control to a large change in load. With the boiler-following mode of operation, the boiler's stored energy is initially used to meet the

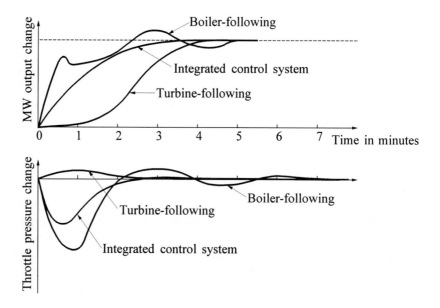

Figure 9.43 Comparison of boiler-turbine control
system response to a large load change

steam demand and the initial response of MW output is rapid. The coupling between throttle pressure and MW output results in a "dip" in the MW output when the boiler pressure is at a minimum. This coupling, along with the boiler time lag, produces an oscillatory response. For large changes, the deviations in boiler variables may be excessive. In contrast, with the turbine-following mode, no use is made of stored energy in the boiler; hence, steam flow and MW output closely follow steam production in the boiler. While the unit can be manoeuvred in a well-controlled manner, the response rate is limited by the slow response of the boiler.

The *integrated* or *coordinated boiler-turbine* control provides an adjustable blend of both boiler-following and turbine-following modes of control. The improvement in unit response achieved through integrated control is demonstrated in Figure 9.43. It is evident that the integrated control strikes a compromise between fast response and boiler safety.

In the *sliding pressure mode* of control, the throttle pressure setpoint is made a function of unit load rather than a constant value. The control valves are left wide open (provided the drum pressure is above a minimum level) and the turbine power output is controlled by controlling the throttle pressure through manipulation of the boiler controls. It is thus essentially a turbine-following mode of operation. A major advantage of this mode of control is that no change in throttling action occurs during load manoeuvring; therefore, temperature in the HP turbine can remain nearly constant.

The *process parameter controls* include systems that regulate unit MW output, main steam pressure, feed-water, and air/fuel.

Protection systems

The function of the protection system is to monitor important plant process parameters and initiate a trip and/or derating of processes (such as furnace and turbine) or auxiliary equipment (such as ID/FD fans) under adverse conditions.

A typical boiler protection monitors the following in determining if the furnace is to be tripped: furnace pressure, status of ID/FD fans, air flow, drum level, feed-water flow, black furnace (loss of flame), and main steam pressure. Each protection is usually time-conditioned.

The protection of auxiliary motors is generally based on voltage, frequency, or current supplied to the motor.

9.3.2 Nuclear-Based Energy Systems [28]

Two basic types of nuclear power plant reactors are in common use in North America: the *pressurized water reactor* (PWR) and the *boiling water reactor* (BWR). A variation of the PWR type is the *Canadian deuterium uranium* (CANDU) reactor.

The pressurized water reactor (PWR)

Figure 9.44 shows a simplified representation of a PWR unit. It uses water under pressure as a heat transport medium which absorbs heat from the reactor core and exchanges it with the shell-side feed-water in a *steam generator* which produces the steam used to drive the turbine. The steam produced is saturated, requiring wet-steam turbines.

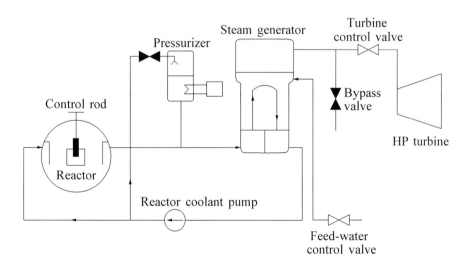

Figure 9.44 Schematic diagram of a PWR

The coolant circuit in a PWR acts as a heat transport medium as well as a nuclear reaction moderating medium. As a moderator, the coolant is used to slow down high energy neutrons to thermal energy levels, thus facilitating the nuclear fission reaction. This function provides a mechanism for self-regulation of the reactor process. An increase of reactor power results in an increase of the coolant temperature and a corresponding decrease of the fluid density. This lowers the nuclear fission reaction rate, thereby acting as a negative feedback in the reactor process dynamics. The coolant pressure control is facilitated by the use of a *pressurizer*, which comprises a liquid-filled tank feeding the coolant circuit. A steam space at the top of the tank is used to maintain the pressure at setpoint by means of electric water heaters and water sprays. The liquid level in the tank is controlled by using a make-up and let-down system of valves. The fluid flow in the reactor is sustained by means of large electrically driven *reactor coolant pumps*. The control of the reactor power is achieved by means of neutron-absorbing *control rods* which are inserted to a variable depth in the reactor core. The controller directs the movement of preselected groups of control rod clusters to increase or decrease reactor power as required to maintain the average coolant temperature at a programmed setpoint. A neutron flux signal and a turbine power signal are used to enhance the controller's response to load variations. Long-term regulation of core reactivity is accomplished by adjusting the concentration of neutron-absorbing boric acid in the reactor coolant.

The most critical pumps in a nuclear power plant are those associated with the reactor systems. Low-frequency operation has been cited as a cause of low reactor coolant flow in PWR units. If reactor power remains constant as coolant flow decreases, the heat content per unit of coolant-flowing increases. Steam bubbles can no longer be swept from the surface of the fuel cladding sufficiently quickly, and a blanketing of the hot surfaces with a steam film results. This condition is termed *departure from nucleate boiling* (DNB). At DNB the heat transfer rate is drastically reduced, and the resultant fuel and cladding temperature increases, leading to cladding failure and the release of radioactive fission products to the coolant. To prevent this condition from occurring, minimum reactor coolant pump-flow limits and motor undervoltage and underfrequency limits are established. Should these limits be violated, the reactor is automatically tripped.

In the conventional PWR design, the coolant temperature is a very important variable - it is used for both reactor control and steam dump operations.

The operation of the PWR is best visualized by considering its response to an increase in load demand. The control valves open to let more steam into the turbine to meet the increased demand. This in turn causes the water level in the steam generator to drop. The feed-water flow then increases to restore the water level. The mismatch between the reactor power level and the steam generator load results in a decrease of the reactor's coolant temperature and pressure, and a corresponding decrease in the moderating fluid's density. This change in density results in an increase of the neutron multiplication rate which provides the initial rise in the reactor power level. In response to the decreased coolant temperature, the reactor control system initiates a control absorber rod withdrawal to raise reactor power. The coolant pressure control system responds to the pressure change by operating the pressurizer heaters. Finally, the pressurizer water level is restored by the make-up and let-down system. A new steady-state condition is established when the equilibrium between the generator electrical power output and the reactor power level is reached.

The response to a decrease in load demand is similar to the above but in reverse. Typically, a PWR is capable of a load manoeuvring rate of ±5% reactor full power (RFP) per minute, and a step change of ±10% RFP.

Canadian deuterium uranium (CANDU) reactors

A CANDU reactor is a variation on the PWR design which uses heavy water in the coolant circuit for the heat transport function but assigns the moderating function to another independently controlled circuit - the moderator system. The moderator temperature is controlled by regulating the service-water flow through cooling heat exchangers. While the conventional PWR design relies on the thermal feedback of the moderating fluid to provide the initial load-following reactor response and steady-state reactor power level fine-tuning, the CANDU design utilizes a separate mechanism to directly control reactivity by means of a series of light water-filled tubes (zones) distributed in the reactor core. This design offers a faster response to reactor power demand change and a decoupling of the reactor controls from the steam generator dynamics. The steam generator pressure is also directly controlled to be maintained at a pressure setpoint. Feedforward action is also added to improve response and maintain the balance between reactor power and turbine load. The coolant circuit of some CANDU plants is not equipped with a pressurizer and relies solely on a make-up and let-down system to control coolant pressure. In these cases, the pressure setpoint is made a function of reactor power level in an effort to minimize swell and shrink in the coolant circuit. Despite this effort, the pressure swings in response to system electrical disturbances are still more pronounced than in a pressurizer-equipped system and may determine the outcome of the reactor protection system's reaction to the disturbance.

CANDU plants can be operated in two basic control modes: *reactor-following-turbine* (RFT) and *turbine-following-reactor* (TFR). Steam generator pressure control plays a pivotal role in these schemes. In TFR mode the reactor power is held constant at a level set by the operator and the turbine governor control valves are used to regulate steam pressure. This mode is suitable for unit operation within a large system network where it is used to supply the base loads while other units are used to supply peak loads. Since reactor conditions remain constant, this mode also constitutes a safer state of operation from the nuclear plant perspective. The control system reverts to this mode of operation during plant upsets. In RFT mode, the turbine load is set by the operator and the reactor is used to regulate steam pressure. In both TFR and RFT modes, the turbine is also used to regulate speed and assist the system in meeting short-term changes in electrical load demand. In RFT mode, the governor control valves also attempt to maintain a constant generator output.

In CANDU units, the steam dump valves are modulated in proportion to the amount of pressure error in excess of a preset value (the pressure error offset) with a power mismatch feedforward term added. Some CANDU units permit a small amount of steam dump (5%) into the condenser without initiating a reactor power reduction. Beyond that, the reactor power is set back at the rate of 0.05%/s until no more than 5% steam dump is admitted. In the case of large load rejections (greater than 40%) the pressure offset term is removed from the bypass valve control equation, and the reactor is stepped back to 60% RFP by a free-fall drop of the control rods into the reactor core. In addition, a fast-action scheme is put in effect whereby the bypass valves are opened fully for a short period of time and the control program sampling rate is increased.

In CANDU reactor design, the coolant temperature is less critical than in the conventional PWR design except for its heat transfer implication. The reactor is controlled based on flux measurements, with thermal measurements serving as correction factors only. The CANDU reactor power level is so tightly controlled that it is virtually decoupled from the turbine side of the plant. Therefore, the steam generator pressure control is more critical than the reactor control in the simulation of CANDU units.

The boiling water reactor (BWR)

Figure 9.45 shows a schematic diagram of a BWR. The feed-water enters directly into the reactor core and heat is transferred to the boiling water. The efficiency of heat removal is thus improved by use of the latent heat of evaporation. The steam produced at the reactor outlet header is used to drive the turbine. As in the case of a PWR, saturated steam is produced, requiring wet-steam turbines.

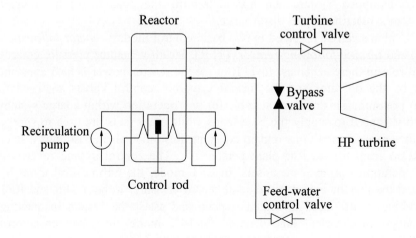

Figure 9.45 Schematic diagram of a BWR

The BWR relies on the moderating effect of the steam generated in the reactor to adjust the neutron multiplication rate. The power level is determined by the balance established between the amount of fuel in the core, the amount of neutron-absorbing material within the control rods in the core, and the core steam/water volume ratio (void content). In addition to the feed-water circuit, which acts as the main source of water into the reactor, a recirculation circuit also contributes to the water flow through the reactor core. Increasing either the feed-water or recirculation flow decreases the reactor steam void and thus raises reactor power. The control rod position establishes the relationship between reactor power level and core flow. Control rod withdrawal/ insertion is used for quick manoeuvring of the reactor. The recirculation flow is sustained by using variable-speed electrically driven pumps and the flow is directed into the reactor core by means of an array of jet pumps. Feed-water flow is also sustained by using electrically driven pumps and adjusted by means of control valves to maintain the reactor water level at setpoint. Steam pressure is maintained by means of the turbine governor control valves.

To illustrate the dynamics and control of a BWR, let us consider its response to an increase in load demand. The governing system responds by opening the control valve to increase the steam flow and, in anticipation of this increase, the master controller increases recirculation flow. The steam pressure setpoint is momentarily decreased to permit the governor control valves to open. The steam pressure drops as the steam outflow is increased. An increase in recirculation flow decreases the steam void in the reactor core (i.e., moderator density increases) and the reactor power level rises. With the increase in reactor power, steam pressure is restored and the steam pressure setpoint is returned to its original value. The reactor water level decreases following the increased steam outflow, but the water inventory is soon restored by an increase in feed-water flow. The BWR maximum load manoeuvring rate is dictated by the recirculation controller response and by limits set on changes to reactor water level and steam pressure.

A BWR is capable of load following over the entire reactor operating range. Newer units, such as the BWR 6 design, are capable of step changes of up to ±25% RFP.

A BWR unit can withstand a decrease in load of up to almost 100% without tripping. For decreases in load of more than 25%, the turbine bypass valves operate to temporarily relieve excess steam which is supplied by the reactor but not delivered to the turbine. The bypass valves close automatically as reactor power is reduced by a combination of flow control and control rod insertion.

9.3.3 Modelling of Thermal Energy Systems

Dynamics of thermal energy systems play an important role in the long-term dynamic response of power systems. In addition to the process physics of the individual elements of these systems, the associated auxiliaries, protections, and controls need to be appropriately represented in the study of long-term stability.

Models for representation of thermal energy systems in system studies are still evolving, and standard models for common use by the utility industry have not yet been developed. Reference 30 presents detailed nuclear and thermal power plant models recently developed by Ontario Hydro and CRIEPI (Japan) under a joint EPRI/CRIEPI study.

REFERENCES

[1]   H.M. Rustebakke (editor), *Electric Utility Systems and Practices*, John Wiley & Sons, 1983.

[2]   G.R. Rich, *Hydraulic Transients*, Dover Publications, Inc., 1963.

[3]   V.L. Streeter and E.B. Wylie, *Fluid Mechanics,* McGraw-Hill, 1975.

[4]   C. Concordia and L.K. Kirchmayer, "Tie-Line Power and Frequency Control of Electric Power Systems," *AIEE Trans.,* Vol. PAS-73, pp. 131-146, 1954.

[5]   Discussion of reference 4 by H.M. Paynter.

[6]   R. Oldenburger and J. Donelson, Jr., "Dynamic Response of a Hydroelectric Plant," *AIEE Trans.,* Vol. PAS-81, Part III, pp. 403-419, October 1962.

[7]   L.M. Hovey, "Optimum Adjustment of Hydro Governors on Manitoba Hydro System," *AIEE Trans.,* Vol. PAS-81, Part III, pp. 581-587, December 1962.

[8]   J.L. Woodward, "Hydraulic-Turbine Transfer Function for Use in Governing Studies," *Proc. IEE,* Vol. 115, No. 3, pp. 424-426, March 1968.

[9]   D.G. Ramey and J.W. Skooglund, "Detailed Hydrogovernor Representation for System Stability Studies," *IEEE Trans.,* Vol. PAS-89, pp. 106-112, January 1970.

[10]   IEEE Committee Report, "Dynamic Models for Steam and Hydro Turbines in Power System Studies," *IEEE Trans.,* Vol. PAS-92, pp. 1904-1915, November/December 1973.

[11]   IEEE Working Group Report, "Hydraulic Turbine and Turbine Control Models for System Dynamic Studies," *IEEE Trans.,* Vol. PWRS-7, No. 1, pp. 167-179, February 1992.

[12]   J.M. Undrill and J.L. Woodward, "Nonlinear Hydro Governing Model and Improved Calculation for Determining Temporary Droop," *IEEE Trans.,* Vol.

PAS-86, pp. 443-453, April 1967.

[13]   F.R. Schleif and A.B. Wilbor, "The Coordination of Hydraulic Turbine Governors for Power System Operation," *IEEE Trans.,* Vol. PAS-85, pp. 750-758, July 1966.

[14]   P.L. Dandeno, P. Kundur, and J.P. Bayne, "Hydraulic Unit Dynamic Performance under Normal and Islanding Conditions - Analysis and Validation," *IEEE Trans.,* Vol. PAS-97, pp. 2134-2143, November/December 1978.

[15]   M. Leum, "The Development and Field Experience of a Transistor Electric Governor for Hydro Turbines," *IEEE Trans.,* Vol. PAS-85, pp. 393-400, April 1966.

[16]   R.E. Goodson, "Distributed System Simulation Using Infinite Product Expansions," *Simulation,* Vol. 15, No. 6, pp. 255-263, December 1970.

[17]   P. Kundur and J.P. Bayne, "A Study of Early Valve Actuation Using Detailed Prime Mover and Power System Simulation," *IEEE Trans.,* Vol. PAS-94, pp. 1275-1287, July/August 1975.

[18]   IEEE Working Group Report, "Dynamic Models for Fossil Fueled Steam Units in Power System Studies," *IEEE Trans.,* Vol. PWRS-6, No. 2, pp. 753-761, May 1991.

[19]   E.F. Church, *Steam Turbines,* McGraw Hill, pp. 70-74, 1950.

[20]   T.D. Younkins and L.H. Johnson, "Steam Turbine Overspeed Control and Behaviour during System Disturbances," *IEEE Trans.,* Vol. PAS-100, pp. 2504-2511, May 1981.

[21]   M.A. Eggenberger, "An Advanced Electrohydraulic Control System for Reheat Steam Turbines," American Power Conference, Chicago, April 1965.

[22]   M.A. Eggenberger, "A Simplified Analysis of the No-Load Stability of Mechanical-Hydraulic Speed Control Systems for Steam Turbines," ASME Paper 60-WA-34, December 1960.

[23]   M. Birnbaum and E.G. Noyes, "Electro Hydraulic Control for Improved Availability and Operation of Large Steam Turbines," ASME-IEEE National Power Conference, Albany, September 1965.

[24]   P. Kundur, D.C. Lee, and J.P. Bayne, "Impact of Turbine Generator Overspeed

Controls on Unit Performance under System Disturbance Conditions," *IEEE Trans.*, Vol. PAS-104, pp. 1262-1267, June 1985.

[25] M.S. Baldwin and D.P. McFadden, "Power Systems Performance as Affected by Turbine-Generator Controls Response during Frequency Disturbances," *IEEE Trans.*, Vol. PAS-100, pp. 2486-2494, May 1981.

[26] H.T. Akers, J.D. Dickinson, and J.W. Skooglund, "Operation and Protection of Large Steam Turbine Generators under Abnormal Conditions," *IEEE Trans.*, Vol. PAS-87, pp. 1180-1188, April 1968.

[27] ANSI/IEEE Standard C37.106-1987, *IEEE Guide for Abnormal Frequency Protection for Power Generating Plants*.

[28] EPRI Report EL-6627, "Long-Term Dynamics Simulation: Modelling Requirements," Final Report of Project 2473-22, Prepared by Ontario Hydro, December 1989.

[29] O.W. Durrant and R.P. Siegfried, "Operation of Drum and Once-Through Boilers during Electrical Power System Emergencies," Paper presented at the ASME-IEEE Joint Power Generation Conference, Detroit, Mich., September 1967.

[30] EPRI Report EL-6627, "Long-Term Dynamic Simulation: Nuclear and Thermal Power Plant Models," Final Report of Project 3144-01, Prepared by Ontario Hydro and CRIEPI (Japan), September 1992.

Chapter 10

High-Voltage Direct-Current Transmission

High-voltage direct-current (HVDC) transmission has advantages over ac transmission in special situations. The first commercial application of HVDC transmission was between the Swedish mainland and the island of Gotland in 1954. This system used mercury-arc valves and provided a 20 MW underwater link of 90 km. Since then, there has been a steady increase in the application of HVDC transmission.

With the advent of thyristor valve converters, HVDC transmission became even more attractive. The first HVDC system using thyristor valves was the Eel River scheme commissioned in 1972, forming a 320 MW back-to-back dc interconnection between the power systems of the Canadian provinces of New Brunswick and Quebec. Thyristor valves have now become standard equipment for dc converter stations. Recent developments in conversion equipment have reduced their size and cost, and improved their reliability. These developments have resulted in a more widespread use of HVDC transmission. In North America, the total capacity of HVDC links in 1987 was over 14,000 MW [1]. There are more links under construction.

The following are the types of applications for which HVDC transmission has been used:

1. Underwater cables longer than about 30 km. AC transmission is impractical for such distances because of the high capacitance of the cable requiring intermediate compensation stations.

2.   Asynchronous link between two ac systems where ac ties would not be feasible because of system stability problems or a difference in nominal frequencies of the two systems.

3.   Transmission of large amounts of power over long distances by overhead lines. HVDC transmission is a competitive alternative to ac transmission for distances in excess of about 600 km.

HVDC systems have the ability to rapidly control the transmitted power. Therefore, they have a significant impact on the stability of the associated ac power systems. An understanding of the characteristics of the HVDC systems is essential for the study of the stability of the power system. More importantly, proper design of the HVDC controls is essential to ensure satisfactory performance of the overall ac/dc system.

This chapter will provide a general introduction to the basic principles of operation and control of HVDC systems and describe their modelling for power-flow and stability studies. Two terminal systems will be considered in detail, followed by a brief discussion of multiterminal systems.

For additional general information on HVDC transmission, the reader may refer to references 2 to 8. Reference 9 provides information related to specification of HVDC systems.

10.1 HVDC SYSTEM CONFIGURATIONS AND COMPONENTS

10.1.1 Classification of HVDC Links

HVDC links may be broadly classified into the following categories:

- Monopolar links
- Bipolar links
- Homopolar links

The basic configuration of a *monopolar link* is shown in Figure 10.1. It uses one conductor, usually of negative polarity. The return path is provided by ground or water. Cost considerations often lead to the use of such systems, particularly for cable transmission. This type of configuration may also be the first stage in the development of a bipolar system.

Instead of ground return, a metallic return may be used in situations where the earth resistivity is too high or possible interference with underground/underwater metallic structures is objectionable. The conductor forming the metallic return is at low voltage.

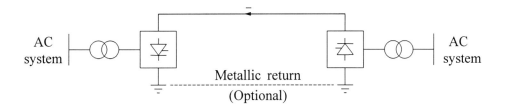

Figure 10.1 Monopolar HVDC link

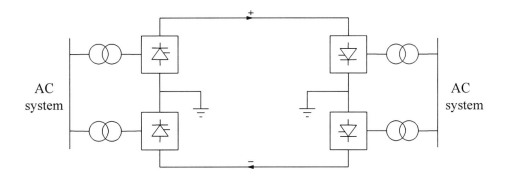

Figure 10.2 Bipolar HVDC link

The *bipolar link* configuration is shown in Figure 10.2. It has two conductors, one positive and the other negative. Each terminal has two converters of equal rated voltage, connected in series on the dc side. The junctions between the converters is grounded. Normally, the currents in the two poles are equal, and there is no ground current. The two poles can operate independently. If one pole is isolated due to a fault on its conductor, the other pole can operate with ground and thus carry half the rated load or more by using the overload capabilities of its converters and line.

From the viewpoint of lightning performance, a bipolar HVDC line is considered to be effectively equivalent to a double-circuit ac transmission line. Under normal operation, it will cause considerably less harmonic interference on nearby facilities than the monopolar system. Reversal of power-flow direction is achieved by changing the polarities of the two poles through controls (no mechanical switching is required).

In situations where ground currents are not tolerable or when a ground electrode is not feasible for reasons such as high earth resistivity, a third conductor is used as a metallic neutral. It serves as the return path when one pole is out of service or when there is imbalance during bipolar operation. The third conductor requires low insulation and may also serve as a shield wire for overhead lines. If it is fully insulated, it can serve as a spare.

The *homopolar link*, whose configuration is shown in Figure 10.3, has two or more conductors, all having the same polarity. Usually a negative polarity is preferred because it causes less radio interference due to corona. The return path for such a system is through ground. When there is a fault on one conductor, the entire converter is available for feeding the remaining conductor(s) which, having some overload capability, can carry more than the normal power. In contrast, for a bipolar scheme reconnection of the whole converter to one pole of the line is more complicated and usually not feasible. Homopolar configuration offers an advantage in this regard in situations where continuous ground current is acceptable.

The ground current can have side effects on gas or oil pipe lines that lie within a few miles of the system electrodes. Pipelines act as conductors for the ground current which can cause corrosion of the metal. Therefore, configurations using ground return may not always be acceptable.

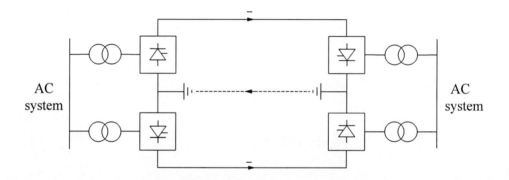

Figure 10.3 Homopolar HVDC link

Each of the above HVDC system configurations usually has cascaded groups of several converters, each having a transformer bank and a group of valves. The converters are connected in parallel on the ac side (transformer) and in series on the dc side (valve) to give the desired level of voltage from pole to ground.

Back-to-back HVDC systems (used for asynchronous ties) may be designed for monopolar or bipolar operation with a different number of valve groups per pole, depending on the purpose of the interconnection and the desired reliability.

Most point-to-point (two-terminal) HVDC links involving lines are bipolar, with monopolar operation used only during contingencies. They are normally designed to provide maximum independence between poles to avoid bipolar shutdowns.

A multiterminal HVDC system is formed when the dc system is to be connected to more than two nodes on the ac network. The possible configurations of multiterminal systems will be discussed in Section 10.8.

10.1.2 Components of HVDC Transmission System

The main components associated with an HVDC system are shown in Figure 10.4, using a bipolar system as an example. The components for other configurations are essentially the same as those shown in the figure. The following is a brief description of each component.

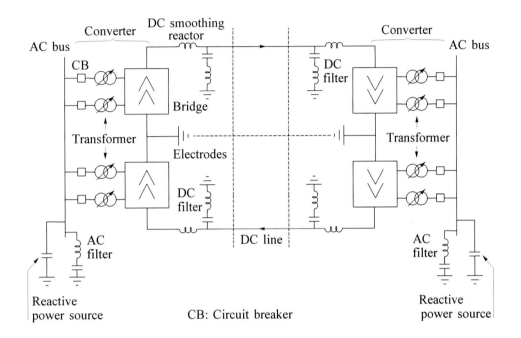

Figure 10.4 A schematic of a bipolar HVDC system
identifying main components

Converters. They perform ac/dc and dc/ac conversion, and consist of valve bridges and transformers with tap changers. The valve bridges consist of high-voltage valves connected in a 6-pulse or 12-pulse arrangement as described in Section 10.2. The converter transformers provide ungrounded three-phase voltage source of appropriate level to the valve bridge. With the valve side of the transformer ungrounded, the dc system will be able to establish its own reference to ground, usually by grounding the positive or negative end of the valve converter.

Smoothing Reactors. These are large reactors having inductance as high as 1.0 H connected in series with each pole of each converter station. They serve the following purposes:

- Decrease harmonic voltages and currents in the dc line.
- Prevent commutation failure in inverters.
- Prevent current from being discontinuous at light load.
- Limit the crest current in the rectifier during short-circuit on the dc line.

Harmonic Filters. Converters generate harmonic voltages and currents on both ac and dc sides. These harmonics may cause overheating of capacitors and nearby generators, and interference with telecommunication systems. Filters are therefore used on both ac and dc sides.

Reactive Power Supplies. As we will see in Section 10.2, dc converters inherently absorb reactive power. Under steady-state conditions, the reactive power consumed is about 50% of active power transferred. Under transient conditions, the consumption of reactive power may be much higher. Reactive power sources are therefore provided near the converters. For strong ac systems, these are usually in the form of shunt capacitors. Depending on the demands placed on the dc link and on the ac system, part of the reactive power source may be in the form of synchronous condensers or static var compensators. The capacitors associated with the ac filters also provide part of the reactive power required.

Electrodes. Most dc links are designed to use earth as a neutral conductor for at least brief periods of time. The connection to the earth requires a large-surface-area conductor to minimize current densities and surface voltage gradients. This conductor is referred to as an electrode. As discussed earlier, if it is necessary to restrict the current flow through the earth, a metallic return conductor may be provided as part of the dc line.

DC Lines. They may be overhead lines or cables. Except for the number of conductors and spacing required, dc lines are very similar to ac lines.

AC Circuit Breakers. For clearing faults in the transformer and for taking the dc link out of service, circuit-breakers are used on the ac side. They are not used for clearing dc faults, since these faults can be cleared more rapidly by converter control.

10.2 CONVERTER THEORY AND PERFORMANCE EQUATIONS

A converter performs ac/dc conversion and provides a means of controlling the power flow through the HVDC link. The major elements of the converter are the valve bridge and converter transformer. The valve bridge is an array of high-voltage switches or valves that sequentially connect the three-phase alternating voltage to the dc terminals so that the desired conversion and control of power are achieved. The converter transformer provides the appropriate interface between the ac and dc systems.

In this section, we will describe the structure and operation of practical converter circuits. In addition, we will develop equations relating dc quantities and fundamental frequency ac quantities.

10.2.1 Valve Characteristics

The valve in an HVDC converter is a controlled electronic switch. It normally conducts in only one direction, the forward direction, from anode to cathode. When it is conducting, there is only a small voltage drop across it. In the reverse direction, when the voltage applied across the valve is such that the cathode is positive relative to the anode, the valve blocks the current.

The early HVDC systems used mercury-arc valves. Mercury-arc valves with rated currents of the order of 1,000 to 2,000 A and rated peak inverse voltage of 50 to 150 kV have been built and used. Among the disadvantages of mercury-arc valves are their large size and tendency to conduct in the reverse direction.

All HVDC systems built since the mid-1970s have used thyristor valves. Thyristor valves rated at 2,500 to 3,000 A and 3 to 5 kV have been developed. The thyristors are connected in series to achieve the desired system voltage. They are available in various designs: air cooled, air insulated; oil cooled, oil insulated; water cooled, air insulated; and freon cooled, SF_6 insulated. The valves can be designed for indoor or outdoor installation.

For the valve to conduct, it is necessary for the anode to be positive relative to the cathode. In a mercury-arc valve, with the control grid at sufficiently negative voltage with respect to the cathode, the valve is prevented from conducting, although the anode may be positive. The instant of firing can be controlled by the grid.

Similarly, a thyristor valve will conduct only when the anode is positive with respect to the cathode and when there is a positive voltage applied to the gate. Conduction may be initiated by applying a momentary or sustained current pulse of proper polarity to the gate.

Once conduction is initiated, the current through the valve continues until current drops to zero and a reverse voltage bias appears across the valve. In the forward direction, the current is blocked until a control pulse is applied to the gate. When not conducting, the valve should be capable of withstanding the forward or reverse bias voltages appearing between its cathode and anode.

Figure 10.5 shows the symbol used to represent a controlled valve (mercury-arc or thyristor).

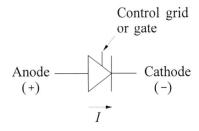

Figure 10.5 Symbol for controlled valve

10.2.2 Converter Circuits

The basic module of an HVDC converter is the three-phase, full-wave bridge circuit shown in Figure 10.6. This circuit is also known as a Graetz bridge. Although there are several alternative configurations possible, the Graetz bridge has been universally used for HVDC converters as it provides better utilization of the converter transformer and a lower voltage across the valve when not conducting [2,6]. The latter is referred to as the peak inverse voltage and is an important factor that determines the rating of the valves.

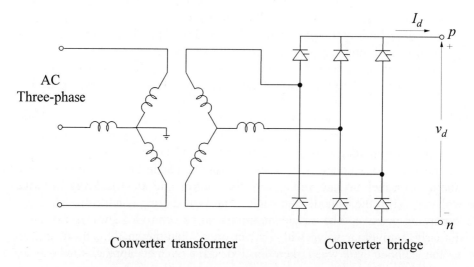

Figure 10.6 Three-phase, full-wave bridge circuit

The converter transformer has on-load taps on the ac side for voltage control. The ac side windings of the transformer are usually star-connected with grounded neutral; the valve side-windings are delta-connected or star-connected with ungrounded neutral.

Analysis of three-phase, full-wave bridge circuit

For purposes of analysis, we will make the following assumptions:

(a)   The ac system, including the converter transformer, may be represented by an ideal source of constant voltage and frequency in series with a lossless inductance (representing primarily the transformer leakage inductance).

(b)   The direct current (I_d) is constant and ripple-free; this is justified because of the large smoothing reactor (L_d) used on the dc side.

(c)   The valves are ideal switches with zero resistance when conducting, and infinite resistance when not conducting.

Based on the above assumptions, the bridge converter of Figure 10.6 may be represented by the equivalent circuit shown in Figure 10.7.

Let the instantaneous line-to-neutral source voltages be

$$e_a = E_m \cos(\omega t + 60°)$$
$$e_b = E_m \cos(\omega t - 60°) \qquad\qquad (10.1)$$
$$e_c = E_m \cos(\omega t - 180°)$$

The line-to-line voltages are then

$$e_{ac} = e_a - e_c = \sqrt{3} E_m \cos(\omega t + 30°)$$
$$e_{ba} = e_b - e_a = \sqrt{3} E_m \cos(\omega t - 90°) \qquad\qquad (10.2)$$
$$e_{cb} = e_c - e_b = \sqrt{3} E_m \cos(\omega t + 150°)$$

Figure 10.8(a) shows the voltage waveforms corresponding to Equations 10.1 and 10.2.

To simplify analysis and help understand the operation of the bridge converter, we will first consider the case with negligible source inductance (i.e., L_c=0) and no ignition delay. After developing a basic understanding of the converter performance, we will extend the analysis to include the effect of delaying the valve ignition through gate/grid control and then the effect of source inductance.

Analysis assuming negligible source inductance

(a) With no ignition delay

In Figure 10.7, the cathodes of valves 1, 3, and 5 of the upper row are connected together. Therefore, when the phase-to-neutral voltage of phase a is more positive than the voltages of the other two phases, valve 1 conducts. The common potential of the cathodes of the three valves is then equal to that of the anode of valve 1. Since the cathodes of valves 3 and 5 are at a higher potential than their anodes, these valves do not conduct.

In the lower row, the anodes of valves 2, 4 and 6 are connected together. Therefore, valve 2 conducts when phase c voltage is more negative than the other two phases.

From the waveforms shown in Figure 10.8(a) we see that valve 1 conducts when ωt is between −120° and 0°, since e_a is greater than e_b or e_c. Valve 2 conducts when ωt is between −60° and 60°, since e_c is more negative than e_a or e_b during this

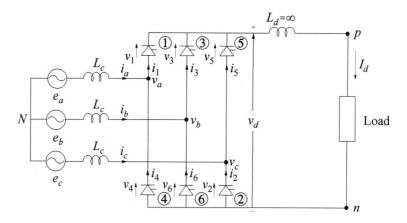

Note: Valves are numbered in order of firing.

Figure 10.7 Equivalent circuit for three-phase full-wave bridge converter

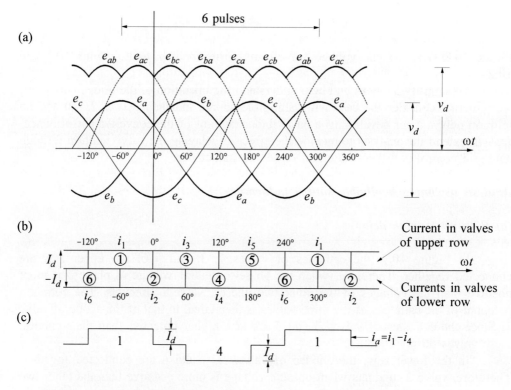

Figure 10.8 Waveforms of voltages and currents of bridge circuit of Figure 10.7
    (a) Source line-to-neutral and line-to-line voltages
    (b) Valve currents and periods of conduction
    (c) Phase current i_a

period. This is shown in Figure 10.8(b), which identifies the period of conduction of each valve, and the magnitude and duration of current in it. Since, by assumption, the direct current I_d is assumed constant, the current in each valve is I_d when conducting and zero when not conducting.

Let us now examine the period when ωt is between 0° and 120°. Just before $\omega t = 0$, valves 1 and 2 are conducting. Just after $\omega t = 0°$, e_b becomes more positive than e_a and valve 3 ignites; valve 1 is extinguished because its cathode is now at a higher potential than its anode. For the next 60°, valves 2 and 3 conduct. At $\omega t = 60°$, e_a is more negative than e_c, causing valve 4 to ignite and valve 2 to extinguish.

At $\omega t = 120°$, e_c is more positive than e_b, resulting in the ignition of valve 5 and extinction of valve 3. Similarly, at $\omega t = 180°$ conduction switches from valve 4 to 6 in the lower row; at $\omega t = 240°$ conduction switches from valve 5 to 1 in the upper row. This completes one cycle, and the sequence continues.

The valve-switching sequence is illustrated in Figure 10.9, which shows only the conducting valves during the six distinct periods of a complete cycle.

Each valve thus conducts for a period of 120°. When it is conducting, the magnitude of valve current is I_d; the valves in the upper row carry positive current and the valves in the lower row carry negative (or return) current.

The current in each phase of the ac source is composed of currents in the two valves connected to that phase. For example, the current in phase a, as shown in Figure 10.8(c), is equal to $i_1 - i_4$. This represents the current in the secondary (valve side) winding of the converter transformer of Figure 10.6.

The transfer of current from one valve to another in the same row is called "commutation." In the above analysis, we have assumed that the source inductance L_c is negligible. Therefore, commutation occurs instantaneously, i.e., without "overlap." The result is that no more than two valves (one from the top row and the other from the bottom row) conduct at any time.

From Figure 10.8(a), we see that the number of pulsations (cycles of ripple) of v_d per cycle of alternating voltage is six. Hence, the bridge circuit of Figure 10.6 is referred to as a "6-pulse bridge circuit."

Average direct voltage:

The instantaneous direct voltage v_d across the bridge (between the cathodes of the upper-row valves and anodes of lower-row valves) is composed of 60° segments of the line-to-line voltages. Therefore, the average direct voltage can be found by integrating the instantaneous values over any 60° period.

Denoting ωt by θ, and considering the period between $\omega t = -60°$ and 0°, the average direct voltage with no ignition delay is given by

$$V_{do} = \frac{3}{\pi} \int_{-60°}^{0} e_{ac} \, d\theta$$

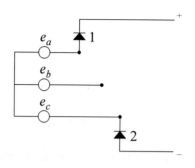

(a) $\omega t = -60°$ to $0°$

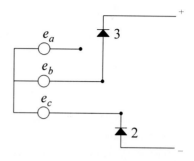

(b) $\omega t = 0°$ to $60°$

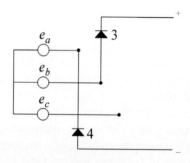

(c) $\omega t = 60°$ to $120°$

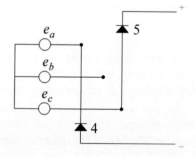

(d) $\omega t = 120°$ to $180°$

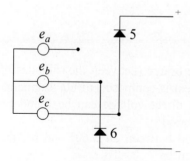

(e) $\omega t = 180°$ to $240°$

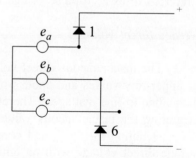

(f) $\omega t = 240°$ to $300°$

Figure 10.9 Valve-switching sequence with no ignition delay and no overlap

Substituting for e_{ac} from Equation 10.2, we get

$$V_{d0} = \frac{3}{\pi} \int_{-60°}^{0} \sqrt{3}E_m \cos(\theta+30°)\,d\theta$$

$$= \frac{3\sqrt{3}}{\pi}E_m \sin(\theta+30°)\Big|_{-60°}^{0} \tag{10.3A}$$

$$= \frac{3\sqrt{3}}{\pi}E_m 2\sin30° = \frac{3\sqrt{3}}{\pi}E_m = 1.65E_m$$

where E_m is the peak value of the line-to-neutral voltage.

In terms of RMS line-to-neutral (E_{LN}) and line-to-line (E_{LL}) voltages, the expression for V_{d0} becomes

$$V_{d0} = \frac{3\sqrt{6}}{\pi}E_{LN} = 2.34\,E_{LN} \tag{10.3B}$$

$$= \frac{3\sqrt{2}}{\pi}E_{LL} = 1.35\,E_{LL} \tag{10.3C}$$

and V_{d0} is called the "ideal no-load direct voltage."

(b) With ignition delay

The grid or gate control can be used to delay the ignition of the valves. The "delay angle" is denoted by α; it corresponds to time delay of α/ω seconds.

With delay, valve 3 ignites when $\omega t = \alpha$ (instead of $\omega t = 0$), valve 4 when $\omega t = \alpha + 60°$, valve 5 when $\omega t = \alpha + 120°$, and so on. This is illustrated in Figure 10.10.

The delay angle is limited to 180°. If α exceeds 180°, the valve fails to ignite. For example, consider the ignition of valve 3. With $\alpha = 0$, valve 3 ignites at $\omega t = 0$. The ignition can be delayed up to $\omega t = 180°$. Beyond this, e_b is no longer greater than e_a, and hence valve 3 will not ignite.

Average direct voltage:

Referring to Figure 10.10, the average direct voltage V_d when the delay angle is equal to α is given by

$$V_d = \frac{3}{\pi} \int_{-(60°-\alpha)}^{\alpha} e_{ac}\,d\theta = \frac{3}{\pi} \int_{\alpha-60°}^{\alpha} \sqrt{3}E_m \cos(\theta+30°)\,d\theta$$

$$= V_{d0} \int_{\alpha-60°}^{\alpha} \cos(\theta+30°)\,d\theta = V_{d0}\sin(\theta+30°)\Big|_{\alpha-60°}^{\alpha} \tag{10.4}$$

$$= V_{d0}[\sin(\alpha+30°)-\sin(\alpha-30°)]$$

$$= V_{d0}(2\sin30°)\cos\alpha = V_{d0}\cos\alpha$$

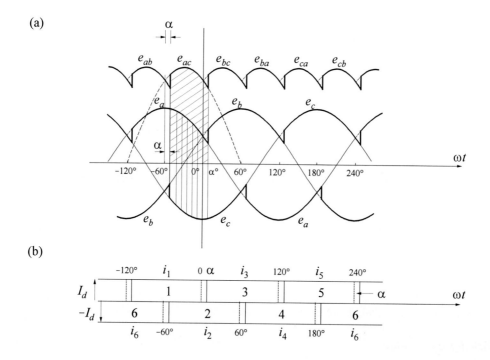

Figure 10.10 Voltage waveforms and valve currents, with ignition delay

The effect of the delayed ignition is thus to reduce the average direct voltage by the factor $\cos \alpha$.

Since α can range from 0° to 180°, $\cos \alpha$ can range from 1 to -1. Therefore, V_d can range from V_{d0} to $-V_{d0}$. Negative V_d, as discussed later in this section, represents inversion as opposed to rectification.

Current and phase relations:

As the ignition delay angle α is increased, the phase displacement between alternating voltage and alternating current in a supply phase also changes. This is illustrated in Figure 10.11 for phase a. The current wave shape, as shown in Figure 10.8(c), is composed of rectangular segments associated with currents in valves 1 and 4.

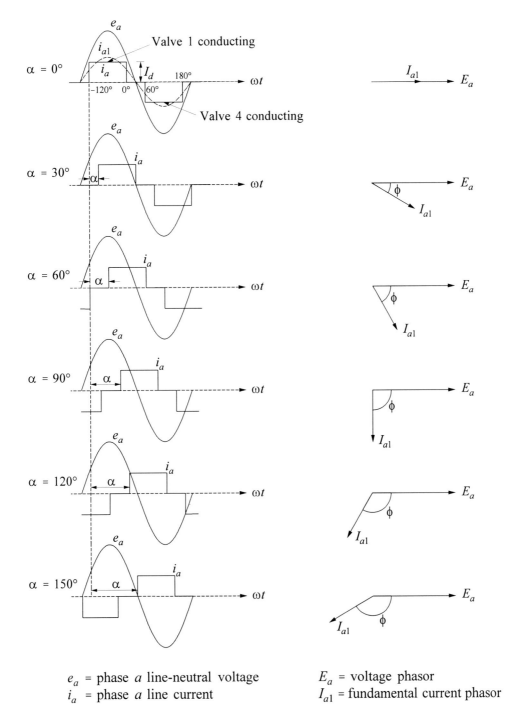

e_a = phase a line-neutral voltage E_a = voltage phasor
i_a = phase a line current I_{a1} = fundamental current phasor

Figure 10.11   Variation of phase displacement between voltage
and current of phase a with delay angle

The direct current I_d is constant by assumption (L_d in Figure 10.7 prevents I_d from changing). Since each valve conducts for a period of 120°, the alternating line currents appear as rectangular pulses of magnitude I_d and duration of 120° or $2\pi/3$ rad. With the assumption that there is no overlap, the shape of the alternating line currents is independent of α. Only the phase displacement changes with α.

The fundamental frequency component of the alternating line current can be determined by Fourier analysis of the current wave shape shown in Figure 10.12.

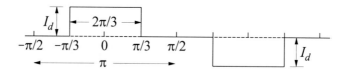

Figure 10.12 Line current waveform

The peak value of the fundamental frequency component of the alternating line current is

$$I_{LM} = \frac{2}{\pi}\int_{-\pi/3}^{\pi/3} I_d\cos x\,dx = \frac{2}{\pi}I_d\sin x\,\Big|_{-60°}^{60°}$$

$$= \frac{2}{\pi}I_d[\sin 60° - \sin(-60°)] \tag{10.5A}$$

$$= \frac{2}{\pi}\sqrt{3}I_d = 1.11 I_d$$

The RMS value of the fundamental frequency component of the alternating line current is

$$I_{L1} = \frac{I_{LM}}{\sqrt{2}} = \frac{2\sqrt{3}}{\pi\sqrt{2}}I_d$$

$$= \frac{\sqrt{6}}{\pi}I_d = 0.78 I_d \tag{10.5B}$$

With losses in the converter neglected, the ac power must equal the dc power. Therefore,

$$3E_{LN}I_{L1}\cos\phi = V_d I_d$$

$$= (V_{d0}\cos\alpha)I_d$$

where

E_{LN} = RMS value of the line-to-neutral voltage

ϕ = angle by which fundamental line current lags the line-to-neutral source voltage as shown in Figure 10.11

Substituting for V_{d0} from Equation 10.3B and for I_{L1} from Equation 10.5B, we have

$$\left(3E_{LN}\frac{\sqrt{6}}{\pi}I_d\right)\cos\phi = \left(\frac{3\sqrt{6}}{\pi}E_{LN}I_d\right)\cos\alpha$$

Hence, the power factor of the fundamental wave is

$$\cos\phi = \cos\alpha \tag{10.6}$$

The term $\cos\phi$ is referred to by some authors as the "vector power factor" or "displacement factor" [2].

The converter thus operates as a device that converts alternating to direct current (or direct to alternating current) so that the current ratio is fixed but the voltage ratio varies with the ignition delay caused by grid or gate control.

The ignition delay α shifts the current wave and its fundamental component by an angle $\phi = \alpha$, as shown in Figure 10.11. With $\alpha = 0°$, the fundamental current component (i_{a1}) is in phase with the phase voltage e_a; the active power ($P_a = E_a I_{a1}\cos\phi$) is positive and the reactive power ($Q_a = E_a I_{a1}\sin\phi$) is zero. As α increases from 0° to 90°, P_a decreases and Q_a increases. At $\alpha = 90°$, P_a is zero and Q_a is maximum. As α increases from 90° to 180°, P_a becomes negative and increases in magnitude; Q_a remains positive and decreases in magnitude. At $\alpha = 180°$, P_a is negative maximum and Q_a is zero. We see that the converter, whether it is acting as a rectifier or as an inverter, draws reactive power from the ac system.

Analysis including commutation overlap

Due to the inductance L_c of the ac source (see Figure 10.7), the phase currents cannot change instantly. Therefore, the transfer of current from one phase to another requires a finite time, called the commutation time or overlap time. The corresponding *overlap* or *commutation angle* is denoted by μ.

In normal operation, the overlap angle is less than 60°; typical full-load values are in the range of 15° to 25°. With $0° < \mu < 60°$, during commutation three valves conduct simultaneously. However, between commutations only two valves conduct. A new commutation begins every 60° and lasts for an angular period of μ. Therefore, the angular period when two valves conduct with no ignition delay (i.e., $\alpha = 0$) is $60° - \mu$, as shown in Figure 10.13. During each commutation period, the current in the incoming valve increases from 0 to I_d, and the current in the outgoing valve reduces from I_d to 0. For simplicity, in Figure 10.13 we have identified only the valve conduction periods, but not the valve currents.

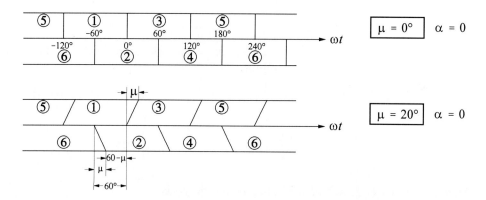

Figure 10.13 Effect of overlap angle on periods of conduction of valves

If $60° \leq \mu < 120°$, an abnormal mode of operation occurs in which alternately three and four valves conduct [2]. Here we will consider only the normal operation when μ is less than $60°$.

Let us analyze the effect of overlap by considering the commutation from valve 1 to valve 3. Figure 10.14 shows the periods of valve conduction, when ignition delay is included. The commutation begins when $\omega t = \alpha$ (delay angle) and ends when $\omega t = \alpha + \mu = \delta$, where δ is *extinction angle* (equal to the sum of the delay angle α and commutation angle μ).

At the beginning of commutation ($\omega t = \alpha$): $i_1 = I_d$ and $i_3 = 0$

At the end of commutation ($\omega t = \alpha + \mu = \delta$): $i_1 = 0$ and $i_3 = I_d$

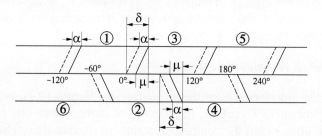

Figure 10.14 Periods of valve conduction with ignition delay

During the period of commutation, valves 1, 2 and 3 are conducting and the effective converter circuit is as shown in Figure 10.15. From the figure, for the loop containing valves 1 and 3, we have

$$e_b - e_a = L_c \frac{di_3}{dt} - L_c \frac{di_1}{dt}$$

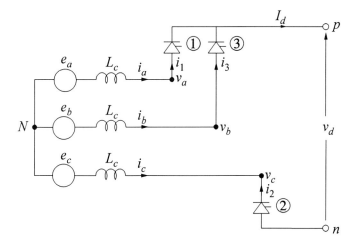

Note: Non-conducting valves not shown.

Figure 10.15 Equivalent circuit during commutation

The voltage $e_b - e_a$ is called the "commutating voltage." From Equation 10.2, it is equal to $\sqrt{3}E_m \sin \omega t$. Therefore,

$$\sqrt{3}E_m \sin \omega t = L_c \frac{di_3}{dt} - L_c \frac{di_1}{dt}$$

Since $i_1 = I_d - i_3$,

$$\frac{di_1}{dt} = 0 - \frac{di_3}{dt}$$

Hence,

$$e_b - e_a = \sqrt{3}E_m \sin \omega t = 2L_c \frac{di_3}{dt} \qquad (10.7A)$$

or

$$\frac{di_3}{dt} = \frac{\sqrt{3}E_m}{2L_c} \sin \omega t \qquad (10.7B)$$

Taking a definite integral with respect to t, with the lower limit corresponding to the beginning of commutation ($\omega t = \alpha$ or $t = \alpha/\omega$) and a running upper limit, we have

$$\int_{0}^{i_3} di_3 = \frac{\sqrt{3}E_m}{2L_c} \int_{\alpha/\omega}^{t} \sin\omega t \, dt$$

Integration of the above equation yields

$$i_3 = \frac{\sqrt{3}E_m}{2\omega L_c}(\cos\alpha - \cos\omega t) \qquad (10.8A)$$

$$= I_{S2}(\cos\alpha - \cos\omega t)$$

where

$$I_{S2} = \frac{\sqrt{3}E_m}{2\omega L_c} \qquad (10.8B)$$

The current i_3 of the incoming valve during commutation consists of a constant term ($I_{S2}\cos\alpha$) and a sinusoidal term ($-I_{S2}\cos\omega t$) lagging the commutating voltage by 90°. This is to be expected because what we have here is a line-to-line short-circuit through an inductance of $2L_c$. The constant component of i_3 depends on α; it serves to make $i_3 = 0$ at the beginning of commutation.

As shown in Figure 10.16, the current during commutation is a segment of a sinusoidal current with a peak value of $I_{S2} = \sqrt{3}E_m/(2\omega L_c)$. The shape of the segment is a function of the control angle α. Therefore, the overlap angle depends on I_d, L_c and α.

During commutation, the shape of i_1 satisfies $i_1 = I_d - i_3$. For α nearly equal to 0° (or 180°), the commutation period or the overlap is the greatest. The overlap is the shortest when $\alpha = 90°$, since i_3 is associated with the segment of the sine wave which is nearly linear. Also, if the source voltage E_m is lowered or if I_d is increased, the overlap increases.

Voltage reduction due to commutation overlap:

During commutation,

$$v_a = v_b = e_b - L_c\frac{di_3}{dt}$$

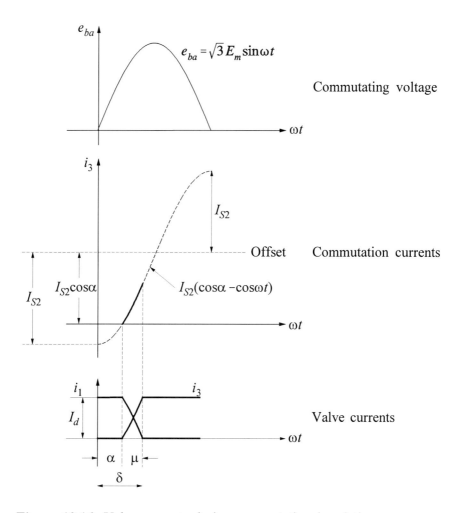

Figure 10.16   Valve currents during commutation in relation
to commutating voltage

From Equation 10.7A,

$$L_c \frac{di_3}{dt} = \frac{e_b - e_a}{2}$$

Hence,

$$v_a = v_b = e_b - \frac{e_b - e_a}{2}$$

$$= \frac{e_a + e_b}{2}$$

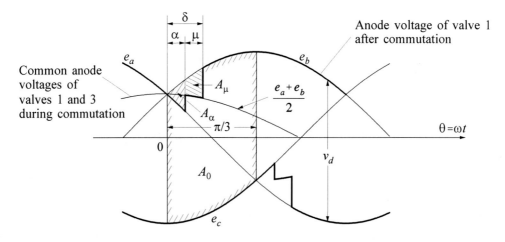

Figure 10.17 Voltage waveforms showing the effect of overlap
during commutation from valve 1 to valve 3

Because of the overlap, the voltage of the terminal p (see Figure 10.15) immediately after $\omega t = \alpha$ recovers to $(e_a + e_b)/2$, instead of e_b. Therefore, as shown in Figure 10.17, the effect of the overlap is measured by subtracting an area A_μ from the area A_0, once every 60° ($\pi/3$ rad).

$$A_\mu = \int_\alpha^\delta \left(e_b - \frac{e_a + e_b}{2} \right) d\theta = \int_\alpha^\delta \frac{e_b - e_a}{2} d\theta$$

$$= \frac{\sqrt{3}E_m}{2} \int_\alpha^\delta \sin\theta\, d\theta = \frac{\sqrt{3}E_m}{2} (\cos\alpha - \cos\delta)$$

The corresponding average voltage drop (due to overlap) is given by

$$\Delta V_d = \frac{A_\mu}{\pi/3} = \frac{3}{\pi} \frac{\sqrt{3}}{2} E_m (\cos\alpha - \cos\delta)$$

$$= \frac{V_{d0}}{2} (\cos\alpha - \cos\delta)$$

(10.9)

where V_{d0} is the ideal no-load voltage given by Equation 10.3. From Equation 10.8A, the current i_3 during commutation is

$$i_3 = \frac{\sqrt{3}E_m}{2\omega L_c} (\cos\alpha - \cos\omega t)$$

Since at the end of the commutation $\omega t = \delta$ and $i_3 = I_d$,

$$I_d = \frac{\sqrt{3}E_m}{2\omega L_c}(\cos\alpha - \cos\delta)$$

(10.10)

$$= I_{s2}(\cos\alpha - \cos\delta)$$

Hence,

$$\frac{\sqrt{3}E_m}{2}(\cos\alpha - \cos\delta) = I_d\omega L_c$$

Substituting in Equation 10.9 gives

$$\Delta V_d = \frac{3}{\pi}I_d\omega L_c$$

With commutation overlap and ignition delay, the reduction in direct voltage is represented by areas A_α and A_μ; the direct voltage is given by

$$V_d = V_{do}\cos\alpha - \Delta V_d$$

(10.11)

$$= V_{do}\cos\alpha - R_c I_d$$

where

$$R_c = \frac{3}{\pi}\omega L_c = \frac{3}{\pi}X_c$$

(10.12)

and R_c is called the "equivalent commutating resistance." It accounts for the voltage drop due to commutation overlap. It does not, however, represent a real resistance and consumes no power.

Rectifier operation

The equivalent circuit of the bridge rectifier based on the above analysis is given in Figure 10.18. The direct voltage and current in the equivalent circuit are the average values. The internal voltage is a function of the ignition angle α. The overlap angle μ does not explicitly appear in the equivalent circuit; the effect of commutation overlap is represented by R_{cr}. The voltage wave shapes and periods of valve conduction for rectifier mode of operation, including commutation overlap, are shown in Figure 10.19(a).

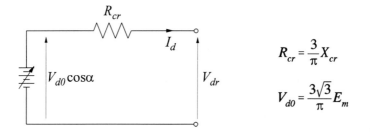

$$R_{cr} = \frac{3}{\pi} X_{cr}$$

$$V_{d0} = \frac{3\sqrt{3}}{\pi} E_m$$

Figure 10.18 Bridge rectifier equivalent circuit

Inverter operation

If there is no commutation overlap, $V_d = V_{d0}\cos\alpha$. Therefore, V_d reverses when $\alpha = 90°$.

With overlap,

$$V_d = V_{d0}\cos\alpha - \Delta V_d$$

Substituting for ΔV_d from Equation 10.9, we have

$$V_d = V_{d0}\cos\alpha - \frac{V_{d0}}{2}(\cos\alpha - \cos\delta)$$

$$= \frac{V_{d0}}{2}(\cos\alpha + \cos\delta) \qquad (10.13)$$

The transitional value of the ignition delay angle, α_t, beyond which inversion takes place is given by

$$\cos\alpha_t + \cos\delta_t = 0$$

or

$$\alpha_t = \pi - \delta_t = \pi - \alpha_t - \mu$$

$$= \frac{\pi - \mu}{2} \qquad (10.14)$$

The effect of the overlap is thus to reduce α_t from 90° to 90° − μ/2.

At first sight, it may seem strange to delay the firing pulse until the actual anode voltage becomes negative. We should, however, realize that commutation is always possible as long as the commutating voltage ($e_{ba} = e_b - e_a$) is positive and as long as the outgoing valve will have reversed voltage applied to it after it extinguishes.

Since valves conduct in only one direction, the current in a converter cannot be reversed. A reversal of V_d results in a reversal of power. An alternating voltage must exist on the primary side of the transformer for inverter operation. The direct voltage of the inverter opposes the current, as in a dc motor, and is called a countervoltage or back voltage. The applied direct voltage from the rectifier forces current through the inverter valves against this back voltage.

Figure 10.19(b) shows the voltage wave shapes and periods of valve conduction for inverter mode of operation. For successful commutation, the changeover from the outgoing valve to the incoming valve must be complete before the commutating voltage becomes negative. For example, commutation from valve 1 to valve 3 is possible only when e_b is more positive than e_a; the current changeover from valve 1 to valve 3 must be complete before e_a becomes more positive than e_b with sufficient margin to allow for valve de-ionization.

For description of rectifier operation, we use the following angles:

α = ignition delay angle
μ = overlap angle
δ = extinction delay angle = $\alpha + \mu$

As illustrated in Figure 10.20, α is the angle by which ignition is delayed from the instant at which the commutating voltage (e_{ba} for valve 3) is zero and increasing.

The inverter operation may also be described in terms of α and δ defined in the same way as for the rectifier, but having values between 90° and 180°. However, the common practice is to use *ignition advance angle* β and *extinction advance angle* γ for describing inverter performance. These angles are defined by their advance with respect to the instant ($\omega t = 180°$ for ignition of valve 3 and extinction of valve 1) when the commutating voltage is zero and decreasing, as shown in Figure 10.20. From the figure, we see that

$\beta = \pi - \alpha$ = ignition advance angle
$\gamma = \pi - \delta$ = extinction advance angle
$\mu = \delta - \alpha = \beta - \gamma$ = overlap

Since $\cos\alpha = -\cos\beta$ and $\cos\delta = -\cos\gamma$, Equations 10.10 and 10.13 may be written in terms of γ and β as follows:

$$I_d = I_{s2}(\cos\gamma - \cos\beta) \tag{10.15}$$

$$V_d = V_{d0}\frac{\cos\gamma + \cos\beta}{2} \tag{10.16}$$

or

$$V_d = V_{d0}\cos\beta + R_c I_d \tag{10.17A}$$

(a) Rectifier mode:

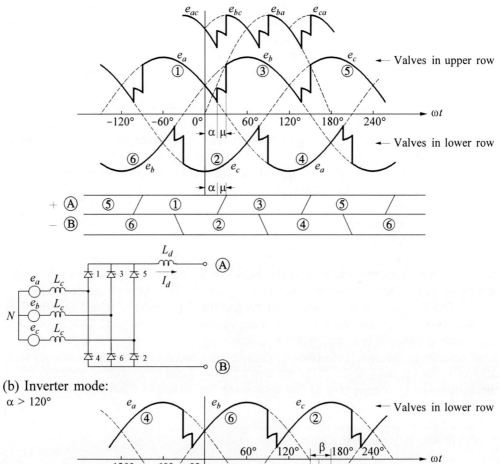

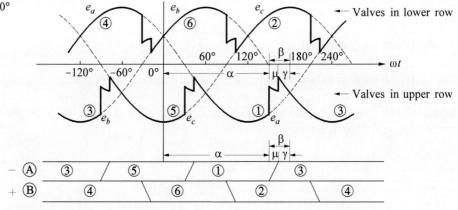

(b) Inverter mode:
$\alpha > 120°$

Figure 10.19 Voltage wave shapes and valve conduction periods

Converter Angle Definitions

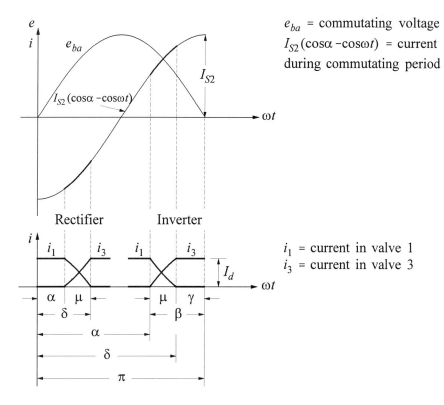

e_{ba} = commutating voltage

$I_{S2}(\cos\alpha - \cos\omega t)$ = current during commutating period

i_1 = current in valve 1

i_3 = current in valve 3

Figure 10.20 Angles used in the description of rectifier and inverter operations

or

$$V_d = V_{d0}\cos\gamma - R_c I_d \qquad (10.17\text{B})$$

The inverter voltage, considered negative in general converter equations, is usually taken as positive when written specifically for an inverter.

Based on the above equations, the inverter may be represented by the two alternative equivalent circuits shown in Figure 10.21.

Relationship between ac and dc quantities

From Equation 10.13, the average direct voltage V_d is given by

$$V_d = V_{d0}\cos\alpha - \Delta V_d = V_{d0}\frac{\cos\alpha + \cos\delta}{2} \qquad (10.18\text{A})$$

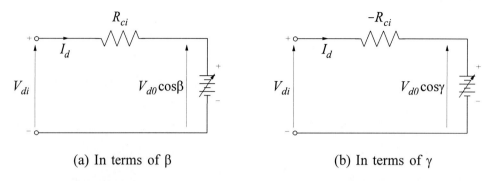

(a) In terms of β   (b) In terms of γ

Figure 10.21 Inverter equivalent circuits (with V_{di} positive)

Substituting for V_{d0} from Equation 10.3B in terms of RMS line-to-neutral voltage E_{LN}, we get

$$V_d = \frac{3\sqrt{6}}{\pi}\frac{\cos\alpha + \cos\delta}{2}E_{LN} \tag{10.18B}$$

With losses neglected, ac power is equal to the dc power:

$$P_{ac} = P_{dc}$$

Hence,

$$3E_{LN}I_{L1}\cos\phi = V_d I_d$$

where

$\quad E_{LN}$ = RMS line-to-neutral voltage
$\quad I_{L1}$ = RMS fundamental frequency current

From Equation 10.18,

$$3E_{LN}I_{L1}\cos\phi = \frac{3\sqrt{6}}{\pi}\frac{\cos\alpha + \cos\delta}{2}E_{LN}I_d$$

Hence,

$$I_{L1}\cos\phi = \left(\frac{\sqrt{6}}{\pi}I_d\right)\left(\frac{\cos\alpha + \cos\delta}{2}\right) \tag{10.19}$$

From Equation 10.5B, with $\mu=0$

$$I_{L1} = \frac{\sqrt{6}}{\pi}I_d$$

Denoting this value of I_{L1} (when $\mu=0$) by I_{L10}, Equation 10.19 can be written as

$$I_{L1}\cos\phi = I_{L10}\left(\frac{\cos\alpha + \cos\delta}{2}\right) \tag{10.20}$$

where

$$I_{L10} = \frac{\sqrt{6}}{\pi}I_d \tag{10.21}$$

Approximate expressions

As an approximation I_{L1} may be considered equal to I_{L10}:

$$I_{L1} \approx I_{L10} = \frac{\sqrt{6}}{\pi}I_d \tag{10.22}$$

The above relationship is exact if $\mu=0°$; with $\mu=60°$ the error is 4.3%, and with $\mu<30°$ (normal value) the error is less than 1.1% [2].

It follows that the power factor is given by

$$\cos\phi \approx \frac{\cos\alpha + \cos\delta}{2} \tag{10.23}$$

As a result of the approximation, from Equation 10.18A,

$$V_d \approx V_{d0}\cos\phi \tag{10.24A}$$

Hence,

$$\cos\phi \approx \frac{V_d}{V_{d0}} \tag{10.24B}$$

From Equation 10.11, $V_d = V_{d0}\cos\alpha - R_c I_d$. Hence,

$$\cos\phi \approx \cos\alpha - \frac{R_c I_d}{V_{d0}} \qquad (10.25)$$

Substituting for V_{d0} from Equation 10.3A in Equation 10.24A, we get

$$V_d \approx \frac{3\sqrt{6}}{\pi} E_{LN} \cos\phi \qquad (10.26)$$

We see from Equation 10.22 that the converter has essentially fixed current ratio I_d/I_{L1}, the variation with load being only a few percent. The power factor $\cos\phi$, as seen from Equation 10.25, depends on load in addition to ignition delay angle α.

10.2.3 Converter Transformer Rating

The RMS value of the transformer secondary current (total and not just the fundamental frequency component) I_{TRMS} is given by

$$I_{TRMS}^2 = \frac{1}{T}\int_0^T i^2(t)\,dt$$

The alternating line-current wave consists of rectangular pulses of amplitude I_d and width $2\pi/3$ rad as shown in Figure 10.12. Therefore,

$$I_{TRMS}^2 = \frac{1}{\pi}\int_{-\pi/2}^{\pi/2} i^2(t)\,dt$$
$$= \frac{1}{\pi}\int_{-\pi/3}^{\pi/3} I_d^2\,dt$$
$$= \frac{2}{3}I_d^2$$

and hence,

$$I_{TRMS} = \sqrt{2/3}\,I_d \qquad (10.27)$$

The RMS value of the line-to-neutral transformer secondary voltage is given by

$$E_{LN} = \frac{\pi}{3\sqrt{6}}V_{d0}$$

Transformer volt-ampere rating is given by

$$3\text{-phase rating} = 3E_{LN}I_{TRMS}$$

$$= 3\left(\frac{\pi}{3\sqrt{6}}\right)V_{do}\sqrt{\frac{2}{3}}I_d \tag{10.28}$$

$$= \frac{\pi}{3}V_{do}I_d$$

$$= 1.0472\,(\text{ideal no-load direct voltage})\,(\text{rated direct current})$$

10.2.4 Multiple-Bridge Converters

Two or more bridges are connected in series to obtain as high a direct voltage as required. The bridges are in series on the dc side and parallel on the ac side. A bank of transformers is connected between the ac source and bridge-connected valves. The ratios of the transformers are adjustable under load.

In practice, multiple-bridge converters have an even number of bridges arranged in pairs so as to result in a 12-pulse arrangement. As shown in Figure 10.22,

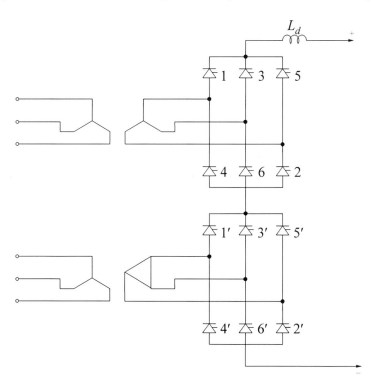

Figure 10.22   12-pulse bridge converter

two banks of transformers, one connected Y-Y and the other Y-Δ, are used to supply each pair of bridges. The three-phase voltages supplied at one bridge are displaced by 30° from those supplied at the other bridge. The ac wave shapes for the two bridges, as illustrated in Figure 10.23, add up so as to produce a wave shape which is more sinusoidal than the current waves for each of the 6-pulse bridges. As we will see in Section 10.6, with a 12-pulse arrangement, *fifth and seventh harmonics are effectively eliminated* on the ac side. This reduces the cost of harmonic filters significantly.

In addition, with a 12-pulse bridge arrangement, the dc voltage ripple is reduced; sixth and eighteenth harmonics are eliminated (6-pulse bridges have multiples of sixth harmonics on the dc side whereas 12-pulse bridges have only multiples of the twelfth harmonic).

For converters having more than two bridges, higher pulse numbers are possible: 18-pulse, three-bridge converter; 24-pulse, four-bridge converters. The transformer connections required are more complex than those for 12-pulse converters. Therefore, it is more practical to use 12-pulse converters and provide the necessary filtering.

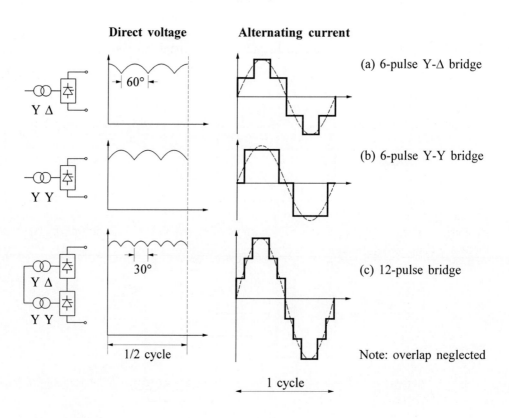

Figure 10.23 Direct voltage and alternating current wave shapes

Relationships between ac and dc quantities, with multiple bridges

Let

B = no. of bridges in series

T = transformer ratio

The ideal no-load voltage (corresponding to Equation 10.3C) is

$$V_{d0} = \frac{3\sqrt{2}}{\pi} BTE_{LL} = 1.3505 BTE_{LL} \qquad (10.29)$$

Since voltage drop per bridge is $I_d(3X_c/\pi)$ and there are B bridges in series, the dc voltage (corresponding to Equations 10.11) is given by

$$V_d = V_{d0}\cos\alpha - I_d B\left(\frac{3}{\pi}X_c\right) \qquad (10.30A)$$

or

$$V_d = V_{d0}\cos\gamma - I_d B\left(\frac{3}{\pi}X_c\right) \qquad (10.30B)$$

The dc voltage in terms of the power factor per Equation 10.24A is given by

$$V_d \approx V_{d0}\cos\phi \qquad (10.31)$$

However, with multiple bridges, V_{d0} is given by Equation 10.29. The average dc output voltage of a 12-pulse bridge is, therefore, twice that of a 6-pulse bridge converter. The RMS value of the fundamental frequency component of the total alternating current (corresponding to Equation 10.22) is given by

$$I_{L1} \approx \frac{\sqrt{6}}{\pi}BTI_d = 0.78BTI_d \qquad (10.32)$$

Example 10.1

A three-phase, 12-pulse rectifier is fed from a transformer with nominal voltage ratings of 220 kV/110 kV.

(a)   If the primary voltage is 230 kV and the effective turns ratio T is 0.48, determine the dc output voltage when the ignition delay angle α is 20° and the commutation angle μ is 18°.

(b)   If the direct current delivered by the rectifier is 2,000 A, calculate the effective commutating reactance X_c, RMS fundamental component of alternating current, power factor $\cos\phi$, and reactive power at the primary side of the transformer.

Solution

(a) A 12-pulse bridge circuit comprises two 6-pulse bridges. Hence $B=2$.

The no-load direct voltage is

$$V_{d0} = \frac{3\sqrt{2}}{\pi}BTE_{LL}$$

$$= 1.3505 \times 2 \times 0.48 \times 230$$

$$= 298.18 \quad \text{kV}$$

The extinction angle is

$$\delta = \alpha + \mu = 20° + 18° = 38°$$

Hence, the reduction in average direct voltage due to commutation overlap is

$$\Delta V_d = V_{d0}\frac{\cos\alpha - \cos\delta}{2}$$

$$= 298.18 \times \frac{\cos 20° - \cos 38°}{2}$$

$$= 22.61 \quad \text{kV}$$

The dc output voltage is

$$V_d = V_{d0}\cos\alpha - \Delta V_d$$

$$= 298.18\cos 20° - 22.61$$

$$= 257.58 \quad \text{kV}$$

Alternatively,

$$V_d = V_{d0}\frac{\cos\alpha + \cos\delta}{2}$$

$$= 298.18 \times \frac{\cos 20° + \cos 38°}{2}$$

$$= 257.58 \quad \text{kV}$$

(b) $\Delta V_d = B R_c I_d$. Hence,

$$R_c = \frac{\Delta V_d}{B I_d} = \frac{22.61}{2 \times 2} = 5.65 \quad \Omega$$

$$X_c = \frac{\pi R_c}{3} = \frac{\pi \times 5.65}{3} = 5.92 \quad \Omega/\text{phase}$$

Fundamental component of alternating current on the primary side is

$$I_{L1} = \frac{\sqrt{6}}{\pi} B T I_d$$

$$= 0.7797 \times 2 \times 0.48 \times 2$$

$$= 1.497 \quad \text{kA}$$

Power factor at the HT bus is

$$\cos\phi \approx \frac{V_d}{V_{d0}} = \frac{257.58}{298.18} = 0.8638$$

Hence, $\phi = \cos^{-1}(0.8638) = 30.25°$

$$P_{ac} = P_{dc} = V_d I_d = 257.58 \times 2 = 515.16 \quad \text{MW}$$

$$Q_{HT} = P_{ac}\tan\phi = 515.16 \times \tan30.25° = 300.43 \quad \text{MVAr}$$

The solution is shown in Figure E10.1.

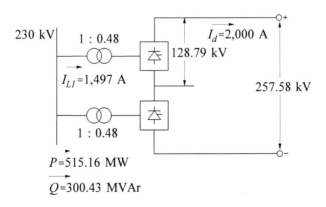

Figure E10.1   ∎

10.3 ABNORMAL OPERATION

10.3.1 Arc-back (Backfire)

Arc-back refers to conduction in the reverse direction and is one of the serious problems associated with mercury-arc valves. Since arc-back is reverse conduction, it can occur only when there is inverse voltage across a valve. In rectification each valve is exposed to inverse voltage during approximately two-thirds of each cycle. Therefore, arc-backs are more common during rectification than during inversion.

Arc-back is a random phenomenon. Among the factors that tend to increase its occurrence are high-peak inverse voltage, overcurrent, high rate of change of current at the end of conduction, condensation of mercury vapour on anodes, and high rate of increase of inverse voltage. The effect of the arc-back is to place a short-circuit across two phases of the secondary of the converter transformers. These short-circuit currents subject the transformers and valves to much greater current than in normal operation. The transformer windings must be firmly braced to withstand more numerous short-circuits than ordinary power transformers; this adds to the cost. The valves require increased maintenance.

To remove an arc-back, current is diverted into a bypass valve. The bypass valve is a separate valve connected across a 6-pulse valve group. This valve has a higher current rating than other valves and is capable of carrying 1 pu direct current for about 60 seconds. The control grid of the bypass valve is normally blocked. When a bridge is to be bypassed, its bypass valve is unblocked and the main valves are simultaneously blocked by discontinuing the transmission of positive pulses to their grids. The direct current shifts from the main valves to the bypass valve in a few milliseconds. By simultaneously unblocking the main valve and blocking the bypass valve, the direct current can be transferred back to the main valves.

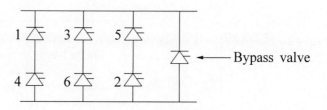

Figure 10.24

Arc-back does not occur in thyristor valves. Thyristors can individually fail in a shorted condition. However, thyristors arrayed into converter valves utilize redundancy and protection to prevent reverse conduction.

10.3.2 Commutation Failure

Failure to complete commutation before the commutating voltage reverses (with sufficient margin for de-ionization) is referred to as commutation failure. It is not due to any misoperation of the valve but to conditions in the circuits outside the valve. Commutation failures are more common with inverters and occur during disturbances such as high direct current or low alternating voltage. A rectifier can have a commutation failure only if the firing circuit fails.

Figure 10.25 illustrates how commutation failure occurs in an inverter. A failure of commutation from valve 1 to valve 3 is considered.

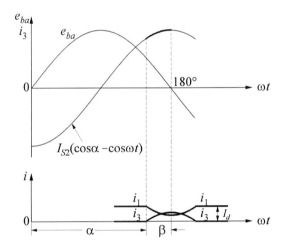

Figure 10.25 Commutation failure in inverters

Because of increased direct current, low alternating voltage (possibly caused by an ac system short-circuit), late ignition, or a combination of these, valve 1 is not extinguished before e_{ba} reverses. Current in valve 3 will decrease to zero and the valve will extinguish. Valve 1 current will return to I_d, and thus valve 1 will continue to conduct.

As shown in Figure 10.26, when valve 4 fires next, because valve 1 is still conducting, a short is placed across the dc side of the bridge. The zero dc voltage keeps the voltage across valve 5 negative so that valve 5 cannot conduct. Valve 4 is extinguished and valve 6 is ignited in the normal fashion.

Valve 1 thus conducts for one full cycle (3 times normal duration) and extinguishes when valve 3 ignites during the next cycle.

For the period when valves 1 and 4 are both conducting (i.e., for 120°), the inverter dc voltage is zero and hence there is no power flow on the dc system.

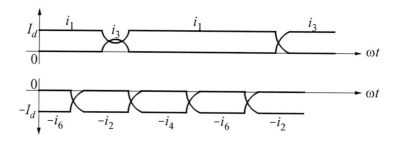

Figure 10.26 Valve currents during commutation failure

Double commutation failure is the failure of two successive commutations in the same cycle. If the unsuccessful commutation from valve 1 to valve 3 were followed by a failure in the commutation from valve 2 to valve 4, valves 1 and 2 would be left conducting until the next cycle when they would be normally conducting again. During the time that only valves 1 and 2 are conducting, the alternating voltages of terminals a and c appear across the dc terminals.

Double commutation failures are very rare. Usually, after one commutation failure, either the firing angle initiating the next commutation is advanced sufficiently by the inverter control, or "double overlap" during the period when valves 1 and 3 as well as valves 2 and 4 are conducting hastens the commutation. Double commutation failure, like the single failure, is self-curing.

10.4 CONTROL OF HVDC SYSTEMS

An HVDC transmission system is highly controllable. Its effective use depends on appropriate utilization of this controllability to ensure desired performance of the power system. With the objectives of providing efficient and stable operation and maximizing flexibility of power control without compromising the safety of equipment, various levels of control are used in a hierarchical manner. In this section, we will describe the principles of operation of these controls, their implementation and their performance during normal and abnormal system conditions.

10.4.1 Basic Principles of Control

Consider the HVDC link shown in Figure 10.27(a). It represents a monopolar link or one pole of a bipolar link. The corresponding equivalent circuit and voltage profile are shown in Figures 10.27(b) and (c), respectively.

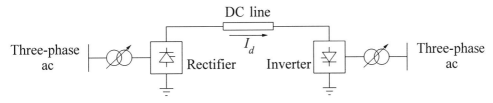

(a) Schematic diagram

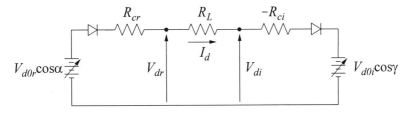

(b) Equivalent circuit

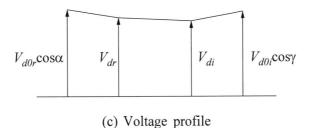

(c) Voltage profile

Figure 10.27 HVDC transmission link

The direct current flowing from the rectifier to the inverter is

$$I_d = \frac{V_{dor}\cos\alpha - V_{doi}\cos\gamma}{R_{cr} + R_L - R_{ci}} \qquad (10.33)$$

The power at the rectifier terminals is

$$P_{dr} = V_{dr}I_d \qquad (10.34)$$

and at the inverter terminal is

$$P_{di} = V_{di}I_d = P_{dr} - R_L I_d^2 \qquad (10.35)$$

Basic means of control

The direct voltage at any point on the line and the current (or power) can be controlled by controlling the internal voltages ($V_{dor}\cos\alpha$) and ($V_{doi}\cos\gamma$). This is accomplished by grid/gate control of the valve ignition angle or control of the ac voltage through tap changing of the converter transformer.

Grid/gate control, which is rapid (1 to 10 ms), and tap changing, which is slow (5 to 6 s per step), are used in a complementary manner. Grid/gate control is used initially for rapid action, followed by tap changing to restore the converter quantities (α for rectifier and γ for inverter) to their normal range.

Power reversal is obtained by reversal of polarity of direct voltages at both ends.

Basis for selection of controls

The following considerations influence the selection of control characteristics:

1.   Prevention of large fluctuations in direct current due to variations in ac system voltage.

2.   Maintaining direct voltage near rated value.

3.   Maintaining power factors at the sending and receiving end that are as high as possible.

4.   Prevention of commutation failure in inverters and arc-back in rectifiers using mercury-arc valves.

Rapid control of the converters to prevent large fluctuations in direct current is an important requirement for satisfactory operation of the HVDC link. Referring to Equation 10.33, the line and converter resistances are small; hence, a small change in V_{dor} or V_{doi} causes a large change in I_d. For example, a 25% change in the voltage at either the rectifier or the inverter could cause direct current to change by as much as 100%. This implies that, if both α_r and γ_i are kept constant, the direct current can vary over a wide range for small changes in the alternating voltage magnitude at either end. Such variations are generally unacceptable for satisfactory performance of the power system. In addition, the resulting current may be high enough to damage the valves and other equipment. Therefore, rapid converter control to prevent fluctuations of direct current is essential for proper operation of the system; without such a control, the HVDC system would be impractical.

For a given power transmitted, the direct voltage profile along the line should be close to the rated value. This minimizes the direct current and thereby the line losses.

There are several reasons for maintaining the power factor high:

(a)   To keep the rated power of the converter as high as possible for given current and voltage ratings of transformer and valve;

(b)   To reduce stresses in the valves;

(c)   To minimize losses and current rating of equipment in the ac system to which the converter is connected;

(d)   To minimize voltage drops at the ac terminals as loading increases; and

(e)   To minimize cost of reactive power supply to the converters.

From Equation 10.23,

$$\cos\phi \approx 0.5[\cos\alpha + \cos(\alpha + \mu)]$$
$$\approx 0.5[\cos\gamma + \cos(\gamma + \mu)]$$

Therefore, to achieve high power factor, α for a rectifier and γ for an inverter should be kept as low as possible.

The rectifier, however, has a *minimum α limit* of about 5° to ensure adequate voltage across the valve before firing. For example, in the case of thyristors, the positive voltage appearing across each thyristor before firing is used to charge the supply circuit providing the firing pulse energy to the thyristor. Therefore, firing cannot occur earlier than about $\alpha = 5°$ [10]. Consequently, the rectifier normally operates at a value of α within the range of 15° to 20° so as to leave some room for increasing rectifier voltage to control dc power flow.

In the case of an inverter, it is necessary to maintain a certain minimum extinction angle to avoid commutation failure. It is important to ensure that commutation is completed with sufficient margin to allow for de-ionization before commutating voltage reverses at $\alpha = 180°$ or $\gamma = 0°$. The extinction angle γ is equal to $\beta - \mu$, with the overlap μ depending on I_d and the commutating voltage. Because of the possibility of changes in direct current and alternating voltage even after commutation has begun, sufficient *commutation margin above the minimum γ limit* must be maintained. Typically, the value of γ with acceptable margin is 15° for 50 Hz systems and 18° for 60 Hz systems.

Control characteristics

Ideal characteristics:

In satisfying the basic requirements identified above, the responsibilities for voltage regulation and current regulation are kept distinct and are assigned to separate terminals. Under normal operation, the rectifier maintains constant current (CC), and the inverter operates with *constant extinction angle*[1] (CEA), maintaining adequate commutation margin. The basis for this control philosophy is best explained by using the steady-state voltage-current (*V-I*) characteristics, shown in Figure 10.28. The voltage V_d and the current I_d forming the coordinates may be measured at some common point on the dc line. In Figure 10.28, we have chosen this to be at the rectifier terminal. The rectifier and inverter characteristics are both measured at the rectifier; the inverter characteristic thus includes the voltage drop across the line.

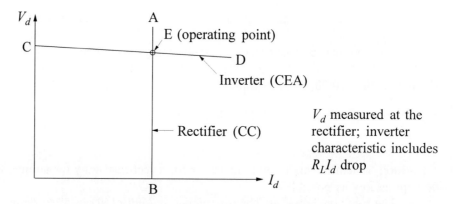

Figure 10.28 Ideal steady-state *V-I* characteristics

With the rectifier maintaining constant current, its *V-I* characteristic, shown as line AB in Figure 10.28, is a vertical line. From Figure 10.27(b),

$$V_d = V_{d0i}\cos\gamma +(R_L-R_{ci})I_d \tag{10.36}$$

This gives the inverter characteristic, with γ maintained at a fixed value. If the commutating resistance R_{ci} is slightly larger than the line resistance R_L, the characteristic of the inverter, shown as line CD in Figure 10.28, has a small negative slope.

---

[1] Constant extinction angle control mode is essentially the same as constant margin angle control. Under normal operation, the commutation margin angle and the extinction angles are equal. The distinction arises during conditions such as operation with a large overlap angle; the valve voltage may become positive earlier than when the sinusoidal portion of the voltage would have crossed zero under normal conditions. Under these conditions the concept of maintaining minimum extinction angle is not meaningful. Therefore, the commutation margin angle (representing the interval between the end of conduction and the instant when the actual voltage across the valve becomes positive) is maintained for safe inverter operation.

Since an operating condition has to satisfy both rectifier and inverter characteristics, it is defined by the intersection of the two characteristics (E).

The rectifier characteristic can be shifted horizontally by adjusting the "current command" or "current order." If measured current is less than the command, the regulator advances the firing by decreasing α.

The inverter characteristic can be raised or lowered by means of its transformer tap changer. When the tap changer is moved, the CEA regulator quickly restores the desired γ. As a result, the direct current changes, which is then quickly restored by the current regulator of the rectifier. The rectifier tap changer acts to bring α into the desired range between 10° and 20° to ensure a high power factor and adequate room for control.

To operate the inverter at a constant γ, the valve firing is controlled by a computer which takes into consideration variations in the instantaneous values of voltage and current. The computer controls the firing times so that the extinction angle γ is larger than the de-ionization angle of the valve.

Actual characteristics:

The rectifier maintains constant current by changing α. However, α cannot be less than its minimum value (α_{min}). Once α_{min} is reached, no further voltage increase is possible, and the rectifier will operate at *constant ignition angle* (CIA). Therefore, the rectifier characteristic has really two segments (AB and FA) as shown in Figure 10.29. The segment FA corresponds to minimum ignition angle and represents the CIA control mode; the segment AB represents the normal constant current (CC) control mode.

In practice, the constant current characteristic may not be truly vertical, depending on the current regulator. With a proportional controller, it has a high negative slope due to the finite gain of the current regulator, as shown below.

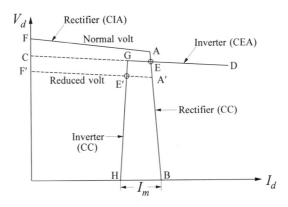

Figure 10.29 Actual converter control steady-state characteristics

With a regulator gain of K, we have

$$V_{d0}\cos\alpha = K(I_{ord} - I_d)$$
$$= V_d + R_{cr}I_d$$

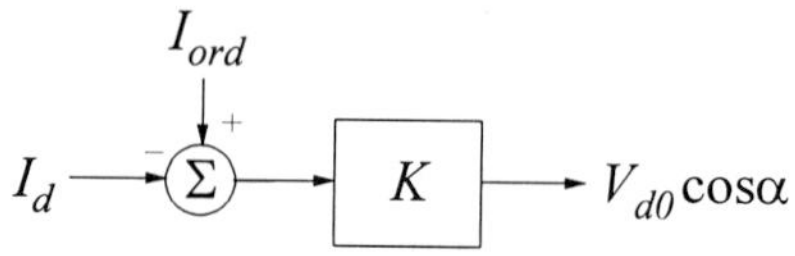

I_{ord} = current order

Figure 10.30 Current regulator

Therefore,

$$V_d = KI_{ord} - (K + R_{cr})I_d$$

In terms of perturbed values,

$$\Delta V_d = -(K + R_{cr})\Delta I_d$$

or

$$\Delta V_d / \Delta I_d = -(K + R_{cr}) \tag{10.37}$$

With a proportional plus integral regulator, the CC characteristic is quite vertical. The complete rectifier characteristic at normal voltage is defined by FAB. At a reduced voltage it shifts, as indicated by F′A′B.

The CEA characteristic of the inverter intersects the rectifier characteristic at E for normal voltage. However, the inverter CEA characteristic (CD) does not intersect the rectifier characteristic at a reduced voltage represented by F′A′B. Therefore, a big reduction in rectifier voltage would cause the current and power to be reduced to zero after a short time depending on the dc reactors. The system would thus run down.

In order to avoid the above problem, the inverter is also provided with a current controller, which is set at a lower value than the current setting for the rectifier. The complete inverter characteristic is given by DGH, consisting of two segments: one of CEA and one of constant current.

The difference between the rectifier current order and the inverter current order is called the *current margin*, denoted by I_m in Figure 10.29. It is usually set at 10 to 15% of the rated current so as to ensure that the two constant current characteristics do not cross each other due to errors in measurement or other causes.

Under normal operating conditions (represented by the intersection point E), the rectifier controls the direct current and the inverter the direct voltage. With a reduced rectifier voltage (possibly caused by a nearby fault), the operating condition is represented by the intersection point E′. The inverter takes over current control and the rectifier establishes the voltage. In this operating mode, the roles of the rectifier and inverter are reversed. The change from one mode to another is referred to as a *mode shift*.

Combined rectifier and inverter characteristics

In most HVDC systems, each converter is required to function as a rectifier

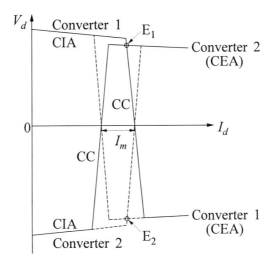

Figure 10.31 Operation with each converter having combined
inverter and converter characteristics

as well as an inverter. Consequently, each converter is provided with a combined
characteristic as shown in Figure 10.31.

The characteristic of each converter consists of three segments: constant
ignition angle (CIA) corresponding to α_{min}, constant current (CC), and constant
extinction angle (CEA).

The power transfer is from converter 1 to converter 2, when the characteristics
are as shown in Figure 10.31 by solid lines. The operating condition in this mode of
operation is represented by point E_1.

The power flow is reversed when the characteristics are as shown by the dotted
lines. This is achieved by reversing the "margin setting," i.e., by making the current
order setting of converter 2 exceed that of converter 1. The operating condition is now
represented by E_2 in the figure; the current I_d is the same as before, but the voltage
polarity has changed.

Alternative inverter control modes

The following are variations to the CEA control mode described above for the
inverter. These variations offer some advantages in special cases.

DC voltage control mode:

Instead of regulating to a fixed γ value (CEA), a closed-loop voltage control
may be used so as to maintain a constant voltage at a desired point on the dc line,

usually the sending end (rectifier). The necessary inverter voltage to maintain the desired dc voltage is estimated by computing the line RI drop. As compared to constant γ control (which has drooping voltage characteristic), the voltage control mode has the advantage that the inverter V-I characteristic is flat as shown in Figure 10.32(a). In addition, the voltage control mode has a slightly higher value of γ and is thus less prone to commutation failure. Normally the voltage control mode maintains a γ of about 18° in conjunction with the tap changers.

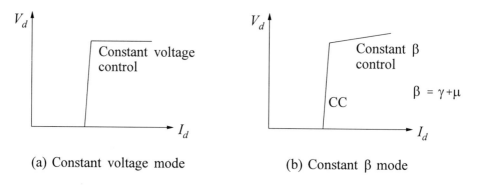

(a) Constant voltage mode   (b) Constant β mode

Figure 10.32 Alternative inverter control modes

Beta (β) control:

The inverter equivalent circuit in terms of ignition advance angle β is as shown in Figure 10.21(a). With constant β, the V-I characteristic of the inverter therefore has a positive slope as shown in Figure 10.32(b). At low loads, constant β gives additional security against commutation failure. However, at higher currents (larger overlap), the minimum γ may be encountered. Constant β control mode is not used for normal operation. It is viewed as a backup type of control mode useful for acting directly upon the firing angle during transient conditions.

Mode stabilization

As shown in Figure 10.33, the intersection of rectifier α_{min} characteristic and the inverter CEA may not be well defined at certain voltage levels near the transition between the inverter CEA and CC characteristics. In this region, a small change in ac voltage will cause a large (10%) change in direct current. There will also be a tendency to "hunt" between modes and tap changing. In order to avoid this problem, a characteristic with positive slope (constant β) at the transition between CEA and CC control characteristic of the inverter is often provided as shown in Figure 10.34(a). Another variation, shown in Figure 10.34(b), controls the direct voltage with a voltage feedback loop.

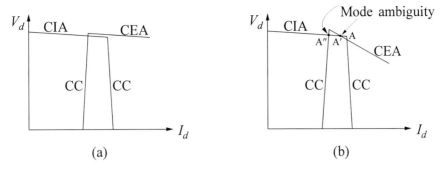

Figure 10.33 Regions of mode ambiguity

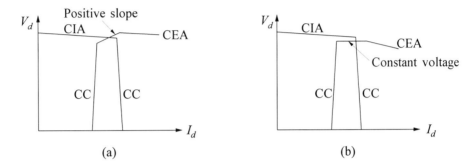

Figure 10.34 Modification of *V-I* characteristic for mode stabilization

Tap changer control

Tap changer control is used to keep the converter firing angles within the desired range, whenever α (for rectifier) or γ (for inverter) exceeds this range for more than a few seconds.

Normally, the inverter operates at constant extinction angle, thus fixing the line voltage with superimposed voltage control by the tap changer. The rectifier operates in current control mode with superimposed $\alpha = \alpha_{nominal}$ control by the tap changer.

Tap changers are usually sized to allow for minimum and maximum steady-state voltage variations, and for minimum to maximum power flow under worst-case steady-state voltage conditions. Unnecessary tap movements during transient conditions are prevented by using time delays. Hunting of the tap changer is avoided by having a dead band wider than the tap-step size.

Current limits

The following limits have to be recognized in establishing the current order.

(a) Maximum current limit:

The maximum short-time current is usually limited to 1.2 to 1.3 times normal full-load current, to avoid thermal damage to valves.

(b) Minimum current limit:

At low values of current, the ripple in the current may cause it to be discontinuous or intermittent. In a 12-pulse operation, the current is then interrupted 12 times per cycle. This is objectionable because of the high voltages (Ldi/dt) induced in the transformer windings and the dc reactor by the high rate of change of current at the instants of interruption.

At low values of direct current, the overlap is small. Operation is objectionable even with continuous current if the overlap is too small. With a very small overlap, the two jumps in direct voltage at the beginning and end of commutation merge to form one jump twice as large, resulting in an increased stress on the valves. It may also cause flashover of protective gaps placed across the terminals of each bridge [2].

(c) Voltage-dependent current-order limit[1] (VDCOL):

Under low voltage conditions, it may not be desirable or possible to maintain rated direct current or power for the following reasons:

(a)   When voltage at one converter drops by more than about 30%, the reactive power demand of the remote converter increases, and this may have an adverse effect on the ac system. A higher α or γ at the remote converter necessary to control the current causes the increase in reactive power. The reduced ac system voltage levels also significantly decrease the reactive power supplied by the filters and capacitors, which often supply much of the reactive power absorbed by the converters.

(b)   At reduced voltages, there are also risks of commutation failure and voltage instability.

These problems associated with operation under low-voltage conditions may be prevented by using a "voltage-dependent current-order limit" (VDCOL). This limit reduces the maximum allowable direct current when the voltage drops below a predetermined value. The VDCOL characteristics may be a function of the ac commutating voltage or the dc voltage. The two types of VDCOLs are illustrated in Figure 10.35.

The rectifier inverter static *V-I* characteristics, including VDCOL, are shown in Figure 10.36. The inverter characteristic matches the rectifier VDCOL to preserve the current margin. The general practice is to transiently reduce the current order through voltage-dependent current limit. For VDCOL operation, the measured direct

---

[1] This is also referred to as *voltage dependent current limit* (VDCL).

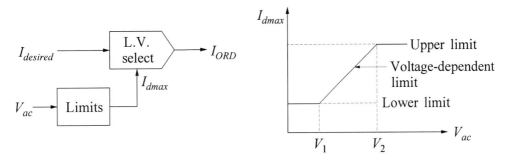

(a) Current limit as a function of alternating voltage

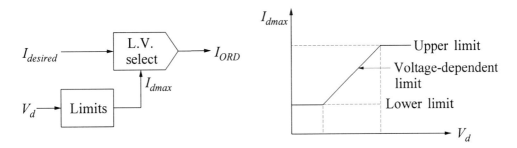

(b) Current limit as a function of direct voltage

Figure 10.35 Voltage-dependent current limits

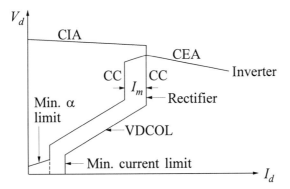

Figure 10.36 Steady-state V-I characteristic with VDCOL,
minimum current limit and firing angle limits

voltage is passed through a first-order time lag element. Generally, this time lag is different for increasing and decreasing voltage conditions. While the voltage is going down, fast VDCOL action is required; hence, the time lag is small. If the same short time lag is used during voltage recovery, it may lead to oscillations and possibly instability. To prevent this, the larger time lag is used when the direct voltage is recovering.

Minimum firing angle limit

As shown in Figure 10.31, power transfer over the dc line can be controlled by manipulation of current order and current margin. These signals are transmitted to the converter stations through a telecommunication link. In the event that the communication fails or in case of a dc line fault, there is a possibility that the inverter station may switch to the rectification mode. This would result in a reversal of power flow. To prevent this from occurring, the inverter control is provided with a minimum-α limit, as indicated by the lowest portion of the inverter V-I characteristic in Figure 10.36. This restricts the firing angle of the inverter to a value greater than $90°$, typically in the range of $95°$ to $110°$. The rectifier is, however, allowed to operate in the inverter region to assist the system under certain fault conditions. As a consequence, the maximum limit imposed on the rectifier firing angle is typically between $90°$ and $140°$.

Power control

Usually, an HVDC link is required to transmit a scheduled power. In such an application, the corresponding current order is determined as being equal to the power order (P_0) divided by the measured direct voltage:

$$I_{ord} = P_0/V_d$$

The current order so computed is used as input signal to the current control. However, high-speed constant power control may have an adverse effect on ac system stability. From the viewpoint of system stability, high-speed constant current control with a superimposed slow power control is preferable. This is acceptable since the dispatcher is not interested in a very high speed power control. Thus, from the stability viewpoint, the HVDC system control performs as a constant current control, but for the dispatcher it appears as constant power control.

Auxiliary controls for ac system

To enhance ac system performance, auxiliary signals derived from the ac system quantities may be used to control the converters. The control strategy could include modulation of either direct current or direct voltage, or both. In addition, special control measures may be used to assist recovery of dc systems from faults. These higher-level controls will be discussed in more detail later in this chapter.

Summary of basic control principles

The HVDC system is basically constant-current controlled for the following two important reasons:

- To limit overcurrent and minimize damage due to faults

- To prevent the system from running down due to fluctuations of the ac voltages

It is because of the high-speed constant current control characteristic that the HVDC system operation is very stable.

The following are the significant aspects of the basic control system:

(a)   The rectifier is provided with a *current control* and an *α-limit control*. The minimum α reference is set at about 5° so that sufficient positive voltage across the valve exists at the time of firing, to ensure successful commutation. In the current control mode, a closed-loop regulator controls the firing angle and hence the dc voltage to maintain the direct current equal to the current order. Tap changer control of the converter transformer brings α within the range of 10° to 20°. A time delay is used to prevent unnecessary tap movements during transient excursions of α.

(b)   The inverter is provided with a *constant extinction angle* (CEA) control and a *current control*. In the CEA control mode, γ is regulated to a value of about 15°. This value represents a trade-off between acceptable var consumption and a low risk of commutation failure. While CEA control is the norm, there are variations which include *voltage control* and *β control*. Tap changer control is used to bring the value of γ close to the desired range of 15° to 20°.

(c)   Under normal conditions, the rectifier is on current control mode and the inverter is on CEA control mode. If there is a reduction in ac voltage at the rectifier end, the rectifier firing angle decreases until it hits the α_{min} limit. At this point, the rectifier switches to α_{min} control and the inverter will assume current control.

(d)   To ensure satisfactory operation and equipment safety, several limits are recognized in establishing the current order: maximum current limit, minimum current limit, and voltage-dependent current limit.

(e)   Higher-level controls may be used, in addition to the above basic controls, to improve ac/dc system interaction and enhance ac system performance.

All schemes used to date have used the above modes of operation for the rectifier and the inverter. However, there are some situations that may warrant serious investigation of a control scheme in which the inverter is operated continuously in current control mode and the rectifier in α-minimum control mode. Enhanced performance into weak systems may be one case.

10.4.2 Control Implementation

In Section 10.4.1, we considered the basic principles of control of one pole. Figure 10.37 illustrates a general scheme for implementing the controls. In many cases there are two to four converter bridges per pole connected in series. Usually, two bridges with Y-Y and Y-Δ connected transformers are considered as one 12-pulse unit. Thus, the smallest unit to be controlled is a 6-pulse or a 12-pulse bridge unit.

The control hierarchy varies from one dc system to another, but the general concepts are common. Figure 10.38 illustrates the control hierarchy of a typical bipolar HVDC system. The control scheme is divided into four levels: bridge or converter unit control, pole control, master control and overall control.

The *bridge* or *converter unit control* determines the firing instants of the valves within a bridge and defines α_{min} and γ_{min} limits. This has the fastest response within the control hierarchy.

The *pole control* coordinates the control of bridges in a pole. The conversion of current order to a firing angle order, tap changer control, and certain protection sequences are handled by the pole control. This includes coordination of starting up, deblocking, and balancing of bridge controls.

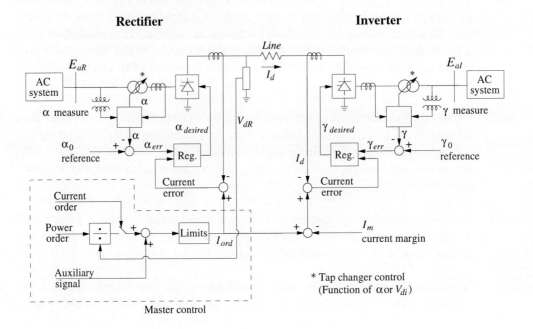

Figure 10.37 Basic control scheme

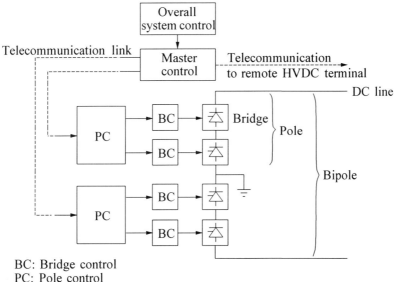

BC: Bridge control
PC: Pole control

Figure 10.38 Hierarchy of different levels of HVDC system controls

The *master control* determines the current order and provides coordinated current order signals to all the poles. It interprets the broader demands for controlling the HVDC system by providing an interface between pole controls and the *overall system control*. This includes power-flow scheduling determined by the control centre and ac system stabilization.

The basic control functions are similar for most applications. However, higher-level control functions are determined by the specific performance objectives of individual systems. For reliable operation of the HVDC system, each pole should function as independently as possible. The control and protection functions should be segregated and implemented at the lowest possible level of hierarchy.

The control of HVDC systems clearly requires communication between terminals for proper operation. In the case of rapid changes in power level, high-speed communication is required to maintain consistent current settings at the two terminals. Change of power direction requires communication to transfer the current margin setting from one terminal to the other. The starting and stopping of the terminals require coordination of the operations at the two terminals. Communication is also required for transmittal of general status information needed by the operators. Protection may also require communication between the terminals for detection of some faults.

There are several alternative transmission media available for the telecommunications: direct wires via private lines or telephone networks, power-line carrier, microwave systems, and fibre optics. The choice of the telecommunication medium will depend on the amount of information to be transmitted and the required speed of response, degree of security, and noise immunity.

10.4.3 Converter Firing-Control Systems

The converter firing-control system establishes the firing instants for the converter valves so that the converter operates in the required mode of control: constant current (CC), constant ignition angle (CIA), or constant extinction angle (CEA).

Two basic types of controls have been used for the generation of converter firing pulses:

- Individual phase control (IPC)
- Equidistant pulse control (EPC)

The implementation of the above basic forms of converter firing control has evolved over the years. There are several different versions in existence depending on the manufacturer and the vintage of equipment. Their detailed description is beyond the scope of this book. The following descriptions illustrate the principles of operation of the two forms of converter firing-control systems.

Individual phase control system [11,12]

This system was widely used in the early HVDC installations. Its main feature is that the firing pulses are generated individually for each valve, determined by the zero crossing of its commutation voltage.

The commutation process in a three-phase bridge circuit was analyzed in Section 10.2. Referring to Figure 10.15 and Equation 10.7A, the commutation voltage is given by

$$e_{ba} = e_b - e_a$$

$$= \sqrt{3} E_m \sin\omega t = 2L_c \frac{di_3}{dt}$$

As depicted in Figure 10.39, commutation begins when ωt equals the ignition angle (α) and ends when ωt equals the normal extinction angle ($\pi - \gamma$).

If $t = t_1$ at the beginning of commutation, and $t = t_2$ at the end of commutation, we may write

$$\int_{t_1}^{t_2} \sqrt{3} E_m \sin\omega t \, dt = 2L_c \int_0^{I_d} di_3$$

or

$$-\sqrt{3} \frac{E_m}{\omega} (\cos\omega t_2 - \cos\omega t_1) = 2L_c I_d$$

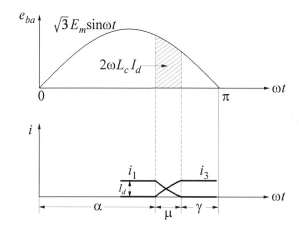

Figure 10.39 Commutation voltage and valve currents

or

$$-\sqrt{3}E_m[\cos(\pi-\gamma)-\cos\omega t_1] = 2\omega L_c I_d$$

Thus the control has to satisfy the condition that the voltage integral $2\omega L_c I_d$ is available between $\omega t=\omega t_1=\alpha$ and $\omega t=\omega t_2=\pi-\gamma$. The equation to be satisfied may be written as

$$-\sqrt{3}E_m\cos\omega t_1-\sqrt{3}E_m\cos\gamma+2X_c I_d = 0 \qquad (10.38)$$

CEA control with IPC:

In Equation 10.38, the direct current (I_d) and the commutating voltage vary with changes in operating conditions. The grid control, therefore, senses these two quantities to determine the instant of firing $(t=t_1)$ so as to satisfy Equation 10.38. With extinction angle equal to a set value γ_c, and $X_c=\omega L_c$, we have

$$-\sqrt{3}E_m\cos\omega t_1-\sqrt{3}E_m\cos\gamma_c+2X_c I_d = 0 \qquad (10.39)$$

The required ignition time t_1 can be found from the solution of Equation 10.39. This is accomplished by using analog circuits in the early converter control applications. The control system consists of three units: the first unit giving a dc output proportional to the direct current I_d; the second giving an output proportional to $E_m\cos\gamma_c$; and the third giving an alternating voltage proportional to the commutating voltage but with a phase lag of 90° (i.e., $E_m\cos\omega t$).

The three outputs are added, and a firing pulse is generated when the sum passes through zero. Under steady-state conditions, such a system controls each valve with constant commutation margin, irrespective of load and voltage variations or unbalance.

Constant current control with IPC:

In this case, an additional signal V_{cc} is added to Equation 10.39 as follows:

$$-\sqrt{3}\,E_m\cos\omega t_1 - \sqrt{3}\,E_m\cos\gamma_c + 2X_c I_d + V_{cc} = 0 \qquad (10.40)$$

where

$$V_{cc} = K(I_o - I_d)$$
$$I_o \; = \text{current order}$$
$$I_d \; = \text{actual direct line current}$$
$$K \; = \text{gain of CC control}$$

The same circuit may be used for both CEA and CC control operations [12]. The error $(I_o - I_d)$ is amplified only when I_d is less than I_o. When I_d is greater than I_o, the amplifier output is clamped to zero, and the converter operates on CEA control. When $(I_o - I_d) > 0$, it operates on constant-current control.

The converter characteristics for the full range of inverter and rectifier operation is shown in Figure 10.40. The realization of CIA control ($\alpha = \alpha_c$) is similar to that of CEA control.

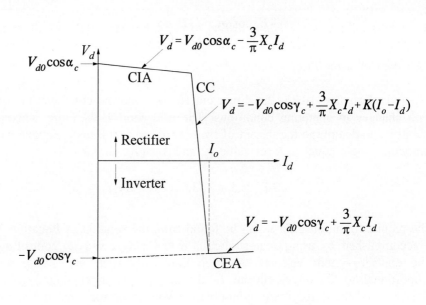

Figure 10.40 Converter characteristics with IPC

The IPC system has the advantage of being able to achieve the highest possible direct voltage under unsymmetrical or distorted supply waveforms since the firing instant for each valve is determined independently. However, the IPC system in effect has a voltage feedback, since the control signal is derived from the alternating line voltage. Any deviation from the ideal voltage waveforms will disturb the symmetry of the current waveforms. This will in turn cause additional waveform distortions, thus introducing non-characteristic harmonics (see Section 10.5). If the ac network to which the converter is connected is weak (i.e., has high impedance), the feedback effect may further distort the altering voltage and thereby lead to harmonic instability.

The harmonic instability problem can be reduced by altering the harmonic characteristics of the ac network (for example, by using additional filters) or adding filters in the control circuit. Alternatively, a firing control system independent of the ac system quantities may be used. This leads to the EPC system described next.

Equidistant pulse control system [10,13]

In this system, valves are ignited at equal time intervals, and the ignition angles of all valves are retarded or advanced equally so as to obtain the desired control mode. There is only indirect synchronization to the ac system voltage.

An EPC system using a phase-locked oscillator to generate the firing pulses was first suggested in reference 13. Since the late 1960s, all manufacturers of HVDC equipment have used this system for converter firing control.

Figure 10.41 shows an EPC-based constant-current control system. The basic components of the system are a voltage-controlled oscillator (VCO) and a ring counter. The VCO delivers pulses at a frequency directly proportional to the input control voltage. The train of pulses is fed to the ring counter, which has six or twelve stages (depending on the pulse number of the converter). Only one stage is on at any time, with the pulse train output of the VCO changing the on stage of the ring counter in cyclic manner. As each stage turns on, it produces a short output pulse once per

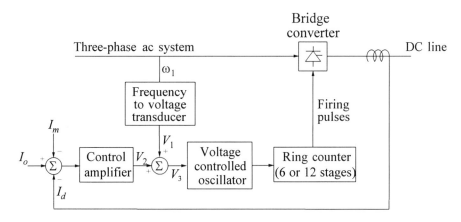

Figure 10.41 Current control system with equidistant pulse control

cycle. Over one full cycle, a complete set of 6 (or 12) output pulses is produced by the ring counter at equal intervals. These pulses are transferred by the firing-pulse generator to the appropriate valves of the converter bridge.

Under steady-state conditions, V_2 is zero and the voltage V_1 is proportional to the ac line frequency ω_1. This generates pulses at line frequency, and maintains a constant firing delay angle α. If there is a change in current order I_o, margin setting I_m, or line frequency ω_1, a change in V_3 occurs which in turn results in a change in the frequency of the firing pulses. A change in firing delay angle ($\Delta\alpha$) results from the *time integral* of the differences between line and firing pulse frequencies. It is apparent that this equidistant pulse control firing scheme is based on pulse frequency control.

An alternative equidistant pulse control firing scheme based on pulse phase control is proposed in reference 14. In this scheme, a step change in control signal causes the spacing of only one pulse to change; this results in a shift of phase only.

For CEA control, the basic circuits of Figure 10.41, illustrated for CC control, must be supplemented by additional circuitry. Since the extinction angle ($\gamma = \pi - \alpha - \mu$) cannot be controlled directly, either a predictive or a feedback control has to be used. In the scheme described in reference 10, a predictive method is used to ensure that adequate commutation voltage-time area (see Figure 10.39) is available at the instant of firing for successful commutation. The firing angle is based on calculation of the overlap angle (μ) from measured values of current and voltage. Reference 13 uses a feedback method to achieve this.

These schemes provide equal pulse spacing in the steady state. Symmetry is maintained relative to the most vulnerable control angle. For example, the smallest γ becomes the set angle in the presence of finite ac voltage unbalance.

The equidistant firing control results in a lower level of non-characteristic harmonics and stable control performance when used with weak ac systems. However, when the ac network asymmetry is large, it results in a lower direct voltage and power than the individual phase control.

Firing system

In modern converters, the valve firing and valve monitoring are provided through an *optical* interface. Light guides are used to carry the firing pulse to each thyristor. Each thyristor is provided with a special control unit that changes a light pulse to an electrical pulse to the gate input on the thyristor. Information about the condition of thyristors, required for protection and supervision of valves, is also transmitted by a light guide system from each thyristor.

At present many manufacturers are developing thyristors that are triggered directly by fibre optics.

10.4.4 Valve Blocking and Bypassing

Valve blocking (stopping) is achieved by interruption of positive pulses to the gates of all the valves in a bridge. However, this may result in overvoltages due to current extinction. In some instances, blocking of the valves at the inverter can lead to continuous conduction through previously conducting phases placing the ac voltage on the dc line and direct current on the converter transformer.

It is, therefore, necessary to bypass the bridge when the valves are blocked. This is achieved by using a bypass valve and bypass switch, as shown in Figure 10.42.

The valve currents are commutated into the bypass valve and then the bypass switch is closed to relieve the bypass valve from carrying current continuously.

Bypass pair operation

In converters using thyristors, the use of a separate bypass valve per bridge has been discontinued. Instead, as shown in Figure 10.43, bypass is implemented by firing a valve to establish a series pair on the same phase as an already conducting valve. A bypass switch closes to relieve the valves during a sustained blocking.

The logic for bypass operation is made part of the converter control.

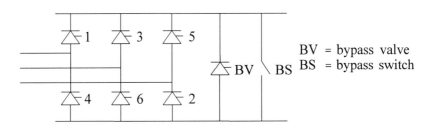

Figure 10.42

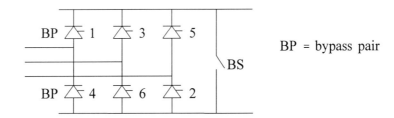

Figure 10.43

10.4.5 Starting, Stopping, and Power-Flow Reversal

The sequences that are used for starting and stopping HVDC systems vary depending on the manufacturer and equipment capability. The rate of rise of voltage, current, and power is tailored to the individual application. The following are typical steps and procedures involved [15].

Normal starting (deblocking) sequence

1. Either the inverter or the rectifier may start first. The converter that is started first establishes valve firing and conduction. Voltage is held low with the deblocking firing angle being in the range of 60° to 70°.

2. Following a communication delay, the other converter also establishes firing. A fairly low voltage is maintained with a firing angle of 60° to 70° and a starting current on the order of 0.2 to 0.3 pu (optimized to equipment or system conditions) is circulated.

3. After a successful start has been established, voltages are increased according to the relaxation rate on firing angle (α or β). The initial current order of 0.2 to 0.3 pu is maintained until the voltage has reached a setpoint of 40% to 80%. The current order is then released to the desired value.

4. When the current is established and can be maintained by the rectifier, the inverter goes into voltage/margin angle control mode.

The entire procedure can take as short a time as a fraction of a second or as long as several minutes, depending on the power order and limitations imposed by the ac system. The load may be increased exponentially or in small steps.

Normal stopping (blocking) sequence

Unlike in ac systems where a circuit-breaker is operated to isolate a line, the dc link is shut down gradually through controls. The blocking of a pole is achieved by reducing the voltage and current to zero as follows:

1. The current and voltage are ramped down within 100 to 300 ms. Then the rectifier is operated in or near the inverter region. This removes any stored energy from the dc system.

2. Bypass switches, if provided, are closed.

If one of two valve groups is to be blocked, it cannot involve reduction of current flow to zero. Therefore, the firing angle of the valve group is ramped to 90°, a bypass is formed with a bypass switch, and the valves are blocked. The ramp rate may be limited to avoid regulator mode changes.

Reversal of power flow

HVDC systems are inherently capable of power flow in either direction. Most schemes have full control features that permit bidirectional power flow.

Power reversal can be smoothly executed by following a prescribed series of ramps or it can be very fast with or without blocking of the firing of the valves.

Control techniques for power reversal may include the following [15]:

1.   Reduction of current to 0.1 to 0.5 pu via a step or ramp.

2.   Decrease/increase of voltage via ramp or exponential function followed by a current ramp to reach the required level.

The basis for deciding the sequence to be followed is normally the ability of the ac system to survive the resulting disturbance and the need for power reversal within a specified time. Typically, fast power reversal times would be on the order of 20 to 30 ms, although ac system limitations, dc cable design constraints, or power dispatch conditions may increase it to several seconds. HVDC controls can meet this entire range of requirements.

10.4.6 Controls for Enhancement of AC System Performance

In a dc transmission system, the basic controlled quantity is the direct current, controlled by the action of the rectifier with the direct voltage maintained by the inverter. A dc link controlled in this manner buffers one ac system from disturbances on the other. However, it does not allow the flow of synchronizing power which assists in maintaining stability of the ac systems. The converters in effect appear to the ac systems as frequency-insensitive loads and this may contribute to negative damping of system swings. Further, the dc links may contribute to voltage collapse during swings by drawing excessive reactive power.

Supplementary controls are therefore often required to exploit the controllability of dc links for enhancing the ac system dynamic performance. There are a variety of such higher level controls used in practice. Their performance objectives vary depending on the characteristics of the associated ac systems. The following are the major reasons for using supplementary control of dc links:

•   Improvement of damping of ac system electromechanical oscillations.

•   Improvement of transient stability.

•   Isolation of system disturbance.

•   Frequency control of small isolated systems.

•   Reactive power regulation and dynamic voltage support.

References 18 to 21 provide descriptions of supplementary controls used in a number of HVDC transmission systems for enhancement of ac system performance. The controls used tend to be unique to each system. To date, no attempt has been made to develop generalized control schemes applicable to all systems.

The supplementary controls use signals derived from the ac systems to modulate the dc quantities. The modulating signals can be frequency, voltage magnitude and angle, and line flows. The particular choice depends on the system characteristics and the desired results.

The principles of dc modulation schemes and details of their application for enhancement of ac system performance are discussed further in Chapter 17.

10.5 HARMONICS AND FILTERS

Converters generate harmonic voltages and currents on both ac and dc sides. In this section we will briefly describe the types of harmonics produced by the converters and the characteristics of filters used to minimize their adverse effects.

10.5.1 AC Side Harmonics

Figure 10.12 shows the wave shape of the alternating current under the "ideal" condition with no commutation overlap, ripple-free direct current, balanced purely sinusoidal commutating voltages, and equally-spaced converter firing pulses. The current may be expressed as a Fourier series.

For a 6-pulse bridge with Y-Y transformer connection, the Fourier series expansion for the alternating current is

$$i = \frac{2\sqrt{3}}{\pi}I_d(\sin\omega t - \frac{1}{5}\sin5\omega t - \frac{1}{7}\sin7\omega t + \frac{1}{11}\sin11\omega t + \frac{1}{13}\sin13\omega t - \cdots) \tag{10.41}$$

For a Δ-Y transformer connection, the current is

$$i = \frac{2\sqrt{3}}{\pi}I_d(\sin\omega t + \frac{1}{5}\sin5\omega t + \frac{1}{7}\sin7\omega t + \frac{1}{11}\sin11\omega t + \frac{1}{13}\sin13\omega t + \cdots) \tag{10.42}$$

The second harmonic and all even harmonics are absent in the above because there are two current pulses of equal size and opposite polarity per cycle. Since the current pulse width is one-third of a cycle, third and all triple-n harmonics are also absent. The remaining harmonics are on the order of $6n\pm1$, where n is any positive integer.

In a 12-pulse bridge, there are two 6-pulse bridges with two transformers, one with Y-Y connection and the other with Y-Δ connection (see Figure 10.23). The harmonics of odd values of n cancel out. Hence,

$$i = \frac{2\sqrt{3}}{\pi}2I_d(\sin\omega t + \frac{1}{11}\sin11\omega t + \frac{1}{13}\sin13\omega t + \frac{1}{23}\sin23\omega t + \frac{1}{25}\sin25\omega t + \cdots) \quad (10.43)$$

The remaining harmonics which have the order $12n\pm1$ (i.e., 11th, 13th, 23th, 25th, etc.) flow into the ac system. Their magnitudes decrease with increasing order; an hth order harmonic has magnitude $1/h$ times the fundamental.

When the commutating reactance is considered, the overlap angle during commutation rounds off the square edges of the current waves, and this reduces the magnitude of harmonic components. The reduction factor of the harmonic components is given by [3]

$$\frac{i_h}{i_{ho}} = \frac{\sqrt{H^2+K^2-2HK\cos(2\alpha+\mu)}}{\cos\alpha-\cos(\alpha+\mu)} \quad (10.44)$$

where

i_h = harmonic current
i_{h0} = harmonic current with no overlap
h = harmonic order
$H = [\sin(h+1)\mu/2]/(h+1)$
$K = [\sin(h-1)\mu/2]/(h-1)$

It is apparent that the harmonics produced on the ac system are a function of the operating conditions. As μ increases, the harmonic component decreases, with the reduction being more pronounced at higher harmonics. Under typical full load conditions, μ is about $20°$ and α is about $15°$; the 11th and 13th harmonics are about 30 to 40% of those shown in the basic equations, which neglected overlap. During faults, however, α reaches nearly $90°$, the overlap angle is reduced, and for a given line current the ac harmonics will increase.

The above discussions consider only balanced conditions. The harmonics produced under such ideal conditions are referred to as "characteristic harmonics."

Various unbalances, such as non-equally spaced firing pulses, bus voltage unbalances and unbalances in the commutating reactance between phases will produce additional harmonics which are referred to as "non-characteristic harmonics." Transformer excitation current also contributes to these harmonics.

The converter manufacturers attempt to minimize these harmonics in the design of the terminals. With modern day equal-spaced firing, the biggest sources of non-characteristic harmonics are bus voltage unbalance, transformer impedance unbalance and the transformer excitation current. Unbalances in ac system voltages depend on the operating conditions and are determined by the design and operating practices of the system. Unbalances in the reactances between the phases of the transformer are usually less than 1% of the phase values. With bus voltage unbalance of less than 1% and normal excitation current levels, non-characteristic harmonics are not significant, unless a resonant condition exists at a particular harmonic frequency.

AC harmonic filters

Harmonics have to be filtered out sufficiently at the terminal so that the harmonics entering the ac system are small and distortion of the ac voltage caused by the harmonic currents is within limits. Harmonics, if not reduced by filters, can produce undesirable effects such as telephone interference, higher losses and heating in ac equipment (machines, capacitors, etc.), or resonance problems which could produce overvoltages and/or overcurrents.

The penetration of the harmonics into the ac system and resonance conditions depends on the harmonic impedance of the ac network, which is difficult to determine. It is constantly varying as the circuits are being added or switched out and the system operating conditions are varying. In filter design, quite elaborate studies are undertaken that consider various factors including possible system resonance conditions.

A typical filter system for a 12-pulse converter terminal is shown in Figure 10.44. The filter impedance is minimum at the 11th and 13th harmonics resulting from the two series-resonant tuned branches. The high pass filter maintains a low impedance for higher harmonic frequencies.

Part of the capacitors required for reactive power compensation is provided by filters. The additional cost of converting the required reactive power capacitors to filters is not very high. Of the 50% reactive power compensation required, about 30% would be in filter form divided among the eleventh, thirteenth, and high pass filters.

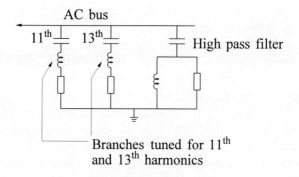

Figure 10.44   Typical filter system configuration

10.5.2 DC Side Harmonics

An ideal 6-pulse bridge converter, with repetitive switching every 60°, produces a direct voltage waveform as shown in Figure 10.19. Fourier analysis of the voltage waveform shows that it contains harmonics of order $6n$ (i.e., 6th, 12th, 18th, etc.). The magnitude of the effective harmonic voltage varies widely over the operating range of α; operation at α near 90° produces higher levels of harmonics than at smaller values of α. The overlap angle μ also has a significant impact on the magnitude of the harmonics.

In a bipole system consisting of two 6-pulse bridges (one in each pole), the transformers would be connected Y-Y and Δ-Y because the 30° phase shift produces cancellation of low order harmonic on the ac bus. This would also have a beneficial effect on the dc side. The 6th, 18th, 30th, ... harmonics are out-of-phase in the two bridges, while the 12th, 24th, 36th, ... harmonics are in phase. The out-of-phase harmonic voltages produce ground return mode currents in the dc line, whereas the in-phase components of harmonic voltages produce line-to-line mode currents.

In the case of a 12-pulse bridge, the out-of-phase components of harmonic voltages will cancel within a 12-pulse bridge; only the in-phase components will produce harmonic currents in the line. For a balanced condition, the significant harmonics in the voltage produced by a dc terminal are therefore of order 12 and its integral multiples.

The non-characteristic harmonics due to unbalances would be small in a well designed system. The dc side harmonics are reduced by the smoothing reactor and filters. Most of the harmonic voltages are dropped across the smoothing reactor. The dc filters are designed to shunt a major portion of the direct harmonic currents so that harmonic currents flowing into the dc line are within the permissible limits.

The smoothing reactor and the rest of the dc system beyond the reactor (dc filters, dc line and remote terminals) act as voltage dividers for the harmonic voltages. In general, a larger value of smoothing reactor will require less dc filtering. However, the smoothing reactor size is influenced by other considerations associated with converter terminal design.

The following are some of the considerations influencing the selection of the smoothing reactor size:

1.   The size of the smoothing reactor has a dominant effect on the ripple in the bridge current; the ripple current is less for higher values of the reactor. As discussed in Section 10.4.1, the ripple current determines the minimum current operating point for the terminal.

2.   The smoothing reactor, filters, and dc lines combine to produce an impedance to the dc bridge which can be resonant at some generated voltage harmonics. Generally, this is not a problem for the higher harmonics (12th, 24th, ...), but the non-characteristic harmonics generated at low frequency (2nd, 4th, 6th, ...) can produce high harmonic currents if the effective impedance is low at these

frequencies. The high harmonic currents at low frequencies can cause problems with control circuitry and interference with communication circuits. Therefore, the smoothing reactor is selected to avoid a low impedance at 2nd or 4th harmonic frequency. In addition, low frequency harmonic resonance must be avoided for other reasons. Repeated commutation failures on an inverter may introduce fundamental frequency pulses on the dc line (see Section 10.3.2). If the input impedance affecting the bridge were near resonance at the fundamental frequency, high transient voltages would occur.

3.   The smoothing reactor size has an influence on the likelihood of commutation failures during a dip in ac voltages and on the likelihood of consequent commutation failure.

4.   A higher smoothing reactor limits the fault current near the rectifier.

10.6 INFLUENCE OF AC SYSTEM STRENGTH ON AC/DC SYSTEM INTERACTION

The nature of ac/dc system interactions and the associated problems are very much dependent on the strength of the ac system relative to the capacity of the dc link. The ac system can be considered as "weak" from two aspects: (a) ac system impedance may be high, (b) ac system mechanical inertia may be low [18]. In this section we will discuss problems associated with dc systems connected to weak ac systems and methods of dealing with such problems.

10.6.1 Short-Circuit Ratio

Since the ac system strength has a very significant impact in the ac/dc system interactions, it is useful to have a simple means of measuring and comparing relative strengths of ac systems. The *short-circuit ratio* (SCR) has evolved as such a measure. It is defined as

$$\text{SCR} = \frac{\text{short-circuit MVA of ac system}}{\text{dc converter MW rating}}$$

The short-circuit MVA is given by

$$\text{SC MVA} = \frac{E_{ac}^2}{Z_{th}}$$

where E_{ac} is the commutation bus voltage at rated dc power and Z_{th} is the Thevenin equivalent impedance of the ac system.

The basic SCR gives the inherent strength of the ac system. From the viewpoint of the HVDC system performance, it is more meaningful to consider the *effective short-circuit ratio* (ESCR), which includes the effects of ac side equipment associated with the dc link: filters, shunt capacitors, synchronous condensers, etc.

HVDC controls play an important role in most ac/dc system interaction phenomena, and must be taken into consideration in assessing acceptable levels of ac system strength. Traditionally, the ac system strength has been classified as follows [16]:

• High, if ESCR is greater than 5;

• Moderate, if ESCR is between 3 and 5; and

• Low, if ESCR is less than 3

With refinements in dc and ac system controls, these classifications change. Reference 18 recommends the following classification:

• High, if ESCR is greater than 3;

• Low, if ESCR is between 2 and 3; and

• Very low, if ESCR is less than 2.

The above classification of ac system strength provides a means for preliminary assessment of potential ac/dc interaction problems. Detailed studies are, however, necessary for proper evaluation of the problems. In addition to the short-circuit ratio, the phase angle of the Thevenin equivalent impedance Z_{th} has an impact on the ac/dc system interaction. It is termed the "damping angle" and has an impact on the dc system control stability. Local resistive loads, while not having a significant effect on ESCR, improve damping of the system. Typical values of the damping angle are in the range of 75° to 85°.

10.6.2 Reactive Power and AC System Strength

From Equations 10.12 and 10.25, we have

$$\cos\phi \approx \cos\alpha - \frac{I_d}{V_{do}}\frac{3}{\pi}X_c$$

Therefore, each converter consumes reactive power which increases with increased power. With normally accepted rectifier ignition delay angle (α) and inverter extinction advance angle (γ) of 15° to 18° and commutating reactance (X_c) of 15%, a converter consumes 50 to 60% reactive power (i.e., if P_{dc}=1.0 pu, Q absorbed by each converter is 0.5 to 0.6 pu). Generally, this has to be provided at the converter site to prevent large reactive power flow through ac lines. Part of the reactive power required is provided by the capacitors associated with the ac filter banks.

The least expensive way to provide the required reactive power is to use shunt capacitor banks. Since the reactive power varies with the dc power transmitted, capacitors must be provided in appropriate sizes of switchable banks, so that steady-state ac voltage is held within an acceptable range (usually ±5%) at all load levels. This is also influenced by the strength of the ac system. The stronger the ac system, the larger the switchable bank size can be for an acceptable voltage change.

Generators, if present near the dc terminal, can be very helpful in handling some of the reactive power demands and in maintaining steady-state voltage within an acceptable range. For weak ac systems, it may be necessary to provide reactive compensation in the form of *static var compensators* (SVCs) or synchronous condensers.

10.6.3 Problems with Low ESCR Systems [16-18]

The following are the problems associated with the operation of a dc system when connected to a weak ac system:

- High dynamic overvoltages,

- Voltage instability,

- Harmonic resonance, and

- Objectionable voltage flicker.

Dynamic overvoltage

When there is an interruption to the dc power transfer, the reactive power absorption of the HVDC converters drops to zero. With a low ESCR system, the resulting increase in alternating voltage due to shunt capacitors and harmonic filters could be excessive. This will require a high insulation level of terminal equipment, thus imposing an economic penalty. It may also cause damage of local customer equipment. Special schemes may be necessary to protect the thyristors in case of restart delays [17].

Voltage stability [16]

With dc systems connected to weak ac systems, particularly on the inverter side, the alternating as well as direct voltages are very sensitive to changes in loading.

An increase in direct current is accompanied by a fall of alternating voltage. Consequently, the actual increase in power may be small or negligible. Control of voltage and recovery from disturbances become difficult. The dc system response may even contribute to collapse of the ac system. The sensitivities increase with large amounts of shunt capacitors.

In such a system, the dc system controls may contribute to voltage instability by responding to a reduction in alternating voltage as follows:

• Power control increases direct current to restore power.

• Inverter γ may increase to maintain volt-second commutation margin.

• Inverter draws more VARs; with reduced voltage, shunt capacitors, however, produce fewer VARs.

• The alternating voltage is reduced, further aggravating the situation.

The process thus leads to progressive fall of voltage.

Harmonic resonance

Most of the problems of harmonic resonance are due to parallel resonance between ac capacitors, filters, and the ac system at lower harmonics.

Capacitors tend to lower the natural resonant frequencies of the ac system, while inductive elements (machines and lines) tend to increase the frequencies. If large numbers of capacitors are added, the natural frequency seen by the commutation bus may drop to 4th, 3rd or even 2nd harmonic. If a resonance at one of these frequencies occurs, there can be a high impedance parallel resonance between the inductive elements and capacitive elements on the commutation bus. A low-impedance series resonance condition could arise in remote points in the system. Harmonic voltages from remote points would tend to be amplified. The avoidance of low-order harmonic resonance is extremely important to reduce transient overvoltages.

Voltage flicker

Another characteristic of a weak ac system is that switching of shunt capacitors and reactors causes unacceptably large voltage changes in the vicinity of compensation equipment. The transient voltage flicker due to frequently switched reactive devices increases with higher levels of dc power transfer.

10.6.4 Solutions to Problems Associated with Weak Systems

The traditional approach to the solution of weak system ac/dc interaction problems is to use synchronous condensers or SVCs. In addition, HVDC controls which switch to current control from power control and reduce direct current for low alternating voltage (for example, VDCOL) will help the situation.

The use of synchronous condensers also reduces the effective system impedance and hence shifts the parallel resonance frequency to higher frequencies at which the system damping is usually better. With 12-pulse bridge circuits, there are no large filters below 11th harmonic; the possibility of excitation of parallel resonance at lower harmonics is therefore low.

An alternative solution to the reactive power and voltage problem is to control the dc converter itself so that the reactive power is modulated in response to voltage variations in a manner similar to an SVC. The shunt capacitors and filters provide the required reactive power; the converter firing angle control stabilizes the ac voltage. Artificial or forced commutation, discussed in Section 10.6.6, provides considerably more freedom in controlling reactive power.

Reference 17 describes five dc links (all back-to-back) connected to weak ac systems, and special techniques used to achieve their satisfactory performance.

10.6.5 Effective Inertia Constant

The ability of the ac system to maintain the required voltage and frequency depends on the rotational inertia of the ac system. For satisfactory performance, the ac system should have a minimum inertia relative to the size of the dc links. A measure of the relative rotational inertia is the effective dc inertia constant, defined as follows [18]:

$$H_{dc} = \frac{\text{total rotational inertia of ac system, MW·s}}{\text{MW rating of dc link}}$$

An effective inertia constant, H_{dc}, of at least 2.0 to 3.0 s is required for satisfactory operation.

For ac systems with very low or no generation, synchronous condensers have to be used to increase the inertia and assist in satisfactory operation of line commutated inverters.

10.6.6 Forced Commutation

The converter bridge circuits we have discussed so far rely on the natural voltage of the ac system for commutation. Such a commutation is known as "natural commutation," and requires the ignition angle (α) to be in the range of 0° to 180° (2° to 165° in practice for reliable turn-on and turn-off of the valves) for successful commutation. As a consequence, the converter absorbs reactive power from the ac system, while operating as a rectifier as well as an inverter.

If our objective is to have the converters supply reactive power to the ac system when so desired, we need to be able to force commutation at any desired point on the ac cycle. This can be achieved by superimposing harmonics or by use of special capacitor circuits to modify the voltages across the valves appropriately in relation to ac system voltages. Such a commutation is called "forced commutation." A system using this form of commutation will also allow feeding into a system without any generation.

A converter bridge circuit with a forced commutation scheme is shown in Figure 10.45 [3,5]. Forced commutation is initiated at the desired times by firing the auxiliary valves (A1 to A6) with the associated capacitors pre-charged. Such capacitive commutations, however, cause considerable stress on the valves and other converter equipment.

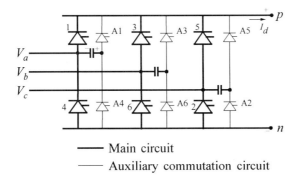

Figure 10.45   Bridge circuit with forced commutation

While converter circuits with forced commutation are feasible, they are very expensive. To date they have not found application in commercial HVDC converters.

Self-commutated voltage sourced inverters using gate turn-off (GTO) thyristors have been used in industrial applications, for example, variable speed drives, uninterruptible power supplies, and battery systems. The voltage and current ratings of these devices are increasing. It is likely that their application for HVDC transmission will become economically practical in the future.

10.7 RESPONSES TO DC AND AC SYSTEM FAULTS

The operation of HVDC transmission is affected by faults on the dc line, converters, or the ac system. The impact of the fault is reflected through the action of converter controls. In ac systems, relays and circuit-breakers are used to detect and remove faults. In contrast, most faults associated with dc systems either are self-clearing or are cleared through action of converter controls. Only in some cases is it necessary to take a bridge or an entire pole out of service. The converter controls thus play a vital role in the satisfactory response of HVDC systems to faults on the dc as well as the ac systems.

10.7.1 DC Line Faults

Faults on dc lines are almost always pole-to-ground faults. A pole-to-pole fault is uncommon since it requires considerable physical damage to bring conductors of two poles together. Lightning never causes a bipolar fault [7].

A pole-to-ground fault blocks power transfer on that pole; the remaining pole is virtually unaffected. As discussed below, the impact of a dc line fault on the connected ac systems is not as disruptive as that of ac faults.

Response of normal control action [2]

A short-circuit momentarily causes the rectifier current to increase (since the rectifier is feeding a low impedance fault rather than the high back voltage of the inverter) and the inverter current to decrease.

The current control of the rectifier acts to reduce the direct voltage and bring back the current to its normal set value (I_{ord}). At the inverter, the current becomes less than its current controller reference setting ($I_{ord}-I_m$). Consequently, the inverter mode of operation changes from CEA control to CC control. This causes the inverter voltage to run down to zero and then reverse polarity as shown by curve 2 of Figure 10.46. The voltages are equal to the RI drop in the line from each converter to the fault. The rectifier current is I_{ord} and inverter current is $I_{ord}-I_m$ in the opposite direction. The steady-state fault current is thus equal to the current margin I_m, which is only about 15% of the rated current.

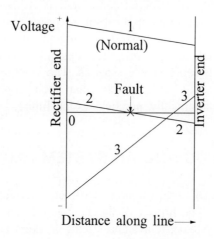

Figure 10.46   Voltage profile of a dc line:
1. Under normal operation
2. With dc line fault and normal control
3. With fast-acting line protection

Fast-acting line protection [7]

The normal control action of the converters, while limiting the fault current to I_m, does not extinguish the fault arc. Therefore, additional control is used to reduce the fault current and recovery voltage across the fault path to zero.

The fault is detected by a collapse in dc voltage usually at the rectifier and by a decrease in the current at the inverter. Both magnitude of voltage drop and rate of change of voltage may be used to detect the fault. Faults on the ac system beyond the dc link do not produce such rapid voltage changes.

To clear the fault, the inverter is kept in inversion and the rectifier is also driven to inversion. To establish terminal voltages of correct polarity for fault clearing, angle β of the inverter is given a maximum limit of about 80° (which allows the inverter voltage to run down to low value but not to reverse) and the rectifier ignition delay angle α is shifted considerably beyond 90°, to about 140°. The resulting voltage profile is as indicated by curve 3 in Figure 10.46. The current in the pole attempts to reverse direction. However, the current in the rectifier cannot reverse because of the unidirectional current characteristic of the valves. Therefore, the current is reduced to zero rapidly (in about 10 ms). This fault clearing process is called "forced retard."

DC overhead lines are restarted after allowing for de-ionization of the air surrounding the arc (60 to 200 ms). If the fault is temporary and the restart is successful, the voltage and current are ramped up. Typically, the total time for fault clearing and return to rated power transfer is on the order of 200 to 300 ms. The recovery time is higher for dc links connected to weak ac systems.

Automatic restarts are not attempted for wholly cable systems because cable faults are nearly always permanent faults.

10.7.2 Converter Faults

Most dc power circuit faults in the converter station will require either the valve group or the pole to be shut down.

A valve group fault, unless it is of a minor nature, will require the entire pole to cease transmission of power. Usually a very fast current reduction to zero is ordered. Coincidentally, the firing angle at the rectifier is shifted to at least 90° and possibly well into the inverter region. The current in the pole can be brought to zero in less than 30 ms.

An isolation sequence will follow which may take several seconds to execute, depending on the type of valve group isolators used. Then the remaining valve groups in the pole may be restarted in the normal manner.

10.7.3 AC System Faults

For ac system transient disturbances, the dc system's response is generally very much faster than that of the ac system. The dc system either rides through the disturbance with temporary reduction of power or shuts down until the ac system recovers sufficiently to allow restarting and restoration of power. Commutation failures and recovery from ac system faults represent important aspects of dc system operation.

Rectifier side ac system faults

For remote three-phase faults, the rectifier commutation voltage drops slightly. This results in a reduction of rectifier direct voltage and hence the current. The current regulator decreases α to restore current by increasing voltage. If α hits the α_{min} limit, the rectifier switches to CIA mode of control. This transfers current control to the inverter whose current order is less than that of the rectifier by an amount equal to the current margin (I_m). If the low voltage persists, the tap changers will operate to restore the direct voltage and current to normal. Depending on how low the voltage drops, the VDCOL may regulate current and power transfers. For close-in three-phase faults, the rectifier commutation voltage drops significantly. The dc system shuts down under VDCOL control until the fault is cleared.

In theory, dc power may be transferred at very low rectifier voltages. This would require the inverter to assume current control by lowering its voltage and greatly increasing β. The resulting increased consumption of reactive power may be more detrimental to the ac system performance than briefly shutting down the dc system.

Remote single-phase and phase-to-phase faults do not usually result in shutting down of the dc link. The average of the alternating voltages is higher than that for three-phase faults. If the resulting direct voltage is sufficiently high, the dc system is likely to ride through the disturbance without any noticeable effect. If, on the other hand, the reduction in direct voltage is significant, the response is similar to that for remote three-phase faults.

For close-in unbalanced faults, the harmonic ripple in the direct voltage may be higher than normal. This may produce ripples in the direct current with a large second harmonic component. The line reactors and filters, designed for smoothing out normal characteristic harmonics, are not effective in reducing the second harmonics. The high ripple current could result in current extinction. Depending on the type of valve-firing system used, this may require blocking of the dc link [7].

Inverter side ac system faults

For remote three-phase faults resulting in small voltage dips at the inverter, an increase in direct current occurs. The rectifier CC and the inverter CEA (or constant voltage) controls respond to the changes. Tap changes will occur to restore converter firing angle and direct voltage, if the low alternating voltage persists.

If the voltage dip is significant, the reduction in commutating voltages may lead to temporary commutation failure at the inverter, prior to any corrective control action. With inverter operation at a γ of 18°, it is likely that a voltage reduction by 10% to 15% will cause commutation failure. It takes about 1 or 2 cycles to clear the commutation failure. Following this, some power may be transmitted with the rectifier direct voltage reduced to match the reduction in inverter direct voltage. The resulting increase in reactive power may necessitate reduction of direct current. The VDCOL function (see Section 10.4.1) normally provided by the dc control system will cause this reduction of direct current. During extremely low voltage conditions, repeated commutation failures cannot be avoided. Therefore, it may be necessary to block and bypass the valves until the ac voltage recovers.

Unbalanced faults (both remote and close-in) may lead to commutation failure, partly due to phase shifts in the timing of the phase voltage crossings. For a severe contingency, it may be necessary to block and then restart the inverter.

When the fault has cleared, the allowable rate of restoration is dependent on the strength of the ac system. The controls are adjusted to provide the desired rate of power buildup. The performance of the overall power system following any system disturbance depends largely on the ac/dc system interaction as discussed below. It is also influenced by the subtle design features and response adjustments associated with the converter controls. These tend to vary with manufacturers. Special control strategies may be helpful in specific cases.

Recovery from ac system faults [18]

The post-fault system performance for ac system faults is far more sensitive to system parameters than for dc system faults. Recovery after an ac system fault is easier and can be more rapid with a strong ac system. Weak ac systems may have difficulty providing sufficient reactive power at the rate required for fast dc system restoration. Such systems also exhibit high temporary overvoltages and severe voltage distortion due to harmonics caused by inrushing magnetizing currents. These may cause subsequent commutation failures. Consequently the rate of recovery has to be slow.

The time for the dc system to recover to 90% of its pre-fault power is typically in the range of 100 ms to 500 ms, depending on the dc and ac system characteristics and the control strategy used. The dc system characteristics which influence the allowable rate of recovery are the line inductance and capacitance (particularly for cables), size of dc reactor, resonant harmonic frequencies of the line, converter transformer and filter characteristics. The significant ac system characteristics are ESCR, impedance at low order (2nd to 4th) harmonics, damping characteristics of loads near the dc system, system inertia, and method of voltage control near the converter bus.

Control strategies that assist in satisfactory dc system recovery (without post-fault commutation failures) include delayed or slow ramp recovery, and at reduced current level during recovery.

The VDCOL function can play a significant role in determining the recovery from faults. It limits the current order as a function of either the direct voltage or the alternating voltage. Consequently, reactive power demand is reduced during periods of depressed voltage. This helps prevent further deterioration of ac system voltage. Following fault clearing, the current order limit imposed by the VDCOL may be removed after a delay and gradually increased at a desired ramping rate.

From the viewpoint of ac system stability and minimization of dc power interruption, too slow a recovery is undesirable. The control strategies must therefore be tailored to meet specific needs of an application so as to maximize the recovery rate without compromising secure recovery of the dc system. Such strategies should be based on a detailed study of the individual system.

Special measures to assist recovery

Special measures may be used to assist HVDC system recovery from disturbances and to protect the valves. These measures act on either the firing angle directly or the current order. They depend on the requirements of a specific installation. The following are examples of such measures:

- Circuits that increase α_{min} from approximately 5° to over 30° during ac undervoltage conditions.

- Circuits in the inverter which transiently increase γ should γ slip below the γ_{min} limit. Then γ is increased on all succeeding firings for a prescribed period of time.

- Circuits that advance the firing angle limits by about 10° during start-up of the second valve group in a pole to prevent commutation failures.

- Circuits that increase β immediately following commutation failure so that the next valve is fired early to aid the recovery and to lessen the likelihood of subsequent commutation failure. Typically, β might be moved to 40°, 60°, and 70° on successive commutation failures. Should commutation failure cease, β would gradually retreat to its original value over a few hundred milliseconds. However, if it does not cease, a temporary bypass is usually ordered.

10.8 MULTITERMINAL HVDC SYSTEMS

In the previous sections, we considered the performance and application of point-to-point dc links, i.e., two-terminal dc systems. Successful application of such systems worldwide has led power system planners to consider the use of dc systems with more than two terminals. It is increasingly being realized that multiterminal dc (MTDC) systems may be more attractive in many cases to fully exploit the economic and technical advantages of HVDC technology.

The first MTDC system designed for continuous operation is the Sardinia-Corsica-Italy scheme. This is an expansion of the Sardinia-Italy two-terminal dc system built in 1967; a third terminal tap was added at Corsica in 1991. The two-terminal dc system between Des Cantons in Quebec and Comerford in New Hampshire built in 1986 is being extended to a three-terminal and then possibly to a five-terminal scheme [22]. Similarly, other MTDC systems are likely to evolve from the expansion of existing two-terminal schemes.

In this section, we will describe alternative MTDC system configurations and basic principles of controlling such systems. References 8, 23, 24, and 25 are recommended for further reading.

10.8.1 MTDC Network Configurations

There are two possible connection schemes for MTDC systems:

- Constant voltage parallel scheme

- Constant current series scheme

In the parallel scheme, the converters are connected in parallel and operate at a common voltage. The connections can be either radial or mesh. Figures 10.47 and 10.48 depict the two types of connections.

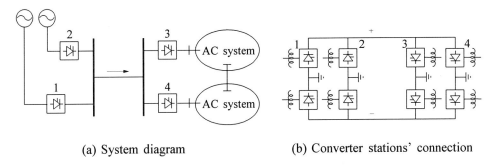

(a) System diagram   (b) Converter stations' connection

Figure 10.47 Parallel MTDC bipolar scheme with radial dc network

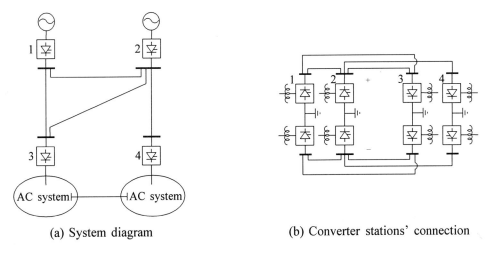

(a) System diagram   (b) Converter stations' connection

Figure 10.48 Parallel MTDC bipolar scheme with mesh type connection

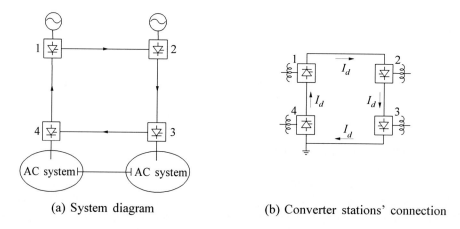

(a) System diagram   (b) Converter stations' connection

Figure 10.49   Series MTDC scheme

In the series scheme, the converters are connected in series with a common direct current flowing through all terminals. The dc line is grounded at one location. Figure 10.49 illustrates the series connection.

It is also conceivable to have a hybrid MTDC system involving both series and parallel-connected converter stations. The availability of dc circuit breakers [25] will add to the flexibility of MTDC systems and influence their selection.

The majority of studies and proposed applications of MTDC have considered parallel configuration with radial-type connection. The mesh connection, although offering more redundancy, requires greater length of dc lines.

Consideration of series connected schemes has been generally confined to applications with small power taps (less than 20%) where it may be more economical to operate at a higher current and lower voltage than for a full voltage tap at full voltage and reduced current. In a series tap, the voltage rating is proportional to the power capacity of the tap. However, the converter transformer must have full dc network voltage insulation. Flexibility of the power transfer could require a wide range for the transformer taps of the series stations.

In any specific MTDC system application, its special needs will determine the preferred network configuration. In general, the parallel scheme is widely accepted as the most practical scheme with fewest operational problems. Compared to the series-connected scheme, it results in fewer line losses, is easier to control, and offers more flexibility for future extension.

10.8.2 Control of MTDC Systems

The basic control principle for MTDC systems is a generalization of that for two-terminal systems. The control characteristic for each converter is composed of segments representing constant-current control and constant-firing angle control (CEA for inverter and CIA for rectifier). In addition, an optional constant-voltage segment may be included.

The converter characteristics, together with the dc network conditions, establish the operating point of the system. For a common point to exist, converter control characteristics must intersect.

For MTDC systems, there is considerable room for providing flexibility of options to meet the requirements of individual systems. References 23 and 25 discuss different proposed control strategies.

The following is a general discussion of significant aspects related to control of parallel- and series-connected systems.

Parallel-connected systems

In a parallel-connected system, one of the terminals establishes the operating voltage of the dc system, and all other terminals operate on constant-current (CC) control. The voltage setting terminal is the one with the smallest ceiling voltage. This may be either a rectifier on CIA control or an inverter on CEA control.

The *V-I* characteristics for a four-terminal dc system are shown in Figure 10.50. The individual converter characteristics are shown in Figure 10.50(a), and the combined characteristics are shown in Figure 10.50(b). It is assumed that two of the terminals are operating as rectifiers and the other two terminals as inverters. The characteristics shown are for one pole. For the sake of simplicity, VDCOLs are not shown. It is assumed that each terminal has only two modes of control (CC, and CEA or CIA); the voltage control option is not considered.

In Figure 10.50, it is assumed that rectifier 1 is the voltage-setting terminal (CIA mode). Tap changers keep angles within the desired range. To maintain stable control operation, a positive-current margin must be maintained.

If an inverter is the voltage-controlling station, it is vulnerable to inadvertent overloading. It is unable to control the current at its terminals in the event of a system disturbance or load change. Disconnection of a current-controlled inverter will require reallocation of rectifier current settings to prevent overloading the voltage-controlled inverter.

On the other hand, if a rectifier defines the system voltage, the operation is more stable. All inverters control current, thereby avoiding operation in the less stable CEA control mode. The voltage-controlling rectifier is capable of protecting itself without causing overloading of other stations. The system is less dependent on high-speed communication and hence is more secure. In general, voltage control at a large rectifier terminal should provide better performance.

The following are the main drawbacks of parallel-connected MTDC systems:

- Any disturbance on the dc system (line fault or commutation failure) affects the entire dc system.

- Reversal of power at any terminal requires mechanical switch operation.

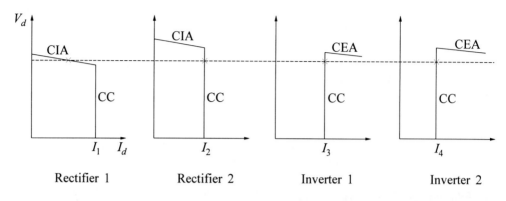

(a) Individual converter characteristics

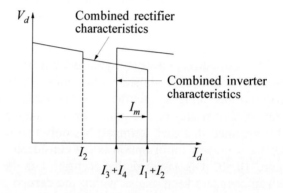

(b) Overall characteristics

Figure 10.50 Control characteristics of parallel-connected MTDC systems

- Blocking of a single bridge, in a converter station consisting of two or more series-connected bridges, requires either operation of the whole system at reduced voltage or disconnection of the affected station.

- Commutation failure at an inverter can draw current from other terminals and this may affect recovery.

Series-connected systems

In a series-connected system, current is controlled by one terminal, and all other terminals either operate at constant-angle (α or γ) control or regulate voltage. Figure 10.51 illustrates the control strategy usually considered for series systems.

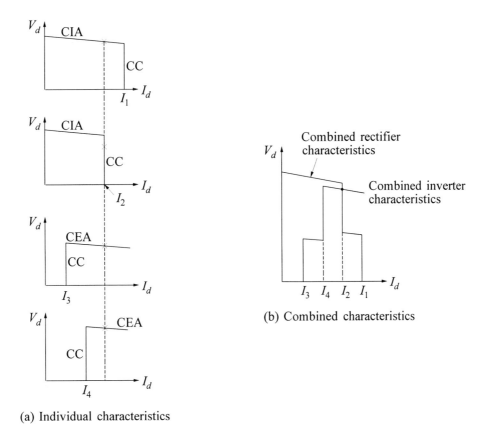

(a) Individual characteristics

(b) Combined characteristics

Figure 10.51 Control characteristics of series systems

Current control is assumed by a rectifier if the sum of the rectifier voltages at the ordered current is greater than the sum of the inverter voltages; the rectifier with the lowest current order assumes the current control. On the other hand, if the sum of the inverter voltages is greater, the inverter with higher current order assumes current control.

For series systems, the voltage references must be balanced, whereas for parallel systems current references must be coordinated. However, for series systems the coordination problem is not as critical as it could be with parallel systems.

The series systems allow high-speed reversal of power at any terminal without the need for switching operations. Bridges and terminals can be taken out of service without affecting the rest of the system. Communication between terminals is required for controlling line loadings to minimize losses; this can be achieved with a relatively slow communication.

The operation of converters in series requires converter operation with high firing angles. This can be minimized by tap-changer control and backing off one bridge against another.

The following are the main drawbacks of the series-connected systems:

- As the voltage to ground is different in various parts of the system, insulation coordination is complex and expensive. Losses are higher in portions with lower voltage.

- A permanent line fault causes interruption of the entire system.

- Flexibility for future extension is limited.

10.9 MODELLING OF HVDC SYSTEMS

In this section, we will discuss the modelling of HVDC systems in power-flow and stability studies. The representation of the dc systems requires consideration of the following:

- Converter model

- DC transmission line/network model

- Interface between ac and dc systems

- DC system controls model

The representation of the converters is based on the following basic assumptions:

(a)   The direct current I_d is ripple-free.

(b)   The ac systems at the inverter and the rectifier consist of perfectly sinusoidal, constant frequency, balanced voltage sources behind balanced impedances. This assumes that all harmonic currents and voltages introduced by the commutation system do not propagate into the ac system because of filtering.

(c)   The converter transformers do not saturate.

The validity of making the first two assumptions for power-flow and stability studies has been demonstrated in reference 27.

10.9.1 Representation for Power-Flow Solution

From the analysis presented in Section 10.2, the converter equations may be summarized as follows:

$$V_{do} = \frac{3\sqrt{2}}{\pi} BTE_{ac}$$

$$V_d = V_{do}\cos\alpha - \frac{3}{\pi}X_c I_d B$$

or

$$V_d = V_{do}\cos\gamma - \frac{3}{\pi}X_c I_d B$$

$$\phi \approx \cos^{-1}(V_d/V_{do})$$

(10.45)

$$P = V_d I_d = P_{ac}$$

$$Q = P\tan\phi$$

where

E_{ac} = RMS line-to-line voltage on HT bus
T = transformer turns ratio
B = no. of bridges in series
P = active power
Q = reactive power
$X_c = \omega L_c$ = commutating reactance
       per bridge/phase
V_d, I_d = direct voltage and current per pole

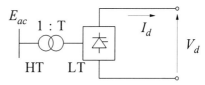

The equation used above for determining the power factor angle (ϕ) is approximate. It simplifies analysis significantly and gives results of acceptable accuracy, consistent with the level of accuracy associated with iterative solution techniques used for power-flow analysis. For specific applications requiring greater accuracy, exact relationships between ac and dc quantities derived in Section 10.2.2 may be used.

For the purpose of illustration, we will consider a two-terminal dc link. Using the subscripts r and i to denote rectifier and inverter quantities, respectively, the equation for a dc line having a resistance R_L is given by

$$V_{dr} = V_{di} + R_L I_d$$

(10.46)

(a) AC/DC interface at the HT bus

Power-flow analysis requires joint solution of the dc and ac system equations. One approach is to solve the two sets of equations iteratively, as illustrated in Figure 10.52, with the converter transformer HT bus (ac side) providing the interface between the ac and dc equations.

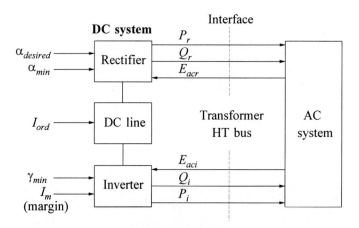

Figure 10.52

Here E_{acr} and E_{aci} are considered to be input quantities for the solution of dc system equations. They are known from the previous step in ac solution.

Variables P_r, Q_r, P_i and Q_i are considered to be the outputs from the solution of the dc system equations. They are used in the next iteration for solving the ac system equations.

The dependent and independent variables in the solution of dc equations depend on rectifier and inverter control modes. The three possible modes of operation are:

    Mode 1: Rectifier on CC control; inverter on CEA control
    Mode 2: Inverter on CC control; rectifier on CIA control
    Mode 3: Rectifier on CIA control; inverter on modified characteristic

In mode 1, alternative inverter control functions are constant voltage control and constant-β control (see Section 10.4.1). For purposes of illustration we will consider here only the CEA control mode.

(1) Mode 1: rectifier on CC control and inverter on CEA control:

In mode 1, we have

- Inverter firing angle adjusted to give $\gamma = \gamma_{min}$.

- Rectifier firing angle adjusted to give $I_d = I_{ord}$.

- Rectifier transformer tap adjusted to give α within a desired range.

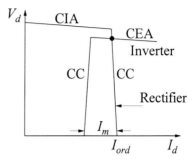

- Inverter transformer tap adjusted to give desired voltage.

Figure 10.53

From Equation 10.45, with $I_d = I_{ord}$, we may write inverter equations as follows:

$$V_{doi} = \frac{3\sqrt{2}}{\pi} B_i T_i E_{aci}$$

$$V_{di} = V_{doi}(\cos\gamma_{min}) - \frac{3}{\pi} X_{ci} B_i I_{ord}$$

$$\phi_i = \cos^{-1}(V_{di}/V_{doi})$$

$$P_i = V_{di} I_{ord}$$

$$Q_i = P_i \tan\phi_i$$

(10.47)

Since γ_{min} and I_{ord} are known and E_{aci} is given by the previous ac solution, V_{di}, P_i and Q_i can be computed. The transformer tap can be adjusted to give V_{di} within the desired range.

The rectifier equations are

$$V_{dr} = V_{di} + R_L I_{ord}$$

$$V_{dor} = \frac{3\sqrt{2}}{\pi} E_{acr} B_r T_r$$

$$\alpha = \cos^{-1}\left(\frac{V_{dr}}{V_{dor}} + \frac{X_{cr} I_{ord}}{\sqrt{2} E_{acr} T_r} \right)$$

(10.48)

In the above equations voltage E_{acr} is known from the previous ac solution. The turns ratio T_r is adjusted to give α within the desired range.

$$\phi_r = \cos^{-1}(V_{dr}/V_{dor})$$

$$P_r = V_{dr}I_{ord} \qquad (10.49)$$

$$Q_r = P_r\tan\phi_r$$

Here P_i, P_r, Q_i and Q_r are outputs to be used in the next iteration of the ac solution.

(2) Mode 2: inverter on CC control and rectifier on CIA control

In mode 2, we have

- Rectifier firing angle $\alpha = \alpha_{min}$.

- Inverter firing angle adjusted to give $I_d = I_{ord} - I_m$.

- Rectifier transformer tap adjusted to maximize dc voltage.

- Inverter transformer tap adjusted so that $\gamma > \gamma_{min}$ and var consumption is minimized.

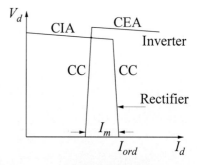

Figure 10.54

With $I_d = I_{ord} - I_m$, the rectifier equations are

$$V_{dor} = \frac{3\sqrt{2}}{\pi}B_r T_r E_{acr}$$

$$V_{dr} = V_{dor}\cos\alpha_{min} - \frac{3}{\pi}X_{cr}(I_{ord}-I_m)B_r \qquad (10.50)$$

$$\phi_r = \cos^{-1}(V_{dr}/V_{dor})$$

$$P_r = V_{dr}(I_{ord}-I_m)$$

$$Q_r = P_r\tan\phi_r$$

In the above equations, E_{acr} is known from the previous ac solution, and I_d is held at $I_{ord}-I_m$ by the inverter. Turns ratio T_r may be adjusted to maximize V_{dr}.

With E_{aci} known from the ac solution, the inverter equations may be solved as follows:

$$V_{doi} = \frac{3\sqrt{2}}{\pi} E_{aci} B_i T_i$$

$$V_{di} = V_{dr} - R_L I_d$$

$$= V_{dr} - R_L (I_{ord} - I_m)$$

$$\gamma = \cos^{-1}\left[\frac{V_{di}}{V_{doi}} + \frac{X_{ci}(I_{ord} - I_m)}{\sqrt{2} E_{aci} T_i} \right] \qquad (10.51)$$

$$\phi_i = \cos^{-1}(V_{di}/V_{doi})$$

$$P_i = V_{di} I_d = V_{di}(I_{ord} - I_m)$$

$$Q_i = P_i \tan\phi_i$$

Turns ratio T_i may be adjusted to ensure $\gamma > \gamma_{min}$ and minimize var consumption. Variables P_r, P_i, Q_r and Q_i are outputs of the above calculations to be used for the next iteration of the ac solution.

(3) Mode 3: rectifier on CIA control and inverter on modified characteristic

For power-flow studies, it is usually sufficient to consider modes 1 and 2. However, for power-flow solutions (solution of network algebraic equations) associated with stability studies, it is necessary to consider the transition between modes 1 and 2. For reasons given in Section 10.4.1, the inverter characteristic is modified as shown in Figure 10.55. The segment JK, with a positive slope, provides a more stable control mode than the segment FK. One method of realizing this modified characteristic is to operate the inverter in the constant-β mode.

In mode 3, we have

- Rectifier ignition delay angle = α_{min}.

- Inverter ignition advance angle = β_c.

- $I_d = I'_d$ such that $I_{ord} > I'_d > (I_{ord} - I_m)$.

The dc system equations are solved as follows to compute the line current.

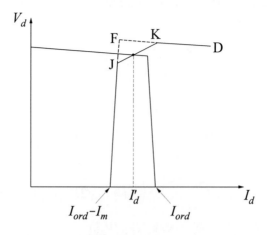

Figure 10.55

$$V_{dor} = \frac{3\sqrt{2}}{\pi} B_r T_r E_{acr}$$

$$V_{doi} = \frac{3\sqrt{2}}{\pi} B_i T_i E_{aci}$$

(10.52A)

Variables E_{acr} and E_{aci} are known from the previous ac solution.

$$V_{dr} = V_{dor}\cos\alpha_{min} - \frac{3}{\pi} I_d' X_{cr} B_r$$

$$V_{di} = V_{doi}\cos\beta_c + \frac{3}{\pi} I_d' X_{ci} B_i$$

(10.52B)

From line equation, we have

$$V_{dr} = V_{di} + R_L I_d'$$

Hence,

$$I_d' = \frac{V_{dr} - V_{di}}{R_L}$$

$$= \frac{1}{R_L}\left[V_{dor}\cos\alpha_{min} - V_{doi}\cos\beta_c - \frac{3}{\pi} I_d'(X_{cr}B_r + X_{ci}B_i) \right]$$

Rearranging, we have the following expression for I'_d in terms of V_{dor}, V_{doi}, α_{min} and β_c, whose values are known.

$$I'_d = \frac{V_{dor}\cos\alpha_{min} - V_{doi}\cos\beta_c}{R_L + \dfrac{3}{\pi}X_{cr}B_r + \dfrac{3}{\pi}X_{ci}B_i} \tag{10.53}$$

With the value of I'_d known, V_{dr} and V_{di} are calculated by using Equation 10.52B. The ac quantities can then be calculated as follows:

$$\begin{aligned}
\phi_r &= \cos^{-1}(V_{dr}/V_{dor}) \\[4pt]
\phi_i &= \cos^{-1}(V_{di}/V_{doi}) \\[4pt]
P_r &= V_{dr}I'_d, \quad Q_r = P_r\tan\phi_r \\[4pt]
P_i &= V_{di}I'_d, \quad Q_i = P_i\tan\phi_i
\end{aligned} \tag{10.54}$$

For transient stability simulations, the tap-changer action is too slow and hence not considered.

For any given system condition, the rectifier and inverter modes of operation may not be known prior to the solution of system equations. Therefore, the following procedure may be used to establish operating modes and solve the ac and dc equations.

1. Solve for ac equations; output E_{acr}, E_{aci}.

2. (a) Solve mode 1 dc equations.
   If $\alpha > \alpha_{min}$, mode 1 condition is satisfied; go to step 3.

   (b) If $\alpha \leq \alpha_{min}$, solve for mode 2 dc equations.
   If $\gamma > \gamma_{min}$, mode 2 condition is satisfied; go to step 3.

   (c) If $\gamma \leq \gamma_{min}$, solve for mode 3 equations.

3. Calculate P_i, Q_i, P_r and Q_r. If mismatch is greater than tolerance, go back to step 1 and solve ac equations.

4. If mismatch is less than tolerance, solution is complete.

(b) AC/DC interface at the LT bus

In the representation discussed above, the ac/dc interface is at the HT side of the commutating transformer. An alternative representation is to have the ac/dc interface at the LT side (the valve side) of the commutating transformer.

An advantage of using the ac/dc interface at the LT bus is that it allows the commutation reactance to be different from the leakage reactance of the commutating transformer. Ideally, it should be the leakage reactance plus the equivalent system reactance at the HT bus. In most cases, the system reactance is small compared to the transformer reactance and, therefore, the LT bus representation may not be essential. In weak systems, this may not be true and the flexibility offered by the so-called LT representation is useful. The LT bus representation also offers flexibility in modelling SVCs, synchronous condensers, and filters connected to the tertiary winding of the converter transformers.

In the LT bus interface approach, the ac system representation includes converter transformers, and the ac solution includes computation of LT bus voltages. The LT voltages are used as input to the solution of dc equations. The HT bus voltage (or more precisely the commutating voltage) is computed from the LT voltage, and used in the solution of dc equations, which are essentially the same as those for the HT bus interface approach. The output of the solution of dc equations, for use in the next iteration of ac solution, is P and Q at the LT bus.

The following are the details of the calculations. Since an equivalent HT bus is used, there is no need for a transformer tap ratio in the dc equations.

Calculation of equivalent HT bus line voltage:

Let

E_{LT} = RMS line-to-line voltage at LT bus
E_{ac} = RMS line-to-line voltage behind X_c
X_c = equivalent commutating reactance per phase
ϕ_{LT} = angle between LT bus phase voltage and line current

Using the LT bus line-to-neutral voltage ($E_{LT}/\sqrt{3}$) as reference phasor, we have

$$\frac{E_{ac}}{\sqrt{3}} = \frac{E_{LT}}{\sqrt{3}} + j\frac{X_c}{B}I_{ac}(\cos\phi_{LT} - j\sin\phi_{LT})$$

$$E_{ac} = E_{LT} + j\sqrt{3}\frac{X_c}{B}\frac{\sqrt{6}I_d B}{\pi}(\cos\phi_{LT} - j\sin\phi_{LT}) \qquad (10.55)$$

$$= \left(E_{LT} + \frac{6X_c}{\sqrt{2}\pi}I_d\sin\phi_{LT}\right) + j\frac{6X_c}{\sqrt{2}\pi}I_d\cos\phi_{LT}$$

Calculation of reactive power at the LT bus:

As before, active power is

$$P = V_d I_d = P_{ac}$$

and the reactive power at the equivalent HT bus is

$$Q_{HT} = P\tan\phi_{HT}$$

where

$$\phi_{HT} = \cos^{-1}(V_d/V_{do})$$

To find the reactive power at the LT bus, Q_{HT} is reduced by the three-phase $X_c I^2$ loss. Since there are B bridges in parallel,

$$Q_{LT} = Q_{HT} - 3\frac{X_c}{B}I_{ac}^2$$

$$= Q_{HT} - 3\frac{X_c}{B}\left(\frac{\sqrt{6}}{\pi}I_d B\right)^2 \qquad (10.56)$$

$$= Q_{HT} - BX_c\frac{18}{\pi^2}I_d^2$$

Reference 28 uses the above approach.

Inclusion of converter station losses

The dc system equations used so far did not consider converter station losses. There are several losses associated with a converter station such as those of converter transformers, filters, valves and valve auxiliaries, and smoothing reactors. The transformer copper loss may be represented by a series resistance, while other ac side losses are neglected. Inclusion of this resistance in the commutation overlap equation, however, results in considerable complexity in the converter equations; this is not usually justified.

The losses associated with the valves and their auxiliaries may be combined with the smoothing reactor and explicitly represented as equivalent series resistance (R_{eq}) on the dc side. There is also a small valve forward voltage drop (V_{drop}) due to arc drop in mercury-arc valves or forward voltage drop in thyristors. Their effects may be included by modifying the converter equations as follows:

$$V_d = V_{do}\cos\alpha - R_c I_d - R_{eq} I_d - V_{drop} \tag{10.57}$$

where

R_c = equivalent commutating resistance = $(3/\pi)X_c$
R_{eq} = equivalent resistance representing the losses in the valves and auxiliaries
V_{drop} = voltage drop across the valve

The above equation is applicable to a rectifier as well as an inverter. However, for inverter calculations we prefer to write the equations in terms of γ:

$$V_d = V_{do}\cos\gamma - R_c I_d + R_{eq} I_d + V_{drop} \tag{10.58}$$

Figure 10.56 shows the corresponding equivalent circuits. These modifications to the original equations present little complexity and are easy to incorporate.

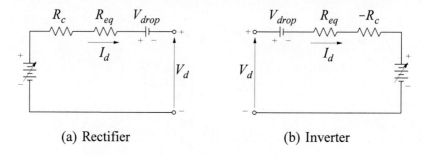

(a) Rectifier   (b) Inverter

Figure 10.56   Converter equivalent circuits

The effect of neglecting losses is to introduce a small error in the computed value of active power and a relatively significant error in the value of reactive power [27]. These effects could be important for the case of back-to-back links.

Multiterminal DC systems

The formulation of power-flow equations developed above applies to two-terminal dc systems. The method can be readily extended to multiterminal systems.

Two approaches have evolved for the solution of power-flow equations. One is the sequential solution approach [29] in which the ac and dc equations are solved separately at each iteration. The other is the unified solution approach [30] in which the ac and dc systems are combined and solved as one set of equations during each iteration.

Example 10.2

Figure E10.2 shows a bipolar dc link with a rating of 1,000 MW, ±250 kV. The line resistance is 10 Ω/line. Each converter has a 12-pulse bridge with $R_c = (3/\pi)X_c = 12\ \Omega$ (6 Ω for each of the 6-pulse bridges).

The performance of the bipolar link is to be analyzed by considering it to be a monopolar link of +500 kV. The rectifier ignition delay angle limit (α_{min}) is equal to 5°. The effects of converter station losses and forward voltage drop across the valves may be neglected.

The dc link is initially operating with the rectifier on CC control with $\alpha_0 = 18°10'$, and the inverter on CEA control with $\gamma_0 = 18°10'$. The current margin I_m is set at 15%, and the transformer turns ratio at each converter is 0.5. At the inverter, the dc power is 1,000 MW, and the dc voltage is 500 kV (for the equivalent monopolar link).

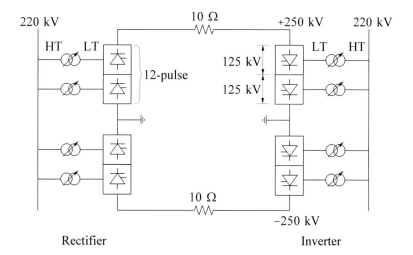

Figure E10.2

(a)   For the above operating condition, compute the following:

   (i)   Power factor and the reactive power at the inverter HT bus.

   (ii)   Inverter commutation overlap angle μ.

   (iii)   RMS values of the line-to-line alternating voltage, fundamental
            component of the line current and the reactive power at rectifier HT
            bus.

(b)   If the rectifier side HT bus ac voltage drops by 20%, compute the following:

   (i)   DC voltages at the rectifier and inverter terminals.

   (ii)   Rectifier angle α, and inverter angles γ and μ.

   (iii)   Active and reactive power at the inverter and rectifier HT buses.

   Assume that the transformer taps have not changed and that the inverter side
   ac voltage is maintained constant.

(c)   If the inverter side HT bus ac voltage drops by 15% and the rectifier side ac
       voltage remains at its initial value, determine the following after the tap
       changer action:

   (i)   The dc voltage at the rectifier and inverter terminals.

   (ii)   Rectifier angle α and the inverter angle γ.

   (iii)   Active and reactive power at the rectifier and inverter terminals.

   The rectifier transformer control action attempts to hold α between 15° and
   20°, and the inverter transformer control action attempts to hold the inverter
   dc voltage within the range of 500 to 510 kV. Assume that the maximum and
   minimum tap positions are 1.2 pu and 0.8 pu (corresponding to turns ratios
   of 0.6 and 0.4) respectively, with a tap step size of 0.01 pu.

Solution

Figure E10.3 shows the equivalent circuit of the bipolar link represented as a
monopolar link of +500 kV.

(a) The initial operating condition is as follows:

$$\alpha_0 = \gamma_0 = 18.167° \qquad\qquad P_i = 1,000 \text{ MW}$$
$$T_r = T_i = 0.5 \qquad\qquad V_{di} = 500 \text{ kV}$$

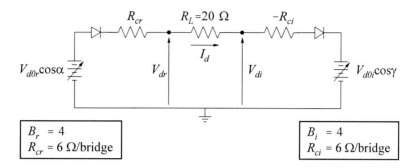

Figure E10.3 Equivalent circuit

The direct current is

$$I_d = \frac{P_i}{V_d} = \frac{1000}{500} = 2 \quad kA$$

The inverter ideal no-load voltage is

$$V_{d0i} = \frac{V_{di} + B_i R_{ci} I_d}{\cos\gamma_0}$$

$$= \frac{500 + 4 \times 6 \times 2}{\cos 18.167°}$$

$$= 576.75 \quad kA$$

(i) The power factor at the inverter HT bus is

$$\cos\phi_i \approx \frac{V_{di}}{V_{doi}} = \frac{500}{576.75} = 0.8669$$

and

$$\phi_i = 29.896°$$

The reactive power at the inverter HT bus is

$$Q_i = P_i \tan\phi_i = 1000 \tan 29.896° = 574.94 \quad MVAr$$

(ii) Since $V_{di} = V_{d0i}(\cos\gamma + \cos\beta)/2$,

$$\cos\beta = 2\frac{V_{di}}{V_{d0i}} - \cos\gamma_0$$

$$= 2\times\frac{500}{576.75} - \cos 18.167°$$

$$= 0.7837$$

and

$$\beta = 38.399°$$

Hence, the inverter commutation angle is

$$\mu_i = \beta - \gamma_0 = 38.399° - 18.167° = 20.232°$$

(iii) At the rectifier, we have

$$V_{dr} = V_{di} + R_L I_d = 500 + 2\times20 = 540 \quad kV$$

and

$$V_{d0r} = \frac{V_{dr} + B_r R_{cr} I_d}{\cos\alpha_0}$$

$$= \frac{540 + 4\times6\times2}{\cos 18.167°}$$

$$= 618.85 \quad kV$$

The RMS line-to-line ac voltage at the HT bus is

$$E_{acr} = \frac{V_{dor}}{1.3505 B_r T_r}$$

$$= \frac{618.85}{1.3505\times4\times0.5}$$

$$= 229.12 \quad kV$$

The sum of the RMS fundamental line currents in the four transformers is

$$I_{L1r} \approx \frac{\sqrt{6}}{\pi} B_r T_r I_d$$

$$= 0.7797\times4\times0.5\times2$$

$$= 3.119 \quad kA$$

The dc power at the rectifier is

$$P_r = V_{dr}I_d = 540 \times 2 = 1080 \quad \text{MW}$$

The power factor at the rectifier HT bus is

$$\cos\phi_r \approx \frac{V_{dr}}{V_{d0r}} = \frac{540}{618.85} = 0.8726$$

and

$$\phi_r = 29.24°$$

The reactive power at the rectifier HT bus is

$$Q_r = P_r \tan\phi_r = 1080 \tan 29.24° = 604.57 \quad \text{MVAr}$$

(b)   With the transformer taps unchanged, V_{d0} is directly proportional to E_{ac}. Therefore, when the rectifier HT bus voltage drops by 20%, we have

$$V_{d0r} = 0.8 \times 618.85 = 495.08 \quad \text{kV}$$

We will first assume that the dc link continues to operate in mode 1: rectifier on CC control with $I_d = I_{ord} = 2$ kA, and the inverter on CEA control with $V_{di} = 500$ kV. The corresponding direct voltage at the rectifier is

$$V_{dr} = V_{di} + R_L I_d = 500 + 20 \times 2 = 540 \quad \text{kV}$$

Therefore,

$$\cos\alpha = \frac{V_{dr} + B_r R_{cr} I_d}{V_{dor}}$$

$$= \frac{540 + 4 \times 6 \times 2}{495.08} > 1.0$$

Therefore, mode 1 condition is not satisfied. The controls switch to mode 2: rectifier on CIA control with $\alpha = \alpha_{min} = 5°$, and the inverter on CC control with

$$I_d = I_{ord} - I_m = (1.0 - 0.15) \times 2 = 1.7 \quad \text{kA}$$

(i) The rectifier and inverter direct voltages now are

$$V_{dr} = V_{d0r}\cos\alpha - B_r R_{cr} I_d$$

$$= 495.08 \cos 5° - 4 \times 6 \times 1.7$$

$$= 452.39 \quad \text{kV}$$

$$V_{di} = V_{dr} - R_L I_d$$

$$= 452.39 - 20 \times 1.7$$

$$= 418.39 \quad \text{kV}$$

(ii) Since the inverter ac bus voltage has not changed, $V_{d0i} = 576.75$ kV. With $I_d = 1.7$ kA, we have

$$\cos\gamma = \frac{V_{di} + B_i R_{ci} I_d}{V_{d0i}}$$

$$= \frac{418.39 + 4 \times 6 \times 1.7}{576.75}$$

$$= 0.796$$

and

$$\gamma = 37.23°$$

Now,

$$V_{di} = V_{d0i} \frac{\cos\gamma + \cos\beta}{2}$$

Hence,

$$\cos\beta = \frac{2V_{di}}{V_{d0i}} - \cos\gamma$$

$$= \frac{2 \times 418.39}{576.75} - 0.796$$

$$= 0.655$$

and

$$\beta = 49.10°$$

The inverter commutation angle is

$$\mu_i = \beta - \gamma = 49.10° - 37.23° = 11.87°$$

(iii) The dc power at the inverter is

$$P_i = V_{di} I_d = 418.39 \times 1.7 = 711.26 \quad \text{MW}$$

The power factor at the inverter HT bus is

$$\cos\phi_i \approx \frac{V_{di}}{V_{d0i}} = \frac{418.39}{576.75} = 0.725$$

and

$$\phi_i = 43.49°$$

The reactive power at the inverter HT bus is

$$Q_i = P_i \tan\phi_i = 711.26 \tan 43.49° = 674.85 \quad \text{MVAr}$$

The dc power at the rectifier is

$$P_r = V_{dr}I_d = 452.39 \times 1.7 = 769.06 \quad \text{MW}$$

The power factor at the rectifier HT bus is

$$\cos\phi_r \approx \frac{V_{dr}}{V_{d0r}} = \frac{452.39}{495.08} = 0.914$$

and

$$\phi_r = 23.97°$$

The reactive power at the rectifier HT bus is

$$Q_r = P_r \tan\phi_r = 769.06 \tan 23.97° = 341.87 \quad \text{MVAr}$$

(c) When the inverter side ac voltage drops by 15% and the rectifier voltage remains at its normal value, mode 1 operation is possible. Hence, the rectifier is on CC control with

$$I_d = I_{ord} = 2 \quad \text{kA}$$

and the inverter is on CEA control with

$$\gamma = 18.167°$$

Due to the reduction in ac voltage, the inverter dc voltage drops. The rectifier α increases and the dc voltage decreases so that I_d is maintained constant. The rectifier transformer tap changer acts to hold α between 15° and 20°, and the inverter transformer tap changer acts to hold V_{di} between 500 kV and 510 kV.

The ideal no-load voltage of a converter is directly proportional to the ac voltage and the transformer turns ratio. In (a), we computed the rectifier and inverter ideal no-load voltages under normal ac voltages and 1.0 pu tap position (turns ratio of 0.5) to be

$$V_{d0r} = 618.85 \quad \text{kV}$$

and

$$V_{d0i} = 576.75 \quad \text{kV}$$

With normal ac voltage and tap changer action, for the rectifier, we have

$$V_{d0r} = 618.85 T_r'$$

where T_r' is the pu tap position.

Similarly, for the inverter with ac voltage reduced by 15%, we have

$$V_{d0i} = 576.75 (0.85) T_i' = 490.24 T_i'$$

Hence, the inverter dc voltage is

$$
\begin{aligned}
V_{di} &= V_{d0i} \cos\gamma - B_i R_{ci} I_d \\
&= 490.24 T_i' \cos 18.167° - 4 \times 6 \times 2 \\
&= 465.72 T_i' - 48 \quad \text{kV}
\end{aligned}
\tag{E10.1}
$$

The rectifier direct voltage required to maintain I_d at 2 kA is

$$V_{dr} = V_{di} + R_L I_d = V_{di} + 20 \times 2 = V_{di} + 40 \quad \text{kV} \tag{E10.2}$$

This should be equal to

$$
\begin{aligned}
V_{dr} &= V_{d0r} \cos\alpha - B_r R_{cr} I_d \\
&= 618.85 T_r' \cos\alpha - 48
\end{aligned}
\tag{E10.3}
$$

From Equations E10.2 and E10.3 we have

$$\cos\alpha = \frac{V_{di} + 88}{618.85 T_r'} \tag{E10.4}$$

Table E10.1 shows the variations in V_{di} and α as T_i' and T_r' change from their initial values to satisfy the control requirements.

Table E10.1

T_i'	T_r'	V_{di} (kV)	α (degrees)
1.0	1.0	417.7	35.2
1.01	0.99	422.4	33.6
$\vdots$	$\vdots$	$\vdots$	$\vdots$
1.07	0.93	450.3	20.7
1.08	0.93	455.0	19.4
$\vdots$	$\vdots$	$\vdots$	$\vdots$
1.10	0.93	464.3	16.3
1.11	0.93	468.9	14.6
1.12	0.94	473.6	15.1
1.13	0.95	478.3	15.6
$\vdots$	$\vdots$	$\vdots$	$\vdots$
1.17	0.98	496.9	15.3
1.18	0.99	501.5	15.8

Notes: V_{di} computed using Equation E10.1
$\quad\quad\quad$ α computed using Equation E10.4

From the table we see that the inverter tap position increases until $T_i' = 1.18$ pu (corresponding to a turns ratio of 0.59) which results in a V_{di} of 501.55 kV.

The rectifier pu tap position T_r' which meets the control requirements is 0.99 (turns ratio of 0.495); the corresponding α is 15.8° and the direct voltage from Equation E10.3 is

$$V_{dr} = 618.85 \times 0.99 \times \cos 15.8° - 48 = 541.51 \quad \text{kV}$$

At the rectifier we have

$$P_r = V_{dr}I_d = 541.51 \times 2 = 1083.02 \quad \text{MW}$$

$$\cos\phi_r \approx \frac{V_{dr}}{V_{d0r}} = \frac{541.51}{618.85 \times 0.99} = 0.884$$

$$Q_r = P_r \tan\phi_r = 1083.02 \tan\phi_r = 573.2 \quad \text{MVAr}$$

At the inverter we have

$$P_i = V_{di}I_d = 501.55 \times 2 = 1003.1 \quad \text{MW}$$

$$\cos\phi_i \approx \frac{V_{di}}{V_{d0i}} = \frac{501.55}{490.24 \times 1.18} = 0.867$$

$$Q_i = P_i \tan\phi_i = 1003.1 \tan\phi_i = 576.7 \quad \text{MVAr}$$ ∎

10.9.2 Per Unit System for DC Quantities

A convenient per unit system for the dc quantities has the following base values:

$$V_{dc\,base} = B\frac{3\sqrt{2}}{\pi}V_{ac\,base} = V_{do}$$

$$I_{dc\,base} = I_{dc\,rated} \qquad\qquad (10.59)$$

$$Z_{dc\,base} = V_{dc\,base}/I_{dc\,base}$$

$$P_{dc\,base} = V_{dc\,base}I_{dc\,base}$$

where

B = number of bridges in series in the dc converter

$V_{ac\,base}$ = line-to-line ac base voltage referred to the LT side of the commutating transformer

If the commutating reactance per bridge expressed in ohms is X, then the per unit total commutating reactance for the B bridges in series is

$$\bar{X}_{dc} = \frac{BX}{Z_{dc\,base}} \qquad\qquad (10.60)$$

Relationship between per unit quantities in the dc and ac systems

The base power and base impedance for the ac system are

$$P_{ac\,base} = \text{MVA}_{base} = \sqrt{3}\,V_{ac\,base}I_{ac\,base}$$

$$Z_{ac\,base} = V_{ac\,base}/(\sqrt{3}\,I_{ac\,base})$$

In the ac solution, the commutating reactance is represented by the parallel combination of B individual transformer reactances. The per unit value of X_c in the ac per unit system is

$$\overline{X}_{ac} = \frac{X}{B Z_{ac\,base}}$$

Therefore, the ratio of the per unit values of X_c in the two systems is

$$\frac{\overline{X}_{dc}}{\overline{X}_{ac}} = \frac{BX}{Z_{dc\,base}} \times \frac{B Z_{ac\,base}}{X}$$

$$= B^2 \left(\frac{Z_{ac\,base}}{Z_{dc\,base}}\right) = B^2 \left(\frac{V_{ac\,base}}{V_{dc\,base}}\right)^2 \frac{P_{dc\,base}}{P_{ac\,base}} \qquad (10.61)$$

$$= \left(\frac{3\sqrt{2}}{\pi}\right)^2 \frac{P_{dc\,base}}{P_{ac\,base}}$$

$$= \frac{18}{\pi^2} \frac{P_{dc\,base}}{P_{ac\,base}}$$

$$\frac{\overline{I}_{dc}}{\overline{I}_{ac}} = \frac{I_{dc}}{I_{dc\,base}} \frac{I_{ac\,base}}{I_{ac}} = \frac{I_{dc}}{B\frac{\sqrt{6}}{\pi}I_{dc}} \frac{I_{ac\,base}}{I_{dc\,base}}$$

$$= \frac{\pi}{B\sqrt{6}} \frac{P_{ac\,base}}{P_{dc\,base}} \frac{V_{dc\,base}}{\sqrt{3}V_{ac\,base}} = \frac{P_{ac\,base}}{P_{dc\,base}} \frac{\pi}{\sqrt{6}} \frac{B3\sqrt{2}}{B\pi\sqrt{3}} \qquad (10.62)$$

$$= \frac{P_{ac\,base}}{P_{dc\,base}}$$

Usually, $P_{dc\,base}$ is chosen to be equal to $P_{ac\,base}$. Alternatively, $P_{dc\,base}$ may be chosen to be the nominal rating of the dc line. It is obvious that the use of the per unit system for the dc quantities offers no particular advantage. The dc quantities may in fact be handled in terms of their natural units, and many computer programs do so.

10.9.3 Representation for Stability Studies

In a stability program, the ac network equations are represented in terms of positive-sequence quantities. This imposes a fundamental limitation on the modelling of the dc systems. In particular, commutation failures cannot be accurately predicted. Commutation failure may result from a severe three-phase fault near the inverter, unbalanced faults on the inverter side ac system, or saturation of converter transformers during dynamic overvoltage conditions.

Notwithstanding the above limitation, models of various degrees of detail have been effectively used to represent dc systems in stability studies [28,31-37]. A general functional block diagram model of dc systems is shown in Figure 10.57.

Some of the early efforts to incorporate HVDC system models into stability programs used detailed representation which accounted for the dynamics of the line and the converter controls [28,31,32]. In recent years, the trend has been toward simpler models [33]. Such models are adequate for general purpose stability studies of systems in which the dc link is connected to strong parts of the ac system. However, for weak ac system applications requiring complex dc system controls and for multiterminal dc systems, detailed models are required. Therefore, the trend is reversing and the preference is to have flexible modelling capability with a wide range of detail [8,34,35,36]. The required degree of detail depends on the purpose of the study and the particular dc system.

Each dc system tends to have unique characteristics tailored to meet the specific needs of its application. Therefore, standard models of fixed structures have not been developed for representation of dc systems in stability studies. Instead, three categories of models have evolved: (a) simple model, (b) response or performance model, and (c) detailed model with flexible modelling capability.

(a) Simple models

For remote dc links, which do not have a significant impact on the results of the stability analysis, very simple models are usually adequate. The dc links may be represented as constant active and reactive power injections at the ac terminals of the converters. Where more realistic models are required, the dc link is represented by the static converter equations and functional effects of the controls. The models appear as algebraic equations and the interface between ac and dc systems is treated in a manner similar to that described for power-flow analysis in Section 10.9.1.

(b) Response models

For general purpose stability studies the dynamics of the dc line and pole controls may be neglected. The pole control action is assumed to be instantaneous and the lines are represented by their resistances.

Many of the control functions are represented in terms of their net effects, rather than actual characteristics of the hardware. The following are the features included in a typical response type model.

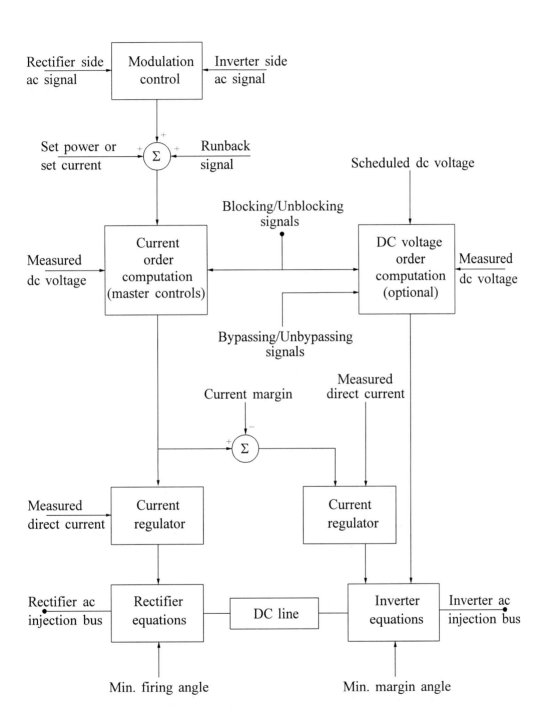

Figure 10.57 Functional block diagram of an HVDC system model

Converter and line equations:

The three modes of control, as identified before, are

- Rectifier on CC and inverter on CEA control

- Rectifier on CIA and inverter on CC control

- Rectifier on CIA and inverter on constant-β control (transition mode)

The control logic associated with the three modes of control may be incorporated into the stability solution as described in Section 10.9.1. In this case, however, transformer taps are not adjusted as they are not fast enough to be effective during the period of interest.

Current-control order with limits:

The current order is determined so as to provide either current control or power control as desired. Constraints are imposed on the current-order level to keep the current within maximum and minimum limits. The maximum current is determined by the voltage-dependent current-order limit (VDCOL) as shown in Figure 10.58. The VDCOL may be given a time delay to assist in riding through ac system faults (see Section 10.7).

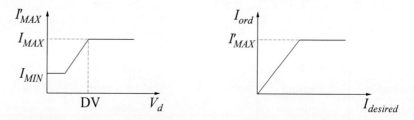

Figure 10.58 Voltage-dependent current-order limits

Control actions during ac faults:

It is necessary to have adequate representation of the actions of controls during ac faults. The following is an example of the logic that may be used to account for the control actions.

If the ac voltage on either side falls below a certain value for longer than a specified time, the direct-current order is set to zero. A ramp limit restricts the rate of decrease of direct current. The line is shut off when the current falls below a specified minimum value.

The direct line current is restored after the ac voltage recovers to an acceptable level. If the voltage recovers before the direct line current has reached its minimum value, the desired current is immediately restored to its original value. The line current increases at a specified maximum rate. If the voltage recovery occurs after the line has been shut down, the recovery is delayed by a specified time. After this, the direct line current is restored to its original value at a specified maximum rate.

The following are examples of two alternative types of recovery procedures:

(i)   The current is increased by controlling rectifier α, with the inverter firing angle fixed at 90°. When the current reaches the desired value minus the current margin, the inverter extinction angle is ramped down to a specified value (normally original value) at a specified rate.

(ii)   The current is increased with the maximum possible dc voltage (with $\gamma = \gamma_0$ or with $\alpha = \alpha_{min}$).

Option (i) ensures that during recovery the maximum reactive power is drawn from the ac system and may be used to control the overvoltages. Option (ii) ensures that maximum possible power is transmitted through the dc link.

The mode of operation and dc blocking and deblocking sequences are system dependent. The optimum sequence is established by experimentation.

A typical dc controller representation is shown in Figure 10.59. The dc shutoff recovery sequence for ac system faults is illustrated in Figure 10.60.

Commutation failure checks:

The dc model usually includes a logic for shutting off the dc line for commutation failure, detected by monitoring commutation voltage or converter margin angle.

Power/Current order modulation:

Dynamics of controls used for ac system stabilization are represented accurately, consistent with the representation used for other forms of stability controls (for example, power system stabilizers).

(c) Detailed flexible model

A wide range of dc system-modelling detail is required in stability studies. Either separate models each representing a dc link in detail or a single detailed model with facilities for simplification should be provided to achieve this flexibility. Because of the variety of controls associated with dc links, "user-defined" control models are likely to be required in addition to the basic ac/dc interface model.

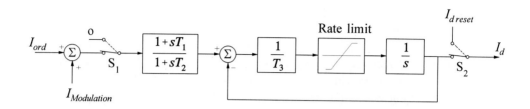

Note: 1. S_1 is switched to o if $E_{acr} < E_{min\,r}$ or $E_{aci} < E_{min\,i}$ for a period $\Delta T > T_{D1}$
        2. S_2 is switched to $I_{d\,reset}$ if line shuts off and E_{acr} and E_{aci} are greater than $E_{recovery}$ and $\Delta T > T_{D2}$
        3. T_3 is the response time of current control loop

Figure 10.59 Current-control block diagram

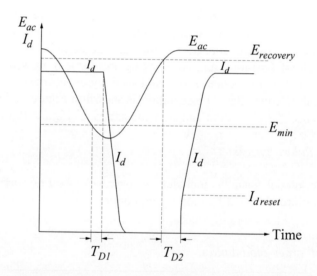

Figure 10.60 Shutoff and recovery sequence

At the highest level of detail, the following modelling features should be provided:

- DC line — a dynamic model which represents the resistance, inductive and capacitive effects of the dc line. The capacitive effects may be particularly important for cables.

- Converter controls represented by appropriate dynamic models for

—   Master controls
—   VDCOL including dynamics
—   Current/power modulation
—   Fast power change logic, including blocking/deblocking
—   Pole controls capable of representing different control options such as CC, CEA, constant ac voltage, constant dc voltage, etc.
—   Commutation failure simulation logic
—   Special controls used to assist HVDC system recovery from disturbances, prevent commutation failure, etc.

- AC/DC interface — correct representation of the commutating voltage, commutating reactance, and transformer tertiary bus to which SVCs and synchronous condensers and other devices are connected.

In such models, the converter angle limits are embedded in the dc controls so that no special mode shift algorithm is required. However, the converter characteristics are still represented by equations relating average values of dc quantities and RMS values of ac fundamental components. This model may be referred to as "quasi steady-state model." It would accurately represent the HVDC system performance in stability studies for analysis of balanced operation, consistent with the representation used for other elements of the power system.

Since the waveforms of the voltages and currents are not represented in the model, only some general predictions regarding commutation failure can be made. These can be based on the magnitude of commutating voltage or inverter margin (extinction) angle.

A flexible quasi steady-state model should have facilities for simplifications so that models with a level of detail that is appropriate for the purpose and scope of the stability study may be used.

Detailed HVDC system models include dynamics which are usually much faster than those associated with the ac system models. In stability studies involving time-domain simulations, very small integration time steps are required to solve the dc equations. Hence, care should be exercised in integrating the ac and dc system equations. One approach often used in such situations is to solve dc equations by using time steps which are submultiples of time steps used for ac equations.

Detailed three-phase representation:

The detailed model described above, based on positive-sequence phasor representation of ac system quantities, is not accurate for analysis of unbalanced faults and for prediction of commutation failure. Accurate simulation of such conditions requires a detailed three-phase, cycle-by-cycle representation including dynamics of the ac line, filters and converter controls, during the disturbance and initial recovery. Thus, the two types of simulation, one using detailed three-phase representation of a small part of the system near the dc link and the other a single-phase quasi steady-state representation of the complete power system, are used in a complementary manner. The first type of simulation can be performed by using an electromagnetic transients program (EMTP) [38] or a dc simulator. It is conceivable that in the future this type of simulation could be incorporated into a transient stability program [6,39].

An example of detailed model

A converter control model used by some early stability programs [28, 31] with detailed representation of dc links is shown in Figure 10.61. It represents controls used for the Pacific DC Intertie and other similar systems.

In these systems, *individual phase control* is used to generate the converter firing pulses (see Section 10.4.3). This is reflected in the model. The heart of the control system is the "delay angle computer," which is represented by block 7 of Figure 10.61. It is based on Equations 10.39 and 10.40. As discussed in Section 10.4.3, the control system consists of three units: the first provides an output proportional to direct current I_d, the second provides an output proportional to $E_m \cos\gamma_c$, and the third provides an alternating voltage proportional to $E_m \cos\omega t$ [12].

The control system uses the ac voltage to establish zero firing point reference. It then adds a direct voltage proportional to the error current to establish the proper firing point. The delay angle computer adds a bias signal $E_0 \cos\gamma_{min}$ to the ac voltage. The input E_0, the commutating voltage, in reality consists of three sinusoidal voltages, one for each pair of elements connected to an ac terminal. The output $\cos\alpha$ actually is a signal suitable for producing a firing pulse at the proper time, i.e., delayed by an angle α. The following is the basis for the "delay angle computer" block shown in Figure 10.61.

From Equation 10.39, the instant of firing $\alpha = \omega t_1$ is given by

$$\sqrt{3}\,E_m \cos\omega t_1 = 2I_d X_c - \sqrt{3}\,E_m \cos\gamma$$

Hence,

$$\cos\alpha = \cos\omega t_1$$

$$= \frac{2X_c I_d}{\sqrt{3}\,E_m} - \cos\gamma$$

$$= \frac{2R_c I_d}{\sqrt{3}\,E_m}\frac{\pi}{3} - \cos\gamma \qquad (10.63)$$

$$= \frac{V_c}{E_0} - \cos\gamma$$

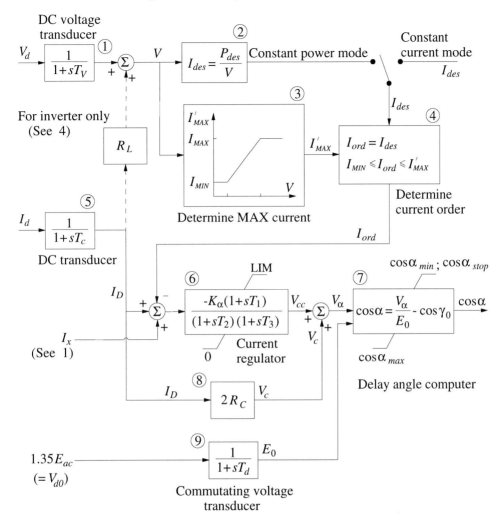

Figure 10.61 A detailed HVDC converter model

Notes:

1. $I_x = I_{ref} - I_{mod}$ for rectifier; $I_x = I_{margin}$ for inverter,

     where I_{mod} is modulating signal if used, and I_{ref} is computed to give required $\cos\alpha$ under steady state.

2.   Limit on current regulator output: LIM $= E_0(\cos\gamma_0 + \cos\alpha_{min}) - V_c$

3. γ_0 = initial inverter extinction angle.

4. P_{des} is specified at rectifier end. Therefore, to determine I_{ord} for inverter, the voltage drop due to line resistance has to be taken into consideration.

where

$$V_c = 2R_c I_d$$
$$E_0 = \text{output of commutating voltage transducer}$$

$$= V_{d0} = \frac{3\sqrt{3}\,E_m}{\pi} = \frac{3\sqrt{2}\,E_{ac}}{\pi} = 1.35 E_{ac}$$

To include current control, an additional signal V_{cc} is added so that

$$\cos\alpha = \frac{V_c}{E_0} + \frac{V_{cc}}{E_0} - \cos\gamma$$

$$= \frac{V_\alpha}{E_0} - \cos\gamma \qquad (10.64)$$

The upper part of Figure 10.61 shows how the current order is determined. Either constant power mode or constant current mode may be specified. The current order is limited by the maximum and minimum current limits. The maximum limit is voltage dependent (VDCOL).

There are three possible modes of operation:

- Constant ignition angle, with $\alpha = \alpha_{min}$;

- Constant current; and

- Constant extinction angle, with $\gamma = \gamma_0$.

(a) CIA mode:

The mode of operation exists when $V_{cc} = \text{LIM}$. The corresponding V_α is

$$V_\alpha = V_{cc} + V_c = \text{LIM} + V_c$$

$$= [E_0(\cos\gamma_0 + \cos\alpha_{min}) - V_c] + V_c$$

Hence,

$$\cos\alpha = \frac{V_\alpha}{E_0} - \cos\gamma_0 = \frac{E_0}{E_0}(\cos\gamma_0 + \cos\alpha_{min}) - \cos\gamma_0$$

$$= \cos\alpha_{min}$$

(b) CC mode:

This is the normal mode for the rectifier. In this mode V_{cc} lies between 0 and LIM. Initially I_{ref} is computed so as to give the required $\cos\alpha$ to satisfy steady-state conditions.

(c) CEA mode:

This is the normal mode for the inverter. For this mode, V_{cc} is equal to its lower limit of zero. From Figure 10.61 and Equation 10.64, with $V_{cc}=0$ and $\gamma=\gamma_0$, we have

$$\cos\alpha = \frac{V_\alpha}{E_0} - \cos\gamma_0 = \frac{2R_c I_d}{E_0} - \cos\gamma_0$$

Therefore, the control will ensure that $\cos\alpha$ corresponds to a condition with $\gamma=\gamma_0$.

Limits on the output block:

The limits ensure that the firing angle is limited to the desired values. Usually, $\alpha_{min}=5°$ and $\alpha_{stop}=110°$.

We see that the above model reflects the dynamic performance of the converter control hardware closely. In contrast, a response model uses logic to represent many of the functions.

Chapter 17 (Section 17.2.3) gives an additional example of a dc link model with a detailed representation of pole and master controls.

Guidelines for selection of modelling detail

The modelling requirements of dc systems are influenced by the following factors:

- Nature and scope of the study,

- Type of disturbance considered, and

- Strength of the associated ac systems.

The following provide general guidelines for selection of modelling detail.

1.   For studies involving disturbances remote from dc links, simple algebraic models may be used unless very low frequency interarea oscillations are excited by the fault. In such a case, response models which include the dc link

modulation controls should be used.

2. For studies associated with preliminary planning of a dc link, response models are usually adequate.

3. For studies involving dc links connected to weak ac systems, response models may be used for initial planning studies provided that they represent the relevant control action adequately. Detailed models are necessary if special dc link controls are to be studied.

4. For studies associated with planning and design specifications of equipment close to dc links, detailed models are required.

5. For studies involving multiterminal HVDC systems, detailed models are required to ensure that the coordination of the converter controls is correct. Convergence difficulties are often experienced with the use of response models for such systems. In initial studies, simple pole controls may be used to minimize the modelling requirement. Special controls necessary for multiterminal systems such as current balances must be modelled.

6. For determination of the effects of dc modulation controls, a response model is usually adequate.

7. For studies involving worst-case disturbances specified by planning/operating criteria, the modelling requirements depend on the disturbance as follows:

   (a) Bipolar outage with no restart (say, due to high-level control malfunction) or with unsuccessful restart on both poles — A simple model could be used, since the dc power would be either zero or relatively low.

   (b) Rectifier side single-line-ground ac fault with breaker and pole outage (the remaining pole would be ramped to cover the loss, subject to overload limits) — A response model is normally adequate. The zero sequence impedance of the Y-Δ converter transformer should be included in the fault shunt.

   (c) Three-phase ac faults at critical locations near the rectifier, in which case either some dc power can be transmitted during the fault, or the dc shuts down and must be restarted rapidly — The restart or recovery characteristic is important. For example, a very fast recovery could be counterproductive if the ac system is weak. Generally, a detailed model is required. In early planning stages, however, a response model including restart may be used.

(d)   Inverter side single-phase ac faults which block dc power during the fault because of commutation failure — A detailed model is required since recovery time and characteristic can be critical (and multiple dc links may be affected).

(e)   Various ac disturbances where dc special stability controls (modulation or fast power changes) are used — A response model is normally needed. More detailed representation of controls would be required in the case of weak ac systems. Normal power modulation can be counter-productive, and it may be necessary to model controls which decrease dc power during voltage dips (desynchronizing effect).

8.   For studies involving unbalanced ac disturbances and unbalances caused by the dc system, depending on the purpose of the study, an EMTP/transient stability combination model may be desirable.

REFERENCES

[1]   "Compendium of HVDC Schemes throughout the World," International Conference on Large High Voltage Electric Systems, CIGRE WG 04 of SC 14, 1987.

[2]   E.W. Kimbark, *Direct Current Transmission*, Wiley-Interscience, 1971.

[3]   E.Uhlmann, *Power Transmission by Direct Current*, Springer-Verlag, 1975.

[4]   C. Adamson and N.G. Hingorani, *High Voltage Direct Current Power Transmission*, Garraway Limited, London, 1960.

[5]   B.J. Cory (editor), *High Voltage Direct Current Converters and Systems*, Macdonald, London, 1965.

[6]   J. Arrillaga, *High Voltage Direct Current Transmission*, IEE Power Engineering Series 6, Peter Peregrinus Ltd., 1983.

[7]   EPRI Report EL-3004, "Methodology for Integration of HVDC Links in Large AC Systems - Phase 1: Reference Manual," Prepared by Ebasco Services Inc., March 1983.

[8]   EPRI Report EL-4365, "Methodology for the Integration of HVDC Links in Large AC Systems - Phase 2: Advanced Concepts," Prepared by Institut de Recherche d'Hydro-Quebec, April 1987.

[9] ANSI/IEEE Standard 1030-1987, *IEEE Guide for Specification of High-Voltage Direct Current Systems - Part I Steady-State Performance.*

[10] A. Ekstrom and G. Liss, "A Refined HVDC Control System," *IEEE Trans.*, Vol. PAS-89, pp. 723-732, May/June 1970.

[11] N.G. Hingorani and P. Chadwick, "A New Constant Extinction Angle Control for AC/DC/AC Static Converters," *IEEE Trans.*, Vol. PAS-87, pp. 866-872, March 1968.

[12] P.G. Engstrom, "Operation and Control of HVDC Transmission," *IEEE Trans.*, Vol. PAS-83, pp. 71-77, January 1964.

[13] J.D. Ainsworth, "The Phase-Locked Oscillator: A New Control System for Controlled Static Convertors," *IEEE Trans.*, Vol. PAS-87, pp. 857-865, March 1968.

[14] E. Rumpf and S. Ranade, "Comparison of Suitable Control Systems for HVDC Stations Connected to Weak AC Systems, Parts I and II," *IEEE Trans.*, Vol. PAS-91, pp. 549-564, March/April 1972.

[15] "High Voltage Direct Current Controls Guide," prepared by the HVDC Controls Committee, Canadian Electrical Association, October 1984.

[16] C.W. Taylor, *Power System Voltage Stability*, McGraw-Hill, 1993.

[17] IEEE Committee Report, "DC Transmission Terminating at Low Short Circuit Ratio Locations," *IEEE Trans.*, Vol. PWRD-1, No. 3, pp. 308-318, July 1986.

[18] CIGRE and IEEE Joint Task Force Report, "Guide for Planning DC Links Terminating at AC Locations Having Low Short-Circuit Capacities, Part I: AC/DC Interaction Phenomena," CIGRE Publication 68, June 1992.

[19] IEEE Committee Report, "Dynamic Performance Characteristics of North American HVDC Systems for Transient and Dynamic Stability Evaluations," *IEEE Trans.*, Vol. PAS-100, pp. 3356-3364, July 1981.

[20] IEEE Committee Report, "HVDC Controls for System Dynamic Performance," *IEEE Trans.*, Vol. PWRS-6, No. 2, pp. 743-752, May 1991.

[21] R. Jötten, J.P. Bowles, G. Liss, C.J.B. Martin, and E. Rumpf, "Control in HVDC Systems, The State of the Art, Part I: Two Terminal Systems," CIGRE Paper 14-10, 1978.

[22]   J. Hegi, M. Bahrman, G. Scott, and G. Liss, "Control of the Quebec-New England Multi-Terminal HVDC System," CIGRE Paper 14-04, Paris 1988.

[23]   J. Reeve, "Multiterminal HVDC Power Systems," *IEEE Trans.*, Vol. PAS-99, pp. 729-737, March/April 1980.

[24]   W.F. Long, J. Reeve, J.R. McNicol, R.E. Harrison, and J.P. Bowels, "Considerations for Implementing Multiterminal DC Systems," *IEEE Trans.*, Vol. PAS-104, pp. 2521-2530, September 1985.

[25]   F. Nozari, C.E. Grund, and R.L. Hauth, "Current Order Coordination in Multiterminal DC Systems," *IEEE Trans.*, Vol. PAS-100, pp. 4629-4636, November 1981.

[26]   J.J. Vithayathil, A.L. Courts, W.G. Peterson, N.G. Hingorani, S. Nilsson, and J.W. Porter, "HVDC Circuit Breaker Development and Field Tests," *IEEE Trans.*, Vol. PAS-104, pp. 2693-2705, October 1985.

[27]   B.K. Johnson, F.P. DeMello, and J.M. Undrill, "Comparing Fundamental Frequency and Differential Equation Representation of AC/DC," *IEEE Trans.*, Vol. PAS-101, pp. 3379-3384, September 1982.

[28]   G.D. Breuer, J.F. Luini, and C.C. Young, "Studies of Large AC/DC Systems on the Digital Computer," *IEEE Trans.*, Vol. PAS-85, pp. 1107-1115, November 1966.

[29]   J. Reeve, G. Fahmy, and B. Stott, "Versatile Load Flow Method for Multiterminal HVDC Systems," *IEEE Trans.*, Vol. PAS-96, pp. 925-933, May/June 1977.

[30]   M.M. El Marsafawy and R.M. Mathur, "A New, Fast Technique for Load Flow Solution of Integrated Multiterminal DC/AC Systems," *IEEE Trans.*, Vol. PAS-99, pp. 246-255, January/February 1980.

[31]   J.F. Clifford and A.H. Schmidt, "Digital Representation of a DC Transmission System and Its Controls," *IEEE Trans.*, Vol. PAS-89, pp. 97-105, January 1970.

[32]   N. Sato, N.V. Dravid, S.M. Chan, A.L. Burns, and J.J. Vithayathil, "Multiterminal HVDC System Representation in a Transient Stability Program," *IEEE Trans.*, Vol. PAS-99, pp. 1927-1936, September/October 1980.

[33]   G.K. Carter, C.E. Grund, H.H. Happ, and R.V. Pohl, "The Dynamics of

AC/DC Systems with Controlled Multiterminal HVDC Transmission," *IEEE Trans.*, Vol. PAS-96, pp. 402-413, March/April 1977.

[34]   R. Proulx, S. Lefebvre, A. Valette, D. Soulier, and A. Venne, "DC Control Modelling in a Transient Stability Program," *Proceedings of International Conference on DC Power Transmission*, Montreal, Canada, pp. 16-22, June 4-8, 1984.

[35]   T. Adielson, "Modelling of an HVDC System for Digital Simulation of AC/DC Transmission Interactions," Paper 100-02, *CIGRE Symposium on AC/DC Interactions and Comparisons*, Boston, September 28-30, 1987.

[36]   D.G. Chapman, J.B. Davies, F.L. Alvarado, R.H. Lasseter, S. Lefebvre, P. Krishnayya, J. Reeve, and N.J. Balu, "Programs for the Study of HVDC Systems," *IEEE Trans.*, Vol. PWRD-3, pp. 1182-1188, July 1988.

[37]   S. Arabi, G.J. Rogers, D.Y. Wong, P. Kundur, and M.G. Lauby, "Small Signal Stability Program Analysis of SVC and HVDC in AC Power Systems," *IEEE Trans.*, Vol. PWRS-6, No. 3, pp. 1147-1153, August 1991.

[38]   H.W. Dommel, "Digital Computer Solution of Electromagnetic Transients in Single- and Multi-Phase Networks," *IEEE Trans.*, Vol. PAS-88, pp. 388-399, April 1969.

[39]   J. Reeve and R. Adapa, "Advances in AC Transient Stability Studies Including DC Modelling," Fourth International Conference on AC and DC Transmission, London, September 23-26, 1985.

Chapter 11

Control of Active Power and Reactive Power

So far this book has concentrated on the characteristics and modelling of individual elements of the power systems. This chapter will examine control of active power and reactive power by considering the system as an entity. In addition, the characteristics and modelling of equipment used for control will be described.

The flows of active power and reactive power in a transmission network are fairly independent of each other and are influenced by different control actions (see Section 6.3). Hence, they may be studied separately for a large class of problems. Active power control is closely related to frequency control, and reactive power control is closely related to voltage control. As constancy of frequency and voltage are important factors in determining the quality of power supply, the control of active power and reactive power is vital to the satisfactory performance of power systems.

11.1 ACTIVE POWER AND FREQUENCY CONTROL

For satisfactory operation of a power system, the frequency should remain nearly constant. Relatively close control of frequency ensures constancy of speed of induction and synchronous motors. Constancy of speed of motor drives is particularly important for satisfactory performance of generating units as they are highly dependent on the performance of all the auxiliary drives associated with the fuel, the feed-water and the combustion air supply systems. In a network, considerable drop in frequency could result in high magnetizing currents in induction motors and transformers. The extensive use of electric clocks and the use of frequency for other timing purposes require accurate maintenance of synchronous time which is proportional to integral of frequency. As a consequence, it is necessary to regulate not only the frequency itself but also its integral.

The frequency of a system is dependent on active power balance. As frequency is a common factor throughout the system, a change in active power demand at one point is reflected throughout the system by a change in frequency. Because there are many generators supplying power into the system, some means must be provided to allocate change in demand to the generators. A speed governor on each generating unit provides the primary speed control function, while supplementary control originating at a central control centre allocates generation.

In an interconnected system with two or more independently controlled areas, in addition to control of frequency, the generation within each area has to be controlled so as to maintain scheduled power interchange. The control of generation and frequency is commonly referred to as *load-frequency control* (LFC).

We will first review the requirements for primary speed governing and then discuss supplementary control.

11.1.1 Fundamentals of Speed Governing

The basic concepts of speed governing are best illustrated by considering an isolated generating unit supplying a local load as shown in Figure 11.1.

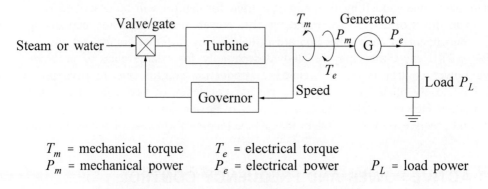

T_m = mechanical torque T_e = electrical torque
P_m = mechanical power P_e = electrical power P_L = load power

Figure 11.1 Generator supplying isolated load

Generator response to load change

When there is a load change, it is reflected instantaneously as a change in the electrical torque output T_e of the generator. This causes a mismatch between the mechanical torque T_m and the electrical torque T_e which in turn results in speed variations as determined by the equation of motion. As shown in Chapter 3 (Section 3.9), the following transfer function represents the relationship between rotor speed as a function of the electrical and mechanical torques.

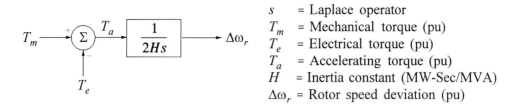

$$
\begin{aligned}
s &= \text{Laplace operator} \\
T_m &= \text{Mechanical torque (pu)} \\
T_e &= \text{Electrical torque (pu)} \\
T_a &= \text{Accelerating torque (pu)} \\
H &= \text{Inertia constant (MW-Sec/MVA)} \\
\Delta\omega_r &= \text{Rotor speed deviation (pu)}
\end{aligned}
$$

Figure 11.2 Transfer function relating speed and torques

For load-frequency studies, it is preferable to express the above relationship in terms of mechanical and electrical power rather than torque. The relationship between power P and torque T is given by

$$P = \omega_r T \tag{11.1}$$

By considering a small deviation (denoted by prefix Δ) from initial values (denoted by subscript 0), we may write

$$
\begin{aligned}
P &= P_0 + \Delta P \\
T &= T_0 + \Delta T \\
\omega_r &= \omega_0 + \Delta\omega_r
\end{aligned} \tag{11.2}
$$

From Equation 11.1,

$$P_0 + \Delta P = (\omega_0 + \Delta\omega_r)(T_0 + \Delta T)$$

The relationship between the perturbed values, with higher-order terms neglected, is given by

$$\Delta P = \omega_0 \Delta T + T_0 \Delta\omega_r \tag{11.3}$$

Therefore,

$$\Delta P_m - \Delta P_e = \omega_0(\Delta T_m - \Delta T_e) + (T_{m0} - T_{e0})\Delta\omega_r \tag{11.4}$$

Since, in the steady state, electrical and mechanical torques are equal, $T_{m0} = T_{e0}$. With speed expressed in pu, $\omega_0 = 1$. Hence,

$$\Delta P_m - \Delta P_e \ = \ \Delta T_m - \Delta T_e \tag{11.5}$$

Figure 11.2 can now be expressed in terms of ΔP_m and ΔP_e as follows:

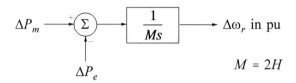

Figure 11.3 Transfer function relating speed and power

Within the range of speed variations with which we are concerned, the turbine mechanical power is essentially a function of valve or gate position and independent of frequency.

Load response to frequency deviation

In general, power system loads are a composite of a variety of electrical devices. For resistive loads, such as lighting and heating loads, the electrical power is independent of frequency. In the case of motor loads, such as fans and pumps, the electrical power changes with frequency due to changes in motor speed. The overall frequency-dependent characteristic of a composite load may be expressed as

$$\Delta P_e \ = \ \Delta P_L + D\Delta\omega_r \tag{11.6}$$

where

ΔP_L = non-frequency-sensitive load change
$D\Delta\omega_r$ = frequency-sensitive load change
D = load-damping constant

The damping constant is expressed as a percent change in load for one percent change in frequency. Typical values of D are 1 to 2 percent. A value of $D=2$ means that a 1% change in frequency would cause a 2% change in load.

The system block diagram including the effect of the load damping is shown in Figure 11.4.

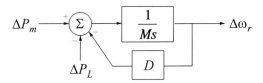

Figure 11.4

This may be reduced to the form shown in Figure 11.5.

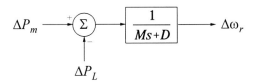

Figure 11.5

In the absence of a speed governor, the system response to a load change is determined by the inertia constant and the damping constant. The steady-state speed deviation is such that the change in load is exactly compensated by the variation in load due to frequency sensitivity. This is illustrated in the following example.

Example 11.1

A small system consists of 4 identical 500 MVA generating units feeding a total load of 1,020 MW. The inertia constant H of each unit is 5.0 on 500 MVA base. The load varies by 1.5% for a 1% change in frequency. When there is a sudden drop in load by 20 MW,

(a)   Determine the system block diagram with constants H and D expressed on 2,000 MVA base.

(b)   Find the frequency deviation, assuming that there is no speed-governing action.

Solution

(a) For 4 units on 2,000 MVA base, $H = 5.0 \times (500/2000) \times 4 = 5.0$. Hence,

$$M = 2H = 10.0 \text{ s}$$

Expressing D for the remaining load ($1020-20=1000$ MW) on 2,000 MVA base,

$$D = 1.5 \times 1000/\ 2000 = 0.75\%$$

(b) With $\Delta P_m = 0$ (no speed governing), the system block diagram with parameters expressed in pu on 2,000 MVA is

$$-\Delta P_L \longrightarrow \boxed{\dfrac{1}{Ms+D}} \longrightarrow \Delta\omega_r$$

This may be expressed in the standard form in terms of a gain and a time constant:

$$-\Delta P_L \longrightarrow \boxed{\dfrac{K}{1+sT}} \longrightarrow \Delta\omega_r$$

where

$$K = \frac{1}{D} = \frac{1}{0.75} = 1.33$$

$$T = \frac{M}{D} = \frac{10}{0.75} = 13.33 \ \ \text{s}$$

The load change is

$$\Delta P_L = -20 \ \ \text{MW}$$

$$= \frac{-20}{2000} = -0.01 \ \ \text{pu}$$

For a step reduction in load by 0.01 pu, Laplace transform of the change in load is

$$\Delta P_L(s) = \frac{-0.01}{s}$$

Hence, from the block diagram

$$\Delta\omega_r(s) = -\left(\frac{-0.01}{s}\right)\left(\frac{K}{1+sT}\right)$$

Taking the inverse transform,

$$\Delta\omega_r(t) = -0.01Ke^{-\frac{t}{T}} + 0.01K$$

$$= -0.01 \times 1.33\, e^{-\frac{t}{13.33}} + 0.01 \times 1.33$$

$$= -0.0133\, e^{-0.075t} + 0.0133$$

The pu speed deviation as function of time is shown in the following figure.

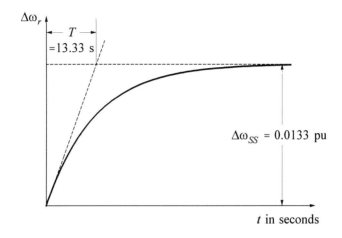

The time constant T is 13.33 s and the steady-state speed deviation is

$$\Delta\omega_{ss} = -\frac{\Delta P_L}{D} = 0.0133 \text{ pu}$$

$$= 0.0133 \times 60 = 0.8 \text{ Hz} \qquad\blacksquare$$

Isochronous governor [1,2]

The adjective isochronous means constant speed. An isochronous governor adjusts the turbine valve/gate to bring the frequency back to the nominal or scheduled value. Figure 11.6 shows the schematic of such a speed-governing system. The measured rotor speed ω_r is compared with reference speed ω_0. The error signal (equal to speed deviation) is amplified and integrated to produce a control signal ΔY which actuates the main steam supply valves in the case of a steam turbine, or gates in the case of a hydraulic turbine. Because of the reset action of this integral controller, ΔY will reach a new steady state only when the speed error $\Delta\omega_r$ is zero.

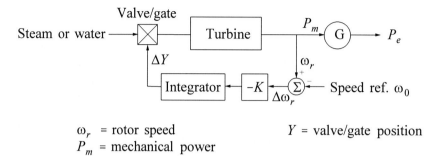

ω_r = rotor speed Y = valve/gate position
P_m = mechanical power

Figure 11.6 Schematic of an isochronous governor

Figure 11.7 shows the time response of a generating unit, with an isochronous governor, when subjected to an increase in load. The increase in P_e causes the frequency to decay at a rate determined by the inertia of rotor. As the speed drops, the turbine mechanical power begins to increase. This in turn causes a reduction in the rate of decrease of speed, and then an increase in speed when the turbine power is in excess of the load power. The speed will ultimately return to its reference value and the steady-state turbine power increases by an amount equal to the additional load.

An isochronous governor works satisfactorily when a generator is supplying an isolated load or when only one generator in a multigenerator system is required to respond to changes in load. For power load sharing between generators connected to the system, speed regulation or droop characteristic must be provided as discussed next.

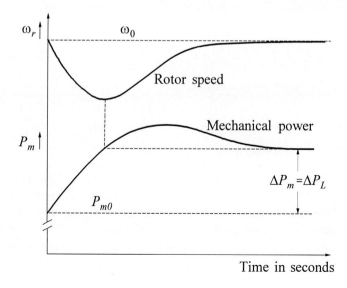

Figure 11.7 Response of generating unit with isochronous governor

Governors with speed-droop characteristic [1,2]

The isochronous governors cannot be used when there are two or more units connected to the same system since each generator would have to have precisely the same speed setting. Otherwise, they would fight each other, each trying to control system frequency to its own setting. For stable load division between two or more units operating in parallel, the governors are provided with a characteristic so that the speed drops as the load is increased.

The speed-droop or regulation characteristic may be obtained by adding a steady-state feedback loop around the integrator as shown in Figure 11.8.

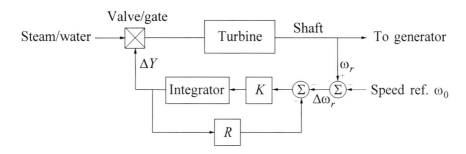

Figure 11.8 Governor with steady-state feedback

The transfer function of the governor of Figure 11.8 reduces to the form shown in Figure 11.9. This type of governor is characterized as a *proportional controller* with a gain of $1/R$.

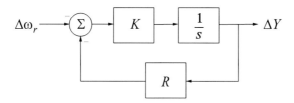

(a) Block diagram with steady-state feedback

$$T_G = \frac{1}{KR}$$

(b) Reduced block diagram

Figure 11.9 Block diagram of a speed governor with droop

Percent speed regulation or droop:

The value of R determines the steady-state speed versus load characteristic of the generating unit as shown in Figure 11.10. The ratio of speed deviation ($\Delta\omega_r$) or frequency deviation (Δf) to change in valve/gate position (ΔY) or power output (ΔP) is equal to R. The parameter R is referred to as speed regulation or droop. It can be expressed in percent as

$$\text{Percent } R = \frac{\text{percent speed or frequency change}}{\text{percent power output change}} \times 100$$

$$= \left(\frac{\omega_{NL} - \omega_{FL}}{\omega_0}\right) \times 100$$

where

ω_{NL} = steady-state speed at no load
ω_{FL} = steady-state speed at full load
ω_0 = nominal or rated speed

For example, a 5% droop or regulation means that a 5% frequency deviation causes 100% change in valve position or power output.

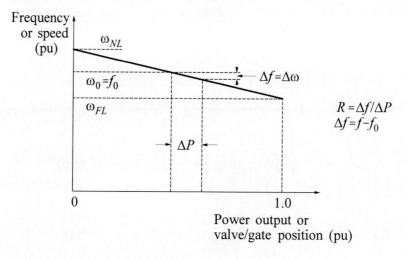

Figure 11.10 Ideal steady-state characteristics
of a governor with speed droop

Load sharing by parallel units:

If two or more generators with drooping governor characteristics are connected to a power system, there will be a unique frequency at which they will share a load change. Consider two units with droop characteristics as shown in Figure 11.11. They are initially at nominal frequency f_0, with outputs P_1 and P_2. When a load increase ΔP_L causes the units to slow down, the governors increase output until they reach a new common operating frequency f'. The amount of load picked up by each unit depends on the droop characteristic:

$$\Delta P_1 = P_1' - P_1 = \frac{\Delta f}{R_1}$$

$$\Delta P_2 = P_2' - P_2 = \frac{\Delta f}{R_2}$$

Hence,

$$\frac{\Delta P_1}{\Delta P_2} = \frac{R_2}{R_1}$$

If the percentages of regulation of the units are nearly equal, the change in the outputs of each unit will be nearly in proportion to its rating.

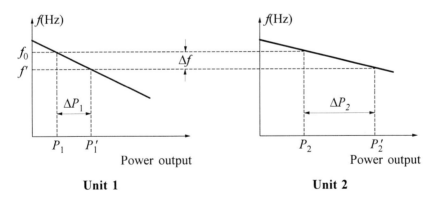

Figure 11.11 Load sharing by parallel units with drooping governor characteristics

Time response

Figure 11.12 shows the time response of a generating unit, with a speed-droop governor, when subjected to an increase in load. Because of the droop characteristic, the increase in power output is accompanied by a steady-state speed or frequency deviation ($\Delta \omega_{SS}$).

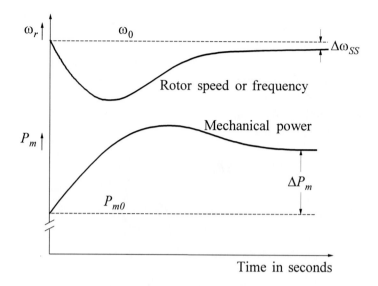

Figure 11.12   Response of a generating unit with a governor
having speed-droop characteristic

11.1.2 Control of Generating Unit Power Output

The relationship between speed and load can be adjusted by changing an input shown as "load reference setpoint" in Figure 11.13.

In practice, the adjustment of load reference setpoint is accomplished by operating the "speed-changer motor." The effect of this adjustment is depicted in Figure 11.14, which shows a family of parallel characteristics for different speed-changer motor settings. The characteristics shown are for a governor associated with a 60 Hz system. Three characteristics are shown representing three load reference settings. At 60 Hz, characteristic A results in zero output, characteristic B results in 50% output, and characteristic C results in 100% output. Thus, the power output of the generating unit at a given speed may be adjusted to any desired value by adjusting the load reference setting through actuation of the speed-changer motor. For each setting, the speed-load characteristic has a 5% droop; that is, a speed change of 5% (3 Hz) causes a 100% change in power output.

When two or more generators are operating in parallel, the speed-droop characteristic (corresponding to a load reference setting) of each generating unit merely establishes the proportion of the load picked up by the unit when a sudden change in system load occurs. The output of each unit at any given system frequency can be varied only by changing its load reference, which in effect moves the speed-droop characteristic up and down.

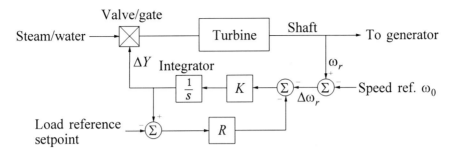

(a) Schematic diagram of governor and turbine

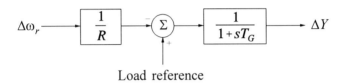

(b) Reduced block diagram of governor

Figure 11.13 Governor with load reference control
for adjusting speed-load relationship

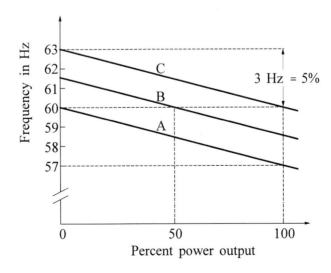

Figure 11.14 Effect of speed-changer setting on governor characteristic

When a generating unit is feeding an isolated load, the adjustment of the speed changer changes the unit speed. However, when the unit is synchronized to a power system, the speed-changer adjustment changes the unit power output; it has only a minor effect on system frequency, depending on the size of the unit relative to that of the total system generation.

Actual speed-droop characteristic [1]

The governor speed-droop characteristic we have considered so far (Figures 11.10 and 11.14) represents the ideal relationship. In actual practice the characteristic departs from the straight-line relationship as depicted in Figure 11.15.

As discussed in Chapter 9, steam turbines have a number of control valves, each having nonlinear flow area versus position characteristic. Hence, they have the speed-droop characteristics of the general nature of curve 2 in Figure 11.15. Each section of curve 2 represents the effect of one control valve. Hydraulic turbines, which have a single gate, tend to have the characteristic similar to curve 3.

The actual speed-droop characteristic may thus exhibit *incremental regulation* ranging from 2% to 12%, depending on the unit output. Modern electrohydraulic governing systems minimize these variations in incremental regulation by using linearizing circuits or first-stage pressure feedback.

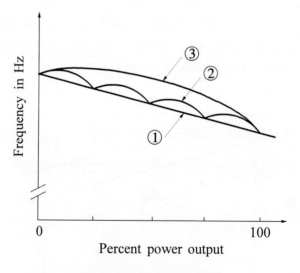

Curve 1: Ideal linear characteristic
Curve 2: Actual characteristic for steam units
Curve 3: Actual characteristic for hydraulic units

Figure 11.15 Actual and ideal governor speed-droop characteristics

11.1.3 Composite Regulating Characteristic of Power Systems

In the analysis of load-frequency controls (LFCs), we are interested in the collective performance of all generators in the system. The intermachine oscillations and transmission system performance are therefore not considered. We tacitly assume the coherent response of all generators to changes in system load and represent them by an equivalent generator. The equivalent generator has an inertia constant M_{eq} equal to the sum of the inertia constants of all the generating units and is driven by the combined mechanical outputs of the individual turbines as illustrated in Figure 11.16. Similarly, the effects of the system loads are lumped into a single damping constant D. The speed of the equivalent generator represents the system frequency, and in per unit the two are equal. We will therefore use rotor speed and frequency interchangeably in our discussion of load-frequency control.

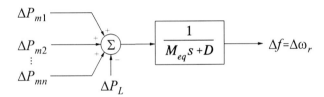

Figure 11.16 System equivalent for LFC analysis

The composite power/frequency characteristic of a power system thus depends on the combined effect of the droops of all generator speed governors. It also depends on the frequency characteristics of all the loads in the system. For a system with n generators and a composite load-damping constant of D, the steady-state frequency deviation following a load change ΔP_L is given by

$$\Delta f_{ss} = \frac{-\Delta P_L}{(1/R_1 + 1/R_2 + \cdots + 1/R_n) + D}$$

$$= \frac{-\Delta P_L}{1/R_{eq} + D} \tag{11.7}$$

where

$$R_{eq} = \frac{1}{1/R_1 + 1/R_2 + \cdots + 1/R_n} \tag{11.8}$$

Thus, the composite frequency response characteristic of the system is

$$\beta = \frac{-\Delta P_L}{\Delta f_{ss}} = \frac{1}{R_{eq}} + D \qquad (11.9)$$

The composite frequency response characteristic β is normally expressed in MW/Hz. It is also sometimes referred to as the *stiffness* of the system. The composite regulating characteristic of the system is equal to $1/\beta$.

The effects of governor speed droop and the frequency sensitivity of load on the net frequency change are illustrated in Figure 11.17, which considers the composite effects of all the generating units and load in the system. An increase of system load by ΔP_L (at nominal frequency) results in a total generation increase of ΔP_G due to governor action and a total system load reduction of ΔP_D due to its frequency-sensitive characteristic.

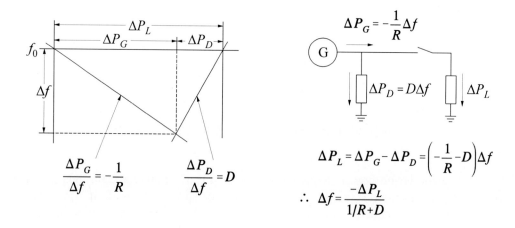

Figure 11.17 Composite governor and load characteristic

Example 11.2

A power system has a total load of 1,260 MW at 60 Hz. The load varies 1.5% for every 1% change in frequency ($D=1.5$). Find the steady-state frequency deviation when a 60 MW load is suddenly tripped, if

(a)   There is no speed control.

(b)   The system has 240 MW of spinning reserve evenly spread among 500 MW of generation capacity with 5% regulation based on this capacity. All other generators are operating with valves wide open. Assume that the effect of governor dead bands is such that only 80% of the governors respond to the reduction in system load.

Neglect the effects of transmission losses.

Solution

Total remaining load is $1260 - 60 = 1200$ MW. The damping constant of remaining load is

$$D = \left(\frac{1.5}{100} \times 1200\right) \times \left(\frac{100}{60 \times 1}\right) = 30 \quad \text{MW/Hz}$$

(a) With no speed control, the resulting increase in steady-state frequency is

$$\Delta f = \frac{-\Delta P_L}{D} = \frac{-(-60) \text{ MW}}{30 \text{ MW/Hz}}$$

$$= 2.0 \quad \text{Hz}$$

(b) Since there is a reduction in system load and an increase in frequency, all generating units (not just those on spinning reserve) respond. However, due to the effects of dead band, only 80% of the total generation contributes to speed regulation.

The total spinning generation capacity is equal to

$$\text{Load} + \text{reserve} = 1260 + 240 = 1500 \text{ MW}$$

Generation contributing to regulation is

$$0.8 \times 1500 = 1200 \text{ MW}$$

A regulation of 5% means that a 5% change in frequency causes a 100% change in power generation. Therefore,

$$\frac{1}{R} = \frac{1200}{(5/100) \times 60} = 400 \quad \text{MW/Hz}$$

The composite system frequency response characteristic is

$$\beta = \frac{1}{R} + D = 400 + 30$$

$$= 430 \quad \text{MW/Hz}$$

Steady-state increase in frequency is

$$\Delta f = \frac{-\Delta P_L}{\beta} = \frac{-(-60) \text{ MW}}{430 \text{ MW/Hz}}$$

$$= 0.1395 \quad \text{Hz} \qquad\qquad\qquad \blacksquare$$

11.1.4 Response Rates of Turbine-Governing Systems

So far, we have analyzed steady-state performance of speed-governing systems. We will now examine the relative response rates of steam and hydraulic turbines and their governing systems.

As discussed in Chapter 9, steam turbines may be of either reheat type or non-reheat type. Figure 11.18 shows the block diagram of a generating unit with a reheat turbine. The block diagram includes representation of the speed governor, turbine, rotating mass and load, appropriate for load-frequency analysis. The turbine representation is based on the simplified transfer function developed in Chapter 9 (Section 9.2.1), and assumes constant boiler pressure.

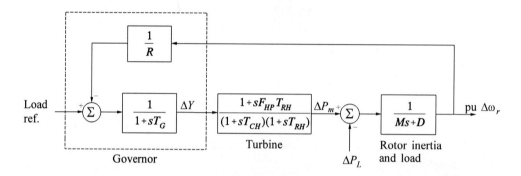

Typical values:

$$R = 0.05 \qquad T_G = 0.2 \text{ s} \qquad F_{HP} = 0.3 \qquad T_{RH} = 7.0 \text{ s}$$
$$T_{CH} = 0.3 \text{ s} \qquad F_{LP} = 0.7 \qquad M = 10.0 \text{ s} \qquad D = 1.0$$

Figure 11.18 Block diagram of a generating unit with a reheat steam turbine

The block diagram of Figure 11.18 is also applicable to a unit with non-reheat turbine. However, in this case $T_{RH}=0$, and the turbine transfer function simplifies to that shown in Figure 11.19.

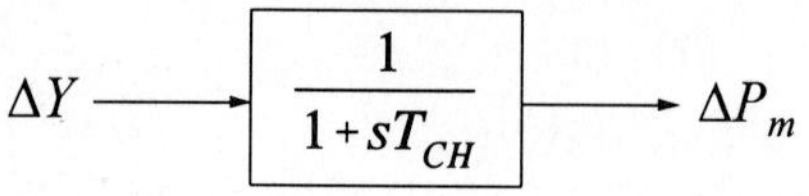

$$\Delta Y \longrightarrow \boxed{\dfrac{1}{1+sT_{CH}}} \longrightarrow \Delta P_m$$

Typical value:

$$T_{CH} = 0.3 \text{ s}$$

Figure 11.19 Non-reheat turbine transfer function

In Chapter 9 (Section 9.1.3), we showed that the governors of hydraulic units require *transient droop compensation* for stable speed control performance. Because a change in the position of the gate at the foot of the penstock produces an initial short-term turbine power change which is opposite to that sought, hydro turbine governors are designed to have relatively large transient droop, with long resetting times. This ensures stable frequency regulation under isolated operating conditions (islanding). Consequently, the response of a hydraulic unit to speed change or to changes in speed-changer setting is relatively slow.

The block diagram of a generating unit with a hydraulic turbine is shown in Figure 11.20. The governor includes transient droop. The transfer functions for the turbine and the speed governor used in the block diagram were developed in Chapter 9. For fast-frequency deviations, the governor exhibits a high regulation (or low gain); for slow changes and in the steady state it has a lower regulation (or high gain).

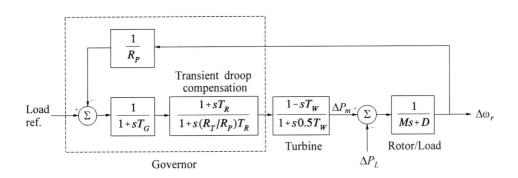

Typical values:

$R_P = 0.05$ $T_G = 0.2$ s $M = 6.0$ s $D = 1.0$
$T_W = 1.0$ s $R_T = 0.38$ $T_R = 5.0$ s

Figure 11.20 Block diagram of hydraulic unit

The nature of the responses of generating units with reheat and non-reheat steam turbines, and hydraulic turbines when subjected to a step change in load (ΔP_L), is illustrated in Figure 11.21. These responses have been computed by using the linearized models and typical parameters shown in Figures 11.18 and 11.19. Constant boiler pressure has been assumed for the steam turbines. Depending on the type of boiler and its controls, and on the size of the load change, the response of the steam turbines may be significantly slower than that shown. On the other hand, a low-head hydraulic unit may have a significantly faster response than the one considered here.

The results presented here demonstrate that, although the steady-state speed deviation is the same for all three units considered, there are significant differences in their transient responses. Unit response characteristics, in fact, vary widely

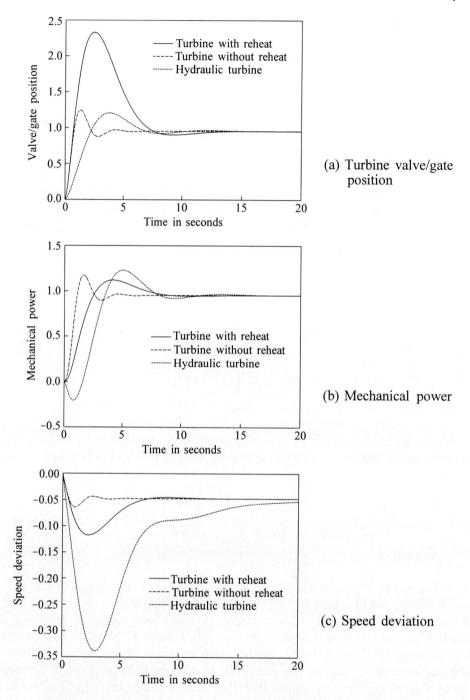

(a) Turbine valve/gate position

(b) Mechanical power

(c) Speed deviation

Figure 11.21 Responses of steam and hydraulic generating units to a small step increase in load demand; values shown are in per unit of the step change.

depending on many factors. These include, in addition to the type of plant, plant controls and mode of operation (e.g., boiler-follow, turbine-follow), and operating point (e.g., nearness to valve point, load limit).

11.1.5 Fundamentals of Automatic Generation Control [3-6]

With primary speed control action, a change in system load will result in a steady-state frequency deviation, depending on the governor droop characteristic and frequency sensitivity of the load. All generating units on speed governing will contribute to the overall change in generation, irrespective of the location of the load change. Restoration of system frequency to nominal value requires supplementary control action which adjusts the load reference setpoint (through the speed-changer motor). Therefore, the basic means of controlling prime-mover power to match variations in system load in a desired manner is through control of the load reference setpoints of selected generating units. As system load is continually changing, it is necessary to change the output of generators automatically.

The primary objectives of *automatic generation control* (AGC) are to regulate frequency to the specified nominal value and to maintain the interchange power between control areas at the scheduled values by adjusting the output of selected generators. This function is commonly referred to as *load-frequency control* (LFC). A secondary objective is to distribute the required change in generation among units to minimize operating costs.

AGC in isolated power systems

In an isolated power system, maintenance of interchange power is not an issue. Therefore, the function of AGC is to restore frequency to the specified nominal value. This is accomplished by adding a reset or integral control which acts on the load reference settings of the governors of units on AGC, as shown in Figure 11.22. The integral control action ensures zero frequency error in the steady state.

The supplementary generation control action is much slower than the primary speed control action. As such it takes effect after the primary speed control (which acts on all units on regulation) has stabilized the system frequency. Thus, AGC adjusts load reference settings of selected units, and hence their output power, to override the effects of the composite frequency regulation characteristics of the power system. In so doing, it restores the generation of all other units not on AGC to scheduled values.

AGC in interconnected power systems

To form the basis for supplementary control of interconnected power systems, let us first look at the performance with primary speed control only.

Consider the interconnected system shown in Figure 11.23(a). It consists of two areas connected by a tie line of reactance X_{tie}. For load-frequency studies, each area may be represented by an equivalent generating unit exhibiting its overall

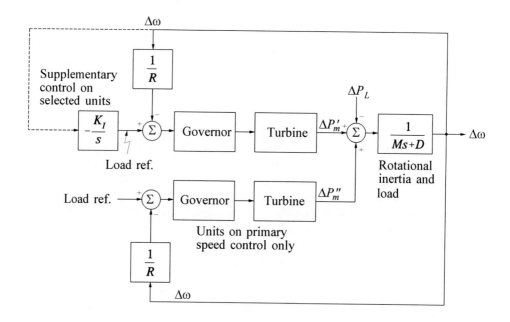

Figure 11.22 Addition of integral control on generating units selected for AGC

performance. Such composite models are acceptable since we are not concerned about intermachine oscillations within each area.

Figure 11.23(b) shows the electrical equivalent of the system, with each area represented by a voltage source behind an equivalent reactance as viewed from the tie bus. The power flow on the tie line from area 1 to area 2 is

$$P_{12} = \frac{E_1 E_2}{X_T} \sin(\delta_1 - \delta_2)$$

Linearizing about an initial operating point represented by $\delta_1 = \delta_{10}$ and $\delta_2 = \delta_{20}$, we have

$$\Delta P_{12} = T \Delta \delta_{12}$$

where $\Delta \delta_{12} = \Delta \delta_1 - \Delta \delta_2$, and T is the synchronizing torque coefficient given by

$$T = \frac{E_1 E_2}{X_T} \cos(\delta_{10} - \delta_{20}) \tag{11.10}$$

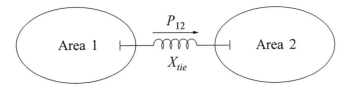

(a) Two-area system

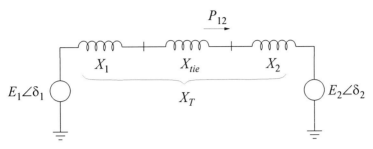

(b) Electrical equivalent

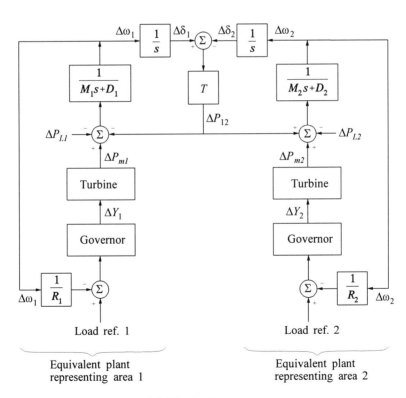

(c) Block diagram

Figure 11.23 Two-area system with only primary speed control

The block diagram representation of the system is shown in Figure 11.23(c) with each area represented by an equivalent inertia M, load-damping constant D, turbine, and governing system with an effective speed droop R. The tie line is represented by the synchronizing torque coefficient T. A positive ΔP_{12} represents an increase in power transfer from area 1 to area 2. This in effect is equivalent to increasing the load of area 1 and decreasing the load of area 2; therefore, feedback of ΔP_{12} has a negative sign for area 1 and a positive sign for area 2.

The steady-state frequency deviation $(f\text{-}f_0)$ is the same for the two areas. For a total load change of ΔP_L,

$$\Delta f = \Delta \omega_1 = \Delta \omega_2 = \frac{-\Delta P_L}{(1/R_1 + 1/R_2) + (D_1 + D_2)} \tag{11.11}$$

Consider the steady-state values following an increase in area 1 load by ΔP_{L1}. For area 1, we have

$$\Delta P_{m1} - \Delta P_{12} - \Delta P_{L1} = \Delta f D_1 \tag{11.12}$$

and for area 2,

$$\Delta P_{m2} + \Delta P_{12} = \Delta f D_2 \tag{11.13}$$

The change in mechanical power depends on regulation. Hence,

$$\Delta P_{m1} = -\frac{\Delta f}{R_1} \tag{11.14}$$

$$\Delta P_{m2} = -\frac{\Delta f}{R_2} \tag{11.15}$$

Substitution of Equation 11.14 in Equation 11.12 and Equation 11.15 in Equation 11.13 yields

$$\Delta f \left(\frac{1}{R_1} + D_1 \right) = -\Delta P_{12} - \Delta P_{L1} \tag{11.16}$$

and

$$\Delta f \left(\frac{1}{R_2} + D_2 \right) = \Delta P_{12} \tag{11.17}$$

Solving Equations 11.16 and 11.17, we get

$$\Delta f = \frac{-\Delta P_{L1}}{(1/R_1 + D_1) + (1/R_2 + D_2)} = \frac{-\Delta P_{L1}}{\beta_1 + \beta_2} \tag{11.18}$$

and

$$\Delta P_{12} = \frac{-\Delta P_{L1}(1/R_2 + D_2)}{(1/R_1 + D_1) + (1/R_2 + D_2)} = \frac{-\Delta P_{L1}\beta_2}{\beta_1 + \beta_2} \tag{11.19}$$

where β_1 and β_2 are the composite frequency response characteristics of areas 1 and 2, respectively. The above relationships are depicted in Figure 11.24.

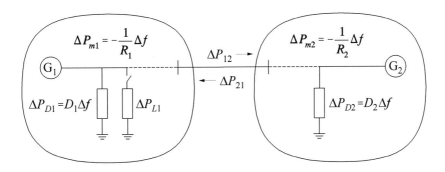

Figure 11.24 Effect of change in area 1 load

An increase in area 1 load by ΔP_{L1} results in a frequency reduction in both areas and a tie line flow of ΔP_{12}. A negative ΔP_{12} is indicative of flow from area 2 to area 1. The tie line flow deviation reflects the contribution of the regulation characteristics $(1/R + D)$ of one area to another.

Similarly, for a change in area 2 load by ΔP_{L2}, we have

$$\Delta f = \frac{-\Delta P_{L2}}{\beta_1 + \beta_2} \tag{11.20}$$

and

$$\Delta P_{12} = -\Delta P_{21} = \frac{\Delta P_{L2}\beta_1}{\beta_1 + \beta_2} \tag{11.21}$$

The above relationships form the basis for the load-frequency control of interconnected systems.

Frequency bias tie line control

The basic objective of supplementary control is to restore balance between each area load and generation. This is met when the control action maintains

- Frequency at the scheduled value
- Net interchange power with neighbouring areas at scheduled values

The supplementary control in a given area should ideally correct only for changes in that area. In other words, if there is a change in area 1 load, there should be supplementary control action only in area 1 and not in area 2.

Examination of Equations 11.18 to 11.21 indicates that a control signal made up of tie line flow deviation added to frequency deviation weighted by a *bias factor* would accomplish the desired objectives. This control signal is known as *area control error* (ACE).

From Equations 11.16 and 11.17, it is apparent that a suitable bias factor for an area is its frequency-response characteristic β. Thus, the area control error for area 2 is

$$ACE_2 = \Delta P_{21} + B_2 \Delta f \qquad (11.22)$$

where

$$B_2 = \beta_2 = \frac{1}{R_2} + D_2 \qquad (11.23)$$

Similarly, for area 1

$$ACE_1 = \Delta P_{12} + B_1 \Delta f \qquad (11.24)$$

where

$$B_1 = \beta_1 = \frac{1}{R_1} + D_1 \qquad (11.25)$$

The ACE represents the required change in area generation, and its unit is

MW.[1] The unit normally used for expressing the frequency bias factor B is MW/0.1 Hz.

The block diagram shown in Figure 11.25 illustrates how supplementary control is implemented. It is applied to selected units in each area and acts on the load reference setpoints.

The area frequency-response characteristic $(1/R+D)$ required for establishing the bias factors can be estimated by examination of chart records following a significant disturbance such as a sudden loss of a large unit.

Basis for selection of bias factor [3-8]

Actually, from steady-state performance considerations, the choice of bias factor is not important. Any combination of area control errors containing components of tie line power deviation and frequency deviation will result in steady-state restoration of the tie flow and frequency since the integral control action ensures that ACE is reduced to zero. In order to illustrate this, consider the following area control error signals applicable to a two-area system:

$$\text{ACE}_1 = A_1 \Delta P_{12} + B_1 \Delta f = 0 \tag{11.26}$$

$$\text{ACE}_2 = A_2 \Delta P_{21} + B_2 \Delta f = 0 \tag{11.27}$$

The above equations result in $\Delta P_{12}=0$ and $\Delta f=0$ for all non-zero values of A_1, A_2, B_1 and B_2.

However, the composition of area control error signals is more important from *dynamic performance* considerations. This can be illustrated by considering the transient response of the AGC system to a sudden increase in area 1 load. The sudden increase in load will result in a decrease in system frequency, followed by governor response (primary speed control of units in both areas) which limits the maximum frequency excursion and subsequently (typically on the order of 10 seconds) brings the frequency deviation back to a value Δf_R determined by the regulation characteristics of both systems:

$$\Delta f_R = \frac{-\Delta P_{L1}}{\beta_1 + \beta_2} \tag{11.28}$$

---

[1] In the literature, several alternative equations have been used to express ACE. A commonly used alternative form is

$$\text{ACE} = \Delta P_{tie} - B\Delta f$$

where $B = -\beta$, a negative number.

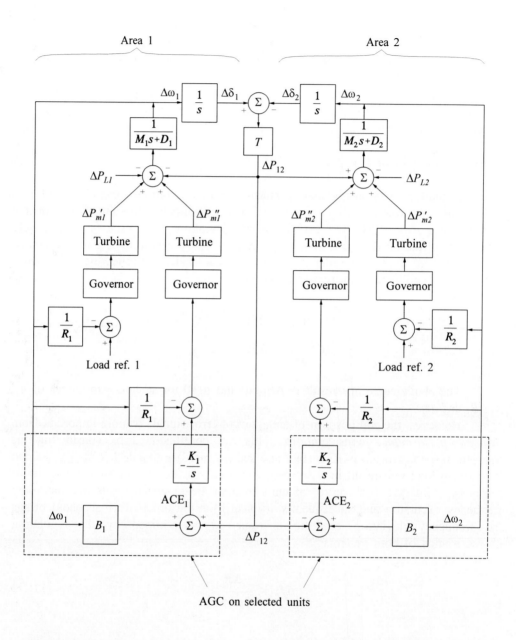

Figure 11.25 Block diagram of two-area system with supplementary control

At this point, there is a deviation of tie line power flow from the scheduled value. The supplementary control, which is much slower than the primary speed control, will now commence responding. Let us examine the performance of the supplementary control for different settings of the area frequency bias factors at the instant when the frequency deviation is Δf_R.

(a)   With B_1 equal to β_1, and B_2 equal to β_2, we have

$$\begin{aligned}
\text{ACE}_1 &= \Delta P_{12} + B_1 \Delta f_R \\
&= \frac{-\Delta P_{L1}}{\beta_1 + \beta_2}(\beta_2 + \beta_1) \\
&= -\Delta P_{L1}
\end{aligned}$$

and

$$\begin{aligned}
\text{ACE}_2 &= -\Delta P_{12} + B_2 \Delta f_R \\
&= \frac{-\Delta P_{L1}}{\beta_1 + \beta_2}(-\beta_2 + \beta_2) \\
&= 0
\end{aligned}$$

Only the supplementary control in area 1 will respond to ΔP_{L1} and change generation so as to bring ACE_1 to zero. The load change in area 1 is thus unobservable to the supplementary control in area 2.

(b)   If, on the other hand, B_1 and B_2 were set to double their respective area frequency-response characteristics,

$$\begin{aligned}
\text{ACE}_1 &= \Delta P_{12} + B_1 \Delta f_R \\
&= \frac{-\Delta P_{L1}}{\beta_1 + \beta_2}(\beta_2 + 2\beta_1) \\
&= -\Delta P_{L1}\left(1 - \frac{1}{\beta_2}\right)
\end{aligned}$$

Similarly,

$$\begin{aligned}
\text{ACE}_2 &= -\Delta P_{12} + 2\beta_2 \Delta f_R \\
&= \frac{-\Delta P_{L1}}{\beta_2}
\end{aligned}$$

Thus, both area 1 and area 2 supplementary controls would respond and correct the frequency deviation twice as fast. However, the generation picked up by area 2 will subsequently reflect itself as a component of ACE_2, and will be backed off again in the steady state.

(c)   If we set the bias factors significantly lower than the respective area βs, a situation opposite to the above would exist. In this case, the supplementary control in area 2 would tend to back off the generation picked up by its generators as a result of primary speed control or governor action. This would result in a degradation of system frequency control.

In addition to the above considerations, a very high value of bias factor is not desirable from the control stability viewpoint. At values significantly higher than the area β, the control action may become unstable.

The appropriateness of setting the frequency bias factor B nearly equal to the area β from dynamic considerations has been examined by a number of investigators [3-8]. A recommendation made in reference 7 to use significantly lower bias settings (B equal to nearly 0.5β) has not gained acceptance. A subsequent optimization study reported in reference 8 showed that $B=β$ is indeed a logical choice.

Systems with more than two areas

The description of the frequency bias tie line control described above applies equally well to systems with more than two areas. The interchange schedule applicable to each area is the algebraic sum of power flows on all the tie lines from that area to the other areas.

When an area is interconnected with more than one additional area, scheduled interchange transfers between them do not necessarily flow directly through the tie lines connecting the respective areas. Actual flows could split over parallel paths through other areas, depending on the relative impedances of the parallel paths. This is illustrated in Figure 11.26 which considers a three-area system.

Performance of AGC under normal and abnormal conditions

Under *normal conditions*, with each area able to carry out its control obligations, steady-state corrective action of AGC is confined to the area where the deficit or excess of generation occurs. Interarea power transfers are maintained at scheduled levels and system frequency is held constant.

Under *abnormal conditions*, one or more areas may be unable to correct for the generation-load mismatch due to insufficient generation reserve on AGC. In such an event, other areas assist by permitting the interarea power transfers to deviate from scheduled values and by allowing system frequency to depart from its pre-disturbance value. Each area participates in frequency regulation in proportion to its available regulating capacity relative to that of the overall system.

The following example illustrates the above aspects of AGC performance.

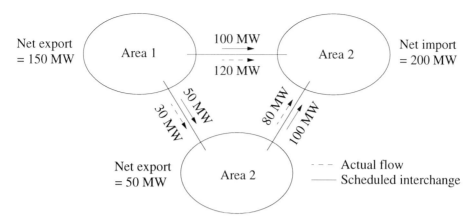

Figure 11.26 Three areas connected by tie lines

Example 11.3

Consider two interconnected areas as follows:

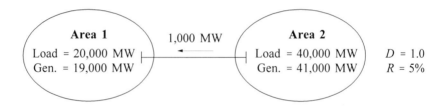

The connected load at 60 Hz is 20,000 MW in area 1 and 40,000 MW in area 2. The load in each area varies 1% for every 1% change in frequency. Area 1 is importing 1,000 MW from area 2. The speed regulation, R, is 5% for all units.

Area 1 is operating with a spinning reserve of 1,000 MW spread uniformly over a generation of 4,000 MW capacity, and area 2 is operating with a spinning reserve of 1,000 MW spread uniformly over a generation of 10,000 MW.

Determine the steady-state frequency, generation and load of each area, and tie line power for the following cases.

(a)   Loss of 1,000 MW load in area 1, assuming that there are no supplementary controls.

(b)   Each of the following contingencies, when the generation carrying spinning reserve in each area is on supplementary control with frequency bias factor settings of 250 MW/0.1 Hz for area 1 and 500 MW/0.1 Hz for area 2.

(i)   Loss of 1,000 MW load in area 1

(ii)   Loss of 500 MW generation, carrying part of the spinning reserve, in area 1

(iii)   Loss of 2,000 MW generation, not carrying spinning reserve, in area 1

(iv)   Tripping of the tie line, assuming that there is no change to the interchange schedule of the supplementary control

(v)   Tripping of the tie line, assuming that the interchange schedule is switched to zero when the ties are lost

Solution

(a) *With no supplementary control.*

Assuming that none of the governors are blocked, all generating units in the two areas respond to the loss of load.

A 5% regulation on 20,000 MW generating capacity (including spinning reserve of 1,000 MW) in area 1 corresponds to

$$\frac{1}{R_1} = \frac{1}{0.05} \times \frac{20,000}{60} = 6,666.67 \quad \text{MW/Hz}$$

Similarly, a 5% regulation on 42,000 MW generating capacity in area 2 corresponds to

$$\frac{1}{R_2} = \frac{1}{0.05} \times \frac{42,000}{60} = 14,000.00 \quad \text{MW/Hz}$$

Total regulation due to 62,000 MW generating capacity in the two areas is

$$\frac{1}{R} = \frac{1}{R_1} + \frac{1}{R_2} = 20,666.67 \quad \text{MW/Hz}$$

Load damping due to 19,000 MW load (remaining after loss of 1,000 MW load) in area 1 is

$$D_1 = 1 \times \frac{19,000}{100} \times \frac{100}{60} = 316.67 \quad \text{MW/Hz}$$

Load damping due to 40,000 MW load in area 2 is

$$D_2 = 1 \times \frac{40,000}{100} \times \frac{100}{60} = 666.67 \quad \text{MW/Hz}$$

Total effective load damping of the two areas is

$$D = D_1 + D_2 = 983.33 \quad \text{MW/Hz}$$

Change in system frequency due to loss of 1,000 MW load in area 1 is

$$\Delta f = \frac{-\Delta P_L}{1/R + D} = \frac{-(-1000)}{20,666.67 + 983.33} = 0.04619 \quad \text{Hz}$$

Load changes in the two areas due to increase in frequency are

$$\Delta P_{D1} = D_1 \Delta f = 316.67 \times 0.04619 = 14.63 \quad \text{MW}$$

$$\Delta P_{D2} = D_2 \Delta f = 666.67 \times 0.04619 = 30.79 \quad \text{MW}$$

Generation changes in the two areas due to speed regulation are

$$\Delta P_{G1} = -\frac{1}{R_1} \Delta f = 6,666.67 \times 0.04619 = -307.93 \quad \text{MW}$$

$$\Delta P_{G2} = -\frac{1}{R_2} \Delta f = 14,000.00 \times 0.04619 = -646.65 \quad \text{MW}$$

The new load, generation and tie line power flows are as follows.

Area 1		Area 2	
Load	= 20,000.00 - 1,000.00 + 14.63 = 19,014.63 MW	Load	= 40,000.00 + 30.79 = 40,030.79 MW
Generation	= 19,000.00 - 307.93 = 18,692.07 MW	Generation	= 41,000.00 - 646.65 = 40,353.35 MW

Tie line power flow from area 2 to area 1 is 322.56 MW. Steady-state frequency is 60.04619 Hz.

(b) *With supplementary control.*

(i) Loss of 1,000 MW load in area 1:

Area 1 has a generating capacity of 4,000 MW on supplementary control, and this will reduce generation so as to bring ACE_1 to zero. Similarly, area 2 generation on supplementary control will keep ACE_2 at zero:

$$ACE_1 = B_1 \Delta f + \Delta P_{12} = 0$$
$$ACE_2 = B_2 \Delta f - \Delta P_{12} = 0$$

Hence,

$$\Delta f = 0 \qquad \Delta P_{12} = 0$$

Area 1 generation and load are reduced by 1,000 MW. There is no steady-state change in area 2 generation and load, or the tie flow.

(ii) Loss of 500 MW generation carrying part of spinning reserve in area 1:

Prior to loss of generation, area 1 had a spinning reserve of 1,000 MW spread uniformly over a generation of 4,000 MW capacity (3,000 MW generation plus 1,000 MW reserve). Spinning reserve lost with generation loss is

$$\frac{500}{3,000} \times 1,000 = 166.67 \quad \text{MW}$$

Spinning reserve remaining is 1,000.00−166.67=833.33 MW. This is sufficient to make up for 500 MW generation loss. Hence, the generation and load in the two areas are restored to their pre-disturbance values. There are no changes in tie line flow or system frequency. However, area 1 spinning reserve is reduced from 1,000 MW to 833.33 MW.

(iii) Loss of 2,000 MW generation in area 1, not carrying spinning reserve:

Half of the generation loss will be made up by the 1,000 MW spinning reserve on supplementary control in area 1. When this limit is reached, area 1 is no longer able to control ACE. Supplementary control in area 2, however, is able to control its ACE. Hence,

$$ACE_2 = B_2 \Delta f - \Delta P_{12} = 0$$

or

$$\Delta P_{12} = B_2 \Delta f = 5,000 \Delta f$$

There is thus a net reduction in system frequency. This causes a reduction in loads due to frequency sensitivity.

Area 1 load damping is

$$D_1 = 1 \times \frac{20,000}{100} \times \frac{100}{60} = 333.33 \quad \text{MW/Hz}$$

The balance of generation loss in area 1 is made up by a reduction in load and tie flow from area 2. Hence,

$$-1,000 = D_1 \Delta f + \Delta P_{12}$$
$$= 333.33 \Delta f + 5,000 \Delta f$$

Solving for Δf, we have

$$\Delta f = \frac{-1,000}{5,000 + 333.33} = -0.1875 \quad \text{Hz}$$

Change in area 1 load is

$$\Delta P_{D1} = D_1 \Delta f = 333.33 \times (-0.1875)$$
$$= -62.5 \quad \text{MW}$$

The tie flow change is

$$\Delta P_{12} = 5,000 \times (-0.1875) = -937.5 \quad \text{MW}$$

Change in area 2 load is

$$\Delta P_{D2} = D_2 \Delta f = 666.67 \times (-0.1875)$$
$$= -125.00 \quad \text{MW}$$

The area load and generation are as follows.

Area 1		Area 2	
Load	= 20,000.0 − 62.5 = 19,937.5 MW	Load	= 40,000.0 − 125.0 = 39,875.0 MW
Generation	= 19,000.0 − 1,000.0 = 18,000.0 MW	Generation	= 41,000.0 − 125.0 + 937.5 = 41,812.5 MW

The steady-state tie line power flow from area 2 to area 1 is 1,937.50 MW, and the system frequency is $60.0 - 0.1875 = 59.8125$ Hz.

(iv) Tripping of the tie line, assuming no change in interchange schedule:

The supplementary control of area 1 attempts to maintain interchange schedule at 1,000 MW. Hence,

$$\text{ACE}_1 = \Delta P_{12} + B_1 \Delta f_1 = 1,000 + 2,500 \Delta f = 0$$

Solving, we find

$$\Delta f_1 = -\frac{1000}{2500} = -0.4 \quad \text{Hz}$$

Change in area 1 load is

$$\Delta P_{D1} = D_1 \Delta f_1 = 333.33 \times (-0.4) = -133.33 \quad \text{MW}$$

Similarly for area 2, we have

$$\Delta f_2 = \frac{1,000}{5,000} = 0.2 \quad \text{Hz}$$

and

$$\Delta P_{D2} = 666.67 \times 0.2 = 133.33 \quad \text{MW}$$

The area load, generation, and frequencies are as follows:

Area 1		Area 2	
Load	= 20,000.00 – 133.33 = 19,866.67 MW	Load	= 40,000.00 + 133.33 = 40,133.33 MW
Generation	= 19,866.67 MW	Generation	= 40,133.33 MW
f_1	= 59.6 Hz	f_2	= 60.2 Hz

(v) Tripping of the tie line, with interchange schedule switched to zero:

With interchange schedule switched to zero, area 1 supplementary control will pick up 1,000 MW generation to make up for loss of import power. Similarly, area 2 supplementary control reduces generation by 1,000 MW to compensate for loss of export. The generation in each area is equal to the respective loads and the area frequencies are equal to 60 Hz.   ∎

Economic allocation of generation

As noted earlier, an important secondary function of automatic generation control is to allocate generation so that each power source is loaded most economically. This function is referred to as *economic dispatch control* (EDC). The theory of economic dispatch is based on the principle of equal incremental costs.

For control of tie line power and frequency, it is necessary to send signals to generating plants to control generation. It is possible to use these signals to control generation to satisfy economic dispatch criteria. Thus, the requirements for EDC can be handled as part of the AGC function.

Since system load is continually changing, economic dispatch calculations have to be made at frequent intervals. The allocation of individual generation output is accomplished by using *base points* and *participation factors* (PFs). The base point represents the most economic output for each generating unit, and the participation factor is the rate of change of the unit output with respect to a change in total generation. The new desired output for each generator is calculated as follows [2]:

$$P_{desired} = P_{base\ point} + \mathrm{PF}(\Delta P_{total})   \qquad (11.29)$$

where

$$\Delta P_{total} = \text{total new generation} - \text{sum of } P_{base\ point} \text{ for all generation}$$

Sum of participation factors of all units is equal to unity.

11.1.6 Implementation of AGC

In modern AGC schemes, the control actions are usually determined for each control area at a central location called the dispatch centre. Information pertaining to tie line flows, system frequency, and unit MW loadings is telemetered to the central location where the control actions are determined by a digital computer. The control signals are transmitted via the same telemetering channels to the generating units on AGC as shown in Figure 11.27. The normal practice is to transmit raise or lower pulses of varying lengths to the units. The control equipment at the plants then changes the reference setpoints of the units up or down in proportion to the pulse length.

Figure 11.27 illustrates the implementation of AGC for one control area (normally the service area of an individual utility). Each control area of an interconnected system is controlled in a similar manner, but independently of the other control areas. That is, the control of generation in the interconnected system is "area-wise decentralized."

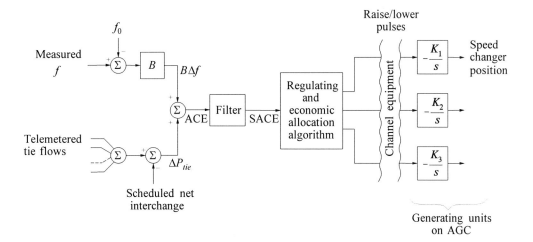

Figure 11.27 AGC control logic for each area

Early AGC systems, developed in the 1950s, were based on analog control equipment. These were gradually superseded by digital systems beginning in the late 1960s. Now, all-digital systems are the universal choice for AGC applications.

Filtering of ACE

Much of the change in ACE is usually due to fast random variations in load to which generating units need not respond. In fact, control action in response to these random components does not reduce ACE but merely causes unnecessary wear and tear on governor motors and turbine valves. Therefore, AGC programs normally use filtering schemes to filter out random variations, and *smoothed ACE* (SACE) is used to control generation.

The conventional approach is to use a low-order filter which reduces noise at the expense of speed of response.

Reference 9 suggests a method of distinguishing random variations and sustained load-change requirements by evaluating whether or not the error is in the direction of correct inadvertent interchange.

Rate limits

In establishing AGC signals, it should be recognized that there is a limit to the rate at which generating unit outputs can be changed. This is particularly true of thermal units where mechanical and thermal stresses are the limiting factors. The maximum loading rate for thermal units is on the order of 2% MCR (*maximum continuous rating*) per minute. For hydro units, the rate is on the order of 100% MCR per minute.

Control performance criteria

The guidelines followed by North American utilities for load-frequency control are developed by the Operating Committee of the North American Electric Reliability Council (NERC). The following criteria specify the minimum control performance standards set by NERC [10].

Under *normal conditions* the following criteria apply:

1.   A1 criterion – The ACE must return to zero within 10 minutes of previously reaching zero. Violations of this criterion count for each subsequent 10-minute period that the ACE fails to return to zero.

2.   A2 criterion – The average ACE for each of the six 10-minute periods during the hour (i.e., for the 10-minute periods ending at 10, 20, 30, 40, 50, and 60 minutes past the hour) must be within specific limits, referred to as L_d, that are determined from the control area's rate of change of demand characteristics:

$$L_d = 5 + 0.025 \, \Delta L \quad \text{MW}$$

where ΔL is the greatest hourly change in the net system load of a control area on the day of its maximum summer or winter peak load.

Under *disturbance conditions* (sudden loss of generation or increase of load) the following criteria apply:

1.   B1 criterion – The ACE must return to zero within 10 minutes following the start of the disturbance.

2.   B2 criterion – The ACE must start to return to zero within 1 minute following the start of the disturbance.

A disturbance is said to have occurred when a sampled value of ACE exceeds $3L_d$.

Time deviation correction

It is the practice of U.S. and Canadian interconnected systems to assign to one area the maintenance of a system time standard. For example, the standard for the eastern systems is maintained by the American Electric Power (AEP) company at Canton, Ohio. Through designated communication channels, information on the status of the system time deviation is relayed to all control areas, and certain periods are designated as time-correction periods. During such periods, all areas are expected to simultaneously offset their frequency schedules by an amount related to accumulated system time deviation.

Frequency of AGC execution

The stability of an AGC system and its ability to react to changing inputs are influenced by phase lags in the input system quantities and in the transmission of its control signals.

With digitally based systems, experience has shown that the execution of AGC once every 2 to 4 seconds results in good performance. This means the ACE is computed and the raise/lower control signals are transmitted to the generating plant once every 2 to 4 seconds.

Figure 11.28 shows the overall function diagram of a typical AGC system.

Frequency bias setting

The general practice in North America is to establish frequency bias in each control area once a year based on the area's natural regulation characteristic $(1/R+D)$ corresponding to the forecasted peak load of the coming year. The average frequency bias setting employed is about 2% per 0.1 Hz based on the estimated peak load and spinning reserve. The bias factor remains fixed through the year for all load levels.

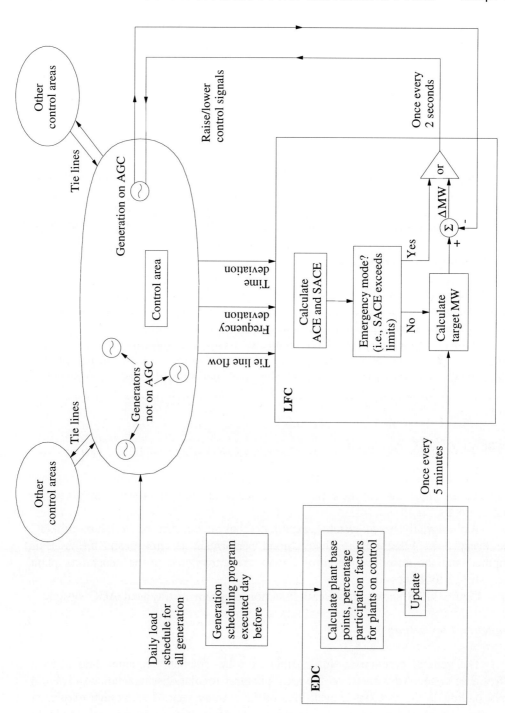

Figure 11.28   Functional diagram of a typical AGC system

A result of the above practice appears to be that, at times of low loads, the bias factor is much higher than the actual frequency regulation characteristic of the area. This is suspected of sometimes causing unstable control action resulting in limit cycle oscillations, large inadvertent interchange power accumulation and time deviation [3].

Reference 11 describes a scheme using a variable, nonlinear frequency bias, which results in improved control performance.

AGC tuning and performance [18,19]

The overriding consideration in the design and tuning of the AGC system is the impact on the power plants that are controlled. The characteristics of power plants vary widely and a number of physical constraints exist with regard to their manoeuvrability. A simple system that results in smooth control and a fairly well-damped system is preferable to a rapid control that attempts to bring ACE to zero rapidly. The plant characteristics in relation to variations in ACE are such that it is impossible to perfectly match generation and load continuously. The realization of the control function is limited by the amount of stored energy in the generating units and the rapidity with which generation can be changed. Therefore, the control attempts to match the average of generation and load over a time.

The control strategy should include the following objectives:

- To minimize fuel cost

- To avoid sustained operation of the generating units in the undesirable ranges (e.g., valve points for steam units)

- To minimize equipment wear and tear by avoiding unnecessary manoeuvring of generating units

Practical AGC systems achieve the above objectives by keeping the control strategies simple, robust, and reliable.

The stability of the control system and its ability to respond to the ACE signals are influenced by the phase lags associated with the measurement and transmission of control signals.

The ideal way to determine the AGC system parameters is by means of simulations. An important parameter that has an influence on the stability of AGC system is the overall loop gain; its value should be finalized based on field tests.

Emergency mode operation

For major system upsets causing splitting of the interconnected system into separate islands or opening of tie lines, AGC is suspended. This may be based on detection of very large changes in frequency or ACE. In some cases, AGC may be suspended for intentional tripping of load/generation so that AGC may not defeat the purpose of such tripping.

Effect of speed-governor dead band

The dead band associated with a speed governor is defined as "the total magnitude of the change in steady-state speed within which there is no resulting measurable change in the position of the governor-controlled valves or gates" [12]. Dead band is expressed in percent of the rated speed.

The effect of the dead band on the governor speed-droop characteristic is depicted in Figure 11.29. The speed-droop characteristic appears as a band rather than a line.

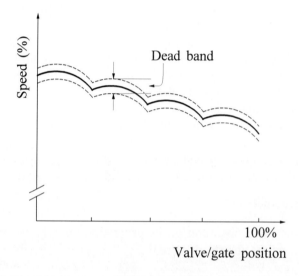

Figure 11.29

Dead band is caused by coulomb friction and backlash effects in various governor linkages, and by valve overlap in the hydraulic relays [1].

The current IEEE standards specify a maximum dead band of 0.06% (0.036 Hz) for governors of large steam turbines [12]. For governors of hydraulic turbines, the standards specify a maximum speed dead band of 0.02% and a maximum blade control dead band of 1.0% [13]. Dead bands of modern electrohydraulic governors are in fact much smaller. In some cases, special measures have been taken to practically eliminate the dead band [14]. However, there are many older units in operation with significant governor dead band.

The effect of the dead band on the speed governor response depends on the magnitude of the frequency deviation. If the deviation is small, it may remain entirely within the dead band; consequently, the speed control will be inactive. In any given situation, however, the entire dead band width will not have to be overcome. The position of each governor would be randomly distributed within its dead band; hence, for small changes in input signal the response of the individual generating units will tend to be random [15]. Speed governor dead bands result in random frequency fluctuations. In large power systems, these random fluctuations have a magnitude on the order of 0.01 Hz.

Results of investigations of the effects of governor dead bands on the performance of AGC systems are reported in references 5, 16 and 17. One effect of the dead band is to reduce the area frequency response characteristic β (i.e., to increase the effective frequency regulation). This requires a reduction in the frequency bias setting to ensure satisfactory AGC performance [16]. Reference 3 provides an in-depth analysis of AGC system dynamic performance and hypothesizes that governor dead bands may cause AGC system limit cycling with periods ranging from 30 seconds to 90 seconds.

Further reading on AGC

The state-of-the-art AGC systems have evolved from the early analog systems to the present digital systems. The result is a simple, yet robust, decentralized system that controls a complex, highly nonlinear, and continuously changing power system. There is a large amount of literature on the subject, some of which has attempted to use modern control theory in an effort to achieve optimal performance. However, these techniques have not been applied to practical AGC systems.

Reference 20 provides a good review of the properties of the conventional AGC systems and of some of the systems proposed in the literature that are based on modern control theory. It identifies potential areas in which the conventional approach may be improved by using concepts based on modern control theory.

Reference 21 presents a general approach to the analysis and synthesis of AGC systems based on multivariable control theory. It proposes two criteria for the design of the load-frequency controllers: (a) non-interaction between controls of frequency and tie line powers; (b) autonomy of controls of each area in taking care of its load variations during steady-state as well as transient conditions.

References 20 and 21 provide additional insight into the requirements and performance of AGC systems. They are both recommended for further reading.

11.1.7 Underfrequency Load Shedding

Severe system disturbances can result in cascading outages and isolation of areas, causing formation of electrical islands. If such an islanded area is undergenerated, it will experience a frequency decline. Unless sufficient generation with ability to rapidly increase output is available, the decline in frequency will be largely determined by frequency sensitive characteristics of loads. In many situations, the frequency decline may reach levels that could lead to tripping of steam turbine generating units by underfrequency protective relays, thus aggravating the situation further. To prevent extended operation of separated areas at lower than normal frequency, load-shedding schemes are employed to reduce the connected load to a level that can be safely supplied by available generation.

Hazards of underfrequency operation

There are two main problems associated with the operation of a power system at low frequency, both related to thermal generating units.

The first problem is concerned with the vibratory stress on the long low-pressure turbine blades. As discussed in Chapter 9 (Section 9.2.3), operation of steam turbines below 58.5 Hz is severely restricted. Since the effects of vibratory stress are cumulative with time, restoration of normal frequency operation as soon as possible is essential.

The second problem is concerned with the performance of plant auxiliaries driven by induction motors. At frequencies below 57 Hz, the plant capability may be severely reduced because of the reduced output of boiler feed pumps or fans supplying combustion air [22]. In the case of nuclear power plants, the reactors may overheat due to reduced flow of coolant as the frequency declines. Low-flow protections or underfrequency relays are normally used to guard against this condition. If the frequency decline is excessive, the generating units may be tripped off the power system.

In addition to avoiding the above consequences, it is necessary to restore normal frequency as soon as possible so that the affected area may be reconnected to the main power system.

Limitations of prime mover systems

The prime movers have several limitations which affect their ability to control frequency decay:

1.   The generation can be increased only to the limits of available spinning reserve within each affected area.

2.   The load that can be picked up by a thermal unit is limited due to thermal stress in the turbine. Initially, about 10% of turbine rated output can be picked up quickly without causing damage by too rapid heating. This is followed by a slow increase of about 2% per minute.

3.   The ability of a boiler to pick up a significant amount of load is limited. An increase in steam flow when the turbine valves open results in a pressure drop. An increase in fuel input to the boiler is required to restore pressure. This takes several minutes and is of little use in limiting frequency drop.

4.   The speed governors have a time delay of 3 to 5 seconds.

As a consequence of the above, generation reserve available for control of frequency is limited to a fraction of the remaining generation.

Factors influencing frequency decay

Assuming that the separated area has negligible spinning reserve with speed governing, the extent to which the frequency of the separated area will decrease and the rate of decay depend mainly on three factors: the magnitude of the overload ΔL (i.e., generation deficiency), load-damping constant D applicable to the area load, and the inertia constant M representing the total rotational inertia of the generators in the area.

Based on the results of Example 11.1, we may write the following expression for the frequency decay:

$$\Delta f = -\Delta L (1 - e^{-\frac{t}{T}}) K \tag{11.30}$$

where $K = 1/D$ and $T = M/D$.

For example, with $D = 1.0$ and $M = 10$ s, the frequency reduction as a function of time t is

$$\Delta f = -\Delta L (1 - e^{-\frac{t}{10}}) \quad \text{pu}$$

$$= -\Delta L (1 - e^{-\frac{t}{10}}) 60 \quad \text{Hz}$$

Figure 11.30 shows the frequency decay for four values of ΔL.

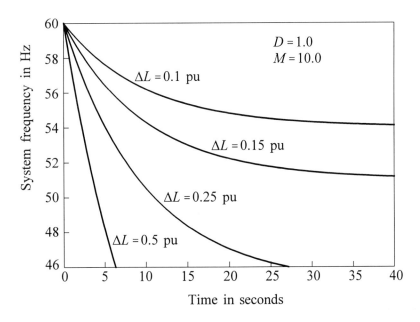

Figure 11.30 Frequency decay due to generation deficiency

Basis for selection of load-shedding schemes

Results such as those presented in Figure 11.30 are helpful as a first step in establishing the basis for load-shedding schemes. Considerations in the selection of the scheme include the maximum generation deficiency for which protection is required, the minimum permissible frequency, likely areas of system separation, and the range of inertia constant M and load-damping constant D.

Reference 23 describes a procedure for developing a load-shedding scheme. A typical scheme shedding load in three steps is as follows:

- 10% load is shed when frequency drops to 59.2 Hz
- 15% additional load is shed when frequency drops to 58.8 Hz
- 20% additional load is shed when frequency reaches 58.0 Hz

Typical operating time with solid-state relays is in the range of 0.1 to 0.2 second.

A scheme based on frequency drop alone is generally acceptable for generation deficiencies up to 25%. For greater generation deficiencies, a scheme which takes into account both frequency drop and rate of change of frequency provides increased selectivity by preventing unnecessary tripping of load [24]. Ontario Hydro uses such a relay to trip appropriate amounts of load in an area. The relay is referred to as a *frequency trend relay* (FTR), and its tripping logic is shown in Figure 11.31. The relay sheds up to 50% of area load. The system within Ontario is divided into 22 areas, and the load shedding is applied to several stations in each area to maintain the integrity of the area under disturbance conditions.

The procedure described above provides a simple means of determining the settings for underfrequency load shedding. After selection of possible areas of separation and specific load blocks for shedding, *detailed dynamic simulation should be carried out to ensure satisfactory system performance, with due regard to power plant and network protections/controls and to system voltages* (see Chapter 16, Section 16.5.2).

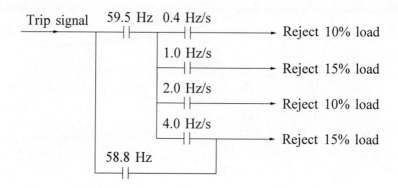

Figure 11.31 Tripping logic for frequency trend relay

11.2 REACTIVE POWER AND VOLTAGE CONTROL

For efficient and reliable operation of power systems, the control of voltage and reactive power should satisfy the following objectives:

(a)   Voltages at the terminals of all equipment in the system are within acceptable limits. Both utility equipment and customer equipment are designed to operate at a certain voltage rating. Prolonged operation of the equipment at voltages outside the allowable range could adversely affect their performance and possibly cause them damage.

(b)   System stability is enhanced to maximize utilization of the transmission system. As we will see later in this section, and in Chapters 12 to 14, voltage and reactive power control have a significant impact on system stability.

(c)   The reactive power flow is minimized so as to reduce RI^2 and XI^2 losses to a practical minimum (see Chapter 6, Section 6.3). This ensures that the transmission system operates efficiently, i.e., mainly for active power transfer.

The problem of maintaining voltages within the required limits is complicated by the fact that the power system supplies power to a vast number of loads and is fed from many generating units. As loads vary, the reactive power requirements of the transmission system vary. This is abundantly clear from the performance characteristics of transmission lines discussed in Chapter 6. Since reactive power cannot be transmitted over long distances, voltage control has to be effected by using special devices dispersed throughout the system. This is in contrast to the control of frequency which depends on the overall system active power balance. The proper selection and coordination of equipment for controlling reactive power and voltage are among the major challenges of power system engineering.

We will first briefly review the characteristics of power system components from the viewpoint of reactive power and then we will discuss methods of voltage control.

11.2.1 Production and Absorption of Reactive Power

Synchronous generators can generate or absorb reactive power depending on the excitation. When overexcited they supply reactive power, and when underexcited they absorb reactive power. The capability to continuously supply or absorb reactive power is, however, limited by the field current, armature current, and end-region heating limits, as discussed in Chapter 5 (Section 5.6). Synchronous generators are normally equipped with automatic voltage regulators which continually adjust the excitation so as to control the armature voltage.

Overhead lines, depending on the load current, either absorb or supply reactive power. At loads below the natural (surge impedance) load, the lines produce net reactive power; at loads above the natural load, the lines absorb reactive power. The reactive power characteristics of transmission lines are discussed in detail in Chapter 6.

Underground cables, owing to their high capacitance, have high natural loads. They are always loaded below their natural loads, and hence generate reactive power under all operating conditions.

Transformers always absorb reactive power regardless of their loading; at no load, the shunt magnetizing reactance effects predominate; and at full load, the series leakage inductance effects predominate.

Loads normally absorb reactive power. A typical load bus supplied by a power system is composed of a large number of devices. The composition changes depending on the day, season, and weather conditions. The composite characteristics are normally such that a load bus absorbs reactive power. Both active power and reactive power of the composite loads vary as a function of voltage magnitudes. Loads at low-lagging power factors cause excessive voltage drops in the transmission network and are uneconomical to supply. Industrial consumers are normally charged for reactive as well as active power; this gives them an incentive to improve the load power factor by using shunt capacitors.

Compensating devices are usually added to supply or absorb reactive power and thereby control the reactive power balance in a desired manner. In what follows, we will discuss the characteristics of these devices and the principles of application.

11.2.2 Methods of Voltage Control

The control of voltage levels is accomplished by controlling the production, absorption, and flow of reactive power at all levels in the system. The generating units provide the basic means of voltage control; the automatic voltage regulators control field excitation to maintain a scheduled voltage level at the terminals of the generators. Additional means are usually required to control voltage throughout the system. The devices used for this purpose may be classified as follows:

(a)   Sources or sinks of reactive power, such as shunt capacitors, shunt reactors, synchronous condensers, and static var compensators (SVCs).

(b)   Line reactance compensators, such as series capacitors.

(c)   Regulating transformers, such as tap-changing transformers and boosters.

Shunt capacitors and reactors, and series capacitors provide *passive* compensation. They are either permanently connected to the transmission and distribution system, or switched. They contribute to voltage control by modifying the network characteristics.

Synchronous condensers and SVCs provide *active* compensation; the reactive power absorbed/supplied by them is automatically adjusted so as to maintain voltages of the buses to which they are connected. Together with the generating units, they establish voltages at specific points in the system. Voltages at other locations in the system are determined by active and reactive power flows through various circuit elements, including the passive compensating devices.

The following is a description of the basic characteristics and forms of application of devices commonly used for voltage and reactive power control.

11.2.3 Shunt Reactors

Shunt reactors are used to compensate for the effects of line capacitance, particularly to limit voltage rise on open circuit or light load.

They are usually required for EHV overhead lines longer than 200 km. A shorter overhead line may also require shunt reactors if the line is supplied from a weak system (low short-circuit capacity) as shown in Figure 11.32. When the far end of the line is opened, the capacitive line-charging current flowing through the large source inductive reactance (X_S) will cause a rise in voltage E_S at the sending end of the line. The "Ferranti" effect (see Chapter 6, Section 6.1) will cause a further rise in receiving-end voltage E_R.

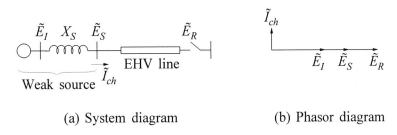

(a) System diagram   (b) Phasor diagram

Figure 11.32 EHV line connected to a weak system

A shunt reactor of sufficient size must be permanently connected to the line to limit fundamental-frequency temporary overvoltages to about 1.5 pu for a duration of less than 1 second. Such line-connected reactors also serve to limit energization overvoltages (switching transients). Additional shunt reactors required to maintain normal voltage under light-load conditions may be connected to the EHV bus as shown in Figure 11.33, or to the tertiary windings of adjacent transformers as shown in Figure 11.34. During heavy loading conditions some of the reactors may have to be disconnected. This is achieved by switching reactors using circuit-breakers.

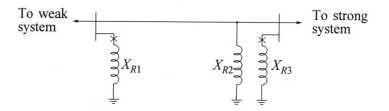

X_{R2} – permanently line-connected reactor
X_{R1}, X_{R3} – switchable bus-connected reactor

Figure 11.33 Line and bus-connected EHV reactors

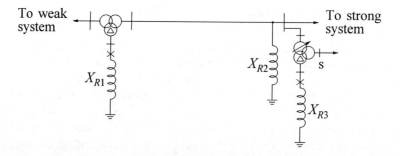

X_{R2} – permanently line-connected reactor
X_{R1}, X_{R3} – switchable reactors connected to tertiary
                windings of transformers

Figure 11.34 Line and transformer-connected reactors

For shorter lines supplied from strong systems, there may not be a need for reactors connected to the line permanently. In such cases, all the reactors used may be switchable, connected either to the tertiary windings of transformers or to the EHV bus. In some applications, tapped reactors with on-voltage tap-change control facilities have been used, as shown in Figure 11.35, to allow variation of the reactance value.

Shunt reactors are similar in construction to transformers, but have a single winding (per phase) on an iron core with air-gaps and immersed in oil. They may be of either single-phase or three-phase construction.

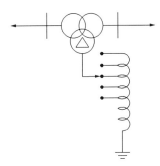

Figure 11.35 Tapped shunt reactor

11.2.4 Shunt Capacitors

Shunt capacitors supply reactive power and boost local voltages. They are used throughout the system and are applied in a wide range of sizes.

Shunt capacitors were first used in the mid-1910s for power factor correction. The early capacitors employed oil as the dielectric. Because of their large size and weight, and high cost, their use at the time was limited. In the 1930s, the introduction of cheaper dielectric materials and other improvements in capacitor construction brought about significant reductions in price and size. The use of shunt capacitors has increased phenomenally since the late 1930s. Today, they are a very economical means of supplying reactive power. The principal advantages of shunt capacitors are their low cost and their flexibility of installation and operation. They are readily applied at various points in the system, thereby contributing to efficiency of power transmission and distribution. The principal disadvantage of shunt capacitors is that their reactive power output is proportional to the square of the voltage. Consequently, the reactive power output is reduced at low voltages when it is likely to be needed most.

Application to distribution systems [25,26]

Shunt capacitors are used extensively in distribution systems for power-factor correction and feeder voltage control. Distribution capacitors are usually switched by automatic means responding to simple time clocks, or to voltage or current-sensing relays.

The objective of *power-factor correction* is to provide reactive power close to the point where it is being consumed, rather than supply it from remote sources. Most loads absorb reactive power; that is, they have lagging power factors. Table 11.1 gives typical power factors and voltage-dependent characteristics of some common types of loads.

Table 11.1 Typical characteristics of individual loads

Type of load	Power factor (lag)	Voltage dependence	
		P	Q
Large industrial motor	0.89	$V^{0.05}$	$V^{0.5}$
Small industrial motor	0.83	$V^{0.1}$	$V^{0.6}$
Refrigerator	0.84	$V^{0.8}$	$V^{2.5}$
Heat pump (cool/heat)	0.81/0.84	$V^{0.2}$	$V^{2.5}$
Dishwasher	0.99	$V^{1.8}$	$V^{3.5}$
Clothes washer	0.65	$V^{0.08}$	$V^{1.6}$
Clothes dryer	0.99	$V^{2.0}$	$V^{3.3}$
Colour TV	0.77	$V^{2.0}$	$V^{5.0}$
Fluorescent lighting	0.90	$V^{1.0}$	$V^{3.0}$
Incandescent lighting	1.00	$V^{1.55}$	-
Range, water or space heat	1.00	$V^{2.0}$	-

Power-factor correction is provided by means of fixed (permanently connected) and switched shunt capacitors at various voltage levels throughout the distribution systems. Low voltage banks are used for large customers and medium voltage banks are used at intermediate switching stations. For large industrial plants, as shown in Figure 11.36, power factor correction is applied at different levels: (i) individual motors, (ii) groups of motors, and (iii) the overall plant.

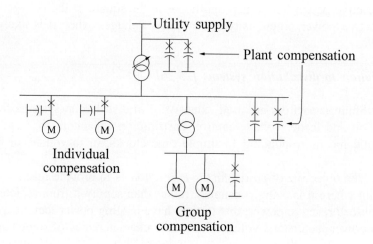

Figure 11.36 Power factor correction in industrial plants [26]

Switched shunt capacitors are also used extensively for *feeder voltage control*. They are installed at appropriate locations along the length of the feeder to ensure that voltages at all points remain within the allowable maximum and minimum limits as the loads vary. As discussed in Section 11.2.10, the application of shunt capacitors is coordinated with that of feeder voltage regulators or booster transformers.

Application to transmission system

Shunt capacitors are used to compensate for the XI^2 losses in transmission systems and to ensure satisfactory voltage levels during heavy loading conditions. Capacitor banks of appropriate sizes are connected either directly to the high voltage bus or to the tertiary winding of the main transformer, as shown in Figure 11.37. They are breaker-switched either automatically by a voltage relay or manually. Switching of capacitor banks provides a convenient means of controlling transmission system voltages. They are normally distributed throughout the transmission system so as to minimize losses and voltage drops. Detailed power-flow studies are performed to determine the size and location of capacitor banks to meet the system design criteria which specify maximum allowable voltage drop following specified contingencies. Procedures for power-flow analysis are discussed in Section 11.3.

The principles of application of shunt capacitors and other forms of transmission system compensation are presented in Section 11.2.8.

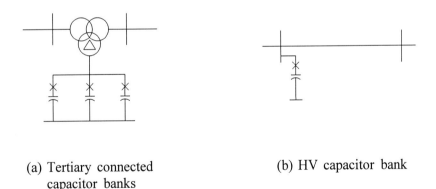

(a) Tertiary connected capacitor banks

(b) HV capacitor bank

Figure 11.37 Capacitor bank connections

11.2.5 Series Capacitors

Series capacitors are connected in series with the line conductors to compensate for the inductive reactance of the line. This reduces the transfer reactance between the buses to which the line is connected, increases maximum power that can be transmitted, and reduces the effective reactive power (XI^2) loss. Although series capacitors are not usually installed for voltage control as such, they do contribute to improved voltage control and reactive power balance. The reactive power produced by a series capacitor increases with increasing power transfer; a series capacitor is self-regulating in this regard.

Application to distribution feeders

Series capacitors have been applied to improve voltage regulation of distribution and industrial feeders since the 1930s. Welders and arc furnaces are typical of loads with poor power factor and intermittent demand. A series capacitor not only reduces voltage drop in the steady state, but it responds almost instantaneously to changes in load current. The series capacitor, by reducing the impedance between the bulk power source and the fluctuating load, is effective in solving light-flicker problems.

There are a number of problems associated with the application of series capacitors to industrial feeders that need careful attention [27-29]:

- Self-excitation of large induction and synchronous motors during starting. The motor may lock in at a fraction of synchronous (subsynchronous) speed due to resonance conditions.

  The most common remedy is to connect, during starting, a suitable resistance in parallel with the series capacitor.

- Hunting of synchronous motors (in some cases induction motors) at light load, due to the high R/X ratio of the feeder.

- Ferroresonance between transformers and series capacitors resulting in harmonic overvoltages. This may occur when energizing an unloaded transformer or when suddenly removing a load.

Because of the above problems and the difficulty in protecting the capacitors from system fault currents, series capacitors are not very widely used in today's distribution systems. They are, however, used in subtransmission systems to modify load division between parallel lines and to improve voltage regulation.

Application to EHV transmission system

Because series capacitors permit economical loading of long transmission lines, their application to EHV transmission has grown. They have been primarily used to improve system stability and to obtain the desired load division among parallel lines.

Complete compensation of the line is never considered. At 100% compensation, the effective line reactance would be zero, and the line current and power flow would be extremely sensitive to changes in the relative angles of terminal voltages. In addition, the circuit would be series resonant at the fundamental frequency. High compensation levels also increase the complexity of protective relaying and the probability of subsynchronous resonance. A practical upper limit to the degree of series compensation is about 80%.

It is not practical to distribute the capacitance in small units along the line. Therefore, lumped capacitors are installed at a few locations along the line. The use of lumped series capacitors results in an uneven voltage profile.

Series capacitors operate at line potential; hence, they must be insulated from ground. A widely accepted practice is to mount the capacitors on platforms insulated from ground. Alternatively, ground-base capacitor banks consisting of capacitor cans placed inside oil-insulated tanks may be used.

The following are some of the key considerations in the application of series-capacitor banks:

(a)   *Voltage rise due to reactive current.* Voltage rise on one side of the capacitor may be excessive when the line reactive-current flow is high, as might occur during power swings or heavy power transfers. This may impose unacceptable stress on equipment on the side of the bank experiencing high voltage. The system design must limit the voltage to acceptable levels, or the equipment must be rated to withstand the highest voltage that might occur.

(b)   *Bypassing and reinsertion.* The series capacitors are normally subjected to a voltage which is on the order of the regulation of the line, i.e., only a few percent of the rated line voltage. If, however, the line is short-circuited by a fault beyond the capacitor, a voltage on the order of the line voltage will appear across the capacitor. It would not be economical to design the capacitor for this voltage, since both size and cost of the capacitor increase with the square of the voltage. Therefore, provision is made for bypassing the capacitor during faults and reinsertion after fault clearing. Speed of reinsertion may be an important factor in maintaining transient stability.

Traditionally, bypassing was provided by a spark gap across the bank or each module of the bank. However, the present trend is to use nonlinear resistors of zinc oxide which have the advantage that reinsertion is essentially instantaneous. Figure 11.38 shows alternative bypass schemes [30]. The scheme shown in Figure 11.38(a) consists of a single spark gap (G) which bypasses the capacitor bank when the capacitor voltage exceeds a set value, usually about three to four times the capacitor rated voltage. The damping circuit (D) limits the discharge current and absorbs the capacitor energy. Upon detection of gap current, the bypass breaker (S) is closed, diverting the current from the gap. When the current returns to normal, the breaker is opened, thereby reinserting the capacitor into the line. This scheme is designed to provide a reinsertion time of 200 to 400 ms.

A dual-gap scheme with reinsertion time on the order of 80 ms is shown in Figure 11.38(b). It has an extra spark gap (G2) which is set low so that it will spark over first. Breaker S2 is normally closed. In the event of a fault, gap

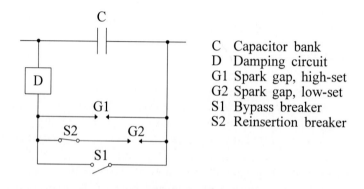

C Capacitor bank
D Damping circuit
G Spark gap
S Bypass breaker

(a) Single-gap protective scheme

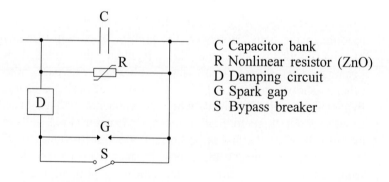

C Capacitor bank
D Damping circuit
G1 Spark gap, high-set
G2 Spark gap, low-set
S1 Bypass breaker
S2 Reinsertion breaker

(b) Dual-gap protective scheme

C Capacitor bank
R Nonlinear resistor (ZnO)
D Damping circuit
G Spark gap
S Bypass breaker

(c) Zinc-oxide protective scheme

Figure 11.38 Series capacitor bypass protective schemes [30]

G2 bypasses the capacitor bank. Breaker S2 opens immediately after the line fault has been cleared and reinserts the capacitor bank into the line. As a result, reinsertion is not delayed due to the de-ionization time. The other gap G1, which is high-set, and the bypass breaker S1 serve as backup protection.

In the scheme shown in Figure 11.38(c), a nonlinear resistor of zinc oxide (ZnO) limits the voltage across the capacitor bank during a fault and reinserts the bank immediately on termination of the fault current. The energy is absorbed by the ZnO resistor without the need to fire the spark gap (G). The spark gap is provided as a backup overvoltage protection for the resistor. The capacitor bank and the ZnO resistor remain in circuit during the fault, with the resistor bypassing most of the current; reinsertion takes place automatically without delay when the fault is cleared.

(c)   *Location.* A series-capacitor bank can theoretically be located anywhere along the line. Factors influencing choice of location include cost, accessibility, fault level, protective relaying considerations, voltage profile and effectiveness in improving power transfer capability.

The following are the usual locations considered:

- Midpoint of the line
- Line terminals
- 1/3 or 1/4 points of the line

In practice all of the above arrangements have been used.

The midpoint location has the advantage that the relaying requirements are less complicated if compensation is less than 50%. In addition, short-circuit current is lower. However, it is not very convenient in terms of access for maintenance, monitoring, security, etc.

Splitting of the compensation into two parts, with one at each end of the line, provides more accessability and availability of station service and other auxiliaries. The disadvantages are higher fault current, complicated relaying, and higher rating of the compensation.

The effectiveness of the compensation scheme depends on the location of the series capacitors and the associated shunt reactors. References 31 and 32 present results of comprehensive studies evaluating the effectiveness of different capacitor locations.

The choice of configuration of the compensation scheme for any particular application requires a detailed study with regard to the overall economy and system reliability. The study should take into account voltage profiles, compensation effectiveness, effect on transmission losses, overvoltages, and proximity to an attended station. References 33 and 34 provide details of engineering considerations related to series-capacitor applications on two power systems.

Adding a capacitor in series with the inductance of a transmission line forms a series-resonant circuit. The natural frequency of the resonant circuit so formed, for the usual range of compensation (20 to 70% of line reactance), is below the power frequency. The transmission network thus has a natural frequency in the subsynchronous range. Consequently, transient current components of subharmonic frequency are excited during any disturbance and are superimposed on the power-frequency currents. The subharmonic currents are usually damped rapidly within a few cycles, due to the resistances of the line and of any connected equipment such as loads. Therefore, the subharmonic natural mode introduced by the use of a series capacitor is rarely troublesome. One notable exception is the possible interaction with a natural frequency of the shaft mechanical system of nearby steam turbine generating units. It may lead to a buildup of torsional oscillations, either spontaneously or after a disturbance. This phenomenon is known as *subsynchronous resonance* (SSR). A detailed discussion of the SSR problem and measures available to counter it are presented in Chapter 15.

11.2.6 Synchronous Condensers

A synchronous condenser is a synchronous machine running without a prime mover or a mechanical load. By controlling the field excitation, it can be made to either generate or absorb reactive power. With a voltage regulator, it can automatically adjust the reactive power output to maintain constant terminal voltage. It draws a small amount of active power from the power system to supply losses.

Synchronous condensers have been used since the 1930s for voltage and reactive power control at both transmission and subtransmission levels. They are often connected to the tertiary windings of transformers. They fall into the category of *active shunt compensators*. Because of their high purchase and operating costs, they have been largely superseded by static var compensators (discussed next). Recent applications of synchronous condensers have been mostly at HVDC converter stations connected to weak systems [35]. There are many old synchronous condensers still in operation, and they serve as an excellent form of voltage and reactive power control devices.

Synchronous compensators have several advantages over static compensators. Synchronous compensators contribute to system short-circuit capacity. Their reactive power production is not affected by the system voltage. During power swings (electromechanical oscillations) there is an exchange of kinetic energy between a synchronous condenser and the power system. During such power swings, a synchronous condenser can supply a large amount of reactive power, perhaps twice its continuous rating. It has about 10 to 20% overload capability for up to 30 minutes. Unlike other forms of shunt compensation, it has an internal voltage source and is better able to cope with low system voltage conditions.

Some combustion turbine peaking units can be operated as synchronous condensers if required. Such units are often equipped with clutches which can be used to disconnect the turbine from the generator when active power is not required from them.

11.2.7 Static Var Systems [36-38]

Terminology [36]

Static var compensators (SVCs) are shunt-connected static generators and/or absorbers whose outputs are varied so as to control specific parameters of the electric power system. The term "static" is used to indicate that SVCs, unlike synchronous compensators, have no moving or rotating main components. Thus an SVC consists of static var generator (SVG) or absorber devices and a suitable control device.

A static var system (SVS) is an aggregation of SVCs and mechanically switched capacitors (MSCs) or reactors (MSRs) whose outputs are coordinated.

Types of SVC

The following are the basic types of reactive power control elements which make up all or part of any static var system:

- Saturated reactor (SR)

- Thyristor-controlled reactor (TCR)

- Thyristor-switched capacitor (TSC)

- Thyristor-switched reactor (TSR)

- Thyristor-controlled transformer (TCT)

- Self- or line-commutated converter (SCC/LCC)

A number of different SVS configurations made up of a combination of one or more of the basic types of SVC and fixed capacitor (FC) banks (i.e., capacitors not switched via local automatic control) have been used in practice for transmission system compensation.

We will first discuss the general principle of SVS operation in an HVAC system using a somewhat idealized compensator and then examine the characteristics of specific configurations.

Static var systems are capable of controlling individual phase voltages of the buses to which they are connected. They can therefore be used for control of negative-sequence as well as positive-sequence voltage deviations. However, we are interested here in the balanced fundamental frequency performance of power systems and therefore our analysis will consider only this aspect of SVS performance.

Fundamental frequency performance of an SVS [36,38]

Characteristic of an ideal SVS:

From the viewpoint of power system operation, an SVS is equivalent to a shunt capacitor and a shunt inductor, both of which can be adjusted to control voltage and reactive power at its terminals (or a nearby bus) in a prescribed manner (see Figure 11.39).

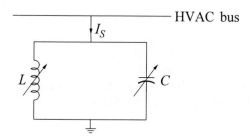

Figure 11.39 Idealized static var system

Ideally, an SVS should hold constant voltage (assuming that this is the desired objective), possess unlimited var generation/absorption capability with no active and reactive power losses and provide instantaneous response. The performance of the SVS can be visualized on a graph of controlled ac bus voltage (V) plotted against the SVS reactive current (I_S). The V/I characteristic of an ideal SVS is shown in Figure 11.40. It represents the steady-state and quasi steady-state characteristics of the SVS.

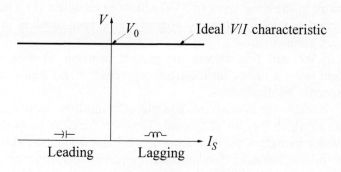

Figure 11.40 *V/I* characteristic of ideal compensator

Characteristic of a realistic SVS:

We consider an SVS composed of a controllable reactor and a fixed capacitor. The resulting characteristics are sufficiently general and are applicable to a wide range of practical SVS configurations.

Figure 11.41 illustrates the derivation of the characteristic of an SVS consisting of a controllable reactor and a fixed capacitor. The composite characteristic is derived by adding the individual characteristics of the components. The characteristic shown in Figure 11.41(a) is representative of the characteristics of practical controllable reactors.

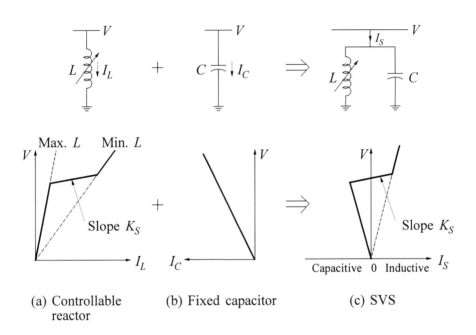

(a) Controllable   (b) Fixed capacitor   (c) SVS
    reactor

Figure 11.41 Composite characteristics of an SVS

Power system characteristic:

In order to examine how the SVS performs when applied to a power system, the characteristics of the SVS and the power system need to be examined together. The system V/I characteristic may be determined by considering the Thevenin equivalent circuit as viewed from the bus whose voltage is to be regulated by the SVS. This is illustrated in Figure 11.42. The Thevenin impedance in Figure 11.42(a) is predominantly an inductive reactance. The corresponding voltage versus reactive current characteristic is shown in Figure 11.42(b). The voltage V increases linearly with capacitive load current and decreases linearly with inductive load current.

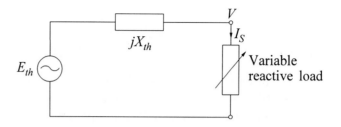

(a) Thevenin equivalent circuit of HVAC network

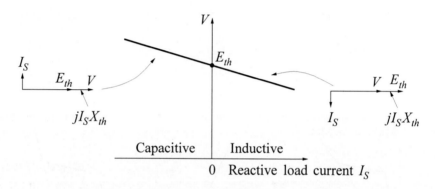

(b) Voltage-reactive current characteristic

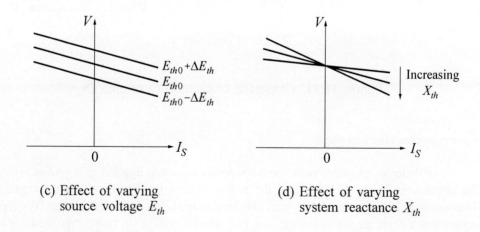

(c) Effect of varying   (d) Effect of varying
source voltage E_{th} system reactance X_{th}

Figure 11.42 Power system voltage versus reactive current characteristic [38]

For each network condition, an equivalent circuit such as that shown in Figure 11.42(a) can be defined. Figures 11.42(c) and (d) show how the network V/I characteristic is affected by changes in source voltage E_{th} and the system equivalent reactance X_{th}, respectively.

Composite SVS - power system characteristic:

The system characteristic may be expressed as

$$V = E_{th} - X_{th}I_S \tag{11.31}$$

The SVS characteristic, within the control range defined by the slope reactance X_{SL}, is given by

$$V = V_0 + X_{SL}I_S \tag{11.32}$$

For voltages outside the control range, the ratio V/I_S is equal to the slopes of the two extreme segments of Figure 11.41(c). These are determined by the ratings of the inductor and capacitor.

The solution of SVS and power system characteristic equations is graphically illustrated in Figure 11.43. Three system characteristics are considered in the figure, corresponding to three values of the source voltage. The middle characteristic

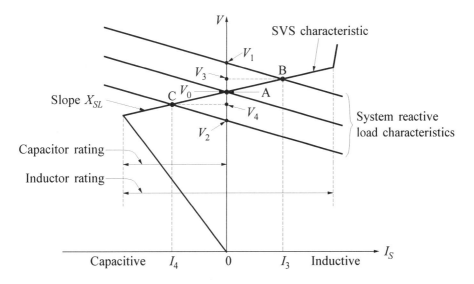

Figure 11.43 Graphical solution of SVS operating point
for given system conditions

represents nominal system conditions, and is assumed to intersect the SVS characteristic at point A where $V=V_0$ and $I_S=0$.

If the system voltage increases by ΔE_{th} (for example, due to a decrease in system load level), V will increase to V_1, without an SVS. With the SVS, however, the operating point moves to B; by absorbing inductive current I_3, the SVS holds the voltage at V_3. Similarly, if the source voltage decreases (due to increase in system load level), the SVS holds the voltage at V_4, instead of at V_2 without the SVS. If the slope K_S of the SVS characteristic were zero, the voltage would have been held at V_0 for both cases considered above.

Effect of using switched capacitors:

In the example considered in Figure 11.43, the SVS control range would be exceeded for larger variations in system conditions. The use of switched capacitor banks can extend the continuous control range of the SVS. This is illustrated in Figure 11.44, which considers three capacitor banks, two of which are switchable. Either thyristors or mechanical switches may be used for switching the capacitors in and out automatically by local voltage-sensing controls. In the figure the unswitched capacitor includes a reactor for filtering harmonics.

We see that an SVS is not a source of voltage as is a synchronous condenser. Instead, it alters the system voltage at the point of connection by varying the reactive current drawn or supplied to the system. In effect, the SVS acts as a variable reactive load which is adjusted so as to keep the ac voltage nearly constant.

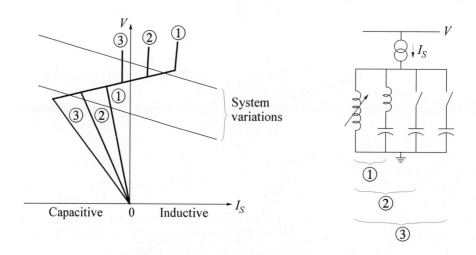

Figure 11.44 Use of switched capacitors to extend continuous control range

In general, the elements of an SVS operate on the principle of adjustable susceptance. The controlled susceptance is either a reactor or a capacitor. We will discuss next the operation of the more commonly used elements: TCR, TSC, and MSC. For a description of other forms of static compensators readers may consult references 36 and 37.

Thyristor-controlled reactor (TCR) [36,37]

Principle of operation:

The basic elements of a TCR are a reactor in series with a bidirectional thyristor switch as shown in Figure 11.45(a).

The thyristors conduct on alternate half-cycles of the supply frequency depending on the firing angle α, which is measured from a zero crossing of voltage. Full conduction is obtained with a firing angle of 90°. The current is essentially reactive and sinusoidal. Partial conduction is obtained with firing angles between 90° and 180°, as shown in Figure 11.45(b). Firing angles between 0 and 90° are not allowed as they produce asymmetrical currents with a dc component.

Let σ be the conduction angle, related to α by

$$\sigma = 2(\pi - \alpha) \qquad (11.33)$$

The instantaneous current i is given by

$$i = \begin{cases} \dfrac{\sqrt{2}V}{X_L}(\cos\alpha - \cos\omega t) & \text{for } \alpha < \omega t < \alpha + \sigma \\[2ex] 0 & \text{for } \alpha + \sigma < \omega t < \alpha + \pi \end{cases} \qquad (11.34)$$

Fourier analysis of the current waveform gives the fundamental component:

$$I_1 = \frac{V}{X_L}\frac{\sigma - \sin\sigma}{\pi} \qquad (11.35)$$

where I_1 and V are RMS values, and X_L is the reactance of the reactor at fundamental frequency.

The effect of increasing α (i.e., decreasing σ) is to reduce the fundamental component I_1. This is equivalent to increasing the effective inductance of the reactor.

In effect, so far as the fundamental frequency current component is concerned, the TCR is a controllable susceptance. The effective susceptance as a function of the firing angle α is

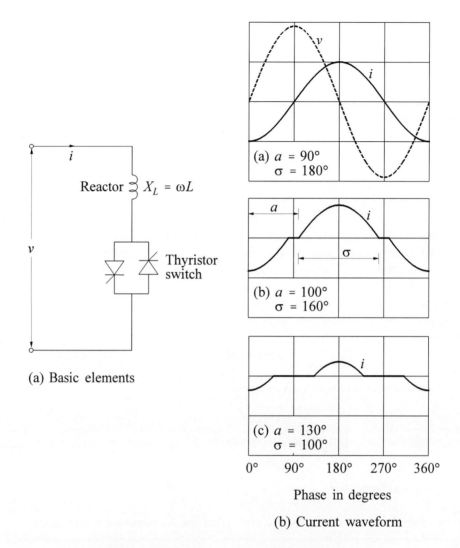

(a) Basic elements

(a) $a = 90°$
$\sigma = 180°$

(b) $a = 100°$
$\sigma = 160°$

(c) $a = 130°$
$\sigma = 100°$

0°   90°   180°   270°   360°

Phase in degrees

(b) Current waveform

Figure 11.45 Thyristor-controlled reactor

$$B(\alpha) = \frac{I_1}{V} = \frac{\sigma - \sin\sigma}{\pi X_L}$$

$$= \frac{2(\pi - \alpha) + \sin 2\alpha}{\pi X_L} \tag{11.36}$$

The maximum value of the effective susceptance is at full conduction ($\alpha = 90°$, $\sigma = 180°$), and is equal to $1/X_L$; the minimum value is zero, obtained with $\alpha = 180°$ or $\sigma = 0°$.

This susceptance control principle is known as *phase control*. The susceptance is switched into the system for a controllable fraction of every half cycle. The variation in susceptance as well as the TCR current is smooth or continuous.

The TCR requires a control system which determines the firing instants (i.e., firing angle α) measured from the last zero crossing of the voltage (synchronization of firing angles). In some designs the control system responds to a signal that directly represents the desired susceptance. In others, the control responds to error signals such as voltage deviation, auxiliary stabilizing signals, etc. The result is a steady-state *V/I* characteristic shown in Figure 11.46, which can be described by

$$V = V_{ref} + X_{SL}I_1 \qquad\qquad (11.37)$$

where X_{SL} is the slope reactance determined by the control system gain.

As illustrated in Figure 11.44, the TCR voltage control characteristic can be extended into the capacitive region by adding in parallel a fixed capacitor bank or switched capacitor banks.

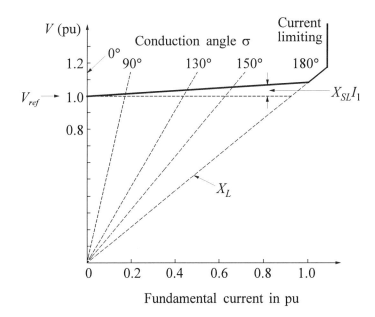

Figure 11.46 Fundamental voltage-current characteristic of TCR

Harmonics:

As α is increased from 90° to 180°, the current waveform becomes less and less sinusoidal; in other words, the TCR generates harmonics. For the single-phase device considered so far, if the firing of the thyristors is symmetrical (equal for both thyristors), only odd harmonics are generated. For a three-phase system, the preferred arrangement is to have the three single-phase TCR elements connected in delta (6-pulse TCR) as shown in Figure 11.47(a). For balanced conditions, all triple (3, 9, ...) harmonics circulate within the closed delta and are therefore absent from the line currents. Filters are often used to remove harmonic currents.

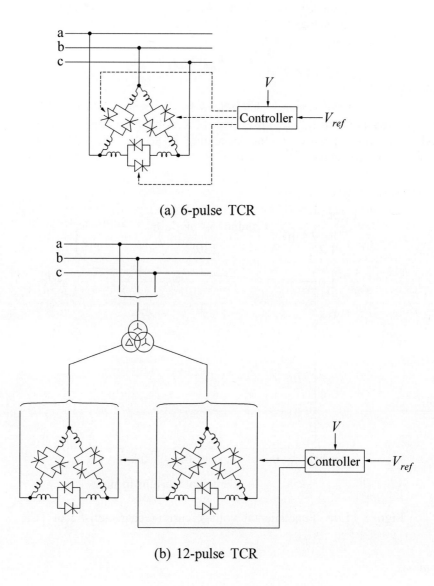

(a) 6-pulse TCR

(b) 12-pulse TCR

Figure 11.47 Three-phase TCR arrangements

Elimination of 5th and 7th harmonics can be achieved by using two 6-pulse TCRs of equal rating, fed from two secondary windings of the step-down transformers, one connected in Y and the other in Δ as shown in Figure 11.47(b). Since the voltages applied to the TCRs have a phase difference of 30°, 5th and 7th harmonics are eliminated from the primary-side line current. This is known as a 12-pulse arrangement because there are 12 thyristor firings every cycle of the three-phase line voltage. With the 12-pulse scheme, the lowest-order characteristic harmonics are the 11th and 13th. These can be filtered with a simple capacitor bank.

Dynamic response:

The TCR responds in about 5 to 10 ms, but delays are introduced by measurement and control circuits. To ensure control loop stability the response rate may have to be limited. For these reasons response times are typically around 1 to 5 cycles of supply frequency.

Thyristor-switched capacitor (TSC) [26,36,37]

Principle of operation:

A thyristor-switched capacitor scheme consists of a capacitor bank split up into appropriately sized units, each of which is switched on and off by using thyristor switches. Each single-phase unit consists of a capacitor (C) in series with a bidirectional thyristor switch and a small inductor (L) as shown in Figure 11.48(a). The purpose of the inductor is to limit switching transients, to damp inrush currents, and to prevent resonance with the network. In three-phase applications, the basic units are connected in Δ as shown in Figure 11.48(b).

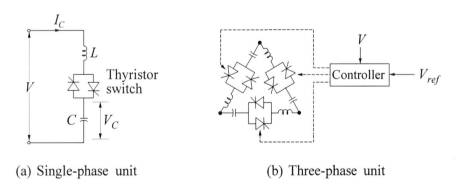

(a) Single-phase unit   (b) Three-phase unit

Figure 11.48 Thyristor-switched capacitor (TSC)

The switching of capacitors excites transients which may be large or small depending on the resonant frequency of the capacitors with the external system. The thyristor firing controls are designed to minimize the switching transients. This is achieved by choosing the switching instant when the voltage across the thyristor switch is at a minimum, ideally zero. Figure 11.49 illustrates the operating principle. The switching-on instant (t_1) is chosen so that the bus voltage V is at its maximum and of the same polarity as the capacitor voltage; this ensures a transient-free switching. The switching-off instant (t_2) corresponds to a current zero. The capacitor will then remain charged to a peak voltage, either positive or negative, ready for the next switch-on operation.

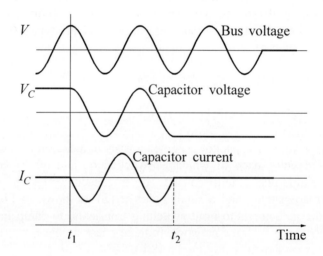

Figure 11.49 Switch operation of a TSC

The susceptance control principle used by a TSC is known as the *integral cycle control*; the susceptance is switched in for an integral number of exact half cycles. The susceptance is divided into several parallel units, and the susceptance is varied by controlling the number of units in conduction. A change can be made every half cycle. This form of control does not generate harmonics.

Figure 11.50 shows the basic scheme of a TSC consisting of parallel Δ connected TSC elements and a controller. When the bus voltage deviates from the reference value (V_{ref}) beyond the dead band in either direction, the control switches in (or out) one or more capacitor banks until the voltage returns inside the dead band, provided that not all the banks have been switched in (or out).

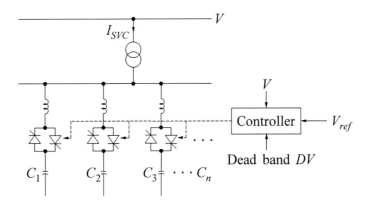

Figure 11.50   TSC scheme

Dynamic response:

The *V/I* characteristic of a TSC compensator is shown in Figure 11.51. We see that the voltage control provided is discontinuous or stepwise. It is determined by the rating and number of parallel connected units. In high voltage applications, the number of shunt capacitor banks is limited because of the high cost of thyristors. The power system *V/I* characteristics, as system conditions change, intersect the TSC *V/I* characteristics at discrete points. The bus voltage *V* is controlled within the range $V_{ref} \pm DV/2$, where *DV* is the dead band. When the system is operating so that its

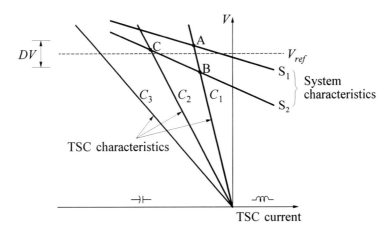

Figure 11.51   *V/I* characteristics of a TSC and power system

characteristic is represented by line S_1, then capacitor C_1 will be switched in and operating point A prevails. If the system characteristic suddenly changes to S_2, the bus voltage drops initially to a value represented by operating point B. The TSC control switches in bank C_2 to change the operating point to C, bringing the voltage within the desired range. Thus the compensator current can change in discrete steps. The time taken for executing a command from the controller ranges from one-half cycle to one cycle.

Mechanically switched capacitor (MSC)

Typically, an MSC scheme consists of one or more capacitor units connected to the power system by a circuit-breaker. A small reactor might be connected in series with the capacitor to damp energizing transients and reduce harmonics. Prestrike and restrike-free circuit-breakers have to be used to avoid system overvoltages due to capacitor-switching transients.

The *V/I* characteristic is linear and similar to that of a TSC. The response time is equal to the switching time of the circuit-breaker arrangement which is on the order of 100 ms following the initiation of switching operation instruction. Frequent switching is not possible unless discharge devices are provided.

Practical static var systems

A static var compensation scheme with any desired control range can be formed by using combinations of the elements described above. Several SVS configurations have been successfully applied to meet differing system requirements. The required speed of response, size range, flexibility, losses, and cost are among the important considerations in selecting a configuration for any particular application.

Figure 11.52 shows a typical SVS scheme consisting of a TCR, a three-unit TSC, and harmonic filters (for filtering TCR-generated harmonics). At power frequency, the filters are capacitive and produce reactive power of about 10 to 30% of TCR MVAr rating. In order to ensure a smooth control characteristic, the TCR current rating should be slightly larger than that of one TSC unit; otherwise dead bands arise (see reference 37, Chapter 4).

The steady-state *V/I* characteristic of the SVS is shown in Figure 11.53(a), and the corresponding *V/Q* characteristic is shown in Figure 11.53(b). The linear control range lies within the limits determined by the maximum susceptance (B_{LMX}) of the reactor, the total capacitive susceptance (B_C) as determined by the capacitor banks in service and the filter capacitance. If the voltage drops below a certain level (typically 0.3 pu) for an extended period, control power and thyristor gating energy can be lost, requiring a shutdown of the SVS. The SVS can restart as soon as the voltage recovers. However, the voltage may drop to low values for short periods, such as during transient faults, without causing the SVS to trip.

Within the linear control range, the SVS is equivalent to a voltage source V_{ref} in series with a reactance of X_{SL}. As is evident from Figure 11.43, the slope reactance

X_{SL} has a significant effect on the performance of the SVS. A large value of X_{SL} makes the SVS less responsive, i.e., changes in system conditions cause large voltage variations at the SVS high voltage bus. The value of X_{SL} is determined by the steady-state gain of the controller (voltage regulator). It may also be effected by a current feedback (with PI controller). Its choice should be based on detailed power-flow and stability studies. Typically, the slope is set within the range of 1 to 5%, depending on the ac system strength.

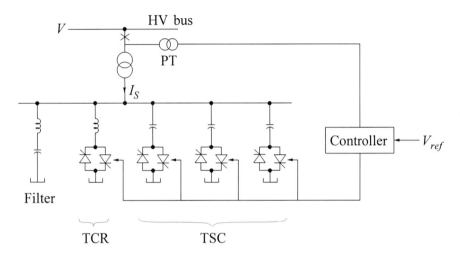

Figure 11.52 A typical static var system

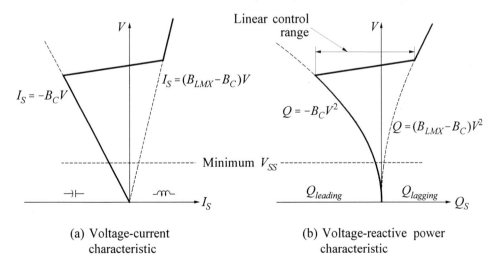

(a) Voltage-current characteristic

(b) Voltage-reactive power characteristic

Figure 11.53 SVS steady-state characteristics

Application of static var compensators

Since their first application in the late 1970s, the use of SVCs in transmission systems has been increasing steadily. By virtue of their ability to provide continuous and rapid control of reactive power and voltage, SVCs can enhance several aspects of transmission system performance. Applications to date include the following:

• Control of temporary (power frequency) overvoltages

• Prevention of voltage collapse

• Enhancement of transient stability

• Enhancement of damping of system oscillations

References 39 to 42 provide details of SVC applications on a number of power systems.

At the subtransmission and distribution system levels, SVCs are used for balancing the three phases of systems supplying unbalanced loads. They are also used to minimize fluctuations in supply voltage caused by repetitive-impact loads such as dragline loads of mining plants, rolling mills, and arc furnaces.

Arc furnaces are a special case of impact loads. They cause voltage fluctuations with frequencies randomly varying between 2 and 10 Hz. These result in flicker of filament lamps in the adjacent load areas. Some electronic equipment and television receivers may also be adversely affected. The term "voltage flicker" is used to describe such rapid fluctuations in voltage. To minimize the adverse effects on the adjacent load areas, the voltage fluctuations must be kept below the acceptable minimum level (typically 0.3%). SVCs provide an effective and economical means of eliminating the voltage flicker problems and have been widely used for such applications since the early 1970s.

11.2.8 Principles of Transmission System Compensation

In the previous sections we have described various devices used for reactive compensation of transmission systems. A well-planned and coordinated application of these devices is essential for the economical design and operation of a reliable system. As reactive compensation affects the steady-state as well as the dynamic performance of the system, detailed power-flow and stability studies are required for establishing appropriate compensation schemes. We will describe the power-flow studies in Section 11.3. Stability studies will be covered in Chapters 12 to 17 in which we consider different aspects of system stability.

In addition to detailed simulations, an understanding of the principles of reactive compensation in transmission systems is invaluable for proper selection and application of the compensating devices. This subject is covered very well by T.J.E.

Miller in reference 37. Here we will briefly review these principles by considering how different forms of compensation affect the performance of a transmission line.

We will first consider the ideal cases of uniformly distributed compensation as they lead to simple relationships which help us to understand the fundamental nature of each type of compensation. We will then consider specific configurations of lumped or concentrated compensation to see how they compare with the ideal cases.

Uniformly distributed fixed series and shunt compensation

In Chapter 6 (Section 6.1) we saw that the line performance is determined by the characteristic impedance Z_C and the electrical length (also referred to as line angle) θ. The objective of compensation is to modify these parameters so as to result in the desired voltage and power transfer characteristics.

Without compensation, assuming a lossless line, the expressions for the two line parameters are

$$Z_C = \sqrt{\frac{L}{C}} = \sqrt{\frac{x_L}{b_C}} = \sqrt{\frac{X_L}{B_C}} \tag{11.38}$$

$$\theta = \beta l \tag{11.39}$$

with the phase constant β given by

$$\beta = \omega\sqrt{LC} = \sqrt{x_L b_C} = \frac{\sqrt{X_L B_C}}{l} \tag{11.40}$$

where

$\quad L \quad$ = series inductance per unit length
$\quad C \quad$ = shunt capacitance per unit length
$\quad x_L$ = series inductive reactance per unit length
$\quad b_C$ = shunt capacitive susceptance per unit length
$\quad X_L$ = total series inductive reactance
$\quad B_C$ = total shunt susceptance
$\quad l \quad$ = line length

Let us designate the corresponding quantities *with compensation* by using a superscript prime (').

With a uniformly distributed *shunt compensation* having a susceptance of b_{sh} per unit length, the effective shunt susceptance is given by

$$b_C' = b_C - b_{sh} = b_C(1 - k_{sh}) \tag{11.41}$$

where k_{sh} is the *degree of shunt compensation* defined as follows:

$$k_{sh} = \frac{b_{sh}}{b_C} \tag{11.42}$$

It is positive for inductive shunt compensation and is negative for capacitive shunt compensation.

The effective values of the characteristic impedance and phase constant with shunt compensation are related to the uncompensated values as follows:

$$Z'_C = \sqrt{\frac{x_L}{b'_C}} = \frac{Z_C}{\sqrt{1-k_{sh}}} \tag{11.43}$$

and

$$\beta' = \beta\sqrt{1-k_{sh}} \tag{11.44}$$

Shunt capacitive compensation in effect decreases Z_C and increases β, whereas shunt inductive compensation increases Z_C and decreases β.

With a uniformly distributed *series capacitive compensation* of C_{se} per unit length, the effective series reactance is

$$
\begin{aligned}
x'_L &= x_L - \frac{1}{\omega C_{se}} = x_L - x_{Cse} \\
&= x_L(1-k_{se})
\end{aligned} \tag{11.45}
$$

where k_{se} is the *degree of series capacitive compensation* defined as follows:

$$k_{se} = \frac{x_{Cse}}{x_L} \tag{11.46}$$

It is positive for capacitive series compensation.

The effective values of characteristic impedance and phase constant with series compensation are given by

$$Z'_C = \sqrt{\frac{x'_L}{b_C}} = Z_C\sqrt{1-k_{se}} \tag{11.47}$$

and

$$\beta' = \beta\sqrt{1-k_{se}} \tag{11.48}$$

Series (capacitive) compensation decreases both Z_C and β.

With *both shunt and series compensation*, the combined effects are as follows:

$$Z_C' = Z_C\sqrt{\frac{1-k_{se}}{1-k_{sh}}} \tag{11.49}$$

$$\beta' = \beta\sqrt{(1-k_{sh})(1-k_{se})} \tag{11.50}$$

The effective line angle (θ') and natural load (P_0') are given by

$$\theta' = \theta\sqrt{(1-k_{sh})(1-k_{se})} \tag{11.51}$$

$$P_0' = P_0\sqrt{\frac{1-k_{sh}}{1-k_{se}}} \tag{11.52}$$

Effect of compensation on line voltage:

Under light load conditions, a flat voltage profile is achieved by inductive shunt compensation. For example, with $k_{sh}=1$ (100% inductive compensation), θ' and P_0' are reduced to zero and Z_C' is increased to infinity; this results in a flat voltage at zero load.

Under heavy load conditions, a flat voltage can be achieved by adding shunt capacitive compensation. For example, in order to transmit $1.4P_0$ with a flat voltage profile, a shunt capacitive compensation of $k_{sh}=-0.96$ is required.

Series capacitive compensation may, in theory, be used instead of shunt compensation to give a flat voltage profile, under heavy loading. For example, a flat voltage profile can be achieved at a load of $1.4P_0$ with a distributed series compensation of $k_{se}=0.49$. In practice, lumped series capacitors are not suitable for obtaining a smooth voltage profile along the line. Obviously, step changes in voltage occur at points where the series capacitors are applied. They do, however, improve voltage regulation at any given point, i.e., voltage changes with load are reduced.

Effect of compensation on maximum power:

In Chapter 6 we developed the expression (Equation 6.51) for power transferred by a line. With compensation, this expression becomes

$$P_R = \frac{E_S E_R}{Z'_C \sin\theta'} \sin\delta \qquad (11.53)$$

The maximum power (corresponding to $\delta = 90°$) can be increased by decreasing either Z'_C, or θ', or both.

The characteristic impedance Z'_C can be decreased with capacitive shunt compensation, but it is accompanied by an increase in the electrical length θ'. On the other hand, inductive shunt compensation decreases θ', but increases Z'_C. Only series capacitor compensation contributes to the decrease of both Z'_C and θ'.

We should, however, recognize that compensation is not required in all cases to satisfy both objectives: (i) increasing P'_0, the power level at which the voltage profile is flat, and (ii) decreasing electrical length in order to improve stability. Short lines may require voltage support, i.e., an increase in P'_0, even though the inherent electrical length is small. This may be achieved by shunt capacitors, provided that θ' does not become excessive as a result. On the other hand, as we saw in Chapter 6 (see Figure 6.13), lines longer than about 500 km cannot be loaded even up to P_0 because of excessive θ; in such cases, reduction of θ' is the first priority.

Illustrative example

For purposes of illustration, we will consider a lossless 500 kV line having the following parameters:

$$\beta = 0.0013 \text{ rad/km} \qquad Z_C = 250 \ \Omega \ (P_0 = 1,000 \text{ MW})$$
$$x_L = 0.325 \ \Omega/\text{km} \qquad b_C = 5.2 \ \mu\text{S/km}$$

The line is 600 km long and transfers power between two sources as shown in Figure 11.54. The magnitudes of the source voltages are held at 1.0 pu. Our objective is to examine the line performance without and with compensation. We will consider shunt capacitor and series capacitor compensations chosen so as to maintain 1.0 pu midpoint voltage when the power transferred (P) is equal to $1.4P_0$.

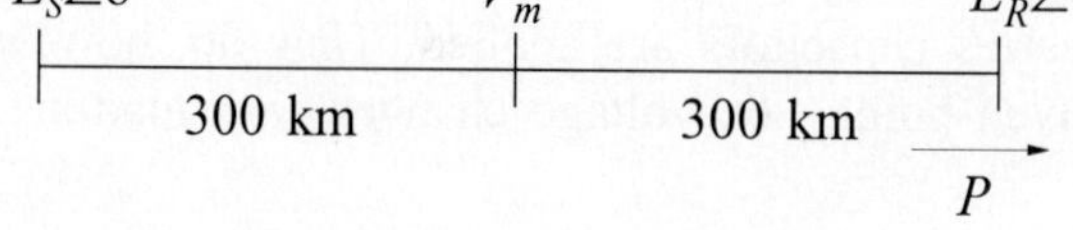

$E_S \angle \delta$ V_m $E_R \angle 0$

300 km   300 km

P

$E_S = E_R = 1.0$ pu
$\theta = \beta l = 0.0013 \times 600$
$= 0.78$ rad $= 44.7°$

Figure 11.54

(a) With *no compensation*, the power-angle relationship is

$$P = \frac{E_S E_R}{Z_C \sin\theta} \sin\delta$$

With E_S and E_R at rated values,

$$\frac{P}{P_0} = \frac{1}{\sin\theta} \sin\delta = \frac{1}{\sin 44.7°} \sin\delta = 1.42 \sin\delta$$

Also, considering one half of the symmetrical line, P may be expressed in terms of V_m as

$$P = \frac{E_S V_m}{Z_C \sin(\theta/2)} \sin(\delta/2)$$

$$= P_0 \frac{V_m \sin(\delta/2)}{\sin(44.7°/2)}$$

Hence, the per unit value of midpoint voltage as a function of P is given by

$$V_m = \frac{P}{P_0} \frac{0.38}{\sin(\delta/2)}$$

(b) With *uniformly distributed fixed shunt compensation*, to maintain V_m at 1.0 pu when $P=1.4P_0$, we have

$$1.4 P_0 = P_0' = P_0 \sqrt{1-k_{sh}}$$

Therefore, $k_{sh}=-0.96$. This will, in fact, result in 1.0 pu voltage throughout the line length at $P=1.4P_0$. The corresponding values of Z_C' and θ' are

$$Z_C' = \frac{Z_C}{\sqrt{1-k_{sh}}} = \frac{250}{\sqrt{1+0.96}} = 178.57 \ \Omega$$

$$\theta' = \theta\sqrt{1-k_{sh}} = 10.92 \ \text{rad} = 62.57°$$

The power transferred is given by

$$P = \frac{E_S E_R}{Z_C' \sin\theta'} \sin\delta$$

Hence,

$$\frac{P}{P_0} = \frac{Z_C \sin\delta}{Z_C' \sin\theta'} = 1.58 \sin\delta$$

The midpoint voltage is now given by

$$V_m = \frac{P}{P_0} \frac{Z_C'}{Z_C} \frac{\sin(\theta'/2)}{\sin(\delta/2)} = \frac{P}{P_0} \frac{0.371}{\sin(\delta/2)}$$

(c) With *uniformly distributed fixed series compensation*, to maintain V_m at 1.0 pu when $P = 1.4P_0$, we have

$$1.4P_0 = P_0' = P_0 \Big/ \sqrt{1 - k_{se}}$$

Therefore, $k_{se} = 0.49$. The line parameters change to

$$Z_C' = Z_C \sqrt{1 - k_{se}} = 250\sqrt{1 - 0.49} = 178.57 \quad \Omega$$

$$\theta' = \theta \sqrt{1 - k_{se}} = 0.557 \quad \text{rad} = 31.92°$$

The power transfer and midpoint voltage equations now become

$$\frac{P}{P_0} = \frac{Z_C \sin\delta}{Z_C' \sin\theta'} = 2.65 \sin\delta$$

and

$$V_m = \frac{P}{P_0} \frac{Z_C'}{Z_C} \frac{\sin(\theta'/2)}{\sin(\delta/2)} = \frac{P}{P_0} \frac{0.1964}{\sin(\delta/2)}$$

Next, we consider the performance of lumped shunt capacitor and series capacitor compensations. With lumped compensation, the expression for P-δ and V_m-P relationships cannot be obtained by a simple modification of the corresponding equations for an uncompensated line. We will, therefore, use a power-flow program to analyze the performance of the compensated line.

(d) With *lumped midpoint shunt compensation*, a capacitor bank of 412 MVAr is required to maintain V_m at 1.0 pu when the power transferred is $1.4P_0$.

A distributed shunt compensation of -0.96 represents a total MVAr of about 749. This is nearly twice the MVAr rating of the lumped shunt compensation. The two sources at the ends of the line supply about half of the reactive power compensation required when lumped compensation is used.

(e) With *lumped series compensation*, split into four units (two at the ends and two in the middle of the line), as shown in Figure 11.55, the degree of compensation k_{se} required is 0.505 to maintain V_m at 1.0 pu when the power transferred is $1.4P_0$.

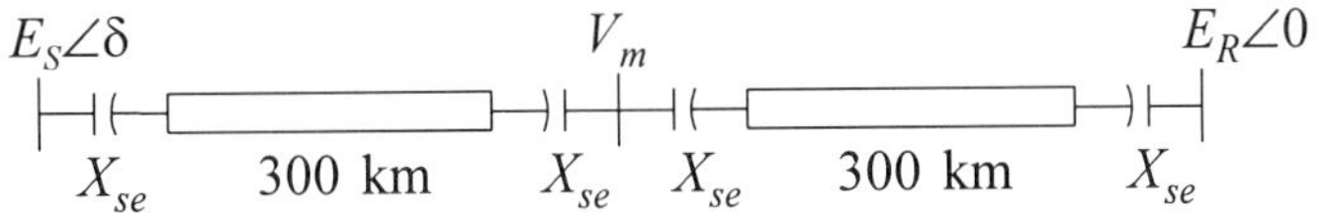

$$X_{se} = 24.625 \ \Omega \qquad k_{se} = \frac{4 \times 24.625}{600 \times 0.325} = 0.505$$

Figure 11.55 Lumped series compensation

Figure 11.56(a) shows plots of power transferred as a function of transmission angle δ for all the five cases we have considered: (a) no compensation, (b) distributed shunt compensation, (c) distributed series compensation, (d) lumped shunt compensation, and (e) lumped series compensation. Plots of midpoint voltage as a function of power transferred for the five cases are shown in Figure 11.56(b). We have considered here a 500 kV line. With line losses neglected and power expressed in per unit of natural power (P_0), the only parameter that has an impact on line performance is β. As noted in Chapter 6, β is practically the same for lines of all voltage levels. Hence, the results presented are generally applicable to a 600 km line of any voltage level.

The P-δ and V_m-P characteristics with each type of lumped compensation, for the system considered, are practically identical to the characteristics computed with the corresponding uniformly distributed compensation.

From the results presented in Figure 11.56, we can make some general observations:

1.   The shunt capacitor compensation (chosen to keep midpoint voltage at 1.0 pu when $P=1.4P_0$) increases the maximum power capability to 1.58 pu of natural power. This represents an increase of 0.16 pu over the limit of 1.42 pu for the uncompensated case.

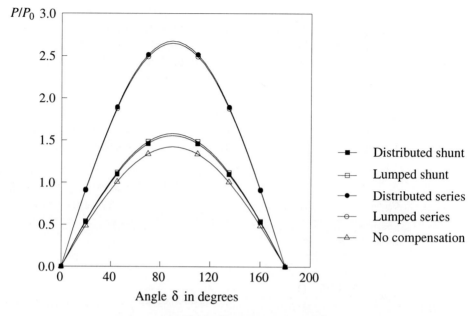

(a) Power transfer as a function of transmission angle δ

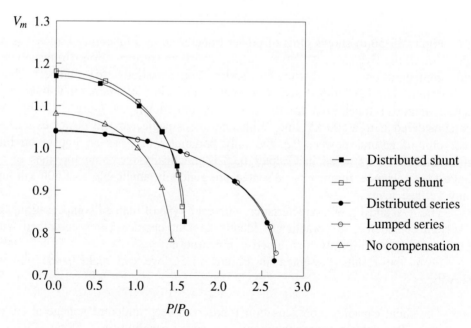

(b) Midpoint voltage as a function of power transfer

Figure 11.56   Performance of 600 km line with and without compensation

While the shunt compensation has helped maintain the midpoint voltage at the rated values when P is equal to 1.4 pu, the voltage magnitude is very sensitive to variations in power transfer level. For example, when the shunt compensated line is operating near the effective natural load ($1.4P_0$), a small change in power changes the voltage magnitude significantly.

We also see that, when operating at a power transfer level corresponding to V_m of 1.0 pu, the shunt compensated line is much closer to the stability limit than the uncompensated line. Thus, shunt compensation has increased the effective natural load at the expense of stability margin and voltage regulation.

2.   The series capacitor compensation (chosen so as to keep midpoint voltage at 1.0 pu when $P=1.4P_0$) increases the maximum transfer capability to 2.65 pu. In addition to nearly doubling the stability limit, the voltage regulation is significantly improved; for example, at $P=1.4P_0$, a large variation in P causes a very small change in V_m.

Uniformly distributed regulated shunt compensation

Consider a shunt compensation in which k_{sh} is continuously *regulated* so that the effective natural load (P_0') is equal to the power transmitted (P) at all times. From Equation 11.53,

$$P = \frac{P_0'}{\sin\theta'}\sin\delta$$

Therefore, when a line is operating at natural load, the transmission angle is equal to the line angle (see Chapter 6, Section 6.1.9). As a result, with continuously regulated shunt compensation, $\theta'=\delta$ at all times. Hence,

$$\frac{P}{\delta} = \frac{P_0'}{\theta'} = \frac{P_0\sqrt{1-k_{sh}}}{\theta\sqrt{1-k_{sh}}} = \frac{P_0}{\theta} = \text{constant} \tag{11.54}$$

We thus have a *linear* relationship between P and δ, instead of the sinusoidal relationship with fixed or no compensation. This means that the small-signal (steady-state) stability limit with the regulated compensation is infinite. With V_0 denoting the rated line voltage, the slope of the P-δ characteristic is given by

$$\frac{P_0}{\theta} = \frac{V_0^2}{Z_c\theta} = \frac{V_0^2}{X_L} \tag{11.55}$$

This is equal to the peak of the sinusoidal $P\text{-}\delta$ characteristic with 100% shunt inductive compensation $(k_{sh}=1)$.

Let us consider the 600 km, 500 kV line for illustration. From Equation 11.54, with δ and θ expressed in radians,

$$\frac{P/P_0}{\delta} = \frac{1}{\theta} = \frac{1}{0.78} = 1.282 \quad \text{pu/rad}$$

This is illustrated in Figure 11.57, which shows $P\text{-}\delta$ characteristics with uniformly distributed regulated compensation and with fixed distributed shunt compensation of $k_{sh}=1.0$.

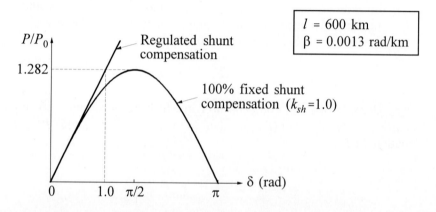

Figure 11.57 Power-angle characteristics of a 600-km line
with regulated and shunt inductive compensation

The regulated shunt compensation effectively changes k_{sh} continuously so that $P_0'=P$. For each value of k_{sh}, there is a sinusoidal $P\text{-}\delta$ characteristic. As the transmitted power P changes, the regulator effectively adjusts k_{sh} so that the operating point shifts from one curve to another in such a way that it lies on a straight line with a positive slope. For satisfactory operation at high power transfer levels, the regulator must be rapid and continuous to prevent movement along the sinusoidal characteristic corresponding to the current value of k_{sh} before moving to the new characteristic.

In practice, this form of compensation can be nearly achieved by placing active compensators, such as synchronous condensers or static var systems, at discrete intervals along the line. The compensators would hold constant voltage equal to V_0 at many points along the line for all load levels; this nearly satisfies the requirement for the effective natural load (P_0') to be equal to the current value of power (P) being transmitted.

The regulators must, however, have sufficient capacity to satisfy the reactive power requirements for maintaining constant voltage at all possible load levels. For the ideal case, the total reactive power supplied or absorbed by the constant voltage regulators is given by

$$Q_V = (V^2 \omega C - I^2 \omega L) l$$

$$= P_0 \theta \left[1 - (P/P_0)^2 \right]$$

(11.56)

For P greater than P_0, the reactive power required is capacitive. The reactive power increases as the square of the power transmitted.

Satisfactory performance of such a scheme depends on the ability of all regulators to maintain constant voltage along the line. If one regulator fails or if it reaches a reactive power limit, the stability of the entire system may be affected.

Hydro Quebec's James Bay transmission system [39] uses this form of compensation. It is of historical interest to note that the concept of long distance ac power transmission using distributed regulated shunt compensation was proposed by F.G. Baum as far back as 1921 [43]. He envisioned a 220 kV line having its voltage maintained by synchronous condensers at intervals of about 100 mi. He showed that with equal voltage (within 3%) at all compensator stations, there was virtually no limit to transmission distance.

Regulated compensation at discrete intervals

We will now examine how regulated compensation applied at discrete intervals approximates the performance of uniformly distributed compensation. We will first consider a midpoint compensator as shown in Figure 11.58, and then we extend the analysis to an arbitrary number of compensators.

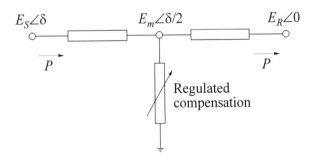

Figure 11.58 Line with midpoint regulated compensation

Midpoint regulated compensation:

The compensating susceptance is continuously varied such that E_m is constant. For simplicity, we will assume that $E_S = E_R = E_m = E$. The line may be considered to be made up of two *independent sections*, each section having a line angle of $\theta/2$, where θ is line angle of the entire line. By applying Equation 11.53 to each section, the expression for power transmitted is

$$P = \frac{E^2}{Z_C \sin(\theta/2)} \sin(\delta/2) \qquad (11.57)$$

If E is equal to rated voltage, the expression for P may be expressed in terms of the natural power P_0 of the uncompensated line as follows:

$$P = \frac{P_0}{\sin(\theta/2)} \sin(\delta/2) \qquad (11.58)$$

For an uncompensated line, the power transmitted is given by

$$P = \frac{P_0}{\sin\theta} \sin\delta \qquad (11.59)$$

The ratio of maximum power that can be transmitted with and without midpoint regulated compensation is

$$\frac{P'_{max}}{P_{max}} = \frac{\sin\theta}{\sin(\theta/2)} \qquad (11.60)$$

Table 11.2 lists the values of the above ratio for different line lengths (assuming $\beta = 0.0013$ rad/km).

Figure 11.59 shows plots of power versus angle characteristics with and without midpoint regulated compensation for a line length of 600 km ($\theta = 44.7°$).

If the regulated compensation has unlimited reactive power capability, it will hold the midpoint voltage through any load level as seen above. On the other hand, if it has a finite size, it can regulate only up to its maximum capacitive output. Its performance beyond this limit depends on the type of device.

Table 11.2

Line Length l (km)	Line Angle θ (degrees)	P'_{max}/P_{max}
200	14.9	1.98
400	29.8	1.93
600	44.7	1.85
800	59.6	1.73
1000	74.5	1.59
1200	89.4	1.42

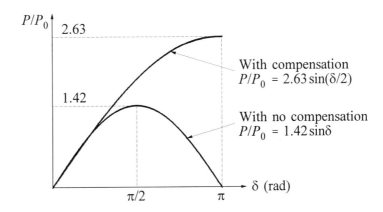

Figure 11.59 Power-angle relationship with and without midpoint regulated compensation

As an example, let us consider an SVS applied to the midpoint of the 600 km line considered above. The SVS performs as a regulated compensation until the reactive power output limit is reached. If the SVS has unlimited capacity, the P-δ characteristic is shown as curve 1 in Figure 11.60. If the SVS has a maximum capacitive rating of 750 MVAr at rated voltage, when the capacitive limit is reached the SVS behaves as a simple capacitor. Thus point A in Figure 11.60 represents the transition from a variable susceptance voltage control operation to a constant-B_{sh} operation represented by curve 2.

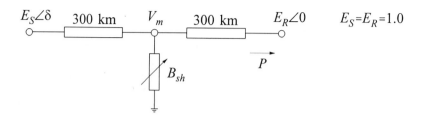

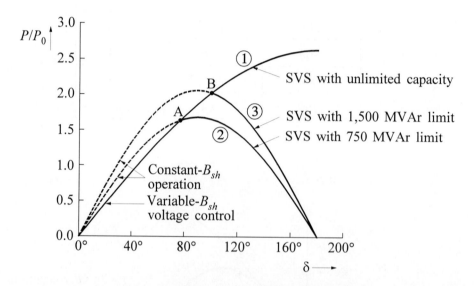

Figure 11.60 Performance of a 600 km line with
an SVS regulating midpoint voltage

If the SVS capacitive limit is 1,500 MVAr, the corresponding constant-B_{sh} characteristic is represented by curve 3. In this case, the transition point (B) is on the unstable part (beyond $\delta = 90°$) of the power-angle curve. The operation would be unstable immediately following the transition from the variable susceptance mode to the fixed capacitor mode.

With an SVS of unlimited capacity, the maximum power that can be transmitted is $2.63P_0$. However, this is achieved by the SVS supplying very large amounts of reactive power at high values of P. Figure 11.61 shows a plot of the reactive power (Q) supplied (or absorbed) by the SVS to maintain 1.0 pu midpoint voltage as the transmitted power P is varied. The SVS absorbs Q when P is less than P_0, and supplies Q when P is greater than P_0. The Q supplied is not excessive for P less than about $1.4P_0$. Beyond a certain limit, it may not be economical to provide an SVS with the required reactive power capability; it may be cheaper to achieve the desired transmission capability by an alternative means.

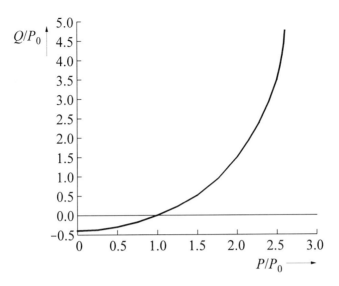

Figure 11.61 Reactive power supplied by SVS
as a function of transmitted power

In our analysis, we have neglected the effect of the SVS droop characteristic; that is, we have assumed the slope reactance X_{SL} to be zero. The effect of a finite X_{SL} is to reduce the maximum power and the reactive power requirements.

Arbitrary number of regulated compensators:

We will next consider the general case where $n-1$ regulated compensators are applied to the line at regular intervals so that the line is divided into n independent sections. The power transmitted is now given by

$$P = \frac{P_0}{\sin(\theta/n)}\sin(\delta/n) \tag{11.61}$$

The ratio of maximum power that can be transmitted with and without the compensation is

$$\frac{P'_{max}}{P_{max}} = \frac{\sin\theta}{\sin(\theta/n)} \tag{11.62}$$

Table 11.3 lists the above ratio for different values of n for a line length of 600 km (assuming $\beta = 0.0013$ rad/km).

Table 11.3

n	θ/n (degrees)	P'_{max}/P_{max}
1	44.70	1.00
2	22.35	1.85
3	14.90	2.74
4	11.17	3.63
6	7.45	5.42
8	5.59	7.22
10	4.47	9.03

Figure 11.62 shows plots of P-δ characteristics for different values of n.

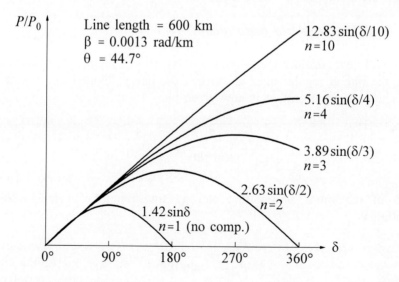

Figure 11.62 Power-angle relationships with regulated compensation at discrete intervals dividing line into n independent sections

Comparative summary of alternative forms of compensation [44-46]

1.   Switched shunt capacitor compensation generally provides the most economical reactive power source for voltage control. It is ideally suited for compensating transmission lines if reduction of the effective characteristic impedance (Z'_C), rather than reduction of the effective line angle (θ') is the primary consideration.

      However, heavy use of shunt capacitor compensation could lead to reduction of small-signal (steady-state) stability margin and poor voltage regulation.

2.   Series capacitor compensation is self-regulating, i.e., its reactive power output increases with line loading. It is ideally suited for applications where reduction of the effective line angle (θ') is the primary consideration. It increases the effective natural load as well as the small-signal stability limit and it improves voltage regulation. It is normally used to improve system stability and to obtain the desired load division among parallel lines.

      Series capacitor compensation could cause subsynchronous resonance problems requiring special solution measures. In addition, protection of lines with series capacitors requires special attention.

3.   A combination of series and shunt capacitors may provide the ideal form of compensation in some cases. This allows independent control of the effective characteristic impedance and the load angle δ. An example of such an application is a long line requiring compensation which, in addition to increasing the effective SIL, causes the phase angle across the line to take a desired value so as not to adversely affect loading patterns on parallel lines.

4.   A static var system is ideally suited for applications requiring direct and rapid control of voltage. It has a distinct advantage over series capacitors where compensation is required to prevent voltage sag at a bus involving multiple lines. Since shunt compensation is connected to the bus and not to particular lines, the total cost of the regulated shunt compensation may be substantially less than that for series compensation of each of the lines.

      When an SVS is used to permit a high power transfer over a long distance, the possibility of instability when the SVS is pushed to its limit must be recognized. When operating at its capacitive limit, the SVS becomes a simple capacitor; it offers no voltage control and its reactive power drops with the square of the voltage. Systems heavily dependent on shunt compensation may experience nearly instantaneous collapse when loadings exceed the levels for which the SVS is sized. The ratings of the SVS should be based on very thorough studies which define its total MVAr and the switched and dynamically controlled portions. An SVS has limited overload capability and has higher losses than series capacitor compensation.

11.2.9 Modelling of Reactive Compensating Devices

In power-flow and stability studies, the passive compensating devices (shunt capacitors, shunt reactors, and series capacitors) are modelled as admittance elements of fixed values. They are included in the network admittance matrix (see Chapter 6, Section 6.4.1) along with the other transmission network passive elements.

Synchronous condensers are modelled as synchronous generators but with no steady-state active power output. Except for the fact that there is no prime mover, the representation of a synchronous condenser and the associated excitation system is similar to that for a synchronous generator.

However, special models are required for representing static var compensators.

Modelling of static var systems [36,38,47]

Figure 11.63 shows the schematic diagram of a typical SVS. It consists of a TCR and switched (thyristor or mechanical) capacitors, and has the steady-state characteristic shown in Figure 11.64. The characteristics and modelling of SVSs with different configurations comprising other types of SVCs are conceptually the same.

Representation of SVS in power-flow studies:

From Figure 11.64 we see that the SVS has three possible modes of operation. The corresponding equivalent circuits of the SVS as seen from the HV bus are shown in Figure 11.65.

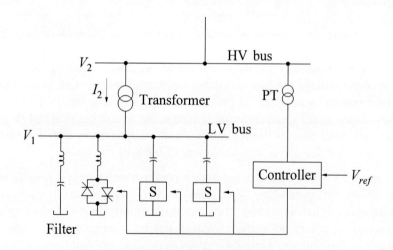

S = thyristor or mechanically controlled switch

Figure 11.63 Schematic of a typical SVS

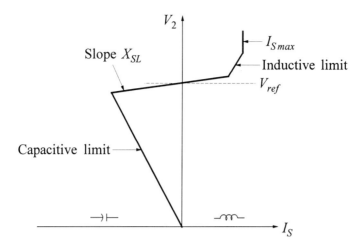

Figure 11.64   Steady-state *V-I* characteristic

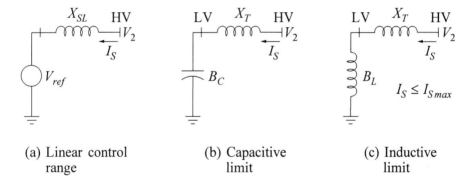

(a) Linear control
     range

(b) Capacitive
     limit

(c) Inductive
     limit

Figure 11.65   Steady-state equivalent circuits of an SVS

The normal mode of operation is in the linear control range. In this mode, the SVS as seen from the HV bus is equivalent to a voltage source V_{ref} in series with the slope reactance X_{SL}. When the SVS operation reaches the capacitive limit, it becomes a fixed capacitive susceptance (B_C) connected to the LV bus. The reactance between the LV bus and the HV bus is the transformer leakage reactance X_T. Similarly, when the SVS operation hits the inductive limit, it becomes a fixed susceptance (B_L) whose net value is inductive. The maximum value of the SVS current is limited to $I_{S\,max}$.

For power-flow analysis, the network configuration shown in Figure 11.66 may be used to represent the SVS. The transformer reactance X_T is split into X_{SL} and X_T-X_{SL}, thereby creating a phantom bus (M). This allows proper representation of the SVS while operating in the linear control mode, including the effect of the slope reactance. In this mode, the SVS is represented as a *PV* bus (with $P=0$) remotely controlling the bus M voltage equal to V_{ref}.

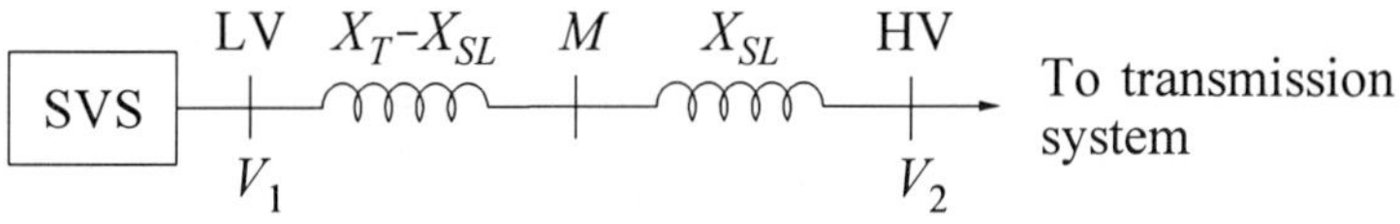

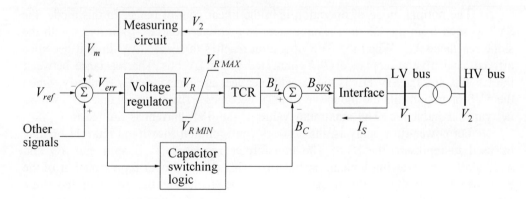

Figure 11.66 SVS representation in power-flow studies

When either of the reactive limits is reached, the SVS becomes a simple susceptance of fixed value connected to the LV bus; bus M voltage is no longer controlled. The SVS current is limited to $I_{S\,max}$.

The above representation accounts for the SVS performance appropriately in all three modes. In addition, it retains the identity of the LV bus to which the SVS is connected.

Representation in stability studies:

Static var systems are usually configured to meet individual system requirements. The control techniques used vary depending on the equipment suppliers and on the vintage of equipment. Therefore, standard stability models capable of representing in detail the wide range of SVSs in use have not been developed. Instead, basic models appropriate for general purpose studies have been recommended by CIGRE [36]. A new set of basic models covering recent developments has been prepared by the IEEE [47]. Such basic models are adequate for general purpose studies in which the special features specific to individual installations do not affect stability analysis. These models may also be used for preliminary studies involving new installations. For detailed studies, however, accurate models reflecting actual controls may be necessary.

Here, we will describe the general approach to modelling of an SVS of the type shown in Figure 11.63. A functional block diagram representation of the SVS is shown in Figure 11.67. The stability model may be developed by identifying the mathematical model for each functional block.

Figure 11.67 SVS functional block diagram

The *thyristor-controlled reactor* (TCR) block represents the variation of reactor susceptance as a function of the firing angle, thyristor firing circuits (Equation 11.36), and linearizing circuit used for compensating the nonlinear relationship between the susceptance B_L and the conduction angle σ. Figure 11.68 shows a model of the TCR block with parameters expressed in per unit with the MVA rating (Q_{TCR}) and kV rating of the TCR as base values. The conduction angle σ is in per unit on a base of 180°.

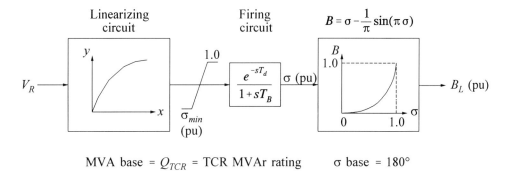

MVA base = Q_{TCR} = TCR MVAr rating σ base = 180°

Figure 11.68 Model of TCR block

The limits shown in the firing circuit block represent the limits on the conduction angle, which in turn determine the limiting values of B_L. In per unit, the maximum value of B_L is 1.0. Typically, the minimum value of B_L for a TCR is 0.02 pu; it may be assumed to be zero. The parameter T_d is the gating transport delay; it has a value of about 1 ms and is normally neglected, making $e^{-sT_d} \approx 1.0$. The time constant T_B associated with the thyristor firing sequence control has a value of about 5 ms; it may also be neglected for most studies.

If the nonlinear relationship between σ and B_L is assumed to be perfectly compensated and T_d and T_B are neglected, the TCR block may be represented by a unit gain with limits as shown in Figure 11.69.

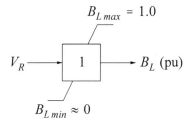

Figure 11.69 Simplified model of TCR block

The structure of the *voltage regulator* block in Figure 11.67 is very much dependent on the particular application as it provides a means of optimizing the SVS dynamic performance. Figure 11.70 shows models of two commonly used forms of regulators.

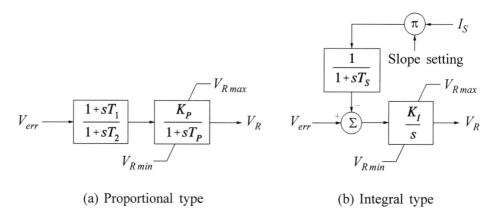

(a) Proportional type   (b) Integral type

Figure 11.70 Voltage regulator models

With the proportional type regulator shown in Figure 11.70(a), the gain K_P is the reciprocal of the droop characteristic, i.e., the slope reactance X_{SL}. The time constant T_P is usually in the range of 50 ms to 100 ms. Figure 11.70(b) shows an integral type regulator. The droop characteristic is obtained through feedback of SVS current. The limiters associated with both forms of regulators are of the non-windup type.

The *measuring circuit* block in Figure 11.67 includes instrument transformers, A/D converters, and rectifiers. It contains a transport delay and time constants which are very small. Consequently, the measuring circuit may be represented by a simple time constant on the order of 5 ms or as a unit gain.

The *capacitor-switching logic* may be represented by a switching sequence depending on the means used for switching, i.e., mechanical or thyristor switches. The intelligence used for switching depends on the requirements of the particular application.

Figure 11.71 shows a model suitable for representing a thyristor switched capacitor (TSC) with parameters expressed in per unit with the MVA rating (Q_{TSC}) and kV rating of the TSC as base values. In the model, T_C is the time constant associated with the thyristor sequence control.

In the model shown, N_T is the total number of individually switched capacitor bank units (assumed to be of equal size), N_C is the number of units switched in at any given time, and T_C is a time constant associated with the thyristor firing sequence control. The control signal V_C attempts to maintain V_2 within a narrow band, with the control action coordinated with that of the TCR.

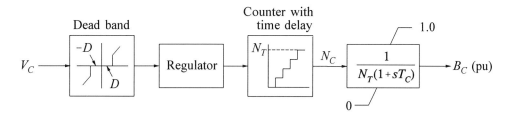

MVA base = Q_{TSC} = TSC MVAr rating

Figure 11.71 TSC model

Figure 11.72 shows the per unit representation of a combined TCR and TSC system. TCR and TSC are each modelled by using a per unit system with their respective MVAr ratings as base MVA. The susceptances B_L and B_C are converted to the common per unit system (usually with 100 MVA base) used to model the entire power system. In Figure 11.72, the net susceptance B_{SVS} and the total SVS current are expressed in the common per unit system. The net susceptance B_{SVS} is inductive if positive, and is capacitive if negative. The current into the SVS represented by I_{SVS} is positive when inductive (load convention). This is consistent with the convention commonly used for representing the V-I characteristic of the SVS (Figure 11.64).

For most system studies, it is not necessary to represent a TSC explicitly as shown in Figure 11.72. The TSC response may be assumed to be instantaneous, and the SVS is represented by a TCR and an FC with the MVAr ratings appropriately selected.

With a fixed capacitor there is no switching logic; the susceptance B_C has a fixed value. Figure 11.73 shows a simplified model of an SVC consisting of a TCR and a fixed capacitor. The TCR is assumed to have a proportional type regulator. The parameters are in per unit with TCR ratings as base values.

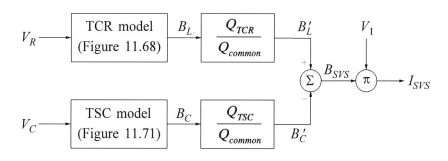

Q_{TCR} = MVAr rating of TCR Q_{common} = common MVA base
Q_{TSC} = MVAr rating of TSC V_1 = LV bus voltage

Figure 11.72 Combined TCR and TSC representation

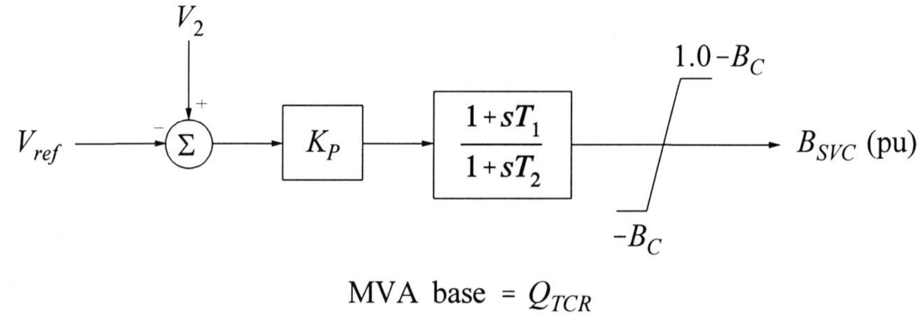

$$\text{MVA base } = Q_{TCR}$$

Figure 11.73 Simplified model of an SVC comprising a TCR and an FC

In addition to the feedback of HV bus voltage to control the SVS, additional signals may be used to enhance system stability.

11.2.10 Application of Tap-Changing Transformers to Transmission Systems

Transformers with tap-changing facilities constitute an important means of controlling voltage throughout the system at all voltage levels. In Chapter 6, we discussed the principle of operation and modelling of tap-changing transformers. Here we will consider how they are used to control voltage and reactive power at the transmission system level.

Autotransformers used to change voltage from one subsystem to another (for example, 500 kV to 230 kV) are often furnished with under-load tap-changing (ULTC) facilities. These may be controlled either automatically or manually. Usually there are many such transformers throughout the network interconnecting transmission systems of different levels. The taps on these transformers provide a convenient means of controlling reactive power flow between subsystems. This in turn can be used to control the voltage profiles, and minimize active and reactive power losses.

The control of a single transformer will cause changes in voltages at its terminals. In addition, it influences the reactive power flow through the transformer. The resulting effect on the voltages at other buses will depend on the network configuration and load/generation distribution. Coordinated control of the tap changers of all the transformers interconnecting the subsystems is required if the general level of voltage is to be changed.

During high system load conditions, the network voltages are kept at the highest practical level to minimize reactive power requirements and increase the effectiveness of shunt capacitors and line charging. The highest allowable operating voltage of the transmission network is governed by the requirement that insulation levels of equipment not be exceeded, taking into consideration possible switching operations and outage conditions. During light load conditions, it is usually required to lower the network voltages to reduce line charging and avoid underexcited operation of generators.

Transformers with off-load tap-changing facilities can also help maintain satisfactory voltage profiles. While transformers with ULTC can be used to take care of daily, hourly, and minute-by-minute variations in system conditions, settings of off-load tap-changing transformers have to be carefully chosen depending on long-term variations due to system expansion, load growth, or seasonal changes. Optimal power-flow analysis provides a convenient method of determining appropriate tap settings with either type of tap-changing facility [49-51].

Utility practices with regard to application of ULTC transformers to transmission systems vary widely. For example, Ontario Hydro's practice is to provide ULTC facilities on most 500/230 kV autotransformers and on all "area supply" transformers stepping down from 230 kV or 115 kV to 44 kV, 27.6 kV, or 13.8 kV. All 230/115 kV autotransformers and generator step-up transformers are provided with off-load tap-changing facilities. Figure 11.74 illustrates the general arrangement.

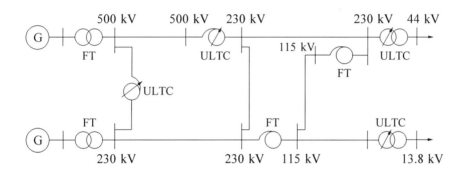

FT = Fixed tap or off-load tap changing
ULTC = Under-load tap changing

Figure 11.74 Single-line diagram of transmission network
illustrating transformer tap change facilities

As an additional example, the practice in the United Kingdom is to provide ULTC on all generator step-up transformers and on autotransformers connecting a 400 kV or 275 kV "supergrid network" to a 132 kV or 66 kV "secondary network." Autotransformers connecting the 400 kV and 275 kV networks have fixed ratios.

Many utilities, on the other hand, do not provide for under-load tap changing of transmission network autotransformers.

11.2.11 Distribution System Voltage Regulation [25,48]

Automatic voltage regulation of distribution systems is provided by using one or more of the following methods:

- Bus regulation at the substation

- Individual feeder regulation in the substation

- Supplementary regulation along the feeders

Substation bus regulation

A distribution substation transformer is usually equipped with ULTC equipment that automatically controls the secondary voltage. Alternatively, the substation may have a separate voltage regulator that regulates the secondary side bus voltage.

Bus regulation generally employs three-phase units, although single-phase regulators could be used in applications where phase voltages have a significant unbalance.

Feeder regulation

Feeder voltage regulators control the voltage of each feeder. Either single-phase units or three-phase units may be used, the former being more common. Single-phase regulators are necessary when the individual phases serve diverse forms of loads. Three-phase regulators can provide the same high quality performance as the single-phase regulators, when the phases are similarly loaded and when the supply voltage is balanced.

If there are several feeders supplied by a substation, feeders with similar characteristics may be grouped, and a common regulator may be used to control the voltages of each group of feeders.

In applications where the feeders are very long, additional regulators and shunt capacitors located at selected points on the feeders provide supplementary regulation.

Feeder regulators are also sometimes referred to as booster transformers. A conventional transformer with ULTC performs two functions: voltage transformation and voltage control. Feeder voltage regulators perform only the latter function; that is, they buck or boost the voltage without changing the basic voltage level.

Types of feeder regulators

There are two basic types of feeder voltage regulators: the induction type and the step type. The following is a brief description of their operating principles.

Induction voltage regulator:

Figure 11.75 shows a schematic of an induction voltage regulator. It consists of two sets of windings: (a) primary winding wound on the rotor and connected across the line, and (b) regulating winding wound on the stator and connected in series with the line.

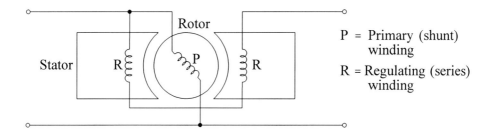

Figure 11.75 Schematic of an induction regulator

The voltage induced in the regulating series winding is added to the primary winding to give the output voltage. The magnitude and polarity of the induced voltage depend on the relative orientation of the regulating winding with respect to the primary winding. By changing the rotor position, the output voltage can be varied between the maximum and minimum limits. The position of the rotor is controlled by an electric motor which responds to a control signal.

The induction regulator provides accurate and continuous control, and performs reliably. Many of the older regulators in service are of this type. It is, however, costly and has been largely superseded by the step type regulator.

Step voltage regulator (SVR):

The step voltage regulator is basically an autotransformer with taps or steps in the series winding, as shown in Figure 11.76. However, it is purely a voltage control device; that is, it is not used for voltage transformation.

The voltage induced in the series winding is either added to or subtracted from the primary voltage depending on the polarity of the series winding. A reversing switch (RS) is provided to change this polarity. The series-winding output voltage magnitude is varied by changing the tap position, which can be done under load.

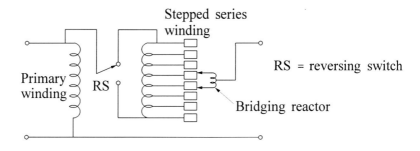

Figure 11.76 Schematic of a step voltage regulator

Typically, the SVR has provision for correcting the voltage by ±10% in 32 steps, each step representing a 5/8% change in voltage. This is achieved by tapping the series winding into eight equal parts, with each part providing one-eighth of the 10% change in voltage. The output terminal is connected to the centre tap of the bridging reactor associated with the tap-changing mechanism. This in effect further divides each step into two equal parts, giving a total of 16 steps of 5/8% each. The reversing switch allows the regulator to raise as well as lower the output voltage, covering a range of plus or minus 10% voltage regulation in a total of 32 steps.

Figure 11.77 shows the major elements of the SVR control mechanism. The SVR is set to hold a constant voltage (within a narrow range) at its secondary terminals or at some selected point out on the feeder as determined by the settings (R and X) of the *line drop compensator*. The voltage sensor compares the input voltage to a preset voltage level. If the input voltage deviates from the setpoint beyond a tolerance or spread for a certain time, the tap-changing motor operates the tap-changing mechanism in a direction so as to bring the voltage back to within a narrow range. This range is called the "bandwidth," and is typically ±2% around the setpoint. The time delay, which is adjustable, prevents the regulator from responding to temporary or self-correcting voltage variations. The time delay setting for the first tap movement can range from 30 to 60 seconds; a 30 seconds setting is typical [48].

The time taken by the tap-changing mechanism for each additional tap movement is in the 2 to 8 seconds range, with 6 seconds being a typical value. Thus, a change from the neutral tap position to full boost or buck takes about 2 minutes.

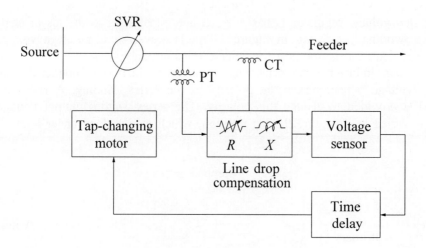

Figure 11.77 SVR control mechanism

Application of voltage regulators and capacitors

Figure 11.78 illustrates the application of regulators and shunt capacitors for control of voltage profile along a feeder. Curve 1 shows the voltage profile with a fairly evenly distributed load along the line, without any regulator or capacitor. The voltage for most parts of the feeder is seen to be below the permissible minimum value. The addition of a voltage regulator (R_1) moves the voltage profile up, as shown by curve 2. A capacitor bank (C), located at approximately two-thirds of the feeder length from the substation, moves the voltage profile to curve 3. The addition of a supplementary regulator (R_2) at approximately one-third of the feeder length from the substation will bring the voltage profile along the entire length of the feeder (from the first consumer to the last) to within the maximum and minimum permissible limits.

For very long feeders, it may be necessary to use two regulators in cascade. One regulator is placed at approximately the middle of the feeder and the other regulator at about one-fifth the distance from the station end. In such cases, it is necessary to ensure proper sequence of operation of the two regulators by setting the time delay of the regulator farther from the station longer than that for the closer regulator. Typical settings are 30 seconds for the regulator closer to the station and 40 seconds for the other regulator.

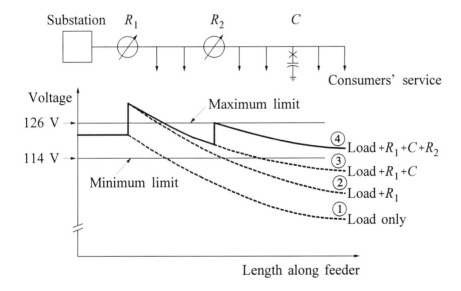

Figure 11.78 Voltage profile of a feeder with a station regulator (R_1), supplementary regulator (R_2) and a shunt capacitor bank (C)

11.2.12 Modelling of Transformer ULTC Control Systems

Modelling of transformers with tap-changing facilities is described in Chapter 6 (Section 6.2). Here we will focus on models for the control systems used for automatically changing transformer taps under load.

The functional block diagram of the control system is shown in Figure 11.79. It consists of the following basic elements [52]:

(a)   Tap-changing mechanism driven by a motor unit

(b)   Voltage regulator consisting of a measuring element and a time-delay element

(c)   Line drop compensator

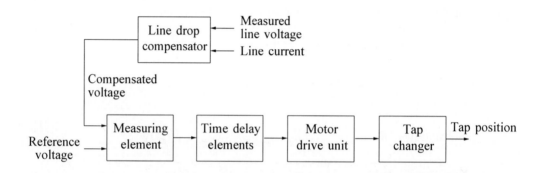

Figure 11.79   Functional block diagram of control system for automatic changing of transformer taps

Figure 11.80 shows the block diagram of the ULTC control system suitable for system studies.

The function of the *line drop compensator*, as discussed in the previous section, is to regulate voltage at a remote point along the line or feeder. The voltage at the remote point is simulated by computing the voltage across the compensator impedance (R_C+jX_C). The magnitude of the compensated voltage is given by

$$V_C = |\tilde{V}_t+(R_C+jX_C)\tilde{I}_t|$$

where $\tilde{V}_t$ is the measured phasor voltage at the transformer secondary side.

The *measuring element* of the voltage regulator consists of an adjustable dead band relay with hysteresis. The input to the regulator is the voltage error.

$$V_{err} = V_{ref}-V_C$$

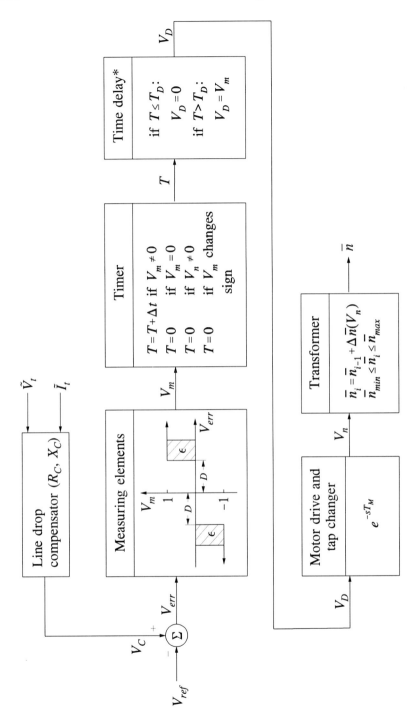

Figure 11.80 ULTC control system model

The output of the measuring element is V_m, which takes a value of 0, 1, or -1, depending on the input V_{err}. With a regulator dead band of D and a hysteresis band of ϵ, the output is

$$
V_m = \begin{cases}
0 & \text{for} \quad -D \le V_{err} \le +D \\
0 & \text{for} \quad D < V_{err} \le D+\epsilon; \quad V_{err} \text{ increasing} \\
0 & \text{for} \quad -D-\epsilon \le V_{err} < -D; \quad V_{err} \text{ decreasing} \\
+1 & \text{for} \quad V_{err} > D+\epsilon \\
+1 & \text{for} \quad D < V_{err} \le D+\epsilon; \quad V_{err} \text{ decreasing} \\
-1 & \text{for} \quad V_{err} < -D-\epsilon \\
-1 & \text{for} \quad -D-\epsilon \le V_{err} < -D; \quad V_{err} \text{ increasing}
\end{cases}
$$

The *time delay element* is used to prevent unnecessary tap changes in response to transient voltage variations and to introduce the desired time delay before a tap movement. The *timer unit* determines the time duration of the error voltage (V_{err}) exceeding the dead band. The timer is advanced if V_{err} is outside the dead band. It is reset if V_{err} is within the dead band, if there is a tap movement ($v_n \ne 0$), or if V_{err} oscillates above and below the dead band. The output V_D of the time delay unit is normally zero. If the accumulated time T of the timer exceeds T_D, then V_D is set to V_m (i.e., 1 or -1), thereby sending a signal to the tap-changer motor to move the tap up or down.

The time delay T_D is equal to T_{D0} for the first tap movement. Some regulators have an inverse time-delay characteristic, in which case the time delay is inversely proportional to the voltage error:

$$
T_D = \frac{T_{D0}}{V_{err}/D}
$$

For the second and subsequent tap movements the time delay T_D is equal to T_{DI}. This allows introduction of intentional time delay between consecutive tap movements, if so desired.

The *motor-drive unit and the tap-changer mechanism* may be represented by a simple time delay T_M inherent to the equipment. The output signal V_n represents an incremental change in tap position, and is equal to 0, 1, or -1.

A change in tap position is reflected in the *transformer* model (see Chapter 6, Section 6.2.1) as an incremental change in per unit turns ratio. The per unit turns ratio after the ith operation is

$$
\bar{n}_i = \bar{n}_{i-1} + \Delta\bar{n}(V_n)
$$

where $\Delta \bar{n}$ represents the per unit turns ratio step corresponding to a change in tap position by one step.

The above assumes that the controlled bus is on the secondary side (see Figure 6.17 of Chapter 6). If the controlled bus is on the primary side, we have

$$\bar{n}_i = \bar{n}_{i-1} - \Delta \bar{n}(V_n)$$

The tap ratio is subject to the maximum and minimum limits $(\bar{n}_{max}$ and $\bar{n}_{min})$.

The following sample data apply to the ULTC control system of the 42 MVA, 110/28.4 kV two-winding transformer considered in Chapter 6, Section 6.2.1:

$$\Delta \bar{n} = 0.0059524 \qquad \bar{n}_{max} = 1.04762 \qquad \bar{n}_{min} = 0.85714$$
$$T_M = 0.5 \text{ s} \qquad\qquad T_{D0} = 30.0 \text{ s} \qquad T_{D1} = 0$$
$$D = 0.00835 \ (0.835\%) \qquad \varepsilon = 0 \qquad\qquad R_C = X_C = 0$$

11.3 POWER-FLOW ANALYSIS PROCEDURES

The main analytical tools used in the planning of reactive power resources are the power-flow and stability programs. The methods of analysis of various aspects of stability are discussed in Chapters 12 to 16.

For reactive power dispatching, in addition to conventional power flows, optimal power-flow programs are being increasingly used. *Optimal power-flow* (OPF) theory was first formulated by Carpentier in 1962 [53]. Since then much research work has been carried out on OPF analysis techniques [49-51,54]. A discussion of the OPF methods and their application is beyond the scope of this book.

In this section, we will limit our discussion to conventional power-flow analysis. Analytical techniques for the solution of the power-flow problem are covered in Chapter 6. We focus on practical considerations and modelling assumptions in power-flow studies of bulk transmission systems. In such studies, the distribution system is not usually represented and loads are represented at substation levels.

The purpose of power-flow analysis is to investigate equipment loadings, power losses, bus voltages and reactive power requirements for the possible range of system operating conditions and contingencies specified by the design criteria. For any given study, the network configuration, load level, and generation schedule are specified. Assumptions regarding equipment modelling depend on the type of power flow: prefault or postfault.

11.3.1 Prefault Power Flows

Prefault power flows generally consider normal system conditions. The basic assumption is that all control actions have taken place and the system is operating in a true steady-state condition. Consequently the system is represented as follows:

- Loads are represented as constant P and Q; this assumes that ULTCs have been successful in holding bus voltages.

- Generator terminal voltages are held at specified values, subject to reactive power outputs being within limits based on capability curves (see Chapter 5, Section 5.4).

- All control actions are accounted for. This includes

  – Transformer tap/phase shift controls
  – Area interchange controls

11.3.2 Postfault Power Flows

Several types of postfault power flows (PFPFs) are generally considered in system studies to ensure satisfactory system performance. These look at *snapshots* of system conditions at various time frames following a disturbance.

Modelling assumptions related to the following should be consistent with the time period considered, purpose of study, and the degree of conservatism required:

- Transmission system transformer ULTCs

- Load characteristics, including effects of distribution system transformer ULTCs and voltage regulators

- Phase shifters

- Tie line power control (AGC)

- Reactive power compensation

- Generator reactive power limits

- Operator actions

For situations that do not involve loss of generation or load, net interchange between AGC control areas does not change significantly; the only changes are due to the voltage sensitive loads and changes in line losses. Therefore, the effect of the tie line power control in such cases is to change the generation on AGC slightly to make up for the changes in the system load and line losses. On the other hand, primary speed controls, frequency dependence of loads, and AGC could be important for contingencies resulting in significant mismatches between generation and load.

Analysis of contingencies not involving loss of generation/load

(a)   For PFPFs representing the time period before automatic control actions of ULTCs and AGC (i.e., several seconds following the disturbance), the following modelling assumptions are made:

- Generator AVRs hold terminal voltages. Normally, there is no automatic control action limiting Q; however, depending on the nature of the study, it may be conservative to assume Q to be within generator capability limits.

- Loads vary as a function of bus voltage (since distribution system ULTCs/regulators have not operated).

- Capacitors/reactors are switched by voltage relays (if applicable).

- Tie line power flows are not controlled.

(b)   For PFPFs representing the time period after automatic control actions, and before operator actions, the following modelling assumptions are made:

- Generator AVRs hold terminal voltages, subject to the generator reactive power output capability limits.

- ULTCs on automatic control operate; load P and Q are restored to their scheduled values.

- Tie line flows are controlled by AGC.

(c)   For PFPFs representing the time period after operator actions, depending on the scope of the study, some or all of the following control actions are assumed to have taken place:

- ULTCs on manual control adjusted.

- Phase shifter angle adjusted.

- Generator Q adjusted to hold HT bus voltage (if applicable), subject to limits on Q.

- Operating reserve generation and exports adjusted to reduce equipment loadings to continuous rating.

Analysis of contingencies resulting in generation-load mismatch

The events that follow loss of generation or load may be categorized as follows:

- Stage 1, representing conditions immediately after the disturbance. During this stage, generator outputs change in approximately inverse proportion to the reactance between the generators and the point where generation or load is lost.

- Stage 2, representing conditions 0.5 s to 2.0 s following the disturbance. The generators accelerate or decelerate due to imbalance between mechanical power input and electrical power output.

  In a small system, all generators accelerate together; the loss of generation/load is shared among the generators in proportion to their inertia.

  In a large system, propagation of the effects of the disturbance to remote parts takes time; there will be time-phase difference between the rotor oscillations of units in different parts of the system.

- Stage 3, representing conditions from 2.0 s to 20.0 s following the disturbance. The speed governors respond and change the turbine outputs. At the end of this stage, the generators throughout the system share the power change in proportion to their capacity, available reserve, and the droop setting. System loads change, depending on their frequency and voltage sensitivity.

- Stage 4, representing conditions tens of seconds to several minutes following the disturbance. The AGC system attempts to correct for deviations in tie line flows and frequency. Depending on the amount of generation reserve on AGC, the tie line flows and frequency are restored. This is followed by manual action by operators.

Special features can be added to power-flow programs representing stages 3 and 4 described above. The corresponding power-flow analyses are referred to as *governor response power flow* and *AGC response power flow*, respectively. Some programs also provide facilities for the *inertial power flow*, which is intended to represent stage 2 identified above. However, such an analysis is meaningful only for small isolated systems in which all generators swing together during the concerned period, sharing generation-load mismatch in proportion to their inertia. For large interconnected systems, this stage can be analyzed only through dynamic simulations such as those used for transient stability analysis.

REFERENCES

[1]   L.K. Kirchmayer, *Economic Control of Interconnected Systems*, John Wiley & Sons, 1959.

[2]   A.J. Wood and B.F. Wollenberg, *Power Generation, Operation, and Control*, John Wiley & Sons, 1984.

[3]   D.N. Ewart, "Automatic Generation Control - Performance under Normal Conditions," *Systems Engineering for Power: Status and Prospects*, CONF. 750867, U.S. Energy Research and Development Administration, Henniker, N.H., August 17-22, 1975.

[4]   C. Concordia and L.K. Kirchmayer, "Tie line Power and Frequency Control of Electric Power Systems," Part I: *AIEE Trans.*, Vol. 72, Part III, p. 562, 1953; Part II: *AIEE Trans.*, Vol. 73, Part III-A, p. 133, 1954.

[5]   N. Cohn, "Some Aspects of Tie line Bias Control on Interconnected Power Systems," *AIEE Trans.*, Vol. 75, Part III, pp. 1415-1436, February 1957.

[6]   O.I. Elgerd, *Electric Energy Systems Theory: An Introduction*, McGraw-Hill, 1971.

[7]   O.I. Elgerd and C.E. Fosha, "Optimum Megawatt-Frequency Control of Multi-Area Electric Energy Systems," *IEEE Trans.*, Vol. PAS-89, pp. 556-563, April 1970.

[8]   J.L. Willems, "Sensitivity Analysis of the Optimum Performance of Conventional Load-Frequency Control," *IEEE Trans.*, Vol. PAS-93, pp. 1287-1291, September/October 1974.

[9]   C.W. Ross, "Error Adaptive Control Computer for Interconnected Power Systems," *IEEE Trans.*, Vol. PAS-85, pp. 742-749, July 1966.

[10]   NERC Operating Guide I: *Systems Control*, February 27, 1991.

[11]   T. Kennedy, S.M. Hoyt, and C.F. Abell, "Variable, Non-Linear Tie line Frequency Bias for Interconnected Systems Control," *IEEE Trans.*, Vol. PWRS-3, No. 3, pp. 1244-1253, August 1988.

[12]   IEEE Standard 122-1991, *Recommended Practice for Functional and Performance Characteristics of Control Systems for Steam Turbine-Generator Units*, February 1992.

[13]   IEEE Standard 125-1988, *Recommended Practice for Preparation of Equipment Specifications for Speed-Governing of Hydraulic Turbines Intended to Drive Electric Generators*, October 1988.

[14]   L.E. Eilts and F.R. Schlief, "Governing Features and Performance of the First 600 MW Hydrogenerating Units at Grand Coulee," *IEEE Trans.*, Vol. PAS-96, pp. 457-466, March/April 1977.

[15]   C. Concordia, "Effect of Prime-Mover Speed Control Characteristics on Electric Power System Performance," *IEEE Trans.*, Vol. PAS-88, pp. 752-756, May 1969.

[16]   C. Concordia, L.K. Kirchmayer, and E.A. Szymanski, "Effect of Speed-Governor Dead Band on Tie line Power and Frequency Control Performance," *AIEE Trans.*, Vol. 76, Part III, pp. 429-434, August 1957.

[17]   C.W. Taylor, K.Y. Lee, and D.P. Dave, "Automatic Generation Control Analysis with Governor Deadband Effects," *IEEE Trans.*, Vol. PAS-98, pp. 2030-2036, November/December 1979.

[18]   F.P. de Mello and J.M. Undrill, "Automatic Generation Control," IEEE Tutorial Course 77 TUDO 010-9-PWR.

[19]   IEEE AGC Task Force Report, "Understanding Automatic Generation Control," *IEEE Trans.*, Vol. PWRS-7, No. 3, pp. 1106-1122, August 1992.

[20]   J. Carpentier, "To Be or Not to Be Modern - That Is the Question for Automatic Generation Control (Point of View of a Utility Engineer)," *Electric Power & Energy Systems*, Vol. 7, No. 2, pp. 83-90, April 1985.

[21]   G. Quazza, "Non-interacting Controls of Interconnected Electric Power Systems," *IEEE Trans.*, Vol. PAS-85, pp. 727-741, July 1966.

[22]   H.A. Bauman, G.R. Hahn, and C.N. Metcalf, "The Effect of Frequency Reduction on Plant Capacity and on System Operation," *AIEE Trans.*, Vol. PAS-73, pp. 1632-1637, February 1955.

[23]   H.E. Lokay and V. Burtnyk, "Application of Underfrequency Relays for Automatic Load Shedding," *IEEE Trans.*, Vol. PAS-87, pp. 776-783, March 1968.

[24]   D.H. Berry, R.D. Brown, J.J. Redmond, and W. Watson, "Underfrequency Protection of the Ontario Hydro System," CIGRE Paper 32-14, August/September 1970.

[25]   T. Gönen, *Electric Power Distribution System Engneering*, McGraw Hill, 1986.

[26]   T. Petersson, "Reactive Power Compensation," ASEA Publication NK 02-3005 E, May 1983.

[27]   J.W. Butler and C. Concordia, "Analysis of Series Capacitor Application Problems," *AIEE Trans.*, Vol. 56, p. 975, 1937.

[28]   E.C. Starr and R.D. Evans, "Series Capacitors for Transmission Circuits," *AIEE Trans.*, Vol. 61, pp. 963, 1942.

[29]   C.F. Wagner, "Self-Excitation of Induction Motors with Series Capacitors," *AIEE Trans.*, Vol. 62, p. 58, 1943.

[30]   *An Introduction to ABB Series Capacitors*, ABB Information Publication, September 9, 1991.

[31]   B.S. Ashok Kumar, K. Parthasarathy, F.S. Prabhakara, and H.P. Kincha, "Effectiveness of Series Capacitors in Long Distance Transmission Lines," *IEEE Trans.*, Vol. PAS-89, pp. 941-950, May/June 1970.

[32]   F. Iliceto and E. Cinieri, "Comparative Analysis of Series and Shunt Compensation Schemes for AC Transmission Systems," *IEEE Trans.*, Vol. PAS-96, pp. 1819-1830, November/December 1977.

[33]   J.A. Maneatis, E.J. Hubacher, W.N. Rothenbuhler, and F. Sabath, "500 kV Series Capacitor Installations in California," *IEEE Trans.*, Vol. PAS-90, pp. 1138-1149, May/June 1971.

[34]   G. Jancke, N. Fahlen, and O. Nerf, "Series Capacitors in Power Systems," *IEEE Trans.*, Vol. PAS-94, pp. 915-925, May/June 1975.

[35]   C.V. Thio and J.B. Davies, "New Synchronous Compensators for the Nelson River HVDC System - Planning Requirements and Specifications," *IEEE Trans. on Power Delivery*, Vol. 6, No. 2, pp. 922-928, April 1991.

[36]   CIGRE Report, "Static VAR Compensators," Prepared by WG 38-01, Task Force No. 2, Paris, 1986.

[37]   T.J.E. Miller (editor), *Reactive Power Control in Electric Systems*, John Wiley & Sons, 1982.

[38]   U.S. Department of Energy Report, "Static Reactive Power Compensators for

High Voltage Power Systems," Prepared by General Electric Company, DOE/NBM-1010, March 1982.

[39]   R. Elsliger, Y. Hotte, and J.C. Roy, "Optimization of Hydro-Quebec's 735 kV Shunt-Compensated System Using Static Compensators on a Large Scale," IEEE Conference paper A 78107-5, Presented at the IEEE PES Winter Meeting, New York, January 29 - February 3, 1978.

[40]   R. Gutman, J.J. Keane, M.Ea. Rahman, and Q. Veraas, "Application and Operation of a Static Var System on a Power System - American Electric Experience, Part I: System Studies, Part II: Equipment Design and Installation," *IEEE Trans.*, Vol. PAS-104, pp. 1868-1881, July 1985.

[41]   IEEE Publication, "Static Var Compensators: Planning, Operating, and Maintenance Experiences," IEEE/PES 90TH 320-2 PWR.

[42]   D. Dickmander, B. Thorvaldsson, G. Stromberg, D. Osborn, A. Poitras, and D. Fisher, "Control System Design and Performance Verification for the Chester Maine Static Var Compensator," *IEEE Trans.*, Vol. PWRD-7, No. 3, pp. 1492-1503, July 1992.

[43]   F.G. Baum, "Voltage Regulation and Insulation for Large-Power Long-Distance Transmission Systems," *AIEE Trans.*, Vol. 40, pp. 1017-1032, June 1921.

[44]   H.K. Clark, "Voltage Control and Reactive Supply Problems," IEEE course text on Reactive Power: Basics, Problems, and Solutions, Publication 87EH0262-6-PWR.

[45]   H.K. Clark, "Considerations in the Evaluation of Series and Shunt Compensation Alternatives," Paper presented at the T & D Expo., Chicago, May 14-16, 1985.

[46]   E.W. Kimbark, "A New Look at Shunt Compensation," *IEEE Trans.*, Vol. PAS-102, pp. 212-218, January 1983.

[47]   IEEE Special Stability Controls Working Group Report, "Static Var Compensator Models for Power Flow and Dynamic Performance Simulation," Paper 93WM173-5-PWRS, Presented at the IEEE PES Winter Meeting, Columbus, Ohio, February 1993.

[48]   General Electric Company, *Voltage Regulation*, Omnitext, 1979.

[49]   J.F. Aldrich, R.A. Fernandes, L.W. Wicks, H.H. Happ, and K.A. Wiragu,

"Benefits of Voltage Scheduling in Power Systems," *IEEE Trans.*, Vol. PAS-99, No. 5, pp. 1701-1712, September/October 1980.

[50]   M.A. El-Kady, B.D. Bell, V.F. Carvalho, R.C. Burchett, H.H. Happ, and D.R. Vierath, "Assessment of Real-Time Optional Voltage Control," *IEEE Trans.*, Vol. PWRS-1, No. 2, pp. 98-105, May 1986.

[51]   N. Flatabø, J.A. Foosnaes, and T.O. Berntsen, "Transformer Tap Setting in Optimal Load Flow," *IEEE Trans.*, Vol. PAS-104, No. 6, pp. 1356-1362, June 1985.

[52]   M.S. Ćalović, "Modelling and Analysis of Under-Load Tap-Changing Transformer Control Systems," *IEEE Trans.*, Vol. PAS-103, No. 7, pp. 1909-1915, July 1984.

[53]   J. Carpentier, "Contribution à l'Etude du Dispatching Economique," *Bulletin de la Société Française des Electriciens*, Vol., 3, No. 8, August 1962.

[54]   H.W. Dommel and W.F. Tinney, "Optimal Power Flow Solutions," *IEEE Trans.*, Vol. PAS-87, pp. 1866-1876, October 1968.

PART III

SYSTEM STABILITY:

physical aspects, analysis,
and improvement

Chapter 12

Small-Signal Stability

Chapter 2 provided a general introduction to the power system stability problem, including a discussion of the basic concepts, classification, and definitions of related terms. We will now consider in detail the various categories of system stability, beginning with this chapter on small-signal stability. Knowledge of the characteristics and modelling of individual system components as presented in Chapters 3 to 11 should be helpful in this regard.

Small-signal stability, as defined in Chapter 2, is the ability of the power system to maintain synchronism when subjected to small disturbances. In this context, a disturbance is considered to be small if the equations that describe the resulting response of the system may be linearized for the purpose of analysis. Instability that may result can be of two forms: (i) steady increase in generator rotor angle due to lack of synchronizing torque, or (ii) rotor oscillations of increasing amplitude due to lack of sufficient damping torque. In today's practical power systems, the small-signal stability problem is usually one of insufficient damping of system oscillations. Small-signal analysis using linear techniques provides valuable information about the inherent dynamic characteristics of the power system and assists in its design.

This chapter reviews fundamental aspects of stability of dynamic systems, presents analytical techniques useful in the study of small-signal stability, illustrates the characteristics of small-signal stability problems, and identifies factors influencing them.

12.1 FUNDAMENTAL CONCEPTS OF STABILITY OF DYNAMIC SYSTEMS

12.1.1 State-Space Representation

The behaviour of a dynamic system, such as a power system, may be described by a set of n first order nonlinear ordinary differential equations of the following form:

$$\dot{x}_i = f_i(x_1, x_2, ..., x_n; \ u_1, u_2, ..., u_r; t) \qquad i = 1, 2, ..., n \qquad (12.1)$$

where n is the order of the system and r is the number of inputs. This can be written in the following form by using vector-matrix notation:

$$\dot{\mathbf{x}} = \mathbf{f}(\mathbf{x}, \mathbf{u}, t) \qquad (12.2)$$

where

$$\mathbf{x} = \begin{bmatrix} x_1 \\ x_2 \\ \vdots \\ x_n \end{bmatrix} \qquad \mathbf{u} = \begin{bmatrix} u_1 \\ u_2 \\ \vdots \\ u_r \end{bmatrix} \qquad \mathbf{f} = \begin{bmatrix} f_1 \\ f_2 \\ \vdots \\ f_n \end{bmatrix}$$

The column vector $\mathbf{x}$ is referred to as the *state vector*, and its entries x_i as *state variables*. The column vector $\mathbf{u}$ is the vector of inputs to the system. These are the external signals that influence the performance of the system. Time is denoted by t, and the derivative of a state variable x with respect to time is denoted by $\dot{x}$. If the derivatives of the state variables are not explicit functions of time, the system is said to be *autonomous*. In this case, Equation 12.2 simplifies to

$$\dot{\mathbf{x}} = \mathbf{f}(\mathbf{x}, \mathbf{u}) \qquad (12.3)$$

We are often interested in output variables which can be observed on the system. These may be expressed in terms of the state variables and the input variables in the following form:

$$\mathbf{y} = \mathbf{g}(\mathbf{x}, \mathbf{u}) \qquad (12.4)$$

where

$$\mathbf{y} = \begin{bmatrix} y_1 \\ y_2 \\ \vdots \\ y_m \end{bmatrix} \qquad \mathbf{g} = \begin{bmatrix} g_1 \\ g_2 \\ \vdots \\ g_m \end{bmatrix}$$

The column vector **y** is the vector of outputs, and **g** is a vector of nonlinear functions relating state and input variables to output variables.

The concept of state

The concept of state is fundamental to the state-space approach. The state of a system represents the minimum amount of information about the system at any instant in time t_0 that is necessary so that its future behaviour can be determined without reference to the input before t_0.

Any set of n linearly independent system variables may be used to describe the state of the system. These are referred to as the *state variables*; they form a minimal set of dynamic variables that, along with the inputs to the system, provide a complete description of the system behaviour. Any other system variables may be determined from a knowledge of the state.

The state variables may be physical quantities in a system such as angle, speed, voltage, or they may be abstract mathematical variables associated with the differential equations describing the dynamics of the system. The choice of the state variables is not unique. This does not mean that the state of the system at any time is not unique; only that the means of representing the state information is not unique. Any set of state variables we may choose will provide the same information about the system. If we overspecify the system by defining too many state variables, not all of them will be independent.

The system state may be represented in an n-dimensional Euclidean space called the *state space*. When we select a different set of state variables to describe the system, we are in effect choosing a different coordinate system.

Whenever the system is not in equilibrium or whenever the input is non-zero, the system state will change with time. The set of points traced by the system state in the state space as the system moves is called the *state trajectory.*

Equilibrium (or singular) points

The equilibrium points are those points where all the derivatives $\dot{x}_1, \dot{x}_2, \ldots, \dot{x}_n$ are simultaneously zero; they define the points on the trajectory with zero velocity. The system is accordingly at rest since all the variables are constant and unvarying with time.

The equilibrium or singular point must therefore satisfy the equation

$$\mathbf{f}(\mathbf{x}_0) = 0 \tag{12.5}$$

where $\mathbf{x}_0$ is the state vector **x** at the equilibrium point.

If the functions $f_i (i=1, 2, \ldots, n)$ in Equation 12.3 are linear, then the system is linear. A linear system has only one equilibrium state (if the system matrix is non-singular). For a nonlinear system there may be more than one equilibrium point.

The singular points are truly characteristic of the behaviour of the dynamic system, and therefore we can draw conclusions about stability from their nature.

12.1.2 Stability of a Dynamic System

The stability of a linear system is entirely independent of the input, and the state of a stable system with zero input will always return to the origin of the state space, independent of the finite initial state.

In contrast, the stability of a nonlinear system depends on the type and magnitude of input, and the initial state. These factors have to be taken into account in defining the stability of a nonlinear system.

In control system theory, it is common practice to classify the stability of a nonlinear system into the following categories, depending on the region of state space in which the state vector ranges:

*   Local stability or stability in the small
*   Finite stability
*   Global stability or stability in the large

Local stability

The system is said to be *locally stable* about an equilibrium point if, when subjected to small perturbation, it remains within a small region surrounding the equilibrium point.

If, as t increases, the system returns to the original state, it is said to be *asymptotically stable* in the small.

It should be noted that the general definition of local stability does not require that the state return to the original state and, therefore, includes small limit cycles. In practice, we are normally interested in asymptotic stability.

Local stability (i.e., stability under small disturbance) conditions can be studied by linearizing the nonlinear system equations about the equilibrium point in question. This is illustrated in the next section.

Finite stability

If the state of a system remains within a finite region R, it is said to be stable within R. If, further, the state of the system returns to the original equilibrium point from any point within R, it is asymptotically stable within the finite region R.

Global stability

The system is said to be *globally stable* if R includes the entire finite space.

12.1.3 Linearization

We now describe the procedure for linearizing Equation 12.3. Let $\mathbf{x}_0$ be the initial state vector and $\mathbf{u}_0$ the input vector corresponding to the equilibrium point about which the small-signal performance is to be investigated. Since $\mathbf{x}_0$ and $\mathbf{u}_0$ satisfy Equation 12.3, we have

$$\dot{\mathbf{x}}_0 = \mathbf{f}(\mathbf{x}_0, \mathbf{u}_0) = 0 \tag{12.6}$$

Let us perturb the system from the above state, by letting

$$\mathbf{x} = \mathbf{x}_0 + \Delta\mathbf{x} \qquad \mathbf{u} = \mathbf{u}_0 + \Delta\mathbf{u}$$

where the prefix Δ denotes a small deviation.

The new state must satisfy Equation 12.3. Hence,

$$\dot{\mathbf{x}} = \dot{\mathbf{x}}_0 + \Delta\dot{\mathbf{x}}$$
$$= \mathbf{f}[(\mathbf{x}_0 + \Delta\mathbf{x}), (\mathbf{u}_0 + \Delta\mathbf{u})] \tag{12.7}$$

As the perturbations are assumed to be small, the nonlinear functions $\mathbf{f}(\mathbf{x}, \mathbf{u})$ can be expressed in terms of Taylor's series expansion. With terms involving second and higher order powers of Δx and Δu neglected, we may write

$$\dot{x}_i = \dot{x}_{i0} + \Delta\dot{x}_i = f_i[(\mathbf{x}_0 + \Delta\mathbf{x}), (\mathbf{u}_0 + \Delta\mathbf{u})]$$

$$= f_i(\mathbf{x}_0, \mathbf{u}_0) + \frac{\partial f_i}{\partial x_1}\Delta x_1 + \cdots + \frac{\partial f_i}{\partial x_n}\Delta x_n$$

$$+ \frac{\partial f_i}{\partial u_1}\Delta u_1 + \cdots + \frac{\partial f_i}{\partial u_r}\Delta u_r$$

Since $\dot{x}_{i0} = f_i(\mathbf{x}_0, \mathbf{u}_0)$, we obtain

$$\Delta\dot{x}_i = \frac{\partial f_i}{\partial x_1}\Delta x_1 + \cdots + \frac{\partial f_i}{\partial x_n}\Delta x_n + \frac{\partial f_i}{\partial u_1}\Delta u_1 + \cdots + \frac{\partial f_i}{\partial u_r}\Delta u_r$$

with $i = 1, 2, \ldots, n$. In a like manner, from Equation 12.4, we have

$$\Delta y_j = \frac{\partial g_j}{\partial x_1}\Delta x_1 + \cdots + \frac{\partial g_j}{\partial x_n}\Delta x_n + \frac{\partial g_j}{\partial u_1}\Delta u_1 + \cdots + \frac{\partial g_j}{\partial u_r}\Delta u_r$$

with $j=1,2,...,m$. Therefore, the linearized forms of Equations 12.3 and 12.4 are

$$\Delta \dot{\mathbf{x}} = \mathbf{A} \, \Delta \mathbf{x} + \mathbf{B} \, \Delta \mathbf{u} \qquad (12.8)$$

$$\Delta \mathbf{y} = \mathbf{C} \, \Delta \mathbf{x} + \mathbf{D} \, \Delta \mathbf{u} \qquad (12.9)$$

where

$$
\mathbf{A} = \begin{bmatrix} \dfrac{\partial f_1}{\partial x_1} & \cdots & \dfrac{\partial f_1}{\partial x_n} \\ \cdots & \cdots & \cdots \\ \dfrac{\partial f_n}{\partial x_1} & \cdots & \dfrac{\partial f_n}{\partial x_n} \end{bmatrix}
\qquad
\mathbf{B} = \begin{bmatrix} \dfrac{\partial f_1}{\partial u_1} & \cdots & \dfrac{\partial f_1}{\partial u_r} \\ \cdots & \cdots & \cdots \\ \dfrac{\partial f_n}{\partial u_1} & \cdots & \dfrac{\partial f_n}{\partial u_r} \end{bmatrix}
$$

$$(12.10)$$

$$
\mathbf{C} = \begin{bmatrix} \dfrac{\partial g_1}{\partial x_1} & \cdots & \dfrac{\partial g_1}{\partial x_n} \\ \cdots & \cdots & \cdots \\ \dfrac{\partial g_m}{\partial x_1} & \cdots & \dfrac{\partial g_m}{\partial x_n} \end{bmatrix}
\qquad
\mathbf{D} = \begin{bmatrix} \dfrac{\partial g_1}{\partial u_1} & \cdots & \dfrac{\partial g_1}{\partial u_r} \\ \cdots & \cdots & \cdots \\ \dfrac{\partial g_m}{\partial u_1} & \cdots & \dfrac{\partial g_m}{\partial u_r} \end{bmatrix}
$$

The above partial derivatives are evaluated at the equilibrium point about which the small perturbation is being analyzed.

In Equations 12.8 and 12.9,

$\Delta \mathbf{x}$ is the state vector of dimension n
$\Delta \mathbf{y}$ is the output vector of dimension m
$\Delta \mathbf{u}$ is the input vector of dimension r
$\mathbf{A}$ is the state or plant matrix of size $n \times n$
$\mathbf{B}$ is the control or input matrix of size $n \times r$
$\mathbf{C}$ is the output matrix of size $m \times n$
$\mathbf{D}$ is the (feedforward) matrix which defines the proportion of input which appears directly in the output, size $m \times r$

By taking the Laplace transform of the above equations, we obtain the state equations in the frequency domain:

$$s\Delta\mathbf{x}(s) - \Delta\mathbf{x}(0) = \mathbf{A} \, \Delta\mathbf{x}(s) + \mathbf{B} \, \Delta\mathbf{u}(s) \qquad (12.11)$$

$$\Delta\mathbf{y}(s) = \mathbf{C} \, \Delta\mathbf{x}(s) + \mathbf{D} \, \Delta\mathbf{u}(s) \qquad (12.12)$$

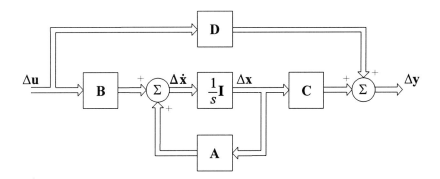

Figure 12.1 Block diagram of the state-space representation

Figure 12.1 shows the block diagram of the state-space representation. Since we are representing the transfer function of the system, the initial conditions $\Delta x(0)$ are assumed to be zero.

A formal solution of the state equations can be obtained by solving for $\Delta x(s)$ and evaluating $\Delta y(s)$, as follows:

Rearranging Equation 12.11, we have

$$(s\mathbf{I}-\mathbf{A})\Delta x(s) \;=\; \Delta x(0) + \mathbf{B}\,\Delta u(s)$$

Hence,

$$\Delta x(s) \;=\; (s\mathbf{I}-\mathbf{A})^{-1}[\,\Delta x(0) + \mathbf{B}\,\Delta u(s)\,]$$

$$\;=\; \frac{\mathrm{adj}(s\mathbf{I}-\mathbf{A})}{\det(s\mathbf{I}-\mathbf{A})}[\,\Delta x(0) + \mathbf{B}\,\Delta u(s)\,] \tag{12.13}$$

and correspondingly,

$$\Delta y(s) \;=\; \mathbf{C}\frac{\mathrm{adj}(s\mathbf{I}-\mathbf{A})}{\det(s\mathbf{I}-\mathbf{A})}[\,\Delta x(0) + \mathbf{B}\,\Delta u(s)\,] + \mathbf{D}\,\Delta u(s) \tag{12.14}$$

The Laplace transforms of Δx and Δy are seen to have two components, one dependent on the initial conditions and the other on the inputs. These are the Laplace transforms of the *free* and *zero-state components* of the state and output vectors.

The poles of $\Delta x(s)$ and $\Delta y(s)$ are the roots of the equation

$$\det(s\mathbf{I}-\mathbf{A}) \;=\; 0 \tag{12.15}$$

The values of s which satisfy the above are known as *eigenvalues* of matrix $\mathbf{A}$, and Equation 12.15 is referred to as the *characteristic equation* of matrix $\mathbf{A}$.

12.1.4 Analysis of Stability

Lyapunov's first method [1]

The *stability in the small* of a nonlinear system is given by the roots of the characteristic equation of the system of first approximations, i.e., by the eigenvalues of $\mathbf{A}$:

(i)   When the eigenvalues have negative real parts, the original system is asymptotically stable.

(ii)   When at least one of the eigenvalues has a positive real part, the original system is unstable.

(iii)   When the eigenvalues have real parts equal to zero, it is not possible on the basis of the first approximation to say anything in the general.

The *stability in the large* may be studied by explicit solution of the nonlinear differential equations using digital or analog computers.

A method that does not require explicit solution of system differential equations is the direct method of Lyapunov.

Lyapunov's second method, or the direct method

The second method attempts to determine stability directly by using suitable functions which are defined in the state space. The sign of the Lyapunov function and the sign of its time derivative with respect to the system state equations are considered.

The equilibrium of Equation 12.3 is *stable* if there exists a positive definite function $V(x_1, x_2, ..., x_n)$ such that its total derivative $\dot{V}$ with respect to Equation 12.3 is not positive.[1]

The equilibrium of Equation 12.3 is *asymptotically stable* if there is a positive definite function $V(x_1, x_2, ..., x_n)$ such that its total derivative with respect to Equation 12.3 is negative definite.

The system is *stable* in that region in which $\dot{V}$ is negative semidefinite, and *asymptotically stable* if $\dot{V}$ is negative definite.[2]

The stability in the large of power systems is the subject of the next chapter. This chapter is concerned with the stability in the small of power systems, and this is given by the eigenvalues of $\mathbf{A}$. As illustrated in the following section, the natural

---

[1] A function is called *definite* in a domain D of state space if it has the same sign for all $\mathbf{x}$ within D and vanishes for $\mathbf{x}=0$. For example, $V(x_1, x_2, x_3) = x_1^2 + x_2^2 + x_3^2$ is positive definite.

[2] A function is called *semidefinite* in a domain D of the state space if it has the same sign or is zero for all $\mathbf{x}$ within D. For example, $V(x_1, x_2, x_3) = (x_1 - x_2)^2 + x_3^2$ is positive semi-definite since it is zero for $x_1 = x_2$, $x_3 = 0$.

modes of system response are related to the eigenvalues. Analysis of the eigenproperties of **A** provides valuable information regarding the stability characteristics of the system.

It is worth recalling that the matrix **A** is the Jacobian matrix whose elements a_{ij} are given by the partial derivatives $\partial f_i/\partial x_j$ evaluated at the equilibrium point about which the small disturbance is being analyzed. This matrix is commonly referred to as the *state matrix* or the *plant matrix*. The term "plant" originates from the area of process control and is entrenched in control engineering vocabulary. It represents that part of the system which is to be controlled.

12.2 EIGENPROPERTIES OF THE STATE MATRIX

12.2.1 Eigenvalues

The eigenvalues of a matrix are given by the values of the scalar parameter λ for which there exist non-trivial solutions (i.e., other than $\boldsymbol{\phi} = 0$) to the equation

$$\mathbf{A}\boldsymbol{\phi} = \lambda\boldsymbol{\phi} \tag{12.16}$$

where

        A is an $n \times n$ matrix (real for a physical system such as a power system)
        $\boldsymbol{\phi}$ is an $n \times 1$ vector

To find the eigenvalues, Equation 12.16 may be written in the form

$$(\mathbf{A} - \lambda\mathbf{I})\boldsymbol{\phi} = 0 \tag{12.17}$$

For a non-trivial solution

$$\det(\mathbf{A} - \lambda\mathbf{I}) = 0 \tag{12.18}$$

Expansion of the determinant gives the *characteristic equation*. The n solutions of $\lambda = \lambda_1, \lambda_2, ..., \lambda_n$ are *eigenvalues* of **A**.

The eigenvalues may be real or complex. If **A** is real, complex eigenvalues always occur in conjugate pairs.

Similar matrices have identical eigenvalues. It can also be readily shown that a matrix and its transpose have the same eigenvalues.

12.2.2 Eigenvectors

For any eigenvalue λ_i, the n-column vector $\boldsymbol{\phi}_i$ which satisfies Equation 12.16 is called the *right eigenvector* of **A** associated with the eigenvalue λ_i. Therefore, we have

$$\mathbf{A}\boldsymbol{\phi}_i = \lambda_i\boldsymbol{\phi}_i \qquad i = 1, 2, ..., n \qquad (12.19)$$

The eigenvector $\boldsymbol{\phi}_i$ has the form

$$\boldsymbol{\phi}_i = \begin{bmatrix} \phi_{1i} \\ \phi_{2i} \\ \vdots \\ \phi_{ni} \end{bmatrix}$$

Since Equation 12.17 is homogeneous, $k\boldsymbol{\phi}_i$ (where k is a scalar) is also a solution. Thus, the eigenvectors are determined only to within a scalar multiplier.

Similarly, the n-row vector $\boldsymbol{\psi}_i$ which satisfies

$$\boldsymbol{\psi}_i\mathbf{A} = \lambda_i\boldsymbol{\psi}_i \qquad i = 1, 2, ..., n \qquad (12.20)$$

is called the *left eigenvector* associated with the eigenvalue λ_i.

The left and right eigenvectors corresponding to different eigenvalues are orthogonal. In other words, if λ_i is not equal to λ_j,

$$\boldsymbol{\psi}_j\boldsymbol{\phi}_i = 0 \qquad (12.21)$$

However, in the case of eigenvectors corresponding to the same eigenvalue,

$$\boldsymbol{\psi}_i\boldsymbol{\phi}_i = C_i \qquad (12.22)$$

where C_i is a non-zero constant.

Since, as noted above, the eigenvectors are determined only to within a scalar multiplier, it is common practice to *normalize* these vectors so that

$$\boldsymbol{\psi}_i\boldsymbol{\phi}_i = 1 \qquad (12.23)$$

12.2.3 Modal Matrices

In order to express the eigenproperties of $\mathbf{A}$ succinctly, it is convenient to introduce the following matrices:

$$\boldsymbol{\Phi} = [\; \boldsymbol{\phi}_1 \quad \boldsymbol{\phi}_2 \quad \cdots \quad \boldsymbol{\phi}_n \;] \qquad (12.24)$$
$$\boldsymbol{\Psi} = [\; \boldsymbol{\psi}_1^T \quad \boldsymbol{\psi}_2^T \quad \cdots \quad \boldsymbol{\psi}_n^T \;]^T \qquad (12.25)$$
$$\boldsymbol{\Lambda} = \text{diagonal matrix, with the eigenvalues } \lambda_1, \lambda_2, ..., \lambda_n$$
$$\text{as diagonal elements} \qquad (12.26)$$

Each of the above matrices is $n \times n$. In terms of these matrices, Equations 12.19 and 12.23 may be expanded as follows.

$$\mathbf{A\Phi} = \mathbf{\Phi\Lambda} \qquad (12.27)$$

$$\mathbf{\Psi\Phi} = \mathbf{I} \qquad \mathbf{\Psi} = \mathbf{\Phi}^{-1} \qquad (12.28)$$

It follows from Equation 12.27

$$\mathbf{\Phi}^{-1}\mathbf{A\Phi} = \mathbf{\Lambda} \qquad (12.29)$$

12.2.4 Free Motion of a Dynamic System

Referring to the state equation 12.9, we see that the free motion (with zero input) is given by

$$\Delta\dot{\mathbf{x}} = \mathbf{A}\,\Delta\mathbf{x} \qquad (12.30)$$

A set of equations of the above form, *derived from physical considerations,* is often not the best means of analytical studies of motion. The problem is that the rate of change of each state variable is a linear combination of all the state variables. As the result of cross-coupling between the states, it is difficult to isolate those parameters that influence the motion in a significant way.

In order to eliminate the cross-coupling between the state variables, consider a new state vector $\mathbf{z}$ related to the original state vector $\Delta\mathbf{x}$ by the transformation

$$\Delta\mathbf{x} = \mathbf{\Phi z} \qquad (12.31)$$

where $\mathbf{\Phi}$ is the modal matrix of $\mathbf{A}$ defined by Equation 12.24. Substituting the above expression for $\Delta\mathbf{x}$ in the state equation (12.30), we have

$$\mathbf{\Phi\dot{z}} = \mathbf{A\Phi z} \qquad (12.32)$$

The new state equation can be written as

$$\dot{\mathbf{z}} = \mathbf{\Phi}^{-1}\mathbf{A\Phi z} \qquad (12.33)$$

In view of Equation 12.29, the above equation becomes

$$\dot{\mathbf{z}} = \mathbf{\Lambda z} \qquad (12.34)$$

The important difference between Equations 12.34 and 12.30 is that $\mathbf{\Lambda}$ is a diagonal matrix whereas $\mathbf{A}$, in general, is non-diagonal.

Equation 12.34 represents n uncoupled first order (scalar) equations:

$$\dot{z}_i = \lambda_i z_i \qquad\qquad i = 1, 2, ..., n \qquad\qquad (12.35)$$

The effect of the transformation (12.31) is, therefore, to *uncouple* the state equations.

Equation 12.35 is a simple first-order differential equation whose solution with respect to time t is given by

$$z_i(t) = z_i(0) e^{\lambda_i t} \qquad\qquad (12.36)$$

where $z_i(0)$ is the initial value of z_i.

Returning to Equation 12.31, the response in terms of the original state vector is given by

$$\Delta \mathbf{x}(t) = \mathbf{\Phi} \mathbf{z}(t)$$

$$= \left[\begin{array}{cccc} \mathbf{\phi}_1 & \mathbf{\phi}_2 & \cdots & \mathbf{\phi}_n \end{array}\right] \begin{bmatrix} z_1(t) \\ z_2(t) \\ \vdots \\ z_n(t) \end{bmatrix} \qquad\qquad (12.37)$$

which, in view of Equation 12.36, implies that

$$\Delta \mathbf{x}(t) = \sum_{i=1}^{n} \mathbf{\phi}_i z_i(0) e^{\lambda_i t} \qquad\qquad (12.38)$$

From Equation 12.37, we have

$$\begin{aligned} \mathbf{z}(t) &= \mathbf{\Phi}^{-1} \Delta \mathbf{x}(t) \\ &= \mathbf{\Psi} \Delta \mathbf{x}(t) \end{aligned} \qquad\qquad (12.39)$$

This implies that

$$z_i(t) = \mathbf{\psi}_i \Delta \mathbf{x}(t) \qquad\qquad (12.40)$$

With $t = 0$, it follows that

$$z_i(0) = \mathbf{\psi}_i \Delta \mathbf{x}(0) \qquad\qquad (12.41)$$

By using c_i to denote the scalar product $\mathbf{\psi}_i \Delta \mathbf{x}(0)$, Equation 12.38 may be written as

$$\Delta \mathbf{x}(t) = \sum_{i=1}^{n} \mathbf{\phi}_i c_i e^{\lambda_i t} \qquad\qquad (12.42)$$

In other words, the time response of the ith state variable is given by

$$\Delta x_i(t) = \phi_{i1}c_1 e^{\lambda_1 t} + \phi_{i2}c_2 e^{\lambda_2 t} + \cdots + \phi_{in}c_n e^{\lambda_n t} \qquad (12.43)$$

The above equation gives the expression for the free motion time response of the system in terms of the eigenvalues, and left and right eigenvectors.

Thus, the free (or initial condition) response is given by a *linear combination of n dynamic modes* corresponding to the n eigenvalues of the state matrix.

The scalar product $c_i = \psi_i \Delta x(0)$ represents the *magnitude of the excitation* of the ith mode resulting from the initial conditions.

If the initial conditions lie along the jth eigenvector, the scalar products $\psi_i \Delta x(0)$ for all $i \neq j$ are identically zero. Therefore, only the jth mode is excited.

If the vector representing the initial condition is not an eigenvector, it can be represented by a linear combination of the n eigenvectors. The response of the system will be the sum of the n responses. If a component along an eigenvector of the initial conditions is zero, the corresponding mode will not be excited (see Example 12.1 for an illustration).

Eigenvalue and stability

The time dependent characteristic of a mode corresponding to an eigenvalue λ_i is given by $e^{\lambda_i t}$. Therefore, the stability of the system is determined by the eigenvalues as follows:

(a)   A *real eigenvalue* corresponds to a non-oscillatory mode. A negative real eigenvalue represents a decaying mode. The larger its magnitude, the faster the decay. A positive real eigenvalue represents aperiodic instability.

The values of c's and the eigenvectors associated with real eigenvalues are also real.

(b)   *Complex eigenvalues* occur in conjugate pairs, and each pair corresponds to an oscillatory mode.

The associated c's and eigenvectors will have appropriate complex values so as to make the entries of $x(t)$ real at every instant of time. For example,

$$(a+jb)e^{(\sigma-j\omega)t} + (a-jb)e^{(\sigma+j\omega)t}$$

has the form

$$e^{\sigma t}\sin(\omega t + \theta)$$

which represents a damped sinusoid for negative σ.

The real component of the eigenvalues gives the damping, and the imaginary component gives the frequency of oscillation. A negative real part represents a damped oscillation whereas a positive real part represents oscillation of increasing amplitude. Thus, for a complex pair of eigenvalues:

$$\lambda = \sigma \pm j\omega \tag{12.44}$$

The frequency of oscillation in Hz is given by

$$f = \frac{\omega}{2\pi} \tag{12.45}$$

This represents the actual or damped frequency. The damping ratio is given by

$$\zeta = \frac{-\sigma}{\sqrt{\sigma^2 + \omega^2}} \tag{12.46}$$

The damping ratio ζ determines the rate of decay of the amplitude of the oscillation. The time constant of amplitude decay is $1/|\sigma|$. In other words, the amplitude decays to $1/e$ or 37% of the initial amplitude in $1/|\sigma|$ seconds or in $1/(2\pi\zeta)$ cycles of oscillation.

Figure 12.2 shows the six different eigenvalue combinations and the corresponding trajectory behaviour around the singular points applicable to a two-dimensional case.

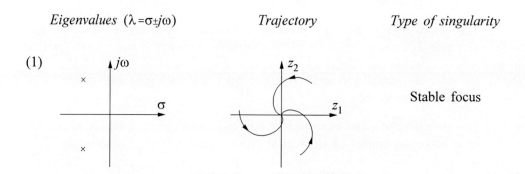

Eigenvalues $(\lambda = \sigma \pm j\omega)$ *Trajectory*   *Type of singularity*

(1)   Stable focus

Figure 12.2 Singular points corresponding to six possible
combinations of eigenvalue pairs
(*Continued on the next page*)

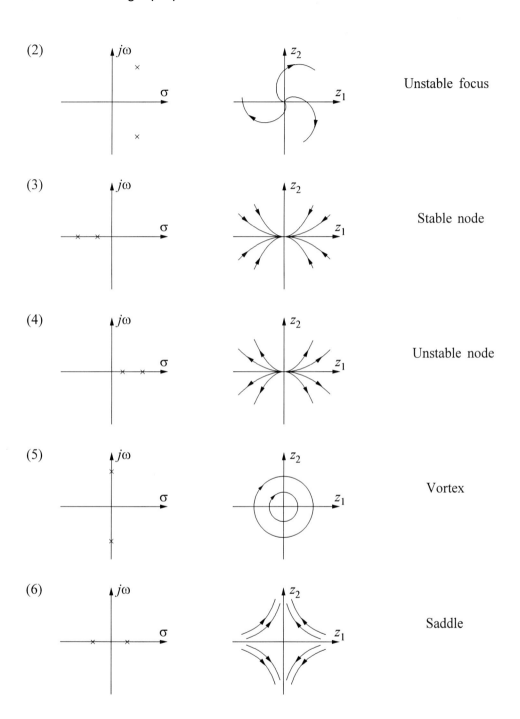

Figure 12.2 (*Continued*) Singular points corresponding to six possible combinations of eigenvalue pairs

Cases (1), (3) and (5) ensure local stability, with (1) and (3) being asymptotically stable.

12.2.5 Mode Shape, Sensitivity, and Participation Factor

(a) Mode shape and eigenvectors

In the previous section, we discussed the system response in terms of the state vectors $\Delta \mathbf{x}$ and $\mathbf{z}$, which are related to each other as follows:

$$\Delta \mathbf{x}(t) = \mathbf{\Phi}\mathbf{z}(t)$$
$$= [\ \mathbf{\phi}_1 \quad \mathbf{\phi}_2 \quad \cdots \quad \mathbf{\phi}_n\ ]\mathbf{z}(t) \tag{12.47A}$$

and

$$\mathbf{z}(t) = \mathbf{\Psi}\Delta \mathbf{x}(t)$$
$$= [\ \mathbf{\psi}_1^T \quad \mathbf{\psi}_2^T \quad \cdots \quad \mathbf{\psi}_n^T\ ]^T \Delta \mathbf{x}(t) \tag{12.47B}$$

The variables $\Delta x_1, \Delta x_2, ..., \Delta x_n$ are the original state variables chosen to represent the dynamic performance of the system. The variables $z_1, z_2, ..., z_n$ are the transformed state variables such that each variable is associated with only one mode. In other words, the transformed variables $\mathbf{z}$ are directly related to the modes.

From Equation 12.47A we see that the *right eigenvector* gives the *mode shape*, i.e., the relative activity of the state variables when a particular mode is excited. For example, the degree of activity of the state variable x_k in the ith mode is given by the element ϕ_{ki} of the right eigenvector $\mathbf{\phi}_i$.

The magnitudes of the elements of $\mathbf{\phi}_i$ give the extents of the activities of the n state variables in the ith mode, and the angles of the elements give phase displacements of the state variables with regard to the mode.

As seen from Equation 12.47B, the left eigenvector $\mathbf{\psi}_i$ identifies which combination of the original state variables displays only the ith mode. Thus the kth element of the right eigenvector $\mathbf{\phi}_i$ measures the activity of the variable x_k in the ith mode, and the kth element of the left eigenvector $\mathbf{\psi}_i$ weighs the contribution of this activity to the ith mode.

(b) Eigenvalue sensitivity

Let us now examine the sensitivity of eigenvalues to the elements of the state matrix. Consider Equation 12.19 which defines the eigenvalues and eigenvectors:

$$\mathbf{A}\mathbf{\phi}_i = \lambda_i \mathbf{\phi}_i$$

Differentiating with respect to a_{kj} (the element of $\mathbf{A}$ in kth row and jth column) yields

$$\frac{\partial \mathbf{A}}{\partial a_{kj}}\boldsymbol{\phi}_i + \mathbf{A}\frac{\partial \boldsymbol{\phi}_i}{\partial a_{kj}} = \frac{\partial \lambda_i}{\partial a_{kj}}\boldsymbol{\phi}_i + \lambda_i\frac{\partial \boldsymbol{\phi}_i}{\partial a_{kj}}$$

Premultiplying by $\boldsymbol{\psi}_i$, and noting that $\boldsymbol{\psi}_i\boldsymbol{\phi}_i = 1$ and $\boldsymbol{\psi}_i(\mathbf{A}-\lambda_i\mathbf{I}) = 0$, we see that the above equation simplifies to

$$\boldsymbol{\psi}_i\frac{\partial \mathbf{A}}{\partial a_{kj}}\boldsymbol{\phi}_i = \frac{\partial \lambda_i}{\partial a_{kj}}$$

All elements of $\partial \mathbf{A}/\partial a_{kj}$ are zero, except for the element in the kth row and jth column which is equal to 1. Hence,

$$\frac{\partial \lambda_i}{\partial a_{kj}} = \psi_{ik}\phi_{ji} \qquad (12.48)$$

Thus the sensitivity of the eigenvalue λ_i to the element a_{kj} of the state matrix is equal to the product of the left eigenvector element ψ_{ik} and the right eigenvector element ϕ_{ji}.

(c) Participation factor

One problem in using right and left eigenvectors individually for identifying the relationship between the states and the modes is that the elements of the eigenvectors are dependent on units and scaling associated with the state variables. As a solution to this problem, a matrix called the *participation matrix* ($\mathbf{P}$), which combines the right and left eigenvectors as follows is proposed in reference 2 as a measure of the association between the state variables and the modes.

$$\mathbf{P} = [\ \mathbf{P}_1 \quad \mathbf{P}_2 \quad \cdots \quad \mathbf{P}_n\ ] \qquad (12.49A)$$

with

$$\mathbf{P}_i = \begin{bmatrix} p_{1i} \\ p_{2i} \\ \vdots \\ p_{ni} \end{bmatrix} = \begin{bmatrix} \phi_{1i}\psi_{i1} \\ \phi_{2i}\psi_{i2} \\ \vdots \\ \phi_{ni}\psi_{in} \end{bmatrix} \qquad (12.49B)$$

where

ϕ_{ki} = the element on the kth row and ith column of the modal matrix Φ
  = kth entry of the right eigenvector ϕ_i
ψ_{ik} = the element on the ith row and kth column of the modal matrix Ψ
  = kth entry of the left eigenvector ψ_i

The element $p_{ki} = \phi_{ki}\psi_{ik}$ is termed the *participation factor* [2]. It is a measure of the relative participation of the kth state variable in the ith mode, and vice versa.

Since ϕ_{ki} measures the *activity* of x_k in the ith mode and ψ_{ik} *weighs* the contribution of this activity to the mode, the product p_{ki} measures the *net participation*. The effect of multiplying the elements of the left and right eigenvectors is also to make p_{ki} dimensionless (i.e., independent of the choice of units).

In view of the eigenvector normalization, the sum of the participation factors associated with any mode ($\sum_{i=1}^{n} p_{ki}$) or with any state variable ($\sum_{k=1}^{n} p_{ki}$) is equal to 1.

From Equation 12.48, we see that the participation factor p_{ki} is actually equal to the sensitivity of the eigenvalue λ_i to the diagonal element a_{kk} of the state matrix $\mathbf{A}$

$$p_{ki} = \frac{\partial \lambda_i}{\partial a_{kk}} \qquad (12.50)$$

As we will see in a number of examples in this chapter, the participation factors are generally indicative of the relative participations of the respective states in the corresponding modes.

12.2.6 Controllability and Observability

In Section 12.1.3 the system response in the presence of input was given as Equations 12.8 and 12.9 and is repeated here for reference.

$$\Delta \dot{\mathbf{x}} = \mathbf{A}\,\Delta\mathbf{x} + \mathbf{B}\,\Delta\mathbf{u} \qquad (12.8)$$

$$\Delta \mathbf{y} = \mathbf{C}\,\Delta\mathbf{x} + \mathbf{D}\,\Delta\mathbf{u} \qquad (12.9)$$

Expressing them in terms of the transformed variables $\mathbf{z}$ defined by Equation 12.31 yields

$$\Phi\dot{\mathbf{z}} = \mathbf{A}\Phi\mathbf{z} + \mathbf{B}\,\Delta\mathbf{u}$$

$$\Delta\mathbf{y} = \mathbf{C}\Phi\mathbf{z} + \mathbf{D}\,\Delta\mathbf{u}$$

The state equations in the "normal form" (decoupled) may therefore be written as

$$\dot{z} = \Lambda z + B' \Delta u \qquad (12.51)$$

$$\Delta y = C' z + D \Delta u \qquad (12.52)$$

where

$$B' = \Phi^{-1} B \qquad (12.53)$$

$$C' = C \Phi \qquad (12.54)$$

Referring to Equation 12.51, if the ith row of matrix B' is zero, the inputs have no effect on the ith mode. In such a case, the ith mode is said to be *uncontrollable*.

From Equation 12.52, we see that the ith column of the matrix C' determines whether or not the variable z_i contributes to the formation of the outputs. If the column is zero, then the corresponding mode is *unobservable*. This explains why some poorly damped modes are sometimes not detected by observing the transient response of a few monitored quantities.

The $n \times r$ matrix $B' = \Phi^{-1} B$ is referred to as the *mode controllability matrix*, and the $m \times n$ matrix $C' = C \Phi$ as the *mode observability matrix*.

By inspecting B' and C' we can classify modes into controllable and observable; controllable and unobservable; uncontrollable and observable; uncontrollable and unobservable.[1]

12.2.7 The Concept of Complex Frequency

Consider the damped sinusoid

$$v = V_m e^{\sigma t} \cos(\omega t + \theta) \qquad (12.55)$$

The unit of ω is radians per second and that of θ is radians. The dimensionless unit neper (Np) is commonly used for σt in honour of the mathematician John Napier (1550-1617) who invented logarithms. Thus the unit of σ is neper per second (Np/s).

For circuits in which the excitations and forced functions are damped sinusoids, such as that given by Equation 12.55, we can use phasor representations of damped sinusoids. This will work as well as the phasors of (undamped) sinusoids normally used in ac circuit analysis because the properties of sinusoids that make the phasors possible are shared by damped sinusoids. That is, the sum or difference of two or more damped sinusoids is a damped sinusoid and the derivative or indefinite

---

[1] This is referred to as *Kalman's canonical structure theorem*, since it was first proposed by R.E. Kalman in 1960.

integral of a damped sinusoid is also a damped sinusoid. In all these cases, V_m and θ may change; ω and σ are fixed.

Analogous to the form of phasor notation used for sinusoids, in the case of damped sinusoids, we may write

$$v = V_m e^{\sigma t} \cos(\omega t + \theta)$$

$$= \mathrm{Re}\left[V_m e^{\sigma t} e^{j(\omega t + \theta)}\right]$$

$$= \mathrm{Re}\left[V_m e^{j\theta} e^{(\sigma + j\omega)t}\right]$$

With $s = \sigma + j\omega$, we have

$$v = \mathrm{Re}\left(\tilde{V} e^{st}\right) \qquad (12.56)$$

where $\tilde{V}$ is the phasor ($V_m \angle \theta$) and is the same for both the undamped and damped sinusoids. Obviously, we may treat the damped sinusoids the same way we do undamped sinusoids by using s instead of $j\omega$.

Since s is a complex number, it is referred to as *complex frequency*, and $V(s)$ is called a *generalized phasor*.

All concepts such as impedance, admittance, Thevenin's and Norton's theorems, superposition, etc., carry over to the damped sinusoidal case.

It follows that, in the s-domain, the phasor current $I(s)$ and voltage $V(s)$, associated with a two-terminal network are related by

$$V(s) = Z(s)I(s)$$

where $Z(s)$ is the generalized impedance.

Similarly, input and output relations of dynamic devices can be expressed as

$$\frac{V_o(s)}{V_i(s)} = G(s) = \frac{b_m s^m + b_{m-1} s^{m-1} + \cdots + b_1 s + b_0}{a_n s^n + a_{n-1} s^{n-1} + \cdots + a_1 s + a_0}$$

In the factored form,

$$G(s) = \frac{b_m (s - z_1)(s - z_2) \cdots (s - z_m)}{a_n (s - p_1)(s - p_2) \cdots (s - p_n)}$$

The numbers $z_1, z_2, ..., z_m$ are called the *zeros* because they are values of s for which $G(s)$ becomes zero. The numbers $p_1, p_2, ..., p_n$ are called the *poles* of $G(s)$. The values of poles and zeros, along with a_n and b_m, uniquely determine the system transfer function $G(s)$. Poles and zeros are useful in considering frequency domain properties of dynamic systems.

12.2.8 Relationship between Eigenproperties and Transfer Functions

The state-space representation is concerned not only with input and output properties of the system but also with its complete internal behaviour. In contrast, the transfer function representation specifies only the input/output behaviour. Hence, one can make an arbitrary selection of state variables when a plant is specified only by a transfer function. On the other hand, if a state-space representation of a system is known, the transfer function is uniquely defined. In this sense, the state-space representation is a more complete description of the system; it is ideally suited for the analysis of multi-variable multi-input and multi-output systems.

For small-signal stability analysis of power systems, we primarily depend on the eigenvalue analysis of the system state matrix. However, for control design we are interested in an open-loop transfer function between specific variables. To see how this is related to the state matrix and to the eigenproperties, let us consider the transfer function between the variables y and u. From Equations 12.8 and 12.9, we may write

$$\Delta \dot{x} = A \Delta x + b \Delta u \qquad (12.57)$$

$$\Delta y = c \Delta x \qquad (12.58)$$

where A is the state matrix, Δx is the state vector, Δu is a single input, Δy is a single output, c is a row vector and b is a column vector. We assume that y is not a direct function of u (i.e., $D = 0$).

The required transfer function is

$$G(s) = \frac{\Delta y(s)}{\Delta u(s)} \qquad (12.59)$$

$$= c(sI - A)^{-1}b$$

This has the general form

$$G(s) = K \frac{N(s)}{D(s)} \qquad (12.60)$$

If $D(s)$ and $N(s)$ can be factored, we may write

$$G(s) = K \frac{(s-z_1)(s-z_2) \cdots (s-z_l)}{(s-p_1)(s-p_2) \cdots (s-p_n)} \qquad (12.61)$$

As discussed in Section 12.2.7, the n values of s, namely, $p_1, p_2, ..., p_n$, which make the denominator polynomial $D(s)$ zero are the *poles* of $G(s)$. The l values of s, namely, $z_1, z_2, ..., z_l$, are the *zeros* of $G(s)$.

Now, $G(s)$ can be expanded in partial fractions as

$$G(s) = \frac{R_1}{s-p_1} + \frac{R_2}{s-p_2} + ... + \frac{R_n}{s-p_n} \qquad (12.62)$$

and R_i is known as the *residue* of $G(s)$ at pole p_i.

To express the transfer function in terms of the eigenvalues and eigenvectors, we express the state variables Δx in terms of the transformed variables z defined by Equation 12.31. Following the procedure used in Section 12.2.4, Equations 12.57 and 12.58 may be written in terms of the transformed variables as

$$\dot{z} = \Phi^{-1}A\Phi z + \Phi^{-1}b\Delta u$$

$$= \Lambda z + \Phi^{-1}b\Delta u \qquad (12.63)$$

and

$$\Delta y = c\Phi z \qquad (12.64)$$

Hence,

$$G(s) = \frac{\Delta y(s)}{\Delta u(s)}$$

$$= c\Phi[sI - \Lambda]^{-1}\Psi b \qquad (12.65)$$

Since Λ is a diagonal matrix, we may write

$$G(s) = \sum_{i=1}^{n} \frac{R_i}{s - \lambda_i} \qquad (12.66)$$

where

$$R_i = c\phi_i\psi_i b \qquad (12.67)$$

We see that the poles of $G(s)$ are given by the eigenvalues of A. Equation 12.67 gives the residues in terms of the eigenvectors. The zeros of $G(s)$ are given by the solution of

$$\sum_{i=1}^{n} \frac{R_i}{s-\lambda_i} = 0 \tag{12.68}$$

Example 12.1

In this example we will study a second-order linear system. Such a system is easy to analyze and is helpful in understanding the behaviour of higher-order systems. The performance of high-order systems is often viewed in terms of a dominant set of second-order poles or eigenvalues. Therefore, a thorough understanding of the characteristics of a second-order system is essential before we study complex systems.

Figure E12.1 shows the familiar RLC circuit, which represents a second-order system. Study the eigenproperties of the state matrix of the system and examine its modal characteristics.

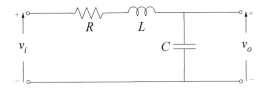

Figure E12.1

Solution

The differential equation relating v_o to v_i is

$$LC\frac{d^2v_o}{dt^2} + RC\frac{dv_o}{dt} + v_o = v_i \tag{E12.1}$$

This may be written in the standard form

$$\frac{d^2v_o}{dt^2} + (2\zeta\omega_n)\frac{dv_o}{dt} + \omega_n^2 v_o = \omega_n^2 v_i \tag{E12.2}$$

where

$$\omega_n = 1/\sqrt{LC} = \text{undamped natural frequency}$$

$$\zeta = (R/2)/\sqrt{L/C} = \text{damping ratio}$$

In order to develop the state-space representation, we define the following state, input and output variables:

$$x_1 = v_o$$

$$x_2 = \frac{dv_o}{dt}$$

$$u = v_i$$

$$y = v_o = x_1$$

(E12.3)

Using the above quantities, Equation E12.2 can be expressed in terms of two first-order equations:

$$\frac{dx_1}{dt} = x_2$$

(E12.4)

$$\frac{dx_2}{dt} = -\omega_n^2 x_1 - (2\zeta\omega_n)x_2 + \omega_n^2 u$$

(E12.5)

In matrix form,

$$\begin{bmatrix} \dot{x}_1 \\ \dot{x}_2 \end{bmatrix} = \begin{bmatrix} 0 & 1 \\ -\omega_n^2 & -2\zeta\omega_n \end{bmatrix} \begin{bmatrix} x_1 \\ x_2 \end{bmatrix} + \begin{bmatrix} 0 \\ \omega_n^2 \end{bmatrix} u$$

(E12.6)

The output variable is given by

$$y = \begin{bmatrix} 1 & 0 \end{bmatrix} \begin{bmatrix} x_1 \\ x_2 \end{bmatrix} + 0u$$

(E12.7)

These have the standard state-space form:

$$\dot{\mathbf{x}} = \mathbf{A}\mathbf{x} + \mathbf{b}u$$

$$y = \mathbf{c}\mathbf{x} + du$$

The eigenvalues of $\mathbf{A}$ are given by

$$\begin{vmatrix} -\lambda & 1 \\ -\omega_n^2 & -2\zeta\omega_n - \lambda \end{vmatrix} = 0$$

Hence,

$$\lambda^2 + 2\zeta\omega_n\lambda + \omega_n^2 = 0$$

(E12.8)

Solving for the eigenvalues, we have

$$\lambda_1 = -\zeta\omega_n + \omega_n\sqrt{\zeta^2 - 1}$$

$$\lambda_2 = -\zeta\omega_n - \omega_n\sqrt{\zeta^2 - 1}$$

(E12.9)

The right eigenvectors are given by

$$(\mathbf{A} - \lambda\mathbf{I})\boldsymbol{\phi} = 0$$

Therefore,

$$\begin{bmatrix} -\lambda_i & 1 \\ -\omega_n^2 & -2\zeta\omega_n - \lambda_i \end{bmatrix} \begin{bmatrix} \phi_{1i} \\ \phi_{2i} \end{bmatrix} = \begin{bmatrix} 0 \\ 0 \end{bmatrix}$$

This may be rewritten as

$$-\lambda_i\phi_{1i} + \phi_{2i} = 0$$

$$-\omega_n^2\phi_{1i} - (2\zeta\omega_n + \lambda_i)\phi_{2i} = 0$$

(E12.10)

If we attempt to solve the above equations for ϕ_{1i} and ϕ_{2i}, we realize that they are not independent. As discussed earlier, this is true in general; for an nth order system, the equation $(\mathbf{A} - \lambda\mathbf{I})\boldsymbol{\phi} = 0$ gives only $n-1$ independent equations for the n components of eigenvectors. One component of the eigenvector may be fixed arbitrarily and then the other components can be determined from the $n-1$ independent equations. It should, however, be noted that the eigenvectors themselves are linearly independent if eigenvalues are distinct.

For the second-order system, we can fix $\phi_{1i} = 1$ and determine ϕ_{2i}, from one of the two relationships in Equation E12.10, for each eigenvalue.

The eigenvector corresponding to λ_1 is

$$\boldsymbol{\phi}_1 = \begin{bmatrix} \phi_{11} \\ \phi_{21} \end{bmatrix} = \begin{bmatrix} 1 \\ \lambda_1 \end{bmatrix} = \begin{bmatrix} 1 \\ -\zeta\omega_n + \omega_n\sqrt{\zeta^2 - 1} \end{bmatrix}$$

(E12.11)

and the eigenvector corresponding to λ_2 is

$$\boldsymbol{\phi}_2 = \begin{bmatrix} \phi_{12} \\ \phi_{22} \end{bmatrix} = \begin{bmatrix} 1 \\ \lambda_2 \end{bmatrix} = \begin{bmatrix} 1 \\ -\zeta\omega_n - \omega_n\sqrt{\zeta^2 - 1} \end{bmatrix}$$

(E12.12)

The nature of the system response depends almost entirely on the damping ratio ζ. The value of ω_n has the effect of simply adjusting the time scale.

If ζ is greater than 1, both eigenvalues are real and negative; if ζ is equal to 1, both eigenvalues are equal to $-\omega_n$; and if ζ is less than 1, eigenvalues are complex conjugates, given as

$$\lambda = -\zeta\omega_n \pm j\omega_n\sqrt{1-\zeta^2}$$

$$= \sigma \pm j\omega$$

(E12.13)

The location of the eigenvalues in the complex plane with respect to ζ and ω_n is indicated in Figure E12.2.

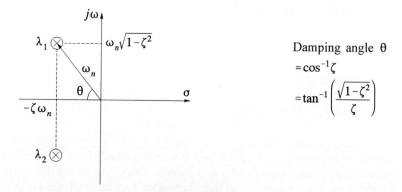

$$\text{Damping angle } \theta$$
$$= \cos^{-1}\zeta$$
$$= \tan^{-1}\left(\frac{\sqrt{1-\zeta^2}}{\zeta}\right)$$

Figure E12.2

We will first examine the singularities of the second-order system and discuss the shape of the state trajectories near the singularity. We will then discuss in detail the case where both eigenvalues are real and negative, with λ_2 greater than λ_1, but with λ_1 and λ_2 not far different.

The state equations in the normal form are given by

$$\dot{z}_1 = \lambda_1 z_1$$

(E12.14)

$$\dot{z}_2 = \lambda_2 z_2$$

(E12.15)

Hence,

$$\frac{\dot{z}_2}{\dot{z}_1} = \frac{dz_2}{dz_1} = \frac{\lambda_2 z_2}{\lambda_1 z_1}$$

(E12.16)

By integration,

$$z_2 = cz_1^{\lambda_2/\lambda_1} \tag{E12.17}$$

where c is the arbitrary constant dependent upon initial conditions. Curves representing the above in the z_1–z_2 plane are generally parabolic, with the exact shape determined by the ratio λ_2/λ_1 and constant c. The slope of the curves is given by

$$\frac{dz_2}{dz_1} = c\frac{\lambda_2}{\lambda_1}z_1^{\lambda_2/\lambda_1-1} \tag{E12.18}$$

Near the origin, $dz_2/dz_1 \to 0$ as $z_1 \to 0$, since $\lambda_2/\lambda_1 > 1$.

Typical trajectories in the phase plane with "normal" coordinates are shown in Figure E12.3.

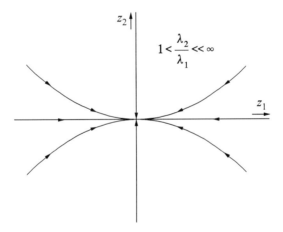

Figure E12.3

The curves represent loci of points determined by the corresponding values of z_1 and z_2. As the independent variable t increases, the point relating instantaneous values of z_1 and z_2 moves along the curve in the direction of the arrowheads. Initial conditions determine the value of c and the quadrant within which a particular solution lies. Since the roots are negative, both z_1 and z_2 decrease and ultimately reach zero as t increases. The singularity is referred to as a *stable node*.

If the initial conditions are such that one of the variables z_1 and z_2 is equal to zero, this variable remains zero, and the solution curve is on either the z_2 or z_1 axis. The axes then represent special cases of solution curves, corresponding to special initial conditions.

The corresponding plots in the x_1–x_2 plane are shown in Figure E12.4. The lines corresponding to z_1 and z_2 on this plane are skewed in accordance with the transformation $\mathbf{x} = \mathbf{\Phi z}$.

If the input v_i is zero, and if the initial conditions are such that (x_1, x_2) is on one of the eigenvectors, the state vector will remain in the same direction but will vary in magnitude by the factor $e^{\lambda_1 t}$ or $e^{\lambda_2 t}$ as the case may be.

If the vector representing the initial condition is not an eigenvector, it can be represented by a linear combination of the two eigenvectors. The response of the circuit will be the sum of the two responses. As time increases, the component in the direction of the eigenvector $\mathbf{\phi}_2$ becomes less significant because $e^{\lambda_2 t}$ decays faster than $e^{\lambda_1 t}$. Thus the trajectories always approach the origin along the $\mathbf{\phi}_1$ direction unless the component of this eigenvector was initially zero. If the eigenvectors are not real, such a simple physical interpretation of eigenvectors is not possible.

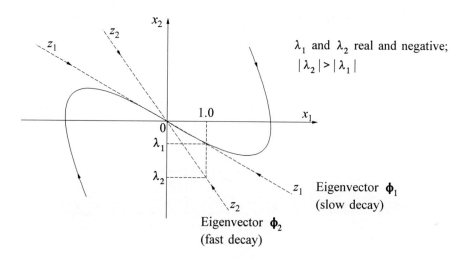

λ_1 and λ_2 real and negative; $|\lambda_2| > |\lambda_1|$

z_1 Eigenvector $\mathbf{\phi}_1$ (slow decay)

Eigenvector $\mathbf{\phi}_2$ (fast decay)

Figure E12.4

12.2.9 Computation of Eigenvalues

In the above example, we computed the eigenvalues by solving the characteristic equation of the system. This was possible because we were analyzing a simple second-order system. For higher-order systems with eigenvalues of widely differing magnitudes, this approach fails. The method that has been widely used for the computation of eigenvalues of real non-symmetrical matrices is the *QR* transformation method originally developed by J.G.F. Francis [3]. The method is numerically stable, robust, and converges rapidly. It is used in a number of very good general purpose commercial codes and has been successfully used for analyzing small-signal stability of power systems with several hundred states. The right eigenvectors may be computed by using the inverse iteration technique. A good description of the *QR* transformation and inverse iteration methods may be found in reference 4.

For large systems involving several thousand states, the QR method cannot be used for computing the eigenvalues. The reason for this and a description of special techniques for eigenvalue analysis of very large systems are presented in Section 12.8.

12.3 SMALL-SIGNAL STABILITY OF A SINGLE-MACHINE INFINITE BUS SYSTEM

In this section we will study the small-signal performance of a single machine connected to a large system through transmission lines. A general system configuration is shown in Figure 12.3(a). Analysis of systems having such simple configurations is extremely useful in understanding basic effects and concepts. After we develop an appreciation for the physical aspects of the phenomena and gain experience with the analytical techniques, using simple low-order systems, we will be in a better position to deal with large complex systems.

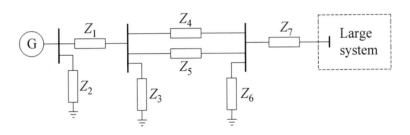

(a) General configuration

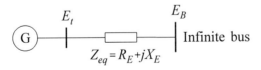

(b) Equivalent system

Figure 12.3 Single machine connected to a large system
through transmission lines

For the purpose of analysis, the system of Figure 12.3(a) may be reduced to the form of Figure 12.3(b) by using Thévenin's equivalent of the transmission network external to the machine and the adjacent transmission. Because of the relative size of the system to which the machine is supplying power, dynamics associated with the machine will cause virtually no change in the voltage and frequency of Thévenin's voltage E_B. Such a voltage source of constant voltage and constant frequency is referred to as an *infinite bus*.

For any given system condition, the magnitude of the infinite bus voltage E_B remains constant when the machine is perturbed. However, as the steady-state system conditions change, the magnitude of E_B may change, representing a changed operating condition of the external network.

In what follows we will analyze the small-signal stability of the system of Figure 12.3(b) with the synchronous machine represented by models of varying degrees of detail. We will begin with the classical model and gradually increase the model detail by accounting for the effects of the dynamics of the field circuit, excitation system, and amortisseurs. In each case, we will develop the expressions for the elements of the state matrix as explicit functions of system parameters. This will help make clear the effects of various factors associated with a synchronous machine on system stability. In addition to the state-space representation and modal analysis, we will use the block diagram representation and torque-angle relationships to analyze the system-stability characteristics. The block diagram approach was first used by Heffron and Phillips [5] and later by deMello and Concordia [6] to analyze the small-signal stability of synchronous machines. While this approach is not suited for a detailed study of large systems, it is useful in gaining a physical insight into the effects of field circuit dynamics and in establishing the basis for methods of enhancing stability through excitation control.

12.3.1 Generator Represented by the Classical Model

With the generator represented by the classical model (see Section 5.3.1) and all resistances neglected, the system representation is as shown in Figure 12.4.

Here E' is the voltage behind X_d'. Its magnitude is assumed to remain constant at the pre-disturbance value. Let δ be the angle by which E' leads the infinite bus voltage E_B. As the rotor oscillates during a disturbance, δ changes.

With E' as reference phasor,

$$\tilde{I}_t = \frac{E' \angle 0° - E_B \angle -\delta}{jX_T} = \frac{E' - E_B(\cos\delta - j\sin\delta)}{jX_T} \qquad (12.69)$$

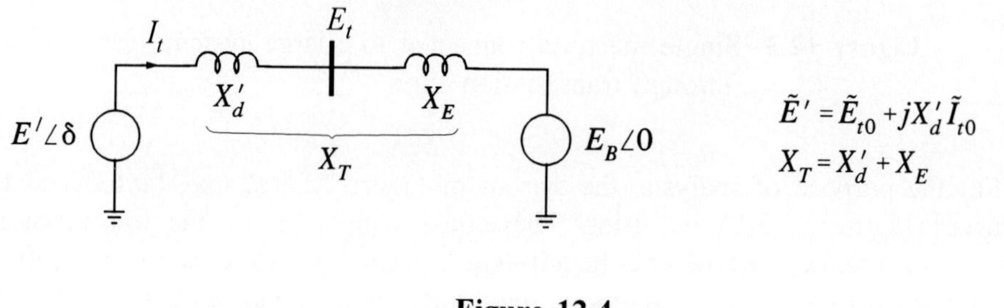

$$\tilde{E}' = \tilde{E}_{t0} + jX_d'\tilde{I}_{t0}$$
$$X_T = X_d' + X_E$$

Figure 12.4

The complex power behind X_d' is given by

$$S' = P + jQ' = \tilde{E}'\tilde{I}_t^*$$

$$= \frac{E'E_B \sin\delta}{X_T} + j\frac{E'(E' - E_B \cos\delta)}{X_T}$$

(12.70)

With stator resistance neglected, the air-gap power (P_e) is equal to the terminal power (P). In per unit, the air-gap torque is equal to the air-gap power (see Section 5.1.2). Hence,

$$T_e = P = \frac{E'E_B}{X_T} \sin\delta$$

(12.71)

Linearizing about an initial operating condition represented by $\delta = \delta_0$ yields

$$\Delta T_e = \frac{\partial T_e}{\partial\delta}\Delta\delta = \frac{E'E_B}{X_T}\cos\delta_0 (\Delta\delta)$$

(12.72)

The equations of motion (Equations 3.209 and 3.210 of Chapter 3) in per unit are

$$p\Delta\omega_r = \frac{1}{2H}(T_m - T_e - K_D\Delta\omega_r)$$

(12.73)

$$p\delta = \omega_0 \Delta\omega_r$$

(12.74)

where $\Delta\omega_r$ is the per unit speed deviation, δ is rotor angle[1] in electrical radians, ω_0 is the base rotor electrical speed in radians per second, and p is the differential operator d/dt with time t in seconds.

Linearizing Equation 12.73 and substituting for ΔT_e given by Equation 12.72, we obtain

$$p\Delta\omega_r = \frac{1}{2H}\left[\Delta T_m - K_S\Delta\delta - K_D\Delta\omega_r\right]$$

(12.75)

where K_S is the synchronizing torque coefficient given by

---

[1] As discussed in Section 5.3.1, for a classical generator model, the angle of E' with respect to a synchronously rotating reference phasor can be used as a measure of the rotor angle. Here we have chosen E_B as the reference, and the rotor angle δ is measured as the angle by which E' leads E_B.

$$K_S = \left(\frac{E'E_B}{X_T}\right)\cos\delta_0 \qquad (12.76)$$

Linearizing Equation 12.74, we have

$$p\Delta\delta = \omega_0\Delta\omega_r \qquad (12.77)$$

Writing Equations 12.75 and 12.77 in the vector-matrix form, we obtain

$$\frac{d}{dt}\begin{bmatrix} \Delta\omega_r \\ \Delta\delta \end{bmatrix} = \begin{bmatrix} -\dfrac{K_D}{2H} & -\dfrac{K_S}{2H} \\ \omega_0 & 0 \end{bmatrix}\begin{bmatrix} \Delta\omega_r \\ \Delta\delta \end{bmatrix} + \begin{bmatrix} \dfrac{1}{2H} \\ 0 \end{bmatrix}\Delta T_m \qquad (12.78)$$

This is of the form $\dot{\mathbf{x}} = \mathbf{A}\mathbf{x} + \mathbf{b}u$. The elements of the state matrix $\mathbf{A}$ are seen to be dependent on the system parameters K_D, H, X_T, and the initial operating condition represented by the values of E' and δ_0. The block diagram representation shown in Figure 12.5 can be used to describe the small-signal performance.

From the block diagram of Figure 12.5, we have

$$\begin{aligned} \Delta\delta &= \frac{\omega_0}{s}\left[\frac{1}{2Hs}\left(-K_S\Delta\delta - K_D\Delta\omega_r + \Delta T_m\right)\right] \\ &= \frac{\omega_0}{s}\left[\frac{1}{2Hs}\left(-K_S\Delta\delta - K_D s\frac{\Delta\delta}{\omega_0} + \Delta T_m\right)\right] \end{aligned} \qquad (12.79)$$

Rearranging, we get

$$s^2(\Delta\delta) + \frac{K_D}{2H}s(\Delta\delta) + \frac{K_S}{2H}\omega_0(\Delta\delta) = \frac{\omega_0}{2H}\Delta T_m$$

Therefore, the characteristic equation is given by

$$s^2 + \frac{K_D}{2H}s + \frac{K_S\omega_0}{2H} = 0 \qquad (12.80)$$

This is of the general form

$$s^2 + 2\zeta\omega_n s + \omega_n^2 = 0$$

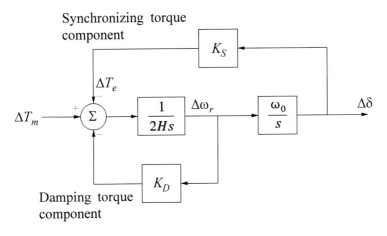

Synchronizing torque component

Damping torque component

Figure 12.5 Block diagram of a single-machine infinite bus system with classical generator model

K_S = synchronizing torque coefficient in pu torque/rad
K_D = damping torque coefficient in pu torque/pu speed deviation
H = inertia constant in MW·s/MVA
$\Delta\omega_r$ = speed deviation in pu = $(\omega_r - \omega_0)/\omega_0$
$\Delta\delta$ = rotor angle deviation in elec. rad
s = Laplace operator
ω_0 = rated speed in elec. rad/s = $2\pi f_0$
      = 377 for a 60 Hz system

Therefore, the undamped natural frequency is

$$\omega_n = \sqrt{K_S \frac{\omega_0}{2H}} \qquad \text{rad/s} \tag{12.81}$$

and the damping ratio is

$$\zeta = \frac{1}{2} \frac{K_D}{2H\omega_n}$$

$$= \frac{1}{2} \frac{K_D}{\sqrt{K_S 2H\omega_0}} \tag{12.82}$$

As the synchronizing torque coefficient K_S increases, the natural frequency increases and the damping ratio decreases. An increase in damping torque coefficient K_D increases the damping ratio, whereas an increase in inertia constant decreases both ω_n and ζ.

Example 12.2

Figure E12.5 shows the system representation applicable to a thermal generating station consisting of four 555 MVA, 24 kV, 60 Hz units.

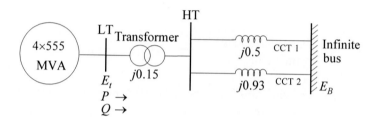

Figure E12.5

The network reactances shown in the figure are in per unit on 2220 MVA, 24 kV base (referred to the LT side of the step-up transformer). Resistances are assumed to be negligible.

The objective of this example is to analyze the small-signal stability characteristics of the system about the steady-state operating condition following the loss of circuit 2. The postfault system condition in per unit on the 2220 MVA, 24 kV base is as follows:

$$P = 0.9 \qquad Q = 0.3 \text{ (overexcited)} \qquad E_t = 1.0\angle36° \qquad E_B = 0.995\angle0°$$

The generators are to be modelled as a single equivalent generator represented by the classical model with the following parameters expressed in per unit on 2220 MVA, 24 kV base:

$$X_d' = 0.3 \qquad\qquad H = 3.5 \text{ MW·s/MVA}$$

(a)   Write the linearized state equations of the system. Determine the eigenvalues, damped frequency of oscillation in Hz, damping ratio and undamped natural frequency for each of the following values of damping coefficient (in pu torque/pu speed):

(i) $K_D = 0$ (ii) $K_D = -10.0$ (iii) $K_D = 10.0$

(b)   For the case with $K_D=10.0$, find the left and right eigenvectors, and participation matrix. Determine the time response if at $t=0$, $\Delta\delta=5°$ and $\Delta\omega=0$.

Solution

(a) Figure E12.6 shows the circuit model representing the postfault steady-state operating condition with all parameters expressed in per unit on 2220 MVA base.

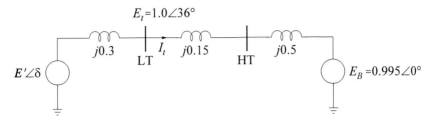

Figure E12.6

With E_t as reference phasor, the generator stator current is given by

$$\tilde{I}_t = \frac{(P+jQ)^*}{\tilde{E}_t^*} = \frac{0.9-j0.3}{1.0}$$

$$= 0.9-j0.3 \quad \text{pu}$$

The voltage behind the transient reactance is

$$\tilde{E}' = \tilde{E}_t + jX_d'\tilde{I}_t$$

$$= 1.0 + j0.3(0.9 - j0.3)$$

$$= 1.09 + j0.27 = 1.123\angle 13.92° \quad \text{pu}$$

The angle by which E' leads E_B is

$$\delta_0 = 13.92° + 36° = 49.92°$$

The total system reactance is

$$X_T = 0.3 + 0.15 + 0.5 = 0.95 \quad \text{pu}$$

The corresponding synchronizing torque coefficient, from Equation 12.76, is

$$K_S = \frac{E'E_B}{X_T}\cos\delta_0$$

$$= \frac{1.123 \times 0.995}{0.95}\cos 49.92°$$

$$= 0.757 \quad \text{pu torque/rad}$$

Linearized system equations are

$$\begin{bmatrix} \Delta\dot{\omega}_r \\ \Delta\dot{\delta} \end{bmatrix} = \begin{bmatrix} -\dfrac{K_D}{2H} & -\dfrac{K_S}{2H} \\ \omega_0 & 0 \end{bmatrix} \begin{bmatrix} \Delta\omega_r \\ \Delta\delta \end{bmatrix} + \begin{bmatrix} \dfrac{1}{2H} \\ 0 \end{bmatrix} \Delta T_m$$

$$= \begin{bmatrix} -0.143K_D & -0.108 \\ 377.0 & 0 \end{bmatrix} \begin{bmatrix} \Delta\omega_r \\ \Delta\delta \end{bmatrix} + \begin{bmatrix} 0.143 \\ 0 \end{bmatrix} \Delta T_m$$

The eigenvalues of the state matrix are given by

$$\begin{vmatrix} -0.143K_D - \lambda & -0.108 \\ 377.0 & -\lambda \end{vmatrix} = 0$$

or

$$\lambda^2 + 0.143K_D\lambda + 40.79 = 0$$

This is of the form

$$\lambda^2 + 2\zeta\omega_n\lambda + \omega_n^2 = 0$$

with

$$\omega_n = \sqrt{40.79} = 6.387 \text{ rad/s} = 1.0165 \text{ Hz}$$
$$\zeta = 0.143K_D/(2\times6.387) = 0.0112K_D$$

The eigenvalues are

$$\lambda_1, \lambda_2 = -\zeta\omega_n \pm \omega_n\sqrt{1-\zeta^2}$$

The damped frequency is

$$\omega_d = \omega_n\sqrt{1-\zeta^2}$$

The following are the required results for different values of K_D.

K_D	0	10	−10
Eigenvalues λ	$0 \pm j6.39$	$-0.714 \pm j6.35$	$0.714 \pm j6.36$
Damped frequency ω_d	1.0165 Hz	1.0101 Hz	1.0101 Hz
Damping ratio ζ	0	0.112	−0.112
Undamped natural frequency ω_n	1.0165 Hz	1.0165 Hz	1.0165 Hz

(b) The right eigenvectors are given by

$$(\mathbf{A} - \lambda \mathbf{I})\boldsymbol{\phi} = 0$$

For the given system, with $K_D = 10$, the above equation becomes

$$\begin{bmatrix} -1.43 - \lambda_i & -0.108 \\ 377.0 & -\lambda_i \end{bmatrix} \begin{bmatrix} \phi_{1i} \\ \phi_{2i} \end{bmatrix} = 0$$

For $\lambda = -0.714 + j6.35$, the corresponding equations are

$$(0.714 + j6.35)\phi_{11} + 0.108\phi_{21} = 0$$

$$377.0\phi_{11} + (0.714 - j6.35)\phi_{21} = 0$$

The above equations are not linearly independent. As discussed in Example 12.1, one of the eigenvectors corresponding to an eigenvalue has to be set arbitrarily. Therefore, let

$$\phi_{21} = 1.0$$

then

$$\phi_{11} = -0.0019 + j0.0168$$

Similarly, eigenvectors corresponding to $\lambda_2 = -0.714 - j6.35$ are

$$\phi_{22} = 1.0 \qquad \phi_{12} = -0.0019 - j0.0168$$

The right eigenvector modal matrix is

$$\boldsymbol{\Phi} = \begin{bmatrix} -0.0019 + j0.0168 & -0.0019 - j0.0168 \\ 1.0 & 1.0 \end{bmatrix}$$

The left eigenvectors normalized so that $\psi_i \phi_i = 1.0$ are given by

$$\Psi = \Phi^{-1} = \frac{\text{adj}(\Phi)}{|\Phi|}$$

$$= \frac{\begin{bmatrix} 1.0 & -1.0 \\ 0.0019+j0.0168 & -0.0019+j0.0168 \end{bmatrix}^T}{(-0.0019+j0.0168+0.0019+j0.0168)}$$

$$= \begin{bmatrix} -j29.76 & 0.5-j0.056 \\ j29.76 & 0.5+j0.056 \end{bmatrix}$$

The participation matrix is

$$P = \begin{bmatrix} \phi_{11}\psi_{11} & \phi_{12}\psi_{21} \\ \phi_{21}\psi_{12} & \phi_{22}\psi_{22} \end{bmatrix}$$

$$= \begin{bmatrix} 0.5+j0.056 & 0.5-j0.056 \\ 0.5-j0.056 & 0.5+j0.056 \end{bmatrix} = \begin{bmatrix} 0.503\angle 6.4° & 0.503\angle -6.4° \\ 0.503\angle -6.4° & 0.503\angle 6.4° \end{bmatrix}$$

The time response is given by

$$\begin{bmatrix} \Delta\omega_r(t) \\ \Delta\delta(t) \end{bmatrix} = \begin{bmatrix} \phi_{11} & \phi_{12} \\ \phi_{21} & \phi_{22} \end{bmatrix} \begin{bmatrix} c_1 e^{\lambda_1 t} \\ c_2 e^{\lambda_2 t} \end{bmatrix}$$

With $\Delta\delta=5°=0.0873$ rad and $\Delta\omega_r=0$ at $t=0$, we have

$$\begin{bmatrix} c_1 \\ c_2 \end{bmatrix} = \begin{bmatrix} \psi_{11} & \psi_{12} \\ \psi_{21} & \psi_{22} \end{bmatrix} \begin{bmatrix} \Delta\omega_r(0) \\ \Delta\delta(0) \end{bmatrix}$$

$$= \begin{bmatrix} -j29.76 & 0.5-j0.056 \\ j29.76 & 0.5-j0.056 \end{bmatrix} \begin{bmatrix} 0 \\ 0.0873 \end{bmatrix}$$

$$= \begin{bmatrix} 0.0436-j0.0049 \\ 0.0436+j0.0049 \end{bmatrix}$$

The time response of speed deviation is

$$\Delta\omega_r(t) = \phi_{11}c_1e^{\lambda_1 t} + \phi_{12}c_2e^{\lambda_2 t}$$

$$= (-0.0019+j0.0168)(0.0436-j0.0049)e^{(-0.714+j6.35)t} +$$

$$(-0.0019-j0.0168)(0.0436+j0.0049)e^{(-0.714-j6.35)t}$$

$$= -0.0015e^{-0.714t}\sin(6.35t) \qquad \text{pu}$$

Similarly, the time response of rotor angle deviation is

$$\Delta\delta(t) = 0.088e^{-0.714t}\cos(6.35t-0.112) \qquad \text{rad}$$

This is a second-order system with an oscillatory mode of response having a damped frequency of 6.35 rad/s or 1.0101 Hz. The oscillations decay with a time constant of $1/0.714$ s. This corresponds to a damping ratio ζ of 0.112. As this is a rotor angle mode, $\Delta\omega_r$ and $\Delta\delta$ participate in it equally. ∎

12.3.2 Effects of Synchronous Machine Field Circuit Dynamics

We now consider the system performance including the effect of field flux variations. The amortisseur effects will be neglected and the field voltage will be assumed constant (manual excitation control).

In what follows, we will develop the state-space model of the system by first reducing the synchronous machine equations to an appropriate form and then combining them with the network equations. We will express time in seconds, angles in electrical radians, and all other variables in per unit.

Synchronous machine equations

As in the case of the classical generator model, the acceleration equations are

$$p\Delta\omega_r = \frac{1}{2H}(T_m - T_e - K_D\Delta\omega_r) \qquad (12.83)$$

$$p\delta = \omega_0\Delta\omega_r \qquad (12.84)$$

where $\omega_0 = 2\pi f_0$ elec. rad/s. In this case, the rotor angle δ is the angle (in elec. rad) by which the q-axis leads the reference E_B. As shown in Figure 12.6, the rotor angle δ is the sum of the internal angle δ_i (see Section 3.6.3) and the angle by which E_t leads E_B. We need a convenient means of identifying the rotor position with respect to an appropriate reference and keeping track of it as the rotor oscillates. As discussed in

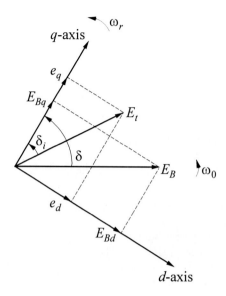

Figure 12.6

Chapter 3 (Section 3.6), the q-axis offers this convenience when the dynamics of rotor circuits are represented in the machine model. The choice of E_B as the reference for measuring rotor angle is convenient from the viewpoint of solution of network equations.

The per unit synchronous machine equations were summarized in Section 3.4.9 and the simplifications essential for large-scale stability studies were discussed in Section 5.1. From Equation 5.10, with time t in seconds instead of per unit, the field circuit dynamic equation is

$$p\psi_{fd} = \omega_0(e_{fd} - R_{fd}i_{fd})$$

$$= \frac{\omega_0 R_{fd}}{L_{adu}}E_{fd} - \omega_0 R_{fd}i_{fd} \tag{12.85}$$

where E_{fd} is the exciter output voltage defined in Section 8.6.1. Equations 12.83 to 12.85 describe the dynamics of the synchronous machine with $\Delta\omega_r$, δ, and ψ_{fd} as the state variables. However, the derivatives of these state variables appear in these equations as functions of T_e and i_{fd}, which are neither state variables nor input variables. In order to develop the complete system equations in the state-space form, we need to express i_{fd} and T_e in terms of the state variables as determined by the machine flux linkage equations and network equations.

With amortisseurs neglected, the equivalent circuits relating the machine flux linkages and currents are as shown in Figure 12.7.

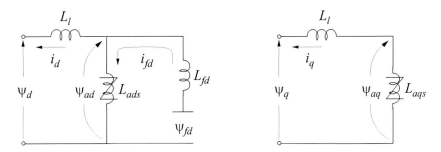

Figure 12.7

The stator and rotor flux linkages are given by

$$\psi_d = -L_l i_d + L_{ads}(-i_d + i_{fd})$$
$$= -L_l i_d + \psi_{ad} \qquad (12.86)$$

$$\psi_q = -L_l i_q + L_{aqs}(-i_q)$$
$$= -L_l i_q + \psi_{aq} \qquad (12.87)$$

$$\psi_{fd} = L_{ads}(-i_d + i_{fd}) + L_{fd} i_{fd}$$
$$= \psi_{ad} + L_{fd} i_{fd} \qquad (12.88)$$

In the above equations ψ_{ad} and ψ_{aq} are the air-gap (mutual) flux linkages, and L_{ads} and L_{aqs} are the saturated values of the mutual inductances.

From Equation 12.88, the field current may be expressed as

$$i_{fd} = \frac{\psi_{fd} - \psi_{ad}}{L_{fd}} \qquad (12.89)$$

The d-axis mutual flux linkage can be written in terms of ψ_{fd} and i_d as follows:

$$\psi_{ad} = -L_{ads} i_d + L_{ads} i_{fd}$$
$$= -L_{ads} i_d + \frac{L_{ads}}{L_{fd}}(\psi_{fd} - \psi_{ad}) \qquad (12.90)$$
$$= L'_{ads}\left(-i_d + \frac{\psi_{fd}}{L_{fd}}\right)$$

where

$$L'_{ads} = \frac{1}{\dfrac{1}{L_{ads}} + \dfrac{1}{L_{fd}}} \tag{12.91}$$

Since there are no rotor circuits considered in the q-axis, the mutual flux linkage is given by

$$\psi_{aq} = -L_{aqs} i_q \tag{12.92}$$

The air-gap torque is

$$\begin{aligned}
T_e &= \psi_d i_q - \psi_q i_d \\
&= \psi_{ad} i_q - \psi_{aq} i_d
\end{aligned} \tag{12.93}$$

With $p\psi$ terms and speed variations neglected as discussed in Section 5.1, the stator voltage equations are

$$\begin{aligned}
e_d &= -R_a i_d - \psi_q \\
&= -R_a i_d + (L_l i_q - \psi_{aq})
\end{aligned} \tag{12.94}$$

$$\begin{aligned}
e_q &= -R_a i_q + \psi_d \\
&= -R_a i_q - (L_l i_d - \psi_{ad})
\end{aligned} \tag{12.95}$$

As a first step, we have expressed i_{fd} and T_e in terms of ψ_{fd}, i_d, i_q, ψ_{ad} and ψ_{aq}. In addition, e_d and e_q have been expressed in terms of these variables and will be used in conjunction with the network equations to provide expressions for i_d and i_q in terms of the state variables.

The advantages of using ψ_{ad} and ψ_{aq} as intermediate variables in the elimination process will be more apparent when we account for the effects of amortisseur circuits in Section 12.6.

Network equations

Since there is only one machine, the machine as well as network equations can be expressed in terms of one reference frame, i.e., the d-q reference frame of the machine. Referring to Figure 12.6, the machine terminal and infinite bus voltages in terms of the d and q components are

$$\tilde{E}_t = e_d + je_q \tag{12.96}$$

$$\tilde{E}_B = E_{Bd} + jE_{Bq} \tag{12.97}$$

The network constraint equation for the system of Figure 12.3(b) is

$$\tilde{E}_t = \tilde{E}_B + (R_E + jX_E)\tilde{I}_t$$

$$(e_d + je_q) = (E_{Bd} + jE_{Bq}) + (R_E + jX_E)(i_d + ji_q) \tag{12.98}$$

Resolving into d and q components gives

$$e_d = R_E i_d - X_E i_q + E_{Bd} \tag{12.99}$$

$$e_q = R_E i_q + X_E i_d + E_{Bq} \tag{12.100}$$

where

$$E_{Bd} = E_B \sin\delta \tag{12.101}$$

$$E_{Bq} = E_B \cos\delta \tag{12.102}$$

Using Equations 12.94 and 12.95 to eliminate e_d, e_q in Equations 12.99 and 12.100, and using the expressions for ψ_{ad} and ψ_{aq} given by Equations 12.90 and 12.92, we obtain the following expressions for i_d and i_q in terms of the state variables ψ_{fd} and δ:

$$i_d = \frac{X_{Tq}\left[\psi_{fd}\left(\dfrac{L_{ads}}{L_{ads}+L_{fd}}\right) - E_B\cos\delta\right] - R_T E_B\sin\delta}{D} \tag{12.103}$$

$$i_q = \frac{R_T\left[\psi_{fd}\left(\dfrac{L_{ads}}{L_{ads}+L_{fd}}\right) - E_B\cos\delta\right] + X_{Td} E_B\sin\delta}{D} \tag{12.104}$$

where

$$R_T = R_a + R_E$$
$$X_{Tq} = X_E + (L_{aqs} + L_l) = X_E + X_{qs}$$
$$X_{Td} = X_E + (L'_{ads} + L_l) = X_E + X'_{ds} \qquad (12.105)$$
$$D = R_T^2 + X_{Tq} X_{Td}$$

The reactances X_{qs} and X'_{ds} are saturated values. In per unit they are equal to the corresponding inductances.

Equations 12.103 and 12.104, together with Equations 12.89, 12.90 and 12.92, can be used to eliminate i_{fd} and T_e from the differential equations 12.83 to 12.85 and express them in terms of the state variables. These equations are nonlinear and have to be linearized for small-signal analysis.

Linearized system equations

Expressing Equations 12.103 and 12.104 in terms of perturbed values, we may write

$$\Delta i_d = m_1 \Delta\delta + m_2 \Delta\psi_{fd} \qquad (12.106)$$

$$\Delta i_q = n_1 \Delta\delta + n_2 \Delta\psi_{fd} \qquad (12.107)$$

where

$$m_1 = \frac{E_B(X_{Tq}\sin\delta_0 - R_T\cos\delta_0)}{D}$$

$$n_1 = \frac{E_B(R_T\sin\delta_0 + X_{Td}\cos\delta_0)}{D}$$

$$\qquad (12.108)$$

$$m_2 = \frac{X_{Tq}}{D} \frac{L_{ads}}{(L_{ads} + L_{fd})}$$

$$n_2 = \frac{R_T}{D} \frac{L_{ads}}{(L_{ads} + L_{fd})}$$

By linearizing Equations 12.90 and 12.92, and substituting in them the above expressions for Δi_d and Δi_q, we get

$$\Delta\psi_{ad} = L'_{ads}\left(-\Delta i_d + \frac{\Delta\psi_{fd}}{L_{fd}}\right)$$

$$= \left(\frac{1}{L_{fd}} - m_2\right)L'_{ads}\,\Delta\psi_{fd} - m_1 L'_{ads}\,\Delta\delta \tag{12.109}$$

$$\Delta\psi_{aq} = -L_{aqs}\,\Delta i_q$$

$$= -n_2 L_{aqs}\,\Delta\psi_{fd} - n_1 L_{aqs}\,\Delta\delta \tag{12.110}$$

Linearizing Equation 12.89 and substituting for $\Delta\psi_{ad}$ from Equation 12.109 gives

$$\Delta i_{fd} = \frac{\Delta\psi_{fd} - \Delta\psi_{ad}}{L_{fd}}$$

$$= \frac{1}{L_{fd}}\left(1 - \frac{L'_{ads}}{L_{fd}} + m_2 L'_{ads}\right)\Delta\psi_{fd} + \frac{1}{L_{fd}}m_1 L'_{ads}\,\Delta\delta \tag{12.111}$$

The linearized form of Equation 12.93 is

$$\Delta T_e = \psi_{ad0}\,\Delta i_q + i_{q0}\,\Delta\psi_{ad} - \psi_{aq0}\,\Delta i_d - i_{d0}\,\Delta\psi_{aq}$$

Substituting for Δi_d, Δi_q, $\Delta\psi_{ad}$, and $\Delta\psi_{aq}$ from Equations 12.106 to 12.110, we obtain

$$\Delta T_e = K_1\Delta\delta + K_2\Delta\psi_{fd} \tag{12.112}$$

where

$$K_1 = n_1(\psi_{ad0} + L_{aqs}i_{d0}) - m_1(\psi_{aq0} + L'_{ads}i_{q0}) \tag{12.113}$$

$$K_2 = n_2(\psi_{ad0} + L_{aqs}i_{d0}) - m_2(\psi_{aq0} + L'_{ads}i_{q0}) + \frac{L'_{ads}}{L_{fd}}i_{q0} \tag{12.114}$$

By linearizing Equations 12.83 to 12.85 and substituting the expressions for Δi_{fd} and ΔT_e given by Equations 12.111 and 12.112, we obtain the system equations in the desired final form:

$$
\begin{bmatrix} \Delta\dot{\omega}_r \\ \Delta\dot{\delta} \\ \Delta\dot{\psi}_{fd} \end{bmatrix} = \begin{bmatrix} a_{11} & a_{12} & a_{13} \\ a_{21} & 0 & 0 \\ 0 & a_{32} & a_{33} \end{bmatrix} \begin{bmatrix} \Delta\omega_r \\ \Delta\delta \\ \Delta\psi_{fd} \end{bmatrix} + \begin{bmatrix} b_{11} & 0 \\ 0 & 0 \\ 0 & b_{32} \end{bmatrix} \begin{bmatrix} \Delta T_m \\ \Delta E_{fd} \end{bmatrix} \qquad (12.115)
$$

where

$$
a_{11} = -\frac{K_D}{2H}
$$

$$
a_{12} = -\frac{K_1}{2H}
$$

$$
a_{13} = -\frac{K_2}{2H}
$$

$$
a_{21} = \omega_0 = 2\pi f_0
$$

$$
a_{32} = -\frac{\omega_0 R_{fd}}{L_{fd}} m_1 L'_{ads} \qquad (12.116)
$$

$$
a_{33} = -\frac{\omega_0 R_{fd}}{L_{fd}}\left[1 - \frac{L'_{ads}}{L_{fd}} + m_2 L'_{ads}\right]
$$

$$
b_{11} = \frac{1}{2H}
$$

$$
b_{32} = \frac{\omega_0 R_{fd}}{L_{adu}}
$$

and ΔT_m and ΔE_{fd} depend on prime-mover and excitation controls. With constant mechanical input torque, $\Delta T_m=0$; with constant exciter output voltage, $\Delta E_{fd}=0$.

It is interesting to compare the above state-space equations with those derived in Section 12.3.1 by assuming the classical generator model (which is equivalent to assuming $R_{fd}=0$, $R_a=0$ and $X_q=X'_d$).

The mutual inductances L_{ads} and L_{aqs} in the above equations are saturated values. The method of accounting for saturation for small-signal analysis is described below.

Representation of saturation in small-signal studies

Since we are expressing small-signal performance in terms of perturbed values of flux linkages and currents, a distinction has to be made between *total saturation* and *incremental saturation.*

Total saturation is associated with total values of flux linkages and currents. The method of accounting for total saturation was discussed in Section 3.8.

Incremental saturation is associated with perturbed values of flux linkages and currents. Therefore, the incremental slope of the saturation curve is used in computing the incremental saturation as shown in Figure 12.8.

Denoting the incremental saturation factor $K_{sd\,(incr)}$, we have

$$L_{ads\,(incr)} = K_{sd\,(incr)}\,L_{adu} \tag{12.117}$$

Based on the definitions of A_{sat}, B_{sat} and ψ_{TI} in Section 3.8.2, we can show that

$$K_{sd\,(incr)} = \frac{1}{1 + B_{sat}A_{sat}\,e^{B_{sat}(\psi_{at0} - \psi_{TI})}} \tag{12.118}$$

A similar treatment applies to *q*-axis saturation.

For computing the initial values of system variables (denoted by subscript 0), total saturation is used. For relating the perturbed values, i.e., in Equations 12.105, 12.108, 12.113, 12.114, and 12.116, the incremental saturation factor is used.

The method of computing the initial steady-state values of machine parameters was described in Section 3.6.5.

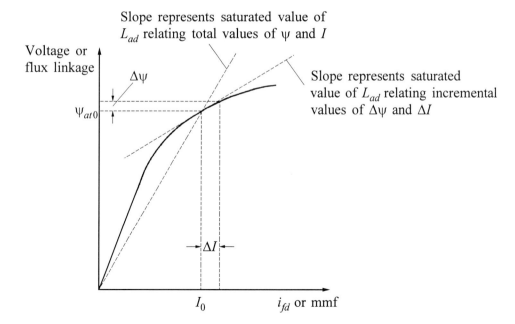

Figure 12.8 Distinction between incremental and total saturation

Summary of procedure for formulating the state matrix

(a)   The following steady-state operating conditions, machine parameters and network parameters are given:

$$P_t \quad Q_t \quad E_t \quad R_E \quad X_E$$
$$L_d \quad L_q \quad L_l \quad R_a \quad L_{fd} \quad R_{fd} \quad A_{sat} \quad B_{sat} \quad \psi_{Tl}$$

Alternatively E_B may be specified instead of Q_t or E_t.

(b)   The first step is to compute the initial steady-state values of system variables:

$I_t,$ power factor angle ϕ

Total saturation factors K_{sd} and K_{sq} (see Section 3.8)

$X_{ds} = L_{ds} = K_{sd}L_{adu}+L_l$

$X_{qs} = L_{qs} = K_{sq}L_{aqu}+L_l$

$$\delta_i = \tan^{-1}\left(\frac{I_t X_{qs}\cos\phi - I_t R_a\sin\phi}{E_t + I_t R_a\cos\phi + I_t X_{qs}\sin\phi}\right)$$

$e_{d0} = E_t\sin\delta_i$

$e_{q0} = E_t\cos\delta_i$

$i_{d0} = I_t\sin(\delta_i+\phi)$

$i_{q0} = I_t\cos(\delta_i+\phi)$

$E_{Bd0} = e_{d0}-R_E i_{d0}+X_E i_{q0}$

$E_{Bq0} = e_{q0}-R_E i_{q0}-X_E i_{d0}$

$$\delta_0 = \tan^{-1}\left(\frac{E_{Bd0}}{E_{Bq0}}\right)$$

$$E_B = \left(E_{Bd0}^2 + E_{Bq0}^2\right)^{1/2}$$

$$i_{fd0} = \frac{e_{q0}+R_a i_{q0}+L_{ds}i_{d0}}{L_{ads}}, \qquad E_{fd0} = L_{adu}i_{fd0}$$

$$\psi_{ad0} = L_{ads}(-i_{d0}+i_{fd0}), \qquad \psi_{aq0} = -L_{aqs}i_{q0}$$

(c)   The next step is to compute incremental saturation factors and the corresponding saturated values of L_{ads}, L_{aqs}, L'_{ads}, and then

R_T, X_{Tq}, X_{Td}, D	from Equation 12.105
m_1, m_2, n_1, n_2	from Equation 12.108
K_1, K_2	from Equations 12.113 and 12.114

(d)   Finally, compute the elements of matrix **A** from Equation 12.116.

Block diagram representation

Figure 12.9 shows the block diagram representation of the small-signal performance of the system. In this representation, the dynamic characteristics of the system are expressed in terms of the so-called K constants [5]. The basis for the block diagram and the expressions for the associated constants are developed below.

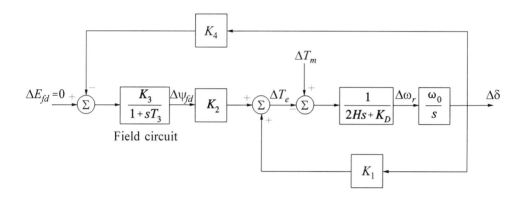

Figure 12.9 Block diagram representation with constant E_{fd}

From Equation 12.112, we may express the change in air-gap torque as a function of $\Delta\delta$ and $\Delta\psi_{fd}$ as follows:

$$\Delta T_e = K_1\Delta\delta + K_2\Delta\psi_{fd}$$

where

$K_1 = \Delta T_e/\Delta\delta$ with constant ψ_{fd}
$K_2 = \Delta T_e/\Delta\psi_{fd}$ with constant rotor angle δ

The expressions for K_1 and K_2 are given by Equations 12.113 and 12.114.

The component of torque given by $K_1\Delta\delta$ is in phase with $\Delta\delta$ and hence represents a synchronizing torque component.

The component of torque resulting from variations in field flux linkage is given by $K_2\Delta\psi_{fd}$.

The variation of ψ_{fd} is determined by the field circuit dynamic equation:

$$p\Delta\psi_{fd} = a_{32}\Delta\delta + a_{33}\Delta\psi_{fd} + b_{32}\Delta E_{fd}$$

By grouping terms involving $\Delta\psi_{fd}$ and rearranging, we get

$$\Delta\psi_{fd} = \frac{K_3}{1+pT_3}[\Delta E_{fd} - K_4\Delta\delta] \tag{12.119}$$

where

$$K_3 = -\frac{b_{32}}{a_{33}}$$

$$K_4 = -\frac{a_{32}}{b_{32}} \tag{12.120}$$

$$T_3 = -\frac{1}{a_{33}} = K_3 T'_{do} \frac{L_{adu}}{L_{ffd}}$$

Equation 12.119, with s replacing p, accounts for the field circuit block in Figure 12.9.

Expression for the K constants in the expanded form

We have expressed the K constants in terms of the elements of matrix **A**. In the literature [5,6], they are usually expressed explicitly in terms of the various system parameters, as summarized below.

The constant K_1 was expressed in Equation 12.113 as

$$K_1 = n_1(\psi_{ad0} + L_{aqs}i_{d0}) - m_1(\psi_{aq0} + L'_{ads}i_{q0})$$

From Equation 12.95, the first term in parentheses in the above expression for K_1 may be written as

$$\psi_{ad0} + L_{aqs}i_{d0} = e_{q0} + R_a i_{q0} + X_{qs}i_{d0} = E_{q0} \tag{12.121}$$

where E_{q0} is the predisturbance value of the voltage behind $R_a + jX_q$. The second term in parentheses in the expression for K_1 may be written as

$$\psi_{aq0} + L'_{ads}i_{q0} = -L_{aqs}i_{q0} + L'_{ads}i_{q0}$$
$$= -(X_q - X'_d)i_{q0} \tag{12.122}$$

Substituting for n_1, m_1 from Equation 12.108 and for the terms given by Equations 12.121 and 12.122 in the expression for K_1, yields

$$K_1 = \frac{E_B E_{q0}}{D}(R_T \sin\delta_0 + X_{Td}\cos\delta_0) +$$
$$\frac{E_B i_{q0}}{D}(X_q - X_d')(X_{Tq}\sin\delta_0 - R_T\cos\delta_0) \tag{12.123}$$

Similarly, the expanded form of the expression for the constant K_2 is

$$K_2 = \frac{L_{ads}}{L_{ads}+L_{fd}}\left[\frac{R_T}{D}E_{q0} + \left(\frac{X_{Tq}(X_q-X_d')}{D}+1\right)i_{q0}\right] \tag{12.124}$$

From Equations 12.91, 12.108, and 12.116, we may write

$$a_{33} = -\omega_0\frac{R_{fd}}{L_{fd}}\left[1 - \frac{L_{ads}}{L_{ads}+L_{fd}} + \frac{X_{Tq}}{D}\frac{L_{ads}}{(L_{ads}+L_{fd})}\frac{L_{ads}L_{fd}}{(L_{ads}+L_{fd})}\right]$$

$$= -\omega_0\frac{R_{fd}}{L_{ads}+L_{fd}}\left[1 + \frac{X_{Tq}}{D}\frac{L_{ads}^2}{(L_{ads}+L_{fd})}\right] \tag{12.125}$$

$$= -\omega_0\frac{R_{fd}}{L_{ads}+L_{fd}}\left[1 + \frac{X_{Tq}}{D}(X_d-X_d')\right]$$

Substitution of the above in the expression for K_3 and T_3 given by Equation 12.120 yields

$$K_3 = \frac{L_{ads}+L_{fd}}{L_{adu}}\frac{1}{1+\frac{X_{Tq}}{D}(X_d-X_d')} \tag{12.126}$$

$$T_3 = \frac{L_{ads}+L_{fd}}{\omega_0 R_{fd}}\frac{1}{1+\frac{X_{Tq}}{D}(X_d-X_d')}$$
$$= \frac{T_{d0s}'}{1+\frac{X_{Tq}}{D}(X_d-X_d')} \tag{12.127}$$

where T_{d0s}' is the saturated value of T_{d0}'. Similarly, from Equations 12.91, 12.108 and

12.116, we may write

$$a_{32} = -\omega_0 \frac{R_{fd}}{L_{fd}} \frac{E_B}{D}(X_{Tq}\sin\delta_0 - R_T\cos\delta_0)\frac{L_{ads}L_{fd}}{L_{ads}+L_{fd}}$$

Substitution of the above in the expression for K_4 given by Equation 12.120 yields

$$K_4 = L_{adu}\frac{L_{ads}}{L_{ads}+L_{fd}}\frac{E_B}{D}(X_{Tq}\sin\delta_0 - R_T\cos\delta_0) \quad (12.128)$$

If the effect of saturation is neglected, this simplifies to

$$K_4 = \frac{E_B}{D}(X_d-X_d')(X_{Tq}\sin\delta_0 - R_T\cos\delta_0) \quad (12.129)$$

If the elements of matrix **A** are available, the K constants may be computed directly from them. The expanded forms are derived here to illustrate the form of expressions used in the literature. An advantage of these expanded forms is that the dependence of the K constants on the various system parameters is more readily apparent. A disadvantage, however, is that some inconsistencies appear in representing saturation effects.

In the literature, $E_q'=(L_{ad}/L_{ffd})\psi_{fd}$ is often used as a state variable instead of ψ_{fd} (see Section 5.2). The effect of this is to remove the $L_{ad}/(L_{ad}+L_{fd})$ term from the expressions for K_2 and K_3. The product K_2K_3 would, however, remain the same.

Effect of field flux linkage variation on system stability

We see from the block diagram of Figure 12.9 that, with constant field voltage ($\Delta E_{fd}=0$), the field flux variations are caused only by feedback of $\Delta\delta$ through the coefficient K_4. This represents the demagnetizing effect of the armature reaction.

The change in air-gap torque due to field flux variations caused by rotor angle changes is given by

$$\left.\frac{\Delta T_e}{\Delta\delta}\right|_{\text{due to }\Delta\psi_{fd}} = -\frac{K_2K_3K_4}{1+sT_3} \quad (12.130)$$

The constants K_2, K_3, and K_4 are usually positive. The contribution of $\Delta\psi_{fd}$ to synchronizing and damping torque components depends on the oscillating frequency as discussed below.

(a)   In the steady state and at very low oscillating frequencies ($s=j\omega \to 0$):

$$\Delta T_e \text{ due to } \Delta \psi_{fd} = -K_2 K_3 K_4 \Delta \delta$$

The field flux variation due to $\Delta\delta$ feedback (i.e., due to armature reaction) introduces a negative synchronizing torque component. The system becomes monotonically unstable when this exceeds $K_1\Delta\delta$. The steady-state stability limit is reached when

$$K_2 K_3 K_4 = K_1$$

(b)   At oscillating frequencies much higher than $1/T_3$:

$$\Delta T_e \approx -\frac{K_2 K_3 K_4}{j\omega T_3}\Delta\delta$$

$$= \frac{K_2 K_3 K_4}{\omega T_3}j\Delta\delta$$

Thus, the component of air-gap torque due to $\Delta\psi_{fd}$ is 90° ahead of $\Delta\delta$ or in phase with $\Delta\omega$. Hence, $\Delta\psi_{fd}$ results in a positive damping torque component.

(c)   At typical machine oscillating frequencies of about 1 Hz (2π rad/s), $\Delta\psi_{fd}$ results in a positive damping torque component and a negative synchronizing torque component. The net effect is to reduce slightly the synchronizing torque component and increase the damping torque component.

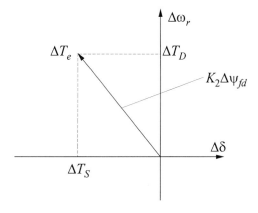

Figure 12.10 Positive damping torque and negative
synchronizing torque due to $K_2\Delta\psi_{fd}$

Special situations with K_4 negative:

The coefficient K_4 is normally positive. As long as it is positive, the effect of field flux variation due to armature reaction ($\Delta\psi_{fd}$ with constant E_{fd}) is to introduce a positive damping torque component. However, there can be situations where K_4 is negative. From the expression given by Equation 12.128, K_4 is negative when $(X_E+X_q)\sin\delta_0-(R_a+R_E)\cos\delta_0$ is negative. This is the situation when a hydraulic generator without damper windings is operating at light load and is connected by a line of relatively high resistance to reactance ratio to a large system. This type of situation was reported in reference 7.

Also K_4 can be negative when a machine is connected to a large local load, supplied partly by the generator and partly by the remote large system [8]. Under such conditions, the torques produced by induced currents in the field due to armature reaction have components out of phase with $\Delta\omega$, and produce negative damping.

Example 12.3

In this example we analyze the small-signal stability of the system of Figure E12.5 (considered in Example 12.2) including the effects of the generator field circuit dynamics. The parameters of each of the four generators of the plant in per unit on its rating are as follows:

$$X_d = 1.81 \qquad X_q = 1.76 \qquad X_d' = 0.3 \qquad X_l = 0.16$$
$$R_a = 0.003 \qquad T_{d0}' = 8.0 \text{ s} \qquad H = 3.5 \qquad K_D = 0$$

The above parameters are unsaturated values. The effect of saturation is to be represented by assuming that d and q axes have similar saturation characteristics with

$$A_{sat} = 0.031 \qquad\qquad B_{sat} = 6.93 \qquad\qquad \psi_{TI} = 0.8$$

The effects of the amortisseurs may be neglected. The excitation system is on manual control (constant E_{fd}), and transmission circuit 2 is out of service.

(a)   If the plant output in per unit on 2220 MVA, 24 kV base is

$$P = 0.9 \qquad\qquad Q = 0.3 \text{ (overexcited)} \qquad\qquad E_t = 1.0$$

compute the following:

(i)   The elements of the state matrix **A** representing the small-signal performance of the system.

(ii)   The constants K_1 to K_4 and T_3 associated with the block diagram representation of Figure 12.9.

(iii)   Eigenvalues of **A** and the corresponding eigenvectors and participation matrix; frequency and damping ratio of the oscillatory mode.

(iv)   Steady-state synchronizing torque coefficient; damping and synchronizing torque coefficients at the rotor oscillating frequency.

(b)   Determine the limiting value of P (within ± 0.025 pu) and the corresponding value of the rotor angle δ beyond which the system is unstable, with

(i)   Saturation effects neglected

(ii)   Saturation effects included

Assume that $Q=P/3$ as P varies and $E_t=1.0$. Comment on the mode of instability and the effect of representing saturation.

Solution

The four units of the plant may be represented by a single generator whose parameters on 2220 MVA base are the same as those of each unit on its rating. The circuit model of the system in per unit on 2220 MVA base is shown in Figure E12.7.

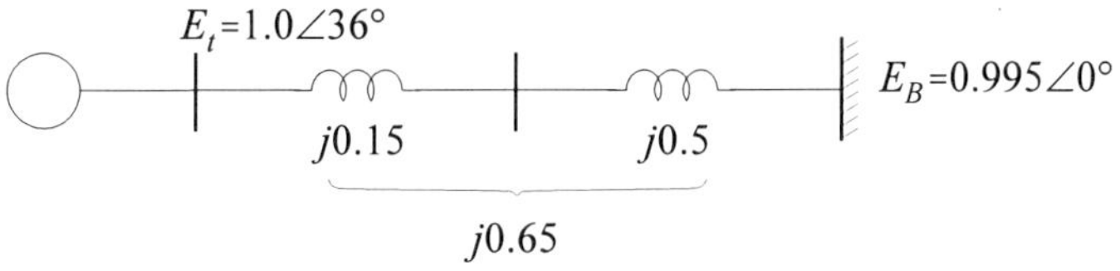

$E_t=1.0\angle 36°$

$j0.15$ $j0.5$ $E_B=0.995\angle 0°$

$j0.65$

Figure E12.7

The generators of this example have the same characteristics as the generator considered in examples of Chapters 3 and 4, except for L_l.

The per unit fundamental parameters (elements of the d- and q-axis equivalent circuits) of the equivalent generator following the procedure used in Example 4.1 are

$$L_{adu} = 1.65 \qquad L_{aqu} = 1.60 \qquad L_l = 0.16$$
$$R_a = 0.003 \qquad R_{fd} = 0.0006 \qquad L_{fd} = 0.153$$

(a) (i) The initial steady-state values of the system variables are computed by using the procedure summarized earlier in this section.

$$K_{sd} = K_{sq} = 0.8491 \qquad\qquad \delta_i = 43.13°$$
$$e_{d0} = 0.6836 \qquad e_{q0} = 0.7298 \qquad i_{d0} = 0.8342 \qquad i_{q0} = 0.4518$$
$$\delta_0 = 79.13° \qquad E_{fd0} = 2.395 \qquad K_{sd(incr)} = K_{sq(incr)} = 0.434$$

From Equation 12.108,

$$m_1 = 1.0437 \qquad m_2 = 0.8802 \qquad n_1 = 0.1268 \qquad n_2 = 0.0018$$

From Equation 12.116,

$$\mathbf{A} = \begin{bmatrix} 0 & -0.1092 & -0.1236 \\ 376.991 & 0 & 0 \\ 0 & -0.1938 & -0.4229 \end{bmatrix}$$

(ii) From Equations 12.113, 12.114, and 12.120, the constants of the block diagram of Figure 12.9 are

$$\begin{aligned} K_1 &= 0.7643 & K_2 &= 0.8649 \\ K_3 &= 0.3230 & K_4 &= 1.4187 & T_3 &= 2.365 \end{aligned}$$

(iii) Eigenvalues computed by using a standard routine based on the QR transformation method are

$$\begin{aligned} \lambda_1, \lambda_2 &= -0.11 \pm j6.41 \quad (\omega_d = 1.02 \text{ Hz}, \zeta = 0.017) \\ \lambda_3 &= -0.204 \end{aligned}$$

The right eigenvector matrix is

$$\mathbf{\Phi} = \begin{bmatrix} 0.0169 - j0.0014 & 0.0169 + j0.0014 & 0.0004 \\ -0.0988 - j0.9945 & -0.0988 + j0.9945 & -0.7490 \\ 0.0301 - j0.0015 & 0.0301 + j0.0015 & 0.6625 \end{bmatrix}$$

The left eigenvector matrix normalized so that $\mathbf{\Psi\Phi} = \mathbf{I}$ is

$$\mathbf{\Psi} = \begin{bmatrix} 29.3884 + j1.9097 & -0.0410 + j0.4992 & -0.0643 + j0.5632 \\ 29.3884 - j1.9097 & -0.0410 - j0.4992 & -0.0643 - j0.5632 \\ -2.6820 & 0.0014 & 1.5126 \end{bmatrix}$$

The participation matrix is

$$\begin{bmatrix} 0.501\angle -1.0° & 0.501\angle 1.0° & 0.001\angle 180° \\ 0.501\angle -1.0° & 0.501\angle 1.0° & 0.001\angle 180° \\ 0.017\angle 94° & 0.017\angle -94° & 1.002\angle 0° \end{bmatrix} \begin{matrix} \Delta\omega_r \\ \Delta\delta \\ \Delta\psi_{fd} \end{matrix}$$
$$\qquad\quad \lambda_1 \qquad\qquad\quad \lambda_2 \qquad\qquad\quad \lambda_3$$

From the participation matrix, we see that $\Delta\omega_r$ and $\Delta\delta$ have a high participation in the oscillatory mode (corresponding to eigenvalues λ_1 and λ_2); the field flux linkage has a high participation in the non-oscillatory mode, represented by the eigenvalue λ_3.

(iv) The steady-state synchronizing torque coefficient due to $\Delta\psi_{fd}$ is

$$-K_2 K_3 K_4 = -0.8649 \times 0.323 \times 1.4187$$

$$= -0.3963$$

The total steady-state synchronizing torque coefficient is

$$K_S = K_1 - K_2 K_3 K_4$$

$$= 0.7643 - 0.3963 = 0.3679 \quad \text{pu torque/rad}$$

From the block diagram of Figure 12.9,

$$\left.\frac{\Delta T_e(s)}{\Delta\delta(s)}\right|_{\text{due to } \Delta\psi_{fd}} = \frac{-K_2 K_3 K_4}{1+sT_3} = \frac{-K_2 K_3 K_4 (1-sT_3)}{1-s^2 T_3^2}$$

Therefore, ΔT_e due to $\Delta\psi_{fd}$ is

$$\Delta T_e\big|_{\Delta\psi_{fd}} = \frac{-K_2 K_3 K_4}{1-s^2 T_3^2}\Delta\delta + \frac{K_2 K_3 K_4 T_3}{1-s^2 T_3^2}s\Delta\delta$$

$$= K_S(\Delta\psi_{fd})\Delta\delta + K_D(\Delta\psi_{fd})\Delta\omega_r \omega_0$$

From the eigenvalues, the complex frequency of rotor oscillation is $-0.11 + j6.41$. Since the real component is much smaller than the imaginary component, we can compute K_S and K_D at the oscillation frequency by setting $s=j6.41$ without loss of much accuracy.

$$K_S(\Delta\psi_{fd}) = \frac{-K_2 K_3 K_4}{1-s^2 T_3^2} = \frac{-0.3963}{1-(j6.41 \times 2.365)^2}$$

$$= -0.00172 \quad \text{pu torque/rad}$$

$$K_D(\Delta\psi_{fd}) = \frac{K_2 K_3 K_4 T_3 \omega_0}{1-s^2 T_3^2} = \frac{0.3963 \times 2.365 \times 377}{1-(j6.41 \times 2.365)^2}$$

$$= 1.53 \quad \text{pu torque/pu speed change}$$

The effect of field flux variation (i.e., armature reaction) is thus to reduce the synchronizing torque slightly and to add a damping torque component.

The net synchronizing torque component is

$$K_S = K_1 + K_S(\Delta \psi_{fd}) = 0.7643 - 0.00172$$

$$= 0.7626 \quad \text{pu torque/rad}$$

The only source of damping is due to field flux variation. Hence, the net damping torque coefficient is

$$K_D = K_D(\Delta \psi_{fd}) = 1.53 \quad \text{pu torque/pu speed change}$$

From Equation 12.81, the undamped natural frequency is

$$\omega_n = \sqrt{\frac{K_S \omega_0}{2H}} = \sqrt{\frac{0.7626 \times 377}{2 \times 3.5}}$$

$$= 6.41 \quad \text{rad/s}$$

and from Equation 12.82, the damping ratio is

$$\zeta = \frac{1}{2} \frac{K_D}{\sqrt{K_S 2H \omega_0}} = \frac{1}{2} \frac{1.53}{\sqrt{0.7626 \times 2 \times 3.5 \times 377}}$$

$$= 0.0171$$

The above values of ω_n and ζ agree with those computed from the eigenvalues.

(b) The stability limit is determined by increasing P with $Q=P/3$ and $E_t=1.0$ pu (E_B is allowed to take appropriate values so as to satisfy the network equations). The results with and without saturation effects are as follows.

(i) With saturation effects:

The limiting P (within ±0.025 pu) and the corresponding system conditions in per unit are

$$
\begin{array}{lll}
P = 1.3875 & Q = 0.4625 & E_t = 1.0 \\
E_B = 1.1413 & \delta = 102.6° & E_{fd} = 3.3
\end{array}
$$

The corresponding K constants are

$$K_1 = 0.4566 \quad K_2 = 0.960 \quad K_3 = 0.3028 \quad K_4 = 1.5749$$

Hence, the steady-state synchronizing torque coefficient is

$$K_S = K_1 - K_2 K_3 K_4 = -0.0014 \quad \text{pu torque/rad}$$

The eigenvalues of the system state matrix are

$$\lambda_1, \lambda_2 = -0.226 \pm j4.95 \quad (\omega_d = 0.79 \text{ Hz}, \zeta = 0.046)$$
$$\lambda_3 = +0.00142$$

The above represents conditions just past the stability limit. The system instability is due to lack of synchronizing torque. This is reflected in the real eigenvalue becoming slightly positive, representing a mode of instability through a non-oscillatory mode.

(ii) Without saturation effects:

The limiting value of P and the corresponding system conditions in this case are

$$\begin{array}{lll}
P = 1.0375 & Q = 0.3458 & E_t = 1.0 \\
E_B = 1.0275 & \delta = 89.6° & E_{fd} = 2.49
\end{array}$$

The K constants are

$$K_1 = 0.6862 \quad K_2 = 0.9956 \quad K_3 = 0.422 \quad K_4 = 1.633$$

The steady-state synchronizing torque coefficient is

$$K_S = K_1 - K_2 K_3 K_4 = 0.0001 \quad \text{pu torque/rad}$$

The eigenvalues are

$$\lambda_1, \lambda_2 = -0.162 \pm j6.08 \quad (\omega_d = 0.97 \text{ Hz}, \zeta = 0.027)$$
$$\lambda_3 = -0.00006$$

The system is on the verge of instability. The limiting rotor angle δ is very close to 90°. With constant E_{fd} and negligible saliency, the limiting rotor angle will be equal to 90° if the values of L_{ad} and L_{aq} used to compute the initial operating condition are the same as the values used to relate incremental flux linkages and currents.

In case (i), when we represented saturation, we made a distinction between total saturation and incremental saturation. Hence, the limiting rotor angle was about 102°, significantly higher than 90°.   ∎

12.4 EFFECTS OF EXCITATION SYSTEM

In this section, we will extend the state-space model and the block diagram developed in the previous section to include the excitation system. We will then examine the effect of the excitation system on the small-signal stability performance of the single-machine infinite bus system under consideration.

The input control signal to the excitation system is normally the generator terminal voltage E_t. In the generator model we implemented in the previous section, E_t is not a state variable. Therefore, E_t has to be expressed in terms of the state variables $\Delta\omega_r$, $\Delta\delta$, and $\Delta\psi_{fd}$.

In Section 3.6.2, we showed that $\tilde{E}_t$ may be expressed in complex form:

$$\tilde{E}_t = e_d + je_q$$

Hence,

$$E_t^2 = e_d^2 + e_q^2$$

Applying a small perturbation, we may write

$$(E_{t0} + \Delta E_t)^2 = (e_{d0} + \Delta e_d)^2 + (e_{q0} + \Delta e_q)^2$$

By neglecting second-order terms involving perturbed values, the above equation reduces to

$$E_{t0}\Delta E_t = e_{d0}\Delta e_d + e_{q0}\Delta e_q$$

Therefore,

$$\Delta E_t = \frac{e_{d0}}{E_{t0}}\Delta e_d + \frac{e_{q0}}{E_{t0}}\Delta e_q \tag{12.131}$$

In terms of the perturbed values, Equations 12.94 and 12.95 may be written as

$$\Delta e_d = -R_a\Delta i_d + L_l\Delta i_q - \Delta\psi_{aq}$$

$$\Delta e_q = -R_a\Delta i_q - L_l\Delta i_d + \Delta\psi_{ad}$$

Use of Equations 12.106, 12.107, 12.109, and 12.110 to eliminate Δi_d, Δi_q, $\Delta\psi_{ad}$ and $\Delta\psi_{aq}$ from the above equations in terms of the state variables and substitution of the resulting expressions for Δe_d and Δe_q in Equation 12.131 yield

$$\Delta E_t = K_5 \Delta \delta + K_6 \Delta \psi_{fd} \tag{12.132}$$

where

$$K_5 = \frac{e_{d0}}{E_{t0}}\left[-R_a m_1 + L_l n_1 + L_{aqs} n_1\right] + \frac{e_{q0}}{E_{t0}}\left[-R_a n_1 - L_l m_1 - L'_{ads} m_1\right] \tag{12.133}$$

$$K_6 = \frac{e_{d0}}{E_{t0}}\left[-R_a m_2 + L_l n_2 + L_{aqs} n_2\right] + \frac{e_{q0}}{E_{t0}}\left[-R_a n_2 - L_l m_2 + L'_{ads}(\frac{1}{L_{fd}} - m_2)\right] \tag{12.134}$$

For the purpose of illustration and examination of the influence on small-signal stability, we will consider the excitation system model shown in Figure 12.11. It is representative of thyristor excitation systems classified as type ST1A in Chapter 8. The model shown in Figure 12.11, however, has been simplified to include only those elements that are considered necessary for representing a specific system. A high exciter gain, without transient gain reduction or derivative feedback, is used. Parameter T_R represents the terminal voltage transducer time constant.

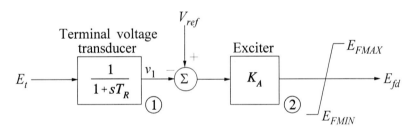

Figure 12.11 Thyristor excitation system with AVR

The only nonlinearity associated with the model is that due to the ceiling on the exciter output voltage represented by E_{FMAX} and E_{FMIN}. For small-disturbance studies, these limits are ignored as we are interested in a linearized model about an operating point such that E_{fd} is within the limits. Limiters and protective circuits (UEL, OXL, V/Hz) are not modelled as they do not affect small-signal stability.

From block 1 of Figure 12.11, using perturbed values, we have

$$\Delta v_1 = \frac{1}{1 + pT_R} \Delta E_t$$

Hence

$$p\Delta v_1 = \frac{1}{T_R}(\Delta E_t - \Delta v_1)$$

Substituting for ΔE_t from Equation 12.132, we get

$$p\Delta v_1 = \frac{K_5}{T_R}\Delta\delta + \frac{K_6}{T_R}\Delta\psi_{fd} - \frac{1}{T_R}\Delta v_1 \qquad (12.135)$$

From block 2 of Figure 12.11,

$$E_{fd} = K_A(V_{ref} - v_1)$$

In terms of perturbed values, we have

$$\Delta E_{fd} = K_A(-\Delta v_1) \qquad (12.136)$$

The field circuit dynamic equation developed in the previous section, with the effect of excitation system included, becomes

$$p\Delta\psi_{fd} = a_{31}\Delta\omega_r + a_{32}\Delta\delta + a_{33}\Delta\psi_{fd} + a_{34}\Delta v_1 \qquad (12.137)$$

where

$$a_{34} = -b_{32}K_A = -\frac{\omega_0 R_{fd}}{L_{adu}}K_A \qquad (12.138)$$

The expressions for a_{31}, a_{32} and a_{33} remain unchanged and are given by Equation 12.116.

Since we have a first-order model for the exciter, the order of the overall system is increased by 1; the new state variable added is Δv_1.

From Equation 12.135,

$$p\Delta v_1 = a_{41}\Delta\omega_r + a_{42}\Delta\delta + a_{43}\Delta\psi_{fd} + a_{44}\Delta v_1 \qquad (12.139)$$

where

$$a_{41} = 0$$
$$a_{42} = \frac{K_5}{T_R}$$
$$a_{43} = \frac{K_6}{T_R} \qquad (12.140)$$
$$a_{44} = -\frac{1}{T_R}$$

and K_5 and K_6 are given by Equations 12.133 and 12.134.

Since $p\Delta\omega_r$ and $p\Delta\delta$ are not directly affected by the exciter,

$$a_{14} = a_{24} = 0$$

The complete state-space model for the power system, including the excitation system of Figure 12.11, has the following form:

$$
\begin{bmatrix} \Delta\dot{\omega}_r \\ \Delta\dot{\delta} \\ \Delta\dot{\psi}_{fd} \\ \Delta\dot{v}_1 \end{bmatrix}
=
\begin{bmatrix} a_{11} & a_{12} & a_{13} & 0 \\ a_{21} & 0 & 0 & 0 \\ 0 & a_{32} & a_{33} & a_{34} \\ 0 & a_{42} & a_{43} & a_{44} \end{bmatrix}
\begin{bmatrix} \Delta\omega_r \\ \Delta\delta \\ \Delta\psi_{fd} \\ \Delta v_1 \end{bmatrix}
+
\begin{bmatrix} b_1 \\ 0 \\ 0 \\ 0 \end{bmatrix}
\Delta T_m
\qquad (12.141)
$$

With constant mechanical torque input,

$$\Delta T_m = 0$$

Block diagram including the excitation system

Figure 12.12 shows the block diagram obtained by extending the diagram of Figure 12.9 to include the voltage transducer and AVR/exciter blocks. The

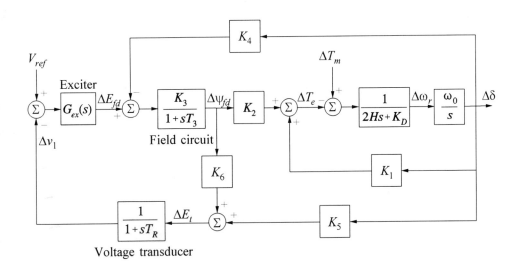

Figure 12.12 Block diagram representation with exciter and AVR

representation is applicable to any type of exciter, with $G_{ex}(s)$ representing the transfer function of the AVR and exciter. For a *thyristor exciter*,

$$G_{ex}(s) = K_A$$

The terminal voltage error signal, which forms the input to the voltage transducer block, is given by Equation 12.132:

$$\Delta E_t = K_5 \Delta \delta + K_6 \Delta \psi_{fd}$$

The coefficient K_6 is always positive, whereas K_5 can be either positive or negative, depending on the operating condition and the external network impedance $R_E + jX_E$. The value of K_5 has a significant bearing on the influence of the AVR on the damping of system oscillations as illustrated below.

Effect of AVR on synchronizing and damping torque components

With automatic voltage regulator action, the field flux variations are caused by the field voltage variations, in addition to the armature reaction. From the block diagram of Figure 12.12, we see that

$$\Delta \psi_{fd} = \frac{K_3}{1+sT_3}\left[-K_4 \Delta \delta - \frac{G_{ex}(s)}{1+sT_R}(K_5 \Delta \delta + K_6 \Delta \psi_{fd})\right] \qquad (12.142)$$

By grouping terms involving $\Delta \psi_{fd}$ and rearranging,

$$\Delta \psi_{fd} = \frac{-K_3[K_4(1+sT_R) + K_5 G_{ex}(s)]}{s^2 T_3 T_R + s(T_3 + T_R) + 1 + K_3 K_6 G_{ex}(s)} \Delta \delta \qquad (12.143)$$

The change in air-gap torque due to change in field flux linkage is

$$\Delta T_e \big|_{\Delta \psi_{fd}} = K_2 \Delta \psi_{fd} \qquad (12.144)$$

As noted before, the constants K_2, K_3, K_4, and K_6 are usually positive; however, K_5 may take either positive or negative values. The effect of the AVR on damping and synchronizing torque components is therefore primarily influenced by K_5 and $G_{ex}(s)$. We will illustrate this by considering a specific case with parameters as follows:

$$
\begin{array}{lllll}
K_1 = 1.591 & K_2 = 1.5 & K_3 = 0.333 & K_4 = 1.8 & T_3 = 1.91 \\
K_5 = -0.12 & K_6 = 0.3 & T_R = 0.02 & G_{ex}(s) = K_A & \\
H = 3.0 & K_D = 0.0 & & &
\end{array}
$$

This represents a system with a thyristor exciter and system conditions such that K_5 is negative.

(a) Steady-state synchronizing torque coefficient:

From Equations 12.143 and 12.144, with $s=j\omega=0$, ΔT_e due to $\Delta\psi_{fd}$ is

$$\Delta T_e|_{\Delta\psi_{fd}} = \frac{-K_2 K_3 (K_4 + K_5 K_A)}{1 + K_3 K_6 K_A} \Delta\delta$$

$$= \frac{-1.5 \times 0.333(1.8 - 0.12 K_A)}{1 + 0.333 \times 0.3 K_A} \Delta\delta$$

$$= \frac{0.06 K_A - 0.9}{1 + 0.1 K_A} \Delta\delta$$

Hence, the synchronizing torque coefficient due to $\Delta\psi_{fd}$ is

$$K_{S(\Delta\psi_{fd})} = \frac{0.06 K_A - 0.9}{1 + 0.1 K_A}$$

We see that the effect of the AVR is to increase the synchronizing torque component at steady state. With $K_A = 0$ (i.e., constant E_{fd}), $K_{S(\Delta\psi_{fd})} = -0.9$. When $K_A = 15$, the AVR compensates exactly for the demagnetizing effect of the armature reaction. With $K_A = 200$, $K_{S(\Delta\psi_{fd})} = 0.529$ and the total synchronizing torque coefficient is

$$K_S = K_1 + K_{S(\Delta\psi_{fd})} = 1.591 + 0.529$$

$$= 2.12 \quad \text{pu torque/rad}$$

Here, we considered a case with K_5 negative. With a positive K_5 the AVR would have an effect opposite to the above; that is, the effect of the AVR would be to reduce the steady-state synchronizing torque component.

Although we have considered a thyristor exciter in our example, the above observations apply to any type of exciter with a steady-state exciter/AVR gain equal to K_A.

(b) Damping and synchronizing torque components at the rotor oscillation frequency:

Substitution of the numerical values applicable to the specific case under consideration in Equation 12.143 yields

$$\Delta \psi_{fd} = \frac{-0.6 - 0.333 K_5 K_A - 0.012 s}{0.0382 s^2 + 1.93 s + 1 + 0.1 K_A} \Delta \delta$$

From Equation 12.144,

$$\Delta T_e \big|_{\Delta \psi_{fd}} = K_2 \Delta \psi_{fd}$$

$$= \frac{1.5(-0.6 - 0.333 K_5 K_A - 0.012 s)}{0.0382 s^2 + 1.93 s + 1 + 0.1 K_A} \Delta \delta$$

We will assume that the rotor oscillation frequency is 10 rad/s (1.6 Hz). With $s = j\omega = j10$,

$$\Delta T_e \big|_{\Delta \psi_{fd}} = \frac{-0.9 - 0.5 K_5 K_A - j0.18}{-2.82 + 0.1 K_A + j19.3} \Delta \delta$$

With $K_5 = -0.12$ and $K_A = 200$,

$$\Delta T_e \big|_{\Delta \psi_{fd}} = \frac{11.1 - j0.18}{17.18 + j19.3} \Delta \delta$$

$$= 0.2804 \, \Delta \delta - 0.3255 (j \Delta \delta)$$

Thus the effect of the AVR is to increase the synchronizing torque component and decrease the damping torque component, when K_5 is negative.

The net synchronizing torque coefficient is

$$K_S = K_1 + K_{S(\Delta \psi_{fd})} = 1.591 + 0.2804$$

$$= 1.8714 \qquad \text{pu torque/rad}$$

The damping torque component due to $\Delta \psi_{fd}$ is

$$K_{D(\Delta \psi_{fd})} = -0.3255 (j \Delta \delta)$$

Since $\Delta \omega_r = s \Delta \delta / \omega_0 = j\omega \Delta \delta / \omega_0$,

$$K_{D(\Delta \psi_{fd})} = -\frac{0.3255 \omega_0}{\omega} \Delta \omega_r$$

With $\omega = 10$ rad/s, the damping torque coefficient is

$$K_{D(\Delta\psi_{fd})} = -12.27 \quad \text{pu torque/pu speed change}$$

In the absence of any other source of damping, the total $K_D = K_{D(\Delta\psi_{fd})}$.

It is readily apparent that, with K_5 positive, the synchronizing and damping torque components due to $\Delta\psi_{fd}$ would be opposite to the above.

For the system under consideration, Table 12.1 summarizes the effect of the AVR on K_S and K_D at $\omega = 10$ rad/s for different values of K_A.

Table 12.1

K_A	$K_{S(\Delta\psi_{fd})}$	$K_S = K_1 + K_{S(\Delta\psi_{fd})}$	$K_{D(\Delta\psi_{fd})}$
0.0	−0.0025	1.5885	1.772
10.0	−0.0079	1.5831	0.614
15.0	−0.0093	1.5817	0.024
25.0	−0.0098	1.5812	−1.166
50.0	0.0029	1.5939	−4.090
100.0	0.0782	1.6692	−8.866
200.0	0.2804	1.8714	−12.272
400.0	0.4874	2.0784	−9.722
1000.0	0.5847	2.1757	−4.448
Infinity	0.6000	2.1910	0.000

With $K_A = 0$, $\Delta\psi_{fd}$ is entirely due to armature reaction. The effect of the AVR is to decrease K_D for all positive values of K_A. The net damping is minimum (most negative) for $K_A = 200$, and is zero for $K_A = \infty$. For low values of K_A, the effect of the AVR is to decrease K_S very slightly, the net K_S being minimum at K_A of about 46. As K_A is increased beyond this value, K_S increases steadily. For infinite value of K_A, the torque due to $\Delta\psi_{fd}$ is in phase with $\Delta\delta$, and hence has no damping component.

We are normally interested in the performances of excitation systems with moderate or high responses. For such excitation systems, we can make the following general observations regarding the effects of the AVR:

- With K_5 *positive*, the effect of the AVR is to introduce a negative synchronizing torque and a positive damping torque component.

  The constant K_5 is positive for low values of external system reactance and low generator outputs.

  The reduction in K_S due to AVR action in such cases is usually of no particular concern, because K_1 is so high that the net K_S is significantly greater than zero.

- With K_5 negative, the AVR action introduces a positive synchronizing torque component and a negative damping torque component. This effect is more pronounced as the exciter response increases.

  For high values of external system reactance and high generator outputs K_5 is negative. In practice, the situation where K_5 is negative are commonly encountered. For such cases, a high response exciter is beneficial in increasing synchronizing torque. However, in so doing it introduces negative damping. We thus have conflicting requirements with regard to exciter response. One possible recourse is to strike a compromise and set the exciter response so that it results in sufficient synchronizing and damping torque components for the expected range of system-operating conditions. This may not always be possible. It may be necessary to use a high-response exciter to provide the required synchronizing torque and transient stability performance. With a very high external system reactance, even with low exciter response the net damping torque coefficient may be negative.

An effective way to meet the conflicting exciter performance requirements with regard to system stability is to provide a power system stabilizer as described in the following section.

12.5 POWER SYSTEM STABILIZER

The basic function of a *power system stabilizer* (PSS) is to add damping to the generator rotor oscillations by controlling its excitation using auxiliary stabilizing signal(s). To provide damping, the stabilizer must produce a component of electrical torque in phase with the rotor speed deviations.

The theoretical basis for a PSS may be illustrated with the aid of the block diagram shown in Figure 12.13. This is an extension of the block diagram of Figure 12.12 and includes the effect of a PSS.

Since the purpose of a PSS is to introduce a damping torque component, a logical signal to use for controlling generator excitation is the speed deviation $\Delta\omega_r$.

If the exciter transfer function $G_{ex}(s)$ and the generator transfer function between ΔE_{fd} and ΔT_e were pure gains, a direct feedback of $\Delta\omega_r$ would result in a damping torque component. However, in practice both the generator and the exciter (depending on its type) exhibit frequency dependent gain and phase characteristics. Therefore, the PSS transfer function, $G_{PSS}(s)$, should have appropriate phase compensation circuits to compensate for the phase lag between the exciter input and the electrical torque. In the ideal case, with the phase characteristic of $G_{PSS}(s)$ being an exact inverse of the exciter and generator phase characteristics to be compensated, the PSS would result in a pure damping torque at all oscillating frequencies.

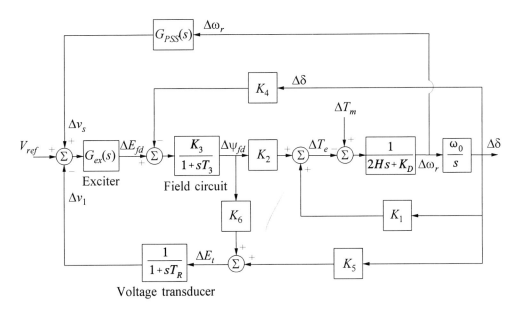

Figure 12.13 Block diagram representation with AVR and PSS

It should be recognized that the generator model assumed in the representation shown in Figure 12.13 neglects amortisseurs to simplify the system model and allow its representation in the form of a block diagram. However, amortisseurs could have a significant effect on the generator phase characteristics and should be considered in establishing the parameters of the PSS. Representation of amortisseur effects will be considered in Section 12.6.

We will illustrate the principle of PSS application by considering the system used in the previous section for examining the effect of the excitation system. The system parameters, as before, are

$$K_1 = 1.591 \quad K_2 = 1.5 \quad K_3 = 0.333 \quad K_D = 0.0 \quad H = 3.0$$
$$T_3 = 1.91 \quad K_5 = -0.12 \quad K_6 = 0.3 \quad G_{ex}(s) = K_A = 200$$

Since T_R is very small in comparison to T_3, we will neglect its effect in examining the PSS performance. This simplifies the analysis without loss of accuracy.

From the block diagram of Figure 12.13, with T_R neglected, $\Delta\psi_{fd}$ due to PSS is given by

$$\Delta\psi_{fd} = \frac{K_3 K_A}{1 + sT_3}(-K_6\Delta\psi_{fd} + \Delta v_s)$$

Therefore,

$$\frac{\Delta\psi_{fd}}{\Delta v_s} = \frac{K_3 K_A}{s T_3 + 1 + K_3 K_6 K_A}$$

$$= \frac{0.333 \times 200}{1.91 s + 1 + 0.333 \times 0.3 \times 200}$$

$$= \frac{66.66}{1.91 s + 21}$$

Let us examine the PSS phase compensation required to produce damping torque at a rotor oscillation frequency of 10 rad/s. With $s = j\omega = j10$,

$$\frac{\Delta\psi_{fd}}{\Delta v_s} = \frac{66.66}{21 + j19.1}$$

$$\Delta T_{PSS} = \Delta T_e \text{ due to PSS} = K_2 (\Delta\psi_{fd} \text{ due to PSS})$$

Therefore, at a frequency of 10 rad/s,

$$\frac{\Delta T_{PSS}}{\Delta v_s} = K_2 \left(\frac{66.66}{21 + j19.1} \right)$$

$$= \frac{1.5 \times 66.66}{21 + j19.1}$$

$$= 3.522 \angle -42.3°$$

If ΔT_{PSS} has to be in phase with $\Delta\omega_r$ (i.e., a purely damping torque), the $\Delta\omega_r$ signal should be processed through a *phase-lead network* so that the signal is advanced by $\theta = 42.3°$ at a frequency of oscillation of 10 rad/s. The amount of damping introduced depends on the gain of PSS transfer function at that frequency. Therefore,

$$\Delta T_{PSS} = (\text{gain of PSS at } \omega = 10)(3.522)(\Delta\omega_r)$$

With the phase-lead network compensating exactly for the phase lag between ΔT_e and Δv_s, the above compensation is purely damping.

The damping torque coefficient due to PSS at $\omega = 10$ rad/s is equal to

$$K_D(\text{PSS}) = (\text{gain of PSS})(3.522)$$

In the previous section, K_D due to AVR action was found to equal -12.27. Therefore, the net K_D including the effects of AVR and PSS is

$$K_D = -12.27 + (\text{gain of PSS at } \omega = 10)(3.522)$$

With a gain of $12.27/3.522 = 3.48$, the PSS produces just enough damping to compensate for the negative damping due to AVR action. As the PSS gain is increased further, the amount of damping increases.

   If the phase-lead network provides more compensation than the phase lag between ΔT_e and Δv_s, the PSS introduces, in addition to a damping component of torque, a negative synchronizing torque component. Conversely, with under-compensation a positive synchronizing torque component is introduced. Usually, the PSS is required to contribute to the damping of rotor oscillations over a range of frequencies, rather than a single frequency.

   We will illustrate the basic structure, modelling, and performance of power system stabilizers by considering a thyristor excitation system. Figure 12.14 shows the block diagram of the excitation system, including the AVR and PSS. Since we are concerned with small-signal performance, stabilizer output limits and exciter output limits are not shown in the figure. The following is a brief description of the basis for the PSS configuration and considerations in the selection of the parameters.

   The PSS representation in Figure 12.14 consists of three blocks: a phase compensation block, a signal washout block, and a gain block.

   The *phase compensation* block provides the appropriate phase-lead characteristic to compensate for the phase lag between the exciter input and the generator electrical (air-gap) torque. The figure shows a single first-order block. In practice, two or more first-order blocks may be used to achieve the desired phase compensation. In some cases, second-order blocks with complex roots have been used.

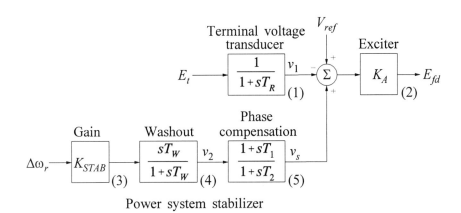

Power system stabilizer

Figure 12.14 Thyristor excitation system with AVR and PSS

Normally, the frequency range of interest is 0.1 to 2.0 Hz, and the phase-lead network should provide compensation over this entire frequency range. The phase characteristic to be compensated changes with system conditions; therefore, a compromise is made and a characteristic acceptable for different system conditions is selected. Generally some undercompensation is desirable so that the PSS, in addition to significantly increasing the damping torque, results in a slight increase of the synchronizing torque.

The *signal washout* block serves as a high-pass filter, with the time constant T_W high enough to allow signals associated with oscillations in ω_r to pass unchanged. Without it, steady changes in speed would modify the terminal voltage. It allows the PSS to respond only to changes in speed. From the viewpoint of the washout function, the value of T_W is not critical and may be in the range of 1 to 20 seconds. The main consideration is that it be long enough to pass stabilizing signals at the frequencies of interest unchanged, but not so long that it leads to undesirable generator voltage excursions during system-islanding conditions.

The stabilizer gain K_{STAB} determines the amount of damping introduced by the PSS. Ideally, the gain should be set at a value corresponding to maximum damping; however, it is often limited by other considerations.

In applying the PSS, care should be taken to ensure that the overall system stability is enhanced, not just the small-signal stability [9]. The performance objectives of the PSS, alternative input signals, and a detailed description of the procedure for selection of its parameters so as to enhance overall system stability are presented in Chapter 17.

System state matrix including PSS

From block 4 of Figure 12.14, using perturbed values, we have

$$\Delta v_2 = \frac{pT_W}{1+pT_W}(K_{STAB}\Delta\omega_r)$$

Hence

$$p\Delta v_2 = K_{STAB}\,p\Delta\omega_r - \frac{1}{T_W}\Delta v_2$$

Substituting for $p\Delta\omega_r$ given by Equation 12.115, we obtain the following expression for $p\Delta v_2$ in terms of the state variables:

$$p\Delta v_2 = K_{STAB}\left[a_{11}\Delta\omega_r + a_{12}\Delta\delta + a_{13}\Delta\psi_{fd} + \frac{1}{2H}\Delta T_m\right] - \frac{1}{T_W}\Delta v_2$$

$$\tag{12.145}$$

$$= a_{51}\Delta\omega_r + a_{52}\Delta\delta + a_{53}\Delta\psi_{fd} + a_{55}\Delta v_2 + \frac{K_{STAB}}{2H}\Delta T_m$$

where

$$a_{51} = K_{STAB}a_{11} \qquad a_{52} = K_{STAB}a_{12}$$

$$a_{53} = K_{STAB}a_{13} \qquad a_{55} = -\frac{1}{T_W}$$

(12.146)

Since $p\Delta v_2$ is not a function of Δv_1 and Δv_3, $a_{54}=a_{56}=0$.
From block 5,

$$\Delta v_s = \Delta v_2 \left(\frac{1+pT_1}{1+pT_2}\right)$$

Hence

$$p\Delta v_s = \frac{T_1}{T_2}p\Delta v_2 + \frac{1}{T_2}\Delta v_2 - \frac{1}{T_2}\Delta v_s$$

Substitution for $p\Delta v_2$, given by Equation 12.145, yields

$$p\Delta v_s = a_{61}\Delta\omega_r + a_{62}\Delta\delta + a_{63}\Delta\psi_{fd} +$$

$$a_{64}\Delta v_1 + a_{65}\Delta v_2 + a_{66}\Delta v_s + \frac{T_1}{T_2}\frac{K_{STAB}}{2H}\Delta T_m$$

(12.147)

where

$$a_{61} = \frac{T_1}{T_2}a_{51} \qquad a_{62} = \frac{T_1}{T_2}a_{52}$$

$$a_{63} = \frac{T_1}{T_2}a_{53} \qquad a_{65} = \frac{T_1}{T_2}a_{55} + \frac{1}{T_2}$$

(12.148)

$$a_{66} = -\frac{1}{T_2}$$

From block 2 of Figure 12.14,

$$\Delta E_{fd} = K_A(\Delta v_s - \Delta v_1)$$

The field circuit equation, with PSS included, becomes

$$p\Delta\psi_{fd} = a_{32}\Delta\delta + a_{33}\Delta\psi_{fd} + a_{34}\Delta v_1 + a_{36}\Delta v_s$$

(12.149)

where

$$a_{36} = \frac{\omega_0 R_{fd}}{L_{adu}} K_A \qquad (12.150)$$

The complete state-space model, including the PSS, has the following form (with $\Delta T_m = 0$):

$$
\begin{bmatrix} \Delta\dot\omega_r \\ \Delta\dot\delta \\ \Delta\dot\psi_{fd} \\ \Delta\dot v_1 \\ \Delta\dot v_2 \\ \Delta\dot v_s \end{bmatrix}
=
\begin{bmatrix}
a_{11} & a_{12} & a_{13} & 0 & 0 & 0 \\
a_{21} & 0 & 0 & 0 & 0 & 0 \\
0 & a_{32} & a_{33} & a_{34} & 0 & a_{36} \\
0 & a_{42} & a_{43} & a_{44} & 0 & 0 \\
a_{51} & a_{52} & a_{53} & 0 & a_{55} & 0 \\
a_{61} & a_{62} & a_{63} & 0 & a_{65} & a_{66}
\end{bmatrix}
\begin{bmatrix} \Delta\omega_r \\ \Delta\delta \\ \Delta\psi_{fd} \\ \Delta v_1 \\ \Delta v_2 \\ \Delta v_s \end{bmatrix}
\qquad (12.151)
$$

Alternative methods of treating blocks 4 and 5:

Block 4 may be considered to be made up of two blocks:

In this case, $\Delta v_2'$ becomes the state variable, with

$$p\Delta v_2' = \frac{1}{T_W}(K_{STAB}\Delta\omega_r - \Delta v_2')$$

and the output Δv_2 of the block is given by

$$\Delta v_2 = T_W p\Delta v_2'$$

$$= K_{STAB}\Delta\omega_r - \Delta v_2'$$

The advantage of this approach is that the expression for the derivative of the input variable to the block is not required. This is important in situations where the input is not a state variable, in which case the expression for its derivative is not readily available. Similarly, block 5 may be treated as follows:

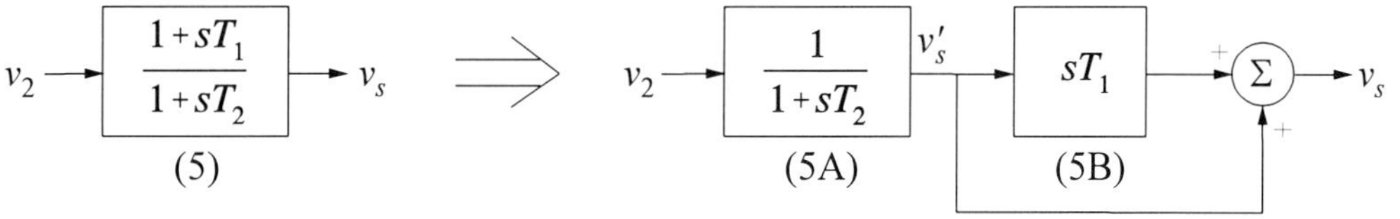

In this case, v'_s is the state variable, with

$$p\Delta v'_s = \frac{1}{T_2}(\Delta v_2 - \Delta v'_s)$$

and the output Δv_s is given by

$$\Delta v_s = T_1 p \Delta v'_s + \Delta v'_s$$

$$= \frac{T_1}{T_2}\Delta v_2 + \left(1 - \frac{1}{T_2}\right)\Delta v'_s$$

The disadvantage of this approach is that the output of the block is not a state variable and hence cannot be monitored directly by computing the state variable.

For excitation system models with higher-order transfer function blocks the following general procedure may be used.

Higher-order transfer function blocks:

Consider an nth order transfer function whose block diagram is shown in Figure 12.15.

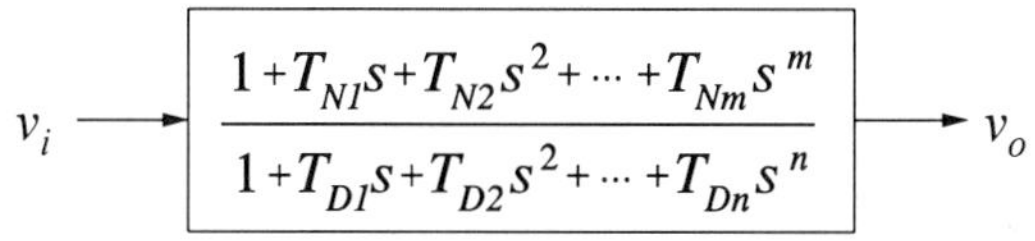

Figure 12.15

The above system has n poles and m zeros ($m<n$). The transfer function of Figure 12.15 may be divided into two parts:

$$G(s) = \frac{v_o(s)}{v_i(s)} = \frac{x_1(s)}{v_i(s)} \frac{v_o(s)}{x_1(s)} \tag{12.152}$$

Hence the nth order transfer function may be represented in terms of two blocks as shown in Figure 12.16.

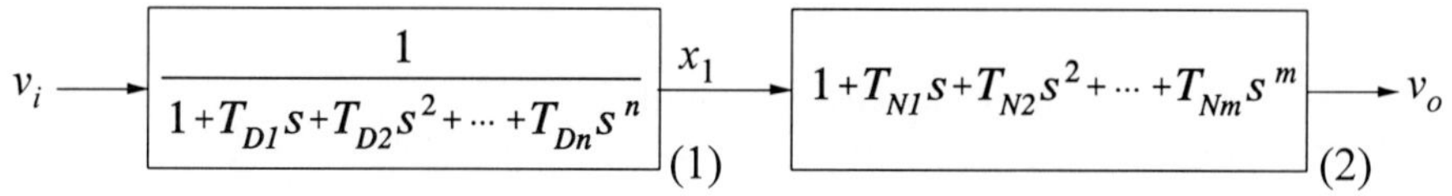

$$v_i \longrightarrow \boxed{\frac{1}{1+T_{D1}s+T_{D2}s^2+\cdots+T_{Dn}s^n}} \underset{(1)}{\overset{x_1}{\longrightarrow}} \boxed{1+T_{N1}s+T_{N2}s^2+\cdots+T_{Nm}s^m} \underset{(2)}{\longrightarrow} v_o$$

Figure 12.16

From block 1 of Figure 12.16,

$$x_1(1+T_{D1}s+T_{D2}s^2+\cdots+T_{Dn}s^n) = v_i \tag{12.153}$$

Let

$$x_2 = \dot{x}_1 = \frac{dx_1}{dt}$$

$$x_3 = \dot{x}_2 = \frac{dx_2}{dt} \tag{12.154}$$

$$\cdots \cdots \cdots \cdots \cdots \cdots$$

$$x_n = \dot{x}_{n-1} = \frac{dx_{n-1}}{dt}$$

From Equation 12.153, in the time domain

$$x_1+T_{D1}x_2+T_{D2}x_3+\cdots+T_{D(n-1)}x_n+T_{Dn}\dot{x}_n = v_i$$

Therefore,

$$\dot{x}_n = \frac{1}{T_{Dn}}\left[v_i-x_1-T_{D1}x_2-\cdots-T_{D(n-1)}x_n\right] \tag{12.155}$$

Combining Equations 12.154 and 12.155, we have

$$
\begin{bmatrix} \dot{x}_1 \\ \dot{x}_2 \\ \dots \\ \dot{x}_{n-1} \\ \dot{x}_n \end{bmatrix} = \begin{bmatrix} 0 & 1 & 0 & \dots & 0 \\ 0 & 0 & 1 & \dots & 0 \\ \dots & \dots & \dots & \dots & \dots \\ 0 & 0 & 0 & \dots & 1 \\ -\dfrac{1}{T_{Dn}} & -\dfrac{T_{D1}}{T_{Dn}} & -\dfrac{T_{D2}}{T_{Dn}} & \dots & -\dfrac{T_{D(n-1)}}{T_{Dn}} \end{bmatrix} \begin{bmatrix} x_1 \\ x_2 \\ \dots \\ x_{n-1} \\ x_n \end{bmatrix} + \begin{bmatrix} 0 \\ 0 \\ \dots \\ 0 \\ \dfrac{1}{T_{Dn}} \end{bmatrix} v_i \qquad (12.156)
$$

From block 2 of Figure 12.16, the expression for the output is

$$
\begin{aligned}
v_o &= x_1 + T_{N1}x_2 + T_{N2}x_3 + \dots + T_{Nm}x_{m+1} \\
&= \begin{bmatrix} 1 & T_{N1} & T_{N2} & \dots & T_{Nm} & 0 & \dots & 0 \end{bmatrix} x + 0 v_i
\end{aligned}
\qquad (12.157)
$$

Equations 12.156 and 12.157 have the state-space form

$$
\dot{x} = Ax + bu
$$

$$
y = cx + du
$$

The variables x_1, x_2, ..., x_n defined above are referred to as *phase variables*. They are a particular set of state variables that consists of one variable and its n-1 derivatives.

Example 12.4

This example is an extension of the previous two examples which analyzed the small-signal stability of the system shown in Figure E12.5. Here we consider the effect of two alternative types of excitation control:

(a)   A thyristor exciter with AVR as shown in Figure 12.11, with

$$K_A = 200 \qquad T_R = 0.02 \text{ s}$$

(b)   The above excitation system with PSS added as shown in Figure 12.13, with

$$K_{STAB} = 9.5 \qquad T_W = 1.4 \text{ s} \qquad T_1 = 0.154 \text{ s} \qquad T_2 = 0.033 \text{ s}$$

The generator parameters and saturation data are as given in Example 12.3. With the transmission circuit 2 out of service, the plant output in per unit on 2220 MVA and 24 kV base is as follows:

$$P = 0.9 \qquad\qquad Q = 0.3 \text{ (overexcited)}$$
$$E_t = 1.0\angle 36° \qquad\qquad E_B = 0.995\angle 0°$$

Analyze the small-signal stability of the system with each of the two types of excitation control, by determining

(i)   Elements of the state matrix

(ii)   The constants of the block diagram of Figure 12.12

(iii)   Eigenvalues of the state matrix, participation matrix, and frequency and damping ratio of each oscillatory mode

(iv)   Synchronizing and damping torque coefficients at the rotor oscillation frequency

Compare with the results of Examples 12.2 and 12.3.

Solution

(i)   From equations developed in Section 12.3.3, the state matrix of the system including the AVR and PSS is

$$
\mathbf{A} =
\begin{array}{c}
\left[
\begin{array}{cccccc}
0 & -0.1092 & -0.1236 & 0 & 0 & 0 \\
376.99 & 0 & 0 & 0 & 0 & 0 \\
0 & -0.1938 & -0.4229 & -27.3172 & 0 & 27.3172 \\
0 & -7.3125 & 20.8391 & -50.0 & 0 & 0 \\
0 & -1.0372 & -1.1738 & 0 & -0.7143 & 0 \\
0 & -4.8404 & -5.4777 & 0 & 26.9697 & -30.3030
\end{array}
\right]
&
\begin{array}{l}
\text{States} \\
\Delta\omega_r \\
\Delta\delta \\
\Delta\psi_{fd} \\
\Delta v_1 \\
\Delta v_2 \\
\Delta v_s
\end{array}
\end{array}
$$

With AVR only (i.e., without PSS), the last two rows and columns disappear.

(ii) The constants of the block diagram are

$$
\begin{array}{lll}
K_1 = 0.7643 & K_2 = 0.8649 & K_3 = 0.3230 \\
K_4 = 1.4187 & K_5 = -0.1463 & K_6 = 0.4168 \\
T_3 = 2.365 & T_R = 0.02 & G_{ex}(s) = 200.0
\end{array}
$$

(iii) (a) With AVR only, the eigenvalues are

$$\lambda_1, \lambda_2 = +0.504 \pm j7.23 \quad (\omega_d = 1.15 \text{ Hz}, \zeta = -0.07)$$
$$\lambda_3 = -20.202$$
$$\lambda_4 = -31.230$$

and the participation matrix is

$$
\begin{bmatrix}
0.474 & 0.474 & 0.077 & 0.024 \\
0.474 & 0.474 & 0.077 & 0.024 \\
0.065 & 0.065 & 2.524 & 1.633 \\
0.010 & 0.010 & 1.677 & 2.681
\end{bmatrix}
\begin{matrix}
\Delta\omega_r \\
\Delta\delta \\
\Delta\psi_{fd} \\
\Delta v_1
\end{matrix}
$$
$$
\quad\;\; \lambda_1 \qquad\;\; \lambda_2 \qquad\;\; \lambda_3 \qquad\;\; \lambda_4
$$

Only the magnitudes of the participation factors are shown above; the angles do not provide any useful information.

With AVR only, the system becomes unstable through an oscillatory mode of 1.15 Hz. From the participation factors, we see that this mode is associated primarily with the rotor angle and speed. The two non-oscillatory modes, both of which decay rapidly, are associated with the AVR and the field circuit.

(b) With AVR and PSS, the eigenvalues are

$$
\begin{aligned}
\lambda_1 &= -39.097 \\
\lambda_2, \lambda_3 &= -1.005 \pm j6.607 \qquad (\omega_d = 1.05 \text{ Hz}, \zeta = 0.15) \\
\lambda_4 &= -0.739 \\
\lambda_5, \lambda_6 &= -19.797 \pm j12.822 \qquad (\omega_d = 2.04 \text{ Hz}, \zeta = 0.84)
\end{aligned}
$$

and the participation matrix (magnitudes only) is

$$
\begin{bmatrix}
0.004 & 0.528 & 0.525 & 0.035 & 0.013 & 0.013 \\
0.004 & 0.528 & 0.528 & 0.035 & 0.013 & 0.013 \\
0.188 & 0.073 & 0.073 & 0.002 & 0.984 & 0.984 \\
0.908 & 0.025 & 0.025 & 0.000 & 0.527 & 0.527 \\
0.012 & 0.160 & 0.160 & 1.072 & 0.094 & 0.094 \\
0.300 & 0.052 & 0.052 & 0.001 & 0.417 & 0.417
\end{bmatrix}
\begin{matrix}
\Delta\omega_r \\
\Delta\delta \\
\Delta\psi_{fd} \\
\Delta v_1 \\
\Delta v_2 \\
\Delta v_s
\end{matrix}
$$
$$
\;\; \lambda_1 \quad\;\; \lambda_2 \quad\;\; \lambda_3 \quad\;\; \lambda_4 \quad\;\; \lambda_5 \quad\;\; \lambda_6
$$

With the addition of the PSS, the system has become very stable. There are two oscillatory modes: one is the rotor angle mode with a frequency of 1.05 Hz; the other has a frequency of 2.04 Hz and is associated with the excitation system and field circuit. The two non-oscillatory modes are associated with the excitation system.

(iv) (a) *Synchronizing and Damping Torque Coefficients with AVR Only*

From the block diagram of Figure 12.12, field flux linkage change due to AVR and armature reaction (AR) is

$$
\Delta\psi_{fd} = \frac{-K_3[K_4(1 + sT_R) + K_A K_5]}{s^2 T_3 T_R + s(T_3 + T_R) + (1 + K_3 K_6 K_A)} \Delta\delta
$$

Substituting the values of the constants summarized in (i) above and simplifying, we get

$$\Delta\psi_{fd} = \frac{8.993 - 0.0092\,s}{0.0473\,s^2 + 2.385\,s + 27.925}\,\Delta\delta$$

From the eigenvalue calculations, the complex frequency of rotor oscillation is $0.504 + j7.23$. With s equal to this complex frequency, the above expression for $\Delta\psi_{fd}$ simplifies to

$$\Delta\psi_{fd} = \frac{8.988 - j0.067}{26.67 + j17.59}\,\Delta\delta$$

Hence, ΔT_e due to $\Delta\psi_{fd}$ is

$$\Delta T_e\big|_{(AVR+AR)} = K_2 \Delta\psi_{fd} = 0.8649\,\Delta\psi_{fd}$$

$$= \frac{7.774 - j0.0573}{26.67 + j17.59}\,\Delta\delta$$

$$= 0.202\,\Delta\delta - 0.1354\,(j\Delta\delta)$$

Now,

$$\text{pu } \Delta\omega_r = \frac{s\,\Delta\delta}{\omega_0}$$

$$= \frac{(0.504 + j7.23)\,\Delta\delta}{377}$$

Hence,

$$j\Delta\delta = \frac{377}{7.23}\,\Delta\omega_r - \frac{0.504}{7.23}\,\Delta\delta$$

$$= 52.14\Delta\omega_r - 0.07\Delta\delta$$

Substituting for $j\Delta\delta$ in the expression for ΔT_e, we get

$$\Delta T_e\big|_{(AVR+AR)} = 0.202\,\Delta\delta - 0.1354\,(52.14\,\Delta\omega_r - 0.07\,\Delta\delta)$$

$$= 0.2115\,\Delta\delta - 7.06\,\Delta\omega_r$$

Therefore,

$$K_{S(AVR+AR)} = 0.2115 \quad \text{pu torque/rad}$$
$$K_{D(AVR+AR)} = -7.06 \quad \text{pu torque/pu speed change}$$

The field flux linkage variation is due to armature reaction and AVR. In Example 12.3, we saw that the effect of the armature reaction alone is to decrease K_S by 0.00172 and increase K_D by 1.53 at the rotor oscillation frequency of 6.41 rad/s. The effect of AVR alone is therefore to increase K_S by nearly 0.2 and to decrease K_D by nearly 8.5.

The total synchronizing and damping torque coefficients are

$$K_S = K_1 + K_{S(\text{AVR}+\text{AR})} = 0.7643 + 0.2115$$

$$= 0.9758 \quad \text{pu torque/rad}$$

and

$$K_D = K_{D(\text{AVR}+\text{AR})} = -7.06 \quad \text{pu torque/pu speed change}$$

From Equation 12.82, the corresponding damping ratio is

$$\zeta = \frac{1}{2}\frac{K_D}{\sqrt{K_S 2H\omega_0}} = \frac{1}{2}\frac{-7.06}{\sqrt{0.9758 \times 2 \times 3.5 \times 377}}$$

$$= -0.07$$

This agrees with ζ computed from the eigenvalues.

(b) *Synchronizing and Damping Torque Coefficients with PSS*

The complex frequency of rotor oscillation in this case is $-1.005 + j6.607$. From the block diagram of Figure 12.13, $\Delta\psi_{fd}$ due to PSS is

$$\Delta\psi_{fd}\big|_{\text{PSS}} = \frac{K_3 K_A}{1 + sT_3}\left(\Delta v_s - \frac{K_6}{1 + sT_R}\Delta\psi_{fd}\right)$$

with

$$\Delta v_s = K_{\text{STAB}}\frac{sT_W}{1 + sT_W}\frac{1 + sT_1}{1 + sT_2}\Delta\omega_r$$

Therefore,

$$\Delta T_e\big|_{\text{PSS}} = K_2 \Delta\psi_{fd}\big|_{\text{PSS}}$$

$$= \frac{K_{\text{STAB}}K_2 K_3 K_A (1 + sT_R)(1 + sT_1)sT_W}{[(1 + sT_3)(1 + sT_R) + K_3 K_6 K_A](1 + sT_2)(1 + sT_W)}\Delta\omega_r$$

Substituting the numerical values of the parameters, setting $s = -1.005 + j6.607$, and simplifying, we get

$$\Delta T_e \big|_{PSS} = (24.0 + j8.089)\, \Delta\omega_r$$

Now,

$$j\Delta\omega_r = j\frac{s}{\omega_0}\Delta\delta = j\frac{-1.005 + j6.607}{377}\Delta\delta$$

$$= -0.01752\,\Delta\delta - \frac{1.005}{377}(j\Delta\delta)$$

$$= -0.01752\,\Delta\delta - \frac{1.005}{377}\left(\frac{377}{6.607}\Delta\omega_r + \frac{1.005}{6.607}\Delta\delta\right)$$

$$= -0.0179\,\Delta\delta - 0.1521\,\Delta\omega_r$$

Hence,

$$\Delta T_e \big|_{PSS} = 24.0\,\Delta\omega_r + 8.089\,(-0.0179\,\Delta\delta - 0.1521\,\Delta\omega_r)$$

$$= -0.145\,\Delta\delta + 22.77\,\Delta\omega_r$$

The synchronizing and damping torque coefficients due to PSS are

$$K_{S(PSS)} = -0.145 \quad \text{pu torque/rad}$$
$$K_{D(PSS)} = 22.77 \quad \text{pu torque/pu speed change}$$

and K_S and K_D due to AVR and armature reaction (AR) must now also be evaluated at $s = -1.005 + j6.607$. Following the same procedure used before, we have

$$K_{S(AVR+AR)} = 0.21 \quad \text{pu torque/rad}$$
$$K_{D(AVR+AR)} = -8.69 \quad \text{pu torque/pu speed change}$$

The total synchronizing and damping torque coefficients are

$$K_S = K_1 + K_{S(AVR+AR)} + K_{S(PSS)}$$

$$= 0.7643 + 0.21 - 0.145$$

$$= 0.8293 \quad \text{pu torque/rad}$$

$$K_D = K_{D(AVR+AR)} + K_{D(PSS)}$$

$$= -8.69 + 22.77$$

$$= 14.08 \quad \text{pu torque/pu speed change}$$

From Equation 12.81, the undamped natural frequency is

$$\omega_n = \sqrt{\frac{K_S \omega_0}{2H}} = \sqrt{\frac{0.8293 \times 377}{2 \times 3.5}} = 6.683 \quad \text{rad/s}$$

$$\zeta = \frac{1}{2}\frac{K_D}{\sqrt{K_S 2 H \omega_0}} = \frac{1}{2}\frac{14.08}{\sqrt{0.8293 \times 7.0 \times 377}} = 0.15$$

$$\omega_d = \omega_n \sqrt{1 - \zeta^2} = 6.683 \sqrt{1 - 0.15^2} = 6.607 \quad \text{rad/s}$$

The above results agree with ζ and ω_d computed from the eigenvalues.

We see that the PSS has increased K_D by 22.77 and reduced K_S by 0.145. The reduction in K_S indicates that phase lead compensation used results in slight overcompensation at the rotor oscillation frequency of 6.61 rad/s. By adjusting T_1 and/or T_2, the compensation may be altered to give zero or even positive synchronizing torque component.

The following is a summary of frequency of rotor oscillation, damping ratio, K_S and K_D with different types of excitation control and with classical model, based on the results of this and the previous two examples.

Model	Classical Model (K_D=0)	Constant E_{fd}	AVR only	AVR and PSS
ω_d	1.02 Hz	1.07 Hz	1.15 Hz	1.05 Hz
ζ	0	0.017	−0.07	0.15
K_S	0.757	0.763	0.976	0.829
K_D	0	1.53	−7.06	14.08

■

12.6 SYSTEM STATE MATRIX WITH AMORTISSEURS

When generator amortisseurs are included, the system equations developed in Section 12.3.2 change as follows. We will assume that the model includes one d-axis amortisseur and two q-axis amortisseurs, as shown in Figure 12.17.

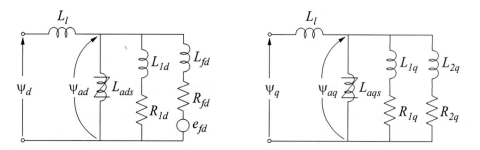

Figure 12.17 Synchronous machine equivalent circuits

Rotor circuit equations

$$p\psi_{fd} = \frac{\omega_0 R_{fd}}{L_{adu}} E_{fd} - \omega_0 R_{fd} i_{fd}$$

$$p\psi_{1d} = -\omega_0 R_{1d} i_{1d}$$

$$p\psi_{1q} = -\omega_0 R_{1q} i_{1q} \qquad (12.158)$$

$$p\psi_{2q} = -\omega_0 R_{2q} i_{2q}$$

The rotor currents are given by

$$i_{fd} = \frac{1}{L_{fd}}(\psi_{fd} - \psi_{ad})$$

$$i_{1d} = \frac{1}{L_{1d}}(\psi_{1d} - \psi_{ad})$$

$$i_{1q} = \frac{1}{L_{1q}}(\psi_{1q} - \psi_{aq}) \qquad (12.159)$$

$$i_{2q} = \frac{1}{L_{2q}}(\psi_{2q} - \psi_{aq})$$

The d- and q-axis mutual flux linkages are given by

$$\psi_{ad} = -L_{ads}i_d + L_{ads}i_{fd} + L_{ads}i_{1d}$$

$$= L''_{ads}\left(-i_d + \frac{\psi_{fd}}{L_{fd}} + \frac{\psi_{1d}}{L_{1d}}\right) \tag{12.160}$$

$$\psi_{aq} = L''_{aqs}\left(-i_q + \frac{\psi_{1q}}{L_{1q}} + \frac{\psi_{2q}}{L_{2q}}\right) \tag{12.161}$$

where

$$L''_{ads} = \frac{1}{\dfrac{1}{L_{ads}} + \dfrac{1}{L_{fd}} + \dfrac{1}{L_{1d}}} \tag{12.162}$$

$$L''_{aqs} = \frac{1}{\dfrac{1}{L_{aqs}} + \dfrac{1}{L_{1q}} + \dfrac{1}{L_{2q}}} \tag{12.163}$$

The expressions for i_d and i_q (corresponding to Equations 12.103 and 12.104) become

$$i_d = \frac{X_{Tq}E_{qN} - R_T E_{dN}}{D}$$

$$i_q = \frac{R_T E_{qN} + X_{Td}E_{dN}}{D} \tag{12.164}$$

where

$$E_{dN} = E''_d + E_B\sin\delta \qquad\qquad E_{qN} = E''_q - E_B\cos\delta$$

$$E''_d = \bar{\omega}L''_{aqs}\left(\frac{\psi_{1q}}{L_{1q}} + \frac{\psi_{2q}}{L_{2q}}\right) \qquad E''_q = \bar{\omega}L''_{ads}\left(\frac{\psi_{fd}}{L_{fd}} + \frac{\psi_{1d}}{L_{1d}}\right)$$

$$X_{Td} = X_E + \bar{\omega}(L''_{ads} + L_l) \qquad X_{Tq} = X_E + \bar{\omega}(L''_{aqs} + L_l)$$

$$= X_E + X''_{ds} \qquad\qquad\qquad = X_E + X''_{qs} \tag{12.165}$$

$$R_T = R_a + R_E \qquad\qquad\qquad D = R_T^2 + X_{Td}X_{Tq}$$

As discussed in Chapter 5, $\bar{\omega}$ is assumed to be 1.0 pu to counterbalance the effects of neglecting stator transients.

Expressing Equations 12.164 and 12.165 in terms of perturbed values, we get

$$\Delta i_d = m_1 \Delta \delta + m_2 \Delta \psi_{fd} + m_3 \Delta \psi_{1d} + m_4 \Delta \psi_{1q} + m_5 \Delta \psi_{2q}$$

$$\Delta i_q = n_1 \Delta \delta + n_2 \Delta \psi_{fd} + n_3 \Delta \psi_{1d} + n_4 \Delta \psi_{1q} + n_5 \Delta \psi_{2q}$$

(12.166)

with

$$m_1 = \frac{E_B}{D}(X_{Tq} \sin \delta_0 - R_T \cos \delta_0)$$

$$m_2 = \frac{X_{Tq}}{D} \frac{L''_{ads}}{L_{fd}} \qquad\qquad m_3 = \frac{X_{Tq}}{D} \frac{L''_{ads}}{L_{1d}}$$

$$m_4 = -\frac{R_T}{D} \frac{L''_{aqs}}{L_{1q}} \qquad\qquad m_5 = -\frac{R_T}{D} \frac{L''_{aqs}}{L_{2q}}$$

$$n_1 = \frac{E_B}{D}(R_T \sin \delta_0 + X_{Td} \cos \delta_0)$$

(12.167)

$$n_2 = \frac{R_T}{D} \frac{L''_{ads}}{L_{fd}} \qquad\qquad n_3 = \frac{R_T}{D} \frac{L''_{ads}}{L_{1d}}$$

$$n_4 = \frac{X_{Td}}{D} \frac{L''_{aqs}}{L_{1q}} \qquad\qquad n_5 = \frac{X_{Td}}{D} \frac{L''_{aqs}}{L_{2q}}$$

The expressions for $\Delta \psi_{ad}$ and $\Delta \psi_{aq}$ are given by

$$\Delta \psi_{ad} = L''_{ads}\left(-\Delta i_d + \frac{\Delta \psi_{fd}}{L_{fd}} + \frac{\Delta \psi_{1d}}{L_{1d}}\right)$$

$$= (-m_1 L''_{ads})\Delta \delta + L''_{ads}\left(\frac{1}{L_{fd}} - m_2\right)\Delta \psi_{fd} + L''_{ads}\left(\frac{1}{L_{1d}} - m_3\right)\Delta \psi_{1d} +$$

(12.168)

$$(-m_4 L''_{ads})\Delta \psi_{1q} + (-m_5 L''_{ads})\Delta \psi_{2q}$$

$$\Delta \psi_{aq} = (-n_1 L''_{aqs})\Delta \delta + (-n_2 L''_{aqs})\Delta \psi_{fd} + (-n_3 L''_{aqs})\Delta \psi_{1d} +$$

(12.169)

$$L''_{aqs}\left(\frac{1}{L_{1q}} - n_4\right)\Delta \psi_{1q} + L''_{aqs}\left(\frac{1}{L_{2q}} - n_5\right)\Delta \psi_{2q}$$

The expression for ΔT_e (corresponding to Equation 12.112) is given by

$$\Delta T_e = \psi_{ad0}\Delta i_q + i_{q0}\Delta\psi_{ad} - \psi_{aq0}\Delta i_d - i_{d0}\Delta\psi_{aq}$$

$$= K_1\Delta\delta + K_2\Delta\psi_{fd} + K_{21}\Delta\psi_{1d} + K_{22}\Delta\psi_{1q} + K_{23}\Delta\psi_{2q} \tag{12.170}$$

where

$$K_1 = n_1(\psi_{ad0} + L''_{aqs}i_{d0}) - m_1(\psi_{aq0} + L''_{ads}i_{q0})$$

$$K_2 = n_2(\psi_{ad0} + L''_{aqs}i_{d0}) - m_2(\psi_{aq0} + L''_{ads}i_{q0}) + \frac{L''_{ads}}{L_{fd}}i_{q0}$$

$$K_{21} = n_3(\psi_{ad0} + L''_{aqs}i_{d0}) - m_3(\psi_{aq0} + L''_{ads}i_{q0}) + \frac{L''_{ads}}{L_{1d}}i_{q0} \tag{12.171}$$

$$K_{22} = n_4(\psi_{ad0} + L''_{aqs}i_{d0}) - m_4(\psi_{aq0} + L''_{ads}i_{q0}) - \frac{L''_{aqs}}{L_{1q}}i_{d0}$$

$$K_{23} = n_5(\psi_{ad0} + L''_{aqs}i_{d0}) - m_5(\psi_{aq0} + L''_{ads}i_{q0}) - \frac{L''_{aqs}}{L_{2q}}i_{d0}$$

From Equations 12.83 and 12.170

$$p\Delta\omega_r = \frac{1}{2H}[\Delta T_M - \Delta T_e - K_D\Delta\omega_r]$$

$$= \frac{1}{2H}[\Delta T_m - K_1\Delta\delta - K_2\Delta\psi_{fd} - K_{21}\Delta\psi_{1d} -$$

$$K_{22}\Delta\psi_{1q} - K_{23}\Delta\psi_{2q} - K_D\Delta\omega_r] \tag{12.172}$$

$$= a_{11}\Delta\omega_r + a_{12}\Delta\delta + a_{13}\Delta\psi_{fd} + a_{14}\Delta\psi_{1d} +$$

$$a_{15}\Delta\psi_{1q} + a_{16}\Delta\psi_{2q} + b_{11}\Delta T_m$$

where

$$a_{11} = -\frac{K_D}{2H} \qquad a_{12} = -\frac{K_1}{2H} \qquad a_{13} = -\frac{K_2}{2H}$$

$$a_{14} = -\frac{K_{21}}{2H} \qquad a_{15} = -\frac{K_{22}}{2H} \qquad a_{16} = -\frac{K_{23}}{2H} \tag{12.173}$$

$$b_{11} = \frac{1}{2H}$$

As before,

$$p\Delta\delta = a_{21}\Delta\omega_r \qquad (12.174)$$

where

$$a_{21} = \omega_0 = 2\pi f_0 \qquad (12.175)$$

From Equations 12.158, 12.159, and 12.168,

$$
\begin{aligned}
p\Delta\psi_{fd} = {}& a_{31}\Delta\omega_r + a_{32}\Delta\delta + a_{33}\Delta\psi_{fd} + a_{34}\Delta\psi_{1d} \\
& a_{35}\Delta\psi_{1q} + a_{36}\Delta\psi_{2q} + b_{32}\Delta e_{fd}
\end{aligned}
\qquad (12.176)
$$

where

$$a_{31} = 0 \qquad\qquad a_{32} = -\frac{\omega_0 R_{fd}}{L_{fd}} m_1 L''_{ads}$$

$$a_{33} = -\frac{\omega_0 R_{fd}}{L_{fd}}\left(1 - \frac{L''_{ads}}{L_{fd}} + m_2 L''_{ads}\right)$$

$$a_{34} = -\frac{\omega_0 R_{fd}}{L_{fd}}\left(m_3 L''_{ads} - \frac{L''_{ads}}{L_{fd}}\right) \qquad (12.177)$$

$$a_{35} = -\frac{\omega_0 R_{fd}}{L_{fd}} m_4 L''_{ads} \qquad a_{36} = -\frac{\omega_0 R_{fd}}{L_{fd}} m_5 L''_{ads}$$

$$b_{32} = \frac{\omega_0 R_{fd}}{L_{adu}}$$

Similarly,

$$
\begin{aligned}
p\Delta\psi_{1d} &= -\omega_0 R_{1d}\Delta i_{1d} \\
&= a_{41}\Delta\omega_r + a_{42}\Delta\delta + a_{43}\Delta\psi_{fd} + a_{44}\Delta\psi_{1d} + a_{45}\Delta\psi_{1q} + a_{46}\Delta\psi_{2q}
\end{aligned}
\qquad (12.178)
$$

where

$$a_{41} = 0 \qquad\qquad a_{42} = -\frac{\omega_0 R_{1d}}{L_{1d}} m_1 L''_{ads}$$

$$a_{43} = -\frac{\omega_0 R_{1d}}{L_{1d}} \left(m_2 L''_{ads} - \frac{L''_{ads}}{L_{fd}} \right)$$

$$a_{44} = -\frac{\omega_0 R_{1d}}{L_{1d}} \left(1 - \frac{L''_{ads}}{L_{fd}} + m_3 L''_{ads} \right)$$

(12.179)

$$a_{45} = -\frac{\omega_0 R_{1d}}{L_{1d}} m_4 L''_{ads} \qquad a_{46} = -\frac{\omega_0 R_{1d}}{L_{1d}} m_5 L''_{ads}$$

$$p\Delta\psi_{1q} = -\omega_0 R_{1q} \Delta i_{1q}$$

$$= a_{51}\Delta\omega_r + a_{52}\Delta\delta + a_{53}\Delta\psi_{fd} + a_{54}\Delta\psi_{1d} + a_{55}\Delta\psi_{1q} + a_{56}\Delta\psi_{2q}$$

(12.180)

where

$$a_{51} = 0 \qquad\qquad a_{52} = -\frac{\omega_0 R_{1q}}{L_{1q}} n_1 L''_{aqs}$$

$$a_{53} = -\frac{\omega_0 R_{1q}}{L_{1q}} n_2 L''_{aqs} \qquad a_{54} = -\frac{\omega_0 R_{1q}}{L_{1q}} n_3 L''_{aqs}$$

$$a_{55} = -\frac{\omega_0 R_{1q}}{L_{1q}} \left(1 - \frac{L''_{aqs}}{L_{1q}} + n_4 L''_{aqs} \right)$$

(12.181)

$$a_{56} = -\frac{\omega_0 R_{1q}}{L_{1q}} \left(n_5 L''_{aqs} - \frac{L''_{aqs}}{L_{2q}} \right)$$

$$p\Delta\psi_{2q} = -\omega_0 R_{2q} \Delta i_{2q}$$

$$= a_{61}\Delta\omega_r + a_{62}\Delta\delta + a_{63}\Delta\psi_{fd} + a_{64}\Delta\psi_{1d} + a_{65}\Delta\psi_{1q} + a_{66}\Delta\psi_{2q}$$

(12.182)

where

$$a_{61} = 0 \qquad\qquad a_{62} = -\frac{\omega_0 R_{2q}}{L_{2q}} n_1 L''_{aqs}$$

$$a_{63} = -\frac{\omega_0 R_{2q}}{L_{2q}} n_2 L''_{aqs} \qquad a_{64} = -\frac{\omega_0 R_{2q}}{L_{2q}} n_3 L''_{aqs}$$

(12.183A)

$$a_{65} = -\frac{\omega_0 R_{2q}}{L_{2q}} \left(n_4 L''_{aqs} - \frac{L''_{aqs}}{L_{1q}} \right)$$

$$a_{66} = -\frac{\omega_0 R_{2q}}{L_{2q}}\left(1 - \frac{L''_{aqs}}{L_{2q}} + n_s L''_{aqs}\right)$$
(12.183B)

The complete state equation is given by

$$
\begin{bmatrix} \Delta\dot\omega_r \\ \Delta\dot\delta \\ \Delta\dot\psi_{fd} \\ \Delta\dot\psi_{1d} \\ \Delta\dot\psi_{1q} \\ \Delta\dot\psi_{2q} \end{bmatrix}
=
\begin{bmatrix}
a_{11} & a_{12} & a_{13} & a_{14} & a_{15} & a_{16} \\
a_{21} & 0 & 0 & 0 & 0 & 0 \\
0 & a_{32} & a_{33} & a_{34} & a_{35} & a_{36} \\
0 & a_{42} & a_{43} & a_{44} & a_{45} & a_{46} \\
0 & a_{52} & a_{53} & a_{54} & a_{55} & a_{56} \\
0 & a_{62} & a_{63} & a_{64} & a_{65} & a_{66}
\end{bmatrix}
\begin{bmatrix} \Delta\omega_r \\ \Delta\delta \\ \Delta\psi_{fd} \\ \Delta\psi_{1d} \\ \Delta\psi_{1q} \\ \Delta\psi_{2q} \end{bmatrix}
+
\begin{bmatrix}
b_{11} & 0 \\
0 & 0 \\
0 & b_{32} \\
0 & 0 \\
0 & 0 \\
0 & 0
\end{bmatrix}
\begin{bmatrix} \Delta T_m \\ \Delta E_{fd} \end{bmatrix}
$$
(12.184)

With constant input torque, $\Delta T_m=0$. Similarly, with constant field voltage, $\Delta E_{fd}=0$.

If AVR action is to be represented, we need the expression for ΔE_t. Following the procedure used in Section 12.3.3,

$$\Delta E_t = \frac{e_{d0}}{E_{t0}}\Delta e_d + \frac{e_{q0}}{E_{t0}}\Delta e_q$$

with

$$\Delta e_d = -R_a \Delta i_d + L_l \Delta i_q - \Delta\psi_{ad}$$

$$\Delta e_q = -R_a \Delta i_q - L_l \Delta i_d + \Delta\psi_{aq}$$

By using Equations 12.166, 12.168, and 12.169, the perturbed value of the terminal voltage may be expressed in the following form:

$$\Delta E_t = K_5 \Delta\delta + K_6 \Delta\psi_{fd} + K_{61}\Delta\psi_{1d} + K_{62}\Delta\psi_{1q} + K_{63}\Delta\psi_{2q}$$
(12.185)

It is left to the reader as an exercise to develop the expressions for the K constants of the above equation in terms of system parameters.

Example 12.5

In this example, we extend the analysis presented in Examples 12.3 and 12.4 by including the effects of generator amortisseurs. In addition, we examine the effects of varying the exciter gain and stabilizer gain to provide additional insight into how the AVR and PSS influence the small-signal dynamic characteristic.

The per unit generator parameters related to the amortisseurs are as follows:

$$X_q' = 0.65 \qquad\qquad X_d'' = 0.23 \qquad\qquad X_q'' = 0.25$$
$$T_{q0}' = 1.0 \text{ s} \qquad\qquad T_{d0}'' = 0.03 \text{ s} \qquad\qquad T_{q0}'' = 0.07 \text{ s}$$

All other data (including generator and excitation system parameters, and system conditions) are as considered in Example 12.4.

(a)   With the excitation system on manual control, determine the state matrix and its eigenvalues. Identify the state variable with the highest participation for each mode.

(b)   With AVR in service and PSS out of service, examine the effect of varying the exciter gain K_A on the eigenvalues. Assume $T_R = 0.02$ s.

(c)   With AVR and PSS in service, examine the effect of varying the stabilizer gain K_{STAB} on the eigenvalues. Assume all other parameters of the excitation system are fixed at values given in Example 12.4.

Solution

(a) The per unit values of the leakage inductances and resistances of the amortisseur circuits are computed following the procedure used in Example 4.1 as follows:

$$L_{1d} = 0.1400 \qquad\qquad L_{1q} = 0.7063 \qquad\qquad L_{2q} = 0.1102$$
$$R_{1d} = 0.0248 \qquad\qquad R_{1q} = 0.0061 \qquad\qquad R_{2q} = 0.0227$$

Based on equations developed in Section 12.6, the state matrix of the system including amortisseurs, but with constant E_{fd}, is

$$
\mathbf{A} = \begin{bmatrix}
0 & -0.121 & -0.069 & -0.076 & -0.003 & -0.023 \\
377 & 0 & 0 & 0 & 0 & 0 \\
0 & -0.109 & -0.883 & 0.645 & 0 & 0.0003 \\
0 & -4.928 & 26.72 & -37.47 & 0.002 & 0.013 \\
0 & -0.058 & -0.0005 & -0.0005 & -2.914 & 2.250 \\
0 & -1.39 & -0.011 & -0.012 & 8.367 & -24.18
\end{bmatrix}
$$

The eigenvalues of **A** and the corresponding state variable with highest participation are

Eigenvalues		Dominant State(s)	ω_d	ζ
λ_1, λ_2	$-0.171 \pm j6.47$	$\Delta\omega_r, \Delta\delta$	1.03 Hz	0.0265
λ_3	-0.200	$\Delta\psi_{fd}$		
λ_4	-2.045	$\Delta\psi_{1q}$		
λ_5	-25.01	$\Delta\psi_{2q}$		
λ_6	-37.85	$\Delta\psi_{1d}$		

Comparing with the results of Example 12.3, we see that the effect of the amortisseurs is to increase the damping ratio from 0.017 to 0.0265 and to increase the frequency very slightly from 1.02 to 1.03 Hz.

(b) Table E12.1 shows the effect of varying K_A on the eigenvalues of the state matrix with PSS out of service. The table also indicates for each mode the state variable with highest participation as well as any other state with significant participation.

Table E12.1

K_A	Eigenvalues						
	λ_1, λ_2	λ_3	λ_4	λ_5	λ_6	λ_7
–	$-0.171 \pm j6.47$	-0.20	-37.85	-25.01	-2.04	–
0.1	$-0.170 \pm j6.47$	-0.20	-37.84	-25.01	-2.05	-50.00
5.0	$-0.112 \pm j6.46$	-0.59	-37.34	-25.02	-2.06	-50.22
10.0	$-0.050 \pm j6.46$	-0.98	-36.83	-25.02	-2.10	-50.42
15.0	$0.011 \pm j6.46$	-1.34	-36.34	-25.02	-2.16	-50.79
30.0	$0.182 \pm j6.50$	-2.94	-34.86	-25.03	-1.86	-51.12
50.0	$0.357 \pm j6.62$	-4.60	-32.87	-25.06	-1.94	-51.70
100.0	$0.552 \pm j6.95$	-9.40	-26.81	-25.54	-1.96	-52.84
110.0	$0.564 \pm j7.01$	-10.64	$-25.47 \pm j0.96$		-1.97	
125.0	$0.573 \pm j7.09$	-13.03	$-24.15 \pm j0.88$		-1.97	-53.30
130.0	$0.574 \pm j7.11$	-14.14	-22.92	-24.18	-1.97	
150.0	$0.570 \pm j7.21$	$-18.09 \pm j5.09$		-24.72	-1.97	-53.71
200.0	$0.528 \pm j7.38$	$-17.63 \pm j11.33$		-24.86	-1.97	-54.41
400.0	$0.315 \pm j7.66$	$-16.40 \pm j22.63$		-24.92	-1.98	-56.38
Dominant States	$\Delta\omega_r, \Delta\delta$	$\Delta\psi_{fd}, \Delta\psi_{1d}$		$\Delta\psi_{2q}$	$\Delta\psi_{1q}$	Δv_1
Significant States		Δv_1				$\Delta\psi_{1d}$

We see that the effect of the AVR is to introduce negative damping. For values of K_A higher than about 14, the system is unstable. The system is most unstable for a K_A of about 125; as K_A is increased further, the system becomes less and less unstable.

With K_A of about 110, two real eigenvalues (λ_4 and λ_5) merge to become a complex pair of eigenvalues. At K_A of about 130, these eigenvalues become real again. For K_A of about 150 and higher, the real eigenvalues λ_3 and λ_4 merge to form a complex pair. The oscillatory mode represented by these eigenvalues is restricted to the field circuit, d-axis amortisseur, and the AVR.

(c) Table E12.2 shows the effect of varying the stabilizer gain K_{STAB}, with K_A=200.

Table E12.2

K_{STAB}	Eigenvalues							
	λ_1, λ_2	λ_3, λ_4	λ_5	λ_6	λ_7	λ_8	λ_9
–	$0.53\pm j7.4$	$-17.6\pm j11.3$	-24.9	-1.97	-54.4	–	–
0.1	$0.51\pm j7.4$	$-17.6\pm j11.4$	-24.9	-1.97	-54.4	-30.4	-0.71
1.0	$0.34\pm j7.4$	$-17.1\pm j11.8$	-24.9	-1.97	-54.4	-30.9	-0.72
3.0	$-0.01\pm j7.4$	$-16.2\pm j12.7$	-24.9	-1.97	-54.5	-31.9	-0.72
5.0	$-0.37\pm j7.2$	$-15.4\pm j13.6$	-24.9	-1.97	-54.5	-32.8	-0.73
9.5	$-1.08\pm j6.8$	$-13.9\pm j15.5$	-25.0	-1.96	-54.5	-34.3	-0.74
15.0	$-1.75\pm j6.2$	$-12.5\pm j17.7$	-25.0	-1.95	-54.6	-35.6	-0.76
21.0	$-2.20\pm j5.5$	$-11.4\pm j20.0$	-25.0	-1.94	-54.7	-36.7	-0.78
Dominant States	$\Delta\omega_r, \Delta\delta$	$\Delta\psi_{fd}, \Delta\psi_{1d}$	$\Delta\psi_{2q}$	$\Delta\psi_{1q}$	Δv_1	Δv_2	Δv_s

From the results summarized in Table E12.2, we see that as K_{STAB} is increased the damping of rotor angle mode increases and the frequency decreases slightly. The reduction in frequency is indicative of reduction in synchronizing torque; this is due to the PSS phase lead circuit providing overcompensation at the rotor oscillating frequency. The frequency of the second oscillatory mode increases and its damping decreases as K_{STAB} is increased. This mode is associated with the field circuit. All other modes are not affected by the variation of K_{STAB}. ∎

12.7 SMALL-SIGNAL STABILITY OF MULTIMACHINE SYSTEMS

Analysis of practical power systems involves the simultaneous solution of equations representing the following:

- Synchronous machines, and the associated excitation systems and prime movers

- Interconnecting transmission network

- Static and dynamic (motor) loads

- Other devices such as HVDC converters, static var compensators

Figure 12.18 shows the general structure of the complete system model.
As discussed in Chapter 5, for system stability studies it is appropriate to neglect the transmission network and machine stator transients. The dynamics of machine rotor circuits, excitation systems, prime mover and other devices are represented by differential equations. The result is that the complete system model consists of a large number of ordinary differential and algebraic equations.

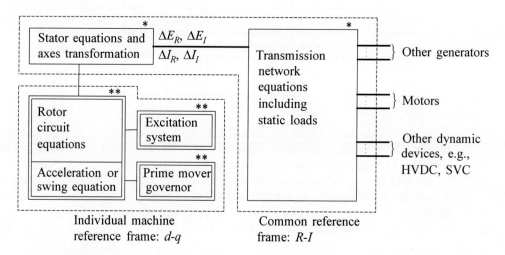

*   Algebraic equations
**   Differential equations

Figure 12.18   Structure of the complete power system model

Each machine model is expressed in its own d-q reference frame which rotates with its rotor. For the solution of interconnecting network equations, all voltages and currents must be expressed in a common reference frame. Usually a reference frame rotating at synchronous speed is used as the common reference. Axis transformation equations are used to transform between the individual machine (d-q) reference frames and the common (R-I) reference frame as shown in Figure 12.19. For convenience in the organization of the complete set of algebraic equations, the machine stator equations are also expressed in the common reference frame.

The R-axis of the common reference frame is usually used as the reference for measuring the machine rotor angle. For a machine represented in detail including dynamics of one or more rotor circuits, the rotor angle δ is defined as the angle by which the machine q-axis leads the R-axis, as shown in Figure 12.19. For a machine represented by a classical model, the rotor angle is the angle by which the voltage E' leads the R-axis. Under dynamic conditions, the angle δ changes with rotor speed.

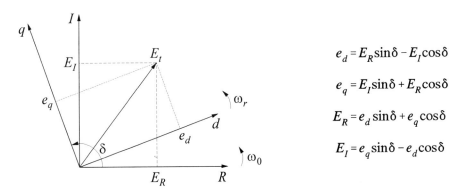

$$e_d = E_R\sin\delta - E_I\cos\delta$$

$$e_q = E_I\sin\delta + E_R\cos\delta$$

$$E_R = e_d\sin\delta + e_q\cos\delta$$

$$E_I = e_q\sin\delta - e_d\cos\delta$$

Figure 12.19 Reference frame transformation

The formulation of the state equations for small-signal analysis involves the development of linearized equations about an operating point and elimination of all variables other than the state variables. The general procedure is similar to that used for a single-machine infinite bus system in the previous sections. However, the need to allow for the representation of extensive transmission networks, loads, a variety of excitation systems and prime mover models, HVDC links, and static var compensators makes the process very complex. Therefore, the formulation of the state equations requires a systematic procedure for treating the wide range of devices. The following is a description of one such procedure [10,11,12].

Formulation of the state equations

The linearized model of each dynamic device is expressed in the following form:

$$\dot{\mathbf{x}}_i = \mathbf{A}_i \mathbf{x}_i + \mathbf{B}_i \Delta \mathbf{v} \tag{12.186}$$

$$\Delta \mathbf{i}_i = \mathbf{C}_i \mathbf{x}_i - \mathbf{Y}_i \Delta \mathbf{v} \tag{12.187}$$

where

        $\mathbf{x}_i$ are the perturbed values of the individual device state variables
        $\mathbf{i}_i$ is the current injection into the network from the device
        $\mathbf{v}$ is the vector of the network bus voltages

In Equations 12.186 and 12.187, $\mathbf{B}_i$ and $\mathbf{Y}_i$ have non-zero elements corresponding only to the terminal voltage of the device and any remote bus voltages used to control the device. The current vector $\mathbf{i}_i$ has two elements corresponding to the real and imaginary components. Similarly, the voltage vector $\mathbf{v}$ has two elements per bus associated with the device. Such state equations for all the dynamic devices in the system may be combined into the form

$$\dot{\mathbf{x}} = \mathbf{A}_D \mathbf{x} + \mathbf{B}_D \Delta \mathbf{v} \tag{12.188}$$

$$\Delta \mathbf{i} = \mathbf{C}_D \mathbf{x} - \mathbf{Y}_D \Delta \mathbf{v} \tag{12.189}$$

where $\mathbf{x}$ is the state vector of the complete system, and $\mathbf{A}_D$ and $\mathbf{C}_D$ are block diagonal matrices composed of $\mathbf{A}_i$ and $\mathbf{C}_i$ associated with the individual devices.

As described in Chapter 6, the interconnecting transmission network is represented by the node equation:

$$\Delta \mathbf{i} = \mathbf{Y}_N \Delta \mathbf{v} \tag{12.190}$$

The elements of $\mathbf{Y}_N$ include the effects of nonlinear static loads as shown later in this section.

Equating Equation 12.189 associated with the devices and Equation 12.190 associated with the network, we obtain

$$\mathbf{C}_D \mathbf{x} - \mathbf{Y}_D \Delta \mathbf{v} = \mathbf{Y}_N \Delta \mathbf{v}$$

Hence,

$$\Delta \mathbf{v} = (\mathbf{Y}_N + \mathbf{Y}_D)^{-1} \mathbf{C}_D \mathbf{x}$$

Substituting the above expression for $\Delta \mathbf{v}$ in Equation 12.188 yields the overall system state equation:

$$\dot{\mathbf{x}} = \mathbf{A_D}\mathbf{x} + \mathbf{B_D}(\mathbf{Y_N} + \mathbf{Y_D})^{-1}\mathbf{C_D}\mathbf{x} \qquad (12.191)$$

$$= \mathbf{A}\mathbf{x}$$

where the state matrix $\mathbf{A}$ of the complete system is given by

$$\mathbf{A} = \mathbf{A_D} + \mathbf{B_D}(\mathbf{Y_N} + \mathbf{Y_D})^{-1}\mathbf{C_D} \qquad (12.192)$$

The method of building $\mathbf{A_i}$, $\mathbf{B_i}$, $\mathbf{C_i}$, and $\mathbf{Y_i}$ matrices for the synchronous machine and the associated controls can follow the general approach described in the previous sections. Motor loads can be treated in a similar way. Reference 12 gives details of treating HVDC links and SVCs.

The MASS (*Multi-Area Small-signal Stability*) computer program described in references 10 and 11 uses the above general approach for formulating the system state matrix. The program computes all eigenvalues of the matrix by using the *QR* transformation method discussed in Section 12.2.9. Since this approach cannot exploit sparsity, it cannot be used for analysis of very large systems. Depending on the capability of the computer used, the maximum system size is limited to a few hundred states. This is satisfactory for the analysis of stability of small systems or local stability problems in large systems. Since all modes of the system are computed, it is also ideally suited for the design and coordination of controls. In the next section, we will describe techniques for the analysis of global problems in very large systems.

Representation of static loads

(a) Constant impedance (linear) load:

The shunt admittance to ground representing the load is computed as

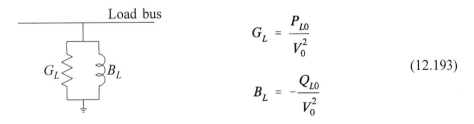

$$G_L = \frac{P_{L0}}{V_0^2}$$

$$(12.193)$$

$$B_L = -\frac{Q_{L0}}{V_0^2}$$

Figure 12.20

where

P_{L0} = initial value of the active component of load
Q_{L0} = initial value of the reactive component of load
V_0 = initial value of the bus voltage magnitude

(b) Nonlinear load:

Let us consider the load whose voltage dependent characteristics are represented as

$$P_L = P_{L0}\left(\frac{V}{V_0}\right)^m$$

$$Q_L = Q_{L0}\left(\frac{V}{V_0}\right)^n \tag{12.194}$$

where V is the magnitude of the bus voltage given by

$$V = \sqrt{v_R^2 + v_I^2}$$

The R and I components of the load current are

$$i_R = P_L \frac{v_R}{V^2} + Q_L \frac{v_I}{V^2}$$

$$i_I = P_L \frac{v_I}{V^2} - Q_L \frac{v_R}{V^2}$$

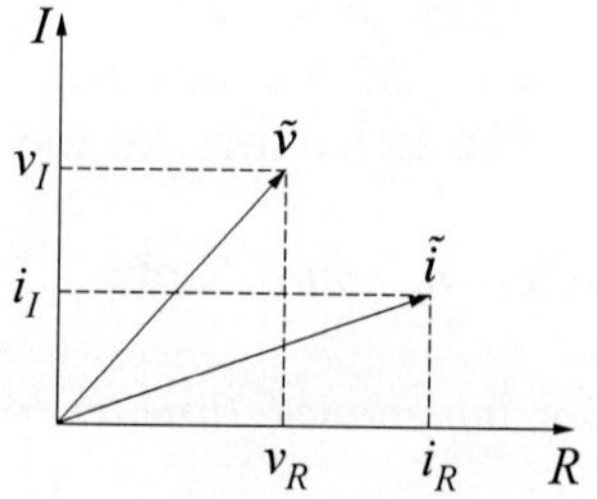

Figure 12.21

Linearizing, we find

$$\Delta i_R = \frac{v_{R0}}{V_0^2}\Delta P_L + \frac{v_{I0}}{V_0^2}\Delta Q_L + \frac{P_{L0}}{V_0^2}\Delta v_R + \frac{Q_{L0}}{V_0^2}\Delta v_I +$$

$$(P_{L0}v_{R0} + Q_{L0}v_{I0})\left(-\frac{2}{V_0^3}\right)\Delta V$$

$$\tag{12.195}$$

$$\Delta i_I = \frac{v_{I0}}{V_0^2}\Delta P_L - \frac{v_{R0}}{V_0^2}\Delta Q_L + \frac{P_{L0}}{V_0^2}\Delta v_I - \frac{Q_{L0}}{V_0^2}\Delta v_R +$$

$$(P_{L0}v_{I0} - Q_{L0}v_{R0})\left(-\frac{2}{V_0^3}\right)\Delta V$$

where

$$\Delta V = \frac{v_{R0}}{V_0}\Delta v_R + \frac{v_{I0}}{V_0}\Delta v_I \qquad (12.196)$$

Now,

$$\Delta P_L = m\frac{P_{L0}}{V_0}\Delta V$$

$$\qquad (12.197)$$

$$\Delta Q_L = n\frac{Q_{L0}}{V_0}\Delta V$$

Substitution of Equations 12.196 and 12.197 in Equation 12.195 yields

$$\begin{bmatrix} \Delta i_R \\ \Delta i_I \end{bmatrix} = \begin{bmatrix} G_{RR} & B_{RI} \\ -B_{IR} & G_{II} \end{bmatrix}\begin{bmatrix} \Delta v_R \\ \Delta v_I \end{bmatrix} \qquad (12.198)$$

where

$$G_{RR} = \frac{P_{L0}}{V_0^2}\left((m-2)\frac{v_{R0}^2}{V_0^2}+1\right)+\frac{Q_{L0}}{V_0^2}\left((n-2)\frac{v_{R0}v_{I0}}{V_0^2}\right)$$

$$B_{RI} = \frac{Q_{L0}}{V_0^2}\left((n-2)\frac{v_{I0}^2}{V_0^2}+1\right)+\frac{P_{L0}}{V_0^2}\left((m-2)\frac{v_{R0}v_{I0}}{V_0^2}\right)$$

$$\qquad (12.199)$$

$$B_{IR} = \frac{Q_{L0}}{V_0^2}\left((n-2)\frac{v_{R0}^2}{V_0^2}+1\right)-\frac{P_{L0}}{V_0^2}\left((m-2)\frac{v_{R0}v_{I0}}{V_0^2}\right)$$

$$G_{II} = \frac{P_{L0}}{V_0^2}\left((m-2)\frac{v_{I0}^2}{V_0^2}+1\right)-\frac{Q_{L0}}{V_0^2}\left((n-2)\frac{v_{R0}v_{I0}}{V_0^2}\right)$$

As an example, consider a load such that

$$P_{L0} = 5.0 \text{ pu} \qquad Q_{L0} = 2.0 \text{ pu} \qquad \tilde{v}_0 = 1.0\angle 20°$$

The real and imaginary components of the initial bus voltage are

$$v_{R0} = 1.0\cos 20° = 0.94 \quad \text{pu}$$

$$v_{I0} = 1.0\sin 20° = 0.342 \quad \text{pu}$$

From Equation 12.199, with constant impedance characteristic $(m=n=2.0)$,

$$
\begin{array}{ll}
G_{RR} = 5.0 & B_{RI} = 2.0 \\
-B_{IR} = -2.0 & G_{II} = 5.0
\end{array}
$$

Similarly, with constant current characteristic $(m=n=1.0)$,

$$
\begin{array}{ll}
G_{RR} = -0.06 & B_{RI} = 0.16 \\
-B_{IR} = -1.84 & G_{II} = 5.06
\end{array}
$$

and with constant MVA characteristic $(m=n=0.0)$,

$$
\begin{array}{ll}
G_{RR} = -5.12 & B_{RI} = -1.68 \\
-B_{IR} = -1.68 & G_{II} = 5.12
\end{array}
$$

The equivalent admittance matrix of Equation 12.198 representing a static load may be directly implemented in the network admittance matrix. However, as evident from the above example, the equivalent admittance matrix representing nonlinear loads is not symmetrical and does not represent a simple shunt admittance to ground as in the case of a constant impedance load.

Redundant state variables

The formulation of the system state equations described above uses absolute changes in machine rotor speed and angle as state variables. With such a formulation, the state matrix of a system which does not contain an infinite bus will have one or two zero eigenvalues [13].

One of these zero eigenvalues is associated with the lack of uniqueness of absolute rotor angle. In other words, if rotor angles of all machines are increased by a constant value, the system stability is not affected. The redundancy in rotor angle states can be eliminated by choosing one of the machines as a reference and expressing angle changes of all other machines with respect to this reference as follows.

For the reference machine R,

$$p\Delta\delta_R = 0$$

for any other machine i $(i=1, ..., n; i \neq R)$,

$$p\Delta\delta_i = (\Delta\omega_r \text{ of machine } i) - (\Delta\omega_r \text{ of machine } R)$$

The second zero eigenvalue exists if all the generator torques are assumed to be independent of speed deviations, i.e., if a damping term represented by K_D is not included in the swing equation and a speed governor is not represented. This zero eigenvalue can also be avoided by measuring speed deviations with respect to that of a reference machine. Mathematically, the process of referring rotor angles or speed deviations to a reference machine is equivalent to a similarity transformation.

The zero eigenvalues, however, may not be computed exactly because of mismatches in the power flow solution and the limited accuracy of eigenvalue calculation routines. They may therefore appear as small eigenvalues.

12.8 SPECIAL TECHNIQUES FOR ANALYSIS OF VERY LARGE SYSTEMS

Analysis of interarea oscillations in a large interconnected power system requires a detailed modelling of the entire system. System representations with as many as 2,000 dynamic devices and 12,000 buses are not uncommon for such studies. With an average of 15 states per device, the number of state variables required for modal analysis may thus be on the order of 30,000. This is well outside the range of the conventional eigenvalue analysis methods. Special techniques have therefore been developed that focus on evaluating a *selected subset of eigenvalues* associated with the complete system response.

One such technique is the AESOPS algorithm originally presented in reference 14. It uses a novel frequency response approach to calculate the eigenvalues associated with the rotor angle modes. References 10 and 15 describe improved implementation of the AESOPS algorithm.

Several powerful methods for the computation of eigenvalues associated with a small number of selected modes of oscillation have been published in the literature on power system stability. Reference 16 describes the application of two sparsity-based eigenvalue techniques: simultaneous iterations and modified Arnoldi methods. The S-method described in reference 17 is particularly suited for finding the unstable modes. The *selective modal analysis* (SMA) approach described in references 2 and 18 computes eigenvalues associated with selected modes of interest by using special techniques to identify variables that are relevant to the selected modes, and then constructing a reduced-order model that involves only the relevant variables.

The PEALS (*Program for Eigenvalue Analysis of Large Systems*) described in references 10 and 11 uses two of these techniques: the AESOPS algorithm and the modified Arnoldi method. These two methods have been found to be efficient and reliable, and they complement each other in meeting the requirements of small-signal stability analysis of large complex power systems. The following is a description of these two techniques:

(a) The AESOPS algorithm

The acronym AESOPS stands for the *Analysis of Essentially Spontaneous Oscillations in Power Systems* [14]. It is based on an ingenious approach developed using sound engineering judgement rather than a rigorous mathematical procedure.

The algorithm computes eigenvalues associated *only with rotor angle modes*, one complex conjugate pair of eigenvalues at a time. The eigenvalues are determined by applying to the rotor of a selected generator an external torque having a complex sinusoidal form, expressed in terms of an initial estimate of an eigenvalue. Under steady-state conditions, all incremental variables of the linear system model will be complex sinusoids similar in form to the external torque. The complex frequency response of the system is calculated by solving an appropriate set of complex *algebraic equations*. A revised estimate of the eigenvalue is determined from the linear system response, and the process is repeated until successive estimates converge within a desired tolerance. The resulting eigenvalue represents a mode of oscillation in which the selected generator participates significantly. If the generator has several dominant modes of rotor oscillations, the eigenvalue computed depends on the initial estimate.

The procedure may be applied to different generators to compute the eigenvalues associated with the modes in which the generators have significant participation.

The AESOPS algorithm is derived from the linearized equation of motion of a generator:

$$2H\frac{d\Delta\omega_r}{dt} = \Delta T_m - \Delta T_e$$

$$= \Delta T_m - (K_S\Delta\delta + K_D\Delta\omega_r)$$

Taking the Laplace transform, we get

$$2Hs\Delta\omega_r = \Delta T_m - \left[K_S(s)\frac{\Delta\omega_r}{s} + K_D(s)\Delta\omega_r\right] \quad (12.200)$$

The above equation recognizes that synchronizing and damping torque coefficients K_S and K_D are functions of the frequency of oscillation.

Rearranging Equation 12.200, we have

$$\Delta T_m = \left[2Hs + K_D(s) + \frac{K_S(s)}{s}\right]\Delta\omega_r \quad (12.201)$$

The characteristic values (eigenvalues) of the system are given by the zeros of

$$2Hs + K_D(s) + \frac{K_S(s)}{s} = 0 \quad (12.202)$$

The characteristic values are thus those values of s which force ΔT_m to be zero, provided $\Delta \omega_r$ is not zero. This is used by the AESOPS algorithm to determine zeros of Equation 12.202.

With a finite external torque, the magnitude of system variables tends to be unbounded at complex frequencies close to an eigenvalue. From a computational viewpoint, it is preferable to limit the extent of system response and allow the magnitude of external torque to go to zero at an eigenvalue. This is achieved by setting the complex speed deviation of the perturbed generator to 1.0+j0.0 per unit. The external torque ΔT_m is determined in the solution process. As iterations converge, the magnitude of the external torque tends to zero and provides an indication of the accuracy of the eigenvalue.

In practice, the model representing the system has poles as well as zeros. Zeros close to a pole prevent the AESOPS algorithm from converging to the corresponding eigenvalues. Poles associated with modes in which the machine has a significant participation are not cancelled by zeros. Thus, for each generator in the system, a subset of the overall system modes may be determined.

Iterative Procedure: The zeros of $\Delta T_m(s)$ may be determined by using Newton's method. This requires the derivative of ΔT_m with respect to s. From Equation 12.201,

$$\frac{\partial \Delta T_m}{\partial s} = \left[2H + \frac{\partial K_D(s)}{\partial s} + \frac{1}{s} \frac{\partial K_S(s)}{\partial s} - \frac{K_S(s)}{s^2} \right] \Delta \omega_r \qquad (12.203)$$

This can be simplified by making the following approximations.

Close to a system eigenvalue, from Equation 12.202,

$$\frac{K_S(s)}{s^2} + \frac{K_D(s)}{s} = -2H \qquad (12.204)$$

If $K_D(s)$ is small,

$$\frac{K_S(s)}{s^2} \approx -2H \qquad (12.205)$$

Therefore,

$$\frac{\partial \Delta T_m}{\partial s} \approx \left[4H + \frac{\partial K_D(s)}{\partial s} + \frac{1}{s} \frac{\partial K_S(s)}{\partial s} \right] \Delta \omega_r \qquad (12.206)$$

Further, $\dfrac{\partial K_D}{\partial s}$ and $\dfrac{\partial K_S}{\partial s}$ are small compared to $4H$. Therefore,

$$\frac{\partial \Delta T_m}{\partial s} \approx 4H \Delta \omega_r \qquad (12.207)$$

The nth iteration of Newton's method for computing the eigenvalues, i.e., the values of s, is given by

$$s_{n+1} = s_n - \frac{\Delta T_m(s)}{\left[\dfrac{\partial \Delta T_m}{\partial s} \right]_{s=s_n}}$$

$$= s_n - \left[\frac{\Delta T_m(s)}{4H \Delta \omega_r} \right]_{s=s_n}$$

The momentum of the disturbed machine is

$$M_n = J \Delta \omega_r = 2H \Delta \omega_r$$

Hence,

$$s_{n+1} = s_n - \left[\frac{\Delta T_m(s)}{2M_n} \right]_{s=s_n} \qquad (12.208)$$

For modes that involve many machines, this makes too large a change in the eigenvalue at each iteration. In such cases, $\partial K_S/\partial s$ and $\partial K_D/\partial s$ are not small. Therefore, a modified momentum is used based on an equivalent inertia defined so that kinetic energy associated with the change in speed at all machines in the system is equal to the equivalent inertia multiplied by the square of the speed change of the disturbed machine. The equivalent inertia is given by

$$H_e = \sum_{i=1}^{N_m} H_i |\Delta \omega_{ri}|^2 \qquad (12.209)$$

with N_m = number of machines. Equation 12.208 is then modified to

$$s_{n+1} = s_n - \left[\frac{\Delta T_m(s)}{4H_e \Delta \omega_r} \right]_{s=s_n}$$

If as discussed earlier $\Delta \omega_r$ is set to $1.0+j0.0$ pu,

$$s_{n+1} = s_n - \frac{\Delta T_m(s_n)}{4H_e} \tag{12.210}$$

Implementation of the AESOPS algorithm in PEALS

The application of the AESOPS algorithm as developed above requires the calculation of

(a)   The torque necessary to keep the speed change of a chosen disturbed machine equal to $1.0+j0.0$ per unit; and

(b)   Speed changes of all the other machines and the equivalent inertia H_e.

These depend on Equation 12.186 of each dynamic device and Equation 12.190 of the interconnecting network.

We will first organize the device equations in a suitable form to perform the above calculations. From Equation 12.186, for any complex frequency s, for each dynamic device, we may write

$$s\mathbf{x}_i = \mathbf{A}_i \mathbf{x}_i + \mathbf{B}_i \Delta \mathbf{v}$$

Therefore,

$$\mathbf{x}_i = (s\mathbf{I} - \mathbf{A}_i)^{-1} \mathbf{B}_i \Delta \mathbf{v} \tag{12.211}$$

Substitution of the above in Equation 12.187, and expressing in the s domain yields

$$\Delta \mathbf{i}_i = \mathbf{C}_i (s\mathbf{I} - \mathbf{A}_i)^{-1} \mathbf{B}_i \Delta \mathbf{v} - \mathbf{Y}_i \Delta \mathbf{v}$$
$$= -\mathbf{Y}_{ie}(s) \Delta \mathbf{v} \tag{12.212}$$

where

$$\mathbf{Y}_{ie}(s) = \left[\mathbf{Y}_i - \mathbf{C}_i (s\mathbf{I} - \mathbf{A}_i)^{-1} \mathbf{B}_i \right]$$
$$= \left[\mathbf{Y}_i - \mathbf{C}_i \mathbf{\Phi}_i (s\mathbf{I} - \mathbf{\Lambda}_i)^{-1} \mathbf{\Phi}_i^{-1} \mathbf{B}_i \right] \tag{12.213}$$

with

$\mathbf{\Lambda}_i$ = diagonal matrix of eigenvalues of $\mathbf{A}_i$ of the device
$\mathbf{\Phi}_i$ = corresponding eigenvector matrix

Computation of $\mathbf{Y}_{ie}(s)$ using Equation 12.213 is simplified due to $s\mathbf{I}-\mathbf{\Lambda}_i$ being diagonal.

For the machine k selected for applying the external torque, the dynamic equations, including the effect of the applied torque ΔT_m, may be written in the partitioned form:

$$\begin{bmatrix} \Delta\dot{\omega}_r \\ \Delta\dot{\delta} \\ \dot{\mathbf{x}}_r \end{bmatrix} = \begin{bmatrix} a_{11} & a_{12} & \mathbf{a}_{1r} \\ \omega_0 & 0 & 0 \\ \mathbf{a}_{r1} & \mathbf{a}_{r2} & \mathbf{A}_{rr} \end{bmatrix} \begin{bmatrix} \Delta\omega_r \\ \Delta\delta \\ \mathbf{x}_r \end{bmatrix} + \begin{bmatrix} \mathbf{b}_1 \\ 0 \\ \mathbf{B}_r \end{bmatrix}\Delta\mathbf{v} + \begin{bmatrix} \dfrac{1}{2H} \\ 0 \\ 0 \end{bmatrix}\Delta T_m \qquad (12.214)$$

$$\Delta\mathbf{i}_k = \begin{bmatrix} \mathbf{c}_1 & \mathbf{c}_2 & \mathbf{C}_r \end{bmatrix}\begin{bmatrix} \Delta\omega_r \\ \Delta\delta \\ \mathbf{x}_r \end{bmatrix} - \mathbf{Y}_k\Delta\mathbf{v} \qquad (12.215)$$

where $\mathbf{x}_r$ is a vector representing all state variables of the machine, except for $\Delta\omega_r$ and $\Delta\delta$. For the disturbed machine, $\Delta\omega_r$ is assumed to be equal to 1.0 pu, and hence $\Delta\delta = \omega_0/s$.

By manipulation and rearrangement of Equations 12.214 and 12.215 to eliminate $\Delta\delta$ and $\mathbf{x}_r$ in terms of $\Delta\omega_r$, and by setting $\Delta\omega_r$ to 1.0 pu, we obtain the following expressions for ΔT_m and $\Delta\mathbf{i}_k$ as functions of $\Delta\mathbf{v}$:

$$\Delta T_M = 2H\left[s - a_{11} - \frac{a_{12}\omega_0}{s} - \mathbf{a}_{1r}(s\mathbf{I}-\mathbf{A}_{rr})^{-1}\left(\mathbf{a}_{r1} + \frac{\mathbf{a}_{r2}\omega_0}{s}\right) \right] -$$
$$2H\left[\mathbf{b}_1 + \mathbf{a}_{1r}(s\mathbf{I}-\mathbf{A}_{rr})^{-1}\mathbf{B}_r \right]\Delta\mathbf{v} \qquad (12.216)$$

$$\Delta\mathbf{i}_k = \Delta\mathbf{I}_{ke}(s) - \mathbf{Y}_{ke}(s)\Delta\mathbf{v} \qquad (12.217)$$

where

$$\Delta\mathbf{I}_{ke}(s) = \left[\mathbf{c}_1 + \frac{\mathbf{c}_2\omega_0}{s} + \mathbf{C}_r(s\mathbf{I}-\mathbf{A}_{rr})^{-1}\left(\mathbf{a}_{r1} + \frac{\mathbf{a}_{r2}\omega_0}{s}\right) \right] \qquad (12.218)$$

$$\mathbf{Y}_{ke}(s) = \mathbf{Y}_k - \mathbf{C}_r(s\mathbf{I}-\mathbf{A}_{rr})^{-1}\mathbf{B}_r \qquad (12.219)$$

As in Equation 12.213, computations associated with the above equations are simplified by expressing $(s\mathbf{I}-\mathbf{A}_{rr})^{-1}$ in diagonal form in terms of the eigenvalues of $\mathbf{A}_{rr}$.

The current injections into the network are given by Equation 12.217 for the disturbed machine and by Equation 12.212 for all other machines and dynamic devices. The equivalent device admittances $\mathbf{Y}_{ie}(s)$ and the equivalent current source $\Delta\mathbf{i}_{ke}(s)$ for the disturbed machine are functions of the parameters of the devices, initial operating condition, and the complex frequency s.

The interconnecting network equations may be written as

$$
\begin{bmatrix} \Delta\mathbf{i}_D \\ 0 \end{bmatrix} = \begin{bmatrix} \mathbf{Y}_{DD} & \mathbf{Y}_{DL} \\ \mathbf{Y}_{LD} & \mathbf{Y}_{LL} \end{bmatrix} \begin{bmatrix} \Delta\mathbf{v}_D \\ \Delta\mathbf{v}_L \end{bmatrix} \tag{12.220}
$$

where subscript D refers to buses with dynamic devices and the subscript L refers to buses with static loads, i.e., buses with no dynamic devices.

Combining the device equations 12.212 and 12.217 and the network equation 12.220 yields

$$
\begin{bmatrix} \Delta\mathbf{I}_{De}(s) \\ 0 \end{bmatrix} = \begin{bmatrix} \mathbf{Y}_{DD}+\mathbf{Y}_{De}(s) & \mathbf{Y}_{DL} \\ \mathbf{Y}_{LD} & \mathbf{Y}_{LL} \end{bmatrix} \begin{bmatrix} \Delta\mathbf{v}_D \\ \Delta\mathbf{v}_L \end{bmatrix} \tag{12.221}
$$

where $\mathbf{Y}_{De}(s)$ is a block diagonal matrix of the device equivalent admittance matrix $\mathbf{Y}_{ie}(s)$, and $\Delta\mathbf{I}_{De}(s)$ is the device current source vector with non-zero element only for the disturbed machine given by Equation 12.218.

Equation 12.221 is solved for bus voltages $\Delta\mathbf{v}$. The network matrix is very large and sparse. PEALS exploits this by using an efficient sparsity-based network solution technique.

The following is a summary of the computation steps associated with the AESOPS algorithm:

1. Form the device models: $\mathbf{A}_i$, $\mathbf{B}_i$, $\mathbf{C}_i$, $\mathbf{Y}_i$; compute eigenvalues and eigenvectors of $\mathbf{A}_i$ using the *QR* transformation method.

2. Set the initial value of complex frequency s equal to the specified initial estimate of the eigenvalue.

3. Compute $\Delta\mathbf{I}_{ke}(s)$ and $\mathbf{Y}_{ke}(s)$ associated with the disturbed machine, using Equations 12.218 and 12.219, and $\mathbf{Y}_{ie}(s)$ for all other dynamic devices, using Equation 12.213.

4. Solve Equation 12.221 to compute bus voltage $\Delta\mathbf{v}$.

5.   Compute ΔT_m for the disturbed machine, using Equation 12.216; $\Delta\omega_r$ for all machines, using Equation 12.211; H_e, using Equation 12.209.

6.   Compute next estimate of s, using Equation 12.210.

7.   If Δs between successive iterations is within the specified tolerance, set eigenvalue to s and stop; otherwise, go to step 3 and repeat the process.

The above approach to the calculation of eigenvalues corresponding to rotor angle modes involves the solution of only the algebraic Equations 12.221. It does not require the formation of the state matrix $\mathbf{A}$ of the overall system. Instead, the state equations for each dynamic device are formed and treated as a small eigenvalue problem to form the equivalent admittance matrix $\mathbf{Y}_{ie}(s)$. The overall admittance matrix of Equation 12.221 is highly sparse. Hence, with the use of efficient sparsity-based solution techniques, this approach has been applied to power systems with over 2,000 dynamic devices and 12,000 buses [10].

One of the limitations of the AESOPS algorithm is that, unless the general characteristics of critical modes are known *a priori*, a considerable amount of searching is required to find all the critical modes. In fact, there is never any assurance that all critical modes have been identified. The method is best suited for tracking changes in specific modes as system conditions change. The modified Arnoldi method described next overcomes these limitations of the AESOPS method.

Calculation of eigenvectors:

In the AESOPS approach, the speed deviations $\Delta\omega_r$ for all the machines computed in the last iteration directly give the corresponding component of the right eigenvector.

The left eigenvector is calculated by using the transposed system dynamic model of Equations 12.188 and 12.189:

$$\dot{\mathbf{y}} = \mathbf{A}_D^T \mathbf{y} + \mathbf{C}_D^T \Delta\mathbf{u}$$

$$\Delta\mathbf{j} = \mathbf{B}_D^T \mathbf{y} - \mathbf{Y}_D^T \Delta\mathbf{u}$$

and the transposed network model of Equation 12.190:

$$\Delta\mathbf{j} = \mathbf{Y}_N^T \Delta\mathbf{u}$$

It can be readily shown that the AESOPS algorithm when applied to the above model will converge to a system eigenvalue and the corresponding left eigenvector. With the initial eigenvalue estimate equal to the correct value, one iteration is usually sufficient for the algorithm to converge to the left eigenvector.

Frequency response calculation:

The frequency response can be calculated directly from the system equations in the form used by the AESOPS algorithm with s replaced by $j\omega$. The overall system has the general form of Equation 12.221. Instead of applying an external torque as input to one of the machines, the input signal may be applied to any device and the outputs monitored at the desired locations. Since the network equations are retained explicitly, the input and outputs may be associated with the network as well as the dynamic devices.

(b) The modified Arnoldi method (MAM)

The Arnoldi method was first presented in reference 19. However, in the original form it had poor numerical properties, the main problems being loss of orthogonality and slow convergence if several dominant eigenvalues are needed. These problems were solved in reference 20 by using complete reorthogonalization and an iterative process. The following is a description of the method as implemented in PEALS [10,11].

The modified Arnoldi method is based on a reduction technique in which the matrix $\mathbf{A}$, whose eigenvalues are to be computed, is reduced to an upper Hessenberg matrix by the recurrence:

$$h_{i+1,i}\mathbf{v}_{i+1} = \mathbf{A}\mathbf{v}_i - \sum_{j=1}^{i} h_{j,i}\mathbf{v}_j \qquad i = 1,...,m \qquad (12.222)$$

where

$$h_{j,i} = \mathbf{v}_j^H \mathbf{A}\mathbf{v}_i \text{ (superscript H denotes conjugate transpose)}$$

and

$\mathbf{v}_1$ is an arbitrary starting vector with $\|\mathbf{v}_1\|_2 = 1$
$h_{i+1,i}$ is a scaling factor to make $\|\mathbf{v}_{i+1}\|_2 = 1$
m is a prespecified order of the reduced Hessenberg matrix

Equation 12.222 can be rearranged and assembled for all m equations to give

$$\mathbf{A}\mathbf{V}_m = \mathbf{V}_m\mathbf{H}_m + h_{m+1,m}\mathbf{v}_{m+1}\mathbf{e}_m^T \qquad (12.223)$$

where

$$\mathbf{V}_m = [\ \mathbf{v}_1 \ \cdots \ \mathbf{v}_m\ ]$$

$$\mathbf{H}_m = \begin{bmatrix} h_{1,1} & h_{1,2} & \cdots & h_{1,m} \\ h_{2,1} & h_{2,2} & \cdots & h_{2,m} \\ \cdots & \cdots & \cdots & \cdots \\ 0 & \cdots & h_{m,m-1} & h_{m,m} \end{bmatrix}$$

$$\mathbf{e}_m^T = [\, 0 \ \cdots \ 0 \ 1 \,]$$

Here $\mathbf{H}_m$ is an upper Hessenberg matrix. We can show that ideally the vector sequence $\mathbf{v}_i$ generated by Equation 12.222 is orthonormal. Therefore, if $m=N$ (the order of $\mathbf{A}$), we have

$$h_{N+1,N} = 0$$

and Equation 12.223 becomes

$$\mathbf{AV}_N = \mathbf{V}_N \mathbf{H}_N \qquad\qquad (12.224)$$

Thus $\mathbf{A}$ is reduced to the upper Hessenberg matrix $\mathbf{H}_N$ whose eigenvalues are the eigenvalues of $\mathbf{A}$. An important feature of the method is that the values of $h_{m+1,m}$, the subdiagonal elements of $\mathbf{H}_m$, decrease very rapidly as m increases; therefore, a good approximation can be made for $m \ll N$ by dropping out the second term in Equation 12.223:

$$\mathbf{AV}_m \approx \mathbf{V}_m \mathbf{H}_m \qquad\qquad (12.225)$$

The eigenvalues of $\mathbf{H}_m$, which is of a very low order, approximate a subset of the eigenvalues of $\mathbf{A}$. The corresponding eigenvectors of $\mathbf{A}$ are given by

$$\mathbf{W} \approx \mathbf{V}_m \mathbf{P} \qquad\qquad (12.226)$$

where $\mathbf{P}$ is the eigenvector matrix of $\mathbf{H}_m$. To improve the accuracy of the eigenvalues of $\mathbf{A}$, the above procedure can be iterated with a new starting vector $\mathbf{v}_1^{new}$ derived from a linear combination of the columns of $\mathbf{V}_m$:

$$\mathbf{v}_1^{new} = \sum_{i=1}^{m} \alpha_i \mathbf{v}_i \qquad\qquad (12.227)$$

where α_i are coefficients calculated from the modal properties of $\mathbf{H}_m$ [16].

Equation 12.225 holds only if the vector sequence $\mathbf{v}_i$ is kept orthogonal at each step of calculation [20]. In practice, the orthogonality may be lost rapidly due to numerical cancellation and roundoff errors. The remedy is to introduce a reorthogonalization process [16] after the calculation of each $\mathbf{v}_i$ from Equation 12.222.

An important property of the method is that the eigenvalues of $\mathbf{H}_m$ in Equation 12.225 usually converge to those eigenvalues of $\mathbf{A}$ which have largest (and smallest) modulus. Thus, if eigenvalues of $\mathbf{A}$ around a specified point λ_t are desired, the transformation

$$\mathbf{A}_t = (\mathbf{A} - \lambda_t \mathbf{I})^{-1} \tag{12.228}$$

can be used to magnify the eigenvalues of $\mathbf{A}$ close to λ_t, since

$$\lambda_{ti} = \frac{1}{\lambda_i - \lambda_t} \tag{12.229}$$

where λ_i is an eigenvalue of $\mathbf{A}$, and λ_{ti} is the corresponding eigenvalue of $\mathbf{A}_t$. Then $\mathbf{A}_t$ is the suitable matrix to which the method can be applied in order to find the set of eigenvalues of $\mathbf{A}$ close to λ_t, which is termed the *shift point*.

With the transformation 12.228, the only computation involving $\mathbf{A}$ in constructing the vector sequence $\mathbf{v}_i$ is the solution of the equation

$$(\mathbf{A} - \lambda_t \mathbf{I})\mathbf{u}_i = \mathbf{v}_i \tag{12.230}$$

Equation 12.222 may be written in terms of $\mathbf{u}_i$, as

$$h_{i+1,i}\mathbf{v}_{i+1} = \mathbf{u}_i - \sum_{j=1}^{i} h_{j,i}\mathbf{v}_j \qquad i = 1, \cdots, m \tag{12.231}$$

and the upper triangular elements in $\mathbf{H}_m$ become

$$h_{j,i} = \mathbf{v}_j^H \mathbf{u}_i \qquad j \le i \tag{12.232}$$

Application of the MAM to power system model:

Substituting the expression for the state matrix $\mathbf{A}$ of a power system given by Equation 12.192 in Equation 12.230, we get

$$(\mathbf{A}_D - \lambda_t \mathbf{I})\mathbf{u}_i + \mathbf{B}_D(\mathbf{Y}_N + \mathbf{Y}_D)^{-1}\mathbf{C}_D\mathbf{u}_i = \mathbf{v}_i$$

This may be written as

$$(\mathbf{A}_D - \lambda_t \mathbf{I})\mathbf{u}_i + \mathbf{B}_D\mathbf{q}_i = \mathbf{v}_i \qquad (12.233)$$

where

$$\mathbf{q}_i = (\mathbf{Y}_N + \mathbf{Y}_D)^{-1}\mathbf{C}_D\mathbf{u}_i$$

Rearranging,

$$\mathbf{C}_D\mathbf{u}_i - (\mathbf{Y}_N + \mathbf{Y}_D)\mathbf{q}_i = 0 \qquad (12.234)$$

Equations 12.233 and 12.234 may be combined to give the matrix equation

$$\begin{bmatrix} (\mathbf{A}_D - \lambda_t \mathbf{I}) & \mathbf{B}_D \\ \mathbf{C}_D & -(\mathbf{Y}_N + \mathbf{Y}_D) \end{bmatrix} \begin{bmatrix} \mathbf{u}_i \\ \mathbf{q}_i \end{bmatrix} = \begin{bmatrix} \mathbf{v}_i \\ 0 \end{bmatrix} \qquad (12.235)$$

The following are the computational steps associated with the MAM:

1.   Form the device models $\mathbf{A}_i$, $\mathbf{B}_i$, $\mathbf{C}_i$, $\mathbf{Y}_i$.

2.   Compute the device admittance matrix corresponding to the chosen shift point λ_t:

$$\mathbf{Y}_{De}(\lambda_t) = \mathbf{Y}_D - \mathbf{C}_D(\lambda_t \mathbf{I} - \mathbf{A}_D)^{-1}\mathbf{B}_D \qquad (12.236)$$

3.   Solve for $\mathbf{q}_i$:

$$[\mathbf{Y}_N + \mathbf{Y}_{De}(\lambda_t)]\mathbf{q}_i = -\mathbf{C}_D(\lambda_t \mathbf{I} - \mathbf{A}_D)^{-1}\mathbf{v}_i \qquad (12.237)$$

4.   Calculate $\mathbf{u}_i$:

$$\mathbf{u}_i = (\lambda_t \mathbf{I} - \mathbf{A}_D)^{-1}(\mathbf{B}_D\mathbf{q}_i - \mathbf{v}_i) \qquad (12.238)$$

5.   Compute v_{i+1} using Equation 12.231 and the related elements in $\mathbf{H}_m$ using Equation 12.232, with the order m equal to the specified value. Repeat steps 3 to 5 to form the complete $\mathbf{V}_m$ and $\mathbf{H}_m$.

6.   Compute eigenvalues of $\mathbf{H}_m$ using the QR transformation method. If the desired set of eigenvalues has converged within the specified tolerance, stop; otherwise, compute a new starting vector v_1^{new} using Equation 12.227, and go to step 3 for the next iteration.

The factorization of the network equivalent admittance matrix $\mathbf{Y}_N + \mathbf{Y}_{De}(\lambda_t)$ in Equation 12.237 needs to be performed only once since λ_t does not change during the iteration process; this adds to the efficiency of the method. As the admittance matrix is very sparse, the method can be applied to very large systems by using sparsity-based techniques for solution of algebraic equations.

The MAM, unlike the AESOPS algorithm, can compute eigenvalues associated with any system mode, not just the rotor angle mode. In addition, it does not require *a priori* knowledge about the mode characteristics of the system. Normally, the shift point λ_t is given in order to compute a particular set of eigenvalues close to it. By slightly modifying the algorithm, the method can be adopted to provide the capability for scanning eigenvalues over a frequency range on the complex plane with assurance that all critical eigenvalues within the specified range have been computed.

Calculation of eigenvectors:

The right eigenvectors are given directly by Equation 12.226 from the last iteration. The left eigenvectors are calculated by inverse iteration. This requires the solution of Equation 12.235 transposed as follows:

$$\begin{bmatrix} \mathbf{A}_D^T - \lambda_t \mathbf{I} & \mathbf{C}_D^T \\ \mathbf{B}_D^T & -(\mathbf{Y}_N + \mathbf{Y}_D)^T \end{bmatrix} \begin{bmatrix} \mathbf{y} \\ \mathbf{q} \end{bmatrix} = \begin{bmatrix} \mathbf{y}_0 \\ 0 \end{bmatrix}$$

with the computed eigenvalue λ_t and an arbitrary vector $\mathbf{y}_0$. Normally, one iteration is sufficient to give the left eigenvector.

Comparison of results

Table 12.2 compares the eigenvalue calculations using the QR transformation, AESOPS, and MAM. The test system consisted of 94 generators and about 1,400 buses. Eighteen of the generators were represented by detailed models and the other generators were represented by the classical model. This resulted in a system model with 366 states. The MASS and PEALS referred to above were used to compute the eigenvalues.

Table 12.2

Mode	QR Method	AESOPS	MAM
A	$-0.122\pm j1.650$	$-0.121\pm j1.650$	$-0.121\pm j1.650$
B	$-0.117\pm j3.073$	$-0.117\pm j3.073$	$-0.117\pm j3.073$
C	$-0.276\pm j5.212$	$-0.276\pm j5.212$	$-0.276\pm j5.212$

All eigenvalues of the state matrix and the corresponding participation factors were first computed using the MASS program, which uses the QR transformation method. Then three pairs of complex eigenvalues associated with rotor angle modes were selected for comparison with AESOPS and MAM. The initial estimates of the eigenvalues for both methods were set so that they deviated from the true values by about 10%. From Table 12.2, we see that all three methods give practically identical results. This is particularly reassuring in view of the widely differing theories on which the three methods are based.

As a further validation of the above results, frequency response was computed using PEALS with the change in mechanical torque of a large generator (with high participation in the three modes) as the input signal and the speed change of the same generator as the output signal. Figure 12.22 shows a plot of the computed frequency response, which exhibits resonant frequencies at about 0.26 Hz, 0.49 Hz, and 0.83 Hz. These are in agreement with the frequencies of the selected modes.

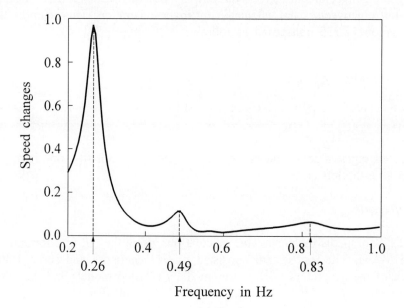

Figure 12.22 Frequency response

Example 12.6

In this example, we will analyze the small-signal stability of a simple two-area system shown in Figure E12.8. This system is similar in structure to the one used in references 23 and 24 to study the fundamental nature of interarea oscillations.

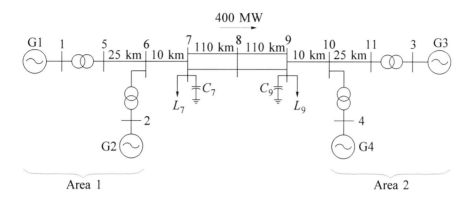

Figure E12.8 A simple two-area system

The system consists of two similar areas connected by a weak tie. Each area consists of two coupled units, each having a rating of 900 MVA and 20 kV. The generator parameters in per unit on the rated MVA and kV base are as follows:

$$
\begin{array}{lllll}
X_d = 1.8 & X_q = 1.7 & X_l = 0.2 & X_d' = 0.3 & X_q' = 0.55 \\
X_d'' = 0.25 & X_q'' = 0.25 & R_a = 0.0025 & T_{d0}' = 8.0 \text{ s} & T_{q0}' = 0.4 \text{ s} \\
T_{d0}'' = 0.03 \text{ s} & T_{q0}'' = 0.05 \text{ s} & A_{Sat} = 0.015 & B_{Sat} = 9.6 & \psi_{T1} = 0.9 \\
H = 6.5 \text{ (for G1 and G2)} & & H = 6.175 \text{ (for G3 and G4)} & & K_D = 0
\end{array}
$$

Each step-up transformer has an impedance of $0+j0.15$ per unit on 900 MVA and 20/230 kV base, and has an off-nominal ratio of 1.0.

The transmission system nominal voltage is 230 kV. The line lengths are identified in Figure E12.8. The parameters of the lines in per unit on 100 MVA, 230 kV base are

$$r = 0.0001 \text{ pu/km} \qquad x_L = 0.001 \text{ pu/km} \qquad b_C = 0.00175 \text{ pu/km}$$

The system is operating with area 1 exporting 400 MW to area 2, and the generating units are loaded as follows:

G1:	$P = 700$ MW,	$Q = 185$ MVAr,	$E_t = 1.03\angle 20.2°$
G2:	$P = 700$ MW,	$Q = 235$ MVAr,	$E_t = 1.01\angle 10.5°$
G3:	$P = 719$ MW,	$Q = 176$ MVAr,	$E_t = 1.03\angle -6.8°$
G4:	$P = 700$ MW,	$Q = 202$ MVAr,	$E_t = 1.01\angle -17.0°$

The loads and reactive power supplied (Q_C) by the shunt capacitors at buses 7 and 9 are as follows:

Bus 7: P_L = 967 MW, Q_L = 100 MVAr, Q_C = 200 MVAr
Bus 9: P_L = 1,767 MW, Q_L = 100 MVAr, Q_C = 350 MVAr

(a)   If all four generators are on manual excitation control (constant E_{fd}), compute the eigenvalues of the system state matrix representing the small-signal performance of the system about the initial operating condition. For each eigenvalue, identify the system state variables with high participation. Determine the frequencies, damping ratios, and mode shapes of the rotor oscillation modes. Assume that the active components of loads have constant current characteristics, and reactive components of loads have constant impedance characteristics.

(b)   Determine the eigenvalues, frequencies, and damping ratios of rotor oscillation modes when all four generators are equipped with the following types of excitation control:

   (i)   Self-excited dc exciter (see Figures 8.38 and 8.40):

   K_A = 20.0 T_A = 0.055 T_E = 0.36 K_F = 0.125
   T_F = 1.8 A_{ex} = 0.0056 B_{ex} = 1.075 T_R = 0.05
   T_B, T_C, R_C, and X_C are not used

   (ii)   Thyristor exciter with a high transient gain:

   K_A = 200.0 T_R = 0.01

   (iii)   Thyristor exciter with a transient gain reduction (TGR):

   K_A = 200.0 T_R = 0.01 T_A = 1.0 T_B = 10.0

   (iv)   Thyristor exciter with high transient gain and PSS:

   K_A = 200.0 T_R = 0.01 K_{STAB} = 20.0 T_W = 10.0
   T_1 = 0.05 T_2 = 0.02 T_3 = 3.0 T_4 = 5.4

   The block diagram of thyristor excitation system with PSS is shown in Figure E12.9.

Solution

(a) *System modes with generators on manual excitation control:*

Table E12.3 summarizes the eigenvalues of the system state matrix and the state variables that have high participation in each mode. The first two eigenvalues represent the *zero eigenvalues* due to the redundant state variables. As described in Section 12.7, one of these zero eigenvalues is due to lack of uniqueness of absolute rotor angle (there is no infinite bus, and the rotor angles are referred to a common reference frame). The other zero eigenvalue is due to the assumption that the generator torques are independent of speed deviation (speed governors are not modelled and K_D=0).

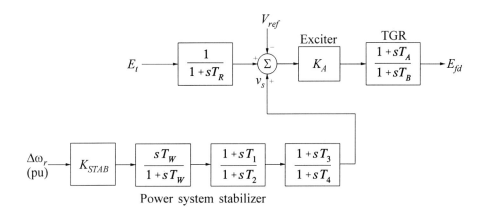

Figure E12.9 Thyristor excitation system with PSS

Table E12.3 System modes with manual excitation control

No.	Eigenvalues Real	Eigenvalues Imaginary	Frequency (Hz)	Damping Ratio	Dominant States
1,2	$-0.76E-3$	$\pm0.22E-2$	0.0003	0.331	$\Delta\omega$ and $\Delta\delta$ of G1, G2, G3, G4
3	$-0.96E-1$	$-$	$-$	$-$	"
4,5	-0.111	±3.43	0.545	0.032	"
6	-0.117	$-$	$-$	$-$	"
7	-0.265	$-$	$-$	$-$	$\Delta\psi_{fd}$ of G3 and G4
8	-0.276	$-$	$-$	$-$	$\Delta\psi_{fd}$ of G1 and G2
9,10	-0.492	±6.82	1.087	0.072	$\Delta\omega$ and $\Delta\delta$ of G1 and G2
11,12	-0.506	±7.02	1.117	0.072	$\Delta\omega$ and $\Delta\delta$ of G3 and G4
13	-3.428	$-$	$-$	$-$	
14	-4.139	$-$	$-$	$-$	
15	-5.287	$-$	$-$	$-$	
16	-5.303	$-$	$-$	$-$	
17	-31.03	$-$	$-$	$-$	d and q axis amortisseur flux linkages
18	-32.45	$-$	$-$	$-$	
19	-34.07	$-$	$-$	$-$	
20	-35.53	$-$	$-$	$-$	
21,22	-37.89	±0.142	0.023	≈1.0	
23,24	-38.01	$\pm0.38E-1$	0.006	≈1.0	

From the table, we see that the system is stable. There are three rotor angle modes of oscillation. Their mode shapes (normalized eigenvector components corresponding to rotor speeds of the four machines) are shown in Figure E12.10. From the mode shapes we see that the 0.55 Hz mode is the interarea mode, with generators G1 and G2 of area 1 swinging against generators G3 and G4 of area 2. The 1.09 Hz mode is the intermachine oscillation local to area 1, with G1 swinging against G2. The third rotor angle mode, with a frequency of 1.12 Hz, is the intermachine mode local to area 2.

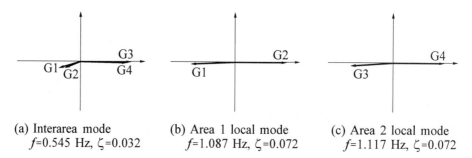

(a) Interarea mode   (b) Area 1 local mode   (c) Area 2 local mode
  f=0.545 Hz, ζ=0.032 f=1.087 Hz, ζ=0.072 f=1.117 Hz, ζ=0.072

Figure E12.10 Mode shapes of rotor angle modes with manual excitation control

(b) *Rotor angle modes with different types of excitation control:*

Table E12.4 summarizes the eigenvalues, frequencies, and damping ratios associated with the rotor oscillation modes for the four alternative forms of excitation control.

We see that the local intermachine modes of oscillation have the same degree of damping with the dc exciter and thyristor exciter (with and without TGR). The interarea mode has a small positive damping with the dc exciter. It is unstable with high gain thyristor exciters. The TGR makes the interarea mode more unstable. The interarea mode as well as the local modes are very well damped when power system stabilizers are added to the thyristor exciters.

Table E12.4 Effects of excitation control on rotor oscillation modes

Type of Excitation Control	Eigenvalue/(Frequency in Hz, Damping Ratio)		
	Interarea Mode	Area 1 Local Mode	Area 2 Local Mode
(i) DC exciter	$-0.018\pm j3.27$ (f=0.52, ζ=0.005)	$-0.485\pm j6.81$ (f=1.08, ζ=0.07)	$-0.500\pm j7.00$ (f=1.11, ζ=0.07)
(ii) Thyristor with high gain	$+0.031\pm j3.84$ (f=0.61, ζ=−0.008)	$-0.490\pm j7.15$ (f=1.14, ζ=0.07)	$-0.496\pm j7.35$ (f=1.17, ζ=0.07)
(iii) Thyristor with TGR	$+0.123\pm j3.46$ (f=0.55, ζ=−0.036)	$-0.450\pm j6.86$ (f=1.09, ζ=0.06)	$-0.462\pm j7.05$ (f=1.12, ζ=0.06)
(iv) Thyristor with PSS	$-0.501\pm j3.77$ (f=0.60, ζ=0.13)	$-1.826\pm j8.05$ (f=1.28, ζ=0.22)	$-1.895\pm j8.35$ (f=1.33, ζ=0.22)

12.9 CHARACTERISTICS OF SMALL-SIGNAL STABILITY PROBLEMS

In large power systems, small-signal stability problems may be either local or global in nature.

Local problems

Local problems involve a small part of the system. They may be associated with rotor angle oscillations of a single generator or a single plant against the rest of the power system. Such oscillations are called *local plant mode oscillations*. The stability problems related to such oscillations are similar to those of a single-machine infinite bus system as studied in Sections 12.3 to 12.6. Most commonly encountered small-signal stability problems are of this category.

Local problems may also be associated with oscillations between the rotors of a few generators close to each other. Such oscillations are called *intermachine or interplant mode oscillations*. Usually, the local plant mode and interplant mode oscillations have frequencies in the range of 0.7 to 2.0 Hz.

Other possible local problems include instability of modes associated with controls of equipment such as generator excitation systems, HVDC converters, and static var compensators. The problems associated with *control modes* are due to inadequate tuning of the control systems [21]. In addition, these controls may interact with the dynamics of the turbine-generator shaft system, causing instability of *torsional mode oscillations* [22]. Torsional oscillation problems are described in Chapter 15.

Analysis of local small-signal stability problems requires a detailed representation of a small portion of the complete interconnected power system. The rest of the system representation may be appropriately simplified by use of simple models and system equivalents. Usually, the complete system may be adequately represented by a model having several hundred states at most.

Global problems

Global small-signal stability problems are caused by interactions among large groups of generators and have widespread effects. They involve oscillations of a group of generators in one area swinging against a group of generators in another area. Such oscillations are called *interarea mode oscillations*.

Large interconnected systems usually have two distinct forms of interarea oscillations:

(a)   A very low frequency mode involving all the generators in the system. The system is essentially split into two parts, with generators in one part swinging against machines in the other part. The frequency of this mode of oscillation is on the order of 0.1 to 0.3 Hz.

(b)   Higher frequency modes involving subgroups of generators swinging against each other. The frequency of these oscillations is typically in the range of 0.4 to 0.7 Hz.

Illustration of global problems

We will consider two large systems with distinct characteristics to illustrate the two forms of interarea oscillations.

Test system A

This test system consists of 2,310 buses, 375 generators, and 3 HVDC links. The 3 dc links and 291 generators are modelled in detail. The remaining 84 generators represent equivalent machines and are modelled by the classical model.

To identify the critical modes of interarea oscillation, eigenvalues of system modes with frequencies in the range of 0.2 to 1.0 Hz are scanned, using the modified Arnoldi method. Table 12.3 lists the eigenvalues, frequencies, and damping ratios of the modes identified by this mode search. Figure 12.23 shows a plot of these eigenvalues on a complex plane. Box A in the figure represents the region on the complex plane within which all eigenvalues have been computed; outside this area, there may exist eigenvalues other than those shown on the complex plane.

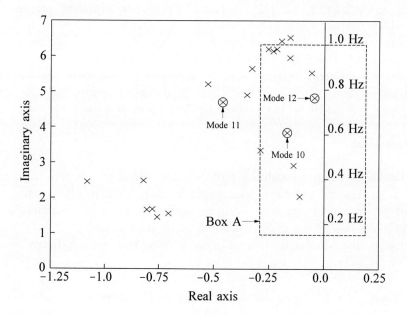

Figure 12.23 Eigenvalues computed by frequency scan using MAM

Table 12.3 Modes of system A computed by using frequency scanning technique

Mode no.	Eigenvalue	Frequency (Hz)	Damping Ratio
1	$-0.7567\pm j1.4163$	0.2254	0.4712
2	$-0.7193\pm j1.4605$	0.2324	0.4419
3	$-0.7999\pm j1.5623$	0.2486	0.4558
4	$-0.7806\pm j1.5776$	0.2511	0.4435
5	$-0.1265\pm j1.9894$	0.3166	0.0634
6	$-0.8248\pm j2.3509$	0.3742	0.3311
7	$-1.0698\pm j2.3642$	0.3763	0.4123
8	$-0.1434\pm j2.7826$	0.4429	0.0515
9	$-0.2871\pm j3.2495$	0.5172	0.0880
10	$-0.1609\pm j3.7797$	0.6016	0.0425
11	$-0.4382\pm j4.6754$	0.7441	0.0933
12	$-0.0470\pm j4.7439$	0.7550	0.0099
13	$-0.3336\pm j4.8262$	0.7681	0.0689
14	$-0.5420\pm j5.2074$	0.8288	0.1035
15	$-0.0551\pm j5.4200$	0.8626	0.0102
16	$-0.3116\pm j5.6159$	0.8938	0.0554
17	$-0.1451\pm j5.8825$	0.9362	0.0247
18	$-0.2385\pm j6.1228$	0.9745	0.0389
19	$-0.2243\pm j6.1745$	0.9827	0.0363
20	$-0.2552\pm j6.1814$	0.9838	0.0412
21	$-0.1961\pm j6.3504$	1.0107	0.0309
22	$-0.1457\pm j6.4356$	1.0243	0.0226

The 22 modes represented by the eigenvalues shown in Figure 12.23 and Table 12.3 can be divided, based on mode shapes, into three categories:

(a)   Interarea modes. Examples are modes 5 and 8 to 12.

(b)   Local plant modes. Examples are modes 13, 21, and 22.

(c)   Control modes. Examples are modes 1 to 4, and 6.

Modes 10, 11, and 12 represent the critical interarea modes. These modes have frequencies in the 0.6 to 0.75 Hz range. Each represents oscillation of a subgroup of generators swinging against another subgroup of generators across a weak or heavily loaded transmission interface. Figure 12.24 shows how the interfaces associated with the three critical interarea modes geographically divide the overall system. The stability characteristics of each interarea mode of oscillation depend on the power transfer across its interface and on the controls associated with the subgroups of generating units on the two sides of the interface.

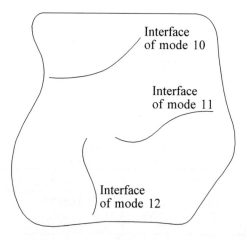

Figure 12.24 Interfaces of critical interarea modes
of oscillation of system A

Test system B

This system consists of nearly 3,000 buses and 300 generators. It has a dominant 0.2 Hz interarea mode in which practically all the generators in the system participate. The mode shape of the interarea mode is such that its interface splits the system into two parts as depicted in Figure 12.25. When the system is perturbed, the generating units on the two sides of the interface swing against each other with a frequency of about 0.2 Hz.

Table 12.4 summarizes the results of analysis carried out to investigate the effects of excitation control of the generators at a large thermal plant having a high participation in the interarea mode. This plant is identified as GS "D" in Figure 12.25 and has a thyristor excitation system.

We see from the results that the control of excitation of generators at the thermal plant has a very significant effect on the stability of the 0.2 Hz interarea mode of oscillation. With a high gain thyristor exciter, for the system conditions considered,

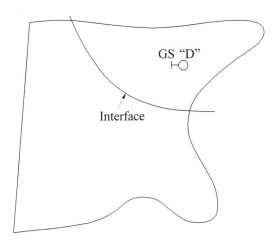

Figure 12.25 Interface of 0.2 Hz interarea
mode of oscillation of system B

Table 12.4 Effect of excitation control of generating units
at GS "D" on the stability of interarea mode

Type of Excitation Control	Interarea Mode	
	Frequency (Hz)	Damping Ratio
(a) Thyristor exciter with high transient gain	0.192	0.009
(b) Thyristor exciter with transient gain reduction	0.187	-0.057
(c) Thyristor exciter with power system stabilizer	0.179	0.122

the interarea mode has a small positive damping ($\zeta=0.009$). With a transient gain reduction, the interarea mode actually becomes unstable ($\zeta=-0.057$). When a power system stabilizer (designed as described in reference 9) is added, the damping increases significantly ($\zeta=0.122$).

Factors influencing interarea modes of oscillation

The characteristics of interarea modes of oscillation are very complex and in some respects significantly differ from the characteristics of local plant modes. Load characteristics, in particular, have a major effect on the stability of interarea modes.

The manner in which excitation systems affect interarea oscillations depends on the types and locations of the exciters, and on the characteristics of loads [23].

Speed-governing systems normally do not have a very significant effect on interarea oscillations. However, if they are not properly tuned, they may decrease damping of the oscillations slightly. In extreme situations, this may be sufficient to aggravate the situation significantly. In the absence of any other convenient means of increasing the damping, adjustment or blocking of the governors may provide some relief [26].

A mode of oscillation in one part of the system may interact with a mode of oscillation in a remote part due to mode coupling. This occurs when the frequencies of the two modes are nearly equal [24,25]. Care should be exercised in interpreting results of analysis in such cases.

The controllability of interarea modes with PSS is a complex function of many factors:

- Location of unit with PSS
- Characteristics and location of loads
- Types of exciters on other units

On some units, the PSS does not have the desired effect on the damping of interarea oscillations. Reference 24 presents results of a detailed study of factors influencing PSS performance in damping interarea and interplant modes of oscillation.

Other effective means of stabilizing interarea modes of oscillation include modulation of HVDC converter controls and static var compensator controls.

Analysis of interarea oscillations requires detailed representation of the entire interconnected power system. Models for excitation systems and loads, in particular, should be accurate, and the same level of modelling detail should be used throughout the system.

Reference 27 provides a detailed account of a comprehensive study of the interarea oscillation problems, including the fundamental nature of the problem, methods of analysis, and control design procedures to mitigate the problem.

Enhancement of small-signal stability is discussed in Chapter 17.

REFERENCES

[1] A.M. Lyapunov, *Stability of Motion*, English translation, Academic Press, Inc., 1967.

[2] G.C. Verghese, I.J. Perez-Arriaga, and F.C. Schweppe, "Selective Modal Analysis with Application to Electric Power Systems, Part I: Heuristic Introduction, Part II: The Dynamic Stability Problem," *IEEE Trans.*, Vol., PAS-101, No. 9, pp. 3117-3134, September 1982.

[3]   J.G.F. Francis, "The QR Transformation - A Unitary Analogue to the LR Transformation," Parts 1 and 2, *The Computer Journal*, Vol. 4, pp. 265-271, 1961; pp. 332-345, 1962.

[4]   J.H. Wilkinson, *The Algebraic Eigenvalue Problem*, Clarendon Press, Oxford, 1965.

[5]   W.G. Heffron and R.A. Phillips, "Effects of Modern Amplidyne Voltage Regulator in Underexcited Operation of Large Turbine Generators," *AIEE Trans.*, Vol. PAS-71, pp. 692-697, August 1952.

[6]   F.P. deMello and C. Corcordia, "Concepts of Synchronous Machine Stability as Affected by Excitation Control," *IEEE Trans.*, Vol. PAS-88, pp. 316-329, April 1969.

[7]   S.B. Crary, *Power System Stability*, Vol. II, John Wiley & Sons, Inc., 1955.

[8]   F.P. DeMello and T.F. Laskoswski, "Concepts of Power System Dynamic Stability," *IEEE Trans.*, Vol. PAS-94, pp. 827-833, May/June 1975.

[9]   P. Kundur, M. Klein, G.J. Rogers, and M.S. Zywno, "Application of Power System Stabilizer for Enhancement of Overall System Stability," *IEEE Trans.*, Vol. PWRS-4, pp. 614-626, May 1989.

[10]   P. Kundur, G.J. Rogers, D.Y. Wong, L. Wang, and M.G. Lauby, "A Comprehensive Computer Program for Small Signal Stability Analysis of Power Systems," *IEEE Trans.*, Vol. PWRS-5, pp. 1076-1083, November 1990.

[11]   EPRI Report EL-5798, "The Small Signal Stability Program Package," Vol. 1, Final Report of Project 2447-1, Prepared by Ontario Hydro, May 1988.

[12]   S. Arabi, G.J. Rogers, D.Y. Wong, P. Kundur, and M.G. Lauby, "Small Signal Stability Program Analysis of SVC and HVDC in AC Power Systems," *IEEE Trans.*, Vol. PWRS-6, pp. 1147-1153, August 1991.

[13]   N. Martins and L.T.G. Lima, "Eigenvalue and Frequency Domain Analysis of Small Signal Electromechanical Stability Problems," *Eigenanalysis and Frequency Domain Methods for System Dynamic Performance*, IEEE publication 90TH0292-3-PWR.

[14]   R.T. Byerly, R.J. Bennon, and D.E. Sherman, "Eigenvalue Analysis of Synchronizing Power Flow Oscillations in Large Electric Power Systems," *IEEE Trans.*, Vol. PAS-101, pp. 235-243, January 1982.

[15]   N. Martins, "Efficient Eigenvalue and Frequency Response Methods Applied to Power System Small-Signal Stability Studies," *IEEE Trans.*, Vol. PWRS-1, pp. 217-225, February 1986.

[16]   L. Wang and A. Semlyen, "Application of Sparse Eigenvalue Techniques to the Small Signal Stability Analysis of Large Power Systems," *IEEE Trans.*, Vol. PWRS-6, pp. 635-642, November 1990.

[17]   N. Uchida and T. Nagao, "A New Eigen-Analysis Method of Steady-State Stability Studies for Large Power Systems: S Matrix Method," *IEEE Trans.*, Vol. PWRS-3, pp. 706-714, May 1988.

[18]   F.L. Pagola, L. Rouco, and I.J. Perez-Arriaga, "Analysis and Control of Small-Signal Stability in Electric Power Systems by Selective Modal Analysis," *Eigenanalysis and Frequency Domain Methods for System Dynamic Performance*, IEEE publication 90TH0292-3-PWR.

[19]   W.E. Arnoldi, "The Principle of Minimized Iterations in the Solution of the Matrix Eigenvalue Problem," *Quart. Appl. Math.*, Vol. 9, pp. 17-29, 1951.

[20]   Y. Saad, "Variations on Arnoldi's Method for Computing Eigenelements of Large Unsymmetric Matrices," *Linear Algebra and Its Applications*, Vol. 34, pp. 184-198, June 1981.

[21]   P. Kundur, D.C. Lee, and H.M. Zein El-Din, "Power System Stabilizers for Thermal Units: Analytical Techniques and On-site Validation," *IEEE Trans.*, Vol. PAS-100, pp. 81-95, January 1981.

[22]   P. Kundur and P.L. Dandeno, "Practical Application of Eigenvalue Techniques in the Analysis of Power System Dynamic Stability Problems," *Proceedings of Fifth Power System Computation Conference*, Cambridge, England, September 1975.

[23]   M. Klein, G.J. Rogers, and P. Kundur, "A Fundamental Study of Inter-Area Oscillations," *IEEE Trans.*, Vol. PWRS-6, No. 3, pp. 914-921, August 1991.

[24]   M. Klein, G.J. Rogers, S. Moorty, and P. Kundur, "Analytical Investigation of Factors Influencing Power System Stabilizers Performance," Paper 92WM016-6EC, Presented at the IEEE PES Winter Meeting, New York, January 1992.

[25]   D.K. Mugwanya and J.E. Van Ness, "Mode Coupling in Power Systems," *IEEE Trans.*, Vol. PWRS-2, pp. 264-270, May 1987.

[26]   O.W. Hanson, C.J. Goodwin, and P.L. Dandeno, "Identification of Excitation and Speed Control Parameters in Stabilizing Inter-System Oscillation," *IEEE Trans.*, Vol. PAS-87, pp. 1306-1313, May 1968.

[27]   Canadian Electrical Association Report, "Investigation of Low Frequency Inter-area Oscillation Problems in Large Interconnected Power Systems," Report of Research Project 294T622, prepared by Ontario Hydro, 1993.

Chapter **13**

Transient Stability

Transient stability, as described in Chapter 2, is the ability of the power system to maintain synchronism when subjected to a severe transient disturbance such as a fault on transmission facilities, loss of generation, or loss of a large load. The system response to such disturbances involves large excursions of generator rotor angles, power flows, bus voltages, and other system variables. Stability is influenced by the nonlinear characteristics of the power system. If the resulting angular separation between the machines in the system remains within certain bounds, the system maintains synchronism. Loss of synchronism because of transient instability, if it occurs, will usually be evident within 2 to 3 seconds of the initial disturbance.

This chapter illustrates the nature of transient stability problems, identifies factors influencing them, and describes modelling considerations and analytical techniques applicable to transient stability analysis.

13.1 AN ELEMENTARY VIEW OF TRANSIENT STABILITY [1-3]

Consider the system shown in Figure 13.1, consisting of a generator delivering power to a large system represented by an infinite bus through two transmission circuits. An infinite bus, as described in Chapter 12 (Section 12.3), represents a voltage source of constant voltage magnitude and constant frequency.

We will present fundamental concepts and principles of transient stability by analyzing the system response to large disturbances, using very simple models. All resistances are neglected. The generator is represented by the classical model (see Chapter 5, Section 5.3.1) and the speed governor effects are neglected. The corresponding system representation is shown in Figure 13.2(a). The voltage behind

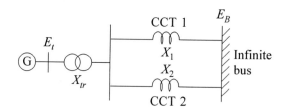

Figure 13.1 Single-machine infinite bus system

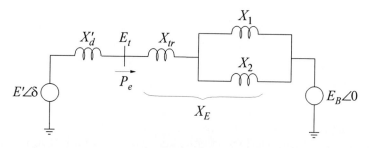

(a) Equivalent circuit

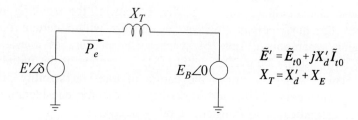

(b) Reduced equivalent circuit

Figure 13.2 System representation with generator represented by classical model

the transient reactance (X_d') is denoted by E'. The rotor angle δ represents the angle by which E' leads E_B. When the system is perturbed, the magnitude of E' remains constant at its predisturbance value and δ changes as the generator rotor speed deviates from synchronous speed ω_0.

The system model can be reduced to the form shown in Figure 13.2(b). It can be analyzed by using simple analytical methods and is helpful in acquiring a basic understanding of the transient stability phenomenon. This model is identical to that shown in Figure 12.4 of Chapter 12. From Equation 12.71, the generator's electrical power output is

$$P_e = \frac{E'E_B}{X_T} \sin\delta = P_{max}\sin\delta \qquad (13.1)$$

where

$$P_{max} = \frac{E'E_B}{X_T} \qquad (13.2)$$

Since we have neglected the stator resistance, P_e represents the air-gap power as well as the terminal power. The power-angle relationship with both transmission circuits in service (I/S) is shown graphically in Figure 13.3 as curve 1. With a mechanical power input of P_m, the steady-state electrical power output P_e is equal to P_m, and the operating condition is represented by point a on the curve. The corresponding rotor angle is δ_a.

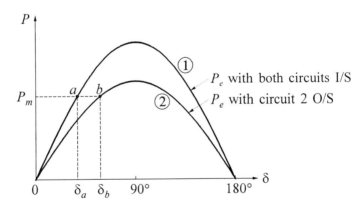

Figure 13.3 Power-angle relationship

If one of the circuits is out of service (O/S), the effective reactance X_T is higher. The power-angle relationship with circuit 2 out of service is shown in Figure 13.3 as curve 2. The maximum power is now lower. With a mechanical power input of P_m, the rotor angle is now δ_b corresponding to the operating point b on curve 2; with a higher reactance, the rotor angle is higher in order to transmit the same steady-state power.

During a disturbance, the oscillation of δ is superimposed on the synchronous speed ω_0, but the speed deviation ($\Delta\omega_r = d\delta/dt$) is very much smaller than ω_0. Therefore, the generator speed is practically equal to ω_0, and the per unit (pu) air-gap torque may be considered to be equal to the pu air-gap power. We will therefore use torque and power interchangeably when referring to the swing equation.

The *equation of motion* or the *swing equation* (see Chapter 3, Section 3.9) may be written as

$$\frac{2H}{\omega_0}\frac{d^2\delta}{dt^2} = P_m - P_{max}\sin\delta \qquad (13.3)$$

where

P_m = mechanical power input, in pu
P_{max} = maximum electrical power output, in pu
H = inertia constant, in MW·s/MVA
δ = rotor angle, in elec. rad
t = time, in s

Response to a step change in P_m

Let us now examine the transient behaviour of the system, with both circuits in service, by considering a sudden increase in the mechanical power input from an initial value of P_{m0} to P_{m1} as shown in Figure 13.4(a). Because of the inertia of the

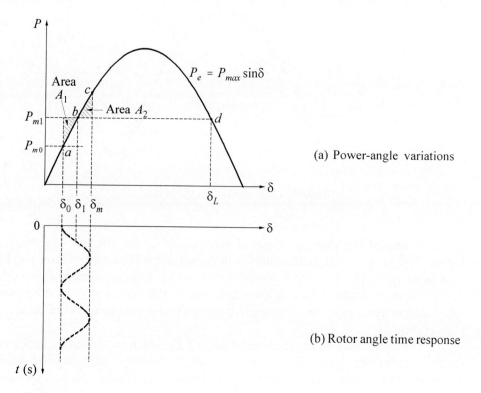

(a) Power-angle variations

(b) Rotor angle time response

Figure 13.4 Response to a step change in mechanical power input

rotor, the rotor angle cannot change instantly from the initial value of δ_0 to δ_1 corresponding to the new equilibrium point b at which $P_e = P_{m1}$. The mechanical power is now in excess of the electrical power. The resulting accelerating torque causes the rotor to accelerate from the initial operating point a toward the new equilibrium point b, tracing the P_e-δ curve at a rate determined by the swing equation. The difference between P_{m1} and P_e at any instant represents the accelerating power.

When point b is reached, the accelerating power is zero, but the rotor speed is higher than the synchronous speed ω_0 (which corresponds to the frequency of the infinite bus voltage). Hence, the rotor angle continues to increase. For values of δ higher than δ_1, P_e is greater than P_{m1} and the rotor decelerates. At some peak value δ_m, the rotor speed recovers to the synchronous value ω_0, but P_e is higher than P_{m1}. The rotor continues to decelerate with the speed dropping below ω_0; the operating point retraces the P_e-δ curve from c to b and then to a. The rotor angle oscillates indefinitely about the new equilibrium angle δ_1 with a constant amplitude as shown by the time plot of δ in Figure 13.4(b).

In our representation of the power system in the above analysis, we have neglected all resistances and the classical model is used to represent the generator. In effect, this neglects all sources of damping. Therefore, the rotor oscillations continue unabated following the perturbation. In practice, as discussed in Chapter 12, there are many sources of positive damping including field flux variations and rotor amortisseur circuits. In a system which is small-signal stable, the oscillations damp out.

Equal-area criterion

For the system model considered above, it is not necessary to formally solve the swing equation to determine whether the rotor angle increases indefinitely or oscillates about an equilibrium position. Information regarding the maximum angle excursion (δ_m) and the stability limit may be obtained graphically by using the power-angle diagram shown in Figure 13.4. Although this method is not applicable to multimachine systems with detailed representation of synchronous machines, it helps in understanding basic factors that influence the transient stability of any system.

From Equation 13.3, we have the following relationship between the rotor angle and the accelerating power:

$$\frac{d^2\delta}{dt^2} = \frac{\omega_0}{2H}(P_m - P_e) \tag{13.4}$$

Now P_e is a nonlinear function of δ, and therefore the above equation cannot be solved directly. If both sides are multiplied by $2d\delta/dt$, then

$$2\frac{d\delta}{dt}\frac{d^2\delta}{dt^2} = \frac{\omega_0(P_m - P_e)}{H}\frac{d\delta}{dt}$$

or

$$\frac{d}{dt}\left[\frac{d\delta}{dt}\right]^2 = \frac{\omega_0(P_m-P_e)}{H}\frac{d\delta}{dt} \tag{13.5}$$

Integrating gives

$$\left[\frac{d\delta}{dt}\right]^2 = \int \frac{\omega_0(P_m-P_e)}{H}d\delta \tag{13.6}$$

The speed deviation $d\delta/dt$ is initially zero. It will change as a result of the disturbance. For stable operation, the deviation of angle δ must be bounded, reaching a maximum value (as at point c in Figure 13.4) and then changing direction. This requires the speed deviation $d\delta/dt$ to become zero at some time after the disturbance. Therefore, from Equation 13.6, as a criterion for stability we may write

$$\int_{\delta_0}^{\delta_m}\frac{\omega_0}{H}(P_m-P_e)d\delta = 0 \tag{13.7}$$

where δ_0 is the initial rotor angle and δ_m is the maximum rotor angle, as illustrated in Figure 13.4. Thus, the area under the function P_m-P_e plotted against δ must be zero if the system is to be stable. In Figure 13.4, this is satisfied when area A_1 is equal to area A_2. Kinetic energy is gained by the rotor during acceleration when δ changes from δ_0 to δ_1. The energy gained is

$$E_1 = \int_{\delta_0}^{\delta_1}(P_m-P_e)d\delta = \text{area } A_1 \tag{13.8}$$

The energy lost during deceleration when δ changes from δ_1 to δ_m is

$$E_2 = \int_{\delta_1}^{\delta_m}(P_e-P_m)d\delta = \text{area } A_2 \tag{13.9}$$

As we have not considered any losses, the energy gained is equal to the energy lost; therefore, area A_1 is equal to area A_2. This forms the basis for the equal-area criterion. It enables us to determine the maximum swing of δ and hence the stability of the system without computing the time response through formal solution of the swing equation.

The criterion can be readily used to determine the maximum permissible increase in P_m for the system of Figure 13.1. The stability is maintained only if an area A_2 at least equal to A_1 can be located above P_{m1}. If A_1 is greater than A_2, then $\delta_m > \delta_L$, and stability will be lost. This is because, for $\delta > \delta_L$, P_{m1} is larger than P_e and the net torque is accelerating rather than decelerating.

We will examine the mechanism of transient instability by considering next the system response to a short-circuit fault on the transmission system, which is a more common form of a disturbance considered in transient stability studies.

Response to a short-circuit fault

Let us consider the response of the system to a three-phase fault at location F on transmission circuit 2, as shown in Figure 13.5(a). The corresponding equivalent circuit, assuming a classical generator model, is shown in Figure 13.5(b). The fault is cleared by opening circuit breakers at both ends of the faulted circuit, the fault-clearing time depending on the relaying time and breaker time (see Section 13.5.2).

If the fault location F is at the sending end (HT bus) of the faulted circuit, no power is transmitted to the infinite bus. The short-circuit current from the generator flows through pure reactances to the fault. Hence, only reactive power flows and the active power P_e and the corresponding electrical torque T_e at the air-gap are zero during the fault. If we had included generator stator and transformer resistances in our model, P_e would have a small value, representing the corresponding resistive losses.

If the fault location F is at some distance away from the sending end as shown in Figures 13.5(a) and (b), some active power is transmitted to the infinite bus while the fault is still on.

Figures 13.5(c) and (d) show P_e-δ plots for the three network conditions: (i) prefault (both circuits in service), (ii) with a three-phase fault on circuit 2 at a location some distance from the sending end, and (iii) postfault (circuit 2 out of service). Figure 13.5(c) considers the system performance with a fault-clearing time of t_{c1} and represents a stable case. Figure 13.5(d) considers a longer fault-clearing time (t_{c2}) such that the system is unstable. In both cases P_m is assumed to be constant.

Let us examine the stable case depicted by Figure 13.5(c). Initially, the system is operating with both circuits in service such that $P_e = P_m$ and $\delta = \delta_0$. When the fault occurs, the operating point suddenly changes from a to b. Owing to inertia, angle δ cannot change instantly. Since P_m is now greater than P_e, the rotor accelerates until the operating point reaches c, when the fault is cleared by isolating circuit 2 from the system. The operating point now suddenly shifts to d. Now P_e is greater than P_m, causing deceleration of the rotor. Since the rotor speed is greater than the synchronous speed ω_0, δ continues to increase until the kinetic energy gained during the period of acceleration (represented by area A_1) is expended by transferring the energy to the system. The operating point moves from d to e, such that area A_2 is equal to area A_1. At point e, the speed is equal to ω_0 and δ has reached its maximum value δ_m. Since P_e is still greater than P_m, the rotor continues to retard, with the speed dropping below ω_0. The rotor angle δ decreases, and the operating point retraces the path from e to *d* and follows the P_e-δ curve for the postfault system farther down. The minimum value of δ is such that it satisfies the equal-area criterion for the postfault system. In the absence of any source of damping, the rotor continues to oscillate with constant amplitude.

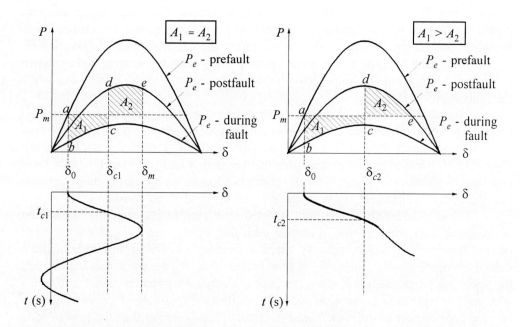

(a) Single-line diagram   (b) Equivalent circuit

(c) Response to a fault cleared   (d) Response to a fault cleared
    in t_{c1} seconds - stable case   in t_{c2} seconds - unstable case

Figure 13.5 Illustration of transient stability phenomenon

With a delayed fault clearing, as shown in Figure 13.5(d), area A_2 above P_m is less than A_1. When the operating point reaches e, the kinetic energy gained during the accelerating period has not yet been completely expended; consequently, the speed is still greater than ω_0 and δ continues to increase. Beyond point e, P_e is less than P_m, and the rotor begins to accelerate again. The rotor speed and angle continue to increase, leading to *loss of synchronism*.

Factors influencing transient stability

From the above discussions, and referring to Figure 13.5, we can conclude that transient stability of the generator is dependent on the following:

(a)   How heavily the generator is loaded.

(b)   The generator output during the fault. This depends on the fault location and type.

(c)   The fault-clearing time.

(d)   The postfault transmission system reactance.

(e)   The generator reactance. A lower reactance increases peak power and reduces initial rotor angle.

(f)   The generator inertia. The higher the inertia, the slower the rate of change in angle. This reduces the kinetic energy gained during fault; i.e., area A_1 is reduced.

(g)   The generator internal voltage magnitude (E'). This depends on the field excitation.

(h)   The infinite bus voltage magnitude E_B.

As a means of introducing basic concepts, we have considered a system having a simple configuration and represented by a simple model. This has enabled the analysis of stability by using a graphical approach. Although rotor angle plots as a function of time are shown in Figures 13.4 and 13.5, we have not actually computed them, and hence the time scales have not been defined for these plots. Practical power systems have complex network structures. Accurate analysis of their transient stability requires detailed models for generating units and other equipment. *At present, the most practical available method of transient stability analysis is time-domain simulation in which the nonlinear differential equations are solved by using step-by-step numerical integration techniques.* This leads us to the next section in which some of the commonly used numerical integration methods are described.

13.2 NUMERICAL INTEGRATION METHODS [4-6]

The differential equations to be solved in power system stability analysis are nonlinear ordinary differential equations with known initial values:

$$\frac{d\mathbf{x}}{dt} = \mathbf{f}(\mathbf{x},t) \tag{13.10A}$$

where $\mathbf{x}$ is the state vector of n dependent variables and t is the independent variable (time). Our objective is to solve $\mathbf{x}$ as a function of t, with the initial values of $\mathbf{x}$ and t equal to $\mathbf{x}_0$ and t_0, respectively.

In this section we provide a general description of numerical integration methods applicable to the solution of equations of the above form. In describing these methods, without loss of generality, we will treat Equation 13.10A as if it were a first-order differential equation. This simplifies presentation and makes it easier for a novice reader to comprehend the special features of each method.

We will first describe the Euler method, which by virtue of its simplicity serves as a good introduction to numerical integration, and then we discuss more advanced methods.

13.2.1 Euler Method

Consider the first-order differential equation

$$\frac{dx}{dt} = f(x,t) \tag{13.10B}$$

with $x=x_0$ at $t=t_0$. Figure 13.6 illustrates the principle of applying the Euler method.

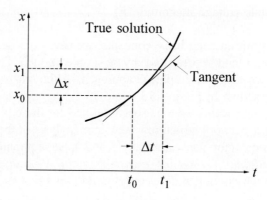

Figure 13.6

At $x=x_0$, $t=t_0$ we can approximate the curve representing the true solution by its tangent having a slope

$$\left.\frac{dx}{dt}\right|_{x=x_0} = f(x_0,t_0)$$

Therefore,

$$\Delta x = \left.\frac{dx}{dt}\right|_{x=x_0} \cdot \Delta t$$

The value of x at $t=t_1=t_0+\Delta t$ is given by

$$x_1 = x_0+\Delta x = x_0+\left.\frac{dx}{dt}\right|_{x=x_0} \cdot \Delta t \tag{13.11}$$

The Euler method is equivalent to using the first two terms of the Taylor series expansion for x around the point (x_0,t_0):

$$x_1 = x_0+\Delta t(\dot{x}_0)+\frac{\Delta t^2}{2!}(\ddot{x}_0)+\frac{\Delta t^3}{3!}(\dddot{x}_0)+\cdots \tag{13.12}$$

After using the Euler technique for determining $x=x_1$ corresponding to $t=t_1$, we can take another short time step Δt and determine x_2 corresponding to $t_2=t_1+\Delta t$ as follows:

$$x_2 = x_1+\left.\frac{dx}{dt}\right|_{x=x_1} \cdot \Delta t \tag{13.13}$$

By applying the technique successively, values of x can be determined corresponding to different values of t.

The method considers only the first derivative of x and is, therefore, referred to as a *first-order* method. To give sufficient accuracy for each step, Δt has to be small. This will increase round-off errors, and the computational effort required will be very high.

In the application of numerical integration methods, it is very important to consider the *propagation of error*, which may cause slight errors made early in the process to be magnified at later steps. *Numerical stability* depends on the propagation of error. If early errors carry through but cause no significant further errors later, the method is said to be numerically stable. If, on the other hand, early errors cause other large errors later, the method is said to be numerically unstable.

13.2.2 Modified Euler Method

The standard Euler method results in inaccuracies because it uses the derivative at the beginning of the interval as though it applied throughout the interval. The modified Euler method tries to overcome this problem by using the average of the derivatives at the two ends.

The modified Euler method consists of the following steps:

(a)   *Predictor step.* By using the derivative at the beginning of the step, the value at the end of the step is *predicted*

$$x_1^p = x_0 + \frac{dx}{dt}\bigg|_{x=x_0} \cdot \Delta t \qquad (13.14)$$

(b)   *Corrector step.* By using the predicted value of x_1^p, the derivative at the end of the step is computed and the average of this derivative and the derivative at the beginning of the step is used to find the *corrected value*

$$x_1^c = x_0 + \frac{1}{2}\left(\frac{dx}{dt}\bigg|_{x=x_0} + \frac{dx}{dt}\bigg|_{x=x_1^p} \right)\Delta t \qquad (13.15)$$

If desired, a more accurate value of the derivative at the end of the step can be calculated, again by using $x=x_1^c$. This derivative can be used to calculate a more accurate value of the average derivative which is in turn used to apply the corrector step again. This process can be used repeatedly until successive steps converge with the desired accuracy.

The modified Euler method is the simplest of *predictor-corrector* (P-C) methods. Among the well known higher order P-C methods are the Adams-Bashforth method, Milne method, and Hamming method [4]. The applicability of these methods to power system stability analysis has been investigated in reference 7 and has been found to suffer from a number of limitations. They are not self-starting, need more computer storage and require smaller time steps than the Runge-Kutta methods described below.

13.2.3 Runge-Kutta (R-K) Methods [4,5]

The R-K methods approximate the Taylor series solution; however, unlike the formal Taylor series solution, the R-K methods do not require explicit evaluation of derivatives higher than the first. The effects of higher derivatives are included by several evaluations of the first derivative. Depending on the number of terms effectively retained in the Taylor series, we have R-K methods of different orders.

Second-order R-K method

Referring to the differential equation 13.10B, the second-order R-K formula for the value of x at $t=t_0+\Delta t$ is

$$x_1 = x_0+\Delta x = x_0+\frac{k_1+k_2}{2}$$

where

$$k_1 = f(x_0,t_0)\Delta t$$

$$k_2 = f(x_0+k_1,t_0+\Delta t)\Delta t$$

This method is equivalent to considering first and second derivative terms in the Taylor series; error is on the order of Δt^3.

A general formula giving the value of x for the $(n+1)^{st}$ step is

$$x_{n+1} = x_n+\frac{k_1+k_2}{2}$$

where

$$k_1 = f(x_n,t_n)\Delta t$$

$$k_2 = f(x_n+k_1,t_n+\Delta t)\Delta t$$

Fourth-order R-K method

The general formula giving the value of x for the $(n+1)^{st}$ step is

$$x_{n+1} = x_n+\frac{1}{6}(k_1+2k_2+2k_3+k_4) \tag{13.16}$$

where

$$k_1 = f(x_n,t_n)\Delta t$$

$$k_2 = f\left(x_n+\frac{k_1}{2},t_n+\frac{\Delta t}{2}\right)\Delta t$$

$$k_3 = f\left(x_n+\frac{k_2}{2},t_n+\frac{\Delta t}{2}\right)\Delta t$$

$$k_4 = f(x_n+k_3,t_n+\Delta t)\Delta t$$

The physical interpretation of the above solution is as follows:

$$k_1 = \text{(slope at the beginning of time step)}\Delta t$$
$$k_2 = \text{(first approximation to slope at midstep)}\Delta t$$
$$k_3 = \text{(second approximation to slope at midstep)}\Delta t$$
$$k_4 = \text{(slope at the end of step)}\Delta t$$
$$\Delta x = 1/6\,(k_1 + 2k_2 + 2k_3 + k_4)$$

Thus Δx is the incremental value of x given by the *weighted average* of estimates based on slopes at the beginning, midpoint, and end of the time step.

This method is equivalent to considering up to fourth derivative terms in the Taylor series expansion; it has an error on the order of Δt^5.

Gill's version of fourth order R-K method (R-K-G)

With x_0 as the initial value of x at the beginning of a step and by using $j=1,2,3$ and 4 to denote four stages, each stage of the Gill method can be described as follows [4]:

$$k_j = a_j[f(x_{j-1},t) - b_j q_{j-1}]$$

$$x_j = x_{j-1} + k_j \Delta t \qquad\qquad (13.17)$$

$$q_j = q_{j-1} + 3k_j - c_j f(x_{j-1},t)$$

The values of the a, b, and c coefficients are as follows:

$$a_1 = 1/2, \quad b_1 = 2, \quad c_1 = a_1; \quad a_2 = 1-\sqrt{0.5}, \quad b_2 = 1, \quad c_2 = a_2$$
$$a_3 = 1+\sqrt{0.5}, \quad b_3 = 1, \quad c_3 = a_3; \quad a_4 = 1/6, \quad b_4 = 2, \quad c_4 = 1/2$$

Solution at the end of a time step is given by x_4. Initially $q_0 = 0$, thereafter in advancing the solution, q_0 for the next step is equal to q_4 of the previous step.

The following are the advantages of Gill's version of the R-K method:

(a)   Roundoff errors are minimized (variable q is used for this purpose).

(b)   Storage requirements are less than for the original R-K method.

Care should be exercised in applying the R-K-G method when discontinuities or sudden changes in the rate of change of variables occur. Within a time step, a variable should not be limited, as a result of sharp nonlinearities; otherwise the variable q takes on incorrect values.

Richardson's formula for accuracy check

The accuracy of results obtained with the above numerical integration methods may be checked by using Richardson's formula which gives the accumulated truncation error *propagated* in the course of integration. The difference between the true value of a variable and the value obtained by a fourth-order Runge-Kutta method using a step length of Δt is given by [4,7]

$$x(\text{true}) - x(\Delta t) = \frac{x(\Delta t) - x(2\Delta t)}{15} \qquad (13.18)$$

where

      $x(\text{true})$ = true value of x
      $x(\Delta t)$ = value computed with a step length of Δt
      $x(2\Delta t)$ = value computed with a step length of $2\Delta t$

13.2.4 Numerical Stability of Explicit Integration Methods

Integration methods described in Sections 13.2.1 to 13.2.3 (i.e., the Euler, predictor-corrector, and R-K methods) are known as *explicit methods*. In these methods the value of the dependent variable x at any value of t is computed from knowledge of the values of x from the previous time steps. In other words, x_{n+1} for $(n+1)^{\text{st}}$ step is calculated explicitly by evaluating $f(x,t)$ with known x. These methods are easy to implement for the solution of a complex set of system state equations.

A significant limitation of the explicit integration methods is that they are not numerically A-stable [6]. Consequently, the length of the integration time step (Δt) is restricted by the small time constants of the system.

Numerical stability is related to the *stiffness* of the set of differential equations representing the system. The stiffness is associated with the range of time constants in the system model. It is measured by the ratio of the largest to smallest time constant, or more precisely by the ratio of the largest to smallest eigenvalues of the linearized system.

Stiffness in a transient stability simulation increases with modelling detail. In the overall system, not all the time constants may be readily apparent. Thus the stiffness may be hidden and can only be established by computing the eigenvalues of the linearized system.

Explicit integration methods have weak stability numerically; with stiff systems, the solution "blows up" unless a small step size is used. Even after the fast modes die out, small time steps continue to be required to maintain numerical stability.

13.2.5 Implicit Integration Methods

Consider the differential equation

$$\frac{dx}{dt} = f(x,t) \qquad \text{with } x = x_0 \text{ at } t = t_0$$

The solution for x at $t = t_1 = t_0 + \Delta t$ may be expressed in integral form as

$$x_1 = x_0 + \int_{t_0}^{t_1} f(x, \tau) d\tau \tag{13.19}$$

Implicit integration methods use interpolation functions for the expression under the integral. Interpolation implies that the functions must pass through the yet unknown points at time t_1.

The simplest implicit integration method is the *trapezoidal rule*. It uses linear interpolation. As shown in Figure 13.7, this implies that the area under the integral of Equation 13.19 is approximated by trapezoids; hence the name - trapezoidal rule.

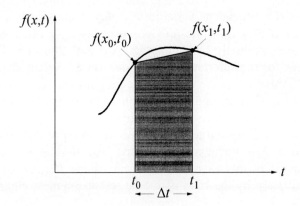

Figure 13.7

The trapezoidal rule for Equation 13.19 is given by

$$x_1 = x_0 + \frac{\Delta t}{2} [f(x_0, t_0) + f(x_1, t_1)] \tag{13.20}$$

A general formula giving the value of x at $t = t_{n+1}$ is

$$x_{n+1} = x_n + \frac{\Delta t}{2}[f(x_n, t_n) + f(x_{n+1}, t_{n+1})] \tag{13.21}$$

We see that x_{n+1} appears on both sides of Equation 13.21. This implies that the variable x is computed as a function of its value at the previous time step as well as the current value (which is unknown). Therefore, an implicit equation must be solved.

The trapezoidal rule is numerically A-stable [6]. The stiffness of the system being analyzed affects accuracy but not numerical stability. With larger time steps, high frequency modes and fast transients are filtered out, and the solutions for the slower modes is accurate. For systems involving simulations in which time steps are limited by numerical stability considerations rather than accuracy, implicit methods are generally better suited than explicit methods.

The trapezoidal rule is a second-order method. Implicit integration methods of higher order have been proposed in the literature on numerical methods. However, they have not been widely used for power system applications since they are more difficult to program and less numerically stable than the trapezoidal rule.

In the above description of different numerical integration methods, for simplicity we have considered a first-order differential equation. When applied to the analysis of power system stability, the system equations are organized as a set of first-order differential equations. The rate of change of each state variable depends on other state variables and is not an explicit function of time t. The following example serves as a simple illustration of the application of numerical integration as well as the equal-area criterion for transient stability analysis.

Example 13.1

In this example, we examine the transient stability of a thermal generating station consisting of four 555 MVA, 24 kV, 60 Hz units supplying power to an infinite bus through two transmission circuits as shown in Figure E13.1. This system is the same as the one considered in Example 12.2 of Chapter 12 in which we examined the small-signal performance.

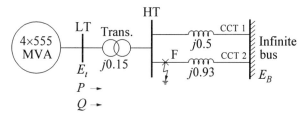

Figure E13.1

The network reactances shown in the figure are in per unit on 2220 MVA, 24 kV base (referred to the LT side of the step-up transformer). Resistances are assumed to be negligible.

The initial system-operating condition, with quantities expressed in per unit on 2220 MVA and 24 kV base, is as follows:

$$P=0.9 \qquad Q=0.436 \text{ (overexcited)} \qquad \tilde{E}_t=1.0\angle 28.34° \qquad \tilde{E}_B=0.90081\angle 0$$

The generators are modelled as a single equivalent generator represented by the classical model with the following parameters expressed in per unit on 2220 MVA, 24 kV base:

$$X'_d=0.3 \qquad H=3.5 \text{ MW·s/MVA} \qquad K_D=0$$

Circuit 2 experiences a solid three-phase fault at point F, and the fault is cleared by isolating the faulted circuit.

(a)   Determine the critical fault-clearing time and the critical clearing angle by computing the time response of the rotor angle, using numerical integration.

(b)   Check the above value of critical clearing angle, using the equal-area criterion.

Solution

With the generator represented by the classical model, the system-equivalent circuit is as shown in Figure E13.2.

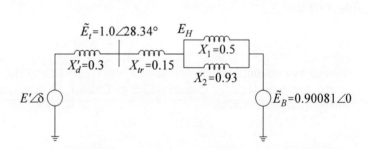

Figure E13.2 Equivalent circuit

For the initial operating condition, the voltage behind X'_d is

$$\tilde{E}' = \tilde{E}_t+jX'_d\tilde{I}_t$$

$$= 1.0\angle 28.34° + \frac{j0.3(0.9-j0.436)}{1.0\angle -28.34°}$$

$$= 1.1626\angle 41.77°$$

Figure E13.3 shows the reduced equivalent circuits representing the three system conditions: (i) prefault, (ii) during fault, and (iii) postfault. Also shown in the figure are the corresponding expressions for the electrical power output as a function of δ.

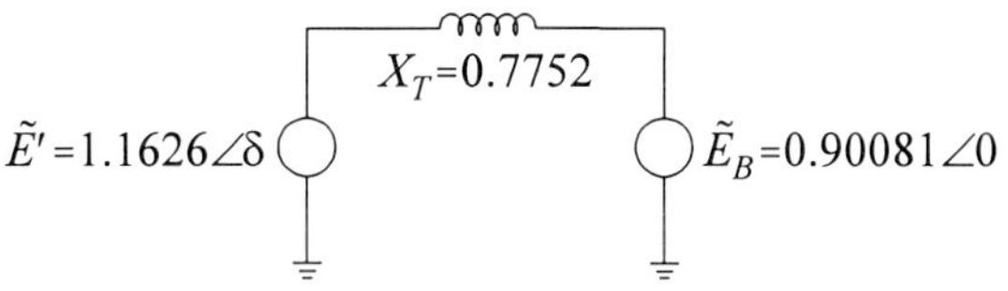

$X_T = 0.7752$

$\tilde{E}' = 1.1626\angle\delta$ $\tilde{E}_B = 0.90081\angle0$

$$P_e = \frac{1.1626\times0.90081}{0.7752}\sin\delta$$
$$= 1.351\sin\delta$$

(a) Prefault

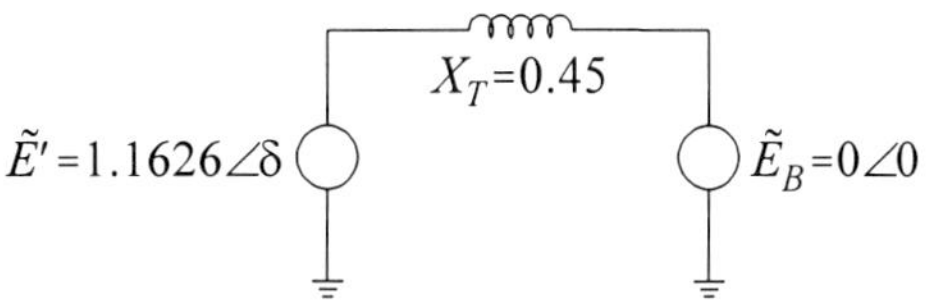

$X_T = 0.45$

$\tilde{E}' = 1.1626\angle\delta$ $\tilde{E}_B = 0\angle0$

$$P_e = 0$$

(b) During fault

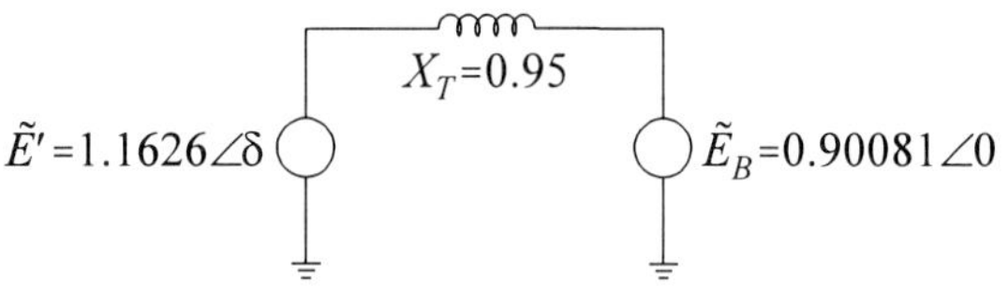

$X_T = 0.95$

$\tilde{E}' = 1.1626\angle\delta$ $\tilde{E}_B = 0.90081\angle0$

$$P_e = \frac{1.1626\times0.90081}{0.95}\sin\delta$$
$$= 1.1024\sin\delta$$

(c) Postfault

Figure E13.3 Reduced equivalent circuits and equations for power output

(a) *Time response using numerical integration:*

Equation 13.3 may be written in terms of the two first-order equations:

$$p(\Delta\omega_r) = \frac{1}{2H}(P_m - P_{max}\sin\delta)$$

$$= \frac{1}{7.0}(0.9 - P_{max}\sin\delta) \qquad (E13.1)$$

$$p(\delta) = \omega_0\Delta\omega_r$$

$$= 377\,\Delta\omega_r \qquad (E13.2)$$

where

$$
P_{max} = \begin{cases} 1.351 & \text{before the fault} \\ 0 & \text{during the fault} \\ 1.1024 & \text{after the fault} \end{cases}
$$

The initial values of δ and $\Delta\omega_r$ are 41.77° and 0 pu, respectively.

Any of the numerical integration methods described in Section 13.2 may be used to solve Equations E13.1 and E13.2. For illustration, let us consider the second-order R-K method. The general formulas giving the values of $\Delta\omega_r$, δ, and t for the $(n+1)^{st}$ step of integration are as follows:

$$
(\Delta\omega_r)_{n+1} = (\Delta\omega_r)_n + \frac{k_1' + k_2'}{2}
$$

$$
\delta_{n+1} = \delta_n + \frac{k_1'' + k_2''}{2}
$$

$$
t_{n+1} = t_n + \Delta t
$$

where

$$
k_1' = \left[0.1286 - \frac{P_{max}}{7.0} \sin(\delta)_n \right] \Delta t
$$

$$
k_1'' = [377(\Delta\omega_r)_n] \Delta t
$$

$$
k_2' = \left[0.1286 - \frac{P_{max}}{7.0} \sin(\delta_n + k_1'') \right] \Delta t
$$

$$
k_2'' = \{ 377[(\Delta\omega_r)_n + k_1'] \} \Delta t
$$

Figure E13.4 shows plots of δ as a function of time, for the three values of fault-clearing time (t_c): 0.07 s, 0.086 s, and 0.087 s. The corresponding values of clearing angle (δ_c) are 48.58°, 52.04°, and 52.30°, respectively. These results were computed using a time step (Δt) of 0.05 s throughout the solution. The time step was, however, adjusted near the fault-clearing time so as to give the exact switching instant.

From the results, we see that the system is stable with $t_c=0.086$ s ($\delta_c=52.04°$), and is unstable with $t_c=0.087$ s ($\delta_c=52.30°$); the critical clearing time is, therefore, 0.0865±0.0005 s, and the critical clearing angle is 52.17°±0.13°.

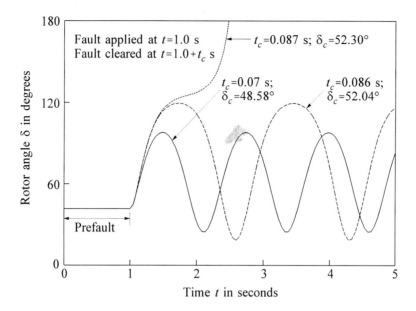

Figure E13.4 Rotor angle response for different values
of fault-clearing time

(b) Equal-area criterion:

The power-angle diagrams for the three network conditions are shown in Figure
E13.5. For the critically stable case, the maximum swing in δ is given by

$$1.1024\sin\delta_m = 0.9 \qquad \text{hence,} \quad \delta_m = 125.27°$$

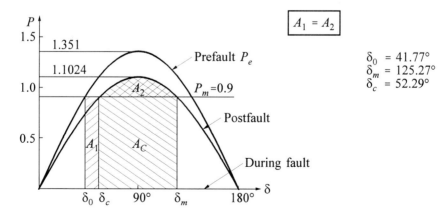

Figure E13.5 Power-angle diagram

From Figure E13.5, for critical clearing,

$$A_1 = A_2$$

or

$$A_1 + A_c = A_2 + A_c$$

Therefore, from the plots, we have

$$0.9(125.27 - 41.77)\frac{\pi}{180} = \int_{\delta_c}^{125.27} 1.1024 \sin\delta \, d\delta$$

$$1.3116 = 1.1024(\cos\delta_c + 0.5781)$$

Thus the critical clearing angle is

$$\delta_c = 52.29°$$

This agrees with the value determined from time responses.   ∎

13.3 SIMULATION OF POWER SYSTEM DYNAMIC RESPONSE

13.3.1 Structure of the Power System Model

Analysis of transient stability of power systems involves the computation of their nonlinear dynamic response to large disturbances, usually a transmission network fault, followed by the isolation of the faulted element by protective relaying.

Figure 13.8 depicts the general structure of the power system model applicable to transient stability analysis. This model structure is similar to that presented in Chapter 12 (Figure 12.18) for small-signal stability analysis. However, for transient stability analysis, nonlinear system equations are solved. In addition, large discontinuities due to faults and network switching, and small discontinuities due to limits on system variables, appear in the system model. Bus voltages, line flows, and performance of protection systems are of interest, in addition to the basic information related to the stability of the system.

As seen in Figure 13.8, the overall power system representation includes models for the following individual components:

•   Synchronous generators, and the associated excitation systems and prime movers

•   Interconnecting transmission network including static loads

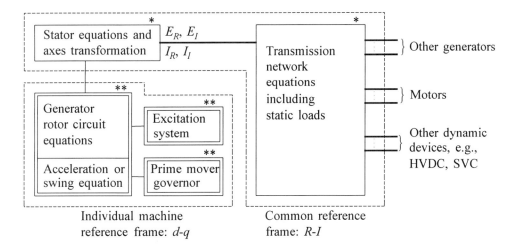

Individual machine   Common reference
reference frame: *d-q*   frame: *R-I*

*   Algebraic equations
**   Differential equations

Figure 13.8 Structure of the complete power system model
for transient stability analysis

- Induction and synchronous motor loads

- Other devices such as HVDC converters and SVCs

The model used for each component should be appropriate for transient stability analysis, and the system equations must be organized in a form suitable for applying numerical methods.

As we will see in what follows, the complete system model consists of a large set of ordinary differential equations and large sparse algebraic equations. The transient stability analysis is thus a differential algebraic initial-value problem.

13.3.2 Synchronous Machine Representation [8]

Synchronous machine modelling is presented in Chapters 3 to 5. The per unit synchronous machine equations are summarized in Section 3.4.9. Modifications of these equations necessary for representation in stability studies are discussed in Section 5.1.

To illustrate the implementation of the generator model for transient stability analysis, we assume that the generator is represented by a model with one *d*-axis and two *q*-axis amortisseurs as shown in Figure 13.9. However, the equations presented here can be readily modified to account for a model with an arbitrary number of amortisseurs.

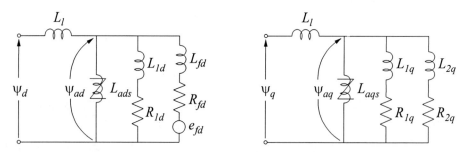

Figure 13.9 Synchronous machine equivalent circuits

The following is a summary of the synchronous machine equations as a set of first order differential equations, with time t in seconds, rotor angle δ in electrical radians, and all other quantities in per unit.

Equations of motion

$$p\Delta\omega_r = \frac{1}{2H}(T_m - T_e - K_D\Delta\omega_r)$$

$$p\delta = \omega_0\Delta\omega_r$$

(13.22)

where

$\omega_0 = 2\pi f_0$ electrical rad/s
$\Delta\omega_r$ = pu rotor speed deviation
p = derivative operator d/dt

Rotor circuit equations

With rotor currents expressed in terms of rotor and mutual flux linkages (see Equation 12.159 of Chapter 12), the rotor circuit dynamic equations are

$$p\psi_{fd} = \omega_0\left[e_{fd} + \frac{(\psi_{ad} - \psi_{fd})R_{fd}}{L_{fd}}\right]$$

$$p\psi_{1d} = \omega_0\left(\frac{\psi_{ad} - \psi_{1d}}{L_{1d}}\right)R_{1d}$$

$$p\psi_{1q} = \omega_0\left(\frac{\psi_{aq} - \psi_{1q}}{L_{1q}}\right)R_{1q}$$

$$p\psi_{2q} = \omega_0\left(\frac{\psi_{aq} - \psi_{2q}}{L_{2q}}\right)R_{2q}$$

(13.23)

The d- and q-axis mutual flux linkages are given by

$$\psi_{ad} = -L_{ads}i_d + L_{ads}i_{fd} + L_{ads}i_{1d}$$
$$= L_{ads}''\left(-i_d + \frac{\psi_{fd}}{L_{fd}} + \frac{\psi_{1d}}{L_{1d}}\right) \tag{13.24}$$

$$\psi_{aq} = L_{aqs}''\left(-i_q + \frac{\psi_{1q}}{L_{1q}} + \frac{\psi_{2q}}{L_{2q}}\right) \tag{13.25}$$

where

$$L_{ads}'' = \cfrac{1}{\cfrac{1}{L_{ads}} + \cfrac{1}{L_{fd}} + \cfrac{1}{L_{1d}}}$$

$$L_{aqs}'' = \cfrac{1}{\cfrac{1}{L_{aqs}} + \cfrac{1}{L_{1q}} + \cfrac{1}{L_{2q}}} \tag{13.26}$$

Here L_{ads} and L_{aqs} are saturated values of the d- and q-axis mutual inductances given by

$$L_{ads} = K_{sd}L_{adu}$$
$$L_{aqs} = K_{sq}L_{aqu} \tag{13.27}$$

and K_{sd} and K_{sq} are computed as a function of the air-gap flux linkage ψ_{at} as described in Chapter 3 (Section 3.8.2).

Stator voltage equations

With stator transients $(p\psi_d, p\psi_q)$ and speed variations (ω/ω_0) neglected as discussed in Chapter 5 (Section 5.1), the stator voltage may be written as follows:

$$e_d = -R_a i_d + (\bar{\omega}L_q'')i_q + E_d''$$
$$e_q = -R_a i_q - (\bar{\omega}L_d'')i_d + E_q'' \tag{13.28}$$

with

$$E_d'' = -\bar{\omega} L_{aqs}'' \left(\frac{\Psi_{1q}}{L_{1q}} + \frac{\Psi_{2q}}{L_{2q}} \right)$$

$$E_q'' = \bar{\omega} L_{ads}'' \left(\frac{\Psi_{fd}}{L_{fd}} + \frac{\Psi_{1d}}{L_{1d}} \right)$$

(13.29)

$$L_d'' = L_l + L_{ads}''$$

$$L_q'' = L_l + L_{aqs}''$$

(13.30)

Since we have neglected the effect of speed variations on the stator voltage, $\bar{\omega} = \omega/\omega_0 = 1.0$ in the above equations. Consequently, $\bar{\omega} L_d'' = X_d''$ and $\bar{\omega} L_q'' = X_q''$. The above equations are in the individual machine d-q reference frame which rotates with the machine's rotor. For the solution of the interconnecting transmission network equations, a synchronously rotating common R-I reference is used. The relationships shown in Figure 13.10 are used to transform variables from one reference frame to the other. The R-axis of the common reference frame also serves as the reference for measuring the rotor angle δ of each machine.

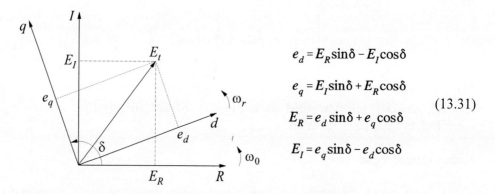

$$e_d = E_R \sin\delta - E_I \cos\delta$$

$$e_q = E_I \sin\delta + E_R \cos\delta$$

(13.31)

$$E_R = e_d \sin\delta + e_q \cos\delta$$

$$E_I = e_q \sin\delta - e_d \cos\delta$$

Figure 13.10 Reference frame transformation
and definition of rotor angle δ

For convenience in the organization of the complete set of algebraic equations, the stator voltage equations are expressed in the common R-I reference frame as indicated in Figure 13.8. Use of Equations 13.31 to transform the stator voltage equations 13.28 yields

$$\begin{bmatrix} E_R \\ E_I \end{bmatrix} = \begin{bmatrix} -R_{RR} & X_{RI} \\ -X_{IR} & -R_{II} \end{bmatrix} \begin{bmatrix} I_R \\ I_I \end{bmatrix} + \begin{bmatrix} E_R'' \\ E_I'' \end{bmatrix} \tag{13.32}$$

The elements of the impedance matrix are given by

$$\begin{aligned} R_{RR} &= (X_d'' - X_q'')\sin\delta\cos\delta + R_a \\ R_{II} &= (X_q'' - X_d'')\sin\delta\cos\delta + R_a \\ X_{RI} &= X_d''\cos^2\delta + X_q''\sin^2\delta \\ X_{IR} &= X_d''\sin^2\delta + X_q''\cos^2\delta \end{aligned} \tag{13.33}$$

As noted earlier, $\bar{\omega}$ is assumed to be equal to 1.0 pu. The internal voltage components are given by

$$\begin{aligned} E_R'' &= E_d''\sin\delta + E_q''\cos\delta \\ E_I'' &= E_q''\sin\delta - E_d''\cos\delta \end{aligned} \tag{13.34}$$

If *subtransient saliency is negligible*, $L_d'' = L_q''$. Then

$$R_{RR} = R_{II} = R_a$$

$$X_{RI} = X_{IR} = \bar{\omega}L_d'' = X_d'' = X_q'' = X''$$

In this case, $E_R'' + jE_I''$ represents the voltage behind the subtransient impedance $R_a + jX''$. For network solution, the generator may be represented by either of the simple equivalent circuits shown in Figure 13.11.

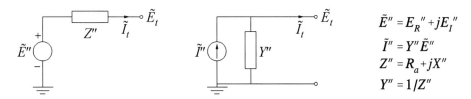

(a) Thevenin's equivalent   (b) Norton's equivalent

$$\begin{aligned} \tilde{E}'' &= E_R'' + jE_I'' \\ \tilde{I}'' &= Y''\tilde{E}'' \\ Z'' &= R_a + jX'' \\ Y'' &= 1/Z'' \end{aligned}$$

Figure 13.11 Synchronous machine equivalent circuits with negligible subtransient saliency

Variations in L_{ads} and L_{aqs} due to saturation may introduce a small amount of subtransient saliency during a transient condition. This is usually insignificant and may be ignored, if it is desired to take advantage of the computational simplicity offered by the above equivalent circuits.

Active power and reactive power at the generator stator terminals are

$$P_t = e_d i_d + e_q i_q \tag{13.35}$$

$$Q_t = e_q i_d - e_d i_q \tag{13.36}$$

The air-gap torque required for the solution of the swing equation (13.22) is

$$\begin{aligned} T_e &= \psi_d i_q - \psi_q i_d \\ &= \psi_{ad} i_q - \psi_{aq} i_d \end{aligned} \tag{13.37A}$$

Since we have assumed $\bar{\omega} = \omega / \omega_0 = 1.0$ pu in the stator voltage equations, in per unit the air-gap torque is equal to the air-gap power (see Chapter 5, Section 5.1.2). Hence,

$$T_e = P_e = P_t + R_a I_t^2 \tag{13.37B}$$

The field current in the reciprocal per unit system is given by

$$i_{fd} = \frac{\psi_{fd} - \psi_{ad}}{L_{fd}} \tag{13.38}$$

The per unit exciter output current I_{fd} (see Chapter 8, Section 8.6) is

$$I_{fd} = L_{adu} i_{fd} \tag{13.39}$$

Here we have considered a generator model with one d-axis and two q-axis amortisseur circuits. For models with a different number of rotor circuits, changes to the above formulation of machine equations are straightforward. However, we have assumed equal mutual inductances between the armature and rotor circuits in each axis. Reference 8 provides a description of the implementation of a model with unequal mutual inductances for stability analysis, using the above general approach.

Initial values of generator variables:

As noted earlier, the transient stability analysis involves the solution of a large set of differential and algebraic equations with known initial values. Prefault power-flow analysis provides the initial values of network variables, including the active and reactive power outputs and voltages at the generator terminals.

The procedure for computing the initial values of the generator quantities for known terminal conditions is presented in Section 3.6.5 of Chapter 3.

Classical generator model and infinite bus

For a machine represented by the classical model, $X_{RI}=X_{IR}=X'_d$, and E'' is replaced by E'. The rotor angle δ is the angle by which E' leads the R-axis. The magnitude of E' is constant throughout the solution. The R and I components of E' change with δ, as determined by the solution of the swing equation.

For a node associated with an infinite bus, both the magnitude and angle of the node voltage remain constant.

13.3.3 Excitation System Representation

Models for different types of excitation systems in use are described in Chapter 8. We will illustrate the method of incorporating these models into a transient stability program by considering the excitation system model shown in Figure 13.12. It represents a bus-fed thyristor excitation system (classified as type ST1A in Chapter 8) with an *automatic voltage regulator* (AVR) and a *power system stabilizer* (PSS). A high exciter gain (K_A) without transient gain reduction or derivative feedback is used. The model of Figure 13.12 is identical to the excitation system model considered in Chapter 12 (see Figure 12.14), except that we also account for the limits on exciter and PSS outputs.

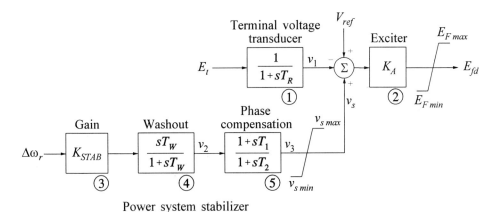

Figure 13.12 Thyristor excitation system with AVR and PSS

For a bus-fed (potential source) thyristor exciter, the voltages vary with the generator terminal voltage (E_t) and exciter output current (I_{fd}):

$$E_{Fmax} = V_{Rmax}E_t - K_c I_{fd}$$
$$E_{Fmin} = V_{Rmin}E_t \tag{13.40}$$

From block 1 of Figure 13.12, we may write

$$pv_1 = \frac{1}{T_R}(E_t - v_1) \tag{13.41}$$

From blocks 3 and 4,

$$pv_2 = K_{STAB}p\Delta\omega_r - \frac{1}{T_W}v_2 \tag{13.42}$$

with $p\Delta\omega_r$ given by Equation 13.22. From block 5,

$$pv_3 = \frac{1}{T_2}[T_1 pv_2 + v_2 - v_3] \tag{13.43}$$

with pv_2 given by Equation 13.42. The stabilizer output v_s is

$$v_s = v_3 \tag{13.44A}$$

with

$$v_{smax} \geq v_s \geq v_{smin} \tag{13.44B}$$

From block 2, the exciter output voltage is

$$E_{fd} = K_A[V_{ref} - v_1 + v_s] \tag{13.45A}$$

with

$$E_{Fmax} \geq E_{fd} \geq E_{Fmin} \tag{13.45B}$$

Both limits considered here are *windup limits*. Modelling of *non-windup limits* is discussed in Chapter 8.

The generator field voltage e_{fd} in the reciprocal per unit system is related to exciter output voltage E_{fd} (see Chapter 8, Section 8.6) as follows:

$$e_{fd} = \frac{R_{fd}}{L_{adu}} E_{fd} \qquad (13.46)$$

An alternative method of treating blocks 4 and 5, in which the derivatives of the input variables are not required, is described in Chapter 12, Section 12.3.4.

Initial values of excitation system variables:

For any given steady-state generator output, the field voltage e_{fd} is determined by the generator equations (see Section 13.3.2). The excitation system quantities are determined as follows:

$$E_{fd} = \frac{L_{adu}}{R_{fd}} e_{fd} \qquad (13.47)$$

$$v_1 = E_t \qquad v_2 = 0 \qquad v_s = 0$$

The AVR reference is

$$V_{ref} = \frac{E_{fd}}{K_A} + v_1 \qquad (13.48)$$

Thus V_{ref} takes a value appropriate to the generator loading condition prior to the disturbance.

Representation of field-shorting circuit [8]

In the case of an ac or static exciter, the exciter current usually cannot be negative. Therefore, a special field-shorting circuit in the form of a "crowbar" or a varistor is provided to bypass the exciter and allow negative field current to flow (see Chapter 8, Section 8.5.8).

The effects of field-shorting circuits are simulated by setting the field voltage to zero and increasing the field resistance (R_{fd}) appropriately during the period when the field current is negative. The resistance added is fixed for a crowbar. For a varistor the resistance is a nonlinear function of the applied voltage. Once the field current becomes positive, the field circuit resistance is restored to its normal value, and the field voltage is established by the exciter output.

13.3.4 Transmission Network and Load Representation

The transients associated with the transmission network decay very rapidly. In fact, network transients will have died out by the time the solution of the swing equation is advanced by one time step. Therefore, it is usually adequate to regard the network, during the electromechanical transient conditions, as though it were passing directly from one steady state to another. The fundamental frequency (50 or 60 Hz) variations may be regarded as microprocesses, and only the variations in the envelopes (amplitude modulation) of current and voltage waveforms are considered for stability analysis. For analysis of balanced conditions, a single-phase representation of the three phases is used. Unbalanced faults are simulated by using symmetrical components as described in Section 13.4.

Without the balanced steady-state representation of the transmission network, stability analysis of large practical power systems would be impractical. For special electromechanical problems requiring inclusion of transmission network and generator stator transients, and three-phase representation, a program such as the electromagnetic transients program (EMTP) may be used [9]. For conventional transient stability analysis, the network representation is similar to that for power-flow analysis. As described in Chapter 6 (Section 6.4.1.), the most convenient form of network representation is in terms of the node admittance matrix. The structure and formulation of this matrix are presented in Section 6.4.1.

The characteristics and modelling of loads are covered in Chapter 7. Dynamic loads are represented as induction or synchronous motors, and their treatment is similar to that of synchronous machines.

Static loads are represented as part of the network equations. Loads with constant impedance characteristics are the simplest to handle and are included in the node admittance matrix. Nonlinear loads are modelled as exponential or polynomial functions of bus voltage magnitude and frequency. The net effect is that a nonlinear static load model is treated as a current injection at the appropriate node in the network equation. The value of the node current from ground into the network is

$$\tilde{I}_L = -\frac{P_L - jQ_L}{\tilde{V}_L^*} \qquad (13.49)$$

where $\tilde{V}_L^*$ is the conjugate of the load bus voltage, and P_L and Q_L are portions of the active and reactive components of the load which vary as nonlinear functions of V_L and frequency deviation. For an inductive load Q_L is positive.

The overall network/load representation comprises a large sparse nodal admittance matrix equation with a structure similar to that of the power-flow problem discussed in Chapter 6, Section 6.4. The network equation is identical to Equation 6.87, which in matrix notation may be written as

$$\tilde{\mathbf{I}} = \mathbf{Y}_N \tilde{\mathbf{V}} \qquad (13.50)$$

The node admittance matrix $\mathbf{Y}_N$ is symmetrical, except for dissymmetry introduced by phase-shifting transformers (see Section 6.2.3). Within the time frame of transient stability simulations, transformer taps and phase-shift angles do not change. Therefore, the elements of the matrix are constant except for changes introduced by network-switching operations.

The effects of generators, nonlinear static loads, dynamic loads, and other devices such as HVDC converters and regulated var compensators are reflected as boundary conditions providing additional relationships between $\tilde{V}$ and $\tilde{I}$ at the respective nodes.

In contrast to power-flow analysis (See Chapter 6), tie line power-flow control, limits on generator reactive power output, and the slack bus make-up for the unknown losses need not be considered in transient stability analysis.

Simulation of faults:

A fault at or near a bus is simulated by appropriately changing the self-admittance of the bus.

Methods of simulating different types of fault will be discussed in Section 13.4.5. Depending on the type of fault, the equivalent negative- and zero-sequence impedances seen at the fault point are computed, combined appropriately, and inserted between the fault point and ground. This simply alters the self-admittance of the node representing the fault bus.

For a three-phase fault, the fault impedance is zero and the faulted bus has the same potential as the ground. This involves placing an infinite shunt admittance. In practice, a sufficiently high shunt admittance (i.e., a very small shunt impedance) is used so that the bus voltage is in effect zero. For example, a conductance G equal to 10^6 pu on 100 MVA base would effectively reduce the bus voltage to zero. The fault is removed by restoring the shunt admittance to the appropriate value depending on the postfault system configuration.

13.3.5 Overall System Equations

In Sections 13.3.2 and 13.3.3 we described the organization of synchronous generator and excitation system equations. Modelling of turbine governing systems is described in Chapter 9. Treatment of these models is similar to that of the excitation system.

Other devices whose dynamics are represented in transient stability studies include motors, HVDC converters and their controls, static var compensators, and synchronous condensers. Models of these devices are described in Chapters 7, 10, and 11.

Equations for each of the generating units and other dynamic devices may be expressed in the following form:

$$\dot{\mathbf{x}}_d = \mathbf{f}_d(\mathbf{x}_d, \mathbf{V}_d) \tag{13.51}$$

$$\mathbf{I}_d = \mathbf{g}_d(\mathbf{x}_d, \mathbf{V}_d) \tag{13.52}$$

where

$\mathbf{x}_d$ = state vector of individual device
$\mathbf{I}_d$ = R and I components of current injection from the device into the network
$\mathbf{V}_d$ = R and I components of bus voltage

The overall system equations, including the differential equations (13.51) for all the devices and the combined algebraic equations for the devices (13.52) and the network (13.50) are expressed in the following general form comprising a set of first-order differential equations

$$\dot{\mathbf{x}} = \mathbf{f}(\mathbf{x}, \mathbf{V}) \tag{13.53}$$

and a set of algebraic equations

$$\mathbf{I}(\mathbf{x}, \mathbf{V}) = \mathbf{Y}_N \mathbf{V} \tag{13.54}$$

with a set of known initial conditions $(\mathbf{x}_0, \mathbf{V}_0)$, where

$\mathbf{x}$ = state vector of the system
$\mathbf{V}$ = bus voltage vector
$\mathbf{I}$ = current injection vector

Time t does not appear explicitly in the above equations. A wide range of approaches has been reported in the literature for solving these equations, depending on the numerical methods and modelling details used. Reference 10 provides a review of these approaches. The many possible schemes for the solution of Equations 13.53 and 13.54 are characterized by the following factors:

(a)   The manner of interface between the differential equations 13.53 and the algebraic equations 13.54. Either a *partitioned* approach or a *simultaneous* approach may be used.

(b)   The integration method used, i.e., implicit method or explicit method.

(c)   The method used for solving the algebraic equations. As in the case of power-flow analysis described in Chapter 6, one of the following methods may be used: (i) the Gauss-Seidel method based on admittance matrix formulation, (ii) a direct solution using sparsity-oriented triangular factorization, and (iii) an iterative solution using the Newton-Raphson method.

All of the above approaches/methods have been successfully used in production-grade transient stability programs. In the next section, two basic solution schemes are described. One scheme uses a partitioned solution and an explicit integration method, and the other uses a simultaneous solution with an implicit integration method.

13.3.6 Solution of Overall System Equations

(a) Partitioned solution with explicit integration

In this approach, the algebraic and differential equations are solved separately. Initially, at $t=0^-$, the values of the state variables $\mathbf{x}$ and the network variables $\mathbf{V}$ and $\mathbf{I}$ are known, and they form a consistent set; that is, the system is in steady state and the time derivatives $\mathbf{f}(\mathbf{x},\mathbf{V})$ are equal to zero.

Following a disturbance, usually a network fault, the state variables $\mathbf{x}$ cannot change instantly. The algebraic equations 13.54 are solved first to give $\mathbf{V}$ and $\mathbf{I}$, and the corresponding power flows and other non-state variables of interest at $t=0^+$. Then the time derivatives $\mathbf{f}(\mathbf{x},\mathbf{V})$ are computed by using the known values of $\mathbf{x}$ and $\mathbf{V}$ in Equations 13.53. These then can be used to initiate the solution of the state variables $\mathbf{x}$ by using any of the explicit integration methods described in Section 13.2.

We will illustrate this by considering Gill's version of the fourth order R-K method. There are four stages ($j=1$ to 4) per time step Δt. Using the values of $\mathbf{f}(\mathbf{x},\mathbf{V})$ computed at the beginning of a time step, $\mathbf{k}_1$, $\mathbf{x}_1$, and $\mathbf{q}_1$ are computed according to Equations 13.17 with $j=1$. Then the algebraic equations 13.54 are solved with $\mathbf{x}=\mathbf{x}_1$ to compute $\mathbf{V}_1$ and $\mathbf{I}_1$. These are in turn used in Equations 13.17 to compute $\mathbf{k}_2$, $\mathbf{x}_2$, and $\mathbf{q}_2$. This process, involving alternating solutions of algebraic and differential equations, is applied successively with $\mathbf{x}_4$ representing the solution at the end of each time step. During a switching operation, the network variables change instantly, but not the state variables.

Since the solution of differential equations requires values of network and state variables only from the previous step/stage, the differential equations may be partitioned in any way desired. For example, differential equations associated with each device may be solved independently. This offers considerable programming flexibility.

For solution of networks associated with large interconnected systems, the most efficient method is to use sparsity-oriented triangular factorization.

The partitioned approach with explicit integration is the traditional approach used widely in production-grade stability programs. Its advantages are programming flexibility and simplicity, reliability, and robustness. Its principal disadvantage is susceptibility to numerical instability. For a stiff system, a small time step is required throughout the solution period, as dictated by the smallest time constant (or eigenvalue).

(b) Simultaneous solution with implicit integration [14]

In this approach, the state variables and the network variables are solved simultaneously. We will illustrate this using the trapezoidal rule.

With $\mathbf{x}=\mathbf{x}_n$ and $\mathbf{V}=\mathbf{V}_n$ at $t=t_n$, the solution of $\mathbf{x}$ at $t=t_{n+1}=t_n+\Delta t$ is given by applying the trapezoidal rule (Equation 13.21) to solve Equation 13.53:

$$\mathbf{x}_{n+1} = \mathbf{x}_n + \frac{\Delta t}{2}[\mathbf{f}(\mathbf{x}_{n+1},\mathbf{V}_{n+1})+\mathbf{f}(\mathbf{x}_n,\mathbf{V}_n)] \tag{13.55}$$

From Equation 13.54, the solution of $\mathbf{V}$ at $t=t_{n+1}$ is

$$\mathbf{I}(\mathbf{x}_{n+1},\mathbf{V}_{n+1}) = \mathbf{Y}_N\mathbf{V}_{n+1} \tag{13.56}$$

The vectors $\mathbf{x}_{n+1}$ and $\mathbf{V}_{n+1}$ are unknown. Let

$$\mathbf{F}(\mathbf{x}_{n+1},\mathbf{V}_{n+1}) = \mathbf{x}_{n+1}-\mathbf{x}_n-\frac{\Delta t}{2}[\mathbf{f}(\mathbf{x}_{n+1},\mathbf{V}_{n+1})+\mathbf{f}(\mathbf{x}_n,\mathbf{V}_n)] \tag{13.57}$$

and

$$\mathbf{G}(\mathbf{x}_{n+1},\mathbf{V}_{n+1}) = \mathbf{Y}_N\mathbf{V}_{n+1}-\mathbf{I}(\mathbf{x}_{n+1},\mathbf{V}_{n+1}) \tag{13.58}$$

At solution,

$$\mathbf{F}(\mathbf{x}_{n+1},\mathbf{V}_{n+1}) = 0 \tag{13.59}$$

$$\mathbf{G}(\mathbf{x}_{n+1},\mathbf{V}_{n+1}) = 0 \tag{13.60}$$

Equations 13.59 and 13.60 are both *nonlinear algebraic* equations. Thus the differential equations have been made algebraic by using an implicit formula. These equations are very sparse; for computational efficiency it is necessary to take advantage of their special structure. Applying the Newton method (see Chapter 6) to solve Equations 13.59 and 13.60, we may write for the $(k+1)^{\text{st}}$ iteration

$$\begin{bmatrix} \mathbf{x}_{n+1}^{k+1} \\ \mathbf{V}_{n+1}^{k+1} \end{bmatrix} = \begin{bmatrix} \mathbf{x}_{n+1}^{k} \\ \mathbf{V}_{n+1}^{k} \end{bmatrix} + \begin{bmatrix} \Delta\mathbf{x}_{n+1}^{k} \\ \Delta\mathbf{V}_{n+1}^{k} \end{bmatrix} \tag{13.61}$$

The following equation is solved to obtain $\Delta\mathbf{x}_{n+1}^{k}$ and $\Delta\mathbf{V}_{n+1}^{k}$:

$$\left[\begin{array}{c} -\mathbf{F}(\mathbf{x}_{n+1}^k, \mathbf{V}_{n+1}^k) \\[2mm] -\mathbf{G}(\mathbf{x}_{n+1}^k, \mathbf{V}_{n+1}^k) \end{array}\right] = \left[\begin{array}{cc} \dfrac{\partial \mathbf{F}}{\partial \mathbf{x}} & \dfrac{\partial \mathbf{F}}{\partial \mathbf{V}} \\[3mm] \dfrac{\partial \mathbf{G}}{\partial \mathbf{x}} & \dfrac{\partial \mathbf{G}}{\partial \mathbf{V}} \end{array}\right] \left[\begin{array}{c} \Delta \mathbf{x}_{n+1}^k \\[2mm] \Delta \mathbf{V}_{n+1}^k \end{array}\right] \tag{13.62}$$

The Jacobian in the above equation is computed at $\mathbf{x} = \mathbf{x}_{n+1}^k$ and $\mathbf{V} = \mathbf{V}_{n+1}^k$. It has the following structure:

$$\mathbf{J} = \left[\begin{array}{cc} \dfrac{\partial \mathbf{F}}{\partial \mathbf{x}} & \dfrac{\partial \mathbf{F}}{\partial \mathbf{V}} \\[3mm] \dfrac{\partial \mathbf{G}}{\partial \mathbf{x}} & \dfrac{\partial \mathbf{G}}{\partial \mathbf{V}} \end{array}\right] = \left[\begin{array}{cc} \mathbf{A_D} & \mathbf{B_D} \\[2mm] \mathbf{C_D} & (\mathbf{Y_N} + \mathbf{Y_D}) \end{array}\right] \tag{13.63}$$

The matrices $\mathbf{A_D}$, $\mathbf{B_D}$, $\mathbf{C_D}$ and $\mathbf{Y_D}$ are associated with the models for the dynamic devices and nonlinear static loads. For a system with m such devices, they have the following structures:

$$\mathbf{A_D} = \left[\begin{array}{cccc} \mathbf{A_{d1}} & 0 & \cdots & 0 \\ 0 & \mathbf{A_{d2}} & \cdots & 0 \\ \vdots & \vdots & \ddots & \vdots \\ 0 & 0 & \cdots & \mathbf{A_{dm}} \end{array}\right] \qquad \mathbf{B_D} = \left[\begin{array}{c} \mathbf{B_{d1}} \\ \mathbf{B_{d2}} \\ \vdots \\ \mathbf{B_{dm}} \end{array}\right] \qquad \mathbf{Y_D} = \left[\begin{array}{cccc} \mathbf{Y_{d1}} & 0 & \cdots & 0 \\ 0 & \mathbf{Y_{d2}} & \cdots & 0 \\ \vdots & \vdots & \ddots & \vdots \\ 0 & 0 & \cdots & \mathbf{Y_{dm}} \end{array}\right]$$

$$\mathbf{C_D} = \left[\begin{array}{cccc} \mathbf{C_{d1}} & \mathbf{C_{d2}} & \cdots & \mathbf{C_{dm}} \end{array}\right]$$

The solutions of Equations 13.59 and 13.60 are given in terms of the above matrices by

$$\mathbf{A_D} \Delta \mathbf{x}_{n+1}^k + \mathbf{B_D} \Delta \mathbf{V}_{n+1}^k = -\mathbf{F}(\mathbf{x}_{n+1}^k, \mathbf{V}_{n+1}^k) \triangleq -\mathbf{F}_{n+1}^k \tag{13.64}$$

$$\mathbf{C_D} \Delta \mathbf{x}_{n+1}^k + (\mathbf{Y_N} + \mathbf{Y_D}) \Delta \mathbf{V}_{n+1}^k = -\mathbf{G}(\mathbf{x}_{n+1}^k, \mathbf{V}_{n+1}^k) \triangleq -\mathbf{G}_{n+1}^k \tag{13.65}$$

In the above equations, k is the iteration counter and good starting values $(\mathbf{x}_{n+1}^0, \mathbf{V}_{n+1}^0)$ are established by extrapolation. Also $\mathbf{F}_{n+1}^k$ and $\mathbf{G}_{n+1}^k$ are the residue vectors of the states and current injections, respectively.

From Equation 13.64, $\Delta \mathbf{x}_{n+1}^k$ can be expressed as a function of $\Delta \mathbf{V}_{n+1}^k$:

$$\Delta \mathbf{x}_{n+1}^k = -\mathbf{A_D}^{-1} [\mathbf{F}_{n+1}^k + \mathbf{B_D} \Delta \mathbf{V}_{n+1}^k] \tag{13.66}$$

Substitution of Equation 13.66 into Equation 13.65 yields

$$(\mathbf{Y}_N + \mathbf{Y}_D - \mathbf{C}_D \mathbf{A}_D^{-1} \mathbf{B}_D) \Delta \mathbf{V}_{n+1}^k = -\mathbf{G}_{n+1}^k + \mathbf{C}_D \mathbf{A}_D^{-1} \mathbf{F}_{n+1}^k \qquad (13.67)$$

Now $\Delta \mathbf{V}_{n+1}^k$ and $\Delta \mathbf{x}_{n+1}^k$ can be calculated by solving Equations 13.66 and 13.67. Then $\mathbf{x}_{n+1}^{k+1}$ and $\mathbf{V}_{n+1}^{k+1}$ are obtained from Equation 13.61.

Treatment of discontinuities:

Equations 13.64 and 13.65 are valid only when the functions given by Equations 13.57 and 13.58 are continuous and differentiable. At points of discontinuity, such as network switching or limits on state variables, the exact formulation by the Newton method or any method requiring derivatives would be complicated [13]. This problem is dealt with in reference 14 as follows:

- For large network discontinuities, such as a network fault or switching operations, the integration method is temporarily changed to the fourth order Runge-Kutta method for one step at the point of discontinuity. This time step has a zero step size and is used only for the calculation of the postfault network conditions (the state vector is not updated). After this, the normal trapezoidal integration is resumed.

- For local non-differentiable functions, such as limits associated with controllers, the device Jacobians are computed by neglecting their effect. This is acceptable since they have only local impact and the overall convergence will not be significantly affected.

Example 13.2

In this example, we analyze the transient stability of the system of Figure E13.1 (considered in Example 13.1) including the effects of rotor circuit dynamics and excitation control. The system diagram is reproduced here as Figure E13.6 for reference.

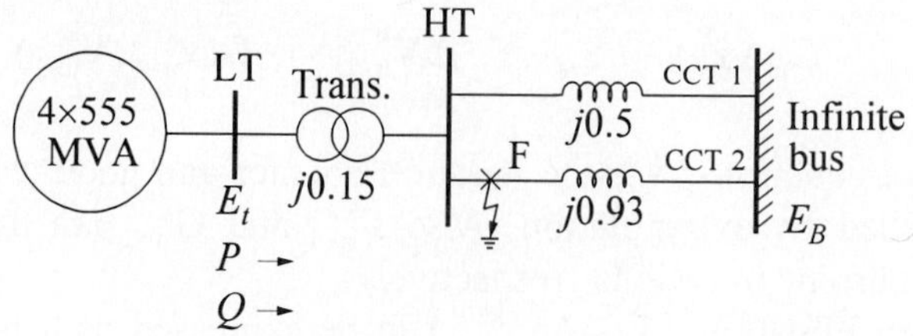

Network reactances in pu on 2220 MVA base

Figure E13.6

Generator parameters:

The four generators of the plant are represented by an equivalent generator whose parameters in per unit on 2220 MVA base are as follows:

$X_d = 1.81$	$X_q = 1.76$	$X'_d = 0.30$	$X'_q = 0.65$
$X''_d = 0.23$	$X''_q = 0.25$	$X_l = 0.15$	$R_a = 0.003$
$T'_{d0} = 8.0$ s	$T'_{q0} = 1.0$ s	$T''_{d0} = 0.03$ s	$T''_{q0} = 0.07$ s
$H = 3.5$	$K_D = 0$		

The above parameters are unsaturated values. The effect of saturation is to be represented by assuming that *d*- and *q*-axes have similar saturation characteristics with

$A_{sat} = 0.031$	$B_{sat} = 6.93$	$\psi_{T1} = 0.8$	$\psi_{T2} = \infty$

Excitation system parameters:

The generators are equipped with thyristor exciters with AVR and PSS as shown in Figure 13.12, with parameters as follows:

$K_A = 200$	$T_R = 0.015$ s	$E_{F max} = 7.0$	$E_{F min} = -6.4$
$K_{STAB} = 9.5$	$T_W = 1.41$ s	$T_1 = 0.154$ s	$T_2 = 0.033$ s
$v_{s max} = 0.2$	$v_{s min} = -0.2$		

The exciter is assumed to be alternator supplied; therefore, $E_{F max}$ and $E_{F min}$ are independent of E_t.

Prefault system condition in pu on 2220 MVA, 24 kV base:

$P = 0.9$	$Q = 0.436$ (overexcited)
$\tilde{E}_t = 1.0\angle 28.34°$	$\tilde{E}_B = 0.90081\angle 0$

Disturbance:

A three-phase fault on circuit 2 at point F cleared by isolating the faulted circuit simultaneously at both ends.

Examine the stability of the system with the following alternative forms of excitation control:

(i)   Manual control, i.e., constant E_{fd}

(ii)   AVR with no PSS

(iii)   AVR with PSS

Consider the following alternative fault-clearing times:

(a)   0.07 s

(b)   0.10 s

Solution

The generators of this example have the same characteristics as the generator considered in examples of Chapters 3 and 4.

The per unit fundamental parameters as determined in Example 4.1 are as follows:

$L_{adu} = 1.66$	$L_{aqu} = 1.61$	$L_l = 0.15$	$R_a = 0.003$
$L_{fd} = 0.165$	$R_{fd} = 0.0006$	$L_{1d} = 0.1713$	$R_{1d} = 0.0284$
$L_{1q} = 0.7252$	$R_{1q} = 0.0062$	$L_{2q} = 0.125$	$R_{2q} = 0.0237$

The initial per unit values of generator variables based on the results of Examples 3.2 and 3.3 are as follows:

$K_{sd} = K_{sq} = 0.835$		$\delta_i = 39.1°$	
$e_d = 0.631$	$e_q = 0.776$	$i_d = 0.906$	$i_q = 0.423$
$i_{fd} = 1.565$	$e_{fd} = 0.000939$		

The initial per unit values of exciter output current and voltages are

$$I_{fd} = L_{adu} i_{fd} = 1.66 \times 1.565 = 2.598$$

$$E_{fd} = (L_{adu}/R_{fd})e_{fd} = (1.66/0.0006) \times 0.000939 = 2.598$$

With E_B assumed to be in phase with the R-axis, the initial value of the rotor angle is

$$\delta = \delta_i + 28.34° = 39.1° + 28.34° = 67.44°$$

(a) Transient response with the fault-clearing time t_c equal to 0.07 s:

The results of time responses computed with the three alternative excitation controls are shown in Figures E13.7(a) to (d), which show time responses of δ, P_e, E_t and E_{fd}, respectively. These results were computed by using Gill's version of the R-K integration method with a time step of 0.02 s.

From Figure E13.7(a), we see that with constant E_{fd}, the system is transiently stable; however, the level of damping of oscillations is low.

With a fast-acting AVR and a high exciter ceiling voltage, the first rotor angle swing is significantly reduced. However, the subsequent swings are negatively damped; the system loses synchronism in the third swing.

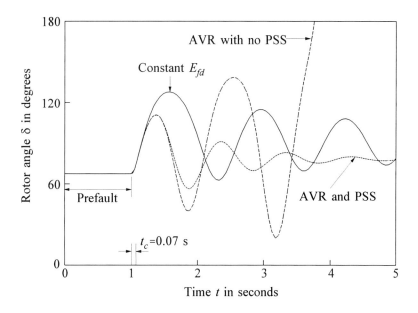

Figure E13.7 (a) Rotor angle response with fault cleared in 0.07 s

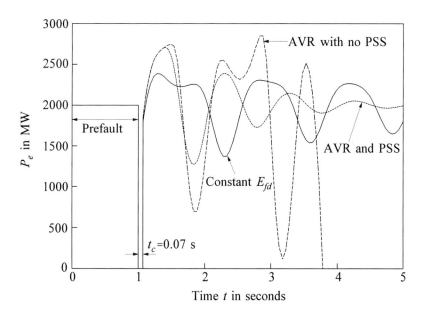

Figure E13.7 (b) Active power response with fault cleared in 0.07 s

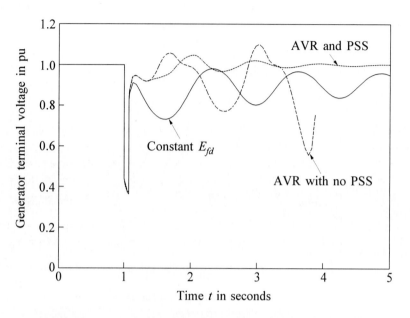

Figure E13.7 (c) Terminal voltage response with fault cleared in 0.07 s

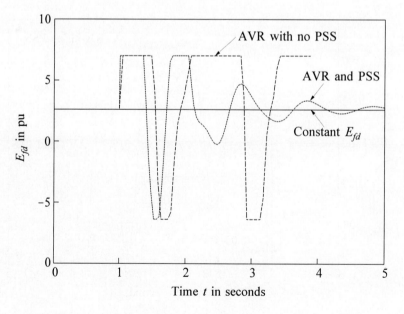

Figure E13.7 (d) Exciter output voltage response with fault cleared in 0.07 s

With the PSS, the rotor oscillations are very well damped without compromising the first-swing stability.

The results with regard to oscillatory stability (i.e., damping of rotor oscillations) are consistent with those of Examples 12.4 and 12.5 of Chapter 12.

(b) Transient response with the fault-clearing time t_c equal to 0.1 s:

The transient responses of rotor angle δ with the three alternative forms of excitation control are shown in Figure E13.8. With constant E_{fd}, the generator is first-swing unstable. With a fast-acting exciter and AVR, the generator maintains first-swing stability but loses synchronism during the second swing. The addition of a PSS contributes to the damping of second and subsequent swings.

It is evident from these results that the use of a fast exciter having a high-ceiling voltage and equipped with a PSS contributes to the enhancement of the overall system stability.

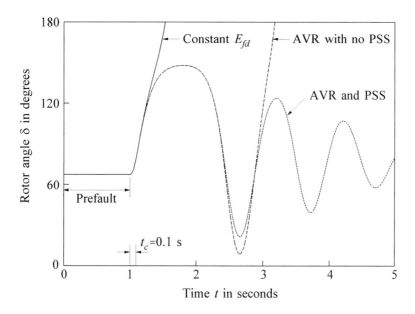

Figure E13.8 Rotor angle response with fault cleared in 0.1 s   ■

Example 13.3

For the system considered in Example 13.2, examine the accuracy and numerical stability of Gill's version of the fourth order R-K method and the trapezoidal rule. Consider the case with AVR and PSS, and the fault cleared in 0.07 s.

Solution

The system considered here includes the effects of rotor circuit dynamics and a fast exciter with a PSS. This represents a very stiff system as is evident from the eigenvalues computed in Example 12.5 of Chapter 12 (see Table E12.2).

R-K method:

Figure E13.9 shows plots of rotor angle transient responses computed by using four different values of time step (Δt): 0.01 s, 0.03 s, 0.04 s, and 0.043 s.

With time steps of 0.01 s and 0.03 s, the results are practically identical. When the time step is increased to 0.04 s, errors become noticeable. With a step size of 0.043 s, the solution is numerically unstable.

In Table E13.1, the accuracy of results obtained with the R-K method is checked by applying Richardson's formula (see Equation 13.18) to the results obtained with step sizes of 0.01 s and 0.02 s. The magnitude of the error is seen never to exceed 0.05°. The table indicates the manner in which the error propagates. It is evident that the accumulated error oscillates, becoming alternately positive and negative.

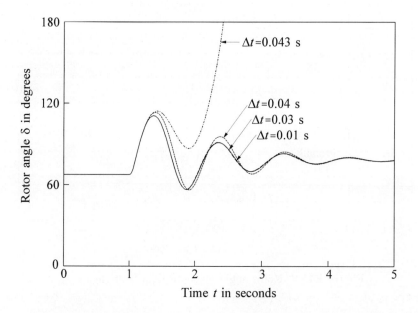

Figure E13.9 Effect of integration step size on the results of R-K method

Table E13.1 Accuracy of results obtained with R-K method

Time (seconds)	Rotor Angle (degrees)		Accumulated Error (degrees)
	$\Delta t = 0.01$ s	$\Delta t = 0.02$ s	
1.4	110.4577	110.4607	−0.0002
1.6	87.2801	87.1575	+0.0082
1.8	58.7696	58.6709	+0.0066
2.0	63.7337	64.1752	−0.0294
2.2	84.8016	85.4362	−0.0423
2.4	90.5653	90.5292	+0.0024
2.6	79.5231	78.9659	+0.0371
2.8	70.0900	69.9465	+0.0096
3.0	73.5296	73.9370	−0.0272
3.2	81.1694	81.4342	−0.0176
3.4	82.5907	82.3862	+0.0136
3.6	78.2763	77.9735	+0.0202
3.8	75.3601	75.3947	−0.0023
4.0	76.9479	77.1938	−0.0164
4.2	79.5847	79.6669	−0.0055
4.4	79.7843	79.6378	+0.0098
4.6	78.1531	78.0250	+0.0085
4.8	77.2835	77.3357	−0.0035
5.0	77.9709	78.0883	−0.0078

$$\text{Accumulated error} = \frac{1}{15}[\delta(\Delta t = 0.01) - \delta(\Delta t = 0.02)]$$

Trapezoidal rule:

The results obtained with step sizes of 0.01 s, 0.059 s, 0.09 s, and 0.093 s are shown in Figure E13.10.

With step sizes of 0.01 s and 0.059 s, the results are practically identical. When the step size is increased to 0.09 s, significant errors result; however, the general characteristics of the overall response are still retained. With a step size of 0.093 s, the errors become so large that the solution blows up after about 1.7 s.

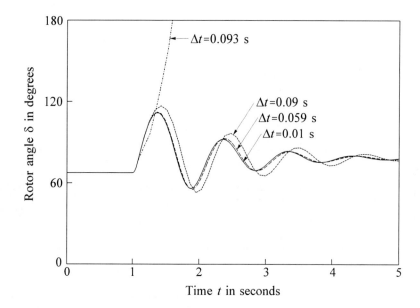

Figure E13.10 Effect of integration step size on the results of trapezoidal rule

◼

13.4 ANALYSIS OF UNBALANCED FAULTS

So far, we have considered only balanced operation of the power system; this allowed for analysis of system performance by solving for the variables in only one of the phases in a manner similar to single-phase circuits. The solution of unbalanced three-phase circuits does not permit this simplification. A three-phase representation can be used instead, but it complicates the problem enormously. The method of symmetrical components is preferred because it offers considerable simplicity in the analysis of power system stability problems involving unbalanced faults.

13.4.1 Introduction to Symmetrical Components [15-18]

The general concept of symmetrical components was first developed by C.L. Fortescue in 1918 [19]. He showed that an unbalanced system of n related phasors can be resolved into n systems of balanced phasors called symmetrical components of the original phasors.

In a three-phase system, a set of three unbalanced phasors can be resolved into three balanced systems of phasors as follows:

(a)   *Positive-sequence components* consisting of a balanced system of three phasors having the same phase sequence as the original phasors.

(b)   *Negative-sequence components* consisting of a balanced system of three phasors having a phase sequence opposite to that of the original phasors.

(c)   *Zero-sequence components* consisting of three phasors equal in magnitude and phase.

Figure 13.13 shows the three sets of balanced phasors that form the symmetrical components of three unbalanced phasors.

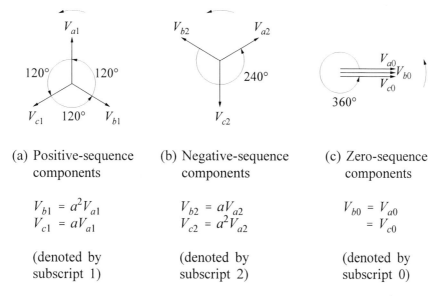

(a) Positive-sequence   (b) Negative-sequence   (c) Zero-sequence
    components   components   components

$$V_{b1} = a^2 V_{a1}$$ $$V_{b2} = a V_{a2}$$ $$V_{b0} = V_{a0}$$
$$V_{c1} = a V_{a1}$$ $$V_{c2} = a^2 V_{a2}$$ $$= V_{c0}$$

(denoted by   (denoted by   (denoted by
subscript 1)   subscript 2)   subscript 0)

Figure 13.13 Symmetrical components of three balanced phasors

Definition of operator a

The operator a denotes 120° phase shift. It is equal to the unit vector $e^{j120°}$. Recalling that $e^{j\theta} = \cos\theta + j\sin\theta$, the unit vector representing a is given by

$$a = e^{j120°} = \cos120° + j\sin120°$$

$$= -0.5 + j0.866$$

The operators a^2 and a^3 represent phase shifts of 240° and 360°, respectively. Thus,

$$a^2 = e^{j240°} = -0.5 - j0.866$$

and

$$a^3 = e^{j360°} = 1.0 + j0.0$$

Relationship between the original unbalanced phasors and their symmetrical components

The total voltage of any phase is equal to the sum of the corresponding components of the different sequences in that phase. Thus,

$$
\begin{aligned}
V_a &= V_{a1} + V_{a2} + V_{a0} \\
V_b &= V_{b1} + V_{b2} + V_{b0} \\
&= a^2 V_{a1} + a V_{a2} + V_{a0} \\
V_c &= V_{c1} + V_{c2} + V_{c0} \\
&= a V_{a1} + a^2 V_{a2} + V_{a0}
\end{aligned}
\tag{13.68}
$$

In matrix form,

$$
\begin{bmatrix} V_a \\ V_b \\ V_c \end{bmatrix} =
\begin{bmatrix} 1 & 1 & 1 \\ 1 & a^2 & a \\ 1 & a & a^2 \end{bmatrix}
\begin{bmatrix} V_{a0} \\ V_{a1} \\ V_{a2} \end{bmatrix}
\tag{13.69}
$$

or

$$\mathbf{V}_{abc} = \mathbf{T}\,\mathbf{V}_{012} \tag{13.70}$$

Figure 13.14 depicts graphically the synthesis of a set of three unbalanced phasors in accord with the above equations.

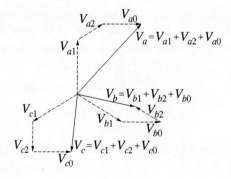

Figure 13.14 Unbalanced set of phasors reconstituted from sequence components

The inverse transformation is given by

$$
\begin{bmatrix} V_{a0} \\ V_{a1} \\ V_{a2} \end{bmatrix} = \frac{1}{3} \begin{bmatrix} 1 & 1 & 1 \\ 1 & a & a^2 \\ 1 & a^2 & a \end{bmatrix} \begin{bmatrix} V_a \\ V_b \\ V_c \end{bmatrix}
\tag{13.71}
$$

Similar transformations apply to currents. The sum of the three line currents is equal to the neutral current. Thus,

$$
I_a + I_b + I_c = I_n
\tag{13.72}
$$

Hence,

$$
I_n = 3I_{a0}
\tag{13.73}
$$

Effect of transformation on impedances

The impedances of system elements satisfy the equation

$$
\mathbf{V}_{abc} = \mathbf{Z}_{abc} \mathbf{I}_{abc}
\tag{13.74}
$$

where $\mathbf{Z}_{abc}$ is the impedance matrix giving self- and mutual impedances in and between phases.

By using the transformation of Equation 13.70, Equation 13.74 can be expressed in terms of the sequence components of voltages and currents:

$$
\mathbf{T}\mathbf{V}_{012} = \mathbf{Z}_{abc}\mathbf{T}\mathbf{I}_{012}
\tag{13.75}
$$

or

$$
\begin{aligned}
\mathbf{V}_{012} &= \mathbf{T}^{-1}\mathbf{Z}_{abc}\mathbf{T}\mathbf{I}_{012} \\
&= \mathbf{Z}_{012}\mathbf{I}_{012}
\end{aligned}
\tag{13.76}
$$

For typical power systems $\mathbf{Z}_{abc}$ is not diagonal but does possess certain *symmetries*. These symmetries are such that the sequence impedance matrix $\mathbf{Z}_{012}$ is *diagonal*, either exactly or approximately. Sequence impedances of different types of system elements will be discussed in Section 13.4.2.

Expression for power in terms of sequence components

Complex three-phase power is given by

$$S_{3\,phase} = V_a I_a^* + V_b I_b^* + V_c I_c^*$$

$$= \mathbf{V}_{abc}^T \mathbf{I}_{abc}^*$$

$$= \left(\mathbf{T}\mathbf{V}_{012}\right)^T \left(\mathbf{T}\mathbf{I}_{012}\right)^* \qquad (13.77)$$

$$= \mathbf{V}_{012}^T \mathbf{T}^T \mathbf{T}^* \mathbf{I}_{012}^*$$

$$= 3\mathbf{V}_{012}^T \mathbf{I}_{012}^*$$

$$= 3\left(V_{a0} I_{a0}^* + V_{a1} I_{a1}^* + V_{a2} I_{a2}^*\right)$$

In the above, the currents are line values and the voltages are line-to-neutral values. It is interesting to note that there are no "cross" terms or components caused by interaction between voltages of one sequence and currents of another sequence. The factor 3 is reasonable when we think in terms of 9 components. This also means that the transformation is not power invariant.

Why symmetrical components?

The resolution of three unbalanced phasors into symmetrical components results in three balanced systems. In symmetrical power systems (systems in which the three phases are identical), currents and voltages of different sequences do not react to each other; that is, currents of one sequence induce voltage drops of like sequence only.

Most of the apparatus used in practice is of the symmetrical type. This results in considerable simplification of many types of unbalanced problems, such as those introduced by unbalanced faults or the opening of one phase.

Since in a symmetrical system there is no interaction between quantities of different sequences, currents of any one sequence may be considered to flow in an independent network associated with that sequence only. The single-phase equivalent circuit composed of the impedances to currents of any one sequence is called the *sequence network* for that particular sequence. To analyze the system performance where there is an unbalanced fault, the sequence networks are interconnected at the fault location to represent the interaction between quantities of different sequences due to unbalance created by the fault.

The resolution of the problem into the sequence components has a further advantage in that it *isolates the quantities into components which represent a better criterion of the controlling factors in certain phenomena.* For example, in stability investigations the synchronizing force between machines is affected principally by the positive-sequence quantities, the heating effects due to unbalanced currents depend on negative-sequence currents, and the ground relays and grounding phenomena are very closely associated with zero-sequence components. Thus, the analysis of many problems is not only simpler, but the results are resolved into factors which may be investigated separately.

Sequence impedances

The impedance of a circuit when positive-sequence currents alone are flowing is called the impedance to positive-sequence currents or *positive-sequence impedance*. Similarly, impedance to negative-sequence currents is called *negative-sequence impedance*, and the impedance to zero-sequence currents the *zero-sequence impedance*.

A knowledge of the sequence impedances of the power system elements is essential for the analysis of system performance during unbalanced faults.

13.4.2 Sequence Impedances of Synchronous Machines [15]

Positive-sequence impedance

In Section 13.3.2, we discussed the detailed representation of synchronous machines in the positive-sequence network appropriate for stability studies. Such a model is accurate for the entire study period, including the subtransient, transient, and steady-state conditions.

Figure 13.15 shows the simplified positive-sequence equivalent circuit of a synchronous machine. As discussed in Chapter 5 (Section 5.3.3), the value of the *positive-sequence reactance* X_1 depends on the time frame of interest: subtransient, transient, and steady state.

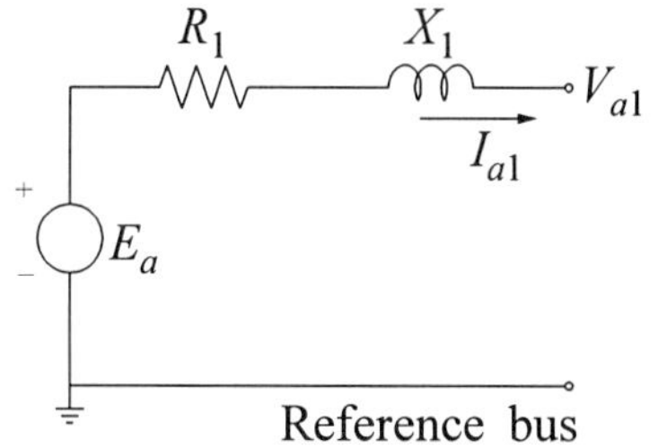

Figure 13.15   Positive-sequence equivalent circuit
of synchronous machine

For the time frame representing

- the initial subtransient condition, $X_1 = X_d''$;

- the initial transient condition, $X_1 = X_d'$;

- the steady-state condition, $X_1 = X_d$.

The above assumes that saliency effects are negligible.

The *positive-sequence resistance* R_1 is the ac resistance of the armature.

The equivalent circuit of Figure 13.15 represents one phase of the balanced positive-sequence circuits. Its reference bus is the neutral of the generator.

Negative-sequence impedance

Negative-sequence reactance (X_2):

When fundamental frequency negative-sequence (reverse phase sequence) currents are applied to the armature windings of a machine, with the rotor running at synchronous speed and the field winding shorted through the exciter, the ratio of the resulting negative-sequence fundamental frequency voltages to the currents gives the negative-sequence impedance. The reactive part of this impedance is the negative-sequence reactance X_2.

In what follows, we will show that the negative-sequence reactance X_2 is equal to the mean between the d- and q-axis subtransient reactances, X_d'' and X_q''. Saliency effects result in somewhat different values of X_2 depending on whether sinusoidal currents are circulated or sinusoidal voltages are applied.

With time t in seconds, the negative-sequence currents may be expressed as

$$i_{a2} = I_{a2}\cos(\omega t + \phi)$$

$$i_{b2} = I_{a2}\cos\left(\omega t + \phi + \frac{2\pi}{3}\right)$$

$$i_{c2} = I_{a2}\cos\left(\omega t + \phi - \frac{2\pi}{3}\right)$$

Transforming into d, q, 0 components (see Chapter 3, Section 3.3), we have

$$i_d = I_{a2}\cos(2\omega t + \phi)$$

$$i_q = -I_{a2}\sin(2\omega t + \phi)$$

$$i_0 = 0$$

The resulting field due to the negative-sequence stator currents rotates at synchronous speed opposite to the direction of the rotor. Therefore, as viewed from the rotor, the stator currents appear to be double frequency currents i_d and i_q. Currents of twice the rated frequency are induced in the rotor circuits, keeping the flux linkages of the rotor circuits almost constant (since resistance is considerably smaller than inductance). The flux due to the stator currents is forced into paths of low permeance which do not link - with any rotor circuit. These flux paths are the same as those encountered in evaluating subtransient inductances. Therefore, the d- and q-axis components of stator currents at 120 Hz encounter subtransient inductances L_d'' and L_q'' (see Chapter 4, Section 4.2). The corresponding per unit stator flux linkages are

$$\psi_d = -L_d'' i_d = -X_d'' i_d$$

$$\psi_q = -L_q'' i_q = -X_q'' i_q$$

Neglecting the stator resistance, with time t in seconds and ω_r equal to 1.0 pu, the per unit stator voltages are given by the following equations:

$$e_d = \frac{1}{\omega_0}\frac{d\psi_d}{dt} - \psi_q$$

$$= \frac{1}{\omega_0}\frac{d}{dt}\left[-X_d'' I_{a2}\cos(2\omega t+\phi)\right] - X_q'' I_{a2}\sin(2\omega t+\phi)$$

$$= 2\frac{\omega}{\omega_0} X_d'' I_{a2}\sin(2\omega t+\phi) - X_q'' I_{a2}\sin(2\omega t+\phi)$$

For fundamental frequency negative-sequence currents, $\omega=\omega_0$. Hence,

$$e_d = (2X_d'' - X_q'')I_{a2}\sin(2\omega_0 t+\phi)$$

Similarly,

$$e_q = (2X_q'' - X_d'')I_{a2}\cos(2\omega_0 t+\phi)$$

and

$$e_0 = 0$$

The corresponding phase voltages (negative sequence) may be determined by applying the inverse Park's transformation (see Chapter 3, Section 3.3):

$$e_a = e_d\cos\theta + e_q(-\sin\theta) + e_0$$

Denoting the negative-sequence stator voltage by e_{a2} and noting that $\theta = \omega t$ at synchronous speed, we have

$$e_{a2} = e_d\cos(\omega t) - e_q\sin(\omega t) + 0$$

$$= \underbrace{\frac{X_d'' + X_q''}{2} I_{a2}\sin(\omega t + \phi)}_{\text{Fundamental frequency}} + \underbrace{\frac{3}{2}(X_d'' - X_q'')I_{a2}\sin(3\omega t + \phi)}_{\text{Third harmonic}}$$

Similarly,

$$e_{b2} = \frac{X_d'' + X_q''}{2} I_{a2}\sin(\omega t + 120° + \phi) + \frac{3}{2}(X_d'' - X_q'')I_{a2}\sin(3\omega t + \phi)$$

$$e_{c2} = \frac{X_d'' + X_q''}{2} I_{a2}\sin(\omega t - 120° + \phi) + \frac{3}{2}(X_d'' - X_q'')I_{a2}\sin(3\omega t + \phi)$$

We see that the voltages have a *third harmonic* negative-sequence component which is a function of subtransient saliency $(X_d'' - X_q'')$. This component is small; unless the system resonates near this harmonic, it may be neglected. As far as the fundamental frequency negative-sequence currents are concerned, the reactance is equal to the average of the subtransient reactances:

$$X_2 = \frac{X_d'' + X_q''}{2} \tag{13.78}$$

Since the mmf wave due to negative-sequence stator currents moves at twice the synchronous speed with respect to the rotor, it alternately meets permeances of the two rotor axes, corresponding to X_d'' and X_q''. Hence the value of X_2 lies between X_d'' and X_q''. *With sinusoidal negative-sequence currents applied*, the negative-sequence reactance is the *arithmetic mean* of X_d'' and X_q''. This would apply when the external circuit has a large reactance in series.

If *sinusoidal negative-sequence voltages are applied* to the stator, the resulting currents in the stator windings can be shown to have a fundamental frequency component proportional to $\frac{1}{2}\left(\frac{1}{X_d''} + \frac{1}{X_q''}\right)$ and a third harmonic component proportional to $\dfrac{3}{2}\left(\dfrac{1}{X_d{}''}-\dfrac{1}{X_q{}''}\right)$. If only the fundamental is considered, the effective negative-sequence reactance is

$$X_2 \;=\; 2\frac{X_d{}''\,X_q{}''}{X_d{}''+X_q{}''} \tag{13.79}$$

When $X_d{}''=X_q{}''$, there is no third harmonic component. In this case, the voltages and currents would have only fundamental frequency components, and $X_2=X_d{}''=X_q{}''$.

Negative-sequence resistance (R_2):

The power associated with the negative-sequence current may be expressed as a resistance times the square of the current. This resistance is designated as the negative-sequence resistance R_2.

As far as the negative-sequence stator currents are concerned, the rotor circuits appear as short-circuited windings. The situation is similar to that of an induction machine. Therefore, the physical nature of the negative-sequence resistance is best visualized by analyzing the performance of an induction motor. Figure 13.16 shows the equivalent circuit of an induction machine with symmetry in both axes. For the case with asymmetry in the two axes, the circuit represents the average of the two axes. With negative-sequence stator currents, slip $s=2$. Therefore, the equivalent resistance representing power transfer from rotor to shaft is $R_r(1-s)/s=-R_r/2$.

The total rotor copper loss is represented by R_r. *Half of the rotor loss is supplied from the stator and the other half by the shaft torque (represented by $-R_r/2$) through the rotor by conversion of mechanical power.*

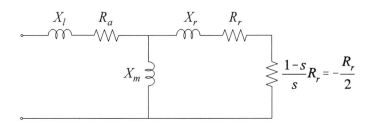

X_m = magnetizing reactance X_l = stator leakage reactance
X_r = rotor leakage reactance R_r = rotor resistance
R_a = stator resistance s = slip

Figure 13.16 Induction motor equivalent circuit

The real part of the equivalent impedance seen from the stator terminals is the negative-sequence resistance R_2. Since X_m is large, it may be neglected. Hence,

$$R_2 \approx R_a + \frac{R_r}{2} \tag{13.80}$$

Figure 13.17 shows the equivalent circuit of a synchronous machine appropriate for use in negative-sequence networks. Since the generated voltages in the machine are of positive sequence only, the negative-sequence equivalent circuit contains no emf, but is represented by the impedance $(R_2 + jX_2)$ offered to the flow of negative-sequence currents. The reference bus of the equivalent circuit is the neutral of the generator.

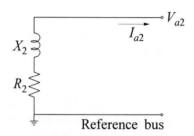

Figure 13.17 Negative-sequence equivalent circuit
of a synchronous machine

Zero-sequence impedance

The machine must be star-connected; otherwise, the term zero-sequence impedance has no significance as no zero-sequence current can flow.

Zero-sequence reactance (X_0):

When zero-sequence currents are applied to the armature, the instantaneous values of currents in the three phases are equal. With the armature windings infinitely distributed so that each phase produces a sinusoidal spatial distribution of the mmf, the net field produced by the three phases with equal instantaneous currents is zero. No flux is produced across the air-gap. The only reactance associated with any phase winding is due to slot leakage flux and end-winding leakage flux. However, in an actual machine, the winding distribution is not perfectly sinusoidal, and the flux due to the three phases will have a small value. Therefore, the actual X_0 is slightly higher than for the ideal case and depends on pitch and breadth factors of the winding.

Zero-sequence resistance (R_0):

The zero-sequence resistance of a synchronous machine is somewhat larger than the positive-sequence resistance for reasons discussed below.

The spatial distribution of stator windings is usually not purely sinusoidal, but consists of higher harmonics. The spatial third harmonic contribution to the mmf due to stator currents $i_a,\ i_b,\ i_c$ is

$$MMF\ =\ K[i_a\cos3\theta+i_b\cos3(\theta-120°)+i_c\cos3(\theta+120°)]$$

When the stator currents form a balanced set, such as the positive- or negative-sequence, the net mmf due to spatial winding distribution harmonics is zero. However, for zero-sequence currents, with $i_a=i_b=i_c=I_m\cos\omega t$ and the rotor travelling at synchronous speed $(\theta=\omega t)$, the mmf due to third harmonic spatial distribution is

$$MMF\ =\ \frac{3}{2}I_m[\cos(3\theta+\omega t)+\cos(3\theta-\omega t)]$$

$$=\ \frac{3}{2}I_m[\cos(4\omega t)+\cos(2\omega t)]$$

Thus the mmf due to third harmonic spatial winding distribution, when zero-sequence currents flow in the stator windings, consists of second and fourth harmonic components. These induce currents in the rotor and cause rotor heating. Therefore, the zero-sequence resistance (R_0) is slightly higher than the positive-sequence resistance (R_1). The difference is usually not significant.

Figure 13.18(a) shows the flow of zero-sequence currents in the three phases and the neutral of the synchronous machine. The corresponding (single-phase) equivalent circuit is shown in Figure 13.18(b).

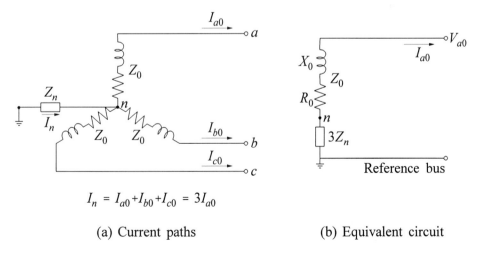

$$I_n\ =\ I_{a0}+I_{b0}+I_{c0}\ =\ 3I_{a0}$$

(a) Current paths   (b) Equivalent circuit

Figure 13.18   Zero-sequence current paths and equivalent circuit of a synchronous machine

The neutral impedance Z_n carries a current equal to the sum of the zero-sequence currents in the three phases. This is reflected as $3Z_n$ in the equivalent circuit of Figure 13.18(b), which represents only one phase. The reference bus for the zero-sequence equivalent circuit is the ground at the generator. If the neutral is isolated from the ground, $Z_n=\infty$ and $I_{ao}=0$.

13.4.3 Sequence Impedances of Transmission Lines

Positive- and negative-sequence impedances

In a symmetrical three-phase static circuit, the positive- and negative-sequence impedances are identical because the impedance of such a circuit is independent of the phase order of the applied voltages. The equivalent circuits of transmission lines developed in Chapter 6 are applicable to the positive-sequence as well as the negative-sequence networks.

Zero-sequence impedance

When only zero-sequence currents flow in a transmission line, the currents in the three phases are identical. The sum of these currents returns through the ground and/or overhead ground wires. Since the three-phase currents are in phase, the magnetic field produced is very different from that produced by positive- or negative-sequence currents. It requires consideration of the current distribution in the earth. The net effect is that the zero-sequence reactance of transmission lines tends to be 2 to 3 times the positive-sequence reactance [15-17].

Zero-sequence currents flowing through ground result in voltage drops due to the impedance of the ground. Evidently, the ground is not necessarily at the same potential at all points. The reference bus of the zero-sequence network does not represent the ground potential at any particular point but is the reference ground for zero-sequence voltages throughout the system. Therefore, the impedances of the ground and ground wires are included in zero-sequence impedances of the transmission lines. This ensures that the zero-sequence voltages at the terminals are correctly given when referred to the reference bus.

13.4.4 Sequence Impedances of Transformers

Positive- and negative-sequence impedances

Transformers, like transmission lines, are non-rotative or static. With the assumption that symmetry exists between different phases, the impedances to balanced three-phase currents do not depend on the phase sequence. Therefore, the equivalent circuits and the associated impedances derived in Chapter 6 for the two-winding and three-winding transformers are equally valid for the positive-sequence and the negative-sequence networks.

Zero-sequence impedances

    The impedance offered by a transformer to the flow of zero-sequence currents depends on the winding connections. Figure 13.19 gives the zero-sequence equivalent circuits for different winding connections applicable to two-winding and three-winding transformers.

    For zero-sequence currents to flow through windings on one side of the transformer and into the connected lines, a return path must exist through which a completed circuit is provided. Additionally, there must be a path for the corresponding - current in the coupled windings on the other side.

    For illustration, let us consider the winding connections for case 1 in Figure 13.19. The Y-connected primary (P) windings with the neutral grounded provide a path to the ground. The Δ-connected secondary (S) windings provide a path for the zero-sequence currents, which circulate within the Δ. This serves to balance the currents in the primary windings; however, the zero-sequence currents circulating within the secondary windings cannot flow in the lines connected to them. The impedance Z_0 represents the total per-phase leakage reactances and resistances of the primary and secondary windings and is equal to the positive-sequence impedance. The neutral of the Y-connected primary is grounded through an impedance Z_n; therefore, the effective zero-sequence impedance as seen from the primary side is Z_0+3Z_n. Since a Δ-connected circuit provides no path for zero-sequence currents flowing in the line, the impedance as seen from the secondary side is infinite.

    If the windings on one side are Y-connected with neutral ungrounded (as in cases 2 and 5), zero-sequence currents cannot flow in the windings on either side. Viewed from either the primary side or the secondary side, the zero-sequence impedance is infinite.

    Cases 6 to 8 in Figure 13.19 consider three-winding transformers where the effects of the leakage reactances and resistances of the three windings are represented by a star-connected equivalent (see Chapter 6). Connection of this equivalent to the primary (P), secondary (S), and tertiary (T) terminals, and the reference bus depends on the winding connections as shown in the figure.

13.4.5 Simulation of Different Types of Faults

    In this section, we consider the simulation of a power system that is essentially symmetrical but is rendered unbalanced by a fault at a particular location on the system. Such faults can be analyzed by appropriately interconnecting the positive-, negative-, and zero-sequence networks at the fault point as described below.

    Figure 13.20 identifies the fault currents and voltages at fault location F. The *currents flowing into the fault* from the three phases *a*, *b*, and *c* are represented by I_a, I_b, and I_c, respectively; V_a, V_b and V_c represent *voltages to ground* at F.

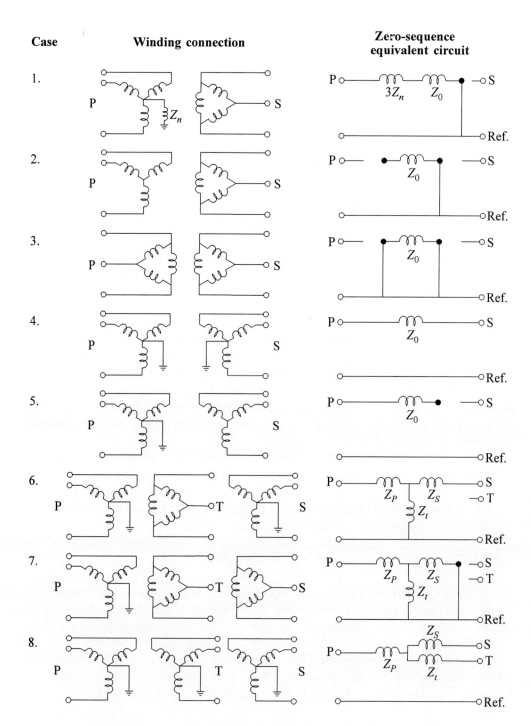

Figure 13.19 Zero-sequence equivalent of three-phase two-
winding and three-winding transformer banks

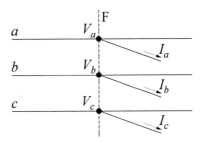

Figure 13.20 Designation of fault currents and voltages

(a) Single line-to-ground fault

Let us assume that phase a is shorted to ground at the fault point F, as shown in Figure 13.21(a). From the figure, we see that

$$V_a = 0 \qquad I_b = I_c = 0$$

In terms of the symmetrical components (see Equations 13.69 and 13.71), we have

$$V_{a1} + V_{a2} + V_{a0} = V_a = 0$$

$$I_{a1} = I_{a2} = I_{a0} = \frac{I_a}{3}$$

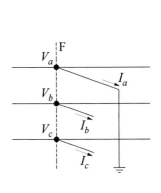

(a) Fault schematic

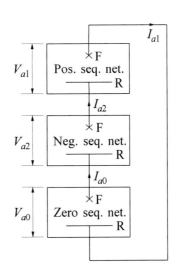

(b) Sequence network connection

Figure 13.21 Simulation of a single line-to-ground fault

The connection of the sequence networks to satisfy the above relationships is shown in Figure 13.21(b). The three sequence networks are connected at the fault point (F) and the reference bus (R) of each sequence network so as to form a series connection.

(b) Line-to-line fault

From Figure 13.22(a),

$$V_b = V_c \qquad I_a = 0 \qquad I_b + I_c = 0$$

In terms of symmetrical components,

$$V_{a1} = V_{a2} \qquad I_{a1} + I_{a2} = 0 \qquad I_{a0} = 0$$

The sequence network connection reflecting the above relationships is shown in Figure 13.22(b). Since the fault does not involve ground, the zero-sequence network is absent.

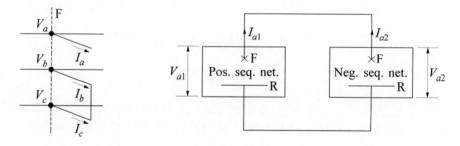

(a) Fault schematic   (b) Sequence network connection

Figure 13.22 Simulation of a line-to-line fault

(c) Double line-to-ground fault

From Figure 13.23(a),

$$V_b = V_c = 0 \qquad I_a = 0$$

In terms of symmetrical components,

$$V_{a1} = V_{a2} = V_{a0}$$
$$I_{a1} + I_{a2} + I_{a0} = 0$$

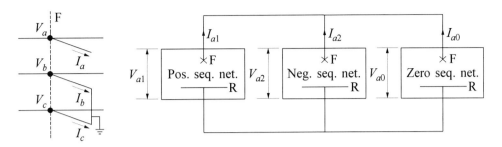

(a) Fault schematic   (b) Sequence network connection

Figure 13.23 Simulation of a double line-to-ground fault

The corresponding sequence network connection is shown in Figure 13.23(b).

(d) Three-phase fault

To complete the picture with regard to simulation of different types of faults in terms of sequence networks, let us examine a symmetrical three-phase fault as shown in Figure 13.24(a).

From Figure 13.24,

$$V_a = V_b = V_c = 0$$

$$I_a + I_b + I_c = 0$$

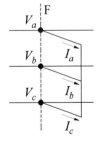 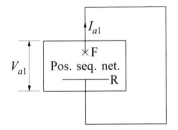

(a) Fault schematic   (b) Sequence network connection

Figure 13.24 Simulation of a three-phase fault

In terms of symmetrical components,

$$I_{a0} = 0$$

$$V_{a1} = V_{a2} = V_{a0} = 0$$

There are no zero-sequence or negative-sequence emf's, and the voltages V_{a0} and V_{a2} at the fault point F are each equal to zero. Therefore, zero- and negative-sequence currents do not flow anywhere in the system. As expected, only the positive-sequence network is involved. The fault is simulated by shorting the fault point F and the reference bus R of the positive-sequence network, as shown in Figure 13.24(b).

Formation of sequence networks

We see from the above analyses that a fault on a power system can be analyzed by appropriately interconnecting the sequence networks of the system. The solution of the resulting network gives the symmetrical components of voltages and currents throughout the system.

The positive-, negative-, and zero-sequence systems represent symmetrical systems. Therefore, each of these systems may be represented by an equivalent single-phase network. Each element of the power system is represented by a model with an appropriate degree of precision.

In each sequence network, the sequence voltages are referred to the reference (or zero potential) bus for the network. In the positive- or negative-sequence systems, all neutral points are at the same potential. Therefore, the neutrals of each of these systems can be connected to a common bus which is the reference or zero-potential bus for its sequence network.

As stated earlier, the ground is not necessarily at the same potential at all points in the system. Therefore, the reference bus for the zero-sequence network is not representative of the ground at any particular point; instead, the reference bus serves as a reference for zero-sequence voltages at all points of the system. The individual equipment models are so constructed that the zero-sequence voltages to ground are correctly represented when referred to the reference bus of the network.

Representation in stability studies

The positive-sequence network is represented in detail with equipment models as described in Chapters 3 to 11.

The negative- and zero-sequence voltages and currents throughout the system are usually not of interest in stability studies. Therefore, it is unnecessary to simulate the complete negative- and zero-sequence networks in system stability simulations. Their effects may be represented by equivalent impedances (Z_2 and Z_0) as viewed at the fault point F. These impedances are combined appropriately depending on the type of fault and inserted as the effective fault impedance Z_{ef} in the positive-sequence network as shown in Figure 13.25.

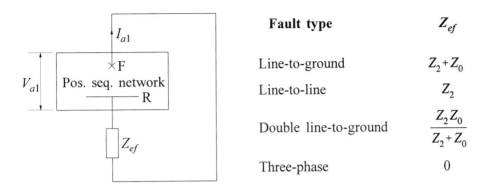

Fault type	Z_{ef}
Line-to-ground	$Z_2 + Z_0$
Line-to-line	Z_2
Double line-to-ground	$\dfrac{Z_2 Z_0}{Z_2 + Z_0}$
Three-phase	0

Figure 13.25 Representation of faults in system stability studies

This representation gives the correct voltages and currents in the positive-sequence network.

There are no negative- or zero-sequence voltages generated in the system. The negative- and zero-sequence currents that flow during unbalanced faults are driven by the positive-sequence voltage sources. The positive-sequence power outputs computed represent the actual power supplied electrically by the machines. This includes all negative- and zero-sequence resistance losses supplied electrically.

Negative-sequence braking torque

As discussed in Section 13.4.2, only half of the negative-sequence rotor losses in synchronous machines are supplied electrically. The other half are supplied mechanically by the shaft power or torque. This represents the component of the total mechanical torque that is converted to supply rotor losses. It acts to decelerate the rotor and is called the negative-sequence braking torque. This effect could be significant for machines close to a fault, particularly those with solid rotors or salient pole rotors with high-resistance damper windings. The method of accounting for this braking torque is described below.

Referring to Figure 13.16, with the magnetizing reactance X_m neglected, the negative-sequence braking torque is

$$T_{b2} = \frac{R_r}{2} I_{a2}^2$$

$$= (R_2 - R_a) I_{a2}^2$$

(13.81)

where R_2 is the negative-sequence resistance of the machine and R_a is its stator resistance.

The machine negative-sequence current I_{a2} can be determined from the negative-sequence network. Depending on the network configuration, I_{a2} can be estimated as a fraction of the machine positive-sequence current [20].

The total accelerating torque acting on the rotor is

$$T_a = T_m - T_e - K_D \Delta \omega_r - T_{b2}$$

The swing equation (Equation 13.22) is appropriately modified to include T_{b2}.

Example 13.4

Figure E13.11 shows the system representation applicable to a 1,000 MVA, 20 kV, 60 Hz generating unit.

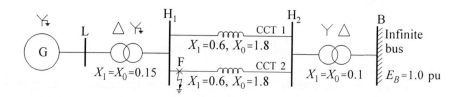

Figure E13.11

The transmission data shown on the figure are in per unit on 1,000 MVA, 20 kV base. Network resistances are assumed to be negligible.

The generator data in per unit on the rating of the unit are as follows:

$X_d = 1.81$	$X_q = 1.76$	$X_d' = 0.3$	$X_q' = 0.65$
$X_d'' = X_q'' = 0.25$	$X_2 = 0.25$	$X_0 = 0.04$	$R_a = 0.003$
$R_2 = 0.063$	$R_0 = 0.005$	$T_{d0}' = 8.0$ s	$T_{q0}' = 1.0$ s
$T_{d0}'' = 0.03$ s	$T_{q0}'' = 0.07$ s	$H = 3.5$	$K_D = 0$

The neutrals of the generator stator and the step-up transformer HT windings are solidly grounded. The neutral of the star-connected winding of the transformer near the infinite bus is ungrounded.

(a)   A double line-to-ground (LLG) fault occurs on circuit 2 at point F as shown in Figure E13.11.

(i) Neglecting all resistances, find the value of the effective fault impedance Z_{ef} which, when inserted in the positive-sequence network, represents the unbalanced fault.

(ii) If the initial generator output conditions are

$$P_0 = 0 \qquad Q_0 = 0 \qquad E_t = 1.0$$

compute the magnitudes of positive-, negative-, and zero-sequence currents throughout the network immediately after the fault occurs, neglecting the effect of generator resistance.

(iii) Compute the negative-sequence braking torque corresponding to the above condition.

(b)   How would the above results change if a line-to-ground fault or a line-to-line fault were to occur at point F?

Solution

For conditions immediately after the fault occurs, the effective positive-sequence machine reactance is $X'' = X''_d = X''_q$.

The three-sequence networks of the system are shown in Figure E13.12. Since $X''_d = X''_q = X_2$, the negative- and positive-sequence reactances are equal for each of the elements in the system.

(a) Double line-to-ground fault

(i) Each sequence network represents a balanced system. The unbalanced effect introduced by the fault at location F is represented by interconnecting the sequence networks as shown in Figure E13.13.

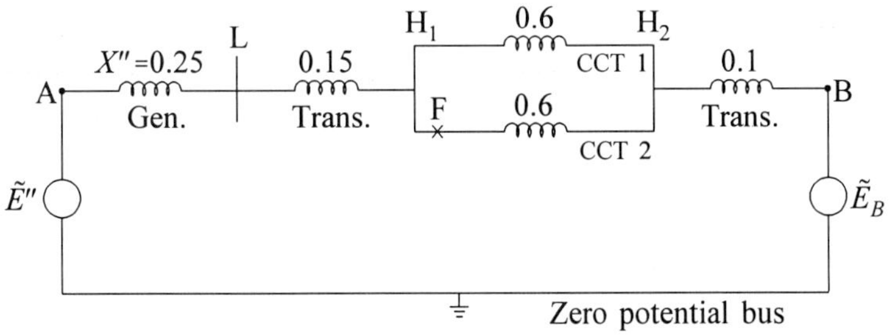

(a) Positive-sequence network

Figure E13.12 (*Continued on the next page*)

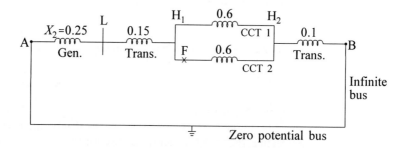

(b) Negative-sequence network

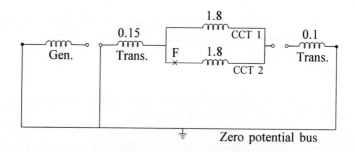

(c) Zero-sequence network

Figure E13.12 (*Continued*)

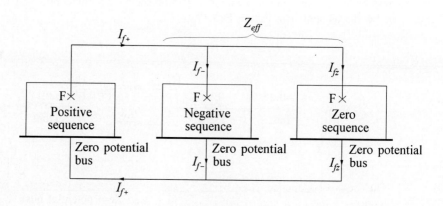

Figure E13.13

The effective negative-sequence and zero-sequence impedances as measured at point F are shown in Figure E13.14.

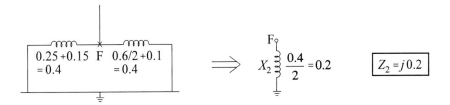

(a) Effective negative-sequence impedance

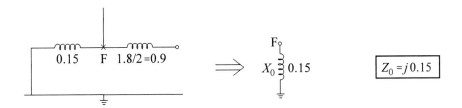

(b) Effective zero-sequence impedance

Figure E13.14

The effective fault impedance representing the L-L-G fault is

$$Z_{ef} = \frac{Z_2 Z_0}{Z_2 + Z_0}$$

$$= \frac{j(0.2 \times 0.15)}{0.35}$$

$$= j0.0857 \quad \text{pu}$$

(ii) Since prior to the fault $P=Q=0$,

$$E'' = E_t = E_B = 1.0 \text{ pu}$$

The positive-sequence network with the effective fault impedance inserted at F reduces to the form shown in Figure E13.15(a).

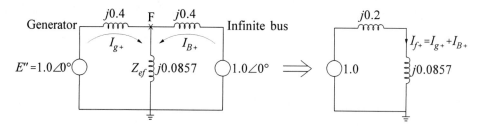

Figure E13.15 (a)

Because of network symmetry the solution is considerably simplified. The positive-sequence component of the fault current is

$$I_{f+} = \frac{1.0}{j0.2857} = -j3.5 \quad \text{pu}$$

The total positive-sequence fault current splits into negative- and zero-sequence components as shown in Figure E13.15(b).

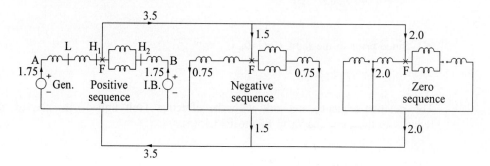

$$I_{f-} = \frac{0.15 \times 3.5}{0.35} = 1.5 \quad \text{pu}$$

$$I_{f0} = I_{f+} - I_{f-} = 3.5 - 1.5 = 2.0 \quad \text{pu}$$

Figure E13.15 (b)

Because of the network symmetry about the point F, one-half of the fault current of each sequence component flows from the generator, and the other half from the infinite bus. Owing to the generator step-up transformer winding connections, the zero-sequence current is restricted to the transformer; no zero-sequence current flows through the generator. Figure E13.16 shows current flows in different parts of the sequence networks.

Figure E13.16

(iii) The negative-sequence braking torque is given by

$$T_{b2} = (R_2 - R_a)I_{g2}^2$$

$$= (0.063 - 0.003)0.75^2 = 0.034 \quad \text{pu}$$

(b) *Line-to-ground fault*

(i) The effective fault impedance is

$$Z_{ef} = Z_2 + Z_0 = j(0.2 + 0.15)$$

$$= j0.35 \quad \text{pu}$$

(ii) Figure E13.17 shows current flows in different parts of the sequence networks.

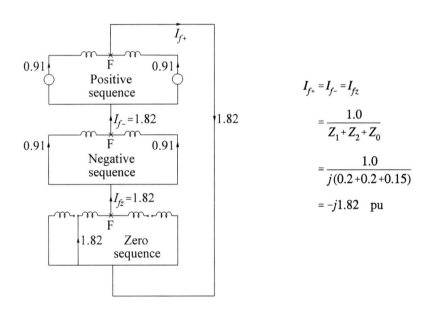

$$I_{f+} = I_{f-} = I_{fz}$$

$$= \frac{1.0}{Z_1 + Z_2 + Z_0}$$

$$= \frac{1.0}{j(0.2 + 0.2 + 0.15)}$$

$$= -j1.82 \quad \text{pu}$$

Figure E13.17

(iii) The negative-sequence braking torque is

$$T_{b2} = (R_2 - R_a)I_2^2$$

$$= (0.063 - 0.003)0.91^2$$

$$= 0.05 \quad \text{pu}$$

Line-to-line fault

(i) The effective fault impedance is

$$Z_{ef} = Z_2 = j0.2$$

(ii) Figure E13.18 shows current flows in the sequence network.

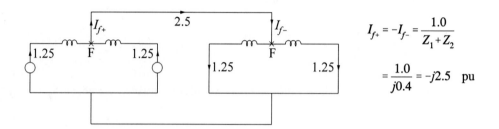

Figure E13.18

(iii) The negative-sequence braking torque is

$$T_{b2} = (R_2 - R_a)I_2^2 = (0.063 - 0.003)1.25^2$$

$$= 0.094 \quad pu \qquad \blacksquare$$

13.4.6 Representation of Open-Conductor Conditions

Open-conductor conditions may be caused by broken conductors or a deliberate single-phase switching operation. Such faults may involve the opening of one phase or two phases of a three-phase circuit.

(a) One open conductor

Figure 13.26(a) shows a section of a three-phase system with phase *a* open between points X and Y. Let v_a, v_b, and v_c be series voltage drops in phases between X and Y in phases *a*, *b*, *c*, respectively; I_a, I_b and I_c are the line currents. From Figure 13.26(a),

$$v_b = v_c = 0 \qquad I_a = 0$$

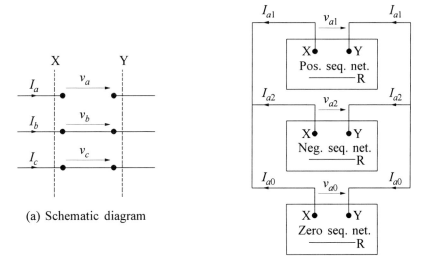

(a) Schematic diagram

(b) Connection of sequence networks

Figure 13.26 Simulation of one open-conductor condition

In terms of symmetrical components,

$$v_{a1} = v_{a2} = v_{a0} = \frac{1}{3}v_a$$

$$I_{a1} + I_{a2} + I_{a0} = 0$$

The connections of sequence networks so as to satisfy the above equations are shown in Figure 13.26(b).

(b) Two open conductors

We now consider the condition with phases b and c open between points X and Y as shown in Figure 13.27(a). The applicable equations are

$$v_a = 0 \qquad I_b = I_c = 0$$

In terms of symmetrical components,

$$v_{a1} + v_{a2} + v_{a0} = 0$$

$$I_{a1} = I_{a2} = I_{a0} = \frac{1}{3}I_a$$

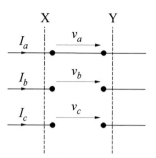

(a) Schematic diagram

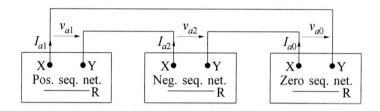

(b) Connection of sequence networks

Figure 13.27 Simulation of two-open-conductors condition

The sequence network connections that satisfy the above equations are shown in Figure 13.27(b).

From Figure 13.26 we see that, for the one-phase open-circuited condition, the three-sequence networks are connected in parallel at points X and Y; therefore, even if the zero-sequence circuit as seen from X and Y has infinite impedance, some power is transferred through the line between X and Y.

For the case with two phases open-circuited as in Figure 13.27, the three-sequence networks are connected in series; unless the zero-sequence circuit provides a path having finite impedance, no power can be transferred through the line.

Example 13.5

The system shown in Figure E13.11 (Example 13.4) is subjected to a single line-to-ground fault at point F on circuit 2. The fault is cleared by single-phase switching which disconnects the faulted conductor at both ends simultaneously. Show how the network may be represented in stability studies during the period when one conductor of circuit 2 is switched out.

Solution

This situation differs slightly from that depicted in Figure 13.26. As shown in Figure E13.19, the impedances of conductors of phases b and c need to be accounted for. This is achieved by including the impedances associated with circuit 2 in the sequence networks.

Figure E13.20 shows the sequence networks of the system of Figure E13.11 and their interconnection to represent the condition with one phase of circuit 2 open at both ends.

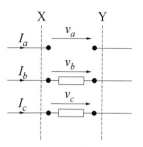

Figure E13.19

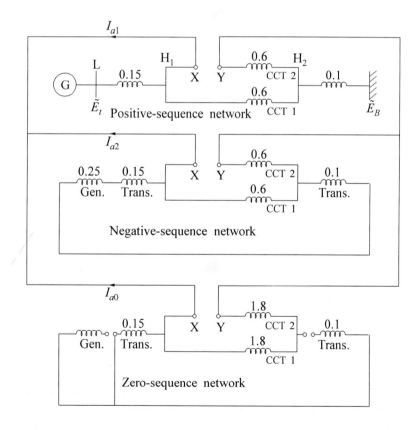

Figure E13.20

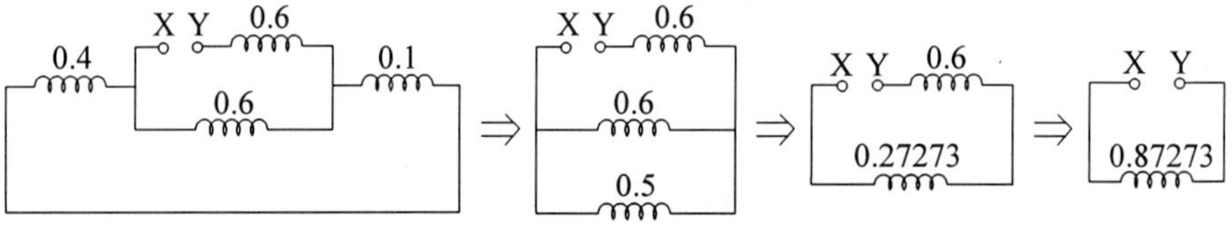

(a) Reduction of negative-sequence network

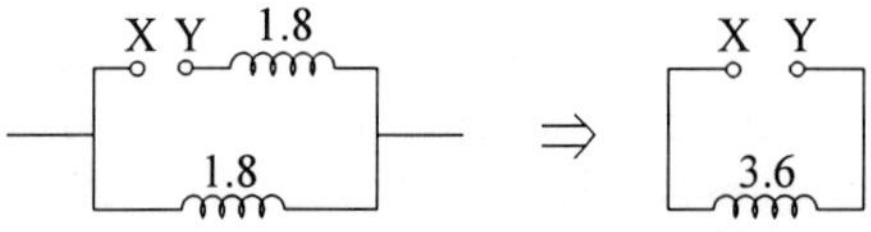

(b) Reduction of zero-sequence network

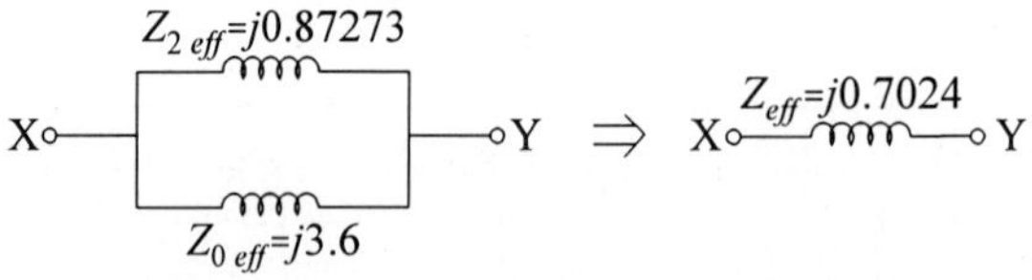

(c) Total effective series impedance

Figure E13.21 Reduction of negative- and zero-sequence networks

As shown in Figure E13.21, the effective negative- and zero-sequence impedances as viewed from points X and Y are

$$Z_{2\,eff} = j0.87273 \quad pu$$

$$Z_{0\,eff} = j3.6 \quad pu$$

and the net effective impedance that has to be inserted between points X and Y of the positive sequence network is

$$Z_{eff} = j0.7024 \quad pu$$

Thus, from the viewpoint of power transfer between the generator and the infinite bus, the effect of opening one phase of circuit 2 is the same as increasing the reactance of the circuit by 0.7024 pu. ■

13.5 PERFORMANCE OF PROTECTIVE RELAYING

Protective relays detect the existence of abnormal system conditions by monitoring appropriate system quantities, determine which circuit breakers should be opened, and energize trip circuits of those breakers. In order to perform their functions satisfactorily, relays should satisfy three basic requirements: selectivity, speed, and reliability.

Since transient stability is concerned with the ability of the power system to maintain synchronism when subjected to a severe disturbance, satisfactory performance of certain protection systems is of paramount importance in ensuring system stability. Protective relays must be able to distinguish among fault conditions, stable power swings and out-of-step conditions. While the relays should initiate circuit-breaker operations to clear faulted elements, it is important to ensure that there are no further relaying operations that cause unnecessary opening of unfaulted elements during stable power swings. Tripping of unfaulted elements would weaken the system further and could lead to system instability.

On the other hand, if the system is unstable we need to know how it separates and whether the point of separation is acceptable. Special relaying may be required to cause separation as desired.

One of the important aspects of transient stability analysis, then, is the evaluation of the performance of protective systems during the transient period, particularly the performance of relaying used for protection of transmission lines and generators.

13.5.1 Transmission Line Protection [21-23]

There are a variety of line protection schemes and practices used by utilities to meet particular system requirements. The following factors influence the choice of the protection scheme:

- Type of circuit: single line, parallel line, multiterminal, magnitude of fault current infeeds, etc.

- Function of line, its effect on service continuity, speed with which fault has to be cleared

- Coordination and matching requirements

Three basic types of relaying schemes are used for line protection: (a) overcurrent relaying, (b) distance relaying, and (c) pilot relaying [22].

(a) Overcurrent relaying

Overcurrent relaying is the simplest and cheapest form of line protection. Two basic forms of relays are used: instantaneous overcurrent relay and time overcurrent relay. They are difficult to apply where coordination, selectivity, and speed are important. Changes to their settings are usually required as the system configuration changes [22].

Overcurrent relaying cannot discriminate between load and fault currents; therefore, when used for phase-fault protection, it is applicable only when the minimum fault current exceeds the full load current.

Overcurrent relaying is used principally on subtransmission systems and radial distribution systems. Since faults on these systems usually do not affect system stability, high-speed protection is not required.

Overcurrent relaying schemes, which utilize communication channels to interchange intelligence between terminal stations, have been developed to overcome the above limitations. The two most common schemes are pilot wire relaying and phase-comparison relaying schemes.

(b) Distance relaying

A distance relay responds to a ratio of measured voltage to measured current. The relay operates if the ratio, which represents the effective impedance of the network, is less than the relay setting. Impedance is a measure of distance along the transmission line and, therefore, the relay is known as a distance relay.

The impedance approach provides an excellent way of obtaining discrimination and selectivity, by limiting relay operation to a certain range of the impedance. The following are the different forms of distance relays:

- Impedance relay

- Reactance relay

- Mho relay

- Modified mho and impedance relays, and combinations thereof

The triggering characteristics of distance relays may be shown conveniently on a complex impedance (R-X) plane [24] as illustrated in Figure 13.28.

The impedance characteristic is a circle on the R-X plane; the reactance characteristic is a horizontal line on the R-X plane; the mho characteristic is a straight line on the G-B (admittance) plane, and a circle through origin on the R-X (impedance) plane. The relay operates when the impedance measured by the relay falls within the characteristic.

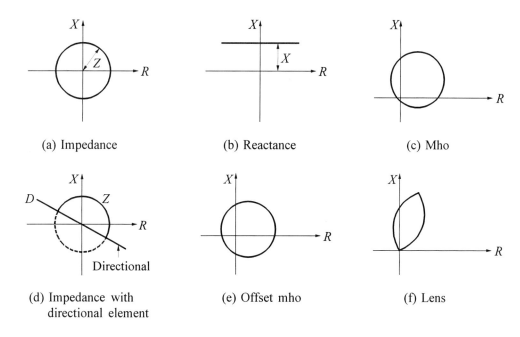

(a) Impedance   (b) Reactance   (c) Mho

(d) Impedance with   (e) Offset mho   (f) Lens
    directional element

Figure 13.28 Distance relay characteristics displayed on a coordinate system
with resistance (R) as the abscissa and reactance (X) as the ordinate

The mho characteristic is inherently directional; i.e., it detects faults only in
one direction. The reactance and impedance characteristics detects faults in all four
quadrants. Therefore, some type of directional supervision is generally used with these
relays. Combinations of similar elements with different settings and/or elements with
different characteristics are used to provide selectivity and coordination.

Distance relaying is the most widely used form of protection for transmission
lines. It is easy to apply and coordinate, provides fast protection, and is less affected
by changes in system conditions.

Figure 13.29 illustrates the application of distance relaying. At each end of the
line, three separate sets of relays are arranged to provide three protective zones as
shown. The first and second protective zones provide *primary protection* for the
protected line, and the third zone provides *remote backup* for the adjacent line.

For example, the first-zone relays at A are set to protect about 80% of line AB
without any time delay. This ensures high-speed protection of the line section and
provides about a 20% margin of safety to avoid incorrect relay operation caused by
equipment errors for faults beyond the end of the line section.

The second-zone relays at A are set to reach beyond the end of the line, a
typical setting being about 120% of the impedance of the protected line AB. This
ensures protection of the remaining 20% of the line not protected by the first-zone
relays. Since the second-zone relays will overreach line AB and operate for faults on

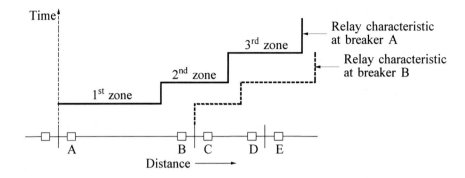

Figure 13.29 Distance relay characteristic

a section of the adjacent line CD, they are set to trip after a time delay (typically 0.3 to 0.5 seconds) thereby providing time coordination with the first-zone relays at C protecting the adjacent line CD.

The third-zone relays at A provide *backup protection* for the adjacent line CD. They are set to reach beyond line CD, so that relay operation is assured for faults on the line. Time coordination with primary protection of line CD is achieved by having a time delay of about 2 seconds for the third-zone protection.

A protection scheme similar to that at A is provided at B, but set to reach in the opposite direction. Thus a fault on line AB opens breakers at both ends of the line.

The number of relays at each end of a line is determined by the requirement that three-phase, phase-to-phase, phase-to-ground, and double phase-to-ground faults be covered. Usually, two sets of relays are provided: one set for phase faults and the other for ground faults.

The reactance type relays are used for short lines, and the mho type relays for long lines. Lens type relays are used when needed to restrict the tripping area of mho units for protection of long lines.

Straight-distance relaying provides adequate protection for many situations. However, it is not satisfactory when simultaneous high-speed tripping of both ends of the line is critical to maintain system stability. Pilot-relaying schemes, described below, are better suited for such applications.

(c) Pilot-relaying schemes

Pilot-relaying schemes utilize communication channels (pilots) between the terminals of the line that they protect. Relays located at each of the terminals determine whether the fault is internal or external to the protected line, and this information is transmitted between the line terminals through the communication channel. For an internal fault, circuit breakers at all terminals of the protected line are tripped; for an external fault the tripping is blocked.

The *communication medium* for transmitting signals between terminals may be pilot wire (metallic wires), power-line carrier, microwave, or fibre optics.

The pilot relay systems can be categorized by the manner in which the transmitted signal is used and by the fault detector principle:

- The transmitted signal may be used for blocking, unblocking, or transfer trip operations. In a *blocking system*, the transmitted signal blocks the tripping function of local relays that are normally set to trip a breaker. In a *transfer trip system*, a signal is sent by relays at one end of the line to either cause or permit the breakers at the other end to trip. An *unblocking system* is a combination of the above two systems. In this system, the tripping function of the local relays is normally blocked; the signal transmitted from the remote terminal unblocks them when tripping is required. In the event that the communication channel is lost for any reason, local relays are permitted to trip, thus protecting against a fault that may occur during this period.

- The detection of the fault location (i.e., internal or external to the protected line) may be based on either the directional comparison or the phase comparison principle, depending on the type of sensing relay used.

  In a scheme using the *directional comparison* principle, distance relays and directional relays are used to detect internal and external faults.

  In a scheme using the *phase comparison* principle, the fault location (i.e., whether the fault is internal or external) is detected by comparison of the relative phase displacement of currents entering and leaving the protected line.

Several different pilot-relaying schemes using a combination of the above categories are possible. We will describe here two such commonly used schemes. For simplicity we will consider only two terminal lines.

Permissive overreaching scheme:

Figure 13.30 illustrates the principle of operation of this scheme.

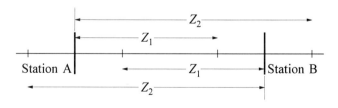

Figure 13.30 Permissive overreaching relay

Each terminal station of the line has the following relays:

- *Underreaching zone 1* phase and ground directional distance relays covering about 80% of the impedance of the protected line. Operation of these relays results in tripping of local breakers. These relays operate instantaneously with no communication required from the remote end.

- *Overreaching zone 2* phase and ground directional distance relays covering about 120% of the impedance of the protected line.

  Operation of these relays results in a permissive signal being sent to the remote end. Provided a permissive trip signal is received from the remote end, a zone 2 operation also results in the local breakers being tripped and a transfer trip signal being sent to the remote end.

  If the apparent impedance remains inside the zone 2 relay characteristic for a fixed time (typically 0.4 seconds), local breakers will be tripped without receiving a permissive trip signal. In this case no transfer trip signal is sent.

Typical zone 1 and zone 2 relay characteristics at each station are shown in Figure 13.31. If the apparent impedance measured by the relays at a station is within zone 1 characteristics, then breakers at that station are tripped instantaneously.

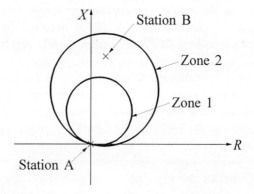

Figure 13.31 Relay characteristic at station A

  If the apparent impedances measured by the relays at both ends are within zone 2 characteristics, then breakers at both ends open at high speed.
  If the apparent impedance is within zone 2 characteristics of the relays at only one station, the breakers at that station are tripped with a time delay of 0.4 seconds, provided the impedance remains within zone 2 for 0.4 seconds.
  To illustrate how this scheme operates, consider the faults F_1, F_2, and F_3 as shown in Figure 13.32.

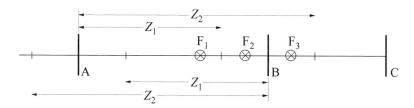

Figure 13.32 Fault locations F_1, F_2 and F_3

For the fault at F_1 within the zone 1 characteristics of the relays at both ends, breakers at the two ends are tripped without intentional time delay. For the fault at F_2 within zone 1 characteristic of the relay at B and zone 2 characteristic (but outside zone 1) of the relay at A, breakers at B trip instantaneously; the zone 2 relays at both ends are picked up and they send a permissive signal to each other. The zone 2 relay at A, upon receiving permissive signal from B (channel time not more than 20 ms), trips the breakers at A.

For the fault at F_3 (outside of the line section), the zone 2 relay at A is picked up. As the zone 2 relay at B is not picked up, no tripping occurs at B and no permissive signal is sent to A. The zone 2 relay at A trips the breakers only if the fault is not cleared by the protective relays of line section BC within 0.4 seconds.

Directional comparison scheme:

The principle of operation of a directional comparison scheme with a blocking-type pilot is illustrated in Figure 13.33. The pilot channel is used to block tripping for faults external to the protected line.

Each terminal station of the line has three sets of relays:

- *Underreaching zone 1* phase and ground directional distance relays covering about 80% of the impedance of the line. These relays operate instantaneously to trip the local breakers.

- *Overreaching zone 2* phase and ground directional distance relays set to reach beyond the remote terminal so as to cover about 120% of the line impedance. This ensures that all internal faults are detected. These relays trip local breakers after a delay of about 25 ms (pickup time of the coordinator timer), if no blocking signal is received from the remote terminal. Zone 2 relays trip local breakers if the fault is not cleared in about 0.4 s, irrespective of any blocking signal received from the remote end.

- *Zone 3 reverse blocking* directional distance relays set in a direction opposite to the protected line. They detect external faults and send a blocking signal to the remote end to prevent tripping.

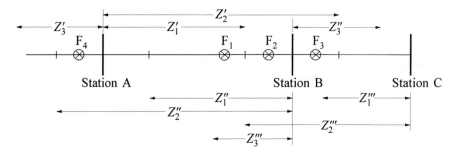

(a) Protective zones of relays at stations A and B, protecting
line AB, and at station C protecting line BC

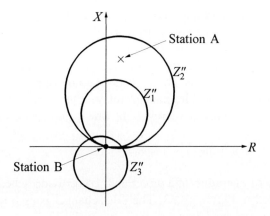

(b) Relay characteristics at station B,
protecting line from B to A

Figure 13.33 Directional comparison relaying scheme

Let us examine the operation of this scheme for four different fault locations
F_1, F_2, F_3 and F_4 identified in Figure 13.33(a). For a fault at location F_1 identified in
Figure 13.33, zone 1 relays at both stations A and B operate and trip breakers at the
two ends with no delay.

For a fault at F_2, the zone 1 relays (Z_1'') at station B trip the local breakers
instantaneously; the zone 2 relays at station A, having received no blocking signal
from the zone 3 relays (Z_3'') at station B, trip the breakers at A after a short time delay
of about 25 ms associated with the coordinating timer. This time delay is the waiting
period in case a blocking signal is sent from the other end. The zone 2 relays (Z_2''')
at station C also detect the fault at F_2, but are prevented from tripping at high speed
by the blocking signal received from the reverse sensing zone 3 relays (Z_3''') at station

B. However, a local timed trip results, if the fault is not cleared and the zone 2 relay at station C continues to detect the fault after about 0.4 s.

For a fault at F_3, the zone 2 relays (Z_2') at station A detect the fault but are prevented from tripping by the blocking signal received from the zone 3 relays (Z_3'') at station B. Similarly, a fault at F_4 is detected by the zone 2 relays (Z_2'') at station B, but is blocked by signals from the zone 3 relays (Z_3') at A.

13.5.2 Fault-Clearing Times

The removal of a faulted element requires a protective relay system to detect that a fault has occurred and to initiate the opening of circuit breakers which will isolate the faulted element from the system. The total fault-clearing time is, therefore, made up of the relay time and breaker-interrupting time. The relay time is the time from the initiation of the short-circuit current to the initiation of the trip signal to the circuit breaker. The interrupting time is the time from the initiation of the trip signal to the interruption of the current through the breaker. On high voltage (HV) and extra-high voltage (EHV) transmission systems, the normal relay times range from 15 to 30 ms (1 to 2 cycles) and circuit-breaker interrupting times range from 30 to 70 ms (2 to 4 cycles).

Local breaker failure backup protection is usually provided on each breaker on critical high-voltage transmission systems. If a breaker fails to operate at a local station, infeed to a fault through the stuck (failed) breaker will be interrupted by sending trip signals to adjacent zone breakers and transferring trip signals to remote end breakers.

An example of a *normally cleared fault* is given in Figure 13.34. Bus A has 2-cycle air-blast breakers, and bus B has 3-cycle oil breakers. The system nominal frequency is 60 Hz and the communication medium is microwave. A fault within zone 1 characteristic of the relay at bus A and within zone 2 (but outside zone 1) characteristic of the relay at bus B is considered. With the typical relaying times shown in Figure 13.34, the fault is cleared from bus A in 64 ms and from bus B in 104 ms.

An example of a *stuck breaker fault* cleared by backup protection is given in Figure 13.35. A fault on line BC within zone 1 characteristic of the relay at bus B and within zone 2 (but outside zone 1) characteristic of the relay at bus C is considered. Breaker 4 at bus B fails to operate, and the fault is cleared from bus B by sending trip signals to breaker 3 at bus B (local backup) and to breakers 1 and 2 at bus A (remote backup). With the typical breaker-interrupting and relaying times shown in Figure 13.35, the fault is cleared from bus C in 104 ms, from bus B in 157 ms, and from bus A in 180 ms. We see from this example that the effect of a breaker-failure is not only to delay fault-clearing, but to isolate from the system additional elements adjacent to the faulted element.

	Local (bus A) breakers 1 and 2	Remote (bus B) breakers 3 and 4
Primary relay time (fault detection)	25 ms	25 ms
Auxiliary relay(s) time	3 ms	9 ms
Communication time	–	17 ms (microwave)
Breaker trip module	3 ms	3 ms
Breaker-clearing time	33 ms (2 cycles)	50 ms (3 cycles)
Total time	64 ms	104 ms

Fault cleared from bus A in 64 ms
Fault cleared from bus B in 104 ms

Notes: (i)   For purposes of illustration, 2-cycle breakers have been assumed at A
              and 3-cycle breakers at B.

       (ii) Communication time depends on channel medium used. With power line
              carrier the time may be longer.

Figure 13.34 Typical fault-clearing times for a normally cleared fault

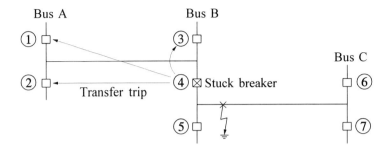

Breaker 4 assumed to be stuck
Breakers 1, 2, 3, 4, and 5 assumed to be 2-cycle air-blast breakers (33 ms)
Breakers 6 and 7 assumed to be 3-cycle oil breakers (50 ms)

	Local breaker 5	Remote breakers 6 and 7	Local backup breaker 3	Remote backup breakers 1 and 2
Primary relay time (at bus B)	25 ms	25 ms	25 ms	25 ms
Auxiliary relay(s) time	3 ms	9 ms	6 ms	12 ms
Communication channel time	–	17 ms	–	17 ms
Breaker failure timer setting	–	–	90 ms	90 ms
Breaker-tripping module time	3 ms	3 ms	3 ms	3 ms
Breaker time	33 ms (2 cyc.)	50 ms (3 cyc.)	33 ms	33 ms
Total time	64 ms	104 ms	157 ms	180 ms

Fault cleared from bus C in 104 ms
Fault cleared from bus B in 157 ms
Fault cleared from bus A in 180 ms

Notes: (i)   Breaker failure timer setting has been assumed to be 90 ms for the 2-cycle
            breaker 4. This could vary from one application to another. For a 3-cycle oil
            breaker, a typical value is 150 ms.

       (ii) Communication channel time depends on channel medium used.

Figure 13.35 Typical fault-clearing times for a stuck breaker fault

13.5.3 Relaying Quantities during Swings

The performance of protective relaying schemes during swings (electromechanical oscillations) and out-of-step conditions may be illustrated by considering the simple two-machine system of Figure 13.36. The machines are assumed to be represented by voltages of constant magnitudes behind their transient impedances (Z_A and Z_B). The effect of swings on relaying quantities is analyzed by considering the voltage, current, and apparent impedance measured by a relay at bus C (terminal of machine A).

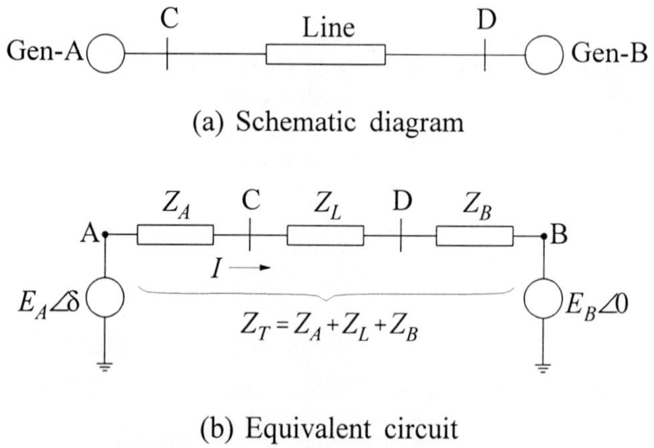
C
Line
D
Gen-A
Gen-B
(a) Schematic diagram
Z_A
C
Z_L
D
Z_B
A
B
I
$E_A\angle\delta$
$E_B\angle 0$
$Z_T = Z_A + Z_L + Z_B$
(b) Equivalent circuit

Figure 13.36 Two-machine system

E_A and E_B are voltages behind the transient impedances of the two machines. E_B is assumed to be the reference phasor, and δ represents the angle by which E_A leads E_B.

The current I is given by

$$\tilde{I} = \frac{E_A\angle\delta - E_B\angle 0}{Z_T} \tag{13.82}$$

The voltage at bus C is

$$\tilde{E}_C = \tilde{E}_A - Z_A\tilde{I} \tag{13.83}$$

The apparent impedance measured by an impedance relay at C protecting line CD is given by

$$Z_C = \frac{\tilde{E}_C}{\tilde{I}} = \frac{\tilde{E}_A - Z_A \tilde{I}}{\tilde{I}}$$

(13.84)

$$= -Z_A + Z_T \frac{E_A \angle \delta}{E_A \angle \delta - E_B \angle 0}$$

If $E_A = E_B = 1.0$ pu,

$$Z_C = -Z_A + \frac{Z_T}{1 \angle 0 - 1 \angle -\delta}$$

$$= -Z_A + Z_T \frac{1 \angle 0 + 1 \angle +\delta}{(1 \angle 0 - 1 \angle -\delta)(1 \angle 0 + 1 \angle +\delta)}$$

$$= -Z_A + Z_T \frac{1 + \cos\delta + j\sin\delta}{2j\sin\delta}$$

(13.85)

$$= -Z_A + Z_T \left[\frac{1}{2} - j \left(\frac{1 + \cos\delta}{2\sin\delta} \right) \right]$$

$$= \left(\frac{Z_T}{2} - Z_A \right) - j \left(\frac{Z_T}{2} \cot\frac{\delta}{2} \right)$$

During a swing, the angle δ changes. Figure 13.37 shows the locus of Z_C as a function of δ on an R-X diagram, when $E_A = E_B$.

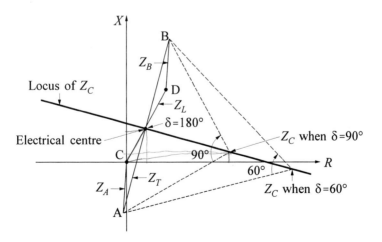

Note: Origin is assumed to be at C, where the relay is located.

Figure 13.37 Locus of Z_C as a function of δ, with $E_A = E_B$

When the voltage magnitudes E_A and E_B are equal, the locus of Z_C is seen to be a straight line which is the perpendicular bisector of the total system impedance between A and B, i.e., of the impedance Z_T. The angle formed by lines from A and B to any point on the locus is equal to the corresponding angle δ.

When $\delta = 0$, the current I is zero and Z_C is infinite. When $\delta = 180°$, the voltage at the *electrical centre* (middle of total system impedance) is zero. Therefore, the relay at C in effect will see a three-phase fault at the electrical centre. The electrical centre and impedance centre coincide in this case.

If E_A is not equal to E_B, the apparent impedance loci are circles with their centres on extensions of the impedance line AB. When $E_A > E_B$, the electrical centre will be above the impedance centre; when $E_A < E_B$, the electrical centre will be below the impedance centre. Figure 13.38 illustrates the shape of the apparent impedance loci for three different values of the ratio E_A/E_B.

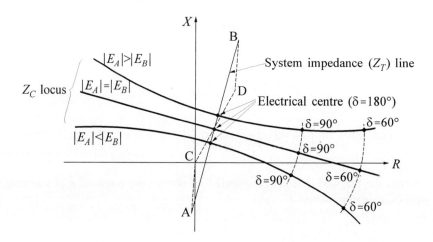

Figure 13.38 Loci of Z_C with different values of E_A/E_B

For generators connected to the main system through a *weak transmission system* (high external impedance), the electrical centre may appear on the transmission line. On the other hand, for the case where a generator is connected to the main system through a *strong transmission system*, the electrical centre will be in the step-up transformer or possibly within the generator itself.

We should recognize that the electrical centres are not fixed points since the effective machine reactance (assumed to be fixed at X_d' in our analysis) and the magnitudes of internal voltages vary during dynamic conditions.

The voltage at the electrical centre drops to zero as δ increases to 180° and then increases in magnitude as δ increases further until it reaches 360°. When δ reaches 180°, the generator will have *slipped a pole*; when δ reaches the initial value where the swing started, one slip cycle will have been completed.

After a system disturbance, the rates of change of the angle, voltage, and current are quite variable. At first the rotor angle changes quickly, slowing down as the peak angle is reached for a *stable* case; then the angle decreases and oscillates with smaller changes until equilibrium is reached. If the system is *unstable*, the angle increases gradually until it reaches 180°, when a pole is slipped. Unless the system is separated by protective systems, this is followed by repeated pole slips in rapid succession. The voltages and apparent impedances at points near the electrical centre oscillate rapidly during pole slipping.

Actual loci of apparent impedances measured by distance relays are more complex than for the idealized case we have considered; they depend on variations in internal machine voltages, voltage regulator and speed governor actions, and interactions between all the machines in the system as influenced by the interconnecting network. Such loci can be readily determined for any given situation by using a transient stability program. However, analysis of idealized cases involving simple-system configurations is helpful in understanding the results obtained with complex simulations.

Generalized impedance diagram

The concept of using the impedance diagrams for the analysis of distance relay performance was initially developed by C.R. Mason and J.H. Neher [25,26]. They considered the "swing line" on the R-X diagram for the case $E_A = E_B$ and analyzed the performance of different relays by plotting their characteristics on the same R-X diagram. This was indeed a pioneering step in understanding the performance of protective systems during power swings.

This concept was subsequently extended by other workers such as Edith Clarke [27] by developing impedance loci for various values of the ratio E_A/E_B. They showed that the impedance locus was a circle for each value of the ratio E_A/E_B, centred on the system impedance line with radii and offset determined by the values of the voltage ratio as shown in Figure 13.39. The specific case of $E_A/E_B = 1$ is the limiting case with infinite radius and offset.

In reference 27, Edith Clarke also developed another family of curves representing apparent impedance loci for constant angular separation (δ). If the angle between E_A and E_B is held constant while the voltage-magnitude ratio is varied, the apparent impedance will trace a portion of a circle which passes through A and B, and whose centre lies on the perpendicular bisector of the system-impedance line. These characteristics are shown in Figure 13.40. Line AB is seen to be a portion of the circle (with infinite radius) which represents 0° and 180° angular separation. It also separates the right and left portions of each constant-δ circle into two characteristics such that the separation angle for the parts differs by 180°. For example, the 90° circle on the right becomes the 270° circle on the left.

The two families of characteristics shown in Figures 13.39 and 13.40 can be combined to form the generalized per-unit impedance diagram [24,27]. These characteristics are helpful in visualizing and analyzing how the apparent impedance varies in the general case when both the angular separation and the ratio of source voltages are varied.

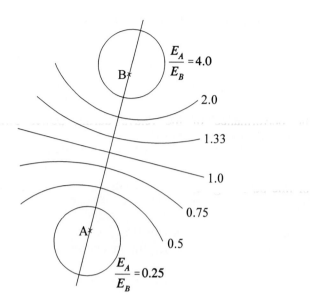

Figure 13.39 Impedance loci for constant voltage ratio

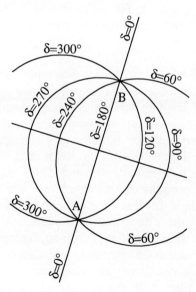

Figure 13.40 Loci of apparent impedance for constant angular separation

13.5.4 Evaluation of Distance Relay Performance during Swings

It is clear from the loci of apparent impedance derived in Section 13.5.3 that the impedance measured by a distance relay during electromechanical oscillations could be within the relay characteristics. The relay should, however, trip only for unstable conditions.

The performance of the relays during power swing conditions can be determined by calculating the impedances measured by the relays during the step-by-step stability simulations and comparing this with the operating characteristics of the relays.

For line pq of Figure 13.41, the apparent impedance measured by a relay at p is given by

$$Z_p = \frac{\tilde{E}_p}{\tilde{I}_{pq}} = \frac{\tilde{E}_p}{y_{pq}(\tilde{E}_p - \tilde{E}_q)}$$

For evaluating the performance of distance relays, the locus of apparent impedance so computed is plotted on the R-X diagram along with the relay characteristics as shown in Figure 13.42.

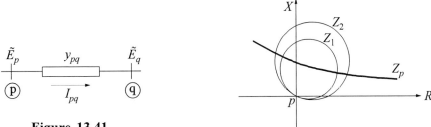

Figure 13.41

Figure 13.42

The performance of protective relaying, using a timed second zone and not using communication between line terminals, does not depend on relaying at the opposite end of the circuit. Tripping occurs if the impedance locus at either end remains in the second-zone characteristic sufficiently long to allow timed operation or if the first-zone characteristic is entered.

Performance of schemes using pilot relaying will depend on impedances measured at both ends.

13.5.5 Prevention of Tripping during Transient Conditions

Requirements for prevention of tripping during swing conditions generally fall into two categories:

(a)   Prevention of tripping during stable swings while allowing tripping for unstable transients.

(b)   Prevention of tripping during unstable transients and forcing of separation at another point.

Prevention of tripping during stable transients

The "mho" relays have smaller tripping characteristics than impedance and reactance relays set to reach the same distance. They may still cover too wide a range to avoid tripping of long lines during swings. If swings up to 120° are to be permitted, the mho relay characteristic is satisfactory for lines whose impedance is less than about half the total impedance of the two generators.

The angular range covered by the mho relay may be reduced by the use of two "ohm" units having linear characteristics as shown in Figure 13.43. The ohm units are similar to reactance units except that they respond to impedances at angles of +160° and −30° instead of +90°. These angles are chosen to make the characteristics of the ohm units parallel to the boundaries of fault area.

The mho unit controls the reach, and the ohm units control the angular range of tripping on swings. The ohm units are referred to as "blinders." If the line flow is only in one direction, it is sufficient to use only one blinder (O_2). Instead of using blinders, relays with shaped characteristics such as the "lens" characteristic shown in Figure 13.28 or the "peanut" characteristic shown in Figure 13.44 may be used for line protection. Such characteristics can be readily realized by using solid-state relays.

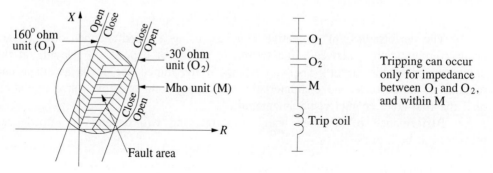

Figure 13.43 Reduction of mho relay angular range

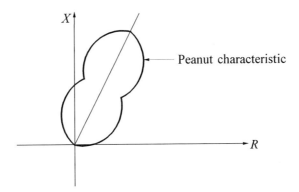

Figure 13.44

Out-of-step blocking and tripping relays

In some cases, it may be desirable to prevent tripping of lines at the natural separation point and choose the separation point so that (a) load and generation are better balanced on both sides, or (b) a critical load is protected, or (c) the separation is at a corporate boundary.

In certain instances, it may be desirable to trip faster in order to prevent the voltage from declining too far. Out-of-step relaying, either blocking or tripping, is used for this purpose.

Principle of out-of-step relaying:

The movement of the apparent impedance under out-of-step conditions is slow compared to its movement when a line fault occurs. Therefore, a transient swing condition can be detected by using two relays having vertical or circular characteristics on an R-X plane as shown in Figure 13.45. If the time required to cross the two characteristics (OOS2, OOS1) exceeds a specified value, the out-of-step function is initiated.

In an *out-of-step tripping scheme*, local breakers would be tripped. Such a scheme could be used to

* Speed up tripping to limit voltage decline

* Ensure that a selected line is tripped instead of other circuits that may be more critical

The out-of-step tripping relays should not operate for stable swings. They must detect all unstable swings and must be set so that normal load conditions are not picked up.

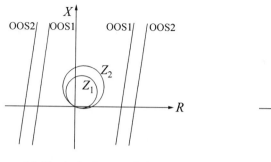 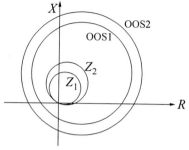

(a) Vertical characteristics   (b) Circular characteristics

Figure 13.45 Out-of-step relaying schemes

In an *out-of-step blocking scheme*, zone 2 and perhaps zone 1 relays are prevented from initiating tripping of the line monitored, and transfer trip signals are sent to open circuits at a remote location in order to cause system separation at a more preferable location.

The out-of-step blocking relays must detect the condition before the line protection operates. To ensure that line relaying is not blocked for fault conditions, the setting of the relays must be such that normal load conditions are not in the blocking area.

13.5.6 Automatic Line Reclosing

The majority (60 to 80%) of transmission line faults are of a transitory nature. An example of such a fault is an insulator flashover due to high transient voltages induced by lightning. After the line is de-energized long enough for the fault source to pass and the fault arc to de-ionize, the line may be reconnected. Therefore, common practice is to reclose the circuit breakers automatically to improve service continuity.

Reclosing may be either single-shot (one attempt) or multishot (several attempts) with time delay between each attempt. If the fault persists after two or three attempts, the operator may attempt to reclose manually after some delay. If this fails, it is indicative of a permanent fault and the line is taken out of service for repair. The first attempt to reclose may be either high-speed or with time delay.

High-speed reclosure refers to the closing of circuit breakers after a time just long enough to permit fault-arc de-ionization. The reclosure can be completed in less than 1 second. The following are the benefits of high-speed reclosure:

- Reduction of the likelihood of system difficulties resulting from several outages in rapid succession, usually caused by severe lightning, wind, or icing, by speeding up restoration of the system to the prior level of security.

- Minimization of the effect of line outage on critical load areas and restoration of normal voltages.

However, high-speed reclosure may not be acceptable in all cases. Reclosure into a permanent fault may cause system instability. Many utilities operate on the premise that all circuit faults are permanent and therefore the system must successfully withstand reclosing into a fault. In such cases, high-speed reclosure does not contribute to increased flexibility of system operation.

In addition, high-speed reclosure of lines close to steam turbine generating units may increase shaft fatigue. This is discussed further in Chapter 15.

In view of the above considerations, *timed and supervised reclosure* is often used on the bulk transmission system. This ensures that specific conditions are satisfied and a preferred sequence of reclosing of circuit breakers is followed. The general procedure is as follows:

- Potential is applied by closing one pre-selected breaker at one terminal after a time delay of 5 to 10 seconds.

- At terminals where out-of-synchronism conditions are a significant probability (e.g., at points in the system with few lines), the other circuit breakers associated with the line are automatically closed after the line has successfully remained on potential with acceptable angles across open breakers (measured by synchrocheck relays) for a period of at least 1 second.

  Where the probability of an out-of-synchronism condition existing between terminals is minimal, the circuit is automatically placed back on load by closing all remaining breakers associated with the circuit after it has successfully remained on potential for about 0.5 seconds.

13.5.7 Generator Out-of-Step Protection

For situations where the electrical centre is out in the transmission system, the detection of an out-of-step condition and the isolation of unstable generator(s) are accomplished by line protection.

However, for situations where the electrical centre is within the generator or step-up transformer, a special relay must be provided at the generator. Such a situation occurs when a generator pulls out of synchronism in a system with strong transmission. A low excitation level on the generator ($E_A < E_B$) also tends to contribute to such a condition.

Effect of generators operating in out-of-step condition

During an out-of-step condition, there are large cyclic variations in currents and voltages of the affected machine with the frequency being a function of the rate of slip of its poles. The high amplitude currents and off-nominal frequency operation could result in winding stresses and pulsating torques that can excite potentially damaging mechanical vibrations. There is also a risk of losing the auxiliaries of the affected unit as well as the auxiliaries of nearby stable units.

In order to avoid these adverse effects on the unit and the rest of the system, it is desirable to have an out-of-step relay that will trip the unit; the tripping should be restricted to disconnection of the machine from the system by tripping the generator breaker, rather than initiating a complete shutdown of the unit. This allows resynchronization of the generator as soon as conditions stabilize.

Relays for out-of-step tripping of generators

The basic schemes available are similar to those used to detect out-of-step conditions on the transmission system. There are, however, no industry standards or commonly used practices with regard to generator out-of-step protection. Reference 28 provides the results of a survey of utility practices and a description of different possible schemes. We will discuss two of the schemes here.

(a) Mho element scheme:

The principle on which this scheme is based is illustrated in Figure 13.46. A mho relay monitors the apparent impedance (looking into the network) at the HT terminal (H) of the unit transformer and is set to reach into the local generator.

The relay will immediately trip the generator when the apparent impedance measured at the HT bus enters the offset mho characteristic. The setting objective (achieved by choosing the size of the circle) is to allow tripping only for unstable swings. Typically, the angle δ_C at the point where the swing impedance enters the relay characteristic is set to about 120°, the maximum angular separation of the machine from the system which may occur without loss of synchronism.

If the relay circle is too large, the protection may trip the generator for stable swings. If the relay circle is too small (δ_C large), the scheme may not trip the generator for unstable swings. In addition, if δ_C is too large the tripping can occur when the angular separation approaches 180°; this should be avoided since it subjects the circuit breaker to the maximum recovery voltage during interruption.

To minimize the probability of false tripping for a loss-of-potential condition, an overcurrent fault detector is sometimes used to supervise the mho characteristic.

(b) Blinder scheme:

This scheme consists of two impedance elements, referred to as blinders, and a supervisory relay with an offset mho characteristic. The characteristics are shown in Figure 13.47.

There are many possible ways of applying blinders. In one scheme, the generator is tripped when the apparent impedance locus is within the mho

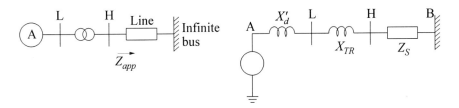

(a) System schematic   (b) System equivalent circuit

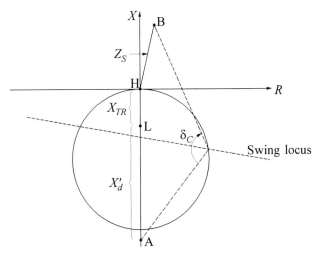

(c) Relay characteristic and swing locus as measured at the HV bus

Figure 13.46 Generator out-of-step protection
using an mho element scheme

characteristic and crosses both blinders, the time for crossing being in excess of a set
value (typically 0.15 s).

The scheme offers more selectivity than the simple mho element scheme. It is
also easy to set the scheme so that circuit breaker operation is effected at a favourable
swing angle. In view of the time delay involved in detecting the out-of-step condition,
it is easy to coordinate with the transmission line protection; this permits the reach to
extend into the system beyond the HT bus (H) of the step-up transformer, as depicted
in Figure 13.47.

In Figure 13.47, the relay is assumed to be located at the HT bus of the step-
up transformer. It may just as conveniently be located at the generator terminals, in
which case the origin of the R-X diagram shifts to point L.

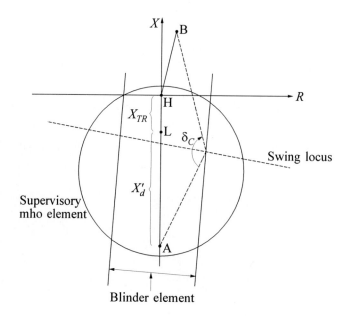

Figure 13.47 Generator out-of-step protection using a blinder scheme

Example 13.6

For the following unstable case simulated in Example 13.2, plot the apparent impedance measured by a distance relay located at the HT bus:

–   Exciter on manual control (constant E_{fd}); fault duration equal to 0.1 s.

Identify the location of the electrical centre and compute the effective machine reactance.

Solution

Figure E13.22 shows the locus of apparent impedance computed during the time-domain simulation.

From the apparent impedance locus plot, we see that the electrical centre (C) lies between the HT bus (H) and the infinite bus (B). The reactance between points H and C, as measured on the plot, is equal to 0.055 pu. Therefore,

$$AC = BC$$

or

$$X_{eff} + 0.15 + 0.055 = 0.5 - 0.055$$

Solving, we get

$$X_{eff} = 0.5 - 0.15 - 2 \times 0.055$$
$$= 0.24 \quad pu$$

This is equal to the average of $X_d'' = 0.23$ and $X_q'' = 0.25$.

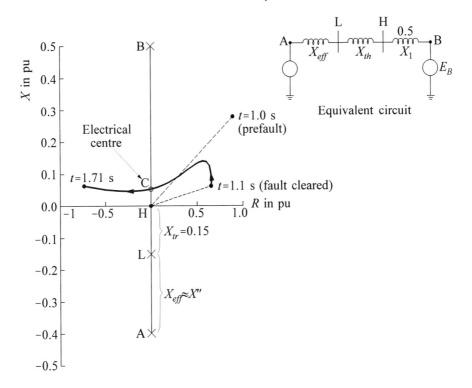

Figure E13.22 Apparent impedance locus   ■

13.5.8 Loss-of-Excitation Protection

Loss of excitation of a synchronous machine can be caused by a wide variety of abnormal field circuit conditions. This may range from a complete short-circuit of the field winding to an effectively open-circuited field winding.

When a generator loses excitation, it operates as an induction generator, running above synchronous speed. The excitation is now supplied from the power system; hence, the machine draws reactive power from the system. This may cause severe system voltage reductions.

Continued operation of the machine as an induction generator may cause damage due to stator and rotor heating. The stator current may be as high as 2 to 3 times rated, depending on slip. The induced slip frequency rotor currents may cause rotor overheating in steam-turbine generators. The time to reach dangerous overheating is on the order of several minutes, depending on the slip.

Loss of generator excitation may cause widespread voltage reduction if the power system to which the generator is connected is weak. This may, in turn, cause loss of load and loss of synchronism of other generators in the system. To prevent this, quick-acting loss-of-excitation relaying should be provided to trip the main and field breakers of the faulty generator.

The older loss-of-excitation protection consisted of undercurrent relays connected in the field circuit. Such relaying could not distinguish between purposely reduced excitation during light load conditions and accidental loss of excitation.

The relay equipment widely used now for loss-of-excitation protection is a directional-distance relay connected to the generator terminals. It is based on the principles developed by C.R. Mason in reference 29. In order to understand the basis for this relay, we need to first understand the loss-of-excitation characteristics of the generator.

Loss-of-excitation (LOE) characteristics [23,24,29]

The response of the machine following the loss of excitation depends on the initial loading condition.

Figure 13.48 shows the general shapes of the paths traced by the apparent impedance as measured at the generator terminal when the field circuit is short-circuited for three different initial power output conditions. The apparent impedance loci have been plotted on an R-X diagram on a per unit machine impedance base. Each characteristic is terminated just before the rotor angle δ is 180°. Beyond this point, the impedance changes become erratic due to pole slipping.

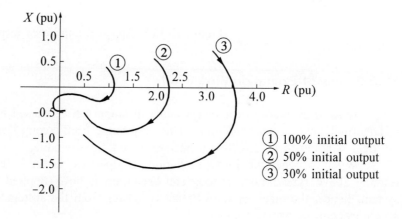

Figure 13.48   Typical LOE characteristics

When excitation is lost, the field flux linkage and the effective internal voltage gradually decrease depending on the field circuit time constant. This results in a gradual reduction in active power output, accompanied by an increase in rotor angle. In addition, the machine absorbs increasing amounts of reactive power as the field flux decays. Referring to the two-machine equivalent circuit of Figure 13.36, what we have is a situation where the internal voltage E_A is decreasing and the rotor angle δ is increasing. The loss-of-excitation characteristic is, therefore, some combination of the two families of curves shown in Figures 13.39 and 13.40. The actual path traced depends on the relative rates of variation in E_A and δ, which in turn depend on the initial generator output, transmission strength, and field circuit time constant. The overall effect on the apparent impedance at the machine terminals is as depicted in Figure 13.48.

With a high initial output, the rotor angle advances more rapidly and the endpoint shown in the figure is reached more quickly, typically within about 10 s. The slip is also relatively high, and therefore the impedance locus terminates at a value corresponding to the average of subtransient d- and q-axis reactances.

On the other hand, with a low initial output, the rotor angle advances slowly and the slip at the endpoint is small. The impedance locus terminates at a value corresponding to the average of d- and q-axis synchronous reactances.

The locus of the endpoints of the loss-of-excitation characteristic is shown in Figure 13.49. It ranges between the subtransient reactances and the synchronous reactances. Although these extreme values are not actually reached, they are helpful in identifying the LOE relay characteristic. By encompassing the locus of the endpoints as shown in Figure 13.49, the relay will operate before the generator first starts to slip poles.

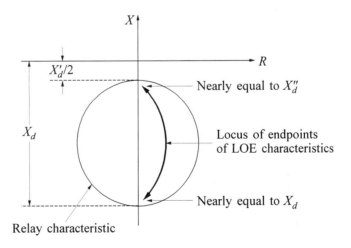

Figure 13.49 LOE relay and system characteristics

Loss-of-excitation relays

The characteristics of an LOE relay proposed by Mason in reference 29 are shown in Figure 13.49. It is an offset mho relay. The offset is equal to one-half of the transient reactance of the generator. The total reach of the relay is equal to the synchronous reactance X_d.

The offsetting of the relay characteristic from the origin provides selectivity against improper operation during power swings or out-of-step conditions. In an extreme case of zero external impedance, the electrical centre (intersection of the swing locus with the system impedance locus) is at the centre of the generator impedance X'_d. Therefore, the LOE relay characteristic is offset by $X'_d/2$.

The offset mho relay with the above characteristic has been widely used and has generally performed satisfactorily.

To provide greater selectivity against false operation during underexcited operation, stable transient swings, or system disturbances causing underfrequency operation, the use of two offset mho relays with characteristics as shown in Figure 13.50 is proposed by J. Berdy in reference 30.

Reference 31 describes a more discriminating relay developed at Ontario Hydro in 1978. This relay can prevent unit trip for recoverable excitation failures and trip the unit more rapidly for unrecoverable failures. The relay uses generator field and terminal voltages as inputs. It operates on the general principle that when the terminal voltage is high the field voltage should be low and, conversely, when the terminal voltage is low the field voltage should be high. For a detected excitation problem, the relay will initially try to transfer voltage control to an alternate AVR/exciter. If this does not rectify the problem, the relay trips the unit. Figure 13.51 shows a block diagram of the relay. The relay also includes selective overexcitation protection as shown by the dashed blocks in the figure.

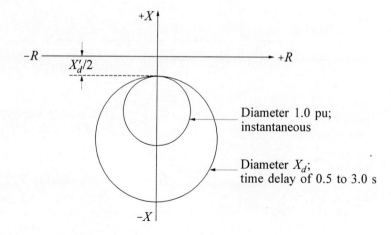

Figure 13.50 LOE protection with two mho relays

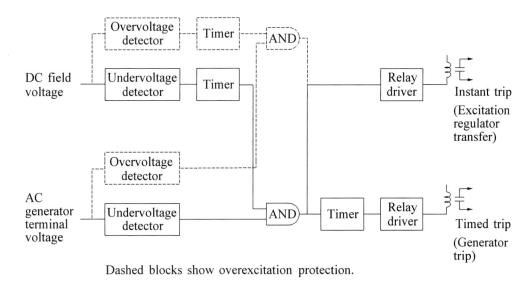

Dashed blocks show overexcitation protection.

Figure 13.51   A discriminating loss-of-excitation relay [31]

Example 13.7

In this example, we examine the response of the system of Example 13.2 to loss of generator excitation.

(a)   If the field of the equivalent generator representing the four units of the plant is suddenly shorted (directly, with zero external resistance), determine the locus of apparent impedance measured by an LOE relay located at the generator terminals. Assume that the LOE relay has a characteristic similar to that shown in Figure 13.49. Show the corresponding time responses of generator terminal voltage, active and reactive power outputs, and rotor angle.

(b)   Determine the locus of apparent impedance if the field of only one of the four generators in the plant is shorted (directly). The other generators operate normally with excitation control.

Assume that in each of the above cases the initial system operating condition is the same as in Example 13.2.

Solution

(a) The apparent impedance locus computed by using a transient stability program is shown in Figure E13.23. The generator is represented in detail including amortisseur circuits, and shorting of the field is simulated by setting the field voltage E_{fd} to zero at $t=1.0$ s.

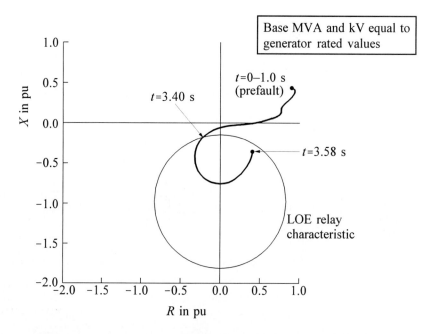

Figure E13.23 Apparent per unit Z measured by an LOE relay
when field windings of all four units are shorted

The corresponding time response plots of E_t, P, Q and δ are shown in Figure E13.24. These plots are helpful in explaining the shape of the apparent impedance locus, if we note that

$$Z_{app} = \frac{\tilde{E}}{\tilde{I}_t} = R + jX$$

where

$$R = \frac{E_t^2 P}{P^2 + Q^2} \qquad\qquad X = \frac{E_t^2 Q}{P^2 + Q^2}$$

From Figure E13.23, we see that the apparent impedance locus enters the LOE characteristic at t=3.4 s, i.e., 2.4 s after the field is shorted. As we see in Figure E13.24(b), this corresponds to the time when the generator has slipped one complete cycle. As the generator continues to slip poles, P, Q, and E_t oscillate in rapid succession.

(b) Figure E13.25 shows the apparent impedance locus when the field of one of the generators is shorted. The other three generators continue to operate normally with AVR and PSS controlling their excitation.

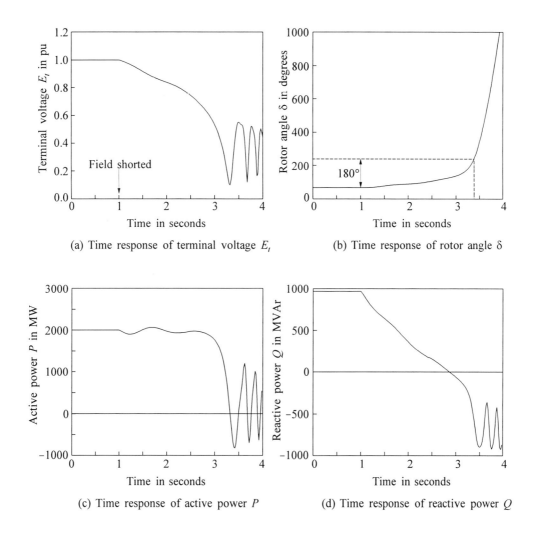

(a) Time response of terminal voltage E_t

(b) Time response of rotor angle δ

(c) Time response of active power P

(d) Time response of reactive power Q

Figure E13.24 Time responses of generator variables
following shorting of field windings

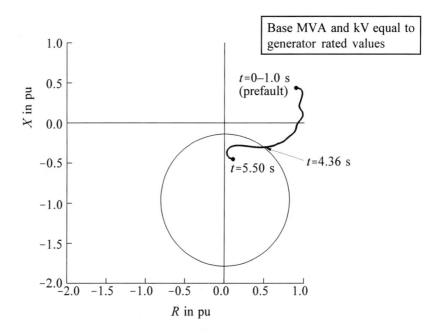

Figure E13.25 Apparent per unit Z measured by an LOE relay when
the field winding of one of the generators is shorted ∎

13.6 CASE STUDY OF TRANSIENT STABILITY OF A LARGE SYSTEM

This case study is intended to demonstrate the analysis of a large practical system for which a disturbance results in transient instability. The results show how the unstable portion of the system is isolated from the rest of the system through the actions of protective relaying. In addition, two possible methods of maintaining stability are illustrated.

The system

The area of interest for the system under study is shown in Figure 13.52. It consists of an 8 unit 7,000 MW nuclear plant connected to the rest of the system through two 500 kV double circuit lines and three 230 kV double circuit lines. The system model, which represents a portion of a large interconnected power system, consists of

- 2279 buses, 467 machines, and 6581 branches
- All machines and associated controls modeled in detail
- Loads modeled as voltage dependent ($P=50\%I+50\%Z$, $Q=100\%Z$)

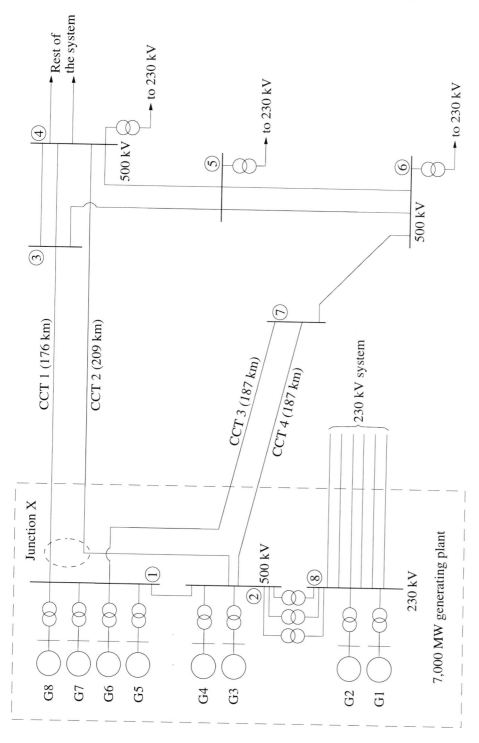

Figure 13.52   Diagram of system in the vicinity of a 7,000 MW plant

The contingency

A double line-to-ground (LLG) fault occurs on the 500 kV double-circuit line comprising circuits 1 and 2 in Figure 13.52. The location of the fault is shown in the figure as junction X, which represents the location nearest the plant where the two circuits share the same tower. The switching sequence is as follows:

Time (ms)	Event
0	No disturbance
100	Apply LLG fault at junction X on circuits 1 and 2
164	Local end clearing:

Open breakers at bus 1 for circuit 1
Open breakers at bus 2 for circuit 2

This occurs 64 ms after the fault is applied, and this time is computed as the sum of fault detection time (25 ms), auxiliary relay time (6 ms), and breaker-clearing time (33 ms = 2 cycles). At this time, the fault remains connected on the ends of circuits 1 and 2 at junction X.

187	Remote end clearing:

Open breakers at bus 4 for circuit 2
Open breakers at bus 3 for circuit 1
Clear fault (the line is isolated)

This occurs 87 ms after the fault is applied, and the time is calculated as the sum of fault detection time (25 ms), auxiliary relay time (12 ms), communication time (17 ms; microwave), and breaker-clearing time (33 ms = 2 cycles)

5000	Terminate simulation

Simulation

The 5-second simulation was conducted using explicit numerical integration (fourth order R-K) and an integration time step of 0.01 second.

Figure 13.53 shows the absolute rotor angle plots for unit G3 and for a unit located in the external system. Absolute angle is measured relative to the synchronously rotating reference frame, and it is clear from plots that the external system unit remains synchronized while the unit G3 *loses synchronism and becomes monotonically unstable.* The responses of the other seven units at the nuclear plant is similar to that of G3. As units G1 to G8 become unstable, the rest of the system

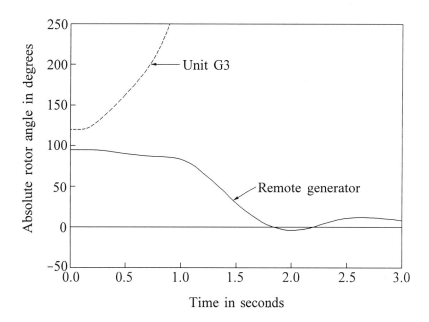

Figure 13.53 Rotor angle time response

becomes generation deficient, and hence the absolute angles of all machines in the system drift slightly. In this regard, using relative angles rather than absolute angles is often a better choice since it permits one to easily observe the relative motion of rotors between machines. The LLG fault results in the loss of one 500 kV double-circuit line, leaving the entire 7,000 MW plant connected to the system through a significantly weaker (i.e., higher impedance) path consisting of one double-circuit 500 kV line and three 230 kV double-circuit lines. The plant is unable to transfer full output through this path, and the resulting accelerating power leads to instability.

How does the system come apart as a result of instability?

Figure 13.54 shows the characteristic for the out-of-step protection on unit G3. The locus of the effective network impedance measured at the unit terminals is superimposed on the circle. We see that the impedance locus *does not enter the circle* except during the brief period when the fault is on; therefore, the protection *will not* trip the units. The point of system separation is clearly beyond the reach of the out-of-step protection.

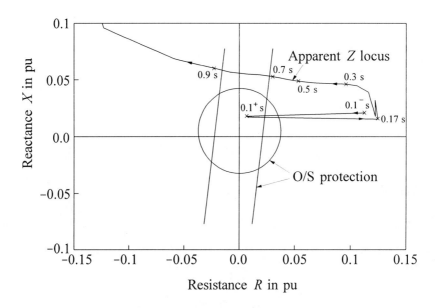

Figure 13.54 Unit G3 out-of-step protection

Figure 13.55 and Figure 13.56 show the line protection performance for relays on circuit 3 at buses 1 and 7, respectively. The apparent impedances measured at each end looking toward the opposite ends are shown on the plots. The relays are mho distance relays and have zone 1 coverage of about 75% of line length and zone 2 overreach of about 125% of line length. These plots show that the apparent impedance enters the zone 2 relays at bus 1 and enters the zone 1 and zone 2 relays at bus 7. The zone 1 relay at bus 7 would trip circuit 3 at bus 7 and send a transfer trip signal to the breakers at bus 1 which would then trip circuit 3 at bus 1. This logic also holds true for the companion 500 kV circuit (#4) which would be tripped in an identical manner. Following the loss of the 500 kV circuits (at approximately 0.8 s), the remaining 230 kV circuits would become extremely overloaded and would be lost through protection actions, thereby completely isolating the unstable plant from the system.

From Figure 13.55, we can determine that the impedance swing *crosses* the circuit at a point about 84% of the line length from bus 1. This point represents the *electrical centre* following the disturbance and is theoretically where separation occurs (although the entire circuit is tripped). Figure 13.57 shows plots of bus voltage measured at buses 1 and 7, and at the electrical centre. The voltage at the electrical centre leads the other voltages in collapsing as instability occurs.

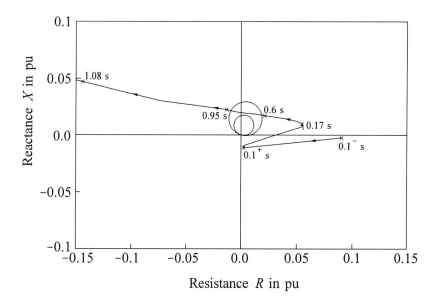

Figure 13.55 Line protection (circuit 3) at bus 1

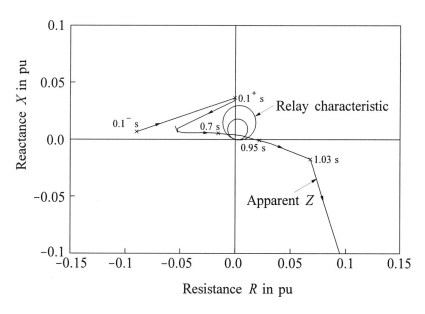

Figure 13.56 Line protection (circuit 3) at bus 7

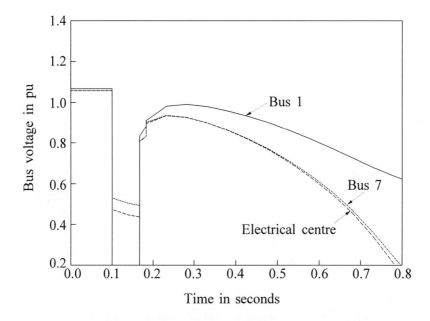

Figure 13.57

Methods of maintaining stability

As discussed in Chapter 17, there exist many possible ways of improving transient stability (including adding transmission facilities, fast-valving, faster fault clearing, etc.). Here, we consider the following two methods:

(a)   *Reduction of the pre-contingency output of the plant.* If one unit is taken off line (with output redispatched to other units in the external system), the total plant output will be reduced sufficiently to ensure stability following the LLG fault. Curve 2 in Figure 13.58 shows the absolute rotor angle for unit G3 for this case. It shows that the plant remains in synchronism. This solution to instability is, however, generally costly because available energy is kept idle or *bottled* in the plant.

(b)   *Rejection of generation following the disturbance.* If the output of the plant can be rapidly reduced following the contingency, the accelerating power can be reduced sufficiently to maintain synchronism. Curve 3 in Figure 13.58 shows the unit G3 rotor angle response with two units (G7 and G8) tripped at 173 ms (73 ms following fault application).

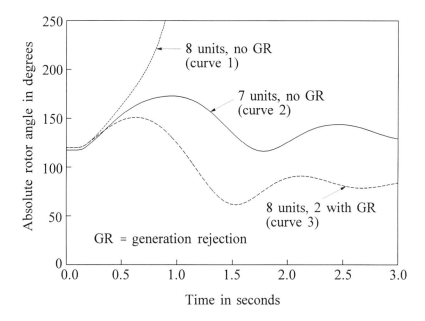

Figure 13.58 Unit G3 rotor angle responses with
and without generation rejection

13.7 DIRECT METHOD OF TRANSIENT STABILITY ANALYSIS

The direct methods determine stability without explicitly solving the system differential equations. This approach is academically appealing and has received considerable attention since the early work of Magnusson [32] and Aylett [33] who used transient energy for assessment of transient stability. The energy-based methods are a special case of the more general Lyapunov's second method or the direct method (see Chapter 12, Section 12.1.4), the energy function being one possible Lyapunov function [34-37].

This section describes the basic concepts on which the direct methods are based. A detailed treatment of the subject is beyond the scope of this book. Interested readers may consult references 38 and 39, books dedicated entirely to the analysis of power system transient stability by the direct methods.

13.7.1 Description of the Transient Energy Function Approach

Rolling ball analogy [40,41]

The transient energy approach can be described by considering a ball rolling on the inner surface of a bowl as depicted in Figure 13.59. The area inside the bowl represents the region of stability, and the area outside is the region of instability. The rim of the bowl is irregular in shape so that different points on the rim have different heights.

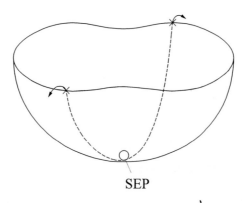

SEP

Figure 13.59 A ball rolling on the inner surface of a bowl

Initially the ball is resting at the bottom of the bowl, and this state is referred to as the *stable equilibrium point* (SEP). When some kinetic energy is injected into the ball, causing it to move in a particular direction, the ball will roll up the inside surface of the bowl along a path determined by the direction of initial motion. The point where the ball will stop is governed by the amount of kinetic energy initially injected. If the ball converts all its kinetic energy into potential energy before reaching the rim, then it will roll back and eventually settle down at the stable equilibrium point again. However, if the kinetic energy injected is high enough to cause the ball to go over the rim, then the ball will enter the region of instability and will not return to the stable equilibrium point. The surface inside the bowl represents the *potential energy surface*, and the rim of the bowl represents the *potential energy boundary surface* (PEBS).

Two quantities are required to determine if the ball will enter the instability region: (a) the initial kinetic energy injected and (b) the height of the rim at the crossing point. The location of the crossing point depends on the direction of the initial motion.

Application to power systems

The basis for the application of the TEF method to the analysis of power system stability is conceptually similar to that for a ball rolling in a bowl. Initially the system is operating at a stable equilibrium point. If a fault occurs, the equilibrium is disturbed and the synchronous machines accelerate. The power system gains kinetic and potential energy during the fault-on period and the system moves away from the SEP. After fault clearing, the kinetic energy is converted into potential energy in the same manner as the ball rolling up the potential energy surface. To avoid instability, the system must be capable of absorbing the kinetic energy at a time when the forces on the generators tend to bring them toward new equilibrium positions. This depends on the potential energy-absorbing capability of the postdisturbance system. For a given postdisturbance network configuration, there is a maximum or critical amount of transient energy that the system can absorb. Consequently, assessment of transient stability requires

(a)   Functions that adequately describe the transient energy responsible for separation of one or more synchronous machines from the rest of the system

(b)   An estimate of the *critical energy* required for the machines to lose synchronism

For a two-machine system the critical energy is uniquely defined, and the TEF analysis is equivalent to the equal-area criterion described in Section 13.1. This is illustrated in Figure 13.60, which shows two plots, both having rotor angle (δ) as the ordinate [40]. The upper plot illustrates the equal-area criterion in which the critical clearing angle (δ_c) is established by equality of areas A_1 and A_2. The lower plot illustrates the transient energy method which can be used to specify the critical clearing angle in terms of potential and kinetic energy. The kinetic energy gained during the fault-on period is added to the potential energy at the corresponding rotor angle, and the sum is compared to the critical potential energy to determine stability.

Given a disturbance, there is a stable equilibrium point for the postfault system. A region of attraction can be defined for this postfault SEP as shown in Figure 13.61. Any postfault system trajectory with the state of the system at fault clearing ($\mathbf{x}_{cl}$) inside this region of attraction will eventually converge to the SEP, and the system is said to be stable. On the other hand, if $\mathbf{x}_{cl}$ lies outside the region of attraction, the postfault system will not converge to the stable equilibrium point, and the system is said to be unstable.

The state of the system at fault clearing ($\mathbf{x}_{cl}$) can be described by the value of the energy function evaluated at $\mathbf{x}_{cl}$, i.e., $V(\mathbf{x}_{cl})$. Hence the direct method solves the stability problem by comparing $V(\mathbf{x}_{cl})$ to the critical energy V_{cr}. The system is stable if $V(\mathbf{x}_{cl})$ is less than V_{cr} and the quantity $V_{cr}-V(\mathbf{x}_{cl})$ is a good measure of system relative stability. This quantity is defined as the *transient energy margin*.

The quantity $V(\mathbf{x}_{cl})$ measures the amount of transient energy injected into the system by the fault while the critical energy measures the strength of the postfault system. More precisely, the critical energy measures the energy-absorbing capability of the postfault system.

Referring to Figure 13.61, if the rotor oscillates within the range δ_{u1} to δ_{u2}, the system will remain transiently stable. If the rotor swings beyond this range, the system will become unstable. Hence the two points δ_{u1} and δ_{u2} on the potential energy curve form a boundary to all stable rotor angle trajectories. This boundary is called the *potential energy boundary surface* (PEBS) and the points on this boundary are local potential energy peaks.

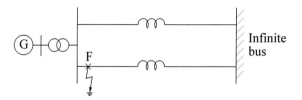

(a) System configuration

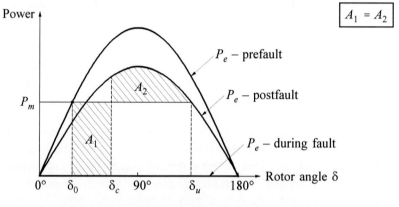

(b) Power-angle relationships

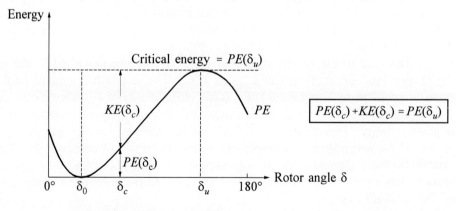

(c) Energy-angle relationships

Figure 13.60 Illustration of equivalence of transient
energy method and equal-area criterion

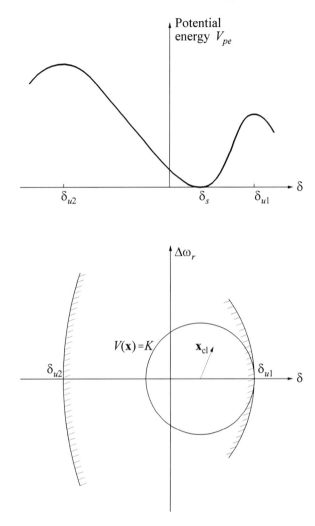

Figure 13.61 Region of stability and its local approximation

The boundary of the stability region is usually approximated locally by a constant energy surface $\{\mathbf{x}\,|\,V(\mathbf{x})=K\}$ as shown in Figure 13.61, where K represents the critical energy V_{cr} of the postfault system.

13.7.2 Analysis of Practical Power Systems [40-49]

The application of the direct method to practical power systems is difficult. A number of simplifying assumptions are necessary. To date, the analysis has been mostly limited to power system representation with the generators represented by the classical model and loads modelled as constant impedances [42-45]. Recently, there have been several attempts to extend the method to include more detailed models [46,47]. This, however, is still at a developmental stage and further research efforts are required to improve the accuracy and reliability of the method with detailed models. Here, we will limit our discussion to the simplified system models.

With the generator transient reactances and load admittances included in the node admittance matrix, we may write

$$\mathbf{I_G} = \mathbf{Y_R}\mathbf{E_G} \tag{13.86}$$

where $\mathbf{Y_R}$ is the reduced admittance matrix with all nodes other than the generator internal nodes eliminated, $\mathbf{E_G}$ is the generator internal source voltage vector, and $\mathbf{I_G}$ is the generator current vector.

Let the internal voltage of the ith generator in phasor notation be given by

$$\tilde{E}_i = E_i \angle \delta_i$$

and the ith row, j^{th} column element of $\mathbf{Y_R}$ by

$$y_{ij} = G_{ij} + jB_{ij}$$

For a system with n machines, the active output of the ith generator is given by

$$
\begin{aligned}
P_i &= \mathrm{Re}(\tilde{E}_i \tilde{I}_i^*) \\
&= \mathrm{Re}(\tilde{E}_i \sum_{j=1}^{n} \tilde{y}_{ij}^* \tilde{E}_j^*) \\
&= E_i^2 G_{ii} + \sum_{\substack{j=1 \\ j \neq i}}^{n} E_i E_j \{ B_{ij}\sin(\delta_i - \delta_j) + G_{ij}\cos(\delta_i - \delta_j) \}
\end{aligned} \tag{13.87}
$$

For the application of the TEF method, it is convenient to describe the transient behaviour of the system with the generator rotor angles expressed with respect to the inertial centre of all the generators. The position of the *centre of inertia* (COI) is defined as

$$\delta_{COI} \triangleq \frac{1}{H_T} \sum_{i=1}^{n} H_i \delta_i \tag{13.88}$$

where H_T is the sum of the inertia constants of all n generators in the system.
The motion of the COI is determined by

$$2H_T p(\Delta\omega_{COI}) = P_{COI} = \sum_{i=1}^{n}(P'_{mi}-P_{ei}) \tag{13.89}$$

and

$$p(\delta_{COI}) = (\Delta\omega_{COI})\omega_0 \tag{13.90}$$

with

$$P'_{mi} = P_{mi}-E_i^2 G_{ii} \tag{13.91}$$

$$P_{ei} = \sum_{\substack{j=1 \\ j \neq i}}^{n}[C_{ij}\sin(\delta_i-\delta_j)+D_{ij}(\delta_i-\delta_j)] \tag{13.92}$$

where

P_{mi} = mechanical input power of the ith machine
C_{ij} = $E_i E_j B_{ij}$
D_{ij} = $E_i E_j G_{ij}$
ω_0 = synchronous speed in elec. rad/s
$\Delta\omega_{COI}$ = per unit speed deviation of COI from synchronous speed

The motion of the generators with respect to the COI can be expressed by defining

$$\theta_i = \delta_i-\delta_{COI} \quad \text{rad} \tag{13.93}$$

and

$$\omega_i = \frac{\dot{\theta}_i}{\omega_0} = \left(\frac{\dot{\delta}_i}{\omega_0}-\Delta\omega_{COI}\right) \quad \text{pu} \tag{13.94}$$

The equations of motion of the ith machine in the COI reference frame are

$$2H_i p(\omega_i) = P'_{mi}-P_{ei}-\frac{H_i}{H_T}P_{COI} \tag{13.95}$$

$$p(\theta_i) = \omega_i\omega_0 \tag{13.96}$$

where ω_i is the pu speed of the ith machine with respect to the COI.

The energy function V describing the total system transient energy for the postdisturbance system is defined as

$$V = \frac{1}{2}\sum_{i=1}^{n} J_i \omega_i^2 - \sum_{i=1}^{n} P'_{mi}(\theta_i - \theta_i^s)$$

$$- \sum_{i=1}^{n-1} \sum_{j=i+1}^{n} \left[C_{ij}(\cos\theta_{ij} - \cos\theta_{ij}^s) - \int_{\theta_i^s + \theta_j^s}^{\theta_i + \theta_j} D_{ij}\cos\theta_{ij}\, d(\theta_i + \theta_j) \right] \tag{13.97}$$

where

θ_i^s = angle of bus i at the postdisturbance SEP

$J_i = 2H_i\omega_0$ = per unit moment of inertia of the ith generator

The transient energy function is composed of the following four terms:

(a) $\frac{1}{2}\Sigma J_i\omega_i^2$: change in rotor kinetic energy of all generators in the COI reference frame

(b) $\Sigma P'_{mi}(\theta_i - \theta_i^s)$: change in rotor potential energy of all generators relative to COI

(c) $\Sigma\Sigma C_{ij}(\cos\theta_{ij} - \cos\theta_{ij}^s)$: change in stored magnetic energy of all branches

(d) $\Sigma\Sigma\int D_{ij}\cos\theta_{ij}\, d(\theta_i + \theta_j)$: change in dissipated energy of all branches

The first term is called the *kinetic energy* (V_{ke}) and is a function of only generator speeds. The sum of terms 2, 3, and 4 is called the *potential energy* (V_{pe}) and is a function of only generator angles.

The transient stability assessment involves the following steps:

1. Calculation of the critical energy V_{cr}

2. Calculation of the total system energy at the instant of fault-clearing V_{cl}

3. Calculation of stability index: $V_{cr} - V_{cl}$. The system is stable if the stability index is positive.

Time-domain simulation is run up to the instant of fault clearing to obtain the angles and speeds of all the generators. These are used to calculate the total system energy (V_{cl}) at fault clearing.

The calculation of V_{cr}, i.e., the boundary of the region of stability, is the most difficult step in applying the TEF method. Three different approaches are briefly described below.

(a) The closest unstable equilibrium point (UEP) approach

Early papers on the application of the TEF method for transient stability analysis used the following approach to determine the smallest V_{cr}:

Step 1:   Determine all the UEPs. This is achieved by solving the postdisturbance system steady-state equations with different initial values of bus angles.

Step 2:   Calculate system potential energy at each of the UEPs obtained in step 1. The critical energy V_{cr} is given by system energy at the UEP which results in the minimum potential energy.

Referring to the rolling ball analogy, the closest UEP is analogous to the lowest point on the rim of the bowl. The critical energy at the closest UEP is equal to the potential energy of the ball when it is at the lowest point on the rim. This assumes that the direction of initial motion is always toward the lowest point on the rim. Similarly, for a power system this approach computes the critical energy by implicitly assuming the worst fault location; hence, the results are very conservative.

(b) The controlling UEP approach [40, 48]

The degree of conservatism introduced by the closest UEP approach is such that the results are usually of little practical value. The controlling UEP approach removes much of this conservativeness by computing the critical energy dependent on the fault location. This approach is based on the observation that the system trajectories for all critically stable cases get near to those UEPs that are closely related to the boundary of system separation. The UEPs are called the controlling or relevant UEPs.

The computation process in this approach consists of two steps: (1) identification of the mode of disturbance and (2) calculation of the controlling UEP.

Mode of disturbance (MOD):

The mode of disturbance identifies the machines that are severely disturbed by a given disturbance. These machines are most likely to lose synchronism with the rest of the system if the disturbance is severe enough to cause system instability. One way to determine the MOD is to perform time-domain simulations and identify which machines go unstable. An automatic procedure for determining the MOD is described in reference 49 based on the normalized V_{pe}. This procedure requires a considerable amount of CPU time when detailed generator models are used.

Controlling UEP calculation:

The SEP or UEP is a solution of the power-flow equations (with $i=1$ to n):

$$f_i = P'_{mi} - P_{ei} - \frac{H_i}{H_T} P_{COI} = 0 \tag{13.98}$$

$$\omega_i = 0 \tag{13.99}$$

There are many solutions to the above equations, one of which is the SEP. All other solutions are UEPs. The controlling UEP is calculated by solving the following minimization problem:

$$\min \left\{ F = \sum_{i=1}^{n} f_i \right\}$$

An approximate UEP based on the MOD is used as a starting point for the solution.

Numerical problems are sometimes encountered when one is solving for the controlling UEP based on the above approach. A robust procedure based on the potential energy boundary surface crossing point suggested in reference 44 is described below.

(c) The boundary of stability-region-based controlling UEP (BCU) method

Earlier CUEP methods face serious convergence problems when solving for the controlling UEP, especially when the system is highly stressed or highly unstressed, or the mode of system instability is complex. These problems usually arise if the starting point for the CUEP solution is not sufficiently close to the exact CUEP. Some of the convergence problems can be overcome by the BCU method which has the capability of producing a much better starting point for the CUEP solution.

In reference 44, the boundary of the stability region is defined as the union of the stable manifolds of the UEPs on the boundary. Any trajectory starting from a point on the boundary will converge to one of the UEPs as time increases. By making use of this property, the BCU method computes the controlling UEP of a power system through computing the CUEP of its associated reduced system.

The dynamic behaviour of a multimachine power system represented by classical generator models and constant impedance loads is described by Equations 13.95 and 13.96. The associated reduced system model for this multimachine power system is defined as

$$\dot{\theta}_i = P'_{mi} - P_{ei} - \frac{H_i}{H_T} P_{COI} = g_i \qquad i = 1, \dots, n \tag{13.100}$$

The equilibrium points of the reduced system are the same as those of the original system. The BCU method computes the CUEP as follows:

(i)   For a given system and a contingency, a sustained fault trajectory in time domain is simulated until it reaches the boundary of stability region of the reduced system (the PEBS). The system state at this time instant is referred to as the *exit point*.

(ii)   Starting with the exit point, the reduced system is simulated in time domain until the quantity $\Sigma|g_i|$ reaches its first local minimum. The rotor angles at this point constitute the starting point for the CUEP solution.

(iii)   The controlling UEP is computed using any robust numerical technique. The UEP so obtained is the controlling UEP relative to the fault-on trajectory.

The BCU method appears to be promising and can be easily extended to accommodate more detailed models. More testing is required to verify its accuracy and convergence characteristics. More importantly, the boundary of stability region for power systems modelled in detail needs to be investigated.

(d) Sustained-fault approach [42]

The calculation of the controlling UEP for large systems is very time-consuming. Under severe system conditions, the solution process may either fail to converge or converge to the wrong UEP. The sustained-fault approach was developed to avoid the calculation of the controlling UEP.

The critical energy is determined as follows:

1.   A time domain simulation with a sustained fault is run until the system crosses the potential energy boundary surface (PEBS). The crossing of the PEBS is indicated by the potential energy reaching its maximum.

2.   The potential energy at the crossing of the PEBS is taken as the critical energy for that particular fault location.

Example 13.8

In this example, we illustrate the use of the TEF method for transient stability analysis by considering the simple system of Example 13.1.

For the single-machine infinite bus system of Figure E13.1,

(a)   Write the dynamic equations for the postdisturbance system corresponding to Equations 13.95 and 13.96.

(b)   Write the expression for the system energy function.

(c)   Calculate the postdisturbance system SEP, UEP, and the critical energy V_{cr}.

(d)   Calculate the energy at fault clearing with t_c=0.07 s, 0.086 s, and 0.087 s. Determine the system stability for each of the three fault durations.

Solution

(a) The postdisturbance reduced equivalent circuit of the system is shown in Figure E13.26.

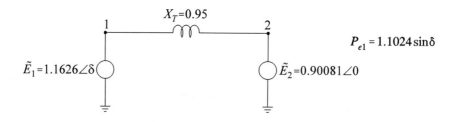

$$X_T = 0.95$$

$$P_{e1} = 1.1024 \sin\delta$$

$$\tilde{E}_1 = 1.1626\angle\delta \qquad \tilde{E}_2 = 0.90081\angle 0$$

Figure E13.26 Postdisturbance equivalent circuit

Because bus 2 represents an infinite bus, $M_T = \infty$ and $\Delta\omega_{COI} = 0$. Consequently, $\theta_1 = \delta$ and $\omega_1 = \dot{\theta}_1/\omega_0 = \Delta\omega_r$.

Since $G_{12}=0$, $D_{12}=0$ and $P'_{m1} = P_{m1} = 0.9$. The postdisturbance dynamic equations of the system, corresponding to Equations 13.95 and 13.96, are

$$2\times 3.5 p(\omega_1) = 0.9 - 1.1024\sin\theta_1$$

$$p(\theta_1) = 377\omega_1$$

(b) From Equation 13.97, the system transient energy function is

$$V = 3.5\times 377\omega_1^2 - 0.9(\theta_1 - \theta_1^s) - 1.1024(\cos\theta_1 - \cos\theta_1^s)$$

(c) At the postdisturbance SEP, we have

$$0.9 - 1.1024\sin\theta_1^s = 0$$

Hence,

$$\theta_1^s = 0.9552 \quad \text{rad} = 54.73°$$

For this simple system there is only one postdisturbance UEP, which is given by

$$\theta_1^u = \pi - \theta_1^s = 2.1864 \quad \text{rad} = 125.27°$$

The critical energy, which equals the system potential energy at the postdisturbance UEP, is

$$V_{cr} = -P'_{m1}(\theta_1^u - \theta_1^s) - C_{12}(\cos\theta_1^u - \cos\theta_1^s)$$

$$= -0.9(2.1864 - 0.9552) - 1.1024(-0.5775 - 0.5775)$$

$$= 0.1651$$

(d)   To compute the energy at fault clearing, we will use the generator angles computed in Example 3.1.

(i) With $t_c = 0.07$ s, $\theta_1^c = 48.58° = 0.8479$ rad, and

$$\omega_c = \frac{0.9}{7} \times 0.07 = 0.009 \quad \text{pu}$$

The total energy at fault clearing is

$$V_{cl} = \frac{7}{2} \times 377 \times 0.09^2 - 0.9(0.8479 - 0.9552) - 1.1024(\cos48.58° - \cos54.73°)$$

$$= 0.1107$$

Since $V_{cl} < V_{cr}$, the system is stable; the stability margin is $V_{cr} - V_{cl} = 0.1651 - 0.1107 = 0.0544$.

(ii) With $t_c = 0.86$ s, $\theta_1^c = 52.04° = 0.9083$ rad, and

$$\omega_c = \frac{0.9}{7} \times 0.086 = 0.011057 \quad \text{pu}$$

The total energy at fault clearing is

$$V_{cl} = \frac{7}{2} \times 377 \times 0.011057^2 - 0.9(0.9083 - 0.9552) - 1.1024(\cos52.04° - \cos54.73°)$$

$$= 0.1620$$

The system is stable; however, the stability margin is now reduced to $0.1651 - 0.1620 = 0.0031$.

(iii) With $t_c = 0.087$ s, $\theta_1^c = 52.3° = 0.9128$ rad, and $\omega_c = 0.011186$ pu. The total energy at fault clearing is

$$V_{cl} = \frac{7}{2} \times 377 \times 0.011186^2 - 0.9(0.9128 - 0.9552) - 1.1024(\cos 52.3° - \cos 54.73°)$$

$$= 0.1657$$

Since $V_{cl} > V_{cr}$, the system is unstable.

The above results are consistent with the time-domain simulation results of Example 13.1. ∎

13.7.3 Limitations of the Direct Methods

In spite of the many significant accomplishments in recent years in the application of the direct methods, modelling limitations and unreliability of computation techniques continue to be major impediments to their widespread practical use. The direct methods are vulnerable to numerical problems when one is solving stressed systems. The use of sophisticated and robust CUEP solution techniques usually imposes a heavy computational burden, making the method slower than the time-domain simulation.

The best course appears to be the use of a hybrid approach in which the transient energy calculation is incorporated into the conventional time-domain simulation [50-54]. This enhances the capability of time-domain simulations by computing the stability margin and minimizing the effort required in determining the stability limits.

REFERENCES

[1]   W.D. Stevenson, *Elements of Power System Analysis*, 3rd Edition, McGraw-Hill, 1975.

[2]   R.T. Byerly and E.W. Kimbark, *Stability of Large Electric Power Systems*, IEEE Press, 1974.

[3]   E.R. Laithwaite and L.L. Freris, *Electric Energy: Its Generation, Transmission and Use*, McGraw-Hill (UK), 1980.

[4]   A. Ralston and H.S. Wilf, *Mathematical Methods for Digital Computers*, John Wiley & Sons, 1962.

[5]   B. Carnahan, H.A. Luther, and J.O. Wilkes, *Applied Numerical Methods*, John Wiley & Sons, 1969.

[6]   C.W. Gear, *Numerical Initial Value Problems in Ordinary Differential Equations*, Prentice-Hall, 1971.

[7]   Kundur Prabhashankar, *Digital Simulation and Analysis of Power System Dynamic Performance*, Ph.D. thesis, University of Toronto, 1967.

[8]   P. Kundur and P.L. Dandeno, "Implementation of Advanced Generator Models into Power System Stability Programs," *IEEE Trans.*, Vol. PAS-102, pp. 2047-2052, July 1983.

[9]   H.W. Dommel, "Digital Computer Solution of Electromagnetic Transients in Single and Multiphase Networks," *IEEE Trans.*, Vol. PAS-88, pp. 388-399, April 1969.

[10]   B. Stott, "Power System Dynamic Response Calculations," *Proc. IEEE*, Vol. 67, pp. 219-241, February 1979.

[11]   P.L. Dandeno and P. Kundur, "A Non-iterative Transient Stability Program Including the Effects of Variable Load-Voltage Characteristics," *IEEE Trans.*, Vol. PAS-92, pp. 1478-1484, September 1973.

[12]   H.W. Dommel and N. Sato, "Fast Transient Stability Solutions," *IEEE Trans.*, Vol. PAS-91, pp. 1643-1650, July/August 1972.

[13]   M. Stubbe, A. Bihain, J. Deuse, and J.C. Baader, "STAG - A New Unified Software Program for the Study of the Dynamic Behaviour of Electrical Power Systems," *IEEE Trans.*, Vol. PWRS-4, No. 1, pp. 129-138, 1989.

[14]   EPRI Report, *Programmers Manual for ETMSP Version 3*, Prepared by Ontario Hydro, 1992.

[15]   C.F. Wagner and R.D. Evans, *Symmetrical Components*, McGraw-Hill, 1933.

[16]   Edith Clarke, *Circuit Analysis of AC Power Systems*, Vol. I, John Wiley & Sons, 1943.

[17]   P.M. Anderson, *Analysis of Faulted Power Systems*, Iowa State Press, Ames, Iowa, 1973.

[18]   Westinghouse Electric Corporation, *Electric Transmission and Distribution Reference Book*, East Pittsburgh, Pa., 1964.

[19]   C.L. Fortescue, "Method of Symmetrical Coordinates Applied to the Solution of Polyphase Networks," *AIEE Trans.*, Vol. 37, Pt. II, pp. 1027-1140, 1918.

[20]   T.M. O'Flaherty and A.S. Aldred, "Synchronous Machine Stability under Unsymmetrical Faults," *Proceedings IEE*, Vol. 109A, p. 431, 1962.

[21] Westinghouse Electric Corporation, *Applied Protective Relaying*, a new "Silent Sentinels" publication, Newark, 1976.

[22] H.M. Rustebakke (editor), *Electric Utility Systems and Practices*, John Wiley & Sons, 1983.

[23] C.R. Mason, *The Art and Science of Protective Relaying*, John Wiley & Sons, 1956.

[24] General Electric Company, *Use of the R-X Diagram in Relay Work*, Philadelphia, Pa., 1966.

[25] C.R. Mason, "Relay Operation during System Oscillations," *AIEE Trans.*, Vol. 56, pp. 823-832, 1937. Discussion by J.H. Neher, pp. 1513-1514. Closure by C.R. Mason, Vol. 57, pp. 111-114, 1938.

[26] J.H. Neher, "A Comprehensive Method of Determining the Performance of Distance Relays," *AIEE Trans.*, Vol. 56, pp. 833-844, 1937.

[27] Edith Clarke, "Impedances Seen by Relays during Power Swings with and without Faults," *AIEE Trans.*, Vol. 64, pp. 372-384, 1945. Discussion by A.J. McConnell, p. 472, June Supplement 1945.

[28] IEEE Working Group Report, "Out of Step Relaying for Generators," *IEEE Trans.*, Vol. PAS-96, pp. 1556-1564, September/October 1977.

[29] C.R. Mason, "A New Loss-of-Excitation Relay for Synchronous Generators," *AIEE Trans.*, Vol. 68, pp. 1240-1245, 1949.

[30] J. Berdy, "Loss of Excitation Protection for Modern Synchronous Generators," *IEEE Trans.*, Vol. PAS-94, pp. 1457-1463, September/October 1975.

[31] D.C. Lee, P. Kundur, and R.D, Brown, "A High Speed, Discriminating Generator Loss of Excitation Protection," *IEEE Trans.*, Vol. PAS-98, pp. 1895-1898, November/December 1979.

[32] P.C. Magnusson, "Transient Energy Method of Calculating Stability," *AIEE Trans.*, Vol. 66, pp. 747-755, 1947.

[33] P.D. Aylett, "The Energy Integral-Criterion of Transient Stability Limits of Power Systems," *Proc. IEE*, Vol. 105c, No. 8, pp. 527-536, September 1958.

[34] G.E. Gless, "Direct Method of Lyapunov Applied to Transient Power System Stability," *IEEE Trans.*, Vol. PAS-85, No. 2, pp. 159-168, February 1966.

[35]   A.H. El-Abiad and K. Nagappan, "Transient Stability Regions of Multimachine Power Systems," *IEEE Trans.*, Vol. PAS-85, No. 2, pp. 169-178, February 1966.

[36]   M. Ribbens-Pavella, "Critical Survey of Transient Stability Studies of Multi-Machine Power Systems by Lyapunov's Direct Method," *Proc. 9th Annual Allerton Conference on Circuits and System Theory*, pp. 751-767, October 1971.

[37]   A.A. Fouad, "Stability Theory-Criteria for Transient Stability," *Proc. Engineering Foundation Conference on System Engineering for Power*, pp. 421-450, Henniker, N.H., 1975.

[38]   M.A. Pai, *Energy Function Analysis for Power System Stability*, Kluwer Academic Press, 1989.

[39]   A.A. Fouad and Vijay Vittal, *Power System Transient Stability Analysis Using the Transient Energy Function Method*, Prentice-Hall, 1992.

[40]   T. Athay, V.R. Sherkat, R. Podmore, S. Virmani, and C. Puech, "Transient Energy Analysis," Final Report of U.S. Department of Energy Contract No. EX-76-C-01-2076, Prepared by Systems Control, Inc., June 1979.

[41]   Chi-Keung Tang, "Evaluation of the Direct Method for Power System Transient Stability Analysis," M.Eng. degree thesis, University of Toronto, August 1984.

[42]   N. Kakimoto, Y. Ohsawa, and M. Hayashi, "Transient Stability Analysis of Electric Power System Via Lure' Type Lyapunov Functions, Parts I and II," *Trans. IEE of Japan*, Vol. 98, No. 516, May/June 1978.

[43]   A.A. Fouad and S.E. Stanton, "Transient Stability of Multimachine Power System, Parts I and II," *IEEE Trans.*, Vol. PAS-100, pp. 3408-3424, July 1981.

[44]   H.D. Chiang, F.F. Wu, and P.P. Varaiya, "Foundations of the Potential Energy Boundary Surface Method for Power System Transient Stability Analysis," *IEEE Trans.*, Vol. CAS-35, pp. 712-728, June 1988.

[45]   M.A. El-Kady, M.M. Abu-Elnaga, R.D. Findlay, and C.G. Bailey, "Sparse Formulation of the Direct Method of Transient Stability for Applications on Micro-Computers," *Proc. 29th Midwest Symposium on Circuits and Systems*, pp. 519-522, Lincoln, Neb., 1986.

[46]   A.A. Fouad, V. Vittal, Y-X. Ni, H.M. Zein-Eldin, E. Vaahedi, H.R. Pota, K. Nodehi, and J. Kim, "Direct Transient Stability Assessment with Excitation Control," *IEEE Trans.*, Vol. PWRS-4, No. 1, pp. 75-82, February 1989.

[47]   V. Vittal, N. Bhatia, A.A. Fouad, G.A. Maria, and H.M. Zein-Eldin, "Incorporation of Nonlinear Load Models in the Transient Energy Function Method," *IEEE Trans.*, Vol. PWRS-4, No. 3, pp. 1031-1036, August 1989.

[48]   T. Athay, R. Podmore, and S. Virmani, "A Practical Method for the Direct Analysis of Transient Stability," *IEEE Trans.*, Vol. PAS-98, No. 2, pp. 573-584, March/April 1979.

[49]   A.A. Fouad, V. Vittal, and Taekyoo Oh, "Critical Energy for Transient Stability Assessment of a Multimachine Power System," *IEEE Trans.*, Vol. PAS-103, pp. 2199-2206, 1984.

[50]   P. Kundur, "A Practical View of the Applicability of the Direct Methods," Minipaper presented at the IEEE Panel Session on Application of Direct Methods to Transient Stability Analysis of Power Systems, *IEEE Trans.*, Vol. PAS-193, pp. 1634-1635, July 1984.

[51]   P. Kundur, "Evaluation of Methods for Studying Power System Stability," *Proc. International Symposium on Power System Stability*, Ames, Iowa, 1985.

[52]   B. Gao, "Assessment of Power System Transient Stability Using Energy Functions," M.A.Sc. thesis, University of Toronto, 1986.

[53]   G. Maria, C. Tang, and J. Kim, "Hybrid Transient Stability Analysis," *IEEE Trans.*, Vol. PWRS-5, No. 2, pp. 384-393, May 1990.

[54]   C.K. Tang, C.E. Graham, M. El-Kady, and R.T.H. Alden, "Transient Stability Index from Conventional Time Domain Simulation," Paper 93SM487-9PWRS, presented at the 1993 IEEE PES Summer Meeting, Vancouver, July 1993.

Chapter 14

Voltage Stability

Voltage control and stability problems are not new to the electric utility industry but are now receiving special attention in many systems. Once associated primarily with weak systems and long lines, voltage problems are now also a source of concern in highly developed networks as a result of heavier loadings. In recent years, voltage instability has been responsible for several major network collapses. The following are some examples [1,2]:

- New York Power Pool disturbances of September 22, 1970

- Florida system disturbance of December 28, 1982

- French system disturbances of December 19, 1978, and January 12, 1987

- Northern Belgium system disturbance of August 4, 1982

- Swedish system disturbance of December 27, 1983

- Japanese system disturbance of July 23, 1987

As a consequence, the terms "voltage instability" and "voltage collapse" are appearing more frequently in the literature and in discussions of system planning and operation.

Although low voltages can be associated with the process of rotor angles going out of step, the type of voltage collapse related to voltage instability can occur where "angle stability" is not an issue. The gradual pulling out of step of machines, as rotor angles between two groups of machines approach or exceed 180°, results in very low voltages at intermediate points in the network (see Chapter 13, Section 13.5.3). However, in such cases the low voltage is a result of the rotors falling out of step rather than a cause of it.

Voltage stability, as described in Chapter 2, is concerned with the ability of a power system to maintain acceptable voltages at all buses in the system under normal conditions and after being subjected to a disturbance. A system enters a state of voltage instability when a disturbance, increase in load demand, or change in system condition causes a progressive and uncontrollable decline in voltage. The main factor causing instability is the inability of the power system to meet the demand for reactive power.

This chapter will review basic concepts related to voltage stability and characterize the "voltage avalanche" phenomenon. The dynamic and static approaches to voltage stability analysis will be described, and methods identified for preventing voltage instability.

14.1 BASIC CONCEPTS RELATED TO VOLTAGE STABILITY

Voltage stability problems normally occur in heavily stressed systems. While the disturbance leading to voltage collapse may be initiated by a variety of causes, the underlying problem is an inherent weakness in the power system. In addition to the strength of transmission network and power transfer levels, the principal factors contributing to voltage collapse are the generator reactive power/voltage control limits, load characteristics, characteristics of reactive compensation devices, and the action of voltage control devices such as transformer under-load tap changers (ULTCs).

This section illustrates the basic concepts related to voltage instability by firstly considering the characteristics of transmission systems and then examining how the phenomenon is influenced by the characteristics of generators, loads, and reactive power compensation devices.

14.1.1 Transmission System Characteristics

The characteristics of interest are the relationships between the transmitted power (P_R), receiving end voltage (V_R), and the reactive power injection (Q_i). Such characteristics were discussed for a simple radial system in Chapter 2 (Section 2.1) and for transmission lines of varying lengths in Chapter 6 (Section 6.1). For complex systems with a large number of voltage sources and load buses, similar characteristics can be determined by using power-flow analysis (see Chapter 6).

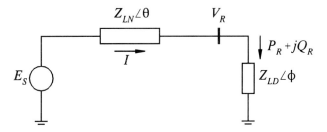

(a) Schematic diagram

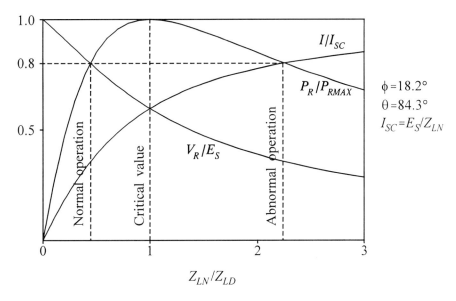

(b) Receiving end voltage, current and power as a function of load demand

Figure 14.1 Characteristics of a simple radial system

Let us briefly review the characteristics of the simple radial system considered in Chapter 2 (Figure 2.4). For reference the schematic diagram of the system is reproduced in Figure 14.1(a). As shown in Section 2.1.2, the current I and receiving end voltage V_R and power P_R are given by the following equations:

$$I = \frac{1}{\sqrt{F}} \frac{E_S}{Z_{LN}} \tag{14.1}$$

$$V_R = \frac{1}{\sqrt{F}} \frac{Z_{LD}}{Z_{LN}} E_S \tag{14.2}$$

$$P_R = \frac{Z_{LD}}{F}\left(\frac{E_S}{Z_{LN}}\right)^2 \cos\phi \qquad (14.3)$$

where

$$F = 1+\left(\frac{Z_{LD}}{Z_{LN}}\right)^2+2\left(\frac{Z_{LD}}{Z_{LN}}\right)\cos(\theta-\phi)$$

Plots of I, V_R, and P_R are shown in Figure 14.1(b) as a function of load demand (Z_{LN}/Z_{LD}), for the case with $\tan\theta=10.0$ and $\cos\phi=0.95$. To make the results applicable to any value of Z_{LN}, the values of I, V_R and P_R are appropriately normalized.

As load demand increases (Z_{LD} decreases), P_R increases rapidly at first and then slowly before reaching a maximum, and finally decreases. There is thus a maximum value of active power that can be transmitted through an impedance from a constant voltage source. The power transmitted is maximum when the voltage drop in the line is equal in magnitude to V_R, i.e., when $Z_{LD}/Z_{LN}=1$. The conditions corresponding to maximum power represent the limits of satisfactory operation. The values of V_R and I corresponding to maximum power are referred to as *critical values*.

For a given value of power P_R delivered ($P_R<P_{RMAX}$), two operating points may be found corresponding to two different values of Z_{LD}. This is shown in Figure 14.1(b) for $P_R=0.8$. The point to the left corresponds to normal operation. At the operating point to the right, I is much larger and V_R much lower than for the point to the left.

For a load demand higher than the maximum power, control of power by varying the load would be unstable, i.e., an increase in load admittance would reduce power. In this region, the load voltage may or may not progressively decrease depending on the load-voltage characteristic. With a constant-admittance load characteristic, the system condition stabilizes at a voltage level that is lower than normal. On the other hand, if the load is supplied by a transformer with ULTC, the tap-changer action will try to raise the load voltage, which has the effect of reducing effective Z_{LD}. This lowers V_R still further and leads to a progressive reduction of voltage. This is the phenomenon of *voltage instability*.

From Equation 14.3, we see that the maximum value of P_R can be increased by increasing the source voltage E_S and/or decreasing ϕ.

A more traditional method of illustrating the phenomenon is to plot the relationship between V_R and P_R, for different values of load power factor with E_S constant as shown in Figure 14.2. The locus of critical operating points is shown by dashed lines in the figure. Only the operating points above the critical points represent satisfactory operating conditions.

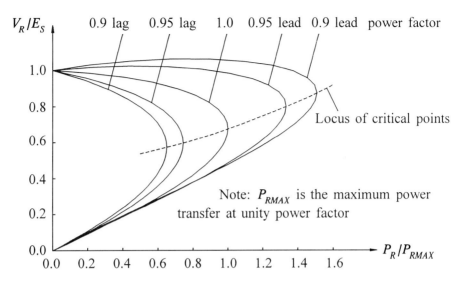

Figure 14.2 The V_R-P_R characteristics of the system of Figure 14.1

   In Chapter 6 (Section 6.1), we developed similar *V-P* characteristics for transmission lines of different lengths. Practical power systems consisting of many voltage sources and load buses also exhibit similar relationships between active power transfer and load bus voltages. We will illustrate this for the system shown in Figure 14.3, consisting of 39 buses with nine generators and one synchronous condenser. Figure 14.4 shows the *V-P* curve for the system; it represents the variation in voltage at bus 530, a critical bus in the load area prone to voltage instability, as a function of total active power load in the shaded area. This curve has been produced by using a series of power-flow solutions for different load levels. The loads in area 1 (shaded) are uniformly scaled up while the power factor is kept constant. The active power outputs of generators are correspondingly increased in proportion to the size of generator. The *P* and *Q* components of each load are assumed to be independent of the bus voltage. At the "knee" of the *V-P* curve, the voltage drops rapidly with an increase in load demand (or nominal voltage load). Power-flow solution fails to converge beyond this limit, which is indicative of instability. Operation at or near the stability limit is impractical and a satisfactory operating condition is ensured by allowing sufficient "power margin."

   We see that complex systems have *V-P* characteristics similar to those of the simple radial system of Figure 14.1. Such characteristics represent the basic property of networks with predominantly inductive elements.

   We have so far considered the *V-P* characteristics with constant load power factor. Voltage stability, in fact, depends on how variations in *Q* as well as *P* in the load area affect the voltages at the load buses. Often, a more useful characteristic for certain aspects of voltage stability analysis is the *Q-V* relationship, which shows the

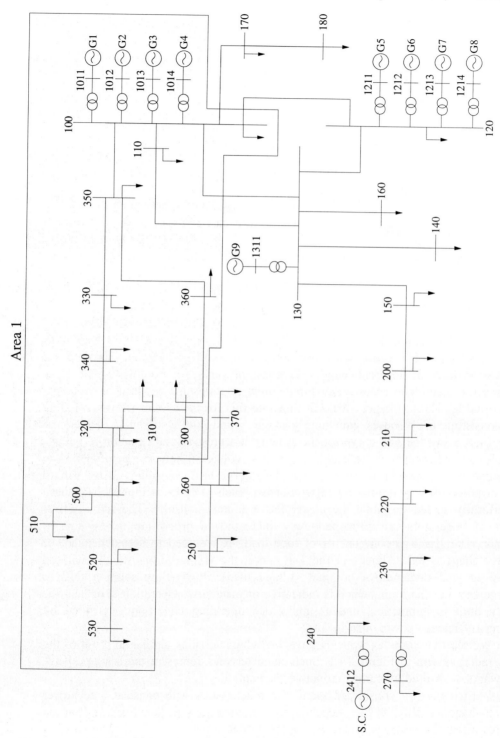

Figure 14.3 A 39-bus, 10-machine test system

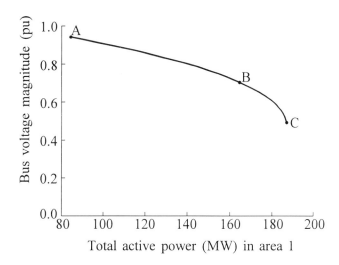

Figure 14.4 The *V-P* curve at bus 530 of the system shown in Figure 14.3

sensitivity and variation of bus voltages with respect to reactive power injections or absorptions. For the simple radial system of Figure 14.1, such characteristics for different values of load power are shown in Figure 2.8 of Chapter 2. These characteristics can be derived more readily than the *V-P* characteristics for systems with a non-radial type structure and are better suited for examining the requirements for reactive power compensation.

Figure 14.5 shows the *Q-V* curves computed at buses 160, 200, 510, and 530 for the three operating conditions represented by points A, B, and C on the *V-P* curve of Figure 14.4. Point A represents the base case, point B a condition near the critical operating point, and point C a condition at the critical operating point. Each of these *Q-V* curves has been produced by successive power-flow calculations with a variable reactive power source at the selected bus and recording its values required to hold different scheduled bus voltages. The bottom of the *Q-V* curve, where the derivative *dQ/dV* is equal to zero, represents the voltage stability limit. Since all reactive power control devices are designed to operate satisfactorily when an increase in *Q* is accompanied by an increase in *V*, operation on the right side of the *Q-V* curve is stable and on the left side is unstable. Also, voltage on the left side may be so low that protective devices may be activated. The bottom of the *Q-V* curve, in addition to identifying the stability limit, defines the minimum reactive power requirement for stable operation [3].

In this section we have examined the characteristics of transmission systems as impacted by the flow of active and reactive power through highly inductive elements. It is evident from the analysis presented here that the following are the principal causes of voltage instability:

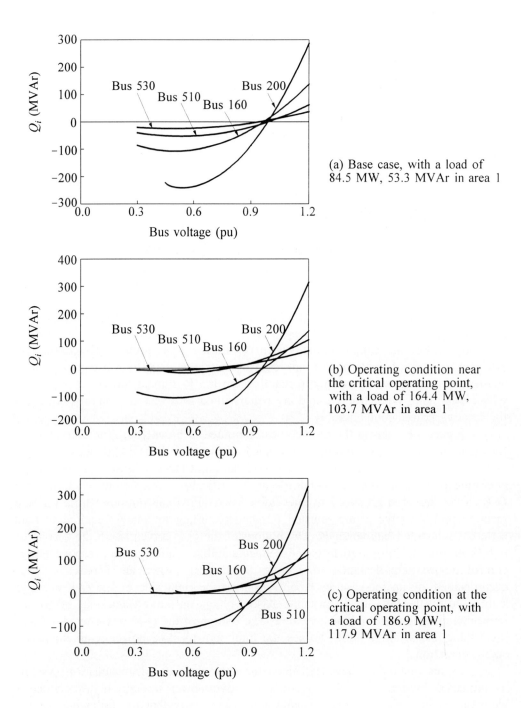

(a) Base case, with a load of 84.5 MW, 53.3 MVAr in area 1

(b) Operating condition near the critical operating point, with a load of 164.4 MW, 103.7 MVAr in area 1

(c) Operating condition at the critical operating point, with a load of 186.9 MW, 117.9 MVAr in area 1

Figure 14.5 The Q-V curves for system shown in Figure 14.3

- The load on the transmission lines is too high.

- The voltage sources are too far from the load centres.

- The source voltages are too low.

- There is insufficient load reactive compensation.

The transmission system V-P and Q-V characteristics have been introduced here primarily to illustrate the basic phenomenon associated with voltage instability. The approach presented for deriving the characteristics by using conventional power-flow programs is, however, not necessarily the most efficient way of studying the voltage stability problem. Methods of analyzing voltage stability are discussed in Section 14.3.

14.1.2 Generator Characteristics

Generator AVRs are the most important means of voltage control in a power system. Under normal conditions the terminal voltages of generators are maintained constant. During conditions of low-system voltages, the reactive power demand on generators may exceed their field current and/or armature current limits (see Chapter 5, Section 5.4). When the reactive power output is limited, the terminal voltage is no longer maintained constant.

The generator field current is automatically limited by an *overexcitation limiter* (OXL). The function and modelling of such limiters are described in Chapter 8 (Sections 8.5 and 8.6). With constant field current, the point of constant voltage is behind the synchronous reactance (see Chapter 3, Figure 3.22). This effectively increases the network reactance significantly, further aggravating the voltage collapse condition.

On most generators, the armature current limit is realized manually by operators responding to alarms. The operator reduces reactive and/or active power output to bring the armature current within safe limits. On some generators, automatic armature current limiters with time delay are used to limit reactive power output through the AVR [2].

To illustrate the impact of loss of generator voltage control capability, consider the system shown in Figure 14.6(a). It consists of a large load supplied radially from an infinite bus, with intermediate generation supplying part of the load and regulating voltage (V_I).

With voltage at the intermediate bus maintained, the V-P characteristic is shown by curve 1 in Figure 14.6(b). When the generating unit at the intermediate point hits its field current limit, the bus voltage (V_I) is no longer maintained and the V-P characteristic is shown by curve 2. An operating condition such as that represented by point A is considerably more stable when on curve 1 than when it is on curve 2. These results demonstrate the importance of maintaining the voltage control capability of generators. In addition, they show that the degree of voltage stability cannot be judged based only on how close the bus voltage is to the normal voltage level.

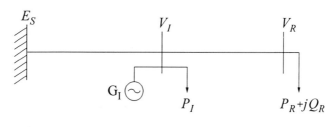

(a) Schematic diagram

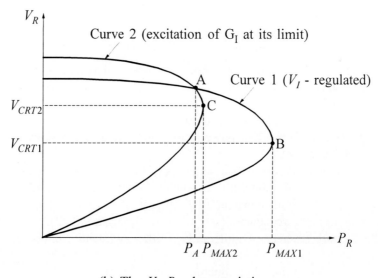

(b) The V_R-P_R characteristics

Figure 14.6 Impact of loss of regulation of intermediate bus voltage

This situation is similar to that which led to voltage collapse in the Brittany region of the French system in December 1965 and in November 1975 [4].

14.1.3 Load Characteristics [1,5]

Load characteristics and distribution system voltage control devices are among the key factors influencing system voltage stability.

The characteristics and modelling of different types of loads are discussed in Chapter 7. Loads whose active and reactive components vary with voltage interact with the transmission characteristics by changing the power flow through the system. The system voltages settle at values determined by the composite characteristic of the transmission system and loads.

Distribution system voltage regulators and substation transformer ULTCs attempt to hold constant voltage at the point of consumption (see Chapter 11, Section 11.2). Within the normal control range, loads appear effectively as constant MVA loads. This may have a destabilizing effect during conditions of voltage collapse. The effect of ULTCs will be discussed further in Section 14.2.

When the ULTCs reach the end of their tap range, distribution system voltages begin to drop. The *residential* active and reactive loads will drop with voltage. This will in turn reduce line loading and, hence, the line reactive losses. The *industrial* loads, with large components of induction motors, will change little. However, the capacitors in the industrial area will supply less reactive power, thereby causing a net increase in the reactive load [1].

When the distribution voltages remain low for a few minutes, thermostats and other load regulation devices, as well as manual controls, tend to restore load. For example, heating-type loads will run longer to bring the temperature to the level called for by the thermostats. Consequently, more such devices will be operating at any given time. Many loads of this type will be restored to their normal full voltage value over a 10 to 15 minute period [5]. As voltage-sensitive controlled loads creep back, transmission and distribution voltages will drop further.

At voltages below 85 to 90% of the nominal value, some induction motors may stall and draw high reactive current. This brings the voltages down further. Industrial and commercial motors are usually controlled by magnetically held contactors; therefore, the voltage drop will cause many motors to drop out. The loss of load will result in the recovery of voltages. After some time, the motors are restored to service. This may cause voltages to drop again if the original cause of the voltage problem still persists.

It is evident from the above discussion that, for accurate analysis of voltage stability, the network representation must include the effects of distribution transformer tap-changer action and capacitors in the distribution systems. Depending on the scope of the study, the representation of load characteristics should take into consideration the effects of thermostats and other load regulation devices. In industrial areas, motors and capacitors may need to be represented explicitly.

14.1.4 Characteristics of Reactive Compensating Devices

In Chapter 11, we considered different types of reactive compensating devices. Here, we briefly describe how these devices influence voltage stability.

(a) Shunt capacitors

By far the most inexpensive means of providing reactive power and voltage support is the use of shunt capacitors. They can be effectively used up to a certain point to extend the voltage stability limits by correcting the receiving end power factor. They can also be used to free up "spinning reactive reserve" in generators and thereby help prevent voltage collapse in many situations.

Shunt capacitors, however, have a number of inherent limitations from the viewpoint of voltage stability and control:

• In heavily shunt capacitor compensated systems, the voltage regulation tends to be poor.

• Beyond a certain level of compensation, stable operation is unattainable with shunt capacitors (this is illustrated in Example 14.1).

• The reactive power generated by a shunt capacitor is proportional to the square of the voltage; during system conditions of low voltage the var support drops, thus compounding the problem.

(b) Regulated shunt compensation

A *static var system (SVS)* of finite size will regulate up to its maximum capacitive output. There are no voltage control or instability problems within the regulating range. When pushed to the limit, an SVS becomes a simple capacitor. The possibility of this leading to voltage instability must be recognized.

A *synchronous condenser*, unlike an SVS, has an internal voltage source. It continues to supply reactive power down to relatively low voltages and contributes to a more stable voltage performance.

(c) Series capacitors

Series capacitors are self-regulating. The reactive power supplied by series capacitors is proportional to square of the line current and is independent of the bus voltages. This has a favourable effect on voltage stability.

Series capacitors are ideally suited for effectively shortening long lines. Unlike shunt capacitors, series capacitors reduce both the characteristic impedance (Z_C) and the electrical length (θ) of the line (see Chapter 11). As a result, both voltage regulation and stability are significantly improved.

Example 14.1

Figure E14.1 shows the system representation applicable to a 322 km (200 mi), 500 kV transmission line supplying a radial load from a strong system. The line parameters are expressed in per unit on 100 MVA and 500 kV base.

(a)   With the sending end voltage (V_1) maintained at 1.0 pu, generate Q-V curves at the receiving end for four different values of receiving end load power: 1300, 1500, 1700, and 1900 MW assumed at unity power factor. Together with the Q-V curves of the transmission system, plot the shunt capacitor Q-V characteristics with the reactive power injection at 1.0 pu voltage being 300, 450, 675, and 950 MVAr, respectively. Examine the effectiveness of shunt

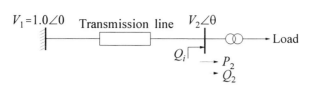

(a) Schematic diagram

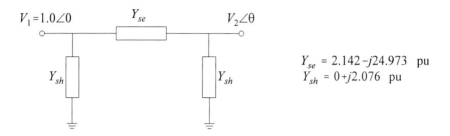

$$Y_{se} = 2.142 - j24.973 \ \text{pu}$$
$$Y_{sh} = 0 + j2.076 \ \text{pu}$$

(b) Equivalent π circuit representation of line

Figure E14.1 A 322 km, 500 kV line supplying a radial load

capacitor compensation as a means of providing reactive power compensation. Assume that the load at the receiving end exhibits a constant MVA steady-state characteristic due to the action of transformer tap changers.

(b)   If the reactive compensation at the receiving end is in the form of an SVC with a capacitive limit of 950 MVAr, examine voltage stability of the system as P_2 is gradually increased from 1300 MW to 1900 MW.

Solution

(a) Figure E14.2 shows the steady-state Q_i-V_2 characteristics of the transmission line and the shunt capacitors.

The transmission line characteristics are shown in solid curves. These curves represent the relationship between voltage at the receiving end bus and injections of reactive power at that bus, each corresponding to a given level of receiving-end power, assumed at unity power factor.

The relationships between voltage and the reactive power produced by shunt capacitors are shown in dashed lines. The intersection between a solid curve and a dashed line establishes the steady-state operating point corresponding to the respective receiving-end power and shunt capacitor rating.

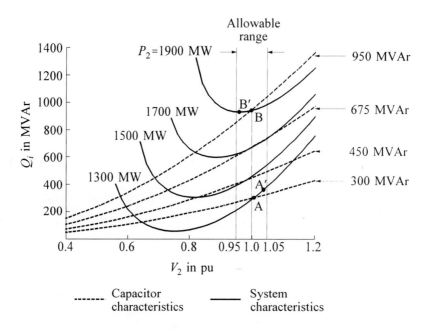

Figure E14.2 System and shunt capacitor steady-state Q-V
characteristics; capacitor MVAr shown at
rated voltage

Let us examine the steady-state system performance with a load level of 1300 MW
and a capacitor bank of 300 MVAr, represented by the operating point A. At this
point the slope $\Delta Q/\Delta V$ of the system is greater than that of the shunt capacitor; this
represents stable operation. When perturbed by a small transient disturbance, the
system returns to operating point A. The addition of a small amount of capacitance
as represented by operating point A′ results in an increase in voltage, a characteristic
normally expected.

The situation is quite different at operating point B, with a receiving power of 1900
MW and a capacitor bank of 950 MVAr. Now, the slope $\Delta Q/\Delta V$ of the system is less
than that of the capacitor. A small perturbation leads to progressive deviation in V_2.
An increase of shunt capacitor by a small amount, as represented by point B′, results
in a decrease in bus voltage.

We thus see that at very high levels of shunt compensation, stable operation is not
possible. The limiting load power level is about 1700 MW requiring a shunt capacitor
of 675 MVAr. At this level, the slope $\Delta Q/\Delta V$ of the system is nearly equal to that of
the shunt capacitor.

In the above analysis we have considered only the steady-state performance with the
load at the receiving end maintaining constant MVA due to transformer ULTC action.

The transient response depends on the inherent load characteristics. For example, with a constant current load characteristic, switching additional load when operating at point B causes a transient reduction of V_2 and P_2. The system is voltage stable in the short term. However, the action of the transformer ULTC, as it attempts to raise the secondary voltage, causes an increase in primary (line) current. This results in a decrease in V_2 and P_2. The voltage V_2 decreases with each tap movement until the tap changer reaches its limit. The system settles at low values of V_2 and P_2.

(b)   Figure E14.3 shows the steady-state Q_i-V_2 characteristics with a *static var compensator* (SVC). The SVC maintains constant voltage V_2 until its maximum capacitive output limit of 950 MVAr is reached. Consequently, for values of P_2 less than 1900 MW, the SVC maintains V_2 at 1.0 pu. When P_2 reaches 1900 MW, the SVC hits its capacitive limit and its characteristic is that of a simple capacitor. This leads to voltage instability.

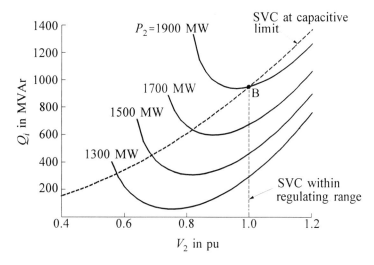

Figure E14.3 System and SVC Q-V characteristics;
SVC capacitive limit 950 MVAr   ■

14.2 VOLTAGE COLLAPSE

Voltage collapse is the process by which the sequence of events accompanying voltage instability leads to a low *unacceptable* voltage profile in a significant part of the power system.

Voltage collapse may be manifested in several different ways. We will describe a typical scenario of voltage collapse, and then provide a general characterization of the phenomenon based on actual incidents of collapse.

14.2.1 Typical Scenario of Voltage Collapse

When a power system is subjected to a sudden increase of reactive power demand following a system contingency, the additional demand is met by the reactive power reserves carried by the generators and compensators. Generally there are sufficient reserves and the system settles to a stable voltage level. However, it is possible, because of a combination of events and system conditions, that the additional reactive power demand may lead to voltage collapse, causing a major breakdown of part or all of the system.

A typical scenario of a voltage collapse would be as follows [6]:

- The power system is experiencing abnormal operating conditions with large generating units near the load centres being out of service. As a result, some EHV lines are heavily loaded and reactive power resources are at a minimum.

- The triggering event is the loss of a heavily loaded line which would cause additional loading on the remaining adjacent lines. This would increase the reactive power losses in the lines (Q absorbed by a line increases rapidly for loads above surge impedance loading), thereby causing a heavy reactive power demand on the system.

- Immediately following the loss of the EHV line, there would be a considerable reduction of voltage at adjacent load centres due to extra reactive power demand. This would cause a load reduction, and the resulting reduction in power flow through the EHV lines would have a stabilizing effect. The generator AVRs would, however, quickly restore terminal voltages by increasing excitation. The resulting additional reactive power flow through the inductances associated with generator transformers and lines would cause increased voltage drop across each of these elements.

  At this stage, generators would likely be within the P-Q output capabilities, i.e., within the armature and field current heating limits. The speed governors would regulate frequency by reducing MW output.

- The EHV level voltage reduction at load centres would be reflected into the distribution system. The ULTCs of substation transformers would restore distribution voltages and loads to prefault levels in about 2 to 4 minutes. With each tap change operation, the resulting increment in load on EHV lines would increase the line XI^2 and RI^2 losses, which in turn would cause a greater drop in EHV levels. If the EHV line is loaded considerably above the SIL, each MVA increase in line flow would cause several MVArs of line losses.

- As a result, with each tap-changing operation, the reactive output of generators throughout the system would increase. Gradually, the generators would hit their reactive power capability limits (imposed by maximum allowable continuous field current) one by one. When the first generator reached its field current limit, its terminal voltage would drop. At the reduced terminal voltage for a fixed MW output, the armature current would increase. This may further limit reactive output to keep the armature current within allowable limits. Its share of reactive loading would be transferred to other generators, leading to overloading of more and more generators. With fewer generators on automatic excitation control, the system would be much more prone to voltage instability. This would likely be compounded by the reduced effectiveness of shunt compensators at low voltages.

The process will eventually lead to voltage collapse or avalanche, possibly leading to loss of synchronism of generating units and a major blackout.

14.2.2 General Characterization Based on Actual Incidents

There have been a number of voltage collapse incidents worldwide (see references 1, 2, 7 and 8 for descriptions). Based on these incidents, voltage collapse may be characterized as follows:

1. The initiating event may be due to a variety of causes: small gradual system changes such as natural increase in system load, or large sudden disturbances such as loss of a generating unit or a heavily loaded line. Sometimes, a seemingly uneventful initial disturbance may lead to successive events that eventually cause system collapse.

2. The heart of the problem is the inability of the system to meet its reactive demands. Usually, but not always, voltage collapse involves system conditions with heavily loaded lines. When transport of reactive power from neighbouring areas is difficult, any change that calls for additional reactive power support may lead to voltage collapse.

3. The voltage collapse generally manifests itself as a slow decay of voltage. It is the result of an accumulative process involving the actions and interactions of many devices, controls, and protective systems. The time frame of collapse in such cases could be on the order of several minutes.

    The duration of voltage collapse dynamics in some situations may be much shorter, being on the order of a few seconds. Such events are usually caused by unfavourable load components such as induction motors or dc converters [8]. The time frame of this class of voltage instability is the same as that of rotor angle instability. In many situations, the distinction between voltage and angle instability may not be clear, and some aspects of both phenomena may exist. This form of voltage instability may be analyzed by conventional transient stability simulations, provided appropriate models are used to represent the devices, particularly induction motor loads, and various controls and protection associated with the generators and transmission equipment.

Based on the above considerations, the companion book by C.W. Taylor [8] classifies voltage stability into *transient* and *longer-term time frames.* Appendix F of the book groups known incidents of voltage instability by these time frames.

4.   Voltage collapse is strongly influenced by system conditions and characteristics. The following are the significant factors contributing to voltage instability/collapse:

   • Large distances between generation and load

   • ULTC action during low voltage conditions

   • Unfavourable load characteristics

   • Poor coordination between various control and protective systems

5.   The voltage collapse problem may be aggravated by excessive use of shunt capacitor compensation. Reactive compensation can be made most effective by the judicious choice of a mixture of shunt capacitors, static var systems, and possibly synchronous condensers.

14.2.3 Classification of Voltage Stability

It is helpful to classify voltage stability into two categories: *large-disturbance* voltage stability and *small-disturbance* voltage stability. These subdivisions essentially decouple phenomena that must be examined by using nonlinear dynamic analysis from those that can be examined by using steady-state analysis. This classification can simplify analytical tool development and application, and it can result in tools that produce *complementary* information.

Large-disturbance voltage stability is concerned with a system's ability to control voltages following large disturbances such as system faults, loss of load, or loss of generation. Determination of this form of stability requires the examination of the dynamic performance of the system over a period of time sufficient to capture the interactions of such devices as ULTCs and generator field current limiters. Large-disturbance voltage stability can be studied by using nonlinear time-domain simulations which include proper modelling.

Large-disturbance voltage stability, as discussed in Section 14.2.2, may be further subdivided into transient and long-term time frames.

Small-disturbance (or small-signal) voltage stability is concerned with a system's ability to control voltages following small perturbations, such as gradual changes in load. This form of stability can be effectively studied with steady-state approaches that use linearization of the system dynamic equations at a given operating point.

Following a disturbance, the system voltages often do not return to the original level. Therefore, it is necessary to define the region of voltage level considered acceptable. The system is then said to have *finite stability* within the specified region of voltage level.

14.3 VOLTAGE STABILITY ANALYSIS

The analysis of voltage stability for a given system state involves the examination of two aspects [9]:

(a)   *Proximity to voltage instability*: How close is the system to voltage instability?

Distance to instability may be measured in terms of physical quantities, such as load level, active power flow through a critical interface, and reactive power reserve. The most appropriate measure for any given situation depends on the specific system and the intended use of the margin; for example, planning versus operating decisions. Consideration must be given to possible contingencies (line outages, loss of a generating unit or a reactive power source, etc.).

(b)   *Mechanism of voltage instability*: How and why does instability occur? What are the key factors contributing to instability? What are the voltage-weak areas? What measures are most effective in improving voltage stability?

Time-domain simulations, in which appropriate modelling is included, capture the events and their chronology leading to instability. However, such simulations are time-consuming and do not readily provide sensitivity information and the degree of stability.

System dynamics influencing voltage stability are usually slow. Therefore, many aspects of the problem can be effectively analyzed by using static methods, which examine the viability of the equilibrium point represented by a specified operating condition of the power system. The static analysis techniques allow examination of a wide range of system conditions and, if appropriately used, can provide much insight into the nature of the problem and identify the key contributing factors. Dynamic analysis, on the other hand, is useful for detailed study of specific voltage collapse situations, coordination of protection and controls, and testing of remedial measures. Dynamic simulations also examine whether and how the steady-state equilibrium point will be reached.

In this section, we discuss static as well as dynamic analysis techniques and we illustrate how the two approaches can be used in a complementary manner.

14.3.1 Modelling Requirements

The following are descriptions of models of power system elements that have a significant impact on voltage stability:

Loads. Load characteristics could be critical in voltage stability analysis. Unlike in conventional transient stability and power-flow analyses, expanded subtransmission system representation in a voltage-weak area may be necessary. This should include transformer ULTC action, reactive power compensation, and voltage regulators in the subtransmission system.

It is important to account for voltage and frequency dependence of loads. It may also be necessary to model induction motors specifically. In some cases, appropriate representation of load characteristics at low voltages may be essential [11].

Generators and their excitation controls. For voltage stability analysis, it may be necessary to account for the droop characteristic of the AVR rather than to assume zero droop. If load (line drop) compensation (see Chapter 8, Section 8.5.4) is provided, its effect should be represented. Field current and armature current limits should be represented specifically rather than as a fixed value of the maximum reactive power limit.

Static var systems (SVSs). When an SVS is operating within the normal voltage control range, it maintains bus voltage with a slight droop characteristic. When operating at the reactive power limits, the SVS becomes a simple capacitor or reactor; this could have a very significant effect on voltage stability. These characteristics of SVS should be represented appropriately in voltage stability studies. Modelling of SVSs is described in Chapter 11 (Section 11.2.9).

Automatic generation control (AGC). For contingencies resulting in a significant mismatch between generation and load, the actions of primary speed control and supplementary tie line bias frequency control can change system generation significantly, sometimes to the detriment of voltage stability. Hence, these functions have to be represented appropriately.

Protection and controls. These include generating unit and transmission network protection and controls. Examples are generator excitation protection, armature over-current protection, transmission line overcurrent protection, capacitor bank controls, phase-shifting regulators, and undervoltage load shedding.

14.3.2 Dynamic Analysis

The general structure of the system model for voltage stability analysis is similar to that for transient stability analysis. As described in Chapter 13 (Section 13.3.5), the overall system equations, comprising a set of first-order differential equations, may be expressed in the following general form:

$$\dot{\mathbf{x}} = \mathbf{f}(\mathbf{x}, \mathbf{V}) \tag{14.4}$$

and a set of algebraic equations

$$\mathbf{I}(\mathbf{x}, \mathbf{V}) = \mathbf{Y}_N \mathbf{V} \tag{14.5}$$

with a set of known initial conditions $(\mathbf{x}_0, \mathbf{V}_0)$, where

$\mathbf{x}$ = state vector of the system
$\mathbf{V}$ = bus voltage vector
$\mathbf{I}$ = current injection vector
$\mathbf{Y}_N$ = network node admittance matrix

Since we include the representation of transformer tap-changer and phase-shift angle controls, the elements of $\mathbf{Y}_N$ change as a function of bus voltages and time. The current injection vector $\mathbf{I}$ is a function of the system states $\mathbf{x}$ and bus voltage vector $\mathbf{V}$, representing the boundary conditions at the terminals of the various devices (generating units, nonlinear static loads, motors, SVSs, HVDC converters, etc.). Due to the time-dependent nature of devices such as field current limiters, the relationship between $\mathbf{I}$ and $\mathbf{x}$ can be a function of time.

Equations 14.4 and 14.5 can be solved in time-domain by using any of the numerical integration methods described in Chapter 13 (Section 13.2) and network power-flow analysis methods described in Chapter 6 (Section 6.4). The study period is typically on the order of several minutes. With the inclusion of special models representing the "slow system dynamics" leading to voltage collapse, the stiffness of the system differential equations is significantly higher than that of transient stability models. Implicit integration methods are ideally suited for such applications. Facilities to automatically change the integration time step, as the solution progresses and fast transients decay, greatly enhance the computational efficiency of such techniques [12].

Example 14.2

In this example we analyze the large-disturbance voltage stability of the system shown in Figure E14.4, using time-domain simulations. The test system considered here is based on the system originally described in reference 25, and used in references 2, 8, and 10 for analysis of the various aspects of voltage stability.

In addition to illustrating the time-domain approach to dynamic analysis of voltage stability, this example shows the effects of generator overexcitation limiter (OXL), transformer tap changer, and load characteristics on voltage stability.

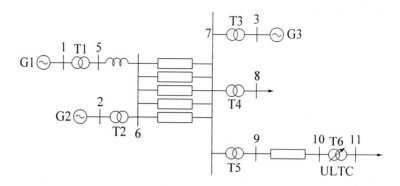

Figure E14.4 Test system

The test system data are as follows:

Transmission lines (R, X and B in pu on 100 MVA base):

Line	R	X	B
5-6	0.0000	0.0040	0.0000
6-7	0.0015	0.0288	1.1730
9-10	0.0010	0.0030	0.0000

Transformers (R and X in pu on 100 MVA base):

Transformer	R	X	Ratio	
T1	0.0000	0.0020	0.8857	
T2	0.0000	0.0045	0.8857	
T3	0.0000	0.0125	0.9024	
T4	0.0000	0.0030	1.0664	
T5	0.0000	0.0026	1.0800	
T6	0.0000	0.0010	0.9750	(load level 1)
			0.9938	(load level 2)
			1.0000	(load level 3)

ULTC for transformer T6 between buses 10 and 11:

Time delay for the first tap movement: 30 s
Time delay for subsequent tap movement: 5 s
Dead band: ±1% pu bus voltage
Tap range: ±16 steps
Step size: 5/8% (=0.00625 pu)

Shunt capacitors:

Bus	MVAr
7	763
8	600
9	1710

Loads:

Bus	P (MW)	Q (MVAr)	
8	3271	1015	(load level 1)
	3320	1030	(load level 2)
	3345	1038	(load level 3)
11	3384	971	(load level 1)
	3435	985	(load level 2)
	3460	993	(load level 3)

Generation:

Bus	P (MW)	V (pu)	
G1	3981	0.9800	(load level 1)
	4094	0.9800	(load level 2)
	4152	0.9800	(load level 3)
G2	1736	0.9646	(load level 1)
	1736	0.9646	(load level 2)
	1736	0.9646	(load level 3)
G3	1154	1.0400	(load level 1)
	1154	1.0400	(load level 2)
	1154	1.0400	(load level 3)

Machine parameters:

Machine 1: Infinite bus
Machine 2: $H = 2.09$, MVA rating = 2200 MVA
Machine 3: $H = 2.33$, MVA rating = 1400 MVA

The following are the parameters of machine 2 and machine 3 on their respective MVA ratings:

$R_a = 0.0046$ $X_d = 2.07$ $X_q = 1.99$

$X_l = 0.155$ $X_d' = 0.28$ $X_q' = 0.49$

$X_d'' = 0.215$ $X_q'' = 0.215$

$$T'_{d0} = 4.10 \qquad\qquad T'_{q0} = 0.56$$
$$T''_{d0} = 0.033 \qquad\qquad T''_{q0} = 0.062$$

Exciters:

Both machine 2 and machine 3 have thyristor exciters with a gain of 400 and the sensing circuit-time constant of 0.02 seconds.

Overexcitation limiter for machine 3:

$$I_{fd\,max1} = 3.02 \text{ pu} \qquad I_{fd\,max2} = 4.60 \text{ pu}$$
$$I_{LIM} = 3.85 \text{ pu} \qquad\quad K_1 = 0.248 \qquad\qquad K_2 = 12.6$$

The OXL is included for generator 3 only. The block diagram of the OXL is shown in Figure E14.5 and its functional characteristic is shown in Figure E14.6.

We will consider three system load levels:

Load level 1:   6655 MW, 1986 MVAr
Load level 2:   6755 MW, 2016 MVAr
Load level 3:   6805 MW, 2031 MVAr

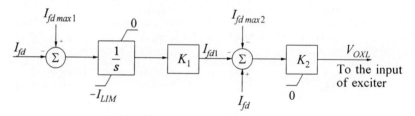

Figure E14.5 Block diagram of OXL

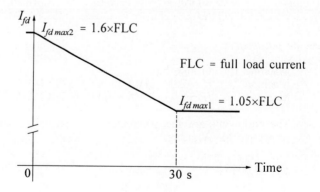

Figure E14.6 OXL characteristic

The disturbance considered is the loss of one of the lines between buses 6 and 7 (without a fault). Investigate the voltage stability of the system with the following representations of loads at buses 8 and 11:

(a)   The load at bus 11 is modelled as 50% constant impedance and 50% constant current for both active and reactive components; the action of the ULTC transformer (T6) supplying this load is modelled in detail. The load at bus 8 is modelled as constant MVA for both active and reactive components. The transformer T4 supplying this load is assumed to have a fixed tap.

(b)   The active power component of load at bus 8 is represented as an equivalent induction motor with the following parameters (see Chapter 7, Section 7.2.1 for description of model):

Motor rating = 3600 MVA, 60 Hz

$X_m = 3.3$ pu $R_s = 0.01$ pu $X_s = 0.145$ pu
$R_r = 0.008$ pu $X_r = 0.145$ pu $H = 0.6$ s
Load torque exponent $m = 2.0$ (that is, $T_L = T_0\omega_r^2$)

All other components are modelled as in (a).

(c)   The load at bus 11 is represented as in (a), with ULTC action of transformer T6. The load at bus 8 is represented as constant MVA for the reactive component and a thermostatically controlled load for the active component with the following parameters (see Chapter 7, Figure 7.4 for description of model):

$K_P = 0.1$ $K_I = 1.0$ $K_1 = 1.0$
$T_1 = 40$ s $T_A = 20$ s $G_{max} = 1.5G_0$

Solution

(a) Figures E14.7(a), (b), and (c) show the time responses of the voltages at buses 11, 10, and 7 following the loss of one of the lines between buses 6 and 7, for each of the three load levels. The corresponding plots of generator G3 field current, reactive power output, and terminal voltage are shown in Figures E14.8(a), (b), and (c).

The effect of the loss of the line is to cause the system voltage to drop initially. For load level 1, the ULTC action of transformer T6 restores bus 11 voltage to nearly its reference value in about 40 seconds. Field current of G3 remains below its continuous limit and the terminal voltage is maintained at the initial value by the AVR. The voltages of buses 10 and 7 settle at values below the predisturbance values. The system is voltage stable.

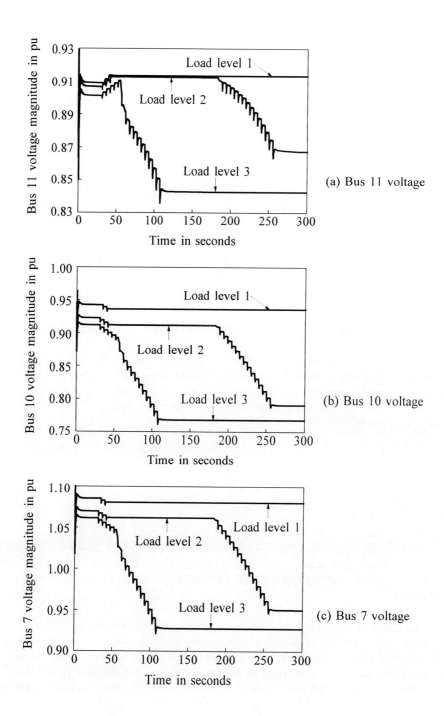

Figure E14.7 Voltages at buses 11, 10, and 7

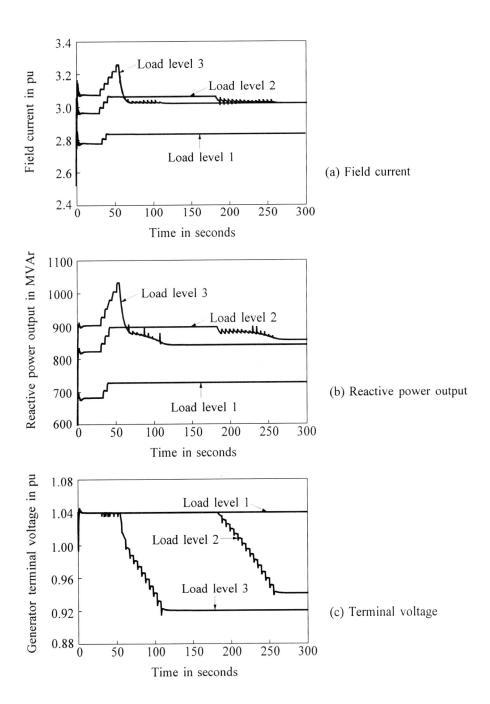

Figure E14.8 Responses of generator G3 variables

With load level 2, the voltage of bus 11 is restored to nearly its reference value in about 40 seconds as in the case of load level 1. However, the reactive power demand on the generators is now higher and the field current of generator G3 exceeds its limit. The overexcitation limiter of G3 starts ramping the field current down at about 180 seconds. This in turn triggers the following chain of events:

- The terminal voltage of G3, which is no longer being controlled by the AVR, drops.

- Voltages at buses 11, 10, and 7 drop.

- The ULTC on transformer T6 operates to restore the voltage and load at bus 11.

- The demand for reactive power on the generators increases. The OXL on G3 continues to hold the field current at its limit and the terminal voltage of G3 continues to drop.

- Voltage at the transmission level (bus 7) drops and this causes further reduction of bus 10 and bus 11 voltages.

- The ULTC on T6 operates again, repeating the above chain of events.

The net effect of each tap movement of transformer T6 is to reduce bus 11 voltage rather than increase it. The voltage at bus 11 falls progressively until the ULTC reaches its upper limit at about 260 seconds. The voltages stabilize at this point. The voltage at bus 11 settles at 0.865 pu.

With load level 3, the demand for reactive power is higher. Consequently, the field current of G3 reaches its limit at about 50 seconds. With loss of voltage control by G3, the voltages at the transmission level and at bus 11 drop with each tap movement of the ULTC. The voltages stabilize when the ULTC reaches its limit at about 110 seconds. The voltage of bus 11 settles at nearly 0.84 pu.

(b) This case differs from the above in that the active power component of load at bus 8 is represented as an induction motor. Only load level 2 is considered. Figures E14.9(a), (b) and (c) show plots of the speed, reactive and active power of the induction motor. The motor stalls at about 65 seconds. This results in a decrease in active power absorbed by the motor. However, the reactive power drawn by the motor increases rapidly. This causes voltage collapse as depicted in Figures E14.10 and E14.11. For comparison, the voltage response with constant MVA load at bus 8, as computed in (a), is also shown in Figure E14.11.

(c) This case considers a thermostatically controlled (TC) load at bus 8, with the system load at level 3. Figures E14.12(a), (b) and (c) show plots of the voltage at bus 8, the conductance of the TC load and active power drawn by the load. Figure E14.13 shows the corresponding plot of voltage at bus 11 (curve 1). For comparison, the response of bus 11 voltage with constant resistance load at bus 8 is also shown in Figure E14.13 (curve 2).

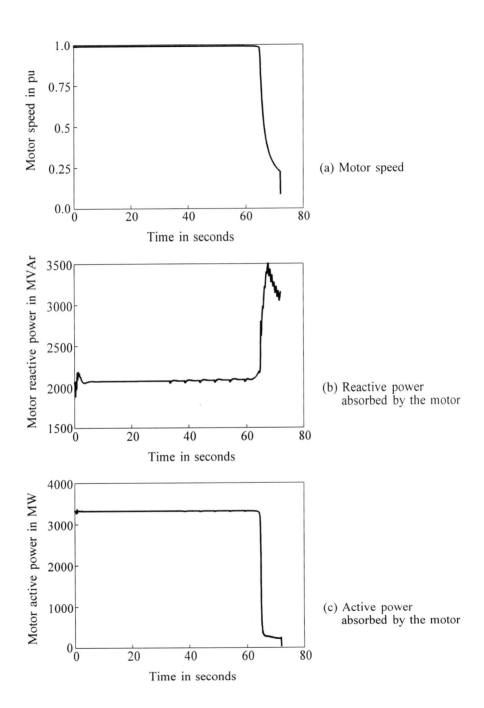

(a) Motor speed

(b) Reactive power
absorbed by the motor

(c) Active power
absorbed by the motor

Figure E14.9 Induction motor response, with load level 2

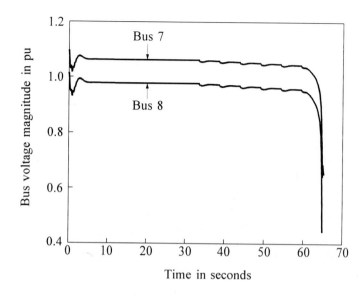

Figure E14.10 Voltage magnitude at bus 7 and bus 8 with induction
motor load at bus 8; system at load level 2

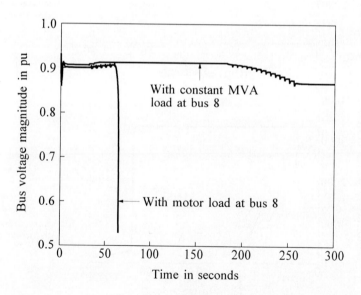

Figure E14.11 Response of voltage magnitude at bus 11 with
(a) constant MVA load and (b) induction motor
load at bus 8; system load at level 2

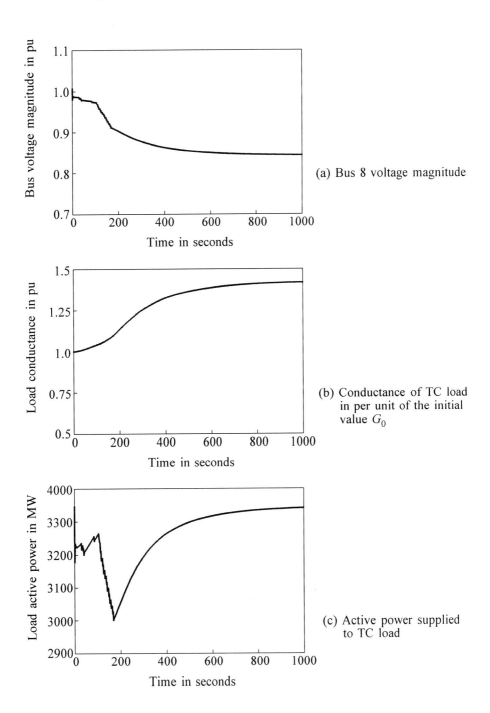

(a) Bus 8 voltage magnitude

(b) Conductance of TC load
in per unit of the initial
value G_0

(c) Active power supplied
to TC load

Figure E14.12 Response of thermostatically controlled load at bus 8

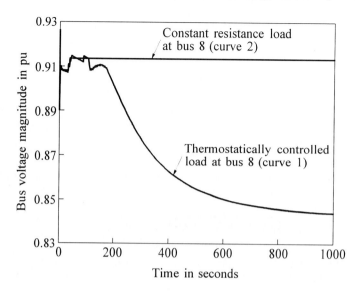

Figure E14.13 Response of voltage magnitude at bus 11 with
TC load and constant resistance load at bus 8

Following the loss of a line between buses 6 and 7, the voltage of bus 11 drops. The ULTC of transformer T6 restores bus 11 voltage to its reference value at about 40 seconds. If the load at bus 8 were a constant resistance load, the voltage would stabilize as depicted by curve 2 of Figure E14.13. With a thermostatically controlled load, bus 11 voltage drops and the ULTC operates again at about 90 seconds. The action of the TC load causes the voltage to drop again. The interaction between the ULTC and the TC load continues until about 170 seconds, when the ULTC reaches its upper limit. The conductance of the TC load and the power continue to increase until the power reaches the predisturbance level of 3345 MW. The voltages at buses 8 and 11 stabilize at about 0.85 pu and 0.845 pu, respectively.   ∎

14.3.3 Static Analysis

The static approach captures *snapshots* of system conditions at various time frames along the time-domain trajectory. At each of these time frames, time derivatives of the state variables (i.e., $\dot{x}$) in Equation 14.4 are assumed to be zero, and the state variables take on values appropriate to the specific time frame. Consequently, the overall system equations reduce to purely algebraic equations allowing the use of static analysis techniques.

In the past, the electric utility industry has largely depended on conventional power-flow programs for static analysis of voltage stability. Stability is determined by computing the *V-P* and *Q-V* curves at selected load buses. Generally, such curves are generated by executing a large number of power flows using conventional models. While such procedures can be automated, they are time-consuming and do not readily provide information useful in gaining insight into causes of stability problems. In addition, these procedures focus on individual buses; that is, the stability characteristics are established by stressing each bus independently. This may unrealistically distort the stability condition of the system. Also, the buses selected for Q-V and V-P analysis must be chosen carefully, and a large number of such curves may be required to obtain complete information. In fact, it may not be possible to generate the Q-V curves completely due to power-flow divergence caused by problems elsewhere in the system.

    A number of special techniques have been proposed in the literature for voltage stability analysis using the static approach. Some of these techniques are described in part 2 of the 1990 IEEE report [1]. In general, these have not found widespread practical application.

    References 13 and 14 describe the practical applications of an approach based on V-Q sensitivity. The modal analysis approach described in references 9 and 10 has also been applied to voltage stability analysis of practical systems. In this section, we describe these two approaches. The advantages of these approaches are that they give voltage stability-related information from a *system-wide perspective* and clearly identify areas that have potential problems. The modal analysis approach has the added advantage that it provides information regarding the *mechanism* of instability. The principal reason for considering the V-Q sensitivity analysis here is that it serves as a good introduction to the modal analysis.

(a) V-Q sensitivity analysis

    As shown in Chapter 6 (Section 6.4.3), the network constraints represented by Equation 14.5 may be expressed in the following linearized form:

$$\begin{bmatrix} \Delta \mathbf{P} \\ \Delta \mathbf{Q} \end{bmatrix} = \begin{bmatrix} \mathbf{J}_{P\theta} & \mathbf{J}_{PV} \\ \mathbf{J}_{Q\theta} & \mathbf{J}_{QV} \end{bmatrix} \begin{bmatrix} \Delta \boldsymbol{\theta} \\ \Delta \mathbf{V} \end{bmatrix} \tag{14.6}$$

where

$\Delta \mathbf{P}$ = incremental change in bus real power
$\Delta \mathbf{Q}$ = incremental change in bus reactive power injection
$\Delta \boldsymbol{\theta}$ = incremental change in bus voltage angle
$\Delta \mathbf{V}$ = incremental change in bus voltage magnitude

The elements of the Jacobian matrix give the sensitivity between power flow and bus voltage changes.

    If the conventional power-flow model is used for voltage stability analysis, the Jacobian matrix in Equation 14.6 is the same as that derived in Chapter 6 for solving the power-flow equations by using the Newton-Raphson technique. With the enhanced device models represented by Equation 14.4, the linear relationship between power and voltage for each device when $\dot{\mathbf{x}} = 0$ may be expressed as follows:

$$\begin{bmatrix} \Delta \mathbf{P}_d \\ \Delta \mathbf{Q}_d \end{bmatrix} = \begin{bmatrix} \mathbf{A}_{11} & \mathbf{A}_{12} \\ \mathbf{A}_{21} & \mathbf{A}_{22} \end{bmatrix} \begin{bmatrix} \Delta \mathbf{V}_d \\ \Delta \boldsymbol{\theta}_d \end{bmatrix} \tag{14.7}$$

where

$\Delta \mathbf{P}_d$ = incremental change in device real power output
$\Delta \mathbf{Q}_d$ = incremental change in device reactive power output
$\Delta \mathbf{V}_d$ = incremental change in device voltage magnitude
$\Delta \boldsymbol{\theta}_d$ = incremental change in device voltage angle

The terms of the network Jacobian matrix in Equation 14.6 associated with each device are modified by $\mathbf{A}_{11}$, $\mathbf{A}_{12}$, $\mathbf{A}_{21}$ and $\mathbf{A}_{22}$ to form the system Jacobian matrix.

System voltage stability is affected by both P and Q. However, at each operating point we may keep P constant and evaluate voltage stability by considering the incremental relationship between Q and V. This is analogous to the Q-V curve approach. Although incremental changes in P are neglected in the formulation, the effects of changes in system load or power transfer level are taken into account by studying the incremental relationship between Q and V at different operating conditions.

Based on the above considerations, in Equation 14.6, let $\Delta P = 0$. Then

$$\Delta \mathbf{Q} = \mathbf{J}_R \Delta \mathbf{V} \tag{14.8}$$

where

$$\mathbf{J}_R = [\mathbf{J}_{QV} - \mathbf{J}_{Q\theta} \mathbf{J}_{P\theta}^{-1} \mathbf{J}_{PV}] \tag{14.9}$$

and $\mathbf{J}_R$ is the reduced Jacobian matrix of the system. From Equation 14.8, we may write

$$\Delta \mathbf{V} = \mathbf{J}_R^{-1} \Delta \mathbf{Q} \tag{14.10}$$

The matrix $\mathbf{J}_R^{-1}$ is the reduced V-Q Jacobian. Its ith diagonal element is the V-Q sensitivity at bus i. For computational efficiency, this matrix is not explicitly formed. The V-Q sensitivities are calculated by solving Equation 14.8.

The V-Q sensitivity at a bus represents the slope of the Q-V curve at the given operating point. A positive V-Q sensitivity is indicative of stable operation; the smaller the sensitivity, the more stable the system. As stability decreases, the magnitude of the sensitivity increases, becoming infinite at the stability limit. Conversely, a negative V-Q sensitivity is indicative of unstable operation. A small negative sensitivity represents a very unstable operation. Because of the nonlinear nature of the V-Q relationships, the magnitudes of the sensitivities for different system conditions do not provide a direct measure of the relative degree of stability.

(b) Q-V modal analysis [9,15]

Voltage stability characteristics of the system can be identified by computing the eigenvalues and eigenvectors (see Chapter 12, Section 12.2) of the reduced Jacobian matrix $\mathbf{J_R}$ defined by Equation 14.9. Let

$$\mathbf{J_R} = \boldsymbol{\xi} \boldsymbol{\Lambda} \boldsymbol{\eta} \tag{14.11}$$

where

$\boldsymbol{\xi}$ = right eigenvector matrix of $\mathbf{J_R}$
$\boldsymbol{\eta}$ = left eigenvector matrix of $\mathbf{J_R}$
$\boldsymbol{\Lambda}$ = diagonal eigenvalue matrix of $\mathbf{J_R}$

From Equation 14.11

$$\mathbf{J_R^{-1}} = \boldsymbol{\xi} \boldsymbol{\Lambda}^{-1} \boldsymbol{\eta} \tag{14.12}$$

Substituting in Equation 14.10 gives

$$\Delta\mathbf{V} = \boldsymbol{\xi} \boldsymbol{\Lambda}^{-1} \boldsymbol{\eta} \Delta\mathbf{Q} \tag{14.13}$$

or

$$\Delta\mathbf{V} = \sum_i \frac{\boldsymbol{\xi}_i \boldsymbol{\eta}_i}{\lambda_i} \Delta\mathbf{Q} \tag{14.14}$$

where $\boldsymbol{\xi}_i$ is the ith column right eigenvector and $\boldsymbol{\eta}_i$ the ith row left eigenvector of $\mathbf{J_R}$.

Each eigenvalue λ_i and the corresponding right and left eigenvectors $\boldsymbol{\xi}_i$ and $\boldsymbol{\eta}_i$ define the ith mode of Q-V response.

Since $\boldsymbol{\xi}^{-1} = \boldsymbol{\eta}$, Equation 14.13 may be written as

$$\boldsymbol{\eta}\Delta\mathbf{V} = \boldsymbol{\Lambda}^{-1} \boldsymbol{\eta} \Delta\mathbf{Q}$$

or

$$\mathbf{v} = \boldsymbol{\Lambda}^{-1} \mathbf{q} \tag{14.15}$$

where

$\mathbf{v} = \boldsymbol{\eta}\Delta\mathbf{V}$ is the vector of modal voltage variations

and

$\mathbf{q} = \boldsymbol{\eta}\Delta\mathbf{Q}$ is the vector of modal reactive power variations

The difference between Equations 14.10 and 14.15 is that $\boldsymbol{\Lambda}^{-1}$ is a diagonal matrix whereas $\mathbf{J}_R^{-1}$, in general, is nondiagonal. Equation 14.15 represents uncoupled first order equations. Thus, for the ith mode we have

$$\mathbf{v}_i = \frac{1}{\lambda_i}\mathbf{q}_i \tag{14.16}$$

If $\lambda_i > 0$, the ith modal voltage and the ith modal reactive power variations are along the same direction, indicating that the system is voltage *stable*. If $\lambda_i < 0$, the ith modal voltage and the ith modal reactive power variation are along opposite directions, indicating that the system is voltage *unstable*. The magnitude of each modal voltage variation equals the inverse of λ_i times the magnitude of the modal reactive power variation. In this sense the magnitude of λ_i determines the *degree of stability* of the ith modal voltage. The smaller the magnitude of positive λ_i, the closer the ith modal voltage is to being unstable. When $\lambda_i = 0$, the ith modal voltage collapses because any change in that modal reactive power causes infinite change in the modal voltage.

Let us now examine the relationship between bus V-Q sensitivities and eigenvalues of $\mathbf{J}_R$. In Equation 14.14, let $\Delta\mathbf{Q} = \mathbf{e}_k$, where $\mathbf{e}_k$ has all zero elements except for the kth element which is equal to 1. Then

$$\Delta\mathbf{V} = \sum_i \frac{\eta_{ik}\boldsymbol{\xi}_i}{\lambda_i}$$

where η_{ik} is the kth element of $\boldsymbol{\eta}_i$.

The V-Q sensitivity at bus k is given by

$$\frac{\partial V_k}{\partial Q_k} = \sum_i \frac{\xi_{ki}\eta_{ik}}{\lambda_i} \tag{14.17}$$

We see from the above equation that the V-Q sensitivities cannot identify individual voltage collapse modes; instead they provide information regarding the combined effects of all modes of voltage-reactive power variations.

If the transmission network resistances are neglected and the node admittance matrix $\mathbf{Y}_N$ is symmetrical, the reduced Jacobian matrix $\mathbf{J}_R$ is also symmetrical [15]. Then the eigenvalues and eigenvectors of $\mathbf{J}_R$ are real. In addition, the right eigenvector and the left eigenvector of an eigenvalue of $\mathbf{J}_R$ are equal.

With phase-shifting transformers (which make the matrix $\mathbf{Y}_N$ unsymmetrical) and line resistances, $\mathbf{J}_R$ is only nearly symmetrical; the eigenvalues of $\mathbf{J}_R$ for all practical purposes are real.

The magnitude of the eigenvalues can provide a relative measure of the *proximity* to instability. Eigenvalues do not, however, provide an absolute measure because of the nonlinearity of the problem. This is analogous to the damping factor in small-signal (angle) stability analysis, which is indicative of the degree of damping but is not an absolute measure of stability margin. If a megawatt distance to voltage instability is required, the system is stressed incrementally until it becomes unstable and modal analysis is applied at each operating point. The application of modal analysis helps in determining how stable the system is and how much extra load or power transfer level should be added. When the system reaches the voltage stability critical point, the modal analysis is helpful in identifying the voltage stability critical areas and elements which participate in each mode.

Example 14.3

For the 500 kV, 322 km line system considered in Example 14.1, write the equations of the power flow from the sending end to the receiving end in the following form:

$$P = f(\theta, V)$$
$$Q = g(\theta, V)$$

Find the expressions for $J_{P\theta}$, J_{PV}, $J_{Q\theta}$, and J_{QV} defined by the linearized load flow equations:

$$\begin{bmatrix} \Delta P \\ \Delta Q \end{bmatrix} = \begin{bmatrix} J_{P\theta} & J_{PV} \\ J_{Q\theta} & J_{QV} \end{bmatrix} \begin{bmatrix} \Delta\theta \\ \Delta V \end{bmatrix}$$

(a)   When $P_2 = 1,500$ MW, calculate the eigenvalues of the reduced Q-V Jacobian matrix and V-Q sensitivities with the following different reactive power injections for each of the corresponding two voltages on the Q-V curve:

(i) $Q_i = 500$ MVAr
(ii) $Q_i = 400$ MVAr
(iii)   Values of Q_i close to the bottom of the Q-V curve

Assume that, for the purpose of analysis, the load and reactive power source have constant P, Q characteristics.

(b)   Determine the voltage stability by computing the eigenvalues of the reduced Q-V Jacobian matrix for the following cases:

(i) $P = 1,500$ MW, $Q_i = 450$ MVAr
(ii) $P = 1,900$ MW, $Q_i = 950$ MVAr

Assume that the reactive power Q_i is supplied by a shunt capacitor.

Solution

From Figure E14.1(b), the admittance matrix of the two-bus system is

$$Y = \begin{bmatrix} 2.142 - j22.897 & -2.142 + j24.973 \\ -2.142 + j24.973 & 2.142 - j22.897 \end{bmatrix} \tag{E14.1}$$

From Equation 6.101 of Chapter 6, the expressions for P and Q at any bus k are given by

$$P_k = V_k \sum_{m=1}^{n} (G_{km} V_m \cos\theta_{km} + B_{km} V_m \sin\theta_{km})$$

$$Q_k = V_k \sum_{m=1}^{n} (G_{km} V_m \sin\theta_{km} - B_{km} V_m \cos\theta_{km}) \tag{E14.2}$$

where $\theta_{km} = \theta_k - \theta_m$.

For the two-bus system, we have

$$P_1 = V_1[(2.142 V_1 \cos\theta_{11} - 22.897 V_1 \sin\theta_{11}) + (-2.142 V_2 \cos\theta_{12} + 24.973 V_2 \sin\theta_{12})]$$

$$P_2 = V_2[(-2.142 V_1 \cos\theta_{21} + 24.973 V_1 \sin\theta_{21}) + (2.142 V_2 \cos\theta_{22} - 22.897 V_2 \sin\theta_{22})]$$

$$Q_1 = V_1[(2.142 V_1 \sin\theta_{11} + 22.897 V_1 \cos\theta_{11}) + (-2.142 V_2 \sin\theta_{12} - 24.973 V_2 \cos\theta_{12})] \tag{E14.3}$$

$$Q_2 = V_2[(-2.142 V_1 \sin\theta_{21} - 24.973 V_1 \cos\theta_{21}) + (2.142 V_2 \sin\theta_{22} + 22.897 V_2 \cos\theta_{22})]$$

where $\theta_{11} = \theta_{22} = 0$, and $\theta_{21} = -\theta_{12} = \theta$.

We are interested only in P_2 and Q_2. With $V_1 = 1.0$,

$$P_2 = -2.142 V_2 \cos\theta + 24.973 V_2 \sin\theta + 2.142 V_2^2$$

$$Q_2 = -2.142 V_2 \sin\theta - 24.973 V_2 \cos\theta + 22.897 V_2^2$$

Hence the expressions for the Jacobian terms are given by

$$J_{P\theta} = \frac{\partial P_2}{\partial\theta} = 2.142 V_2 \sin\theta + 24.973 V_2 \cos\theta$$

$$J_{PV} = \frac{\partial P_2}{\partial V_2} = -2.142 \cos\theta + 24.973 \sin\theta + 4.284 V_2$$

$$J_{Q\theta} = \frac{\partial Q_2}{\partial\theta} = -2.142 V_2 \cos\theta + 24.973 V_2 \sin\theta \tag{E14.4}$$

$$J_{QV} = \frac{\partial Q_2}{\partial V_2} = -2.142 \sin\theta - 24.973 \cos\theta + 45.794 V_2$$

(a) The linearized power-flow equations are

$$\Delta P_2 = J_{P\theta}\Delta\theta + J_{PV}\Delta V_2$$
$$\Delta Q_2 = J_{Q\theta}\Delta\theta + J_{QV}\Delta V_2$$

(E14.5)

With $\Delta P_2 = 0$,

$$\Delta Q_2 = (J_{QV} - J_{Q\theta}J_{P\theta}^{-1}J_{PV})\Delta V_2$$

or

$$\Delta Q_2 = J_R \Delta V_2$$

where

$$J_R = J_{QV} - J_{Q\theta}J_{P\theta}^{-1}J_{PV}$$

The expressions for J_{QV}, $J_{Q\theta}$, $J_{P\theta}$, and J_{PV} are given by Equation E14.4. For this simple system, J_R is a 1×1 matrix. The eigenvalue (λ) of the matrix is same as the matrix itself. The V-Q sensitivity is equal to the inverse of the eigenvalue.

For each value of Q_i, there are two solutions for the receiving end voltage. Table E14.1 summarizes V, θ, λ, and dV/dQ with $P = 1,500$ MW and $Q = 500$, 400, 306, and 305.9 MVAr. For each case, the eigenvalue and V-Q sensitivity are both negative at the low voltage solution and are both positive at the high voltage solution. With $Q = 305.9$ MVAr (close to bottom of the Q-V curve), dV/dQ is large and λ is very small.

Table E14.1

Q_i (MVAr)	High Voltage Solution				Low Voltage Solution			
	V_2	θ	λ	dV/dQ	V_2	θ	λ	dV/dQ
500.0	1.024	-37.3°	17.03	0.059	0.671	-66.7°	-39.87	-0.025
400.0	0.956	-40.1°	12.41	0.081	0.706	-60.3°	-20.96	-0.048
306.0	0.820	-48.2°	0.52	1.923	0.812	-48.8°	-0.95	-1.055
305.9	0.814	-48.7°	0.02	50.1	0.815	-48.6°	-0.70	-1.434

(b)   With a shunt capacitor connected at the receiving end of the line, the self-admittance is

$$Y_{22} = 2.142 - j(22.897 - B_C)$$

(i) With $P=1500$ MW and a 450 MVAr shunt capacitor,

$$V_2=0.981 \qquad \theta=-39.1°$$

Since $B_C=4.5$ pu,

$$Y_{22} = 2.142-j(22.897-4.5) = 2.142-j18.397$$

With this new value of Y_{22}, the reduced Q-V Jacobian matrix, calculated by using Equation E14.4, is

$$J_R = 5.348$$

And J_R is positive, indicating that the system is voltage stable.

(ii) With $P=1900$ MW and a 950 MVAr shunt capacitor,

$$V_2=0.995 \qquad \theta_2=-52.97°$$

Since $B_C=9.5$ pu,

$$Y_{22} = 2.142-j(22.897-9.5) = 2.142-j1.397$$

and the reduced Q-V Jacobian matrix, calculated by using Equation E14.4, is

$$J_R = -13.683$$

And J_R is negative, indicating that the system is voltage unstable.   ■

Bus participation factors

The relative participation of bus k in mode i is given by the *bus participation factor*:

$$P_{ki} = \xi_{ki}\eta_{ik} \tag{14.18}$$

From Equation 14.17, we see that P_{ki} determines the contribution of λ_i to the V-Q sensitivity at bus k.

Bus participation factors determine the areas associated with each mode. The sum of all the bus participations for each mode is equal to unity because the right and left eigenvectors are normalized. The size of bus participation in a given mode indicates the effectiveness of remedial actions applied at that bus in stabilizing the mode.

There are generally two types of modes. The first type has very few buses with large participations and all the other buses with close to zero participations, indicating that the mode is very localized. The second type has many buses with small but similar degrees of participations, and the rest of the buses with close to zero participations; this indicates that the mode is not localized. A typical localized mode occurs if a single load bus is connected to a very strong network through a long transmission line. A typical non-localized mode occurs when a region within a large system is loaded up and the main reactive support for this region is exhausted.

It is impractical and unnecessary to calculate all the eigenvalues of $\mathbf{J}_R$ for a practical system with several thousand buses. On the other hand, calculating only the minimum eigenvalue of $\mathbf{J}_R$ is not sufficient because there is usually more than one weak mode associated with different parts of the system, and the mode associated with the minimum eigenvalue may not be the most troublesome mode as the system is stressed. In practice, it is seldom necessary to compute more than 5 to 10 of the smallest eigenvalues to identify all critical modes.

Branch participation factors

Let us compute the branch participation factor associated with mode i by assuming that the vector of modal reactive power variations $\mathbf{q}$ has all elements equal to zero except for the ith, which equals 1. Then from Equation 14.15, the corresponding vector of bus reactive power variations is

$$\Delta\mathbf{Q}^{(i)} = \mathbf{\eta}^{-1}\mathbf{q} = \xi\mathbf{q} = \xi_i \qquad (14.19)$$

where ξ_i is the ith right eigenvector of $\mathbf{J}_R$. We further assume that all the right eigenvectors are normalized so that

$$\sum_j \xi_{ji}^2 = 1 \qquad (14.20)$$

With the vector of bus reactive power variations equal to $\Delta\mathbf{Q}^{(i)}$, the vector of bus voltage variations, $\Delta\mathbf{V}^{(i)}$, is

$$\Delta\mathbf{V}^{(i)} = \frac{1}{\lambda_i}\Delta\mathbf{Q}^{(i)} \qquad (14.21)$$

and, the corresponding vector of bus angle variation is

$$\Delta\mathbf{\theta}^{(i)} = -\mathbf{J}_{P\theta}^{-1}\mathbf{J}_{PV}\Delta\mathbf{V}^{(i)} \qquad (14.22)$$

With the angle and voltage variations for both the sending end and receiving end known, the linearized change in branch reactive loss can be calculated.

The relative participation of branch j in mode i is given by the participation factor:

$$P_{ji} = \frac{\Delta Q_{loss} \text{ for branch } j}{\text{maximum } \Delta Q_{loss} \text{ for all branches}} \qquad (14.23)$$

Branch participation factors indicate, for each mode, which branches consume the most reactive power in response to an incremental change in reactive load. Branches with high participations are either weak links or are heavily loaded. Branch participations are useful for identifying remedial measures to alleviate voltage stability problems and for contingency selection.

Generator participation factors

As in the case of branch participation, for a given reactive power variation, voltage and angle variations are determined at each machine terminal. These in turn are used to compute the change in reactive power output for each machine.

The relative participation of machine m in mode i is given by the generator participation factor:

$$P_{mi} = \frac{\Delta Q_m \text{ for machine } m}{\text{maximum } \Delta Q \text{ for all machines}} \qquad (14.24)$$

Generator participation factors indicate, for each mode, which generators supply the most reactive power in response to an incremental change in system reactive loading. Generator participations provide important information regarding proper distribution of reactive reserves among all the machines in order to maintain an adequate voltage stability margin.

Calculation of small eigenvalues [15,19]:

Several techniques are available for computing a selected subset of eigenvalues of a real matrix [16-18]. In the EPRI VSTAB program [19], an implicit inverse lopsided simultaneous iteration method is used. This is a combination of the *lopsided simultaneous iteration* (LOPSI) method and the *implicit inverse iteration* (IMPII) method.

LOPSI algorithm

Simultaneous iteration methods are suitable for obtaining dominant eigenvalues and corresponding eigenvectors of real unsymmetrical matrices. These methods involve an accurate eigensolution of a smaller iteration matrix at each iteration cycle, the dimension of which depends on the number of vectors processed simultaneously. A bi-iteration procedure is used to obtain left and right eigenvectors simultaneously. A lop-sided iteration procedure is used to obtain only one set of the eigenvectors, or to obtain the right and left eigenvectors sequentially. A detailed description of the simultaneous iteration method can be found in reference 16. Here, we briefly describe the computational procedure for implementing the LOPSI algorithm for calculating a subset of dominant eigenvalues and corresponding right eigenvectors of a real matrix $\mathbf{A}$. These dominant eigenvalues and corresponding left eigenvectors can be obtained by applying the same procedure to $\mathbf{A}^T$.

Each iteration cycle of the LOPSI algorithm involves a premultiplication and a reorientation, followed by a normalization and a convergence test. Let $\mathbf{A}$ be a real unsymmetrical matrix of order n for which the r dominant eigenvalues and corresponding right eigenvectors are required. Let $\mathbf{R} = [\ \mathbf{R}_1\ \mathbf{R}_2\ \cdots\ \mathbf{R}_r\ ]$ be a set of trial vectors normalized so that the maximum element for each column of $\mathbf{R}$ is 1. Premultiplying $\mathbf{R}$ by $\mathbf{A}$ and denoting the resulting set of vectors $\mathbf{S} = [\ \mathbf{S}_1\ \mathbf{S}_2\ \cdots\ \mathbf{S}_r\ ]$, we have

$$\mathbf{S}\ =\ \mathbf{A}\mathbf{R} \tag{14.25}$$

Let $\mathbf{\Lambda} = [\ \mathbf{\Lambda}_a\ |\ \mathbf{\Lambda}_b\ ] = \mathrm{diag}[\ \lambda_1\ \lambda_2\ \cdots\ \lambda_r\ |\ \lambda_{r+1}\ \cdots\ \lambda_n\ ]$ be a diagonal matrix of the eigenvalues of $\mathbf{A}$ arranged in descending order of absolute magnitude, and let $\mathbf{\Phi} = [\ \mathbf{\Phi}_a\ |\ \mathbf{\Phi}_b\ ] = [\ \phi_1\ \phi_2\ \cdots\ \phi_r\ |\ \phi_{r+1}\ \cdots\ \phi_n\ ]$ be a matrix of the corresponding right eigenvectors of $\mathbf{A}$ so that

$$\mathbf{A}\mathbf{\Phi}_a\ =\ \mathbf{\Phi}_a\mathbf{\Lambda}_a$$

and

$$\mathbf{A}\mathbf{\Phi}_b\ =\ \mathbf{\Phi}_b\mathbf{\Lambda}_b \tag{14.26}$$

The trial vectors $\mathbf{R}$ may be represented as a linear combination of the full set of eigenvectors:

$$\mathbf{R}\ =\ \mathbf{\Phi}_a\mathbf{C}_a + \mathbf{\Phi}_b\mathbf{C}_b \tag{14.27}$$

where $\mathbf{C}_a$ and $\mathbf{C}_b$ are coefficient matrices of size $r{\times}r$ and $(n{-}r){\times}r$, respectively. It follows from Equations 14.25 to 14.27 that

$$\mathbf{S}\ =\ \mathbf{\Phi}_a\mathbf{\Lambda}_a\mathbf{C}_a + \mathbf{\Phi}_b\mathbf{\Lambda}_b\mathbf{C}_b \tag{14.28}$$

The lower eigenvectors contribute relatively less to $\mathbf{S}$ than they do to $\mathbf{R}$. After a number of iterations, due to the washing-out process of earlier iterations, the coefficients of $\mathbf{C}_b$ become much smaller than those of $\mathbf{C}_a$.

The reorientation process involves the complete eigensolution of the $r{\times}r$ iteration matrix $\mathbf{B}$ obtained from the solution of

$$\mathbf{G}\mathbf{B}\ =\ \mathbf{H} \tag{14.29}$$

where $\mathbf{G} = \mathbf{R}^T\mathbf{R}$ and $\mathbf{H} = \mathbf{R}^T\mathbf{S}$.

By substituting the values of $\mathbf{R}$ and $\mathbf{S}$ given by Equations 14.27 and 14.28, and assuming the $\mathbf{C}_b$ coefficients to be negligible by comparison with the $\mathbf{C}_a$ coefficients, we can show that

$$\mathbf{R}^T\mathbf{\Phi}_a\mathbf{C}_a\mathbf{B} \approx \mathbf{R}^T\mathbf{\Phi}_a\mathbf{\Lambda}_a\mathbf{C}_a \tag{14.30}$$

If $\mathbf{R}^T\mathbf{\Phi}_a$ is nonsingular,

$$\mathbf{C}_a\mathbf{B} \approx \mathbf{\Lambda}_a\mathbf{C}_a \tag{14.31}$$

which shows that the matrix of left eigenvectors of $\mathbf{B}$ is an approximation to $\mathbf{C}_a$ and that the eigenvalues of $\mathbf{B}$ are approximations to $\mathbf{\Lambda}_a$. If $\mathbf{T}$ is the $r \times r$ matrix of right eigenvectors of $\mathbf{B}$, then

$$\mathbf{T} \approx \mathbf{C}_a^{-1} \tag{14.32}$$

Hence, the set of vectors obtained by the multiplication

$$\mathbf{W} = \mathbf{S}\mathbf{T} = \mathbf{\Phi}_a\mathbf{\Lambda}_a + \mathbf{\Phi}_a\mathbf{\Lambda}_b\mathbf{C}_b\mathbf{C}_a^{-1} \tag{14.33}$$

gives an improved set of right eigenvectors.

The column vectors in the $\mathbf{W}$ matrix are normalized so that the maximum element for each vector is 1, and $\mathbf{W}$ with its column vectors normalized is denoted by $\mathbf{W}^*$.

Let $\Delta\mathbf{R} = |\mathbf{R}^i - \mathbf{R}^{i-1}|$, with $\mathbf{R}^i$ and $\mathbf{R}^{i-1}$ equal to the eigenvectors solved at iteration i and iteration $i-1$, respectively. The iterative process is continued until the maximum element of $\Delta\mathbf{R}$ is less than a prespecified tolerance.

In summary, the LOPSI procedure to calculate the r dominant eigenvalues and the corresponding right eigenvectors is

a.   Select r initial trial vectors $\mathbf{R} = [\ \mathbf{R}_1\ \mathbf{R}_2\ \cdots\ \mathbf{R}_r\ ]$.

b.   Premultiply $\mathbf{R}$ by $\mathbf{A}$, $\mathbf{S} = \mathbf{A}\mathbf{R}$.

c.   Determine $\mathbf{G} = \mathbf{R}^T\mathbf{R}$, $\mathbf{H} = \mathbf{R}^T\mathbf{S}$.

d.   Solve $\mathbf{G}\mathbf{B} = \mathbf{H}$ for $\mathbf{B}$.

e.   Do full eigensolution of $\mathbf{B}$.

f.   Determine $\mathbf{W} = \mathbf{S}\mathbf{T}$, with $\mathbf{T}$ the right eigenvector matrix of $\mathbf{B}$.

g.   Set $\mathbf{R} = \mathbf{W}^*$, where $\mathbf{W}^*$ is $\mathbf{W}$ normalized such that all the vectors have their largest element equal to 1.

h.   Check convergence; if converged, stop. Otherwise, go back to b.

Among all the r eigenvectors the ones corresponding to the eigenvalues with larger absolute magnitudes will converge first. Therefore, if $j-1$ vectors have already passed the convergence test, they are locked and iteration is carried only on the rest of the r vectors. Such a locking procedure results in savings in computation time, the majority of which is effected in the premultiplication stage.

The convergence of the LOPSI algorithm depends on $|\lambda_r/\lambda_{r+1}|$. If m dominant eigenvalues are required, it is advisable to make r greater than m by including a few extra guard vectors. This not only makes a fast convergence rate probable for the required eigenvalues and eigenvectors but also ensures that the occurrence of complex conjugate eigenvalues in positions λ_m and λ_{m+1} would be of no consequence.

IILSI algorithm

Let us return to the problem of eigensolution of $\mathbf{J}_R$. Since we are interested in the eigenvalues with the smallest absolute magnitudes, which correspond to the largest eigenvalues of $\mathbf{J}_R^{-1}$, the LOPSI algorithm has to be applied to $\mathbf{J}_R^{-1}$. At each iteration, the premultiplication is

$$\mathbf{S} = \mathbf{J}_R^{-1}\mathbf{R} \qquad (14.34)$$

and the LOPSI algorithm becomes an inverse LOPSI algorithm.

Direct solution of Equation 14.34 involves matrix manipulation on $\mathbf{J}_R$. Recall that $\mathbf{J}_R = \mathbf{J}_{QV} - \mathbf{J}_{Q\theta}\mathbf{J}_{P\theta}^{-1}\mathbf{J}_{PV}$, which is not sparse due to reduction. To fully exploit the sparsity of the Jacobian matrix, $\mathbf{S}$ in Equation 14.34 is obtained by solving the following set of sparse linear equations:

$$\begin{bmatrix} \mathbf{J}_{P\theta} & \mathbf{J}_{PV} \\ \mathbf{J}_{Q\theta} & \mathbf{J}_{QV} \end{bmatrix}\begin{bmatrix} \mathbf{Z} \\ \mathbf{S} \end{bmatrix} = \begin{bmatrix} 0 \\ \mathbf{R} \end{bmatrix} \qquad (14.35)$$

It can be readily shown that the $\mathbf{S}$ obtained by solving Equation 14.35 is same as the $\mathbf{S}$ given in Equation 14.34. Since Equation 14.34 is solved implicitly by solving Equation 14.35, the inverse LOPSI algorithm becomes an *implicit inverse lopsided simultaneous iteration* (IILSI) algorithm.

The IILSI algorithm is applied to $\mathbf{J}_R$ and $\mathbf{J}_R^T$ in turn to calculate the smallest eigenvalues and corresponding right and left eigenvectors.

Illustration of modal analysis

To illustrate the application of modal analysis, let us consider the 39-bus, 10-machine system of Figure 14.3. In Section 14.1.1, we computed the *V-P* and *Q-V* characteristics for this system (see Figures 14.4 and 14.5). We now perform modal analysis at the three operating conditions represented by points A, B, and C on the *V-P* curve of Figure 14.4.

Table 14.1 gives the five smallest eigenvalues of $\mathbf{J_R}$ of each of the three operating points. We see that the magnitudes of the eigenvalues decrease as the system approaches instability. At the operating point C, the smallest eigenvalue is 0.0083, indicating that the system is on the verge of instability.

Table 14.2 gives the bus, branch, and generator participations for the least stable mode (λ=0.0083) for operating condition C. Figure 14.7 identifies the area prone to voltage instability (based on bus participations) and branches with higher participations in the least stable mode.

Table 14.1 Five smallest eigenvalues

Operating Point	A	B	C
λ_1	0.3867	0.1446	0.0083
λ_2	1.0271	0.5550	0.3209
λ_3	2.4049	1.5133	0.9334
λ_4	4.1031	2.6280	1.8757
λ_5	4.2699	3.0209	2.3373

Table 14.2 Bus, branch, and generator participations in
the least stable mode for operating point C

Bus Participation		Branch Participation		Generator Participation	
Bus	Participation	Branch	Participation	Bus	Participation
530	0.2638	500-520	1.0000	1311	1.0000
520	0.2091	300-360	0.8414	2412	0.2786
510	0.1025	100-350	0.8175	1011	0.2103
500	0.0941	320-500	0.8093	1014	0.2036
320	0.0482	330-350	0.6534	1013	0.2036
310	0.0319			1012	0.2036
300	0.0296				
340	0.0279				

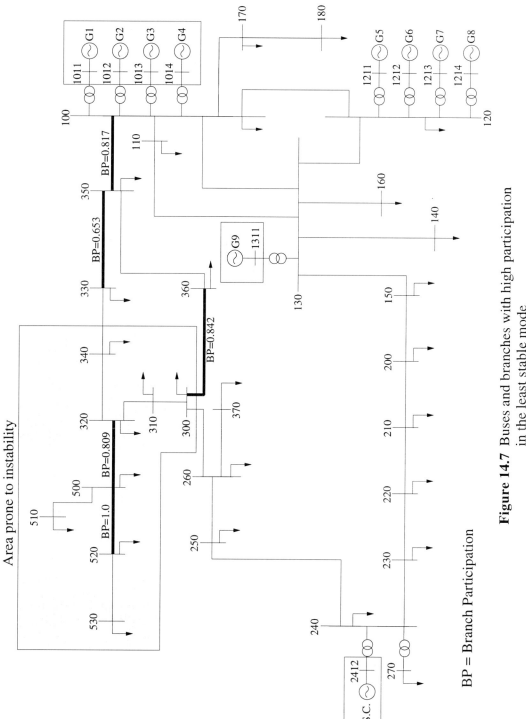

Figure 14.7 Buses and branches with high participation in the least stable mode

BP = Branch Participation

It is of interest to compare these results with the Q-V characteristics shown in Figure 14.5(c). The Q-V curves show that buses 530 and 510 have zero reactive margin. From Table 14.2, we see that these buses have high participation in the mode which is on the verge of becoming unstable. The advantage of modal analysis is that it clearly identifies groups of buses which participate in the instability. The Q-V analysis, being a single-bus approach, is unable to provide any connection between results obtained for individual buses. For large practical systems it may be necessary to compute a very large number of Q-V curves in order to find the problem area. Because of the way the Q-V curve stresses the system, it may not always be possible to complete the curve, in which case the search for a more critical bus must continue. The modal analysis approach eliminates this problem and easily identifies, on a system-wide basis, areas which are potentially troublesome.

Transient state approximations

In the analysis of voltage stability following a contingency, there are several distinct time frames in which different devices act to influence system performance. It is useful to be able to establish the condition of the system during each time frame, particularly if steady state cannot be achieved and we wish to determine what control actions cause instability. This leads to the concept of transient state approximations. Algebraic equations are used to represent system conditions which are approximations to various time frames along the transient trajectory of the system. These snapshots of system conditions represent transient state approximations.

Modal analysis can be applied to each snapshot to investigate the voltage stability of the system.

Proximity to instability

Proximity to small-disturbance voltage instability is determined by increasing load-generation in steps until the system becomes unstable or the power flow fails to converge. The increase in load may be by area, zone, or individual buses. The increase in generation follows a loading order that is usually system-specific. It is essential to capture nonlinearities encountered when moving from one system state to another.

Modal analysis performed at specified operating points provides information regarding areas prone to instability, and bus and generation participation factors. At the point of collapse, the left eigenvector identifies the most effective direction to steer the system to maximize voltage stability. Additionally, such measures as reactive reserves, losses, and bus voltages provide valuable information regarding the mechanism of instability.

References 20 to 25 describe special techniques for determining the point of voltage collapse and proximity to voltage instability. The following sections discuss two of these techniques.

14.3.4 Determination of Shortest Distance to Instability

The distance to voltage instability is normally determined by increasing the system load in a predefined manner, representing the most probable stressing scenario based on historical and forecast data. However, we are also interested in knowing the loading pattern that results in the smallest stability margin. The following is a description of a method of determining the minimum MVA stability margin, based on the technique described in reference 24.

Basic theory

What we want to find is the set of load MW and MVAr increments whose vector sum is a minimum and which, when imposed on the initial operating condition, cause the power-flow Jacobian to be singular. This may be achieved by organizing the power-flow equations in the form

$$\mathbf{f}(\mathbf{x}, \boldsymbol{\rho}) = g \begin{bmatrix} \mathbf{V} \\ \boldsymbol{\theta} \end{bmatrix} - \begin{bmatrix} \mathbf{P} \\ \mathbf{Q} \end{bmatrix} = 0 \tag{14.36}$$

where

$$\mathbf{x} = \begin{bmatrix} \mathbf{V} \\ \boldsymbol{\theta} \end{bmatrix} \qquad \boldsymbol{\rho} = \begin{bmatrix} \mathbf{P} \\ \mathbf{Q} \end{bmatrix}$$

Here $\mathbf{x}$ is the system state vector, and $\boldsymbol{\rho}$ is the parameter vector whose elements are active and reactive load power, and active generator power. Both $\mathbf{x}$ and $\boldsymbol{\rho}$ are $N=2N_{PQ}+N_{PV}$ dimensional vectors, with N_{PQ} the total number of PQ buses and N_{PV} the total number of PV buses. The dimension of the nonlinear vector function $\mathbf{f}$ is also N.

Let $\mathbf{J}_x$ and $\mathbf{J}_\rho$ be the Jacobian matrices of the vector function $\mathbf{f}$ with respect to $\mathbf{x}$ and $\boldsymbol{\rho}$, respectively. Matrix $\mathbf{J}_x$ is the same as the power-flow Jacobian matrix of Equation 14.6. For a given parameter vector $\boldsymbol{\rho}_i$, a system state vector $\mathbf{x}_i$ can be obtained by solving Equation 14.36 using any of the power-flow solution techniques described in Chapter 6. Each parameter vector $\boldsymbol{\rho}_i$ represents a specific system condition in terms of active and reactive loads, and active generation. The system reaches its voltage stability critical point if the parameter vector $\boldsymbol{\rho}_*$ and the corresponding system state vector $\mathbf{x}_*$ are such that the power-flow Jacobian matrix $\mathbf{J}_x$ is singular. Let S denote the hypersurface in the N-dimensional parameter space such that $\mathbf{J}_x(\mathbf{x}_*, \boldsymbol{\rho}_*)$ is singular if $\boldsymbol{\rho}_*$ is a point on S.

Given an initial system operating point $(\mathbf{x}_0, \boldsymbol{\rho}_0)$, we wish to find the parameter vector $\boldsymbol{\rho}_*$ on S such that the distance between $\boldsymbol{\rho}_0$ and $\boldsymbol{\rho}_*$, $k = |\boldsymbol{\rho}_* - \boldsymbol{\rho}_0|$, is a local minimum for the distance between $\boldsymbol{\rho}_0$ and S.

Assuming that S is a smooth hypersurface near $\boldsymbol{\rho}_*$, a normal vector to this hypersurface at $(\mathbf{x}_*, \boldsymbol{\rho}_*)$ is given by

$$\mathbf{\eta}_* = \mathbf{w}_* \mathbf{J}_\rho \tag{14.37}$$

where $\mathbf{w}_*$ is the left eigenvector of $\mathbf{J}_x(\mathbf{x}_*, \mathbf{\rho}_*)$ corresponding to the zero eigenvalue. And $\mathbf{\eta}_*$ is normalized such that $|\mathbf{\eta}_*| = 1$.

Starting from the initial system operating point $(\mathbf{x}_0, \mathbf{\rho}_0)$, the system is stressed by incrementally increasing $\mathbf{\rho}$ along a particular direction. Each time $\mathbf{\rho}$ increases, Equation 14.36 is solved to obtain the system state vector $\mathbf{x}$. And $\mathbf{\rho}$ is continuously increased along the same direction until, at the voltage stability critical point $(\mathbf{x}_*, \mathbf{\rho}_*)$, the power-flow Jacobian matrix $\mathbf{J}_x$ becomes singular; that is,

$$\mathbf{\rho}_* = \mathbf{\rho}_0 + k\mathbf{\eta} \tag{14.38}$$

where k is the distance between the initial system operating point $(\mathbf{x}_0, \mathbf{\rho}_0)$ and the voltage stability critical point $(\mathbf{x}_*, \mathbf{\rho}_*)$ as $k = |\mathbf{\rho}_* - \mathbf{\rho}_0|$.

For a given initial system operating point $(\mathbf{x}_0, \mathbf{\rho}_0)$, $\mathbf{\rho}$ can be increased along different directions. Obviously, the value of k depends on the direction along which $\mathbf{\rho}$ is increased. Our objective is to find the direction of the parameter vector $\mathbf{\rho}$ such that k is the local minimum.

The following procedure determines the vector $\mathbf{\eta}_*$ along which the distance between the initial equilibrium point $(\mathbf{x}_0, \mathbf{\rho}_0)$ and the singular point $(\mathbf{x}_*, \mathbf{\rho}_*)$ is the shortest:

(1)   Let $\mathbf{\eta}_0$ be an initial guess for the direction $\mathbf{\eta}_*$, $|\mathbf{\eta}_0| = 1$.

(2)   Stress the system by incrementally increasing $\mathbf{\rho}$ along the direction of $\mathbf{\eta}_i$ until $\mathbf{J}_x$ becomes singular; that is, determine k_i, $\mathbf{\rho}_i$ and $\mathbf{x}_i$ so that $\mathbf{\rho}_i = \mathbf{\rho}_0 + k_i\mathbf{\eta}_i$ is on the surface S.

(3)   Set $\mathbf{\eta}_{i+1} = \mathbf{w}_i \mathbf{J}_\rho$, and $|\mathbf{\eta}_{i+1}| = 1$.

(4)   Iterate steps 1, 2, 3 until $\mathbf{\eta}_i$ converges to a value $\mathbf{\eta}_*$. Then, $\mathbf{\rho}_* = \mathbf{\rho}_0 + k_*\mathbf{\eta}_*$ is the corresponding equilibrium condition.

A simple radial system example

For illustration, let us consider the radial system shown in Figure 14.8. Corresponding to Equation 14.36, we have

$$\mathbf{f}(\mathbf{x}, \mathbf{\rho}) = \begin{bmatrix} 4V\sin\alpha - P \\ 4V\cos\alpha - 4V^2 - Q \end{bmatrix} = \begin{bmatrix} 0 \\ 0 \end{bmatrix} \tag{14.39}$$

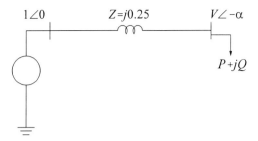

Figure 14.8 A simple radial system

with

$$\mathbf{x} = \begin{bmatrix} V \\ \alpha \end{bmatrix} \qquad \text{and} \qquad \boldsymbol{\rho} = \begin{bmatrix} P \\ Q \end{bmatrix}$$

The Jacobian matrices are

$$\mathbf{J}_x = \begin{bmatrix} 4V\cos\alpha & 4\sin\alpha \\ -4V\sin\alpha & 4\cos\alpha - 8V \end{bmatrix} \tag{14.40}$$

and

$$\mathbf{J}_\rho = \begin{bmatrix} -1 & 0 \\ 0 & -1 \end{bmatrix} \tag{14.41}$$

The determinant of $\mathbf{J}_x$ is

$$\det(\mathbf{J}_x) = 16V - 32V^2\cos\alpha \tag{14.42}$$

On the singular surface S, $\det(\mathbf{J}_x)=0$, that is,

$$16V - 32V^2\cos\alpha = 0$$

or

$$V = \frac{1}{2\cos\alpha} \tag{14.43}$$

Equation 14.43 describes the relationship between V and α, when $\mathbf{J}_x$ is singular. From Equations 14.39 and 14.43, we have the following expression describing the singular surface S in the parameter space:

$$P^2 + 4Q - 4 = 0 \tag{14.44}$$

Let us assume that the initial system operating condition is as follows:

$$P_0 = 0.8 \qquad Q_0 = 0.4 \qquad V_0 = 0.8554 \qquad \alpha_0 = 13.52°$$

Table 14.3 shows the iterative process of finding the point of voltage instability which is closest to the initial operating condition.

Table 14.3 Calculation of the shortest distance to voltage
instability for the system of Figure 14.8

Iteration	Left Eigenvector η_i	Distance to Instability (k_i)	P_i, Q_i
1	$[0.9725 \ -0.2331]^T$	1.0725	1.8430, 0.1500
2	$[0.6776 \ \ 0.7354]^T$	0.4173	1.0828, 0.7069
3	$[0.4869 \ \ 0.8735]^T$	0.4061	0.9977, 0.7541
4	$[0.4443 \ \ 0.8959]^T$	0.4024	0.9788, 0.7605
5	$[0.4405 \ \ 0.8977]^T$	0.4016	0.9769, 0.7605
6	$[0.4378 \ \ 0.8991]^T$	0.4015	0.9758, 0.7610

Figure 14.9 shows the singular surface S in a parameter space with active and reactive components of load as coordinates. On the figure, the point (P_0, Q_0) represents the initial operating condition. The surface S represents the locus of all combinations of P and Q which result in a zero eigenvalue of the Jacobian. All points below S represent voltage stable conditions, and all points above S represent unstable conditions.

General description of the procedure

For any system, the general procedure for finding the minimum distance from an initial load level P_0, Q_0 to S is as follows:

(1)   Increase load from P_0, Q_0 in some direction (the choice of the initial direction will be discussed later) until an eigenvalue of the Jacobian is practically zero. The load level P_1, Q_1 corresponding to this point is the stability limit. This point P_1, Q_1 lies on, or is extremely near, S.

(2)   For the conditions at P_1, Q_1, perform modal analysis and determine the left eigenvector of the *full Jacobian matrix*. The left eigenvector contains elements which provide the increments of MW and MVAr load for each bus. The eigenvector points in the shortest direction to singularity, which is therefore normal to S.

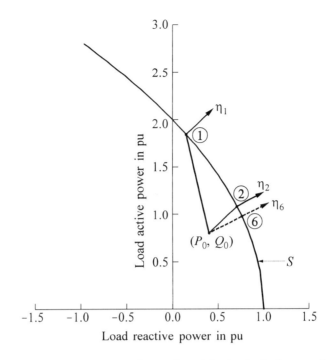

Figure 14.9 The singular surface S in the P-Q plane, and
the convergence of the iterative process

(3)   Go back to the base case load level $\mathbf{P}_0$, $\mathbf{Q}_0$ and load the system again, but this
time in the direction given by the left eigenvector found in (2). When S is
reached, a new left eigenvector is computed.

(4)   Again we return to the base case $\mathbf{P}_0$, $\mathbf{Q}_0$ and load the system in the direction
of the new eigenvector given in (3). This process is repeated until the
computed eigenvector does not change with each new iteration. The process
will then have converged.

When it has converged, the solution gives the minimum vector ($\mathbf{P}$ and $\mathbf{Q}$) distance to
S from $\mathbf{P}_0$, $\mathbf{Q}_0$. This process can be applied to large practical systems; however, S is
not a simple locus, but a hypersurface in a parameter space of dimension $2N$, where
N is the total number of load buses (although we are free to choose the parameter
space we use). The shape of this hypersurface is not known, and therefore we can
expect that this process will find only a *local minimum*. Because of the nonlinearities
encountered in loading the system from $\mathbf{P}_0$, $\mathbf{Q}_0$ to S, the local minimum we find is
dependent on the initial direction we choose for loading. We could use the left
eigenvector corresponding to the minimum eigenvalue of the Jacobian at $\mathbf{P}_0$, $\mathbf{Q}_0$ for
this initial direction, but it may be more appropriate to use uniform loading based on
an expected load pattern.

A large system example

This system has 600 buses. The total load power in the base case is 24,594 MW, 14,112 MVAr. The initial guess for the direction of stressing the system, i.e., η_0 is such that the active and reactive load power in the base case is scaled up uniformly, and the increase in load active power is matched by uniformly scaling up active generation of all the generators. Table 14.4 summarizes the study results. Table 14.5 shows, at each iteration, the elements of the vector η_i associated with the top 20 buses with the largest increase in active power load.

Table 14.4 Summary of iterations to find the shortest distance
to voltage instability of a 600-bus test system

Iteration	Distance to Instability (k_i)	Total Load P and Q' at $(\mathbf{x}_*, \rho_*)$	Smallest Eigenvalue of $\mathbf{J}_x$ at $(\mathbf{x}_*, \rho_*)$
0		24,594 MW, 14,112 MVAr	0.034
1	116	25,729 MW, 14,763 MVAr	0.034
2	38	25,285 MW, 14,046 MVAr	−0.015
3	30	24,788 MW, 14,207 MVAr	−0.014
4	30	24,787 MW, 14,207 MVAr	0.013

From Table 14.4 we see that, with the initial uniform load increase, the stability limit is about 25,729 MW, an increase of about 1,135 MW. The corresponding vector sum of all active and reactive power load increases is 116. The smallest eigenvalue of $\mathbf{J}_x$ is 0.034, indicating that the system is on the verge of instability. The left eigenvector associated with the minimum eigenvalue gives the loading direction for the next iteration. Table 14.5 lists the top 20 entries of the left eigenvector; the load at each bus is increased in proportion to the corresponding eigenvector entries.

We see that the process converges quickly. At convergence, the load increase vector length k_i is only 30, significantly smaller than the initial direction of uniform load increase. The total megawatt load increase for the minimum loading direction is only 193 MW above the base case; this is 942 MW less than the margin found with uniform loading.

14.3.5 The Continuation Power-Flow Analysis

The Jacobian matrix of Equation 14.6 becomes singular at the voltage stability limit. Consequently, conventional power-flow algorithms are prone to convergence problems at operating conditions near the stability limit. The continuation power-flow analysis overcomes this problem by reformulating the power-flow equations so that they remain well-conditioned at all possible loading conditions. This allows the solution of the power-flow problem for stable as well as unstable equilibrium points (that is, for both upper and lower portions of the *V-P* curve).

Table 14.5 Top 20 elements of the vector $\boldsymbol{\eta}_i$ at each iteration

Iteration 1			Iteration 2			Iteration 4		
Bus	E_P	E_Q	Bus	E_P	E_Q	Bus	E_P	E_Q
279	0.146	-0.005	257	0.141	0.001	257	0.138	0.001
278	0.146	-0.005	260	0.140	0.000	260	0.138	0.000
281	0.146	-0.005	283	0.140	0.000	258	0.138	0.008
280	0.145	-0.005	282	0.140	0.000	283	0.137	0.000
293	0.142	-0.004	284	0.140	0.000	282	0.137	0.000
282	0.133	-0.002	258	0.138	0.008	284	0.137	0.000
283	0.133	0.000	256	0.137	0.005	256	0.137	0.005
284	0.132	0.000	246	0.137	0.005	247	0.137	0.008
260	0.131	-0.002	247	0.137	0.008	246	0.137	0.005
257	0.131	-0.002	293	0.136	0.001	248	0.136	0.004
256	0.120	0.000	248	0.136	0.004	294	0.135	0.011
246	0.118	0.000	279	0.135	0.001	265	0.134	0.004
265	0.117	0.000	278	0.135	0.002	293	0.134	0.001
248	0.117	0.000	280	0.135	0.001	261	0.133	0.007
258	0.117	0.002	281	0.135	0.001	267	0.133	0.010
247	0.116	0.002	265	0.134	0.004	250	0.133	0.006
270	0.112	0.000	294	0.134	0.011	269	0.133	0.010
292	0.112	0.000	261	0.133	0.007	281	0.132	0.001
261	0.111	0.001	250	0.132	0.006	268	0.132	0.009
277	0.111	0.000	267	0.132	0.010	279	0.132	0.002

Note: E_P and E_Q are the P and Q terms of the left eigenvector $\boldsymbol{\eta}_i$.

The continuation power-flow method described in this section is based on the approach described in reference 23. It uses a *locally-parameterized continuation method* and belongs to a general class of methods for solving nonlinear algebraic equations known as *path-following methods* [30, 31].

Basic principle

The continuation power-flow analysis uses an iterative process involving predictor and corrector steps as depicted in Figure 14.10. From a known initial solution (A), a tangent predictor is used to estimate the solution (B) for a specified pattern of load increase. The corrector step then determines the exact solution (C) using a conventional power-flow analysis with the system load assumed to be fixed.

The voltages for a further increase in load are then predicted based on a new tangent predictor. If the new estimated load (D) is now beyond the maximum load on the exact solution, a corrector step with loads fixed would not converge; therefore, a corrector step with a fixed voltage at the monitored bus is applied to find the exact solution (E). As the voltage stability limit is reached, to determine the exact maximum load the size of load increase has to be reduced gradually during the successive predictor steps.

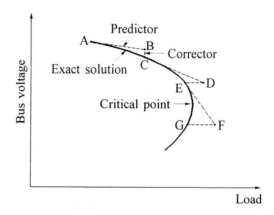

Figure 14.10 A typical sequence of calculations in a continuation power-flow analysis

Mathematical formulation

The basic equations are similar to those of a standard power-flow analysis except that the increase in load is added as a parameter. The reformulated power-flow equations, with provision for increasing generation as the load is increased, may be expressed as

$$\mathbf{F}(\mathbf{\theta},\mathbf{V}) = \lambda\mathbf{K} \qquad (14.45)$$

where
    λ is the load parameter
    $\mathbf{\theta}$ is the vector of bus voltage angles
    $\mathbf{V}$ is the vector of bus voltage magnitudes
    $\mathbf{K}$ is the vector representing percent load change at each bus

The above set of nonlinear equations is solved by specifying a value for λ such that

$$0 \le \lambda \le \lambda_{critical}$$

where $\lambda=0$ represents the base load condition, and $\lambda=\lambda_{critical}$ represents the critical load.

Equation 14.45 may be rearranged as

$$\mathbf{F}(\boldsymbol{\theta},\mathbf{V},\lambda) = 0 \tag{14.46}$$

Predictor step:

In the predictor step, a linear approximation is used to estimate the next solution for a change in one of the state variables (i.e., θ, V, or λ).

Taking the derivatives of both sides of Equation 14.46, with the state variables corresponding to the initial solution, will result in the following set of linear equations:

$$\mathbf{F}_\theta d\boldsymbol{\theta} + \mathbf{F}_\mathbf{V}\, d\mathbf{V} + \mathbf{F}_\lambda d\lambda \;=\; 0$$

or

$$[\;\mathbf{F}_\theta\; \mathbf{F}_\mathbf{V}\; \mathbf{F}_\lambda\;]\begin{bmatrix} d\boldsymbol{\theta} \\ d\mathbf{V} \\ d\lambda \end{bmatrix} = 0 \tag{14.47}$$

Since the insertion of λ in the power-flow equations added an unknown variable, one more equation is needed to solve the above equations. This is satisfied by setting one of the components of the tangent vector to +1 or −1. This component is referred to as the *continuation parameter*. Equation 14.47 now becomes

$$\begin{bmatrix} \mathbf{F}_\theta & \mathbf{F}_\mathbf{V} & \mathbf{F}_\lambda \\ & \mathbf{e}_k & \end{bmatrix}\begin{bmatrix} d\boldsymbol{\theta} \\ d\mathbf{V} \\ d\lambda \end{bmatrix} = \begin{bmatrix} \mathbf{0} \\ \pm 1 \end{bmatrix} \tag{14.48}$$

where $\mathbf{e}_k$ is a row vector with all elements equal to zero except for the kth element (corresponding to the continuation parameter) being equal to 1.

Initially, the load parameter λ is chosen as the continuation parameter and the corresponding component of the tangent vector is set to +1. During the subsequent predictor steps, for reasons given later, the continuation parameter is chosen to be the state variable that has the greatest rate of change near the given solution, and the sign of its slope determines the sign of the corresponding component of the tangent vector. As the maximum load is approached, a voltage will typically be the parameter with the largest change.

Once the tangent vector is found, the prediction for the next solution is given by

$$
\begin{bmatrix} \boldsymbol{\theta} \\ \mathbf{V} \\ \lambda \end{bmatrix} = \begin{bmatrix} \boldsymbol{\theta}_0 \\ \mathbf{V}_0 \\ \lambda_0 \end{bmatrix} + \sigma \begin{bmatrix} d\boldsymbol{\theta} \\ d\mathbf{V} \\ d\lambda \end{bmatrix}
\tag{14.49}
$$

where the subscript "0" identifies the values of the state variables at the beginning of the predictor step. The step size σ is chosen so that a power-flow solution exists with the specified continuation parameter. If for a given step size a solution cannot be found in the corrector step, the step size is reduced and the corrector step is repeated until a successful solution is obtained.

Corrector step:

In the corrector step, the original set of equations $\mathbf{F}(\boldsymbol{\theta}, \mathbf{V}, \lambda) = 0$ is augmented by one equation that specifies the state variable selected as the continuation parameter. Thus the new set of equations is

$$
\begin{bmatrix} \mathbf{F}(\boldsymbol{\theta}, \mathbf{V}, \lambda) \\ x_k - \eta \end{bmatrix} = \begin{bmatrix} \mathbf{0} \end{bmatrix}
\tag{14.50}
$$

In the above, x_k is the state variable selected as the continuation parameter and η is equal to the predicted value of x_k. This set of equations can be solved using a slightly modified Newton-Raphson power-flow method. The introduction of the additional equation specifying x_k makes the Jacobian non-singular at the critical operating point. The continuation power-flow analysis can be continued beyond the critical point and thus obtain solutions corresponding to the lower portion of the *V-P* curve.

The tangent component of λ (i.e., $d\lambda$) is positive for the upper portion of *V-P* curve, is zero at the critical point, and is negative beyond the critical point. Thus the sign of $d\lambda$ will indicate whether or not the critical point has been reached.

If the continuation parameter is the load increase, the corrector will be a vertical line (for example, segment BC in Figure 14.10) on the *V-P* plane. If, on the other hand, a voltage magnitude is the continuation parameter, the corrector will be a horizontal line (for example, segment DE).

Selection of continuation parameter:

The selection of appropriate continuation parameters is particularly important for the corrector steps. A poor choice of the parameter can cause the solution to diverge. For example, the use of load parameter λ as the continuation parameter in the region of critical point can cause the solution to diverge if the estimate exceeds the maximum load. On the other hand, when the voltage magnitude is used as the continuation parameter the solution may diverge if large steps in voltage change are used. A good practice is to choose the continuation parameter as the state variable that has the greatest rate of change near the given solution.

Sensitivity information

In the continuation power-flow analysis, the elements of the tangent vector represent differential changes in the state variables in response to a differential change in system load. Therefore, the dV elements in a given tangent vector are useful in identifying "weak buses", that is, buses which experience large voltage variations in response to a change in load.

Complementary use of conventional and continuation methods

The continuation method of power-flow analysis is robust and flexible. It is ideally suited for solving power-flow problems with convergence difficulties. However, the method is very slow and time-consuming.

The best overall approach for computing power-flow solutions up to and beyond the critical point is to use the conventional and continuation methods of power-flow analysis in a complementary manner. Starting from the base case, Equation 14.45 is solved using a conventional method (Newton-Raphson or Fast Decoupled) to compute power-flow solutions for successively increasing load levels until a solution cannot be obtained. From that point onward, the continuation method is used to obtain the power-flow solutions. Usually, the conventional methods can provide solutions right up to the critical point; the continuation method becomes necessary only if solutions exactly at and past the critical point are required.

Illustration of continuation power-flow application

To illustrate the application of continuation power-flow analysis, let us once again consider the system of Figure 14.3. The voltage stability properties of this system were analyzed in Section 14.1.1 by computing V-P and Q-V characteristics, and in Section 14.3.3 by applying modal analysis. As the power-flow solutions were obtained using conventional methods, the voltage stability analysis had to be limited to only stable equilibrium points, that is, to the upper portion of the V-P curve. At the limiting operating condition considered (point C on the V-P curve of Figure 14.4), as indicated by the eigenvalues of Table 14.1, the system is on the verge of instability but not actually unstable.

We now extend the analysis to operating conditions past the critical point by using continuation power-flow analysis. The load in area 1 of Figure 14.3 (shaded) is increased in the same way as described in Section 14.1.1 with loads assumed to have constant MVA characteristics. Figure 14.11 shows the *complete* V-P curve for the system. The upper portion of the curve (A to C) was computed using a conventional power-flow analysis (Fast Decoupled) and is identical to the characteristic shown in Figure 14.4. The lower portion of the curve (C to B) was, however, computed using the continuation power-flow analysis.

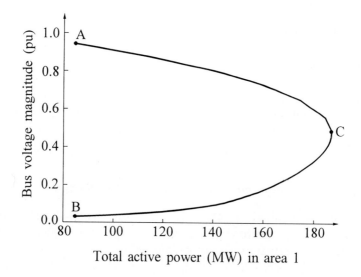

Figure 14.11 *V-P* curve at bus 530 of system shown in Figure 14.3

Table 14.6 Voltage sensitivity and participation factors for critical operating point

Continuation Power-Flow		Modal Analysis	
Bus	Voltage Sensitivity Factor	Bus	Bus Participation Factor
530	0.126	530	0.2654
520	0.112	520	0.2104
510	0.079	510	0.1025
500	0.075	500	0.0941
320	0.054	320	0.0481
310	0.044	310	0.0317
300	0.042	300	0.0294
370	0.041	340	0.0278
260	0.040	370	0.0267
340	0.040	260	0.0263

Note: Voltage sensitivity factor for bus k = $\dfrac{dV \text{ for bus } k}{\text{sum of } dV_s \text{ for all buses}}$

Point C represents the critical operating point, and the corresponding area 1 load is 186.98 MW. For this operating condition, the top 10 entries for the voltage sensitivity factors (dV elements of the tangent vector) are listed in Table 14.6. For comparison, the top 10 entries for the bus participation factors corresponding to the least stable mode (see Section 14.3.3) are also shown in the table. The areas prone to voltage instability identified by the two approaches are practically identical for this system. This is to be expected since, for the loading pattern considered, there is only one voltage weak area.

14.4 PREVENTION OF VOLTAGE COLLAPSE

This section identifies system design and operating measures that can be taken to prevent voltage collapse.

14.4.1 System Design Measures

(a) Application of reactive power-compensating devices

In Section 14.1.4, we discussed the effects of different types of reactive power-compensating devices on voltage stability. Adequate stability margins should be ensured by proper selection of compensation schemes. The selection of the sizes, ratings, and locations of the compensation devices should be based on a detailed study covering the most onerous system conditions for which the system is required to operate satisfactorily.

Design criteria based on maximum allowable voltage drop following a contingency are often not satisfactory from the voltage stability viewpoint. The stability margin should be based on MW and MVAr distances to instability. It is important to recognize voltage control areas and weak transmission boundaries in this regard.

(b) Control of network voltage and generator reactive output

Load (or line drop) compensation of a generator AVR regulates voltage on the high-tension side of, or partway through, the step-up transformer (see Chapter 8, Section 8.5.4). In many situations this has a beneficial effect on voltage stability by moving the point of constant voltage electrically closer to the loads.

Alternatively, secondary outer loop control of generator excitation may be used to regulate network side voltage. This should be much slower than the normal regulation of generator terminal voltage to minimize adverse interaction of the two controls. A response time of about 10 seconds is usually adequate for the outer loop control.

Several utilities are developing special schemes for control of network voltages and reactive power. For example, French and Italian utilities (EDF and ENEL) are developing "secondary voltage control" schemes for centrally controlling network voltages and generator reactive outputs [26,27]. Tokyo Electric Power Company has an adaptive control of reactive power supply [28].

(c) Coordination of protections/controls

As identified in Section 14.2.2, one of the causes of voltage collapse is the lack of coordination between equipment protections/controls and power system requirements. Adequate coordination should be ensured based on dynamic simulation studies.

Tripping of equipment to prevent an overloaded condition should be the last resort. Wherever possible, adequate control measures (automatic or manual) should be provided for relieving the overload condition before isolating the equipment from the system.

(d) Control of transformer tap changers

Tap changers can be controlled, either locally or centrally, so as to reduce the risk of voltage collapse. Where tap changing is detrimental, a simple method is to block tap changing when the source side voltage sags, and unblock when the voltage recovers. Several utilities are using such schemes [1].

There is potential for application of improved ULTC control strategies. Such strategies must be developed based on a knowledge of the load and distribution system characteristics. For example, depressing the distribution voltages in substations which supply predominantly residential loads provides load relief at least temporarily. This will be partially offset eventually when the load is increased by the action of automatic or manually controlled devices. Increasing voltage on industrial loads does not materially affect the load supplied, but increases the reactive power supplied by the capacitors associated with such loads.

Microprocessor-based ULTC controls offer virtually unlimited flexibility for implementing ULTC control strategies so as to take advantage of the load characteristics. Where dropping the downstream voltage offers relief, the voltages may be reduced to a specific level when the primary voltage drops below a threshold. On the other hand, where maintaining secondary voltage is beneficial, normal ULTC controls should be applied. There is even a possibility of actually raising voltages slightly above normal [29]. The best strategy depends on the characteristics of the specific system.

(e) Undervoltage load shedding [1,32]

To cater to unplanned or extreme situations, it may be necessary to use undervoltage load-shedding schemes. This is analogous to underfrequency load shedding, which has become a common utility practice to cater to extreme situations resulting in generation deficiency and underfrequency. Load shedding provides a low-cost means of preventing widespread system collapse. This is particularly true if system conditions and the contingencies leading to voltage instability are of low probability, but would result in serious consequences. The characteristics and locations of the loads to be shed are more important for voltage problems than they are for frequency problems.

Load-shedding schemes should be designed so as to distinguish between faults, transient voltage dips, and low voltage conditions leading to voltage collapse.

14.4.2 System-Operating Measures

(a) Stability margin

The system should be operated with an adequate voltage stability margin by the appropriate scheduling of reactive power resources and voltage profile. There are at present no widely accepted guidelines for selection of the degree of margin and the system parameters to be used as indices. These are likely to be system dependent and may have to be established based on the characteristics of the individual system.

If the required margin cannot be met by using available reactive power resources and voltage control facilities, it may be necessary to limit power transfers and to start up additional generating units to provide voltage support at critical areas.

(b) Spinning reserve

Adequate spinning reactive-power reserve must be ensured by operating generators, if necessary, at moderate or low excitation and switching in shunt capacitors to maintain the desired voltage profile. The required reserve must be identified and maintained within each voltage control area.

(c) Operators' action

Operators must be able to recognize voltage stability-related symptoms and take appropriate remedial actions such as voltage and power transfer controls and, possibly as a last resort, load curtailment. Operating strategies that prevent voltage collapse need to be established. On-line monitoring and analysis to identify potential voltage stability problems and possible remedial measures would be invaluable in this regard.

REFERENCES

[1]   IEEE, Special Publication 90TH0358-2-PWR, *Voltage Stability of Power Systems: Concepts, Analytical Tools, and Industry Experience*, 1990.

[2]   CIGRE Task Force 38-02-10, *Modelling of Voltage Collapse Including Dynamic Phenomena*, 1993.

[3]   CIGRE Task Force 38-01-03, "Planning against Voltage Collapse," *Electra*, No. 111, pp. 55-75, March 1987.

[4]   C. Barbier and J.P. Barret, "An Analysis of Phenomena of Voltage Collapse on Transmission System," *Revue Générale d' Electricité*, pp. 672-690, October 1980.

[5]   H.K. Clark, "Voltage Control and Reactive Supply Problems," IEEE Tutorial Course Text on Reactive Power: Basics, Problems, and Solutions, 87EH0262-6-PWR, pp. 17-27, 1987.

[6]   W.R. Lachs, "Voltage Collapse in EHV Power Systems," Paper No. A78057-2, presented at the 1978 IEEE PES Winter Meeting, New York, January 28-February 3, 1978.

[7]   Y. Mansour and P. Kundur, "Voltage Collapse," *Control and Dynamic Systems*, Vol. 42: *Analysis and Control System Techniques for Electric Power Systems*, pp. 111-162, Academic Press, 1991.

[8]   C.W. Taylor, *Power System Voltage Stability*, McGraw-Hill, 1993.

[9]   B. Gao, G.K. Morison, and P. Kundur, "Voltage Stability Evaluation Using Modal Analysis," *IEEE Trans.*, Vol. PWRS-7, No. 4, pp. 1529-1542, November 1992.

[10]   G.K. Morison, B. Gao, and P. Kundur, "Voltage Stability Analysis Using Static and Dynamic Approaches," Paper 92SM590-0 PWRS, Presented at the IEEE PES Summer Meeting, July 12-16, 1992, Seattle, Washington.

[11]   K. Walve, "Modelling of Power System Components at Severe Disturbances," CIGRE Paper: 38-18, 1986.

[12]   M. Stubbe, A. Bihain, J. Deuse, and J.C. Baader, "STAG - A New Unified Software Program for the Study of the Dynamic Behaviour of Electrical Power Systems," *IEEE Trans.*, Vol. PWRS-4, No. 1, pp. 129-138, 1989.

[13]   N. Flatabø, R. Ognedal, and T. Carlsen, "Voltage Stability Condition in a Power Transmission System Calculated by Sensitivity Methods," *IEEE Trans.*, Vol. PWRS-5, No. 4, pp. 1286-1293, November 1990.

[14]   C. Lemaitre, J.P. Paul, J.M. Tesseron, Y. Harmand, and Y.S. Zhao, "An Indicator of the Risk of Voltage Profile Instability for Real-Time Control Applications," IEEE Summer Meeting 1989, Paper 89SM713-9 PWRS.

[15]   B. Gao, *Voltage Stability Analysis of Large Power Systems*, Ph.D. thesis, University of Toronto, 1992.

[16]   W.J. Stewart and A. Jennings, "A Simultaneous Iteration Algorithm for Real Matrices," *ACM Trans. Mathematical Software*, Vol. 7, No. 2, pp. 184-198, June 1981.

[17]   L. Wang and A. Semlyen, "Application of Sparse Eigenvalue Techniques to the Small Signal Stability Analysis of Large Power Systems," *IEEE Trans.*, Vol. PWRS-5, No. 2, pp. 635-642, May 1990.

[18]   N. Martins, "Efficient Eigenvalue and Frequency Response Methods Applied to Power System Small-Signal Stability Studies," *IEEE Trans.*, Vol. PWRS-1, No. 2, pp. 217-225, February 1986.

[19]   *Voltage Stability Analysis Program Application Guide*, EPRI Project RP3040-1, Prepared by Ontario Hydro, October 1992.

[20]   F.L. Alvarado and T.H. Jung, "Direct Detection of Voltage Collapse Conditions," *Proceedings: Bulk Power System Voltage Phenomena - Voltage Stability and Security*, EPRI EL-6183, pp. 5.23-5.38, January 1989.

[21]   H.D. Chiang, W. Ma, R.J. Thomas, and J.S. Thorp, "A Tool for Analyzing Voltage Collapse in Electric Power Systems," *Proceedings of the Tenth Power System Computation Conference*, Graz, Austria, pp. 1210-1217, August 1990.

[22]   T. Van Cutsem, "A Method to Compute Reactive Power Margins with Respect to Voltage Collapse," *IEEE Trans.*, Vol. PWRS-6, No. 2, pp. 145-156, February 1991.

[23]   V. Ajjarapu and C. Christy, "The Continuation Power Flow: A Tool for Steady State Voltage Stability Analysis," *IEEE PICA Conference Proceedings*, pp. 304-311, May 1991.

[24]   I. Dobson and L. Lu, "Using an Iterative Method to Compute a Closest Saddle Node Bifurcation in the Load Power Parameter Space of an Electric Power

System," Bulk Power System Voltage Phenomena, Voltage Stability and Security NSF Workshop, Deep Creek Lake, Md., August 1991.

[25]   A-P Löf, T. Smed, G. Andersson, and D.J. Hill, "Fast Calculation of a Voltage Stability Index," *IEEE Trans.*, Vol. PWRS-7, No. 1, pp. 54-64, February 1992.

[26]   J.P. Paul, C. Corroyer, P. Jeanell, J.M. Tesseron, F. Maury, and A. Torra, "Improvements in the Organization of Secondary Voltage Control in France," CIGRE 38/39-03, 1990.

[27]   V. Archidiacono, S. Corsi, A. Natale, C. Raffaelli, and V. Mcnditto, "New Developments in the Application of ENEL Transmission System Voltage and Reactive Power Automatic Control," CIGRE 38/39-06, 1990.

[28]   S. Koishikawa, S. Ohsaka, M. Suzuki, T. Michigami, and M. Akimoto, "Adaptive Control of Reactive Power Supply Enhancing Voltage Stability of a Bulk Power Transmission System and a New Scheme of Monitor on Voltage Security," CIGRE 38/39-01, 1990.

[29]   G. Brownell and H. Clark, "Analysis and Solutions for Bulk System Voltage Instability," *IEEE Computer Applications in Power*, pp. 31-35, July 1989.

[30]   R. Seydel, *From Equilibrium to Chaos*, Elsevier, New York, 1988.

[31]   W.C. Rheinboldt, *Numerical Analysis of Parameterized Nonlinear Equations*, John Wiley & Sons, New York, 1986.

[32]   C.W. Taylor, "Concepts of Undervoltage Load Shedding for Voltage Stability," *IEEE Trans. on Power Delivery*, Vol. 7, No. 2, pp. 480-488, April 1992.

Chapter **15**

Subsynchronous Oscillations

In the analysis of power system dynamic performance so far, the rotor of a turbine generator was assumed to be made up of a single mass. Such a representation accounts for the oscillation of the entire turbine-generator rotor with respect to other generators. The frequency of this mode of oscillation is usually in the range of 0.2 to 2 Hz. In reality, a steam turbine-generator rotor has a very complex mechanical structure consisting of several predominant masses (such as rotors of turbine sections, generator rotor, couplings, and exciter rotor) connected by shafts of finite stiffness. Therefore, when the generator is perturbed, torsional oscillations result between different sections of the turbine-generator rotor.

The torsional oscillations in the *subsynchronous* range could, under certain conditions, interact with the electrical system in an adverse manner. Special problems related to torsional oscillations include the following:

(a)   Torsional interaction with power system controls.

(b)   Subsynchronous resonance with series capacitor-compensated transmission lines.

(c)   Torsional fatigue duty due to network switching.

The torsional characteristics of hydro units are such that interaction with power system controls or transmission network has not been a source of concern.

This chapter describes the characteristics and modelling of a turbine-generator shaft system and discusses various problems related to subsynchronous torsional oscillations that require consideration in the design of power systems.

1025

15.1 TURBINE-GENERATOR TORSIONAL CHARACTERISTICS

15.1.1 Shaft System Model [1-6]

The rotor of a thermal generating unit is a complex mechanical system. It may exceed 50 metres in total length and weigh several hundred tons. The rotor contains machined shaft sections of varying sizes and couplings that are either integral or shrunk on and keyed to the rotor. The turbine sections contain disks, blades, and other smaller components. The generator includes coil slots and retaining rings. Such a rotor system has a large number of torsional vibration modes both above and below the rated frequency. A *continuum model* of the rotor system would be required to account for the complete range of torsional oscillations [6]. However, the problem due to interaction between the electrical system and the rotor mechanical system is principally in the subsynchronous frequency range. This allows the representation of the rotor system by a simple *lumped-mass model* for electrical system interaction studies.

Figure 15.1 shows the structure of a typical lumped-mass model of a generating unit driven by a tandem compound reheat turbine. The five torsional masses represent the rotors of the generator, two low-pressure (LP) turbine sections, an intermediate-pressure (IP) turbine section, and a high-pressure (HP) turbine section. The generating unit is assumed to have a static exciter. For a unit with a rotating exciter driven by the same shaft system, there will be an additional mass representing the exciter rotor (see example shown in Figure 15.6).

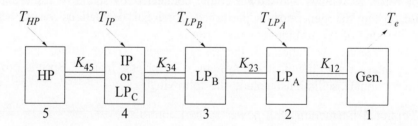

Figure 15.1 Structure of a typical lumped-mass shaft system model

In developing the mathematical equations of the shaft system, we use the following notation:

$$\left. \begin{array}{l} T_{HP},\ T_{IP}, \\ T_{LP_A},\ T_{LP_B} \end{array} \right\} = \text{mechanical torques developed by the respective turbine sections in pu}$$

T_e　= generator air-gap torque in pu

ω_0 = rated speed in electrical rad/s = $2\pi f_0$ = 377 for 60 Hz

ω_{0m} = rated speed in mechanical rad/s = $(2/p_f)\omega_0$

p_f = number of field poles

ω_i = speed of mass i in electrical rad/s

δ_i = angular position of mass i in *electrical radians* with respect to a synchronously rotating reference = $\omega_i t - \omega_0 t + \delta_{i0}$

$\Delta\omega_i$ = speed deviation of mass i in pu = $(\omega_i - \omega_0)/\omega_0$

D = damping coefficient or factor in pu torque/pu speed deviation

H = inertia constant in MW·s/MVA

K = shaft stiffness in pu torque/electrical rad

t = time in seconds

The shaft system dynamic characteristics are defined by three sets of parameters: inertia constant H of the individual masses, torsional stiffness K of shaft sections connecting adjacent masses, and damping coefficient D associated with each mass. The following is a description of each of the parameters.

Inertia constant H

The inertia assigned to each rotor mass includes its share of shaft inertia. Turbine blades are assumed to be rigidly connected to the rotor. If the moment of inertia of a rotor mass is J kg·m^2, the per unit inertia constant H (see Chapter 3, Section 3.9) is given by

$$H = J\frac{1}{2}\frac{\omega_{0m}^2}{VA_{base}} = J\frac{1}{2}\frac{[2\pi\,(r/min)/60]^2}{VA_{base}}$$

Torsional stiffness K

For a shaft of uniform cross-section undergoing elastic strain, the torsional stiffness or spring constant is given by [1]

$$K = \frac{GF}{l} \tag{15.1}$$

where

G = rigidity modulus of shaft material
F = form factor which defines the geometric property
l = length of shaft

For a solid shaft of circular cross-section with diameter d,

$$F = \frac{\pi d^4}{32} \tag{15.2}$$

The torsional stiffness defines the relationship between the torque transmitted and the angular twist between the two ends of the shaft:

$$T = K\theta \tag{15.3}$$

where

   T = torque, N·m
   θ = twist, rad
   K = stiffness, N·m/rad

In a turbine-generator rotor, each shaft span consists of several sections of different diameters. The torsional stiffness of each section is determined and then a single equivalent stiffness is computed as follows:

$$\frac{1}{K_{total}} = \sum \frac{1}{K_{individual\,section}}$$

For the representation of turbine-generator shaft systems in power system studies, angles are expressed in electrical radians (or degrees). With the number of generator field poles equal to p_f,

$$\theta \text{ electrical rad} = (\theta \text{ mechanical rad})\frac{p_f}{2}$$

In system studies, torque is normally expressed in per unit with the base torque equal to

$$T_{base} = \frac{VA_{base}}{\omega_{0m}} = VA_{base}\frac{p_f}{2\omega_0}$$

The torsional stiffness is then given by

$$K \text{ pu torque/electrical rad} = \frac{K \text{ N·m/mechanical rad}}{(p_f/2)\, T_{base}}$$

$$= \frac{K \text{ N·m/mechanical rad}}{VA_{base}} \left(\frac{4\omega_0}{p_f^2} \right)$$

Damping coefficient or factor D [2-6]

There are a number of sources contributing to the damping of torsional oscillations:

(a)   Steam forces on turbine blades. The oscillation of the turbine blades in the steady-state steam flow introduces damping. As an approximation, this may be represented as being proportional to the speed deviation of the respective turbine section.

(b)   Shaft material hysteresis. When the interconnecting shaft sections twist, damping is introduced due to the mechanical hysteresis of the shaft material as it undergoes cyclic stress-strain variations.

(c)   Electrical sources. Generator, exciter, and transmission networks contribute to damping of oscillations.

The damping levels associated with torsional oscillations are very small and are a function of the turbine-generator output. Time constants associated with the decay of torsional oscillations range from 4 to 30 seconds.

The various mechanisms contributing to torsional damping are complex and their contributions are difficult to predict. There is a high degree of variability in the torsional mode damping even among similar units. The best means of determining the actual damping levels is through station tests [3].

In our model of the shaft system, we will assume that all sources of damping may be represented in terms of damping torques proportional to speed deviations of individual masses; that is, damping torques between rotors are assumed to be negligible.

Shaft system equations

We will illustrate the development of equations of individual masses by considering the rotors of the generator and LP_A turbine shown in Figure 15.2.

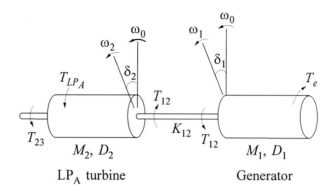

Figure 15.2

The various components of torque associated with the *generator rotor* are as follows:

Input torque $\quad\quad = T_{12} = K_{12}(\delta_2 - \delta_1)$
Output torque $\quad\quad = T_e$
Damping torque $\quad\quad = D_1(\Delta\omega_1)$
Accelerating torque $\quad = T_a = T_{12} - T_e - D_1\Delta\omega_1$

The equations of motion of the generator rotor are

$$2H_1\frac{d(\Delta\omega_1)}{dt} = T_a = K_{12}(\delta_2 - \delta_1) - T_e - D_1(\Delta\omega_1)$$

$$\frac{d(\delta_1)}{dt} = (\Delta\omega_1)\omega_0$$

Similarly, the following are the equations of the LP_A *turbine section*:

Input torque $\quad\quad = T_{LP_A} + T_{23} = T_{LP_A} + K_{23}(\delta_3 - \delta_2)$
Output torque $\quad\quad = T_{12} = K_{12}(\delta_2 - \delta_1)$
Damping torque $\quad = D_2(\Delta\omega_2)$

Equations of motion:

$$2H_2\frac{d(\Delta\omega_2)}{dt} = T_{LP_A} + K_{23}(\delta_3 - \delta_2) - K_{12}(\delta_2 - \delta_1) - D_2(\Delta\omega_2)$$

$$\frac{d\delta_2}{dt} = (\Delta\omega_2)\omega_0$$

The equations of the *complete rotor system* shown in Figure 15.1 may be summarized as follows :

GEN:
$$2H_1\frac{d(\Delta\omega_1)}{dt} = K_{12}(\delta_2 - \delta_1) - T_e - D_1(\Delta\omega_1)$$

$$\frac{d(\delta_1)}{dt} = (\Delta\omega_1)\omega_0$$

LP$_A$:
$$2H_2\frac{d(\Delta\omega_2)}{dt} = T_{LP_A} + K_{23}(\delta_3 - \delta_2) - K_{12}(\delta_2 - \delta_1) - D_2(\Delta\omega_2)$$

$$\frac{d(\delta_2)}{dt} = (\Delta\omega_2)\omega_0$$

LP$_B$:
$$2H_3\frac{d(\Delta\omega_3)}{dt} = T_{LP_B} + K_{34}(\delta_4 - \delta_3) - K_{23}(\delta_3 - \delta_2) - D_3(\Delta\omega_3)$$

$$\frac{d(\delta_3)}{dt} = (\Delta\omega_3)\omega_0$$

(15.4)

IP:
$$2H_4\frac{d(\Delta\omega_4)}{dt} = T_{IP} + K_{45}(\delta_5 - \delta_4) - K_{34}(\delta_4 - \delta_3) - D_4(\Delta\omega_4)$$

$$\frac{d(\delta_4)}{dt} = (\Delta\omega_4)\omega_0$$

HP:
$$2H_5\frac{d(\Delta\omega_5)}{dt} = T_{HP} - K_{45}(\delta_5 - \delta_4) - D_5(\Delta\omega_5)$$

$$\frac{d(\delta_5)}{dt} = (\Delta\omega_5)\omega_0$$

During a transient condition, the generator air-gap torque is determined by the dynamics of the generator and the power system to which it is connected. The torques generated by the individual turbine sections (i.e., T_{HP}, T_{IP}, T_{LP_A}, T_{LP_B}) depend on the dynamics of the steam turbine and its governing system (see Chapter 9).

Example 15.1

The data applicable to a five-mass torsional model (see Figure 15.1) of 960 MVA, 24 kV, 0.9 pf, 1,800 r/min (four poles) nuclear unit with a static exciter is as follows:

Mass No.	Rotor	Power* Fraction	WR^2 lb·ft^2	Shaft Section	Stiffness lb·ft/rad
1	GEN	–	1,114,382	GEN-LP$_A$	279,823,373
2	LP$_A$	0.24	1,831,389	LP$_A$-LP$_B$	235,204,647
3	LP$_B$	0.24	1,830,972	LP$_B$-LP$_C$	207,786,864
4	LP$_C$	0.24	1,830,417	LP$_C$-HP	133,530,219
5	HP	0.28	225,240		

* Fraction of total turbine power generated by the respective turbine
  sections under steady state

(a)   Determine the inertia constant H in MW·s/MVA for each of the five masses and the stiffness K in pu torque/electrical rad for each of the four shaft sections.

(b)   Compute the steady-state value of torque transmitted by each shaft section and the angular displacement between the generator and HP turbine section, when the generator is operating at rated output.

Solution

(a) The moment of inertia of the HP section rotor (mass 5) is

$$J_5 = (WR^2 \text{ lb·ft}^2)\frac{1.356}{32.2}$$

$$= 225,240 \times \frac{1.356}{32.2}$$

$$= 9485.3 \quad \text{kg·m}^2$$

The corresponding inertia constant is

$$H_5 = \frac{1}{2}J\left(\frac{2\pi \text{ r/min}}{60}\right)^2 \frac{1}{\text{VA rating}}$$

$$= \frac{1}{2} \times 9485.3 \times \left(\frac{2\pi \times 1800}{60}\right)^2 \frac{1}{960 \times 10^6}$$

$$= 0.1755 \quad \text{MW·s/MVA}$$

Similarly, the inertia constants of the other masses are

$$LP_C \text{ section} : \quad H_4 = 1.4270$$
$$LP_B \text{ section} : \quad H_3 = 1.4275$$
$$LP_A \text{ section} : \quad H_2 = 1.4278$$
$$\text{Generator} \quad : \quad H_1 = 0.8688$$

The torsional stiffness of the shaft between the HP and LP_C turbine sections is

$$K_{45} = 1.356 K \quad \text{lb·ft/rad}$$

$$= 1.356 \times 133,530,219$$

$$= 181,066,980 \quad \text{N·m/mech. rad}$$

Expressing this in pu torque/electrical rad, we have

$$K_{45} = \frac{181,066,980}{VA_{base}} \left(\frac{4\omega_0}{p_f^2} \right)$$

$$= \frac{181,066,980}{960 \times 10^6} \times \frac{4 \times 377}{4^2}$$

$$= 17.78 \quad \text{pu torque/elec. rad}$$

The torsional stiffnesses of other shaft sections expressed in pu torque/electrical rad are

$$LP_C - LP_B \quad : \quad K_{34} = 27.66$$
$$LP_B - LP_A \quad : \quad K_{23} = 31.31$$
$$LP_A - GEN \quad : \quad K_{12} = 37.25$$

(b) The air-gap torque (T_e) in per unit on 960 MVA base is 0.9.

The torques developed by the different turbine sections are

$$T_{LP_A} = T_{LP_B} = T_{LP_C} = 0.24 \times 0.9 = 0.216 \quad \text{pu}$$

$$T_{HP} = 0.28 \times 0.9 = 0.252 \quad \text{pu}$$

The torque transmitted by the shaft section between the generator and LP_A is

$$T_{12} = T_e = 0.9 \quad \text{pu}$$

Therefore, the angle by which LP_A rotor leads the generator rotor is

$$\delta_{21} = \delta_2 - \delta_1 = T_{12}/K_{12}$$

$$= 0.9/37.25 = 0.02416 \quad \text{elec. rad}$$

The torque transmitted by shaft between LP_A and LP_B turbine sections is

$$T_{23} = T_{12} - T_{LP_A} = 0.9 - 0.216 = 0.684 \quad \text{pu}$$

Hence,

$$\delta_{32} = \delta_3 - \delta_2 = T_{23}/K_{23}$$

$$= 0.684/31.31 = 0.02185 \quad \text{elec. rad}$$

Similarly,

$$\delta_{43} = \delta_4 - \delta_3 = T_{34}/K_{34}$$

$$= (0.684 - 0.216)/27.66 = 0.01692 \quad \text{elec. rad}$$

and

$$\delta_{54} = \delta_5 - \delta_4 = T_{45}/K_{45}$$

$$= 0.252/17.78 = 0.01417 \quad \text{elec. rad}$$

Thus, the HP section rotor leads the generator rotor by

$$\delta_5 - \delta_1 = \delta_{54} + \delta_{43} + \delta_{32} + \delta_{21}$$

$$= 0.01417 + 0.01692 + 0.02185 + 0.02416$$

$$= 0.0771 \quad \text{elec. rad}$$

$$= 4.42 \quad \text{elec. degrees}$$

∎

15.1.2 Torsional Natural Frequencies and Mode Shapes

The natural frequencies and mode shapes can be determined by using the modal analysis technique described in Chapter 12. This requires writing Equation 15.4 in the state-space form:

$$\dot{\mathbf{x}} = \mathbf{A}\mathbf{x} \qquad (15.5)$$

For a small perturbation, the generator air-gap torque may be expressed as

$$T_e = K_S \Delta\delta_1 \tag{15.6}$$

where K_S is the synchronizing torque coefficient (see Chapter 12, Section 12.3).

The linearized equations of the rotor system may be readily written by inspection of Equations 15.4. For example, the generator rotor equations are

$$\frac{d(\Delta\omega_1)}{dt} = -\frac{D_1}{2H_1}(\Delta\omega_1) - \frac{K_{12} + K_S}{2H_1}(\Delta\delta_1) + \frac{K_{12}}{2H_1}(\Delta\delta_2)$$

$$\frac{d(\Delta\delta_1)}{dt} = \omega_0(\Delta\omega_1)$$

and the equations of the LP_A turbine section are

$$\frac{d(\Delta\omega_2)}{dt} = \frac{K_{12}}{2H_2}(\Delta\delta_1) - \frac{D_2}{2H_2}(\Delta\omega_2) - \frac{K_{12} + K_{23}}{2H_2}(\Delta\delta_2) + \frac{K_{23}}{2H_2}(\Delta\delta_3)$$

$$\frac{d(\Delta\delta_2)}{dt} = \omega_0(\Delta\omega_2)$$

Here, we have assumed that the mechanical torque T_{LP_A} developed by the turbine section is constant.

The state variables are thus the speed deviations $\Delta\omega_i$ and the rotor angle $\Delta\delta_i$, with $i=1$ to 5. The elements of the state matrix **A** of the rotor system depend on the torsional stiffness coefficients, the inertia constants of the individual masses, and the generator synchronizing torque coefficient K_S.

The eigenvalues of **A** give the natural frequencies of the shaft system. The corresponding eigenvectors give the "mode shape," i.e., the relative activity of a state variable in a given mode.

Examples of torsional characteristics

Figure 15.3 shows the natural frequencies and mode shapes of the rotor of a 555 MVA, 3,600 r/min fossil-fuel-fired generating unit with a static exciter. Also shown in the figure are the inertia constants of the masses and the stiffnesses of the shaft sections connecting them. The synchronizing torque coefficient (electrical stiffness) K_S is assumed to be 1.98 pu torque/rad. Damping coefficients are assumed to be negligible; this has no effect on the calculation of the natural frequencies or the mode shape.

Since we are considering a rotor with five masses, there are five modes of oscillation. The natural frequencies, as given by the imaginary components of the eigenvalues of the matrix **A**, are 1.67 Hz, 16.35 Hz, 24.1 Hz, 30.3 Hz, and 44.0 Hz.

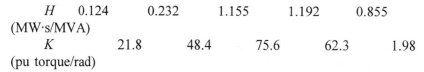

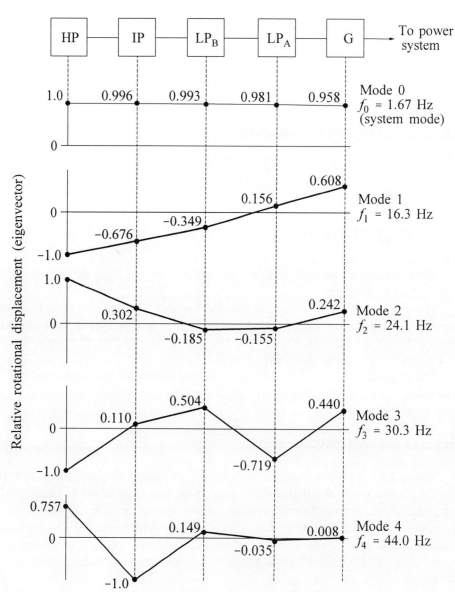

Figure 15.3 Rotor natural frequencies and mode shapes of
a 555 MVA, 3,600 r/min steam turbine generator

The relative rotational displacements of the individual masses for each mode of oscillation are given by the right eigenvector of the corresponding eigenvalue. The elements of the eigenvector associated with either the speed deviations or the angle deviations may be used. In the plots shown in Figure 15.3, each eigenvector has been normalized so that its largest element is equal to 1.0.

The 1.67 Hz mode represents the oscillation of the entire rotor against the power system. This is the mode normally considered in system stability studies. It depends on the synchronizing torque coefficient K_S and the sum of the inertia constants of all the rotor masses. As seen by the mode shape, all five masses participate nearly equally in this mode.

The other four modes represent the torsional modes of oscillation. The first torsional mode has a natural frequency of 16.3 Hz. It has one polarity reversal in the mode shape. The polarities of eigenvector elements associated with the rotors of the generator and the LP_A section are opposite to those associated with the rotors of the LP_B, IP, and HP sections. This indicates that the generator and LP_A rotors oscillate against the other three rotors when this mode is excited.

The second torsional mode has a natural frequency of 24.1 Hz, and its mode shape has two polarity reversals. The third torsional mode, with a natural frequency of 30.3 Hz, has three polarity reversals in its mode shape. Finally, the fourth torsional mode, with a natural frequency of 44.0 Hz, has a mode shape with four polarity reversals.

In Figure 15.3, the rotors of the generator and the LP_A section have very low relative amplitudes of rotational displacement in the fourth torsional mode. This means that this mode cannot be easily excited by applying torques to the generator and LP_A turbine rotors.

In a general case, a rotor with n masses has $n-1$ torsional modes. The ith torsional mode has the ith highest torsional frequency, and its mode shape has i polarity reversals.

Figures 15.4 to 15.6 provide additional examples of torsional characteristics. Only the torsional modes are shown, and the system mode is omitted.

The torsional characteristic shown in Figure 15.4 is that of the 960 MVA, 1,800 r/min (four pole) nuclear unit considered in Example 15.1. This unit has one HP and three LP turbine sections. Its torsional frequencies are 8.4 Hz, 15.2 Hz, 20.3 Hz, and 23.6 Hz.

Figure 15.5 shows the torsional characteristics of a 191 MVA, 3,600 r/min (two pole) coal-fired unit with a static exciter. It has three turbine sections: HP, IP, and one LP. Hence, the rotor is represented by four lumped masses. The principal torsional frequencies are 22.4 Hz, 29.6 Hz, and 52.7 Hz.

| H | 0.176 | 1.427 | 1.428 | 1.428 | 0.869 |
(MW·s/MVA)

| K | | 17.78 | 27.66 | 31.31 | 37.25 |
(pu torque/rad)

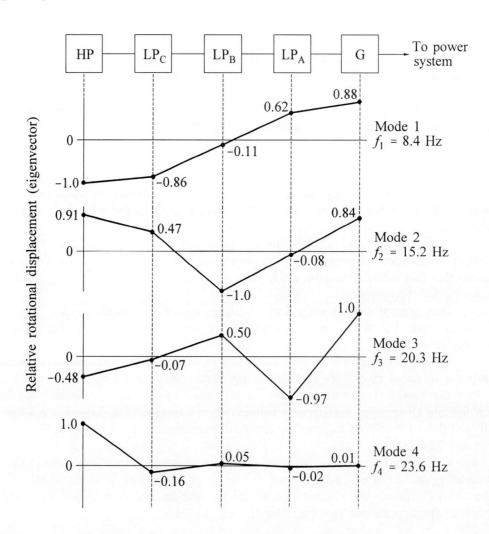

Figure 15.4 Torsional natural frequencies and mode shapes of a 960 MVA, 1,800 r/min turbine generator

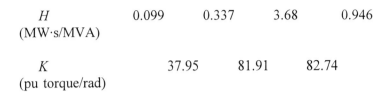

H (MW·s/MVA)	0.099	0.337	3.68	0.946
K (pu torque/rad)		37.95	81.91	82.74

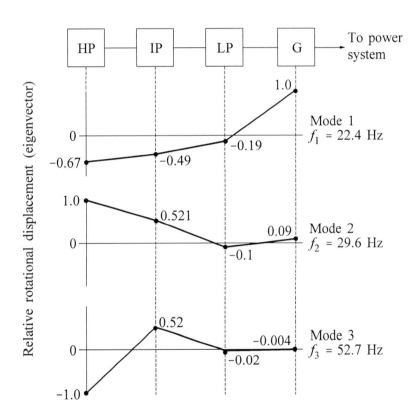

Figure 15.5 Torsional natural frequencies and mode shapes of a 191 MVA, 3,600 r/min turbine generator

Figure 15.6 shows the torsional characteristics of a 635 MVA, 1,800 r/min (four pole) unit with a rotating exciter (EX). The rotor is represented by six lumped masses.

H	0.254	0.983	1.001	1.009	1.035	0.013
(MW·s/MVA)						

K		13.9	18.2	25.2	54.9	5.7
(pu torque/rad)						

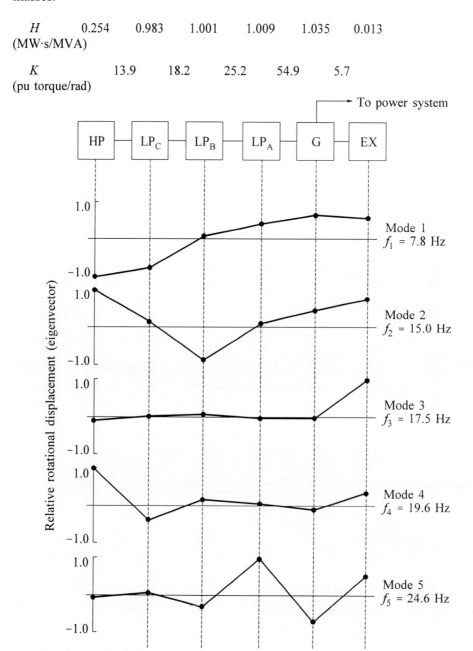

Figure 15.6 Torsional natural frequencies and mode shapes of a 635 MVA, 1,800 r/min turbine generator with a rotating exciter

15.2 TORSIONAL INTERACTION WITH POWER SYSTEM CONTROLS

As discussed in the previous section, torsional oscillations of turbine generators are inherently lightly damped. While the level of damping can vary to some extent with steam conditions and unit output, normally it is not affected by the generating unit or network controls. However, there have been several instances of torsional mode instability due to interactions with the generating unit excitation and prime mover controls, and with nearby HVDC converter controls. In this section, we describe the causes of such torsional instability problems and methods of analyzing and mitigating the problems.

15.2.1 Interaction with Generator Excitation Controls

Torsional mode destabilization by excitation control was first observed in 1969 during application of a power system stabilizer (PSS) on a 555 MVA, 3,600 r/min fossil-fuel-fired unit at the Lambton generation station in Ontario [7,8]. The PSS, which used a stabilizing signal based on speed measured at the generator end of the turbine-generator shaft, was found to excite the lowest torsional mode. The reason for this can be readily seen from the torsional characteristics shown in Figure 15.3, which are similar to those of the Lambton units. Shaft speed measured at the generator end has a high component of the 16 Hz torsional mode. The stabilizer transfer function is designed to provide zero phase shift between the input speed signal and air-gap torque at the system mode frequency of 1.67 Hz, so as to result in a purely damping torque component as shown in Figure 15.7(a) (see Chapter 12). However, the generator characteristics are such that this results in a phase lag of about 135° at 16 Hz [8]. As shown in Figure 15.7(b), the effect is to produce a negative damping torque component and hence instability of the 16 Hz torsional mode.

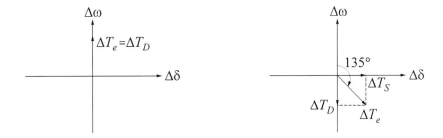

(a) Pure damping torque at 1.67 Hz

(b) Negative damping torque and positive synchronizing torque at 16 Hz

Figure 15.7 Torque components due to PSS

This problem was solved by sensing speed between the two LP turbine sections, close to the "node" of the 16 Hz torsional mode (see Figure 15.3). At this speed pickup location, other torsional modes also have very low amplitude. In addition, an electronic filter, with a notch at 16 Hz and substantial attenuation at other torsional frequencies, was used in the stabilizing path.

The modal analysis approach described in Chapter 12 is ideally suited for investigation of torsional stability problems. This is illustrated in Example 15.2.

Torsional mode instability through excitation control may also be caused by a terminal voltage limiter which uses feedback of terminal voltage to control excitation through a very high gain [8]. This can cause torsional instability, unless some kind of filtering is provided to attenuate high frequency components of the voltage signal.

Example 15.2

This example considers the effect of PSS on the torsional stability of an 889 MVA, 18.5 kV, 1,800 r/min generating unit with a tandem-compound turbine. The shaft system representation is shown in Figure E15.1. It is similar to that of Figure 15.4, except that each double flow LP turbine section is represented by two masses.

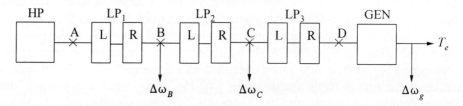

Figure E15.1 Shaft system representation

The shaft system data are as follows:

Mass	Inertia (WR^2 in lb·ft²)	Damping Coefficient (pu torque/pu speed)
Generator	1,081,920	0
LP_{3R} or LP_{3L}	1,054,512	0.125
LP_{2R} or LP_{2L}	1,067,855	0.125
LP_{1R} or LP_{1L}	1,048,380	0.125
HP	258,000	0.125

Shaft Section	Stiffness (lb·ft/rad)	Shaft Section	Stiffness (lb·ft/rad)
Gen - Coupl D	844,391,342	LP_{2L} - Coupl B	625,948,703
Coupl D - LP_{3R}	859,316,929	Coupl B - LP_{1R}	626,684,214
LP_{3R} - LP_{3L}	2,545,306,455	LP_{1R} - LP_{1L}	2,548,322,567
LP_{3L} - Coupl C	627,555,327	LP_{1L} - Coupl A	566,819,519
Coupl C - LP_{2R}	628,294,620	Coupl A - HP	212,716,172
LP_{2R} - LP_{2L}	2,546,764,972		

The mechanical torque developed by each turbine section is assumed to be constant. The system is represented as shown in Figure E15.2 with parameters expressed in per unit on 889 MVA and 18.5 kV.

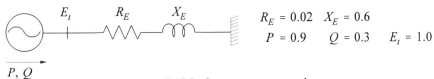

Figure E15.2 System representation

The generator parameters in per unit on 889 MVA and 18.5 kV are as follows:

X_d = 1.72 X'_d = 0.36 X''_d = 0.24 T'_{d0} = 5.7 s T''_{d0} = 0.02 s

X_q = 1.69 X'_q = 0.40 X''_q = 0.23 T'_{q0} = 1.5 s T''_{q0} = 0.04 s

The generator is equipped with a thyristor exciter. Our objective is to examine the system performance with the following alternative forms of PSS:

(a) A delta-omega stabilizer with shaft speed ($\Delta\omega$) as input signal and with speed measured at the generator end, at coupling B, at coupling C, or at both couplings B and C. The excitation system model is shown in Figure E15.3.

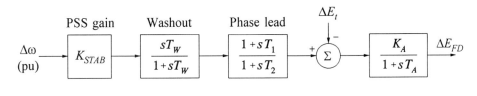

K_A = 200 T_A = 0.005 s T_1 = 0.3 s T_2 = 0.06 s T_W = 1.5 s

Figure E15.3 Thyristor exciter with delta-omega stabilizer

(b) A stabilizer as above, but with a torsional filter having a notch at 9 Hz and substantial attenuation at the higher torsional natural frequencies. The transfer function of the filter is

$$\frac{V_o}{V_i} = \frac{1 + s^2 0.00031258}{(1 + s\,0.022 + s^2 0.000756)(1 + s\,0.012 + s^2 0.009)}$$

(c) A stabilizer with electrical power deviation (delta-P) as stabilizing signal as shown in Figure E15.4. This system is similar to that of Figure E15.3, except that an equivalent speed ($\Delta\omega_{eq}$) derived from electrical power is used instead of actual shaft speed.

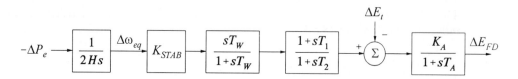

Figure E15.4 Thyristor exciter with delta-P stabilizer

Analysis

(a) *Delta-omega stabilizer.* The effects of PSS are examined by using the modal analysis approach described in Chapter 12. The analysis considers the effects of sensing speed ($\Delta\omega$) at different locations on the shaft. For each case, two or three values of the stabilizer gain K_{STAB} are considered to give an indication of the effect of varying the gain. The results are summarized in Table E15.1, where the six pairs of eigenvalues listed represent six modes which are important from the viewpoint of stabilizer application. The imaginary component of each complex pair of eigenvalues gives the frequency of oscillation in radians per second. The real component gives the rate of decay of the amplitude of oscillation (see Chapter 12).

The first four modes listed in Table E15.1 represent the dominant *torsional modes*. The fifth mode, identified as the *system mode*, is associated with the oscillation of the generator rotor with respect to the rest of the power system. Improving the damping of the system mode is the reason for using the speed signal feedback to the exciter. The sixth mode, identified as the *exciter mode*, is primarily associated with the field voltage and exists only in the presence of speed stabilization. This mode is usually heavily damped and becomes unstable only for high values of K_{STAB}. One of the considerations in choosing the parameters of the speed signal feedback circuit is to achieve maximum damping of the system mode without significantly sacrificing the damping of the exciter mode.

We see from the results that, without the speed signal stabilization ($K_{STAB}=0$), the system mode has a positive real part representing small-signal instability. The eigenvalues associated with the torsional modes, without speed stabilization, give the original level of torsional damping and serve as a reference for purposes of comparison.

The results also show that the effect of sensing speed at the generator end is to cause either instability or decreased damping of all four torsional modes, while significantly damping the system mode.

Sensing speed only at coupling B has an adverse effect on the 23 Hz torsional mode, and sensing only at coupling C has an adverse effect on the 9 Hz torsional mode. Results obtained with the speed signal sensed at other possible locations along the shaft indicated that there is no single location that is acceptable for all torsional modes.

A combination of signals sensed at couplings B and C is seen to have favourable effects on all four torsional modes. The system mode itself is insensitive to speed sensing location and depends only on the stabilizer gain. The exciter mode is heavily damped for the entire range of the stabilizer gain considered, although a higher gain results in a slight reduction of its damping.

(b) *Delta-omega stabilizer with torsional filter.* The results of the analysis with the filter added in the speed signal loop are summarized in Table E15.2. The effect of the filter, as expected, is to make the torsional modes insensitive to the speed signal stabilization. The damping of the system mode with the filter is on the same order as that without the filter. However, the filter has an adverse effect on the exciter mode. This is because the characteristic of the filter circuit is such that it increases the gain in the frequency range associated with the exciter mode. As a consequence, the speed signal gain has to be limited to a value which gives sufficient stability margin for the exciter mode. The limiting value of the gain depends on the external system impedance, generator characteristics, and system-operating conditions [9]. With a lower value of X_E, the exciter mode is less stable.

(c) *Delta-P stabilizer.* The results are summarized in Table E15.3. They show that a delta-P stabilizer is effective without causing instability of the torsional modes. So far as the system and the exciter modes are concerned, there is very little difference between the delta-P stabilizer and the delta-omega stabilizer without a torsional filter.

Table E15.1 Effect of delta-omega stabilization without a torsional filter circuit

K_{STAB}	Speed Sensing Location	Torsional Modes				System Mode	Exciter Mode
		9 Hz	17 Hz	23 Hz	24 Hz		
0.0	-	−0.05±j56.2	−0.07±j105.7	−0.10±j146.1	−0.17±j151.2	+0.23±j5.7	-
9.5	Generator	+0.31±j57.6	+0.20±j106.0	+0.24±j146.4	−0.06±j151.3	−0.73±j5.0	−13.7±j13.1
9.5	Coupl - B	−0.25±j55.3	−0.17±j105.6	−0.01±j146.2	−0.19±j151.2	−0.75±j5.0	−16.7±j13.9
19.0	Coupl - B	−0.47±j54.3	−0.28±j105.5	+0.07±j146.3	−0.20±j151.2	−1.20±j4.2	−14.5±j19.7
9.5	Coupl - C	+0.06±j56.6	−0.26±j105.5	−0.21±j146.1	−0.18±j151.2	−0.74±j5.0	−15.5±j13.7
19.0	Coupl - C	+0.20±j57.0	−0.46±j105.3	−0.31±j146.1	−0.19±j151.2	−1.20±j4.2	−12.8±j18.5
9.5	Both B and C	−0.10±j56.0	−0.22±j105.5	−0.11±j146.1	−0.19±j151.2	−0.74±j5.0	−16.1±j13.8
19.0	Both B and C	−0.16±j55.7	−0.37±j105.4	−0.12±j146.1	−0.20±j151.2	−1.20±j4.2	−13.6±j19.1
28.5	Both B and C	−0.22±j55.5	−0.52±j105.2	−0.13±j146.1	−0.21±j151.2	−1.36±j3.6	−11.9±j22.9

Table E15.2 Effect of delta-omega stabilization with a torsional filter

K_{STAB}	Speed Sensing Location	Torsional Modes				System Mode	Exciter Mode
		9 Hz	17 Hz	23 Hz	24 Hz		
0.0	-	−0.05±j56.2	−0.07±j105.7	−0.10±j146.1	−0.17±j151.2	+0.23±j5.7	-
9.5	Generator	−0.06±j56.2	−0.07±j105.7	−0.11±j146.1	−0.17±j151.2	−0.91±j5.1	−10.4±j15.8
9.5	Coupl - B	−0.05±j56.2	−0.07±j105.7	−0.10±j146.1	−0.17±j151.2	−0.94±j5.0	−8.1±j14.1
19.0	Coupl - B	−0.05±j56.2	−0.06±j105.7	−0.10±j146.1	−0.17±j151.2	−1.42±j4.1	−1.9±j20.5
9.5	Coupl - C	−0.06±j56.2	−0.07±j105.7	−0.10±j146.1	−0.17±j151.2	−0.92±j5.0	−10.5±j20.0
19.0	Coupl - C	−0.06±j56.2	−0.06±j105.7	−0.10±j146.1	−0.17±j151.2	−1.41±j4.2	−3.2±j20.0
9.5	Both B and C	−0.05±j56.2	−0.07±j105.7	−0.10±j146.1	−0.17±j151.2	−0.93±j5.0	−9.4±j21.3
19.0	Both B and C	−0.05±j56.2	−0.06±j105.7	−0.10±j146.1	−0.17±j151.2	−1.42±j4.2	−2.5±j20.4
28.5	Both B and C	−0.05±j56.2	−0.06±j105.7	−0.10±j146.1	−0.17±j151.2	−1.54±j3.5	+0.5±j20.6

Table E15.3 Effect of delta-P stabilization

K_{STAB}	Torsional Modes				System Mode	Exciter Mode
	9 Hz	17 Hz	23 Hz	24 Hz		
0	−0.05±j56.2	−0.07±j105.7	−0.10±j146.1	−0.17±j151.2	+0.23±j5.7	-
9.5	−0.05±j56.2	−0.07±j105.7	−0.10±j146.1	−0.17±j151.2	−0.75±j5.0	−15.6±j13.7
19.0	−0.05±j56.2	−0.07±j105.7	−0.10±j146.1	−0.17±j151.2	−1.20±j4.2	−13.1±j18.6
28.5	−0.05±j56.2	−0.07±j105.7	−0.10±j146.1	−0.17±j151.2	−1.40±j3.6	−11.3±j22.0

15.2.2 Interaction with Speed Governors

The problem of speed-governing systems causing torsional mode instability surfaced in 1983 during the commissioning of a 635 MVA, 1,800 r/min nuclear unit at Ontario Hydro [9]. This unit is equipped with a fast-acting electrohydraulic governing system. The speed governor senses speed at the HP turbine end of the shaft. The torsional mode shape for this unit is similar to that shown in Figure 15.6, which indicates that the 7.8 Hz, 15 Hz, and 19.6 Hz torsional modes will appear in the end-of-shaft speed signal. The governing system uses an electronic linearizing circuit to compensate for the nonlinear flow versus valve position characteristics of the steam valves (see Chapter 9, Section 9.2).

During the commissioning tests, when the unit reached the 100 MW output level, abnormal vibrations were detected in the governor valves. Changes to the phase lead and the valve-linearizing circuits (see Chapter 9, Figure 9.34) partially eliminated the vibrations, and the loading could be increased to 475 MW (nearly 88% of rated output). Beyond this level, severe vibrations of the governor valves were experienced. Unit load was reduced to a safe level until the problem was investigated and satisfactorily solved.

Further field investigations and computer simulations using the modal analysis identified two factors which contributed to the problem:

(a)   Inaccuracies in the valve-linearizing circuit, particularly in the 475 MW region. As a result, the effective governor droop suddenly increased by a factor of about 12, as loading increased.

(b)   Large bandwidth of the electrohydraulic servo system. This allowed significant governor interaction at the higher frequencies associated with the torsional modes.

The problem was solved by providing accurate linearization of valve characteristics which maintained a constant droop of 4% over the entire loading range. As an additional precaution, filters were added to eliminate torsional components from the speed signal. In designing the filters, careful consideration was given to the performance during severe system disturbances and load rejections.

15.2.3 Interaction with Nearby DC Converters

The problem of subsynchronous torsional mode instability due to interaction with controls of nearby HVDC systems first came to light on the Square Butte dc system in North Dakota [10,11]. The Square Butte system consists of a ±250 kV, 500 MW dc link. The rectifier station is located adjacent to the Milton Young generating station consisting of two turbine generators, one rated at 234 MW and the other at 410 MW. The converters employ equidistant firing systems. The normal regulator control modes are constant current control at the rectifier and constant voltage control at the inverter. In addition, a supplementary *frequency sensitive power controller* (FSPC) is provided between the sending end and receiving end systems. Field tests showed that the supplementary damping control destabilized the first torsional mode (11.5 Hz) of the 410 MW generating unit. The tests also showed that the normal constant current control acting without the supplementary damping control could cause instability of the 11.5 Hz torsional mode when some of the ac lines near the rectifier were switched out. Opening of the ac lines reduced the ac system strength which resulted in the instability of the torsional mode. The torsional oscillations stabilized when the dc power level was reduced.

At the time these instabilities were observed, the current control was modified, so as to allow safe operation, with some constraints on the dc power level and the ac system conditions. Subsequently, following a detailed analysis, the control systems were modified so as to allow safe operation under all but extreme system conditions.

This problem has been the subject of investigation of the EPRI research project RP1425-1. The work carried out under this project has led to a broader understanding of the phenomenon and the development of generic solutions [12].

Basic phenomenon

Torsional mode oscillations cause phase and amplitude modulation of the generated alternating voltage waveform. The modulated voltage has frequency components equal to the fundamental frequency minus the torsional frequency (see Figure 15.8). This modulated voltage is impressed on the dc system commutating bus.

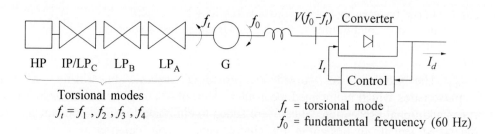

Figure 15.8 Torsional interaction with dc control

With the equidistant firing angle control used in modern HVDC converters, a shift in voltage phase due to a torsional mode causes a similar shift in the firing angle. This firing angle modulation, together with the alternating voltage magnitude modulation, will result in corresponding changes in direct voltage, current, and power. The closed-loop current control responds to correct for these changes. This in turn is reflected as a change in the generator power. If the net phase lag between the variation in shaft speed at the torsional frequency and the resulting change in electrical torque of the generator exceeds 90°, the torsional oscillations become unstable. This situation is similar to that depicted in Figure 15.7(b).

Factors influencing torsional interaction [12,13]

The possibility of torsional instability due to interaction with nearby dc system controls depends on the rating of the dc system in relation to that of the generating unit and on the electrical distance between the two. Reference 12 recommends use of the *unit interaction factor* (UIF) defined below as a measure of the influence of dc system controls on torsional mode stability.

$$\text{UIF} = \frac{\text{MVA}_{dc}}{\text{MVA}_g} \left(1 - \frac{SC_g}{SC_t} \right)^2 \tag{15.7}$$

where

$$\begin{aligned}
\text{MVA}_{dc} &= \text{MVA rating of the dc system} \\
\text{MVA}_g &= \text{MVA rating of the generator} \\
SC_t, \ SC_g &= \text{short-circuit capacity at the dc commutating bus (excluding ac} \\
&\quad \text{filters) with and without the generator, respectively}
\end{aligned}$$

If UIF is less than 0.1, there is little interaction between the torsional oscillation and the dc controls.

Parallel ac lines tend to minimize the possibility of torsional instability. Radial dc lines emanating from isolated thermal-generating stations are most prone to unfavourable torsional interaction.

The nominal firing angle of the rectifier has an impact on the degree of interaction. The tendency to cause torsional instability increases as the nominal firing angle is increased. Therefore, operating modes requiring the dc system voltages to be low may increase the possibility of torsional instability.

Effect of dc system controls [10,11]

In the case of the Square Butte project, the torsional instability was caused through two different control paths: the supplementary damping control and the rectifier current control.

The instability caused by the supplementary damping control was due to high gain and large phase lag in the frequency range at the lower end of the torsional natural frequencies. Only one of the two units near the rectifier had a torsional frequency (11.5 Hz) low enough to interact with the damping control. This problem could be readily solved by use of a notch filter at 11.5 Hz in the supplementary control. The other unit's first torsional frequency was so high as not to cause instability.

The influence of the current or power control depends on the regulator bandwidth. Typically, these controls have a bandwidth in the vicinity of 10 to 20 Hz. Only the torsional modes with frequencies below the upper range of the bandwidth have the potential for becoming unstable.

In the case of Square Butte, the problem due to the current control was solved by modifying the regulator frequency response characteristic. Some HVDC systems may require more than simple modifications of the current regulator to achieve stable operation. In such cases, an additional damping path may be required. In reference 12, a *subsynchronous damping control* (SSDC) has been developed to damp potentially troublesome torsional oscillations.

The problem of torsional interaction with the controls of HVDC systems may be investigated by using the modal analysis approach. Reference 13 describes other methods of studying the problem. Two effects must be accounted for in the simulations: (a) apparent shifts in the bridge firing angle due to shifts in phase of the bus voltage and (b) the speed variation (ω_r/ω_0) effects on the generator voltage [10].

15.3 SUBSYNCHRONOUS RESONANCE

The phenomenon of *subsynchronous resonance* (SSR) occurs mainly in series capacitor-compensated transmission systems. The first SSR problem was experienced in 1970 resulting in the failure of a turbine-generator shaft at the Mohave plant in southern California. It was not until a second shaft failure occurred in 1971 that the real cause of the failure was recognized as subsynchronous resonance [14]. Since these failures, SSR has been a subject of considerable electric utility industry interest.

Before we discuss the SSR phenomenon, let us first examine the characteristics of a series capacitor-compensated system and the possibility of unstable subsynchronous oscillations.

15.3.1 Characteristics of Series Capacitor-Compensated Transmission Systems [15,16]

In an *uncompensated* transmission system, faults and other disturbances result in dc offset components in the generator stator windings (see Chapter 3, Section 3.7.2). These currents, due to the space distribution of armature (stator) windings, result in a component of air-gap torque at the slip frequency of 60 Hz. In the case of unbalanced faults, the negative phase-sequence components of stator currents result in a 120 Hz (slip frequency) component of air-gap torque. Therefore, it is necessary to avoid having any of the torsional frequencies very near either 60 Hz or 120 Hz.

In *series capacitor-compensated* transmission systems, the situation can be very different. Consider, for example, the simple radial system shown in Figure 15.9. In this case, instead of the dc component of fault current, the offset transient current is also alternating current of frequency equal to the natural frequency (ω_n) of the circuit inductance and capacitance:

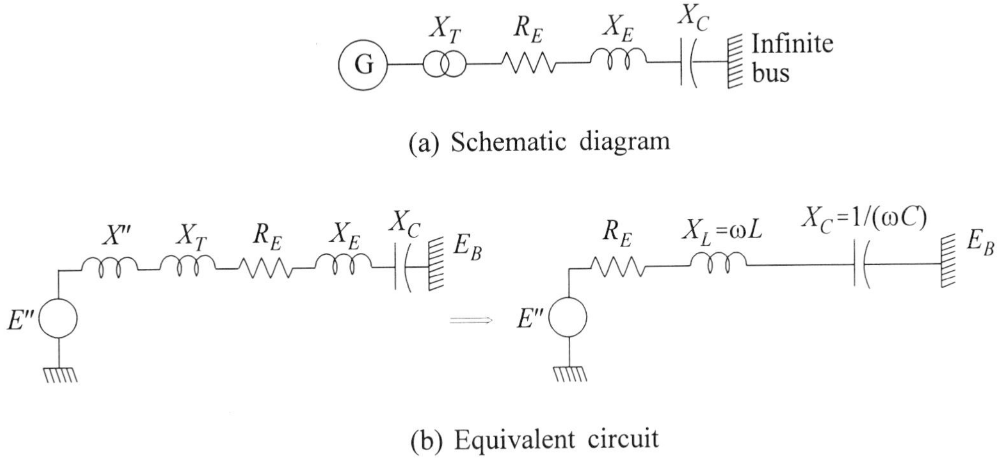
X_T
R_E
X_E
X_C
G
Infinite bus
(a) Schematic diagram
X''
X_T
R_E
X_E
X_C
E_B
E''
R_E
X_L=ωL
X_C=1/(ωC)
E_B
E''
(b) Equivalent circuit

Figure 15.9 Radial series compensated system

$$\omega_n = \frac{1}{\sqrt{LC}} = \frac{\omega_0}{\sqrt{(\omega_0 L)(\omega_0 C)}} = \omega_0 \sqrt{\frac{X_C}{X_L}} \text{ rad/s} \tag{15.8}$$

or

$$f_n = f_0 \sqrt{\frac{X_C}{X_L}} \text{ Hz} \tag{15.9}$$

where f_0 is the synchronous frequency in Hz, and $\omega_0 = 2\pi f_0$ rad/s.

Generator stator current components of frequency f_n induce rotor currents (and hence torque) of slip frequency $(60 - f_n)$ Hz. Table 15.1 shows the natural and slip frequencies as a function of the degree of compensation.

In practice, the frequency-dependent characteristic of the effective impedance of a network is complex and a more detailed system representation is required. A frequency scanning program such as that described in reference 18 can be used to determine this impedance.

The subsynchronous natural frequency (f_n) of the network and the frequency of the associated induced rotor currents (and torque) are said to be complementary, as their sum is equal to synchronous frequency. A series capacitor-compensated transmission network can cause sustained or negatively damped subsynchronous oscillations by two distinctive mechanisms: (a) self-excitation due to induction generator effect and (b) interactions with torsional oscillations.

Table 15.1

Percent Compensation $(X_C/X_L) \times 100$ (%)	Natural Frequency f_n (Hz)	Slip Frequency $60 - f_n$ (Hz)
10	18	42
25	30	30
30	32.6	27.4
40	38	22
50	42.4	17.6

Shunt-compensated transmission systems normally have natural frequencies in the supersynchronous range; therefore, subsynchronous oscillations do not pose a problem. Exceptions are situations involving high degrees of shunt compensation and very long transmission.

15.3.2 Self-Excitation due to Induction Generator Effect

Figure 15.10 is a simplified equivalent circuit of a synchronous machine useful in explaining subsynchronous effects. The effect of saliency has been neglected and the machine has been represented by a circuit similar to that used for induction motors. The effects of the rotor circuits have been lumped and represented by a single circuit with a resistance R_r and a leakage reactance X_r. The power frequency quantities have been omitted and only subsynchronous frequency effects are considered.

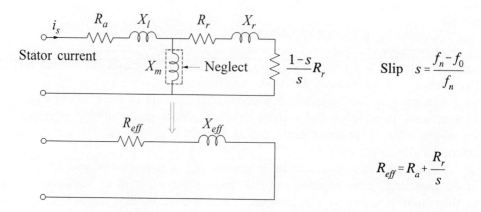

Figure 15.10 Simplified equivalent circuit of a synchronous machine applicable to subsynchronous frequency quantities

Since $f_n < f_0$, the slip s is negative, and the rotor behaves much like that of an induction machine running above synchronous speed. Depending on f_n, the effective resistance R_{eff} can be negative. At high degrees of compensation, this apparent negative resistance may exceed the network resistance, effectively resulting in an RLC circuit with negative resistance. Such a condition will result in self-excitation causing electrical oscillations of intolerable levels. The tendency toward this electrical subsynchronous instability is decreased by increasing the network resistance and by decreasing the resistance of generator rotor circuits (for example, by providing a good pole face damper winding).

This form of self-excitation is purely an electrical phenomenon and is *not* dependent on the shaft torsional characteristics.

15.3.3 Torsional Interaction Resulting in SSR

If the complement of the natural frequency (i.e., synchronous frequency minus the natural frequency) of the network is close to one of the torsional frequencies of the turbine-generator shaft system, torsional oscillations can be excited. This condition is referred to as *subsynchronous resonance* [16]. Under such conditions, a small voltage induced by rotor oscillations can result in large subsynchronous currents; this current will produce an oscillatory component of rotor torque whose phase is such that it enhances the rotor oscillation. When this torque is larger than that resulting from mechanical damping, the coupled electromechanical system will experience growing oscillations.

Subsynchronous resonance is thus a condition where the subsynchronous complement of the natural frequency of a series capacitor-compensated armature circuit of a turbine generator is close to its torsional natural frequencies, resulting in a strong coupling between the electrical and mechanical systems [17].

The consequences of subsynchronous resonance can be dangerous. If the torsional oscillations build up, the turbine-generator shaft will break with disastrous consequences. Even if the oscillations are not unstable, system disturbances can result in *shaft torques of high magnitude, causing loss of fatigue life* of the shaft.

15.3.4 Analytical Methods

Torsional interaction effects involve energy interchange between the generator shaft system and the inductances/capacitances of the network. Therefore, the analysis of SSR problems requires representation of both the electromechanical dynamics of the generating units and the electromagnetic dynamics of the transmission network. The generator stator circuits and the network cannot be represented by steady-state algebraic equations as in system stability studies. Consequently, the system model used for SSR studies is of a higher order and greater stiffness than the model used for stability studies.

Several methods have been used for the study of SSR. The following is a brief description of these methods.

(a)   *Eigenvalue (modal) analysis.* The general approach is the same as that described in Chapter 12. The method is powerful and provides insight into the system characteristics. With the availability of special techniques for the analysis of very large systems (see Section 12.7.2), system size and computational burden are no longer considered as the limitations of this approach.

(b)   *Frequency scanning* [18,19]. This technique computes the equivalent impedance as seen from the internal buses of generators looking into the network, for different values of frequency. It gives information about the natural frequencies of the system and the tendency toward self-excitation and SSR. This approach is particularly suited for preliminary analysis of SSR problems.

(c)   *Frequency response analysis of full system* [20]. The stability of the subsynchronous oscillations is analyzed by using the multidimensional Nyquist criterion. This approach can handle detailed models and large systems.

(d)   *Approximate frequency-domain analysis* [21,22]. This method analyzes the stability of individual torsional modes. It is limited to approximate detection of SSR.

(e)   *Time-domain analysis.* The *electromagnetic transient program* (EMTP) may be used to compute the transient-time response [23]. A full three-phase representation of the system is used. It allows very detailed representation of equipment, including nonlinear effects. It is particularly suited for analyzing transient shaft torques due to SSR. However, this method is limited to the analysis of small systems.

Frequency dependent network equivalents may be used to facilitate the analysis of very large complex systems [24]. Such equivalents preserve the natural resonant characteristics of the network they represent.

Example 15.3

In this example, we illustrate the two forms of instability of subsynchronous oscillations associated with series capacitor-compensated systems: SSR and self-excitation. The test system considered is shown in Figure E15.5. It consists of a 555 MVA, 24 kV, 3,600 r/min turbine generator feeding power through a series capacitor-compensated transmission system to an infinite bus. The shaft system parameters and the torsional characteristics of the generating unit are as shown in Figure 15.3. The damping factors (*D*) of the individual masses are assumed to be negligible.

The generator and network parameters in per unit on 555 MVA, 24 kV base are as follows:

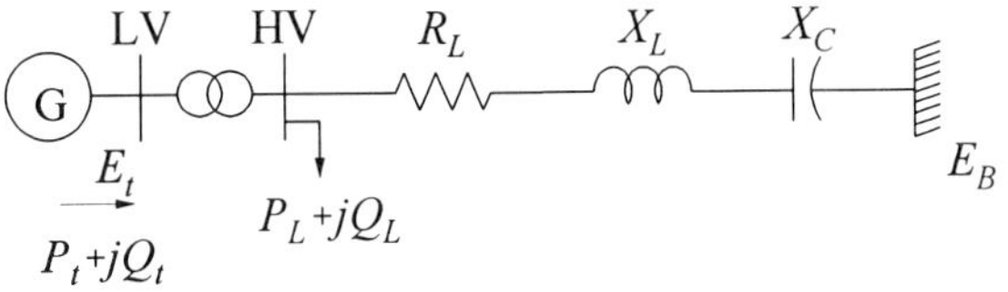

Figure E15.5 Series-compensated system

$X_d = 1.81$	$X_q = 1.76$	$X_l = 0.16$	$X_d' = 0.3$	$X_q' = 0.61$
$X_d'' = 0.217$	$X_q'' = 0.254$	$R_a = 0.002$	$T_{d0}' = 7.8$ s	$T_{q0}' = 0.9$ s
$T_{d0}'' = 0.022$ s	$T_{q0}'' = 0.074$ s	$A_{sat} = 0.008$	$B_{sat} = 9.6$	$\psi_{T1} = 0.8$
$Z_T = 0.008 + j0.13$		$R_L = 0.008$	$X_L = 0.75$	

The power-flow condition is as follows:

$$P_t = 519.5 \text{ MW} \qquad E_t = 1.08 \qquad E_B = 1.0$$

The degree of compensation (i.e., X_C/X_L) is varied and the reactive power output of the generator varies accordingly.

Our focus here is on the interaction between the lowest torsional (16 Hz) mode and the subsynchronous natural frequency oscillation of the network for the following values of load at the HV bus:

(a) $P_L = 166.5$ MW, $Q_L = 0$
(b) $P_L = 0$, $Q_L = 0$

Analysis

(a) With $P_L = 166.5$ MW, $Q_L = 0$:

We use the eigenvalue analysis approach described in Chapter 12 to analyze the modal interaction. However, the system model includes the dynamics of the transmission network and the generator stator circuits. The complete system equations are expressed in the dq reference frame.

Table E15.4 lists the eigenvalues, frequency and damping ratio (ζ) associated with the lowest torsional mode and the network mode as a function of the degree of series compensation (expressed in percent of line reactance X_L). In addition, participation factors associated with the generator speed deviation ($\Delta\omega_1$) and voltage across the series capacitor (Δv_c) are listed. These help identify the extent of interaction between the two modes. Since the network is modelled in the dq reference frame which rotates with the generator rotor, the frequency of the network mode computed is in fact the complement of the network natural frequency (i.e., synchronous frequency minus the natural frequency).

Table E15.4 Torsional and network modes as a function of the degree series compensation

With $P_L = 166.5$ MW, $Q_L = 0$

% Comp. of line 2-3	Torsional Mode			Participation Factors		Network Mode (in d-q reference frame)			Participation Factors	
	Eigenvalue	Freq. (Hz)	ζ	$\Delta\omega_1$	Δv_c	Eigenvalue	Freq. (Hz)	ζ	$\Delta\omega_1$	Δv_c
0.0	$-0.04\pm j102.3$	16.29	0.0004	$0.90\angle0.0°$	—	—	—	—	—	—
25.0	$-0.06\pm j102.6$	16.32	0.0006	$0.90\angle0.0°$	$0.002\angle153°$	$-2.9\pm j221.2$	35.20	0.0130	$0.03\angle-177.0°$	$1.00\angle1.0°$
50.0	$-0.10\pm j103.1$	16.41	0.0010	$0.90\angle0.1°$	$0.01\angle171.3°$	$-4.5\pm j155.5$	24.75	0.0288	$0.24\angle89.3°$	$0.99\angle-21.5°$
60.0	$-0.13\pm j103.7$	16.50	0.0012	$0.90\angle0.0°$	$0.04\angle172.7°$	$-3.2\pm j137.0$	21.80	0.0231	$0.01\angle99.0°$	$1.00\angle4.0°$
65.0	$-0.11\pm j104.4$	16.61	0.0010	$0.91\angle-0.4°$	$0.09\angle171.3°$	$-2.9\pm j126.2$	20.09	0.0230	$0.05\angle164.0°$	$1.00\angle1.7°$
66.0	$-0.10\pm j104.6$	16.64	0.0010	$0.91\angle-0.4°$	$0.10\angle170.0°$	$-2.9\pm j124.1$	19.74	0.0230	$0.07\angle14.0°$	$1.00\angle1.4°$
68.0	$-0.02\pm j105.2$	16.74	0.0002	$0.91\angle-0.4°$	$0.17\angle166.4°$	$-2.9\pm j119.6$	19.04	0.0241	$0.12\angle164.0°$	$1.00\angle0.4°$
70.0	$0.29\pm j106.3$	16.90	-0.0027	$0.91\angle-7.8°$	$0.31\angle152.5°$	$-3.1\pm j114.8$	18.27	0.0273	$0.26\angle150.0°$	$1.00\angle-6.3°$
71.3	$1.38\pm j107.3$	17.08	-0.0128	$0.91\angle-30.0°$	$0.55\angle109.0°$	$-4.2\pm j111.2$	17.70	0.0376	$0.47\angle107.0°$	$0.98\angle-28.0°$
72.0	$2.28\pm j107.5$	17.10	-0.0212	$0.91\angle-37.7°$	$0.64\angle84.5°$	$-5.0\pm j109.9$	17.49	0.0460	$0.55\angle83.0°$	$0.98\angle-36.0°$
74.0	$4.33\pm j106.6$	16.96	-0.0406	$0.92\angle-31.4°$	$0.75\angle48.0°$	$-7.0\pm j107.2$	17.06	0.0658	$0.64\angle47.0°$	$0.96\angle-29.4°$
75.0	$4.97\pm j105.9$	16.85	-0.0468	$0.93\angle-25.5°$	$0.77\angle37.3°$	$-7.6\pm j106.1$	16.88	0.0721	$0.65\angle36.0°$	$0.95\angle-23.2°$
77.0	$5.82\pm j104.4$	16.60	-0.0557	$0.93\angle-13.0°$	$0.77\angle18.0°$	$-8.5\pm j103.8$	16.51	0.0814	$0.66\angle18.0°$	$0.95\angle-9.0°$
78.8	$6.10\pm j103.1$	16.41	-0.0590	$0.93\angle-3.8°$	$0.76\angle5.1°$	$-8.7\pm j102.0$	16.23	0.0850	$0.64\angle6.0°$	$0.95\angle0.8°$
80.0	$6.12\pm j102.3$	16.28	-0.0590	$0.93\angle1.9°$	$0.75\angle-3.2°$	$-8.7\pm j100.9$	16.05	0.0858	$0.63\angle-1.8°$	$0.95\angle7.1°$

With no compensation, the torsional mode has a frequency of 16.29 Hz and a small positive damping due to electrical sources. With 25% compensation, the torsional mode frequency and damping increase very slightly; the network mode has a frequency of 35.2 Hz and a damping ratio of 0.013. There is very little interaction between the two modes as indicated by the participation factors.

As the degree of compensation is increased, the frequency of the torsional mode varies slightly and its damping increases slightly at first and then decreases; the frequency of the network mode decreases and its damping increases. At 70% compensation, the torsional mode is just unstable and there is a noticeable amount of interaction between the two modes. Further increase in the degree of compensation strengthens the coupling between the two modes. This coupling effect tends to pull the two modes closer together in frequency but apart in damping. The torsional mode becomes unstable, while the network mode becomes more stable. With 75% to 80% compensation the two modes have nearly equal frequencies, and the coupling between the two modes is the greatest.

The effect of interaction between the network mode and the torsional mode on the characteristics of the network mode is more evident in Table E15.5. This table provides the frequency and damping ratio of only the network mode with and without the multimass representation of the turbine-generator rotor, that is, with and without the presence of the torsional mode. It is clear that, in this case, the interaction between the two modes increases the damping of the network mode.

(b) With $P_L=Q_L=0$

The results are summarized in Table E15.6. With the load removed, the effective resistance of the network is reduced significantly. Consequently, the network mode becomes unstable due to the induction motor effect. As the frequency of the network mode approaches the torsional mode frequency, the coupling between the two modes increases. In this case, the effect of the interaction is to increase the damping of the torsional mode and to decrease the damping of the network mode.

Table E15.7 compares the damping ratio and frequency of the network mode with and without multimass rotor representation. It is evident that the presence of the torsional mode makes the network mode more unstable. The stronger the interaction between the two modes, the less stable the network mode.

Table E15.5 Comparison of the network mode with and without
multimass representation of the turbine-generator rotor

With P_L = 166.5 MW, Q_L = 0

% Comp. of line 2-3	Network mode with multimass representation of the turbine generator rotor		Network mode with single lumped-mass representation of the turbine-generator rotor	
	Freq. (Hz)	ζ	Freq. (Hz)	ζ
25.0	35.20	0.0130	35.32	0.0133
50.0	24.75	0.0288	25.13	0.0185
60.0	21.80	0.0231	21.82	0.0205
65.0	20.09	0.0230	20.27	0.0215
66.0	19.74	0.0230	19.96	0.0217
68.0	19.04	0.0241	19.37	0.0221
70.0	18.27	0.0273	18.78	0.0224
71.3	17.70	0.0376	18.39	0.0227
72.0	17.49	0.0460	18.20	0.0228
74.0	17.06	0.0658	17.63	0.0231
75.0	16.88	0.0721	17.35	0.0233
77.0	16.51	0.0814	16.74	0.0237
78.0	16.23	0.0850	16.27	0.0239
80.0	16.05	0.0858	15.96	0.0241

Note: Network mode shown above is in the d-q reference frame; the actual network frequency is equal to synchronous frequency (60 Hz) minus the above.

Table E15.6 Torsional and electrical network modes as a function of the degree of series compensation

With $P_L = Q_L = 0$

% Comp.	Torsional Mode					Network Mode				
				Participation Factors					Participation Factors	
	Eigenvalue	Freq. (Hz)	ζ	$\Delta\omega_1$	Δv_c	Eigenvalue	Freq. (Hz)	ζ	$\Delta\omega_1$	Δv_c
0.0	$-0.03\pm j102.3$	16.28	0.0003	$0.896\angle0.0°$	–	–	–	–	–	–
25.0	$-0.04\pm j102.5$	16.32	0.0004	$0.898\angle0.0°$	$0.002\angle153.0°$	$-1.94\pm j221.2$	35.20	0.0088	$0.03\angle180.0°$	$1.00\angle1.4°$
70.0	$-0.53\pm j106.4$	16.94	0.0050	$0.909\angle5.0°$	$0.34\angle-166.0°$	$0.67\pm j114.6$	18.25	-0.0058	$0.33\angle-165.0°$	$1.00\angle6.7°$
75.0	$-6.04\pm j105.5$	16.79	0.0570	$0.940\angle28.0°$	$0.88\angle-30.0°$	$6.51\pm j106.4$	16.94	-0.0611	$0.78\angle-32.0°$	$1.00\angle29.8°$
78.8	$-6.97\pm j102.3$	16.28	0.0680	$0.940\angle3.9°$	$0.88\angle1.7°$	$7.71\pm j102.8$	16.36	-0.0750	$0.79\angle-4.1°$	$1.00\angle4.6°$
80.0	$-6.92\pm j101.4$	16.14	0.0681	$0.930\angle-2.7°$	$0.88\angle10.0°$	$7.74\pm j101.8$	16.20	-0.0758	$0.78\angle33.0°$	$1.00\angle-2.0°$
85.0	$-5.03\pm j97.6$	15.53	0.0500	$0.871\angle-34.°$	$0.83\angle51.0°$	$6.22\pm j97.1$	15.46	-0.0639	$0.71\angle41.3°$	$1.00\angle-32.0°$
93.7	$-0.13\pm j99.5$	15.80	0.0013	$0.870\angle-1.0°$	$0.13\angle-178.0°$	$1.96\pm j81.1$	12.90	-0.0241	$0.07\angle167.1°$	$1.00\angle3.8°$

Note: Network mode shown above is in the d-q reference frame.

Table E15.7 Comparison of the network mode with and without multimass representation of the turbine-generator rotor

With $P_L = Q_L = 0$

% Comp. of line 2-3	Network mode with multimass representation of the turbine generator rotor shafts		Network mode with single lumped-mass representation of the turbine-generator rotor	
	Freq. (Hz)	ζ	Freq. (Hz)	ζ
25.0	35.20	0.0088	35.32	0.0088
70.0	18.25	-0.0058	18.78	-0.0018
75.0	16.94	-0.0611	17.35	-0.0050
78.8	16.36	-0.0750	16.26	-0.0080
80.0	16.20	-0.0758	15.96	-0.0090
85.0	15.46	-0.0639	14.62	-0.0138
93.7	12.90	-0.0241	12.38	-0.0246

Note: Network mode shown above is in the d-q reference frame ∎

15.3.5 Countermeasures to SSR Problems [25,26]

A wide variety of methods have been proposed in the literature for counteracting SSR. These may be broadly classified as follows:

(a) *Static filter*. This can take the form of a blocking filter (parallel-resonance type) in series with the generator, or damping circuits in parallel with the series capacitors.

(b) *Dynamic filter*. This is an active device placed in series with the generator. It picks up a signal derived from rotor motion and produces a voltage in phase opposition so as to compensate for or even exceed the subsynchronous voltage generated in the armature.

(c) *Dynamic stabilizer*. This consists of thyristor modulated shunt reactors connected to the generator terminal. Control of subsynchronous oscillations is achieved by modulating the thyristor switch firing angles, using signals derived from the generator shaft speed.

(d) *Excitation system damper*. Generator excitation control is modulated by using a signal derived from the shaft speed so as to provide increased damping of torsional oscillations.

(e)   *Protective relays.* The SSR condition is detected by a relay and the affected units are tripped. The relay may take several forms: detection of excessive torsional motion by sensing rotor speed or detection of SSR condition by sensing the armature current.

(f)   *NGH scheme [26].* This consists of a linear resistor in series with back-to-back thyristors connected across the series capacitor. The presence of SSR components in the capacitor voltage is detected and reduced by the action of the resistor discharge circuit.

An example of a comprehensive SSR protection scheme is the one at Salt River Project's Navajo Plant [27]. It consists of three forms of protection: (i) a tuned resonant blocking filter in series with the generator, (ii) an excitation system damper to damp subsynchronous oscillations, and (iii) a static relay which trips the units if the transient shaft response exceeds an acceptable level.

15.4 IMPACT OF NETWORK-SWITCHING DISTURBANCES

A generating unit encounters a multitude of switching duties during its lifetime. These include planned as well as unplanned events, such as simple line switching, system faults and fault clearing, reclosing of lines, and faulty synchronizing. Such disturbances can produce high levels of oscillatory shaft torques. The resulting cyclic stress variations undergone by the shaft material may cause loss of fatigue life. This is a cumulative process, with each incident using up a portion of the total fatigue life.

Traditionally, turbine generators have been designed to withstand terminal short circuits in accordance with industry standards [28]. Terminal short circuits are rare events; consequently, a generating unit would not be expected to experience more than a few such events over its entire life. In the early 1970s, it was realized that network-switching operations could contribute to shaft fatigue damage. This recognition came about as a result of studies prompted by the shaft failure of Mohave units due to SSR. Although the SSR phenomenon is of a different nature, it drew attention to other possible transmission network conditions that may lead to shaft damage. The effects of network disturbances and switching events on turbine-generator shafts, independent of SSR conditions, have been examined by a number of investigators [2-6,29,30]. A procedure for estimating torsional fatigue life expenditure has also been developed under an EPRI research project, based on extensive testing of shaft specimens [31].

This problem has been examined by an IEEE working group, and general recommendations have been made to the industry concerning (a) steady-state switching [32] and (b) successive network switching [33].

Steady-state switching

Switching operations involving opening or closing of circuit breakers can result in transient torques which impose severe stress on nearby generators. Usually, the most severe switching operation with the network initially under steady state is the reclosing of transmission lines across a large breaker angle. Depending on the network impedances, the resulting sudden increase in the air-gap torque of a nearby generator could be very large. This causes the mechanical system to respond to the principal torsional natural frequencies. If the resulting torques are high, this may result in loss of shaft fatigue life. For the purpose of explaining the basis used for determining the acceptability of the shaft duty imposed by a switching operation, it is helpful to briefly review the fatigue characteristics of shaft materials.

Torsional fatigue characteristics [2,3,16,33]:

Fatigue is defined as the process of progressive localized permanent structural change occurring in a material subjected to conditions which produce fluctuating stresses and strains at some point or points and which may culminate in cracks or complete fracture after a sufficient number of fluctuations [16].

Fatigue is a cumulative process in which additional events add to previous life expenditure. Observable defects such as cracks will not be formed until all the fatigue life is consumed.

A typical fatigue characteristic of a shaft specimen subjected to a fully reversed cyclic stress is shown in Figure 15.11. The figure shows how the number of cycles to failure is related to the magnitude of the cyclic stress. The figure also defines the

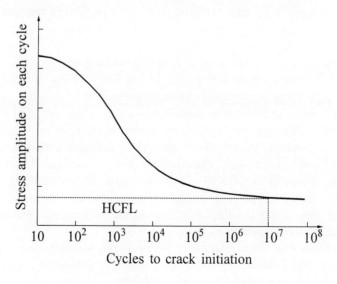

Figure 15.11 A typical fatigue characteristic showing cycle life curve for fully reversed stress

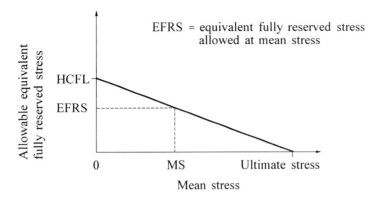

Figure 15.12 Mean stress effect

high-cycle fatigue limit (HCFL) which is the limiting value of the cyclic stress to which the shaft can be subjected such that practically no cumulative fatigue damage occurs. In addition to the alternating stress, the effect of the mean stress should be considered. Figure 15.12 illustrates how the mean stress or the steady-state torque lowers the allowable cyclic torque. The allowable cyclic torque is equal to HCFL when the mean stress is equal to zero, and it is equal to zero when the mean stress is equal to the ultimate stress of the material. For any other value of mean stress between these two extremes, the allowable cyclic torque is determined as shown in Figure 15.12.

Figure 15.13 shows a typical stress-strain curve for a material. The linear region of the curve is known as the region of *elastic strain*. A material which is elastically deformed will return to its original shape and strength after deformation. Values of strain in the nonlinear region are known as *plastic strain*. Plastic strain results in permanent deformation of the material.

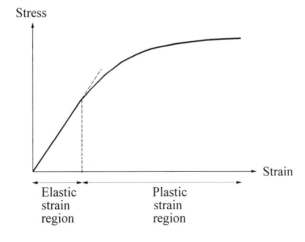

Figure 15.13 Stress-strain curve

High-cycle fatigue is caused by a large number of low amplitude fluctuations; it results in elastic deformation. Conversely, *low-cycle fatigue* is caused by a small number of large amplitude fluctuations; it involves local plastic deformation in regions such as keyways and fillets.

Recommendations:

For planned switching operations, such as a simple line restoration, IEEE guidelines [32] recommend that the switching be conducted so that it does not contribute significantly to cumulative shaft fatigue. This means that the magnitude of the cyclic shaft stress, with mean stress effects included, should be kept mostly below the high cycle fatigue limit (see Figures 15.11 and 15.12). In this way, nearly all of the fatigue capability of the shafts will be preserved to withstand the impact of unplanned and unavoidable disturbances such as faults, fault clearing, reclosing into system faults, and emergency line switching.

Guidelines for screening acceptable line-switching duties:

A detailed investigation of all possible line-switching cases would be impractical. Therefore, the IEEE [32] has suggested general guidelines to permit the utilities to screen switching operations and to determine if any of the cases requires a detailed study. These guidelines assume a *simple line restoration from steady state*; hence, they are applicable only for delayed reclosing, with a reclosing time of about 10 seconds or more. The guidelines are based on detailed studies performed on a number of cases where a large angle across the open breaker indicated that there could be excessive shaft torques.

The breaker angle is not by itself very useful in judging the severity of a switching operation. The circuit impedances play a significant role. Therefore, the severity is measured in terms of the sudden change in the generator power (ΔP), as computed by a conventional transient stability program. A rule-of-thumb limit proposed by the IEEE for ΔP is 0.5 per unit of generator MVA rating. A line-switching case resulting in ΔP of less than 0.5 per unit may be considered safe. Switching operations resulting in ΔP greater than the 0.5 per unit limit have to be studied in detail to assess the shaft duty.

Successive network-switching disturbances

Successive network disturbances, such as automatic high-speed line reclosing following a fault, can result in dangerously high torques. The concern here is for the *compounding effects* of the different switching operations. For example, in the case of high-speed reclosing, the following successive switching disturbances are involved:

- Transmission line fault
- Clearing of fault
- Reclosing of the line, either successful or unsuccessful

The torsional oscillations due to successive impacts may reinforce the initial oscillations. The risk of amplifying vibrations to damaging levels is a function of the type of disturbance and the timing of subsequent switching operations.

In the analysis of the effects of network-switching operations on shaft torques, the 60 Hz component of air-gap torque associated with the dc offset in the stator current is generally found to have a significant effect. Therefore, it is necessary to model the generator stator and network transients. To simulate the circuit-breaker operation adequately, as affected by the current zero crossing in each phase, it is necessary to use a three-phase cycle-by-cycle representation.

The electromagnetic transient program [23] provides a means of representing these effects. It also has facilities for computing shaft torques and determining torsional fatigue life expenditure.

Reference 33 prepared by the IEEE working group gives a summary of the predicted range of fatigue life expenditure for various network disturbances: different types of faults, fault clearing, successful reclosing, and unsuccessful reclosing. Automatic high-speed reclosing of multiple faulted lines near generating plants is identified as posing a significant risk in shaft fatigue life expenditure. Where high-speed reclosing strategies are contemplated, a study to assess the shaft fatigue duty is recommended. The following are possible alternative reclosing strategies with reduced risk of shaft damage:

(a)   Delayed reclosing, with a delay of 10 s or more.

(b)   Sequential reclosing; i.e., automatic reclosing from the end remote from the plant, followed by synchro-check reclosing of the plant end.

(c)   Selective reclosing; i.e., limiting automatic high-speed reclosing to single line-to-ground faults and line-to-line faults.

Reference 30 provides a comparative assessment of the fatigue life expenditure with unrestricted high-speed reclosing and the above alternative reclosing strategies.

15.5 TORSIONAL INTERACTION BETWEEN CLOSELY COUPLED UNITS

A single generator of M masses connected to a power system has M modes of oscillation (for example, see Figure 15.3). One of these modes is the system mode or the rigid-body mode in which all rotor masses move in phase with nearly equal magnitude. The other $M-1$ modes are torsional modes whose frequencies differ from one another and in which the M rotor masses have differing degrees of participation.

If there are N machines operating in parallel, their shaft systems are coupled through the electrical system. The electrical stiffness of such coupling is an order of magnitude smaller than the mechanical stiffness of shafts interconnecting adjacent rotor masses. Therefore, the rigid-body modes of oscillation have frequencies in the range of 0.2 to 3.0 Hz.

   The interaction between the shaft systems of parallel units has an effect on the torsional modes. The need to consider this interaction was first recognized during field tests conducted at the Mohave generating station [4]; a double resonant peak was observed near one of the torsional frequencies with two generating units operating in parallel.

   The effect of interaction between torsional dynamics of two closely coupled units is depicted in Figure 15.14. Each unit is assumed to have the torsional characteristics shown in Figure 15.3.

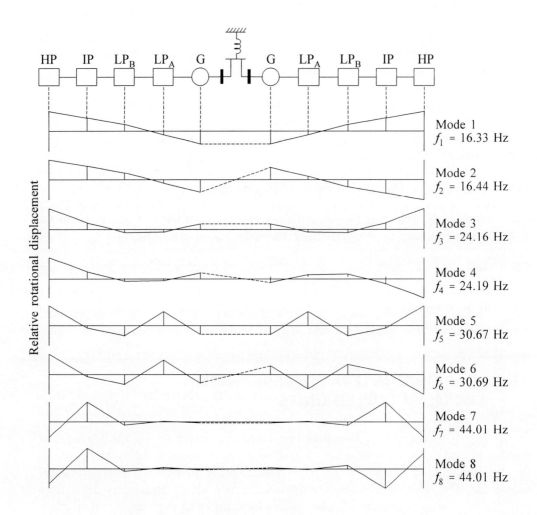

Figure 15.14 Torsional frequencies and mode shapes
of two coupled turbine generators

Modes 1 and 2 represent the lowest torsional modes. In mode 1 (f_1=16.33 Hz) the corresponding masses in the two units oscillate in phase, whereas in mode 2 (f_2=16.44 Hz) the corresponding masses oscillate in antiphase. The difference in frequencies of the two modes is 0.11 Hz. This difference is a function of the stiffness coefficient of the electrical coupling between the two units. With a low coupling, the two frequencies would be equal. As the coupling is increased, the frequency difference increases.

Similarly, there are in-phase and antiphase modes corresponding to each of the other three torsional modes of the individual units.

In a general case of N identical machines (each having M masses) operating in parallel, the following torsional modes exist [34]:

- One group of M modes in which the corresponding rotors in each unit oscillate in unison. These modes are stimulated by symmetrical disturbances on all N units, and are designated *in-phase modes*.

- The M groups of N-1 identical modes representing intermachine dynamics. These modes have linearly independent mode shapes and are designated *antiphase modes*.

The above analysis assumes that the N units are truly identical. If the units are only nominally identical, there will be families of N modes slightly separated in frequency. Each torsional mode family will consist of one in-phase mode, in which the generators participate at slightly different magnitudes, and N-1 antiphase modes with slight differences in frequency. The torsional interaction in this case will be less because of the differences in the individual torsional frequencies of the units.

The possible spread of antiphase torsional modes as additional units are brought on-line is significant in situations where notch filters are used to solve torsional-related problems.

15.6 HYDRO GENERATOR TORSIONAL CHARACTERISTICS

The rotor of a hydraulic generating unit consists of a turbine runner and a generator rotor. If the unit has a shaft-driven exciter, there is an additional rotor mass. Consequently, there are at most two torsional modes of oscillation.

The inertia of the generator rotor is about 10 to 40 times higher than that of the turbine runner (waterwheel). The torsional natural frequencies lie in the range of 6 Hz to 26 Hz [35].

There are no reported cases of adverse dynamic interaction between the hydro generator rotor and the electrical network. The following are the principal reasons for the absence of adverse interaction [35,36]:

1.   The high value of the generator rotor inertia in relation to the inertias of the turbine runner and exciter rotor. This effectively shields the rotor mechanical system. As a result, it is difficult to excite torsional oscillations through disturbances on the generator.

2.   Viscous waterwheel damping. This makes the inherent damping of torsional oscillations for hydro units significantly higher than for steam-turbine generators.

REFERENCES

[1]   J.P. Den Hartog, *Strength of Materials*, Dover, 1961.

[2]   M.C. Jackson, S.D. Umans, R.D. Dunlop, S.H. Horowitz, and A.C. Parikh, "Turbine-Generator Shaft Torques and Fatigue: Part I - Simulation Methods and Fatigue Analysis; Part II - Impact of System Disturbance and High Speed Reclosure," *IEEE Trans.*, Vol. PAS-98, No. 6, pp. 2299-2328, November/December 1979.

[3]   D.N. Walker, S.L. Adams, and R.J. Placek, "Torsional Vibration and Fatigue of Turbine-Generator Shafts," *IEEE Trans.*, Vol. PAS-100, No. 11, pp. 4373-4380, November 1981.

[4]   D.N. Walker, C.E.J. Bowler, R.L. Jackson, and D.A. Hodges, "Results of Subsynchronous Resonance Test at Mohave," *IEEE Trans.*, Vol. PAS-94, No. 5, pp. 1878-1889, September/October 1975.

[5]   R. Quay and A.C. Schwalb, discussion of reference 4, *IEEE Trans.*, Vol. PAS-94, No. 5, pp. 1887-1888, September/October 1975.

[6]   R.G. Ramey, A.C. Sismour, and G.C. Kung, "Important Parameters in Considering Transient Torques on Turbine-Generator Shaft Systems," *IEEE Trans.*, Vol. PAS-99, No. 1, pp. 311-317, January/February 1980.

[7]   W. Watson and M.E. Coultes, "Static Exciter Stability Signal on Large Generators - Mechanical Problems," *IEEE Trans.*, Vol. PAS-92, pp. 205-212, January/February 1973.

[8]   D.C. Lee and P. Kundur, "Advanced Excitation Controls for Power System Stability Enhancement," CIGRE paper 38-01, 1986.

[9]   D.C. Lee, R.E. Beaulieu, and G.J. Rogers, "Effects of Governor Characteristics on Turbo-Generator Shaft Torsionals," *IEEE Trans.,* Vol. PAS-104, No. 6, pp.

1255-1261, June 1985.

[10]   M. Bahrman, E.V. Larsen, R.J. Piwko, and H.S. Patel, "Experience with HVDC - Turbine Generator Torsional Interaction at Square Butte," *IEEE Trans.*, Vol. PAS-99, pp. 966-975, May/June 1980.

[11]   N. Hingorani, S. Nilsson, M. Bahrman, J. Reeve, E.V. Larsen, and R.J. Piwko, "Subsynchronous Frequency Stability Studies of Energy Systems Which Include HVDC Transmission," Paper presented at the Symposium on Incorporating HVDC Power Transmission into System Planning, Phoenix, Ariz., March 24-27, 1980.

[12]   EPRI Report EL-2708, "HVDC System Control for Damping Subsynchronous Oscillations," Final Report of Project RP1425-1, Prepared by General Electric Company, October 1982.

[13]   CIGRE/IEEE guide for "Planning DC Links Terminating at AC Locations Having Low Short-Circuit Capabilities, Part I: AC/DC Interaction Phenomenon," 1992.

[14]   E. Katz, "Subsynchronous Resonance," Paper presented at the panel discussion on Dynamic Stability in the Western Interconnected Power Systems, IEEE PES Summer Meeting, July 18, 1974.

[15]   C. Concordia, J.B. Tice, and C.E.J. Bowler, "Sub-synchronous Torques on Generating Units Feeding Series-Capacitor Compensated Lines," Paper presented at the American Power Conference, Chicago, May 8-10, 1973.

[16]   IEEE Committee Report, "Proposed Terms and Definitions for Subsynchronous Oscillations," *IEEE Trans.*, Vol. PAS-99, No. 2, pp. 506-511, March/April 1980.

[17]   C. Concordia, discussion of reference 16.

[18]   B.L. Agrawal and R.G. Farmer, "Use of Frequency Scanning Techniques for Subsynchronous Resonance Analysis," *IEEE Trans.*, Vol. PAS-98, pp. 341-349, March/April 1979.

[19]   P.M. Anderson, B.L. Agrawal, and J.E. Van Ness, *Subsynchronous Resonance in Power Systems*, IEEE Press, 1990.

[20]   J.M. Undrill and T.E. Kostyniak, "Subsynchronous Oscillations Part I - Comprehensive System Stability Analysis," *IEEE Trans.*, Vol. PAS-95, pp. 1446-1455, July 1976.

[21]   L.A. Kilgore and D.G. Ramey, "Transmission and Generator System Analysis Procedures for Subsynchronous Resonance Problems," tutorial session, IEEE PES Summer Meeting, July 1975.

[22]   EPRI Report EL-2614, "Study of Turbine Generator Shaft Parameters from the Viewpoint of Subsynchronous Resonance," Final Report of Contract TPS81-794, Prepared by Power Technologies Inc., September 1982.

[23]   EPRI/DCG Report EL-7321, "Electromagnetic Transients Program - Version 2: Revised Application Guide," November 1991.

[24]   A.S. Morched, J.H. Ottevangers, and L. Marti, "Multi-port Frequency Dependent Network Equivalents for the EMTP," Paper 92 SM 461-4 PWRD, Presented at the IEEE PES Summer Meeting, Seattle, July 1992.

[25]   IEEE Committee Report, "Countermeasures to Subsynchronous Resonance Problems," *IEEE Trans.*, Vol. PAS-99, No. 5, pp. 1810-1817, September/October 1980.

[26]   N.G. Hingorani, "A New Scheme for Subsynchronous Resonance Damping of Torsional Oscillations and Transient Torque - Part I," *IEEE Trans.*, Vol. PAS-100, pp. 1852-1855, April 1981.

[27]   R.F Wolff, "Stop Subsynchronous T/G - Shaft Damage," *Electrical World*, pp. 129-133, April 1981.

[28]   ANSI Standard C50.13-1977, *Requirements for Cylindrical Rotor Synchronous Generators*.

[29]   J.S. Joyce, T. Kulig, and D. Lambrecht, "The Impact of High-Speed Reclosure of Single and Multi-Phase System Faults on Turbine-Generator Shaft Torsional Fatigue," *IEEE Trans.*, Vol. PAS-99, No. 1, pp. 279-291, January/February 1980.

[30]   C.E.J. Bowler, P.G. Brown, and D.N. Walker, "Evaluation of the Effect of Power Breaker Reclosing Practices on Turbine-Generator Shafts," *IEEE Trans.*, Vol. PAS-99, No. 5, pp. 1764-1779, September/October 1980.

[31]   EPRI Report EL-3083, "Determination of Torsional Fatigue Life of Large Turbine Generator Shafts," Final Report of Project 1531-1, Prepared by General Electric Company, April 1984.

[32]   IEEE Working Group Report, "IEEE Screening Guide for Planned Steady-State Switching Operations to Minimize Harmful Effects on Steam Turbine-

Generators," *IEEE Trans.*, Vol. PAS-99, No. 4, pp. 1519-1521, July/August 1980.

[33]   IEEE Working Group Interim Report, "Effects of Switching Network Disturbances on Turbine-Generator Shaft Systems," *IEEE Trans.*, Vol. PAS-101, No. 9, pp. 3151-3157, September 1982.

[34]   R.T.H. Alden, P.J. Nolan, and J.P. Bayne, "Shaft Dynamics in Closely Coupled Identical Generators," *IEEE Trans.* Vol. PAS-96, pp. 721-728, May/June 1977.

[35]   G. Andersson, R. Almuri, R. Rosenquist, and S. Torseny, "Influence of Hydro Units' Generator-to-Turbine Inertia Ratio on Damping of Subsynchronous Oscillations," *IEEE Trans.*, Vol. PAS-103, pp. 2352-2361, August 1984.

[36]   L.E. Eilts and E. Campbell, "Shaft Torsional Oscillations of Hydrogenerators," U.S. Bureau of Reclamation Report REC-ERC-79-6, Denver, Colo., August 1979.

Chapter 16

Mid-Term and
Long-Term Stability

Mid-term stability and long-term stability are associated with the response of power systems to severe upsets. Severe upsets are those disturbances that result in excursions of frequency, voltage, and power flows either so great or so long-lasting that they invoke the actions of slow processes, protective systems, and controls not modelled in conventional transient stability studies.

This chapter illustrates the nature of system response when subjected to severe upsets and examines the need for distinction between mid-term and long-term stability. It describes modelling considerations and analytical techniques for simulation of long-term dynamic response. In addition, it gives illustrative examples and provides guidelines for enhancing the ability of power systems to cope with severe upsets.

16.1 NATURE OF SYSTEM RESPONSE TO SEVERE UPSETS

In considering the nature of system response to major disturbances, it is helpful to visualize the power system operating conditions in terms of the five states depicted in Figure 16.1. A description of these states and the ways in which the transition takes place from one state to another is provided in Chapter 1 (Section 1.3). This chapter focuses on events and system conditions associated with the *in extremis* state.

1073

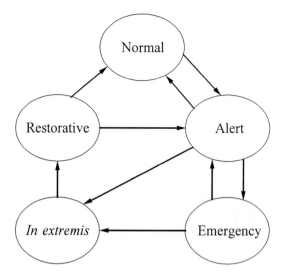

Figure 16.1 Power system operating states

An examination of actual system disturbances [1-6] shows that an underlying pattern exists with regard to the events leading to system failures. This is depicted in Figure 16.2 which shows the events that could cause transition from the alert state to the *in extremis* state. The initiating event could be a disturbance of natural origin, a malfunction of equipment, or a consequence of human factors. Modern power systems are designed and operated to ensure secure operation for the more probable contingencies (see Chapter 1 for a description of the design contingencies). In the vast majority of cases, power systems are able to withstand any single contingency and many multiple contingencies. The protective and control systems act to prevent the propagation of the disturbance to other parts of the network.

Occasionally, however, an unusual combination of circumstances and events causes a portion of the interconnected system to separate completely and form one or more *electrical islands*. The initiating event is usually a contingency more severe than those covered by the normal design criteria: for example, tripping of several transmission lines as the result of a tornado, an ice storm, or a malfunction of communication equipment. The events triggered as a consequence of the initiating event, by stressing the system further, cause uncontrolled cascading outages. Wide variations of frequency (from 58 to 63 Hz) and voltage (from 50% to 120% of pre-disturbance value) may exist during the ensuing system conditions [1]. The system can thus deteriorate to the *in extremis* state, the result being the loss of significant portions of system load. Generally, the actions of control and protective systems dominate the system response during these conditions. The situation is often aggravated by poorly coordinated protection and control systems. Emergency control action should be directed toward saving as much of the system as possible from total collapse.

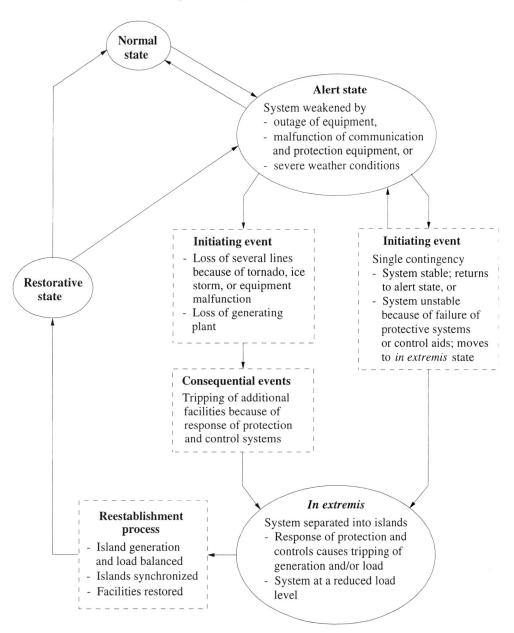

Figure 16.2 Transition of system states during severe upsets

System response to islanding conditions

The system response to an islanding condition is basically a sustained-frequency transient. Therefore, the speed control and the subsequent responses of prime mover and energy supply systems play a major role in determining the nature of the system dynamic performance. Often, the situation is compounded by high- or low-voltage conditions.

Undergenerated islands:

In an island with total initial generation less than the total load, the frequency will decline. If there is sufficient spinning reserve within the island, the system frequency returns, in a few seconds, to a near normal value. If sufficient generation with the ability to rapidly increase output is not available, the frequency may reach levels that could lead to tripping of thermal generating units by underfrequency protective relays, thus aggravating the situation further. Therefore, underfrequency load-shedding schemes are usually employed to reduce the connected load to a level that can be satisfactorily supplied by the available generation (see Chapter 11, Section 11.1.7).

Consequently, in an undergenerated island, the initial transient is dependent on the responses of spinning generation reserve and load-shedding relaying. The minimum value of frequency is reached in a few seconds. The system frequency response beyond this point depends on the prime mover characteristics.

Overgenerated islands:

In an island with an initial excess generation, the frequency will rise, and the speed-governing system will respond by reducing the mechanical power generated by the turbines. The power plants, in effect, experience a "partial load rejection." The performance of the island and its ability to stabilize without loss of load depend on the ability of the power plants to sustain a partial load rejection. This is discussed further in Section 16.3.

Reactive power balance:

The system performance is also influenced by the reactive power balance within the island. A significant mismatch in the total reactive power generated and absorbed could lead to high- or low-voltage conditions. Generator over/underexcitation limiters and controls may be activated. In extreme situations, the response of protective relays could lead to the tripping of generating units. For example, an island with lightly loaded EHV lines and/or cables may cause the generators to absorb high amounts of reactive power; this situation, if not corrected quickly, could lead to tripping of the units by loss-of-excitation protection.

Power plant auxiliaries:

Variations, especially a decrease, in power supply voltage and frequency can degrade the performance of power plant auxiliaries driven by induction motors. For example, pumps associated with circulating water, condensate, heater drain, and feedwater are driven by induction motors. Degraded performance of these pumps may lead to loss of condenser vacuum, high turbine-exhaust temperature, and insufficient condensate/feedwater flow. Many nuclear plants are, therefore, equipped with undervoltage and underfrequency relays set to trip the plant at low voltages (typically, 70% of rated value) and at low frequencies (typically, 59.2 Hz) [8].

Power plant motors commonly use electromagnetic contactors and relays in the starting and protection circuits. The response of these contactors and relays to severe or long voltage dips could disconnect the motors.

System restoration

When the islands reach steady-state conditions, operators take steps to restore the interconnected system. These involve adjusting of generation and load in each island, resynchronizing of islands, and restoration of generating units, loads, and other facilities tripped during the system disturbance. Start-up and reloading of thermal units are constrained by a number of factors, and full power may not be available for several hours. The restoration of power can be speeded up considerably if the thermal units are available quickly for reloading. Many utilities have, therefore, adopted the "tripping to house load" technique. The following are the different methods used to achieve this [9,10]:

(a)   Trip the fire at the moment of disconnection and rely on the thermal storage of the boiler for about 20 minutes.

(b)   As (a), but re-ignite the fire and leave the boiler running on the ignition burners.

(c)   Apply a special firing technique at very low load.

(d)   Leave the boiler at full power or reduce power to minimum load. Dump excess steam into the atmosphere, and draw upon feed-water storage, usually for several hours.

(e)   Leave the boiler at minimum load and bypass excess steam into the condenser.

16.2 DISTINCTION BETWEEN MID-TERM AND LONG-TERM STABILITY

As noted in Chapter 2, the terms *long-term stability* and *mid-term stability* are new in the literature on power system stability. They were introduced as a result of the need to deal with problems associated with the dynamic response of power systems to severe upsets.

Long-term stability analysis as defined in references 11 to 15 assumes that inter-machine synchronizing power oscillations have damped out, the result being uniform system frequency. The focus is on slower and longer duration phenomena that accompany large-scale system upsets and on the resulting sustained mismatches between generation and consumption of active and reactive power. Boiler dynamics of thermal units, penstock and conduit dynamics of hydro units, automatic generation control, power plant and transmission system protections/controls, transformer saturation, and off-nominal effects on loads and the network are likely to be significant.

The term *mid-term stability* was introduced in references 14 and 15. It represents the transition between the transient responses and the long-term responses. In mid-term stability, the focus is on synchronizing power oscillations between machines, as well as on the effects of some of the slower phenomena, and possibly large-voltage and/or frequency excursions.

It is clear, by these definitions, that the distinction between mid-term and long-term stability is weak. What makes long-term stability different is that it assumes that the system frequency is uniform and that fast dynamics are not significant. This is helpful if analytical tools require these assumptions to facilitate simulation. But with today's software that utilizes advanced sparsity and efficient implicit integration techniques, simulating long time frames with fast-dynamic modelling is becoming less of a concern.

There is no well-defined distinction between the mid-term time frame and the long-term time frame so far as modelling requirements are concerned. For time periods beyond the transient period, the choice of models to be included in a simulation should be based on the phenomena being analyzed and the system representation used, rather than on the actual duration of the simulation. For example, neglecting boiler dynamics or steam process control in a "mid-term" simulation may be acceptable for simulation of disturbances in which the driving functions to these dynamics are small. However, neglecting these slower dynamics for more severe disturbances that may invoke protections associated with variables such as steam pressures could have a tremendous impact on the simulation results. The experiences of a number of utilities have identified the need to represent intermachine oscillations and fast transients associated with excitation systems in long-term stability studies [16]. Thus, the distinction between mid-term and long-term stability cannot be satisfactorily based on a fixed time-frame basis, nor on modelling requirements. In light of this, the best approach to classifying stability problems associated with severe system upsets is to eliminate the concept of mid-term as a separate category and to use long-term to encompass all studies beyond the transient time frame.

Long-term stability would then be described as the ability of a power system to reach an acceptable state of operating equilibrium following a severe system disturbance that may or may not have resulted in the system being divided into subsystems. The time frame of interest extends beyond the transient period sufficiently to include, in addition to fast dynamics, the effects of slow dynamics of automatic system controls and protections. Long-term simulations may include severe disturbances, beyond the normal design contingencies, that have resulted in cascading and splitting of the power system into a number of separate islands with the generators in each island remaining in synchronism. *Stability in this case is a question of whether each island will reach an acceptable state of operating equilibrium with minimal disruption to services.* In an extreme case, the system and unit protections may compound the adverse situation and lead to a collapse of the island in whole or in part.

Generally, the long-term stability problems are associated with inadequacies in equipment responses, poor coordination of control and protection equipment, or insufficient active/reactive power reserve.

The characteristic times of the processes and devices activated by the large voltage and frequency shifts will range from a matter of seconds, corresponding to the responses of devices such as generator controls and protections, to several minutes, corresponding to the responses of devices such as prime mover energy supply systems and load-voltage regulators.

From an analytical standpoint, long-term stability programs become extensions of transient stability programs with the desired capability of adjusting the integration time-step according to the dominant transients [17].

One application of long-term stability simulation gaining interest is the dynamic analysis of voltage stability (discussed in Chapter 14) requiring simulation of the effects of transformer tap-changing, generator overexcitation protection and reactive power limits, and thermostatic loads. In this case, intermachine oscillations are not likely to be important and energy supply system transients may not be critical. However, care should be exercised before neglecting some of the fast dynamics.

16.3 POWER PLANT RESPONSE DURING SEVERE UPSETS

16.3.1 Thermal Power Plants

The ability of power plants to survive partial load rejections is of crucial importance in minimizing the impact of a severe upset and quickly restoring normal operation of the power system. References 18 to 21 describe the responses of power plants to partial load rejections and problems experienced in successfully withstanding such disturbances.

Reference 22, prepared by an IEEE Working Group, provides guidelines for enhancing power plant response to partial load rejections. The following is a summary of these guidelines:

(a)   *Overall plant control.* To withstand a partial load rejection, the overall plant control must promptly decrease the input power (fuel flow) to correspond to the output electrical power. As there are time lags in this decrease, the power input must temporarily undershoot the power output. The ideal source of intelligence to determine the input power reduction is the actual power output.

(b)   *Boiler control.* Without a turbine bypass system, a partial load rejection appears to the boiler as a step decrease in steam flow. Prompt reduction of fuel flow is essential, as noted above.

For a *once-through boiler*, prompt reduction of the feedwater flow, tightly coupled to the fuel flow, is also required. However, once-through boilers normally have superheater/turbine bypass systems of limited capacity to protect the furnace tubes and to assist in pressure control. This bypass capability can be used to ease the rate of reduction of fuel and feed-water flow.

For a *drum type boiler*, the reduction of feed-water flow should be delayed because the immediate response of the drum water level is to decrease with steam flow and the resulting rise in drum pressure. In addition, overfeeding is required to obtain the higher-level water inventory required at the lower power level. Adequate water-level control is particularly important for large changes in steam flow. To enhance the ability of water-level control during a partial load rejection, consideration should be given to temporarily increasing the range between high and low water-level trip limits, or delaying the trip for high feedwater flow.

For either type of boiler, a delay in the reduction of air flow is generally desirable since the cooling effect of the excess air flow will tend to compensate for the lags in fuel flow response. However, the air flow to the operating burners must be controlled relative to the fuel flow to maintain stable combustion.

For feed pumps driven by auxiliary steam turbines, the closure of intercept valves will interrupt the steam flow; therefore, it is necessary either to provide an alternate steam source to the auxiliary turbine or to switch to motor-driven pumps.

(c)   *Turbine-generator control.* The turbine overspeed controls are designed to limit overspeed following full-load rejection to about 1% below the overspeed trip settings. This will obviously prevent overspeed trips during partial load rejection.

While the control valves and the intercept valves are closed, there is no steam flow through the turbine. To prevent overheating of the reheater tubes, boiler firing must be tripped off if the interruption of steam flow through the reheater is sustained. However, since the closure of the steam valves is temporary, tripping of the boiler firing can be avoided by proper coordination of boiler protection and fuel control with the turbine controls.

In addition, as recommended in reference 4, the overspeed controls must

(i)   Not interfere with normal speed governing in such a way that the performance of the islanded system is adversely affected

(ii)   Be capable of discriminating between unit rejections and transient system disturbances, e.g., transmission system faults that temporarily reduce unit power

(d)   *Power plant auxiliaries.* The effect of voltage and frequency variations experienced during partial load-rejection conditions should be checked to ensure vital auxiliaries will not trip out.

(e)   *Steam turbine bypasses.* The use of a steam bypass system permits the reduction of boiler power in a controlled manner. A well-designed turbine bypass system significantly enhances the capability of the power plant to withstand a partial load rejection.

Nuclear plants normally have steam turbine bypass systems. In North America, fossil-fuelled power plants with drum type boilers are not usually equipped with turbine bypass systems. Power plants with once-through boilers have turbine and superheater bypass systems; these are installed primarily for start-up and shutdown duty.

16.3.2 Hydro Power Plants

Experience with the operation of hydraulic units under system islanding conditions has demonstrated that governor tuning is critical [23]. The requirement for good stability of speed control under islanding conditions or other isolated modes of operation is in conflict with the governor settings required for fast loading and unloading under normal synchronous operation. For rapid load changes, it is desirable to have a fast governor response. However, the governor settings that result in a fast response usually cause frequency instability under system-islanding conditions. References 23 to 25 provide general guidelines for selection of the governor settings.

As described in Chapter 9 (Section 9.1.3), the governor settings that can be varied to control the dynamic performance of the generating unit are the dashpot reset time T_R, the temporary droop R_T, and the gate servo system gain K_S. A convenient method of analyzing the effects of variations in these parameters on the stability of frequency oscillations is the eigenvalue analysis described in Chapter 12. For a determination of the optimum governor parameters, the stability of the generating unit when supplying an isolated non-synchronous load must be considered. This represents the most severe condition from the viewpoint of frequency control.

Table 16.1 shows the effects of varying the governor parameters on the stability of a hydraulic unit with a mechanical starting time T_M of 6.2 seconds and a water starting time T_W of 1.0 second. The unit is assumed to be supplying rated power to an isolated load having a constant power characteristic (equivalent to having a damping coefficient of –1).

Table 16.1 Effects of governor parameters under isolated operation

Governor Parameters			Eigenvalues		
K_S	R_T	T_R	Mode 1	Mode 2	Mode 3
5.0	0.38	1.5	+0.048±j0.60	−1.058	−5.96
5.0	0.38	2.0	−0.006±j0.59	−0.799	−3.87
5.0	0.38	5.0	−0.173±j0.65	−0.255	−3.77
5.0	0.38	10.0	−0.212±j0.72	−0.104	−3.75
5.0	0.24	5.0	−0.003±j0.75	−0.277	−3.44
5.0	0.30	5.0	−0.081±j0.71	−0.237	−3.58
5.0	0.35	5.0	−0.141±j0.68	−0.248	−3.70
5.0	0.40	5.0	−0.193±j0.64	−0.262	−3.82
2.5	0.38	5.0	−0.107±j0.54	−0.252	−2.86
10.0	0.38	5.0	−0.251±j0.73	−0.256	−5.72
5.0	0.30	1.5	+0.095±j0.65	−0.920	−3.68
5.0	0.50	1.5	−0.010±j0.54	−1.140	−4.32
2.5	0.38	1.5	+0.044±j0.51	−0.954	−2.93
10.0	0.38	1.5	+0.039±j0.68	−1.058	−5.96

Table 16.1 lists eigenvalues associated with three modes that are sensitive to variations in governor parameters. The first mode is an oscillatory mode represented by a pair of complex conjugate eigenvalues having a frequency (given by the imaginary component) on the order of 0.6 rad/s or 0.1 Hz. The stability of this mode is of primary concern for isolated operation and during system-islanding conditions. The other two modes listed are non-oscillatory modes represented by real eigenvalues. The reciprocals of the real eigenvalues represent time constants and are a measure of the responsiveness of the governor under isolated operating conditions.

For isolated operation, selection of governor parameters to ensure that the oscillations are very well damped is of primary concern. It is desirable to achieve this without unduly slowing down the speed of response since speed of response affects the maximum frequency deviation when an island is formed. The results of Table 16.1 show that the variations in the governor parameters have the following effects on the dynamics of the unit under *isolated* operating conditions:

- An increase in dashpot reset time T_R results in an increase in damping, accompanied by a reduction in rate of response.

- An increase in temporary droop R_T results in an increase in damping and a slight decrease in the time to reach steady state.

- An increase in servo system gain K_S results in an improvement in damping as well as in the rate of response. The improvement in the rate of response is primarily through reduction in the time constant associated with mode 3.

The results in Table 16.1 are for a select combination of governor parameters chosen to illustrate the overall nature of the problem. Because a large number of combinations of parameters is possible and because damping of mode 1 is of primary concern, the selection of optimum governor parameters can be better achieved by presenting the results as shown in Figure 16.3. The results are shown in the form of loci of constant damping ratio ζ in an R_T-T_R parameter plane when K_S is held constant. The variations in the natural frequency of oscillation are also shown in the figure.

The results suggest that, for a unit with T_M=6.2 s and T_W=1.0 s, values of R_T less than 0.225, and T_R less than 1.5 s result in unstable frequency oscillations. As T_R is increased from 1.5 s to about 5.0 s, the damping increases significantly. Increasing T_R beyond 5.0 s results in a small improvement in damping at the expense of slowing down the speed of response. This suggests that an appropriate value for T_R is about 5.0 s. The value of R_T can be chosen to give any desired damping ratio. A good value for the damping ratio is 0.25. This can be achieved by setting R_T=0.3 s with T_R=5.0 s. A higher value of R_T would result in higher damping and a slight reduction in time to reach steady state. However, the peak overshoot or maximum frequency deviation would be higher with higher values of R_T.

The effect of the servo system gain K_S is shown in Figure 16.4. These results are an extension of those of Figure 16.3 and show loci of ζ=0 (absolute stability limit) and of ζ=0.25 for three values of K_S. We see that higher values of K_S increase the regions of stability throughout the ranges of R_T and T_R considered.

Similar analyses performed on units having different combinations of water and machine starting times indicate that the optimum choices of the temporary droop R_T and the reset time T_R are represented by the following expressions [23]:

$$R_T = [2.3-(T_W-1.0)0.15]\frac{T_W}{T_M} \tag{16.1}$$

$$T_R = [5.0-(T_W-1.0)0.5]T_W \tag{16.2}$$

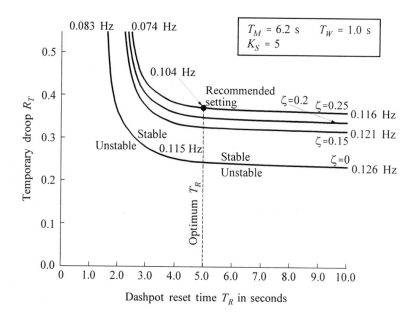

Figure 16.3 Loci of constant damping ratio (ζ) for isolated operation

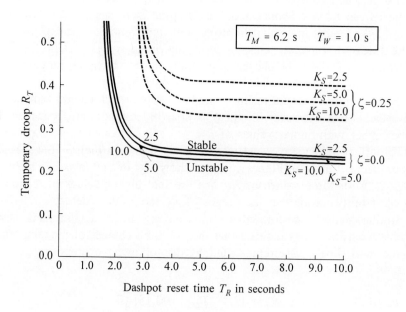

Figure 16.4 Effect of servomotor gain

In addition, the servosystem gain K_S should be set as high as is practically possible. *The above settings ensure good stable performance when the unit is at full load supplying an isolated load; this represents the most severe requirement and ensures stable operation for all situations involving system islanding.*

For loading during normal operation, the above settings result in too slow a response [23]. For satisfactory loading rates, the reset time T_R should be less than 1.0 s, preferably close to 0.5 s.

As discussed in Chapter 9, these conflicting requirements can be met by a dashpot bypass arrangement in which, with the dashpot not bypassed, the settings satisfy the requirements for system islanding conditions or isolated operation; and with the dashpot bypassed, the reset time T_R has a reduced value that results in acceptable loading rates. The dashpot is normally not bypassed so that, in the event of a disturbance leading to an islanding operation, the speed control would be stable. The dashpot is bypassed for brief periods during loading and unloading.

16.4 SIMULATION OF LONG-TERM DYNAMIC RESPONSE

16.4.1 Purpose of Long-Term Dynamic Simulations

The objectives of analysis of long-term dynamics include the following:

(a)   Post-mortem analysis of severe upsets [32]. This will help identify the underlying causes of such incidents and develop corrective measures.

(b)   Evaluation of the ability of power plants to ride through disturbances. This will help identify deficiencies in equipment responses and in the coordination of control and protective systems.

(c)   Examination, at the system planning and design stage, of the system response to extreme contingencies, that is, contingencies that exceed the severity of normal design contingencies. As discussed in Chapter 1 (Section 1.4), the purpose of extreme contingency assessment is to develop measures for reducing the frequency of occurrence of such contingencies and to mitigate the consequences.

(d)   Examination of emergency procedures and provision of operator training.

(e)   Evaluation of load-shedding practices and policies.

16.4.2 Modelling Requirements

The long-term stability simulation program should include, in addition to the models used in conventional transient stability simulations, adequate representation of the prime mover and energy supply systems. It should also include appropriate models for the wide range of protection and control systems that are invoked when the system is in the emergency and *in extremis* states.

An examination of major power system disturbances [1-4,26,27] shows that the following protection and control devices have a significant influence on the system dynamic performance during severe system upsets:

(a)   Generator and excitation system protection and controls:

- Loss-of-excitation relays

- Underexcitation limiters (UEL)

- Overexcitation limiters (OXL)

- V/Hz limiter and protection

Chapter 8 provides descriptions of these devices.

(b)   Electrical network protection and controls:

- Transmission system relays

- Distribution system relays

- Transformer underload tap-changer (ULTC) controls

- Distribution system voltage regulators

- Underfrequency and undervoltage load-shedding relays

Chapters 6 and 11 include descriptions of these devices.

(c)   Prime mover/energy supply system protection and controls:

- Turbine overspeed controls and protections

- Turbine underfrequency protection

- Boiler/reactor controls and protections

- Automatic generation control (AGC)

Chapters 9 and 11 provide descriptions of the above. The representation of prime mover and energy supply systems should include appropriate representation of the power plant auxiliaries. Induction motors driving some of these auxiliaries are affected by off-nominal voltage and/or frequency of the power supply.

For situations involving high network voltage conditions, representation of transformer saturation is important. Where appropriate, the characteristics of motor loads as influenced by off-nominal frequency and voltage variations should be represented.

Off-nominal frequency effects on the representation of the transmission network, synchronous machine stator circuits, and reactive power compensation devices should be accounted for.

The EPRI *long-term stability program* (LTSP) developed under project RP 3144-1 includes most of the above modelling capabilities.

16.4.3 Numerical Integration Techniques

The early long-term dynamic simulation programs, such as LOTDYS [28], assumed uniform frequency and modelled only slow phenomena to achieve computational efficiency. This allowed the use of low order explicit integration techniques, but required special means of initialization. A disadvantage of this approach is that it neglects many important fast transients that influence long-term dynamic response; consequently, many potential system problems are masked.

The present trend is to model fast as well as slow phenomena in long-term dynamic simulations. This is facilitated by the use of implicit integration methods. As discussed in Chapter 13, the implicit integration methods are numerically very stable; stiffness of the differential equations affects accuracy but not numerical stability. With large time-steps, high-frequency modes are filtered out and solutions of slower modes are accurate.

In reference 29, a unified approach to short- and long-term dynamic simulation is implemented by using the trapezoidal integration method. The simulation mode (short- or long-term) is determined by integration step-size. An artificial damping is added to suppress synchronizing power oscillations and to allow larger integration time-steps when simulating the long-term mode. Facilities are provided for switching from one mode to the other by adding or removing the artificial damping and changing the time-step.

Another attractive integration technique for long-term dynamic simulation is the Gear-type backward differentiation formula [30,31]. The second-order Gear method is an implicit self-starting integration algorithm of the predictor-corrector type. The EUROSTAG program described in reference 17 applies this method. The key feature of this application is the automatic adjustment of the integration time-step based on the truncation error. When fast modes are excited, the time-step is reduced to a value below that corresponding to the smallest eigenvalue. As the fast modes decay during the solution process, the time-step is gradually increased. If the fast modes are excited due to a subsequent switching operation, the time-step is automatically reduced. During a long-term dynamic simulation, the time-step may vary between 1 ms and 100 s.

16.5 CASE STUDIES OF SEVERE SYSTEM UPSETS

In this section we consider two case studies: one involving an overgenerated island and the other an undergenerated island.

16.5.1 Case Study Involving an Overgenerated Island

This case study is intended to demonstrate the impact of turbine-generator overspeed controls on the performance of a generation-rich island. It is based on experiences with auxiliary governors (described in Chapter 9, Section 9.2.2) used on some of the generating units in Ontario [4].

An island formed as a result of separation from the rest of the interconnected system of an area consisting of 9,650 MW of generation and 4,750 MW of consumption (load plus losses) is considered (see Figure 16.5). Prior to separation, the area is exporting 4,900 MW; MVAr generation and consumption within the area are nearly equal.

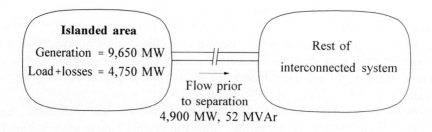

Figure 16.5

The area generation

The generation within the area consists of the following:

- 4,000 MW of nuclear generation at a plant with eight 500 MW units

- 3,850 MW of fossil-fuelled generation at two plants, each with four 500 MW units

- 1,800 MW of hydraulic generation at six plants

Turbine-governing systems

The nuclear units are equipped with *mechanical-hydraulic control* (MHC) governors whose block diagram is shown in Figure 9.31 of Chapter 9. A singular feature of this governing system is the use of an *auxiliary governor*, which becomes operative when the speed exceeds its setting $V_1 = 1\%$. The auxiliary governor acts in parallel with the main governor to effectively increase the gain of the speed control loop by a factor of about 8. This limits overspeed by rapidly closing the control valves (CVs) as well as the intercept valves (IVs).

The fossil-fuelled units have MHC governors whose block diagram is shown in Figure 9.32.

The model for the governing systems of the hydraulic units is shown in Figure 9.10, with settings given by Equations 16.1 and 16.2 so as to result in heavily damped control under islanding conditions.

Figures 9.7 and 9.24 show the representation of hydraulic turbines and steam turbines, respectively.

Simulation

The disturbance simulated is the simultaneous opening of all ties connecting the area to the rest of the system, the result being an island with generation nearly twice the load. The generators and excitation system are represented in detail. Loads are represented as nonlinear functions of voltage and frequency.

The performance of the islanded system is examined with the auxiliary governors of the nuclear units in service and out of service.

(a) With auxiliary governor in service:

Figure 16.6 shows plots of speed deviation, CV and IV positions, and mechanical power of one of the nuclear units. All units in the area swing together; therefore, the speed plot shown is representative of the area frequency.

As a consequence of islanding, the speed increases rapidly to a maximum of about 6.4% above the normal speed of 1,800 r/min and then oscillates with little damping between 3.5% above and 0.7% below the normal speed. The sustained oscillation is due to the action of the auxiliary governors. When the overspeed exceeds the setting V_1 of 1%, the auxiliary governors close the steam valves and reduce the mechanical power of the nuclear units to zero. The deficit in the generated power reduces the speed rapidly, and the valves open again. The resultant increase in mechanical power is such that the speed exceeds the auxiliary governor setting of 1%, and the valves close again. The cycle repeats with a period of about 7 seconds.

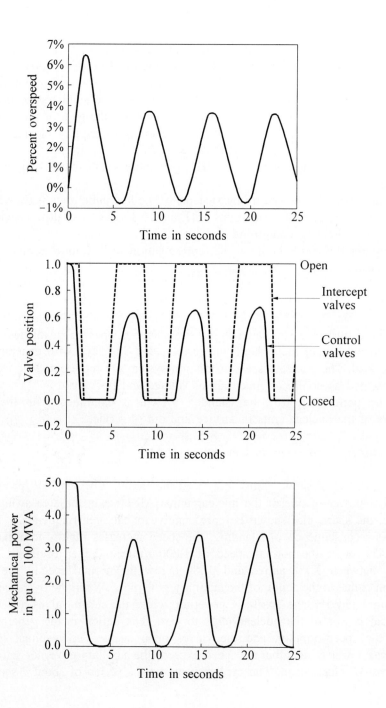

Figure 16.6 Transient response of nuclear units with auxiliary governor

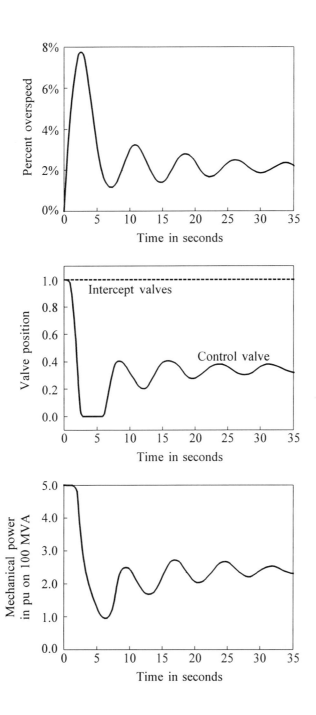

Figure 16.7 Transient response of nuclear units
with auxiliary governor out of service

(b) With auxiliary governor out of service:

Figure 16.7 shows the results with the auxiliary governors of nuclear units out of service. The speed deviation reaches a maximum of about 7.7% and damps to a steady state value of nearly 2% above the nominal. The removal of the auxiliary governor results in stable frequency control. Only the CVs respond to speed changes and the IVs remain fully open because the main governor alone is not enough to overcome the intercept valve-opening bias.

From the above results, it is clear that the auxiliary governors cause instability of the speed control during system-islanding conditions. The other units in the island respond to oscillations of the units with auxiliary governors; the overall effect is to cause oscillations of all units. The resulting movements of steam valves or wicket gates continue until the hydraulic systems of the governors run out of oil and cause unit tripping and possibly a blackout of the island. The power oscillations may also give rise to "priming" of the boilers of the fossil-fuel-fired units that causes water from the boilers to come in contact with the high temperature superheat and HP stages.

A possible solution to this problem is to replace the auxiliary governor with an electronic acceleration detector [4].

This case study has assumed that the steam generation systems perform satisfactorily by following the demands of the turbine. Reference 33 illustrates the types of problems that may be caused by unsatisfactory response of steam generation system protections and controls of a nuclear unit.

16.5.2 Case Study Involving an Undergenerated Island

This case study is intended to illustrate the performance of an undergenerated island as influenced by underfrequency load shedding and power plant control. In addition, it demonstrates the requirement for adequate voltage and reactive control.

Power generation and conditions

An island formed as a result of separation of an area consisting of 2,218 MW generation and 3,283 MW load is considered. Prior to islanding, the area is importing 1,103 MW and exporting 1,590 MVAr, as shown in Figure 16.8.

The generation mix within the area is as follows:

- 1,200 MW nuclear generation supplied by two units in one plant

- 1,018 MW of hydraulic generation supplied by six plants

The two nuclear units are operating at their load limits. The hydraulic generation has a spinning reserve of about 170 MW.

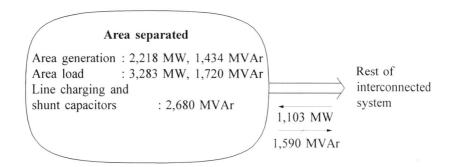

Figure 16.8 Active and reactive power conditions prior to separation

Underfrequency load shedding

Frequency trend relays (see Chapter 11, Section 11.1.7) are used to shed up to 50% load in four blocks depending on the frequency drop and the rate of frequency decline as depicted by Figure 11.31. The load shedding is delayed by about 0.5 second after the relay has operated.

Simulation of islanding condition

The islanding condition is simulated by opening the ties between the area under consideration and the rest of the interconnected system. The detail of representation of the power plants and other elements of the system is similar to that of the case study described in Section 16.5.1. The nuclear units are equipped with thyristor exciters with power system stabilizers and electrohydraulic governing systems. The hydraulic units are equipped with rotating dc exciters and mechanical hydraulic governors. Two cases are considered: one with no capacitor switching and the other with capacitors having a nominal rating of 700 MVAr switched out following islanding. Loads are represented as nonlinear functions of voltage and frequency.

(a) With no capacitor switching:

The switching sequence, including that resulting from load-shedding relays, is as follows:

Time (seconds)	Event
0.5	Island is formed by simultaneous tripping of ties.
0.9	System frequency drops to 59.5 Hz and rate of frequency decline is 1.25 Hz/s; the conditions for tripping the first two blocks of load are satisfied.

1.4	25% of area load is rejected.
1.6	System frequency drops to 58.8 Hz; the condition for tripping the fourth block of load is satisfied.
2.2	Additional 15% of load is rejected.
50.0	Simulation is terminated.

(Note: The rate of frequency does not exceed 2 Hz/s; hence, the third block of load is not shed.)

Figures 16.9 to 16.11 show plots of frequency, terminal voltage and reactive power output of one of the nuclear units, voltage of a load bus, total mechanical power and electrical power of all the generators, and the total active power supplied to the area loads. From Figure 16.9, we see that the system frequency drops below 58.5 Hz in about 2 seconds and remains below this level in spite of shedding about 40% of the area load. The reason for this is evident from Figures 16.10 and 16.11. Because of surplus reactive power in the area, the system voltages increase significantly after islanding, and this keeps the load power high even after shedding load.

The total reactive power absorbed by the nuclear units exceeds the continuous end-region heating limit of the generators (see Chapter 5, Section 5.4.1).

Figure 16.12 shows the apparent impedance seen by a loss-of-excitation (LOE) relay located at the terminals of one of the nuclear units. We see that the impedance locus does not enter the trip characteristic of the relay.

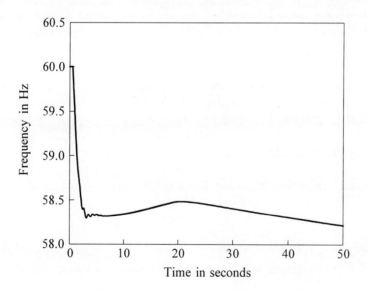

Figure 16.9 System frequency with underfrequency load
shedding and no capacitor switching

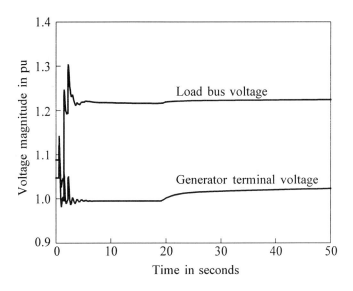

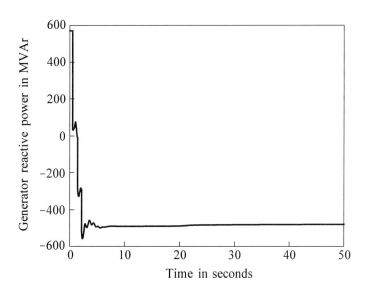

Figure 16.10 Nuclear unit terminal voltage and reactive power and load
bus voltage with load shedding and no capacitor switching

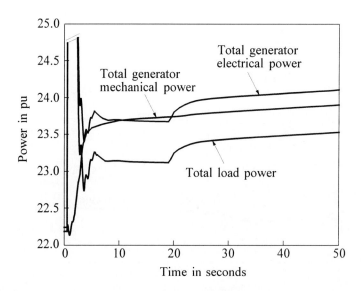

Figure 16.11 Total generated mechanical and electrical power and load
power with load shedding and no capacitor switching

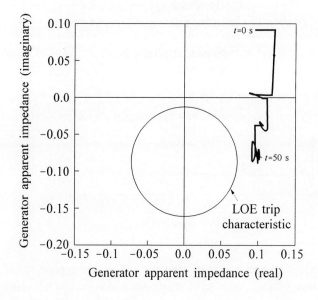

Figure 16.12 Locus of apparent impedance seen by LOE relay;
impedance expressed in per unit on 100 MVA base

(b) With 720 MVAr of shunt capacitors switched out:

The switching sequence in this case is as follows:

Time (seconds)	Event
0.5	Island is formed.
0.9	The conditions for tripping first two blocks of load are satisfied.
1.4	25% of load is shed; 485 MVAr capacitors are switched out.
1.8	Frequency drops below 58.8 Hz; the condition for tripping the fourth block of load is satisfied.
2.3	Additional 15% of load is shed; 235 MVAr capacitors are switched out.
50.0	Simulation is terminated.

Figures 16.13 to 16.15 show the corresponding results. In this case, the system voltages do not rise as high as in the previous case. Following load shedding, the total active power supplied to the loads falls below the available generation. The system frequency recovers to the nominal value (60 Hz) in about 20 seconds and reaches a maximum value of about 60.1 Hz. The steady-state frequency settles at a value slightly above the nominal value as determined by the droop characteristics of the speed governors of all the generating units in the area.

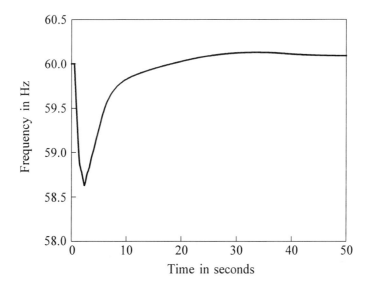

Figure 16.13 System frequency with underfrequency
load shedding and capacitor switching

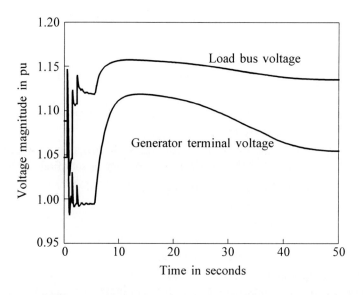

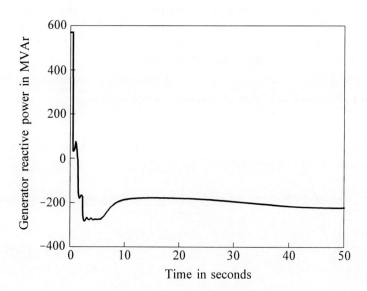

Figure 16.14 Nuclear unit terminal voltage and reactive power and load
bus voltage with load shedding and capacitor switching

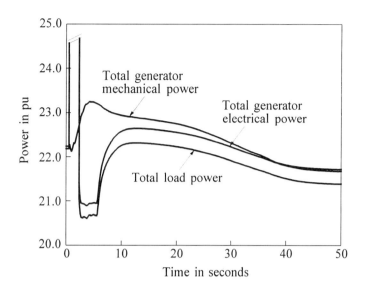

Figure 16.15 Total generated mechanical and electrical power and
load power with load shedding and capacitor switching

Figure 16.14 shows that the terminal voltage of the nuclear unit increases rapidly during the period between 5 and 10 seconds and then gradually drops to about 1.05 pu. This response of the generator voltage is due to the action of the power system stabilizer and can be modified to some extent by changing the parameters of the PSS (see Chapter 17, Section 17.2.1).

With capacitor switching, the reactive power absorbed by the nuclear unit is within the continuous capability of the generator.

REFERENCES

[1]   D.R. Davidson, D.N. Ewart, and L.K. Kirchmayer, "Long-Term Dynamic Response of Power System Disturbance: An Analysis of Major Disturbances," *IEEE Trans.*, Vol. PAS-94, No. 3, pp. 819-826, May/June 1975.

[2]   C. Concordia, D.R. Davidson, D.N. Ewart, L.K. Kirchmayer, and R.P. Schulz, "Long-Term Power System Dynamics - A New Planning Dimension," CIGRE Paper 32-13, 1976.

[3]   U.G. Knight, "International Survey of Major Disturbances," CIGRE SC32/WG01 Report, 1978.

[4]   P. Kundur, D.C. Lee, J.P. Bayne, and P.L. Dandeno, "Impact of Turbine Generator Overspeed Controls on Unit Performance under System Disturbance Conditions," *IEEE Trans.*, Vol. PAS-104, No. 6, pp. 1262-1267, June 1985.

[5]   D.N. Ewart, "Whys and Wherefores of Power System Blackouts," *IEEE Spectrum*, pp. 36-41, April 1978.

[6]   L.H. Fink and K. Carlsen, "Operating under Stress and Strain," *IEEE Spectrum*, pp. 48-53, March 1978.

[7]   R.M. Maleszwski, R.D. Dunlop, and G.L. Wilson, "Frequency Actuated Load Shedding and Restoration," *IEEE Trans.*, Vol. PAS-90, No. 4, pp. 1452-1459, 1971.

[8]   G.A. Weber, M.C. Winter, and S.M. Follen, "Nuclear Plant Response to Grid Electrical Disturbances," EPRI Report NP-2849, February 1983.

[9]   V.F. Carvalho, "Power Plant Control," CIGRE SC39/WG04 Survey Paper D, CIGRE SC39/IFAC Meeting, Florence, September 1983.

[10]   D. Rumpel et al., "Tripping to Houseload - Methods, Application, Experience," CIGRE SC39/IFAC Meeting, Florence, September 1983.

[11]   CIGRE Working Group 32-03, "Tentative Classification and Terminologies Relating to Stability Problems of Power Systems," *Electra*, No. 56, 1978.

[12]   EPRI Report EL-596, "Midterm Simulation of Electric Power Systems," Project RP745, June 1975.

[13]   K. Hemmaplardh, J.W. Manke, W.R. Pauly, and J.W. Lamont, "Considerations for Long Term Dynamic Simulation Program," *IEEE Trans.*, Vol. PWRS-1, pp. 129-135, February 1986.

[14]   V. Converti, D.P. Gelopulos, M. Housley, and G. Steinbrenner, "Long-Term Stability Solution of Interconnected Power Systems," *IEEE Trans.*, Vol. PAS-95, No. 1, pp. 96-104, January/February 1976.

[15]   Discussion of reference 14 by R.P. Schulz and H.H. Happ.

[16]   CIGRE Task Force 38-02-08, "An International Survey of the Present Status and the Perspective of Long Term Dynamics in Power Systems," Final Report, January 1992.

[17]   M. Stubbe, A. Bihain, J. Deuse, and J.C. Badder, "STAG - A New Unified

Software Program for the Study of the Dynamic Behaviour of Electrical Power Systems," *IEEE Trans.*, Vol. PWRS-4, No. 1, pp. 129-138, February 1989.

[18]   P. Kundur, "A Survey of Utility Experiences with Power Plant Response during Partial Load Rejections and System Disturbances," *IEEE Trans.*, Vol. PAS-100, No. 5, pp. 2471-2475, May 1981.

[19]   T.D. Younkins and L.H. Johnson, "Steam Turbine Overspeed Control and Behaviour during System Disturbances," *IEEE Trans.*, Vol. PAS-100, No. 5, pp. 2504-2511, May 1981.

[20]   O.W. Durrant, "Boiler Response to Partial Load Rejections Resulting from System Upsets," IEEE Paper 81-JPGC-922-4, presented at 1981 Joint Power Generation Conference, St. Louis, MO. October 4-8, 1981.

[21]   A.H. Rawdon and F.A. Palacios, "Effects of Partial Load Rejection on Fossil Utility Steam Generators," IEEE Paper 81-JPGC-907-5, presented at 1981 Joint Power Generation Conference, St. Louis, MO, October 4-8, 1981.

[22]   IEEE Working Group Report, "Guidelines for Enhancing Power Plant Response to Partial Load Rejections," *IEEE Trans.*, Vol. PAS-102, No. 6, pp. 1501-1504, June 1983.

[23]   P.L. Dandeno, P. Kundur, and J.P. Bayne, "Hydraulic Unit Dynamic Performance under Normal and Islanding Conditions - Analysis and Validation," *IEEE Trans.*, Vol. PAS-97, No. 6, pp. 2134-2143, November/ December 1978.

[24]   L.M. Hovey, "Optimum Adjustment of Hydro Governors on Manitoba Hydro System," *AIEE Trans.*, Vol. 81, Part III, pp. 581-587, December 1962.

[25]   F.R. Schleif and A.B. Wilbor, "The Coordination of Hydraulic Turbine Governors for Power System Operation," *IEEE Trans.*, Vol. PAS-85, pp. 750-758, July 1966.

[26]   R.P. Schulz, "Capabilities of System Simulation Tools for Analyzing Severe Upsets," *Proceedings of International Symposium on Power System Stability*, Ames, Iowa, pp. 209-215, May 13-15, 1985.

[27]   EPRI Report EL-6627, "Long-Term Dynamic Simulation: Modelling Requirements," Final Report of Project 2473-22, Prepared by Ontario Hydro, December 1989.

[28]   EPRI Report, "Long-Term Power System Dynamics," Final Report of Project

RP 90-7, Prepared by General Electric Company, April 1974.

[29]   R.J. Frowd, J.C. Giri, and R. Podmore, "Transient Stability and Long-Term Dynamics Unified," *IEEE Trans.*, Vol. PAS-101, pp. 3841-3850, October 1982.

[30]   C.W. Gear, "Simultaneous Numerical Solution of Differential-Algebraic Equations," *IEEE Trans.*, Vol. CT-18, No. 1, pp. 89-95, January 1971.

[31]   C.W. Gear, "The Automatic Integration of Ordinary Differential Equations," *Communications of the ACM*, Vol. 14, No. 3, March 1981.

[32]   N.J. Balu, H.W. Mosteller, and R.P. Schulz, "LOTDYS Analysis of August 1973 Gulf Coast Area Power System Disturbance," *IEEE Trans.*, Vol. PAS-100, pp. 2471-2475, May 1981.

[33]   Q.B. Chou, P. Kundur, P.N. Acchione, and B. Lautsch, "Improving Nuclear Generating Station Response for Electrical Grid Islanding," *IEEE Trans.*, Vol. EC-4, No. 3, pp. 406-413, September 1989.

Chapter 17

Methods of Improving Stability

The preceding five chapters have described the physical aspects of the various categories of power system stability, and methods of their analysis. As well, key factors influencing stability problems and methods of mitigation have been covered. This chapter discusses special methods used for enhancing transient and small-signal stability. These two categories of system stability have received considerable attention since the 1960s, and methods for their improvement are further evolved than those for other categories.

For a given system, any one method of improving stability may not be adequate. The best approach is likely to be a combination of several methods judiciously chosen so as to most effectively assist in maintaining stability for different contingencies and system conditions. In applying these methods to the solution of specific stability problems, it is important to keep in mind the overall performance of the power system. Solutions to the stability problem of one category should not be effected at the expense of another category.

Many of the methods for stability enhancement described in this chapter are options normally available for economic design of the system. With proper design and application they should greatly contribute to the flexibility of system operation without compromising other aspects of system performance.

Some of the methods described are, however, somewhat "heroic" in nature and can be justified only in special situations. While improving system stability, they impose duty on some of the equipment. Their application, therefore, has to be based on a careful assessment of the benefits and costs.

17.1 TRANSIENT STABILITY ENHANCEMENT

Methods of improving transient stability try to achieve one or more of the following effects:

(a)   Reduction in the disturbing influence by minimizing the fault severity and duration.

(b)   Increase of the restoring synchronizing forces.

(c)   Reduction of the accelerating torque through control of prime-mover mechanical power.

(d)   Reduction of the accelerating torque by applying artificial load.

The following are various methods of achieving these objectives.

17.1.1 High-Speed Fault Clearing

The amount of kinetic energy gained by the generators during a fault is directly proportional to the fault duration; the quicker the fault is cleared, the less disturbance it causes.

Two-cycle breakers, together with high-speed relays and communication, are now widely used in locations where rapid fault clearing is important.

In special circumstances, even faster clearing may be desirable. Reference 1 describes the development and application of a one-cycle circuit breaker by Bonneville Power Administration (BPA). Combined with a rapid response overcurrent type sensor, which anticipates fault magnitude, nearly one-cycle total fault duration is attained. One-cycle breakers are not yet in widespread use. Reference 2 describes an ultra-high-speed relaying system for EHV lines based on travelling wave detection.

17.1.2 Reduction of Transmission System Reactance

The series inductive reactances of transmission networks are primary determinants of stability limits. The reduction of reactances of various elements of the transmission network improves transient stability by increasing postfault synchronizing power transfers. Obviously, the most direct way of achieving this is by reducing the reactances of transmission circuits, which are determined by the voltage rating, line and conductor configurations, and number of parallel circuits. The following are additional methods of reducing the network reactances:

(a)   Use of transformers with lower leakage reactances.

(b)   Series capacitor compensation of transmission lines.

Typically, the per unit transformer leakage reactance ranges between 0.1 and 0.15. For newer transformers, the minimum acceptable leakage reactance that can be achieved within the normal transformer design practices has to be established in consultation with the manufacturer [3]. In many situations, there may be a significant economic advantage in opting for a transformer with the lowest possible reactance.

Series capacitors directly offset the line series reactance. We discussed the application of series capacitors for reactive power and voltage control in Chapter 11. As shown in Section 11.2.8, the maximum power transfer capability of a transmission line may be significantly increased by the use of series capacitor banks. This directly translates into enhancement of transient stability, depending on the facilities provided for bypassing the capacitor during faults and for reinsertion after fault clearing (see Section 11.2.5). Speed of reinsertion is an important factor in maintaining transient stability [4]. Early designs of protective gaps and bypass switches limited the benefits achievable by series capacitor compensation. However, with the present trend of using nonlinear resistors of zinc oxide, the reinsertion is practically instantaneous.

As discussed in Chapter 15 (Section 15.3), one problem with series capacitor compensation is the possibility of subsynchronous resonance with the nearby turbo alternators. This aspect must be analyzed carefully and appropriate preventative measures must be taken.

Traditionally, series capacitors have been used to compensate for very long overhead lines. Recently, there has been an increasing recognition of the advantages of compensating shorter, but heavily loaded, lines by using series capacitors.

For transient stability applications, the use of *switched series capacitors* offers some advantages. Upon detection of a fault or power swing, a series capacitor bank can be switched in and then removed about 0.5 second later. Such a switched bank can be located in a substation where it can serve several lines.

In reference 5 it is shown that, for a given transient stability limit, the aggregate rating of series capacitors required is less if some are switched than if all are unswitched. The scheme with a portion of the capacitors switched reduces the angular swings of the machines, and this in turn reduces fluctuation of loads, particularly those near the electrical centre.

Protective relaying is made more complex when series compensation is used, particularly if the series capacitors are switched.

17.1.3 Regulated Shunt Compensation

Shunt compensation capable of maintaining voltages at selected points of the transmission system can improve system stability by increasing the flow of synchronizing power among interconnected generators. Synchronous condensers or static var compensators described in Chapter 11 (Sections 11.2.6 to 11.2.8) can be used for this purpose.

As illustrated in Section 11.2.8, regulated shunt compensation increases the maximum power transfer capability of a long transmission line. This clearly enhances transient stability.

17.1.4 Dynamic Braking [6-10]

Dynamic braking uses the concept of applying an artificial electrical load during a transient disturbance to increase the electrical power output of generators and thereby reduce rotor acceleration.

One form of dynamic braking involves the switching in of *shunt resistors* for about 0.5 second following a fault to reduce the accelerating power of nearby generators and remove the kinetic energy gained during the fault. BPA has used such a scheme for enhancing transient stability for faults in the Pacific northwest [8]; the brake consists of a 1400 MW, 240 kV resistor made up of 45,000 ft of 1/2 inch stainless steel wire strung on three towers. Reference 9 provides brief reports of braking resistor applications in Japan, China, Russia, and Australia.

To date, braking resistors have been applied only to hydraulic generating stations remote from load centres. Hydraulic units, in comparison to thermal units, are quite rugged; therefore, they can withstand the sudden shock from the switching in of resistors without any adverse effect on the units.

If braking resistors are applied to thermal units, the effect on shaft fatigue life must be carefully examined (see Chapter 15, Section 15.4). If the switching duty is found unacceptable, the switching in of the resistors may have to be performed in three or four steps spread over one full cycle of the lowest torsional mode [10].

The braking resistors used to date are all shunt devices. Alternatively, *series resistors* may be used to provide the braking effect. In this case, the energy dissipated is proportional to the generator current rather than to the voltage. One way of inserting the resistors in series is to install a star-connected three-phase resistor arrangement with a bypass switch in the neutral of the generator step-up transformer to reduce resistor insulation and switch requirements [10]. The resistor is inserted during a transient disturbance by opening the bypass switch.

Another form of braking resistor application that enhances system stability for unbalanced ground faults only, consists of a resistor connected permanently between the ground and the neutral of the Y-connected high-voltage winding of the generator step-up transformer [30]. Under balanced conditions no current flows through the neutral resistor. When line-to-ground or double line-to-ground faults occur, current flows through the neutral connection and the resistive losses act as a dynamic brake.

With the switched form of braking resistors, the switching times should be based on detailed simulations. If the resistors remain connected too long, there is a possibility of instability on the "backswing."

17.1.5 Reactor Switching

Shunt reactors near generators provide a simple and convenient means of improving transient stability. The reactor normally remains connected to the network. The resulting reactive load increases the generator internal voltage, and this is beneficial to stability. Following a fault, switching out the reactor further improves stability.

17.1.6 Independent-Pole Operation of Circuit Breakers

Independent-pole operation refers to the use of separate mechanisms for each phase of the circuit-beaker so that the three phases are closed and opened independently of each other. As a result, the failure of one pole will not restrict the operation of the other two poles. Although the breaker poles operate independently of each other, the relaying system is normally arranged to trip all three poles for any type of fault.

Independent-pole operation can be used advantageously at locations where the system design criteria include a three-phase fault compounded by breaker failure. Maintaining system stability for the contingency of a three-phase fault with all three poles of a primary circuit-breaker failing to open is extremely difficult. With breakers designed for independent-pole operation, a failure of all three poles is highly improbable. The use of duplicate relay systems, circuit-breaker trip coils, and operating mechanisms practically guarantees that at least two poles will open. Therefore, the independent operation of the failed breaker will reduce a three-phase fault to a single line-to-ground fault when two of the poles open. Thus, the severity of a three-phase fault with a stuck breaker is significantly reduced.

17.1.7 Single-Pole Switching

Single-pole switching uses separate operating mechanisms on each phase; for single line-to-ground faults, the relaying is designed to trip only the faulted phase, followed by fast reclosure within 0.5 to 1.5 seconds. For multiphase faults, all three phases are tripped.

During the period when one phase is open, power is transferred over the remaining two phases.

As most faults on transmission lines are of the single line-to-ground type, opening and reclosing of only the faulted phase results in an improvement in transient stability over three-phase tripping and reclosing.

Single-pole switching is particularly attractive for situations where a single major line connects two systems or where a single major line connects a generating station to the rest of the system. It may also be used on systems with multiple lines to improve system security against multiple contingency disturbances [11].

There are three potential problems that need to be considered in applying single-pole switching:

• Secondary-arc extinction.

• Fatigue duty on turbine-generator shafts and turbine blades.

• Thermal duty on nearby generators due to negative-sequence currents.

Secondary-arc extinction

When one phase of a three-phase line is open at both ends, the faulted phase is capacitively and inductively coupled to the two unfaulted phases, which are still energized (see Figure 17.1). A voltage will be induced in the isolated phase because of the capacitive coupling, and to a lesser extent because of the inductive coupling. The magnitude of the induced voltage is a direct function of the phase-to-phase and phase-to-ground capacitances. The induced voltage may sustain the fault arc for an extended period following the opening of the phase. The arc on the faulted phase conductor after it has been switched off is called the *secondary arc*. The secondary-arc current is primarily determined by the circuit voltage and the length of the line section that is switched out. It is also influenced by the fault location, any shunt reactors connected directly to the open phase, and line transposition.

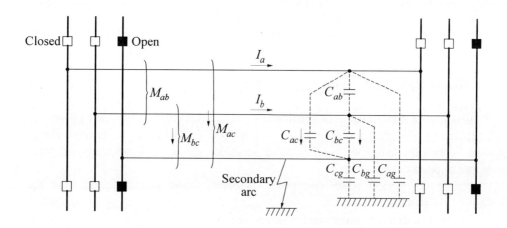

Figure 17.1 Secondary-arc current

The capacitive component of the secondary-arc current is essentially independent of the fault location and the load current. The inductive component of the current is very dependent on the load currents on the two energized phases and on the location of the fault; it is nearly zero for a fault at the middle of the line and is maximum for a fault at either end. Reference 12 gives typical values of the secondary-arc current for 765 kV and 345 kV lines: the capacitive component of the current per 100 mi (160 km) of line length is on the order of 50 A for 765 kV lines and 30 A for 345 kV lines; the maximum inductive component of the current is in the range of 10 to 15 A for 765 kV lines when the power transfer on the two energized phases corresponds to the surge impedance loading.

In reference 13, the results of arc extinction tests carried out on the American Electric Power system are compared with the published results of tests on other systems and laboratory tests. The general conclusion is that the secondary arc should self-extinguish within 500 ms if the arc current is on the order of 40 A or less on lines with shunt reactor compensation and 20 A or less on uncompensated lines.

For applications where the secondary-arc current is higher than the above values, it is necessary to neutralize the capacitive reactance. One method is to use shunt inductive reactors. The four-legged reactor scheme described in references 14 and 15 has been used in a number of single-pole switching applications for reduction of secondary-arc current [11]. In this scheme, a phase reactor is connected between each phase and a neutral point, and a neutral reactor is connected between ground and the neutral point. This method is particularly attractive if shunt reactors are required to compensate the normal line-charging current for voltage control (see Chapter 11, Section 11.2.3). By appropriate connection, these reactors can be made to serve the additional purpose of reducing secondary-arc current.

Another method of secondary-arc suppression is to use high-speed grounding switches as described in reference 16. This method is attractive for lines without shunt reactors.

Fatigue duty on turbine-generator turbine blades and shafts

As discussed in Chapter 9 (Section 9.2.3), one of the design considerations of long low-pressure turbine blades is that the frequencies of the fundamental vibratory modes must not coincide with multiples of running speed. Single-pole switching during the reclosing dead time when one phase is open can excite 120 Hz torque oscillations. These oscillations may resonate with the blade natural-vibration modes and thus damage the blades.

A single-pole switching sequence imposes successive disturbances on the turbine-generator shaft system. The initial disturbance is due to the fault itself, which for a line-to-ground fault is not very severe. When the fault is cleared by opening the faulted phase, a second disturbance is imposed. Reclosing of the opened phase, either successful or unsuccessful, imposes an additional disturbance. As discussed in Chapter 15 (Section 15.4), the shaft torsional oscillations resulting from successive impacts may amplify the transient shaft torques. Each such incident of single-pole switching operation may contribute to a small loss of shaft system fatigue life. If the turbine-generator is subjected to a large number of such incidents, their cumulative effect may cause shaft failure.

The effects of single-pole switching on the fatigue life of shafts and blades have been investigated for specific units in reference 17. The results showed no significant loss of life to either the blades or the shaft sections. In addition, experience to date has not indicated any turbine-generator blade or shaft problems resulting from single-pole switching [11]. However, it would be prudent to study each application individually before implementation.

Thermal duty on nearby generators

System unbalance during the period when one phase is open results in negative-sequence currents in the stators of nearby generators and synchronous condensers. The negative-sequence stator currents induce 120 Hz rotor currents that cause heating. Standards have been established defining generator continuous and short-time unbalanced current capability [18]. The short-time capability is expressed in terms of $I_2^2 t$, the integrated product of machine negative-sequence phase current (I_2) and time (t). The permissible $I_2^2 t$ depends on the type of machine as shown in Table 17.1.

Table 17.1 Short-time unbalanced current capability

Type of Machine	Permissible $I_2^2 t$
Salient pole generator	40
Synchronous condenser	30
Round rotor generators	
- Indirectly cooled	30
- Directly cooled: up to 800 MVA	10
- Directly cooled: 800-1600 MVA	$10 - [(MVA - 800) \times 0.00625]$

I_2 expressed in per unit of rated armature current;
t expressed in seconds.

Studies of specific applications of single-pole switching applied to lines near thermal units have shown the thermal duty to be not limiting (for example, see reference 12). However, each application should be examined individually to be sure that the negative-sequence heating duty is not excessive.

17.1.8 Steam Turbine Fast-Valving

Fast-valving (or early valving, as it is sometimes referred to) is a technique applicable to thermal units to assist in maintaining power system transient stability. It involves rapid closing and opening of steam valves in a prescribed manner to reduce the generator accelerating power following the recognition of a severe transmission system fault.

Although the principle of fast-valving as a stability aid was recognized in the early 1930s [19], the procedure has not been very widely applied for several reasons. Among them are the concerns for any possible adverse effects on the turbine and energy supply system.

Since the mid-1960s, utilities have realized that fast-valving could be an effective method of improving system stability in some situations. A number of technical papers have been published describing the basic concepts and effects of fast-valving [20-23]. Several utilities have tested and implemented fast-valving on some of their units [24-27].

Fast-valving procedures

Chapter 9 discussed steam turbine configurations for fossil-fuel and nuclear-generating units. For illustration of fast-valving applications, let us consider a fossil-fuel generating unit with a tandem-compound single reheat turbine and a nuclear unit. The turbine configurations are shown in Figure 17.2. For these units, the main inlet control valves (CVs) and the reheat intercept valves (IVs) provide a convenient means of controlling the turbine mechanical power. Depending on how these valves are used to control the steam flow, a variety of possibilities exists for the implementation of fast-valving schemes.

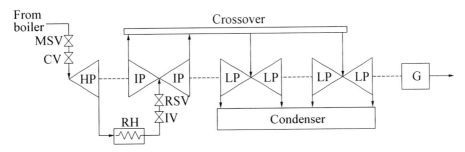

(a) Tandem-compound single reheat steam turbine of a fossil-fuel unit

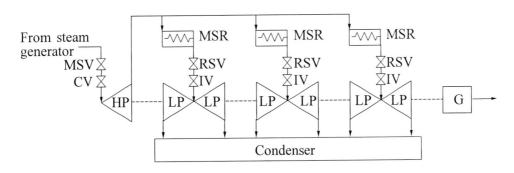

(b) Turbine of a nuclear unit

HP = high-pressure (turbine)   IP = intermediate-pressure (turbine)
LP = low-pressure (turbine)   MSV = main (inlet) safety valve
RSV = reheat safety valve   IV = intercept valve
MSR = moisture separator reheater   CV = control valve   RH = reheater

Figure 17.2 Typical steam turbine configurations

In one commonly used scheme, only the intercept valves are rapidly closed and then fully reopened after a short time delay. Since the intercept valves control nearly 70% of the total unit power, this method results in a fairly significant reduction in turbine power. A more pronounced temporary reduction in turbine power can be achieved through actuation of both control and intercept valves. The procedure of rapid closing and subsequent full opening of the valves is called *momentary fast-valving*.

In some situations where the postfault transmission system is much weaker than the prefault one, it is desirable to have the prime mover power, after being reduced rapidly, return to a level lower than the initial power. One approach is to provide for rapid closure of control and intercept valves, followed by partial opening of control valves and full opening of intercept valves. An alternative approach is to provide for rapid closure and full opening of intercept valves, coupled with partial closure of control valves. This procedure whereby, in addition to a rapid temporary reduction, a sustained reduction in turbine power is achieved is referred to as *sustained fast-valving*.

Improvement of system stability

Fast-valving assists in maintaining system stability following a severe fault by reducing the turbine mechanical power. Results of studies on the effectiveness of fast-valving in enhancing the stability of individual systems have been reported in references 27 to 32. Many of these studies also consider other means of enhancing system stability such as high-response exciters, series capacitor compensation, independent-pole switching, braking resistors, and generator tripping.

Generally, fast-valving has been found to be an effective and economical method of meeting the performance requirements of power systems whose design and operating criteria require stability to be maintained for a three-phase fault with delayed clearing because of a stuck breaker. The transient stability problem in such cases is generally associated with a swing mode local to a generating station and having a period of about 0.5 to 1.2 seconds.

Although not very widely recognized, fast-valving could be very effective in situations where instability occurs through a slow interarea swing having a period of about 2.0 to 4.0 seconds. The problem is usually caused by generators in one area, because of a fault in the area, advancing with respect to the generators in other interconnected areas. This causes heavy power flows through weak ties connecting the areas, which in turn leads to separation of the areas. Fast-valving of one or more units within the area where the fault occurs minimizes the accelerating power and hence reduces the transient power swings through the weak ties. Fast-valving is particularly effective in such situations, since the periods of the power swings are long; this, therefore, allows more time for the reduction in power to be achieved.

The cost of implementing fast-valving is usually small. However, in view of the concerns about possible adverse effects on the turbine and boiler/steam generator, fast-valving should be used only in situations where other less "heroic" measures are not able to maintain system stability. In such situations, often the only other effective means of maintaining system stability is to use generator tripping. Compared to generator tripping, fast-valving has the advantage that the unit remains connected to the system. As a result, the total system inertia is not reduced and full or partial power output is restored within a few seconds. In addition, the resulting stress on the prime mover is believed to be significantly less severe. However, fast-valving is not as effective in aiding stability as generator tripping.

The fast-valving logic can take several forms, depending on the supplier and the intended use of fast-valving. Generally, such logic will contain two main circuits: one for generating the valve control sequence and the other for generating an unloading signal if fast-valving is to be followed by a reduced generator output. The valve control sequence will generate a signal to close the valves at a preset rate, hold them closed for an adjustable period of time and reopen them at a preset rate. Figure 17.3 shows a typical valve closing- and opening-sequence. The allowable valve actuation times are influenced by equipment as well as system considerations.

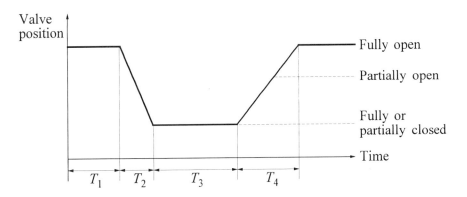

T_1 = delay between the time of initiation and the time
          when the valve begins to close
T_2 = valve-closing time
T_3 = time during which the valve remains closed
T_4 = valve-opening time

Figure 17.3 Typical valve closing- and opening- sequence

The ability of steam valves to rapidly close and reopen depends on the type of governor system used. The electrohydraulic turbine governor system that uses solid-state electronics and a high-pressure hydraulic actuator system is capable of rapid control. A common practice, particularly of North American manufacturers, is to use fast-acting hydraulic valves arranged to dump oil from the spring-loaded actuating cylinders of the intercept and control valves to facilitate rapid valve closure.

This permits complete valve closure in 0.08 to 0.4 second. Reopening after a fast closure is inherently delayed for approximately 0.3 to 1.0 second to allow for restoration of oil to the hydraulic cylinder. The valve-opening time depends on the size of the hydraulic operating cylinder but is usually in the range of 3 to 10 seconds. For some applications, faster valve-reopening may be needed for fast-valving than is used in normal electrohydraulic governing operations. This can be achieved with the aid of accumulators on the hydraulic system [22,33].

In contrast, some European manufacturers tend to use large servo valves with facilities for fast operation; this achieves reopening times of less than 1 second.

Fast-valving can also be applied to units with mechanical-hydraulic turbine governors, but it is less flexible and somewhat more difficult to implement.

For fossil-fuel-fired units, reheater metal protection considerations dictate the intercept valve-opening time. The problem is one of reheater tube heating when the steam flow is shut off and subsequently restored. The time within which reheater steam flow should be reestablished is typically on the order of 10 to 12 seconds.

From a system stability viewpoint, the valves should be closed as rapidly as possible. With momentary fast-valving, it may also be desirable to initiate valve reopening slightly before the first peak of rotor angle is reached. Slow reopening increases the backswing, and this could lead to second-swing instability [22,29]. If the delay in valve opening is significant, it could also accentuate the swinging of generators in one area against the rest of the system, and this could lead to instability [22]. For any specific system, acceptable valve actuation times should therefore be based on detailed stability studies.

Fast-valving action can be initiated by power load unbalance relays, acceleration detectors, or relays that recognize severe transmission faults [6]. References 23 and 33 describe typical fast-valving initiating logics used in the standard circuitry provided by North American turbine-generator manufacturers. The logic is usually a simple extension of the scheme used to limit overspeed and depends on the comparison of mechanical input power and electrical output power. The former is measured by monitoring the reheat steam bowl pressure. In order to avoid unnecessary initiation of fast-valving action, it is usually necessary to introduce additional logic to discriminate against faults that do not require fast-valving. In most cases, since rapid closing of the valves is critical, it is important to ensure that the delay introduced by the selectivity of the logic is minimal.

The time delay introduced by the feedback control logic, such as detection based on power-load imbalances, is on the order of 0.1 second and this may be unacceptable in some cases. A feedforward control logic derived directly from protective relay systems could reduce the initiating time to one or two cycles [22].

The effectiveness of fast-valving in improving transient stability is demonstrated in Figure 17.4. The figure shows the responses with and without fast-valving of a fossil-fuel-fired power plant consisting of two 500 MW units. The fast-valving sequence assumed closing and full reopening of intercept valves only, with T_1=0.1 s, T_2=0.25 s, T_3=0.1 s, and T_4=0.85 s. The units have a thyristor excitation system with power system stabilizers. The disturbance considered is a three-phase fault on a major transmission line close to the station, cleared by primary relaying in 60 ms. Without fast-valving, the generating units exhibit a dominant interarea swing having a period of about 4.5 seconds. Consequently, the fast-valving sequence is very effective in limiting the peak of the first swing in rotor angle. Here, very fast valve closing and opening times have been assumed. With slower valve-opening times, the fast-valving sequence, as shown in reference 29, results in second-swing instability.

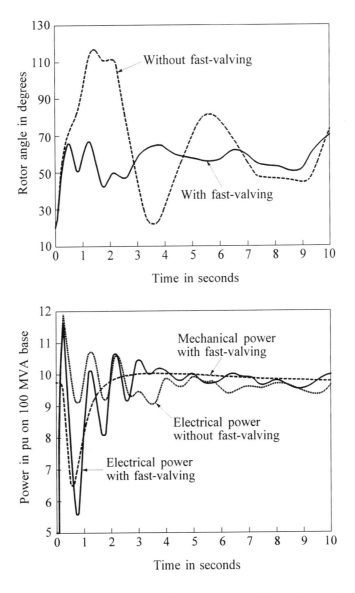

Figure 17.4 Effect of fast-valving on the stability of a fossil-fuel-fired station exhibiting a slow interarea swing

Plant considerations

Fast-valving imposes a relatively severe transient on the turbine and steam generator system. There are several potential problem areas that must be considered in the application of fast-valving. The following is a brief discussion of the equipment considerations that must be addressed.

Fossil-fuel-fired unit steam generator considerations:

Closing the intercept valves results in a temporary increase in reheater pressure that could cause the reheat safety relief valves to lift. The safety valves may not reseat properly following this operation. Also, any noise impact on the environment resulting from the escaping steam must be addressed. Closing both the control and intercept valves reduces the possibility of a lifting of the reheat relief valves but this merely transfers the problem to the main relief valves.

As noted earlier, reheater tube protection is a limiting factor that determines how soon the steam flow should be restored to the reheater. For a typical reheater design, the limiting time is on the order of 11 seconds. If power-operated relief valves are utilized on the reheater outlet and automatically reopened, the problem is minimized in any occurance where only the intercept valves are closed.

Sustained fast-valving appears to the boiler as a partial load rejection. Without a turbine or superheater bypass system, the sudden closure of turbine inlet valves requires venting of the throttled steam to the atmosphere. If this is done by means of spring-loaded safety valves, it may be followed by an outage required to repair these valves. Therefore, power-assisted pressure relief valves and/or a turbine bypass should be used. Under optimum conditions, however, a once-through boiler can be run back quickly enough to withstand sustained fast-valving without lifting the safety valves; this was demonstrated by TVA tests on the Cumberland units [26].

With a drum-type boiler, if a bypass system is not used, sustained fast-valving causes a reduction of drum level. This could result in tripping of the unit by low drum-level protection that is provided to prevent overheating of the furnace firewalls. Improvements in control of feedwater, drum level and rate of reduction of fuel input to the furnace and the use of power-operated relief valves should enhance the ability of drum-type boilers to cope with sustained fast-valving.

Nuclear unit steam generator considerations:

Compared to fossil-fuel-fired units, nuclear units have smaller reheater time constants, and hence closing of only the intercept valves results in a faster rise in reheater pressure. Reheat safety valve lifting is therefore more likely in a nuclear unit. In addition, the probability of a damaged safety valve seat is higher because less superheat in the steam can lead to water-cutting during lifting of the safety valves. The reheater pressure rise can be of greater concern if large bursting disks, rather than safety relief valves, have been used in the reheat system. The bursting of a disk essentially takes the unit out of service and adds to the system upset.

The closing of both the control and intercept valves has the same effect as for the fossil-fuel units; that is, it passes the problem on to the main safety-relief valves. The closing of the control valves results in a rise in steam generator pressure, and steam must be bypassed either to the condenser or to the atmosphere. The boiler pressure-control system uses the turbine or the bypass system to control the boiler pressure.

Turbine considerations:

The integrity of control-valve (CV) seat and stem is of some concern. The dumping of the actuator fluid results in an accelerated closing of the valve that may lead to valve-bouncing or stem-bending. On existing units, it may be necessary to add more solenoid dump valves to achieve the fast-closing time for the CVs.

The combined action of the control and intercept valves necessitates a review of the extraction steam system to determine if there is a likelihood of actuating any start-up bypass valve from the HP turbine to the condenser. The sustained fast-valving cycle may therefore require the addition of a time delay in the control logic for this bypass valve.

The decay of pressure downstream of the intercept valves means that the valve actuators must be capable of reopening the valves against a high-differential pressure. As in the case of the control valves, the control system must have fast-acting dump solenoids and an adequate-control fluid supply for resetting the valves without initiating a low-control oil pressure trip.

The intercept valves are larger than the control valves; thus the effect of higher impact loading on the devices used for stopping the valves at the end of their stroke must be evaluated. Avoiding a full closure would be preferable.

In the case of a nuclear unit with a moisture separator reheater, higher-than-normal steam velocities are created in part of the tubing; the result is higher stress on the tubes. The tubing response to this excitation must be evaluated.

The rapid decay of the pressure in the LP turbine raises the possibility of steam flashing in the feed-water heaters and hence a water induction incident that would result in the feed train being closed by the protection system.

The turbine-generator rotors must not suffer a significant loss of life from the torsional stresses imposed on them by a fast-valving event. Results of simulations of specific applications have shown that the resulting shaft stresses are acceptable [24].

Reference 34 describes fast-valving tests on two units at the Pickering-B nuclear station in Ontario. The tests revealed no obvious adverse effects on equipment. From the plant viewpoint, fast-valving appeared to be a nonevent. From the system viewpoint, the fast-valving sequence was found to be effective in causing a rapid and substantial reduction of turbine power.

17.1.9 Generator Tripping

The selective tripping of generating units for severe transmission system contingencies has been used as a method of improving system stability for many years. The rejection of generation at an appropriate location in the system reduces power to be transferred over the critical transmission interfaces. Since generating units can be tripped rapidly, this is a very effective means of improving transient stability. (See Chapter 13, Section 13.6 for illustration.)

Historically, the practice of generator tripping as a stability aid was confined to hydro plants. Such plants are usually remote from load centres and there is little risk of damage to the unit from a sudden trip. Since the 1970s, this practice has been gradually extended to fossil-fuel-fired and nuclear units as a means of solving severe stability problems.

The scheme used for detection of system conditions requiring unit tripping is often an extension of trip circuits from various remote and local line protections. If the faulted line is restored within minutes, the rejected units are brought back on-line quickly.

Unless special facilities are provided, the rejected unit goes through a standard shut-down and start-up cycle; consequently, full power may not be available for several hours. A practice used by many utilities is to design thermal units so that, after tripping, they continue to run, supplying unit auxiliaries. This permits the units to be resynchronized to the system and restored to full load in about 15 to 30 minutes [35].

The sequence of events that follow generator tripping is as follows: the boiler is tripped; turbine controls limit overspeed (the turbine is not tripped) and the speed is returned to near-rated speed; the unit then operates either at no load or supplies the unit auxiliary load; the boiler is purged and refired in preparation for reloading; the unit is resynchronized to the power system and reloaded at a predetermined rate.

Evidently, the tripping of a unit subjects it to sudden changes in mechanical and electrical loading, with the associated impact on the generator, prime mover, and energy supply system. While thermal units and their controls are designed to withstand such shocks, there is some possibility that controls will not function correctly. Also, thermal units are not designed for frequent full-load rejection; significantly increasing the number of full-load rejections may increase unit maintenance requirements and reduce unit availability. Therefore, this type of control measure for improving system stability should not be used indiscriminately.

The following are the major turbine-generator concerns:

(a)   The overspeed resulting from tripping the generator

(b)   Thermal stresses caused by the rapid load changes

(c)   High levels of shaft torques as the result of successive disturbances

Overspeed considerations

The turbine controls should be capable of full-load rejection without a turbine trip. As discussed in Chapter 9 (Section 9.2.2), the controls of steam turbines are generally designed to limit the overspeed following a full-load rejection to about 1% below the overspeed trip level of 110 to 115% of rated speed. Prevention of turbine trip is essential if the unit has to carry its auxiliary load until it is resynchronized to the power system.

Following the initial overspeed, the unit will reach a steady-state speed determined by the governor droop characteristic. The standard 5% regulation will result in a steady-state speed slightly below 105% of rated speed (corresponding to a frequency of 63 Hz). Continuous operation at this frequency is forbidden as it may cause damage to the low-pressure turbine blades (see Section 9.2.3). On older units with mechanical-hydraulic control (MHC), the operator has to adjust the speed-load changer to return the unit to rated speed. On modern units with electrohydraulic control (EHC), this speed reduction to rated speed is automatic.

If frequent generator tripping is required to maintain system stability, there is a greater risk of turbine-generator runaway because of equipment malfunction. To ensure safe and reliable operation of speed controls, greater emphasis should be placed on preventive maintenance and on periodic testing of the turbine valves and protective system.

Thermal stress considerations

Generator tripping imposes a high thermal stress on the turbine because of the variation of unit output over a wide range. Prior to the disturbance, the unit output is relatively high and the turbine metal temperatures in the critical parts would be near rated conditions. When the unit is tripped, the unit operates for about 20 minutes at very light load or no load; this results in rapid cooling of the turbine metals, followed by resynchronization and reloading of the unit. The turbine metals continue to cool during the initial reloading until the steam temperatures match turbine metal temperatures in the critical areas. An increase in load beyond this point begins to heat the metals, the rate of heating being a function of the loading rate and the degree of cooling that occurred during light-load operation. The parts of the turbine that are critical from thermal stress considerations are high-temperature regions of the HP turbine section.

The thermal stresses resulting from generator tripping can be limited to acceptable levels by appropriate operating procedures developed in consultation with the turbine-generator manufacturer.

Shaft fatigue-life considerations

When a generating unit in a multiunit power plant is tripped, all units in the plant experience successive disturbances that can result in high levels of shaft torques.

As discussed in Chapter 15 (Section 15.4), the torsional oscillations caused by successive impacts may reinforce the initial oscillations. Detailed simulations should be carried out to ensure that there are no adverse effects on the shaft fatigue-life from the compounding effects of a network fault, clearing of the fault, and unit tripping. For the unit that is tripped, the electric torque is suddenly reduced to a very low value. The remaining units initially experience a sudden proportionate increase in the electric torque. If two or more units are tripped, it is very unlikely that they will be tripped at precisely the same instant. Therefore, a multiunit tripping scheme may impose a more severe duty.

References 36 and 37 describe a generation and load rejection scheme implemented by Ontario Hydro at the Bruce Nuclear Power Development Complex having a total generation capacity of about 6000 MW. The scheme, implemented as a stopgap measure because of delays in acquiring transmission right-of-way provided for rejection of up to four 750 MW units and 1500 MW of customer load, depending on total generation at the complex, the number of transmission circuits in service, and the type of fault. The decision to implement the rejection scheme was based on a detailed evaluation of the benefits and impacts. The following factors were given consideration in the selection of the scheme:

(a)   The cost of displacing locked-in nuclear energy with fossil-fuelled generation

(b)   The expected frequency of transmission system faults and the estimated number of units trips

(c)   The impact on generating units with regard to risk of turbine-generator runaway, increased forced and maintenance outage rates, increased cost because of additional wear and tear, and cumulative loss of component life, including shaft fatigue-life

17.1.10 Controlled System Separation and Load Shedding

Controlled separation may be used to prevent a major disturbance in one part of an interconnected system from propagating into the rest of the system and causing a severe system breakup. The initiating disturbance may be the loss of a major transmission line (ac or dc) carrying a large amount of power or loss of a significant amount of generation. The incipient instability in such cases is usually characterized by sudden changes in tie line power. If this is detected in time and the information is used to initiate corrective actions, severe system upsets can be averted.

The impending system instability is detected by monitoring one or more of the following system quantities: sudden change in power flow through specific transmission circuits, change of bus voltage angle, rate of power change, and circuit-breaker auxiliary contacts.

Upon detection of the impeding instability, controlled system separation is initiated by opening the appropriate tie lines before cascading outages can occur. In some instances it may be necessary to shed selected loads to balance generation and load in the separated systems.

In other instances, only load shedding can be initiated so that transient stability is maintained without resorting to system separation.

An example of relaying used for controlled separation is the $\Delta P / \Delta \theta$ relay on the tie lines between Ontario Hydro and Manitoba Hydro. The relay monitors sudden changes in power flow (ΔP) over the tie lines and changes in phase angle ($\Delta \theta$) of the bus voltage at the Ontario end of the tie lines. The $\Delta \theta$ element supervises the ΔP element. The settings of the relay are chosen so that the ties are tripped for disturbances in Manitoba that cause instability in Ontario. For some system conditions, tie line tripping is accompanied by load shedding to ensure satisfactory post-separation system conditions.

Reference 38 describes a local independent detection scheme developed to isolate the U.S. southwest power system from the U.S. northwest system for outages of the Pacific ac intertie. The scheme uses the power rate of change at two locations for detection of conditions requiring northwest/southwest separation. This involves separation of the Utah and Colorado power systems from the Arizona system by opening five ties, and separation of the Nevada system from the California system north of San Francisco by opening three ties.

Reference 39 describes a load-shedding scheme used by the U.S. Northwest Power Pool to maintain transient stability of the Western System Coordinating Council (WSCC) systems.

17.1.11 High-Speed Excitation Systems

Significant improvements in transient stability can be achieved through rapid temporary increase of generator excitation. The increase of generator field voltage during a transient disturbance has the effect of increasing the internal voltage of the machine; this in turn increases the synchronizing power.

During a transient disturbance following a transmission system fault and clearing of the fault by isolating the faulted element, the generator terminal voltage is low. The automatic voltage regulator responds to this condition by increasing the generator field voltage, and this has a beneficial effect on the transient stability. The effectiveness of this type of control depends on the ability of the excitation system to quickly increase the field voltage to the highest possible value. High-initial-response excitation systems with high-ceiling voltages are most effective in this regard. Ceiling voltages are, however, limited by generator rotor insulation considerations. For thermal units, the ceiling voltages are limited to about 2.5 to 3.0 times the rated-load field voltage.

Fast excitation response to terminal voltage variations required for improvement of transient stability often leads to degrading the damping of local plant mode oscillations (see Chapter 12, Sections 12.4 and 12.5). Supplementary excitation control, commonly referred to as *power system stabilizer* (PSS), provides a convenient means of damping system oscillations that enables the high-response excitation system to be used. The use of high-initial-response excitation systems supplemented with PSS is by far the most effective and economical method of enhancing the overall system stability [40]. As an example of such a system, Figure 17.5 shows a general block diagram representation of a thyristor exciter with an automatic voltage regulator (AVR), a PSS, and a terminal voltage limiter. The function of the limiter is to prevent the terminal voltage from exceeding a set level of typically 1.15 pu. The effectiveness of the excitation system in improving the overall system stability depends on proper control design and tuning procedures. This is covered in Section 17.2.

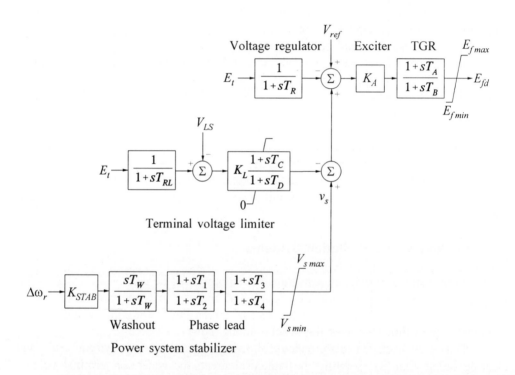

Figure 17.5 Block diagram of thyristor excitation system with PSS

The influence of excitation system response on transient stability is illustrated in Figure 17.6. The figure compares the responses of the fossil-fuel-fired plant considered in Section 17.1.8 with two alternative forms of excitation system: (a) an ac exciter with diode rectifiers, having a response ratio of 2.0, and (b) a bus-fed thyristor exciter with a PSS. The disturbance considered is a three-phase fault on a major transmission line near the power plant, cleared in 60 ms. The system is unstable with the rotating exciter and is stable with the high-initial-response thyristor exciter. The critical fault-clearing time with the ac exciter is 47.5 ms, and 62.5 ms with the thyristor exciter.

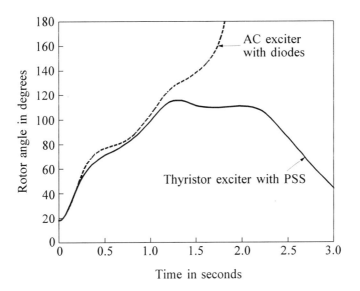

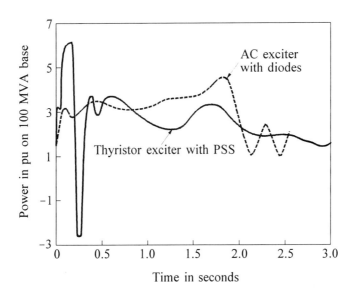

Figure 17.6 Comparison of transient stability with
an ac exciter and a bus-fed thyristor exciter

17.1.12 Discontinuous Excitation Control [40]

A properly applied power system stabilizer provides damping to both local and interarea modes of oscillations. Under large-signal or transient conditions, the stabilizer generally contributes positively to first-swing stability. In the presence of both local and interarea swing modes, however, the normal stabilizer response can allow the excitation to be reduced after the peak of the first local-mode swing and before the highest composite peak of the swing is reached. Additional improvements in transient stability can be realized by keeping the excitation at ceiling, within terminal voltage constraints, until the highest point of the swing is reached.

A discontinuous excitation control scheme referred to as *transient stability excitation control* (TSEC) has been developed by Ontario Hydro to achieve the above [40]. This control improves transient stability by controlling the generator excitation so that the terminal voltage is maintained near the maximum permissible value of about 1.12 to 1.15 pu over the entire positive swing of the rotor angle. The scheme uses a signal proportional to the change in angle of the generator rotor, in addition to the terminal voltage and rotor speed signals. However, the angle signal is used only during the transient period of about 2 seconds following a severe disturbance, since it results in oscillatory instability if used continuously. The angle signal prevents premature reversal of field voltage and hence maintains the terminal voltage at a high level during the positive swing of the rotor angle. Excessive terminal voltage is prevented by the terminal voltage limiter.

Figure 17.7 shows a block diagram of the discontinuous excitation control scheme. The TSEC circuitry is integrated with the PSS circuitry. The speed signal ($\Delta\omega_r$) provides continuous control to maintain small-signal stability under normal operation. The angle signal is derived by integrating the speed signal. The TSEC block shown in the figure is an integrator with a washout. The value of T_{ANG} is such

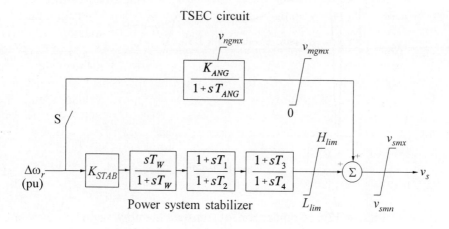

Figure 17.7 Block diagram of TSEC scheme

that, at the frequency range of interest, the output is proportional to the angular deviation. The TSEC is in effect a closed-loop control based on local measurements. The relay contact (S) is closed if the terminal voltage drop exceeds a preset value, field voltage is at positive ceiling, and the speed increase is above a preset value. The relay contact is opened when either the speed drops below a threshold value or the exciter comes out of saturation; the output of the TSEC block then decays exponentially with a time constant T_{ANG}.

The discontinuous excitation control is most effective in improving the transient stability of generating units exhibiting low-frequency interarea swings.

When TSEC is applied to several generating stations in an area, the system voltage level in the entire area is raised. This increases the power consumed by the voltage-dependent loads in the area, thereby contributing to further improvement in transient stability.

To illustrate the performance of TSEC, we once again consider the fossil-fuel-fired power plant used in Sections 17.1.8 and 17.1.11 to demonstrate the effectiveness of fast-valving and high-initial-response excitation system. As seen from the rotor angle plots of Figure 17.6, the generators at this plant exhibit a dominant low-frequency interarea swing. Figure 17.8 shows the responses of the generators with and without TSEC. Clearly, the system transient stability is very significantly improved by TSEC. The critical clearing time, which is 62.5 ms with a thyristor exciter and PSS, is increased to 117.5 ms by the discontinuous excitation control.

Figure 17.9 shows an extended simulation of the system with and without TSEC. For purposes of comparison, the response obtained with fast-valving of intercept valves as described in Section 11.1.8 is also shown in Figure 17.9. The discontinuous excitation control is seen to be as effective as fast-valving for this particular application.

Compared to other methods of improving system stability, such as fast-valving and generator-tripping, TSEC imposes very little duty on the turbine-generator shaft and steam supply system. However, parts of the power system will experience a rise in voltage of up to 15% (depending on the setting of the terminal voltage limiter) for 1 to 2 seconds. This excitation control scheme must be coordinated with other overvoltage protection and control functions. It must also be coordinated with transformer differential protection to ensure that the increased magnetizing current resulting from the elevated voltage level does not cause this protection to operate.

The discontinuous control described above for transient boosting of excitation uses local intelligence to detect a severe system disturbance condition. In some applications, it may be necessary to initiate the transient excitation boosting by using remote telemetered signals. Reference 41 describes such an application.

17.1.13 Control of HVDC Transmission Links

As discussed in Chapter 10 (Section 10.4), an HVDC transmission link is highly controllable. It is possible to take advantage of this unique characteristic of the HVDC link to augment the transient stability of the ac system.

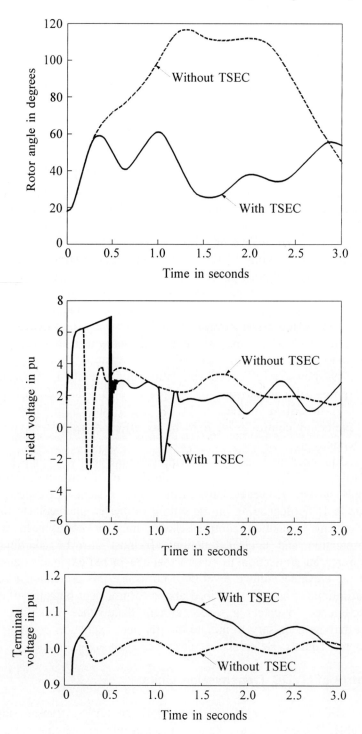

Figure 17.8 Effect of TSEC on transient stability

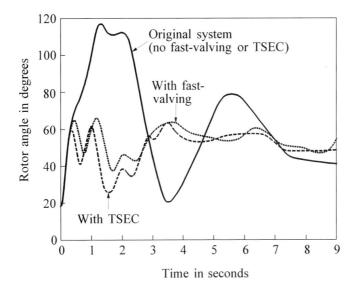

Figure 17.9 Effects of TSEC and fast-valving on extended system response

During a transient disturbance, the dc power can be ramped down rapidly to reduce generation/load unbalance of the ac system on both sides. In some situations, it may be necessary to ramp up the dc power to assist system stability by taking advantage of the short-term overload capability of the HVDC system. From the viewpoint of the ac system performance, the rapid control of dc power has the same beneficial effect as generator/load tripping. References 42 and 43 provide descriptions of special controls used in a number of HVDC installations for transient stability augmentation.

Transient stability augmentation can also be achieved by controlling the HVDC converters so as to provide reactive power and voltage support [44].

17.2 SMALL-SIGNAL STABILITY ENHANCEMENT

The problem of small-signal stability, as described in Chapter 12, is usually one of insufficient damping of system oscillations. The use of power system stabilizers to control generator excitation systems is the most cost-effective method of enhancing the small-signal stability of power systems. Additionally, supplemental stabilizing signals may be used to modulate HVDC converter controls and static var compensator controls to enhance damping of system oscillations.

The controls used for small-signal stability enhancement should perform satisfactorily under severe transient disturbances. Therefore, while the controls are designed using linear techniques, their overall performance is assessed by considering small- as well as large-signal responses.

17.2.1 Power System Stabilizers

Section 12.5 of Chapter 12 developed the theoretical basis for a PSS. The function of a PSS is to add damping to the generator rotor oscillations. This is achieved by modulating the generator excitation so as to develop a component of electrical torque in phase with rotor speed deviations. Shaft speed, integral of power and terminal frequency are among the commonly used input signals to the PSS.

Alternative types of PSS

(a) Stabilizer based on shaft speed signal (delta-omega):

PSS based on shaft speed signal has been used successfully on hydraulic units since the mid-1960s. Reference 45 describes a technique developed to derive a stabilizing signal from measurement of shaft speed of a hydraulic unit. Among the important considerations in the design of equipment for the measurement of speed deviation is the minimization of noise caused by shaft run-out and other causes [45,46]. The allowable level of noise is dependent on its frequency. For noise frequencies below 5 Hz, the level must be less than 0.02%, since significant changes in terminal voltage can be produced by low-frequency changes in the field voltage. A frequency corresponding to shaft rotational speed and resulting from shaft run-out is generally the most important noise component in this range. Lateral movements of the shaft of 0.075 cm are typical at points close to the generator guide bearing. Such low-frequency noise cannot be removed by conventional electric filters; its elimination must be inherent to the method of measuring the speed signal. This is achieved by summing the outputs from several pick-ups around the shaft. At gate positions below 70%, the stabilizing signal is disconnected automatically by an auxiliary (pallet) switch to prevent excessive modulation of the field voltage by vibrations generated in the turbine at partial gate openings.

The application of shaft speed-based stabilizers to thermal units requires a careful consideration of the effects on torsional oscillations. The stabilizer, while damping the rotor oscillations, can cause instability of the torsional modes. This problem is discussed in Chapter 15 (Section 15.2.1). One approach successfully used to circumvent the problem is to sense the speed at a location on the shaft near the nodes of the critical torsional modes [47,48]. In addition, an electronic filter is used in the stabilizing path to attenuate the torsional components.

While stabilizers based on direct measurement of shaft speed have been used on many thermal units, this type of stabilizer has several limitations. The primary disadvantage is the need to use a torsional filter. In attenuating the torsional components of the stabilizing signal, the filter also introduces a phase lag at lower frequencies. As demonstrated in Example 15.2 of Chapter 15, this has a destabilizing effect on the "exciter mode", thus imposing a maximum limit on the allowable stabilizer gain. In many cases, this is too restrictive and limits the overall effectiveness of the stabilizer in damping system oscillations. In addition, the stabilizer has to be custom-designed for each type of generating unit depending on its torsional characteristics. The delta-*P*-omega stabilizer described next was developed to overcome these limitations.

(b) Delta-P-omega stabilizer [40,49,50]:

The principle of this stabilizer is illustrated by the following equation that shows how a signal proportional to rotor speed deviation can be derived from the accelerating power:

$$\Delta\omega_{eq} = \frac{1}{M}\int(\Delta P_m - \Delta P_e)\,dt \qquad (17.1)$$

where

      M = inertia constant, $2H$
      ΔP_m = change in mechanical power input
      ΔP_e = change in electric power output
      $\Delta\omega_{eq}$ = derived or equivalent speed deviation

The objective is to derive the equivalent speed signal $\Delta\omega_{eq}$ so that it does not contain torsional modes. As shown in Chapter 15 (Example 15.2), torsional components are inherently attenuated in the integral of ΔP_e signal. The problem is to measure the integral of ΔP_m free of torsional modes.

In many applications, the ΔP_m component is neglected. This is satisfactory, except when changing load on the unit and other system conditions when the mechanical power changes. Under such conditions, a spurious stabilizer output is produced if ΔP_e alone is used as the stabilizing signal. This in turn results in transient oscillations in voltage and reactive power.

The integral of mechanical power is related to shaft speed and electrical power as follows:

$$\int\Delta P_m\,dt = M\Delta\omega + \int\Delta P_e\,dt \qquad (17.2)$$

The delta-*P*-omega stabilizer makes use of the above relationship to simulate a signal proportional to the integral of mechanical power change by adding signals proportional to shaft-speed change and integral of electrical power change. This signal will contain torsional oscillations unless a filter is used. Because mechanical power changes are relatively slow even for fast-valve movements, the derived integral of the mechanical power signal can be conditioned with a simple low-pass filter to remove torsional frequencies.

The overall transfer function for deriving the equivalent rotor speed deviation signal from shaft speed and electrical power measurements is given by

$$\Delta\omega_{eq}(s) = -\frac{\Delta P_e(s)}{Ms} + G(s)\left[\frac{\Delta P_e(s)}{Ms} + \Delta\omega(s)\right] \qquad (17.3)$$

where $G(s)$ is the transfer function of the torsional filter. A realization of this is shown in the block diagram of Figure 17.10.

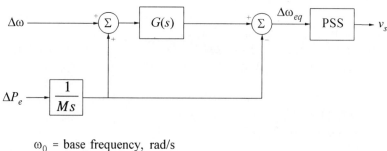

ω_0 = base frequency, rad/s
M = inertia coefficient = $2H$
s = d/dt

Figure 17.10 Block diagram realization of delta-P-omega stabilizer

The delta-P-omega stabilizer has two major advantages over the delta-omega stabilizer:

1. The ΔP_e signal has a high degree of torsional attenuation, and hence there is generally no need for a torsional filter in the main stabilizing path. This eliminates the exciter mode stability problem, thereby permitting a higher stabilizer gain that results in better damping of system oscillations.

2. An end-of-shaft speed-sensing arrangement with a simple torsional filter can be used with electrical power to derive the mechanical power signal. This allows the use of a standard design for all units irrespective of their torsional characteristics.

(c) Frequency-based stabilizers:

Terminal frequency has been used as the stabilizing signal for several PSS applications. Normally, the terminal frequency signal is used directly as the stabilizer input signal. In some cases, terminal voltage and current are used to derive the frequency of a voltage behind a simulated machine reactance so as to approximate the machine rotor speed.

In these systems, as in the case of speed-based stabilizers, care must be taken to filter torsional modes when they are used on steam turbine units.

The sensitivity of the frequency signal to rotor oscillations increases as the external transmission system becomes weaker, tending to offset the reduction in gain from stabilizer output to electrical torque that results from a weaker transmission system [51]. Hence, the gain of a frequency-based stabilizer may be adjusted to obtain the best possible performance under weak ac transmission system conditions where the contribution of the stabilizer is required most.

The frequency signal is more sensitive to modes of oscillation between large areas than to modes involving only individual units, including those between units within a power plant. Thus it seems possible to obtain greater damping contributions to interarea modes of oscillation than would be obtainable with the speed input signal.

The frequency-based stabilizer, however, suffers from several shortcomings:

1.   During a rapid transient, the terminal frequency signal will undergo a sudden phase shift. This results in a spike in the field voltage that is reflected in the generator output quantities.

2.   The frequency signal often contains power system noise caused by large industrial loads such as arc furnaces. In many cases this has prevented the use of frequency as an input signal.

3.   Torsional filtering is required. Hence, a frequency-based stabilizer has the same basic limitation as the delta-omega stabilizer.

(d) Digital stabilizer:

Digital versions of some of the above stabilizers have been developed and are now commercially available [52].

Manufacturers are producing excitation systems with complete digital circuitry. In this environment, if the appropriate inputs are provided, the stabilizer becomes just another program in the excitation control processor.

Excitation control design [40,53]

The parameters of the PSS and other elements of the excitation system are chosen to enhance the overall system stability. Specifically, the following are the objectives of excitation control design:

•   Maximization of the damping of the local plant mode as well as interarea mode oscillations without compromising the stability of other modes.

•   Enhancement of system transient stability.

- Prevention of adverse affects on system performance during major system upsets that cause large frequency excursions.

- Minimization of the consequences of excitation system malfunction because of component failures.

The procedure used for meeting the above objectives is illustrated by considering a thyristor excitation system whose block diagram is shown in Figure 17.5. The input to the PSS may be either the shaft speed deviation ($\Delta\omega_r$) or the equivalent rotor speed deviation ($\Delta\omega_{eq}$). The terminal voltage transducer circuitry is represented by time constants necessary for filtering the rectified terminal voltage waveform. These can usually be reduced to a single time constant (T_R) in the range of 0.01 to 0.02 seconds. Other time constants through to the exciter output, including any associated with the exciter itself, are negligible, and the main path can be represented simply by the gain K_A.

The following is a description of the considerations and procedures used for selection of the various parameters.

Exciter gain:

A high value of K_A is desirable from the viewpoint of transient stability. A suitable value of K_A is about 200 with no *transient gain reduction* (TGR) [40,53].

A common industry practice is to reduce the gain of the exciter at high frequencies by use of TGR [51,52]. Typical values of T_A and T_B used for this purpose are one and ten seconds, respectively. On excitation systems with large values of T_R, TGR is required for satisfactory operation of the generating unit on open-circuit. With T_R on the order of 0.02 seconds, TGR is unnecessary for stable open-circuit operation (see closure of reference 53).

The need to use TGR should be based on a careful assessment of the overall system dynamic performance and on how TGR affects selection of other parameters of the excitation system. References 40 and 53 discuss in detail the impact of using TGR for specific applications.

Phase-lead compensation:

To damp rotor oscillations, the PSS must produce a component of electrical torque in phase with rotor speed deviation. This requires phase-lead circuits to be used to compensate for the lag between the exciter input (i.e., PSS output) and the resulting electrical torque (see Section 12.5). Two first-order phase-compensation blocks are shown in Figure 17.5. If the degree of phase compensation required is small, a single first-order block may be used. For hydraulic generators with low values of d-axis open-circuit time constant (T'_{d0}), the phase compensation required is small and phase-lead circuitry may in fact not be utilized.

The PSS is often required to enhance damping of either a local plant mode or an interarea mode of oscillation. While this mode receives special attention, the phase compensation should be designed so that the PSS contributes to damping over a wide range of frequency covering both interarea and local modes of oscillation.

The first step in determining the phase compensation is to compute the frequency response between the exciter input and the generator electrical torque, using a tool such as the MASS program which was described in Chapter 12. In computing this response, however, the generator speed and rotor angle should remain constant. This is because when the excitation of a generator is modulated, the resulting change in electrical torque causes variations in rotor speed and angle that in turn affect the electrical torque. As we are interested only in the phase characteristic between exciter input and electrical torque, the feedback effect through rotor angle variation should be eliminated by holding the speed constant. (See the block diagram of Figure 12.13 in Chapter 12.) Therefore, the phase characteristic as a function of frequency is obtained with a large inertia assumed for the machine under consideration (say 100 times the actual inertia). This ensures that the speed and angle do not change over the frequency range of importance for stabilizer design (0.1 to 3 Hz).

The required frequency response of any machine is sensitive to the Thevenin equivalent system impedance at its terminals but relatively independent of the dynamics of other machines. It is, therefore, appropriate to assume that all other machines act as infinite buses. This has the effect of eliminating their dynamics from the response calculation while retaining the correct Thevenin impedance at the terminals of the machine under study. The resulting phase characteristic has a relatively simple form free from the effects of natural frequencies of the external machines.

The phase characteristic to be compensated varies to some extent with system conditions. Therefore, a characteristic acceptable for different system conditions is selected. Generally, slight undercompensation is preferable to overcompensation so that the PSS does not contribute to the negative synchronizing torque component (see Section 12.5). An undercompensation by about $10°$ over the entire frequency range of interest provides the required degree of tolerance to allow for uncertainties in machine and system modelling.

Stabilizing signal washout:

The signal washout is a high-pass filter that prevents steady changes in speed from modifying the field voltage. The value of the washout time constant T_W should be high enough to allow signals associated with oscillations in rotor speed to pass unchanged.

From the viewpoint of the "washout function," the value of T_W is not critical and may be anywhere in the range of 1 to 20 seconds. The main considerations are that it should be long enough to pass stabilizing signals at the frequencies of interest relatively unchanged, but not so long that it leads to undesirable generator voltage excursions as a result of stabilizer action during system-islanding conditions. Ideally, the stabilizer should not respond to system-wide frequency variations.

For local mode oscillations in the range of 0.8 to 2.0 Hz, a washout of 1.5 seconds is satisfactory. From the viewpoint of low-frequency interarea oscillations, a washout time constant of 10 seconds or higher is desirable, since lower-time constants result in significant phase lead at low frequencies. Unless this is compensated for elsewhere, it will reduce the synchronizing torque component at interarea frequencies. This desynchronizing effect is detrimental to interarea transient stability as it will cause the areas to swing farther apart following a disturbance.

Stabilizer gain:

The stabilizer gain, K_{STAB}, has an important effect on damping of rotor oscillations. The value of the gain is chosen by examining the effect for a wide range of values. The damping increases with an increase in stabilizer gain up to a certain point beyond which further increase in gain results in a decrease in damping. Ideally, the stabilizer gain should be set at a value corresponding to maximum damping. However, the gain is often limited by other considerations. With a delta-omega stabilizer, as a result of the effect of the torsional filter, the stability of the "exciter mode" becomes an overriding consideration. With a delta-P-omega stabilizer, exciter mode stability is not a problem, and a considerably higher value of gain is acceptable provided that the phase-lead compensation has been chosen to provide satisfactory phase characteristics over a range of frequencies that includes all dominant modes. In such cases, the maximum value of the stabilizer gain is likely to be limited by practical considerations such as the effect on signal noise.

Stabilizer gain is normally set to a value that results in as high a damping of the critical system mode(s) as practical without compromising the stability of other system modes or causing excessive amplification of signal noise.

Stabilizer limits:

The positive output limit of the stabilizer is set at a relatively large value in the range of 0.1 to 0.2 pu. This allows a high level of contribution from the PSS during large swings. With such a high value of stabilizer output limit, it is essential to have a means of limiting the generator terminal voltage to its maximum allowable value, typically in the 1.12 to 1.15 pu range. Therefore, a terminal voltage limiter is used as shown in Figure 17.5. To be effective, the limiter gain K_L must be very high. The terminal voltage signal, however, contains small components of torsional components. Hence, feedback of this signal to the excitation system through a high gain may cause torsional mode instability. Therefore, T_C and T_D are chosen so as to provide high attenuation at torsional frequencies, in addition to ensuring an adequate degree of limiter loop stability.

On the negative side, a limit of -0.05 to -0.1 pu is appropriate. This allows sufficient control range while providing satisfactory transient response. In the unlikely event of the PSS output being held at the negative limit because of a failure of the stabilizer, this will not result in a unit trip.

Check on selected settings:

   The final stage in stabilizer design involves the evaluation of its effect on the overall system performance. First, the effect of the stabilizer on various modes of system oscillations is determined over a wide range of system conditions by using a small-signal stability program. This includes analysis of the effects of the PSS on local plant modes, interarea modes, and control modes. In particular, it is important to ensure that there are no adverse interactions with the controls of other nearby generating units and devices such as HVDC converters and SVCs.
   After checking the PSS performance under small perturbations, it is important to examine its effect on transient stability and long-term stability.
   For systems with voltage problems, the acceptability of the chosen PSS output limits should be carefully assessed. In some situations, it is possible for the machine terminal voltage to fall below the exciter reference level while the speed is also falling. This can lead to the stabilizer overriding the voltage signal to the exciter, causing reduced transient recovery. It is important to limit the stabilizer output to prevent this.
   It is also important to coordinate the performance of the PSS with other protections and controls such as V/Hz limiters and overexcitation/underexcitation protection.
   Reference 53 provides a detailed account of the application of the above procedure for the design of the PSS of a large nuclear power plant.

General comments on excitation control design:

   The excitation control systems, designed as described above, provide effective decentralized controllers for the damping of electromechanical oscillations in power systems. Generally, the resulting design is much more robust than can be achieved through use of other methods such as pole placement techniques and multivariable state space techniques. The overall approach is based on a knowledge of the physical aspects of the power system stabilization problem. The method used for establishing the phase characteristics of the PSS is simple and requires only the dynamic characteristics of the concerned machines to be modelled in detail. Detailed analysis of the performance of the power system is used to establish other parameters and to ensure adequacy of the overall performance of the excitation control. The result is a control that enhances the overall stability of the system under different operating conditions. Since the PSS is tuned to increase the damping torque component for a wide range of frequencies, it contributes to the damping of all system modes in which the respective generator has a high participation. This includes any new mode that may emerge as a result of changing system conditions. It is possible to satisfy the requirements for a wide range of system conditions with fixed parameters; hence, there has been little incentive to date to consider an adaptive control system.

Example of excitation control design

This example illustrates the application of power system stabilizers to a two-unit thermal generating station. Each unit has a rating of 488 MVA and is equipped with a thyristor excitation system. The power system characteristics are such that these units exhibit two dominant rotor oscillation modes: an interarea mode of about 0.5 Hz and a local intermachine mode of about 2.0 Hz. The objective of excitation control design is to enhance the transient as well as small-signal stability of the power system.

For the purpose of comparison, we will also examine the performance of the two thermal units when equipped with slow-rotating exciters.

The power system:

The power system consists of two areas, as shown in Figure 17.11. Area 1 has a peak load of 805 MW and is supplied by the two-unit thermal plant (G1 and G2) for which we are designing the excitation control. Area 2 has a peak load of 360 MW and is supplied by seven smaller power plants.

The parameters of each of the two 488 MVA units of area 1 in per unit on its rating are as follows:

$$
\begin{array}{lllll}
X_d = 1.81 & X_q = 1.76 & X_l = 0.16 & X_d' = 0.3 & X_q' = 0.61 \\
X_d'' = 0.217 & X_q'' = 0.217 & R_a = 0.002 & T_{d0}' = 7.80 \text{ s} & T_{q0}' = 0.90 \text{ s} \\
T_{d0}'' = 0.022 \text{ s} & T_{q0}'' = 0.074 \text{ s} & H = 3.53 & K_D = 0 &
\end{array}
$$

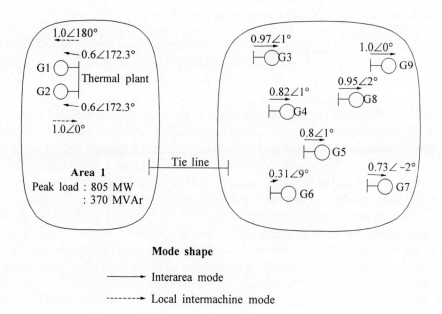

Figure 17.11 Two-area system

Thyristor excitation system:

The block diagram of the thyristor excitation system is shown in Figure 17.5. The parameters of the exciter and AVR are as follows:

$$K_A = 212 \qquad T_R = 0.01 \qquad E_{fmax} = 7.77 \quad E_{fmin} = -6.86$$

A high exciter gain of 212 (with no transient gain reduction) is used to ensure good transient stability performance.

Slow rotating excitation system:

The alternative excitation system considered is a self-excited dc exciter with the following parameters (see block diagram of Figure 8.40, Chapter 8):

$$K_A = 19.2 \qquad T_A = 0.1 \qquad T_E = 0.65 \qquad T_C = 0 \qquad T_B = 0$$
$$A_{ex} = 0.46 \qquad B_{ex} = 0.19 \qquad K_F = 0.05 \qquad T_F = 0.7 \qquad T_R = 0.02$$
$$V_{R\,MAX} = 1.15 \quad V_{R\,MIN} = -1.15 \quad R_C = 0 \qquad X_C = 0$$

Small-signal stability performance:

The frequencies and damping ratios (ζ) of the rotor angle modes associated with units G1 and G2 are summarized in Table 17.2. These results were computed by using the MASS program with a detailed representation of all generating units. Peak system load conditions were considered and a major transmission circuit in area 1 near the generating station was assumed to be out of service. The network was represented by 83 buses, with the active component of loads modelled as 50% constant current and 50% constant impedance, and the reactive component of loads modelled as constant impedance.

With thyristor exciters (no PSS), the interarea mode has a frequency of 0.55 Hz and is just stable with a damping ratio of 0.006; the local intermachine mode has a frequency of 1.823 Hz and a damping ratio of 0.049. The mode shapes are shown in Figure 17.11.

With the rotating exciters, the frequencies of the two modes decrease slightly and the damping ratios increase slightly. The mode shapes are essentially similar to those with thyristor exciters.

Table 17.2 Rotor angle modes of units G1 and G2

Type of Exciter	Local Intermachine Mode		Interarea Mode	
	Frequency	ζ	Frequency	ζ
(a) Thyristor (no PSS)	1.823 Hz	0.049	0.550 Hz	0.006
(b) Rotating exciter	1.793 Hz	0.075	0.498 Hz	0.046

Selection of PSS parameters for the thyristor exciter:

The basis for the selection of the phase-lead compensation is illustrated in Figure 17.12. Curve 1 of the figure shows the phase lag between the exciter input and the generator electrical torque as a function of frequency. This characteristic was computed using the MASS program with G1 and G2 represented in detail as a single equivalent generator having a large inertia, and the generators at all other generating stations as infinite buses. Curve 2 of the figure shows the phase-lead compensation provided by choosing the following PSS parameters (see Figure 17.5):

$$T_1 = 0.06 \qquad T_2 = 0.02 \qquad T_3 = 1.5 \qquad T_4 = 4.0 \qquad T_w = 7.5$$

The effect of supplementing the thyristor excitation system of units G1 and G2 with the PSS on the small-signal stability is shown in Table 17.3 for different values of stabilizer gain K_{STAB}. The damping of both modes of rotor oscillation increases as K_{STAB} is increased. A gain of about 30 is considered satisfactory.

Other parameters of the PSS (including the terminal voltage limiter) that affect the large-signal performance are chosen as follows:

$$V_{s\,max} = 0.2 \qquad V_{s\,min} = -0.05 \qquad V_{LS} = 1.15 \qquad K_L = 17$$
$$T_C = 0.025 \qquad T_D = 1.212 \qquad T_{RL} = 0.01$$

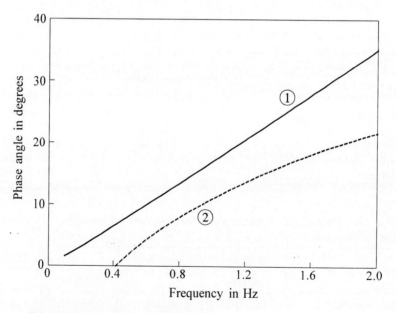

① Phase lag to be compensated
② Phase compensation provided

Figure 17.12 Determination of phase-lead compensation

Table 17.3 Effect of PSS gain on rotor angle modes

K_{STAB}	Local Intermachine Mode		Interarea Mode	
	Frequency	ζ	Frequency	ζ
0	1.823 Hz	0.049	0.550 Hz	0.006
20	2.079 Hz	0.156	0.547 Hz	0.087
30	2.218 Hz	0.197	0.548 Hz	0.124
40	2.366 Hz	0.227	0.533 Hz	0.156

Transient stability performance:

The results of transient stability simulations for a contingency involving a three-phase fault on a circuit close to the thermal station in area 1 are shown in Figures 17.13 and 17.14. The fault is applied when time is equal to 50 cycles and cleared 5 cycles later by opening both ends of the faulted circuit simultaneously. Three types of excitation control are considered for units G1 and G2: the thyristor exciter without and with PSS (K_{STAB}=30), and the slow dc exciter.

The results shown in Figure 17.13 are for the peak system load conditions. The system is transiently unstable with the rotating exciter. It is stable with the thyristor exciter (no PSS), but poorly damped. The PSS, in addition to significantly increasing the damping of system oscillations, improves transient stability performance by reducing the first rotor swing.

The results shown in Figure 17.14 are for system conditions with 94.5% of peak load and the output of the thermal station in area 1 proportionately reduced. With the rotating exciters, units G1 and G2 are in this case first-swing stable but become unstable in the second swing. With thyristor exciters, the stability performance is similar to that for the peak load case. Clearly, thyristor exciters with the PSS result in very good overall stability performance.

Selection of PSS location

In large systems, the selection of units on which to install the PSS to damp interarea oscillations may not be readily apparent. Although the principles of PSS design for damping of local and interarea modes are similar, the mechanisms by which a PSS contributes to the damping of the two types of oscillation are different. A PSS adds damping to an interarea mode largely by modulating system loads, whereas the performance of the PSS with regard to a local mode is only slightly affected by the load characteristics [54,55]. Understanding these mechanisms is essential to the effective application of the PSS.

Participation factors corresponding to speed deviations of generating units are very useful for initial screening of generating units on which to add stabilizers. However, a high participation factor is a necessary, but not a sufficient, condition for a PSS at the unit to effectively damp interarea oscillations. Following the initial screening based on participation factors, a more rigorous evaluation using residues and frequency responses should be carried out to determine appropriate locations for the stabilizers [54].

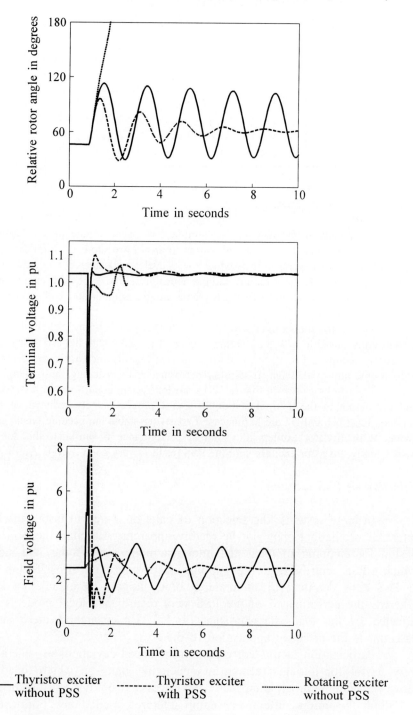

Figure 17.13 Response of unit G1 to a five-cycle
three-phase fault; peak load conditions

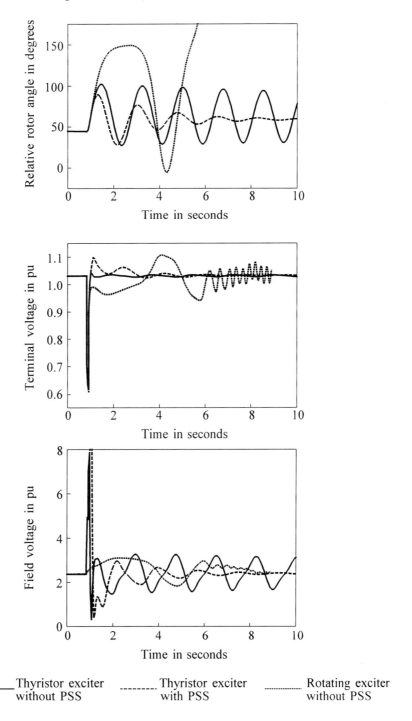

Thyristor exciter
without PSS

Thyristor exciter
with PSS

Rotating exciter
without PSS

Figure 17.14 Response of unit G1 for a five-cycle three-phase
fault; 94.5% peak load conditions

17.2.2 Supplementary Control of Static Var Compensators

In Chapter 11, the characteristics and modelling of static var compensators (SVCs) were described, along with the basic principles of their application. By rapidly controlling the voltage and reactive power, an SVC can contribute to the enhancement of the power system dynamic performance. Normally, voltage regulation is the primary mode of control, and this improves voltage stability and transient stability. However, the contribution of an SVC to the damping of system oscillations resulting from voltage regulation alone is usually small; supplementary control is necessary to achieve significant damping. The effectiveness of an SVC in enhancing small-signal stability depends on the location of the SVC, input signals used, and controller design.

Example of SVC application

The simple two-area system shown in Figure 17.15 is used to illustrate the application of an SVC for the enhancement of system stability. This system is identical to the one used in Example 12.6 of Chapter 12 except for slight differences in power-flow conditions because of additional shunt capacitor compensation at bus 8. All four generators are assumed to have self-excited dc exciters.

System performance without an SVC:

Figure 17.16 shows the system response to a three-phase fault near bus 9 on one of the lines between buses 8 and 9. The fault is assumed to be cleared in 74 ms by isolating the faulted line. We see from the results that the system becomes unstable through growing oscillations of about 0.4 Hz.

Stability enhancement with an SVC:

The first step is to determine a suitable location for the SVC. For this simple system, an obvious choice would be the middle of the interconnection between the two areas, where voltage swings are the greatest without the SVC [56,57].

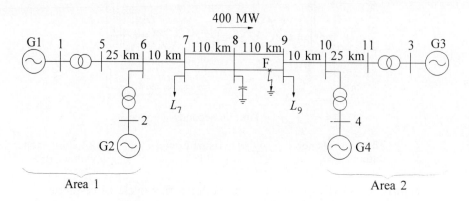

Figure 17.15 A simple two-area system

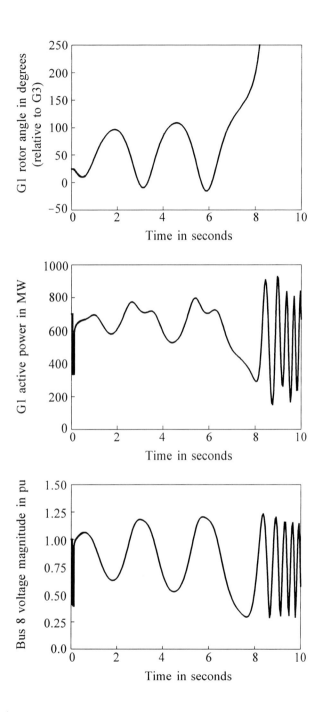

Figure 17.16 System response to a severe disturbance with no SVC

For large complex systems, the best SVC location is not obvious. In such situations, the PEALS (described in Chapter 12) or the VSTAB program (described in Chapter 14) may be used to identify the SVC locations. The bus participations computed by the VSTAB program and the voltage participation factors computed by PEALS serve as useful sensitivity indices for identifying SVC locations. Tables 17.4 and 17.5 give these indices for the prefault system conditions. The VSTAB results show bus 8 to be the best location. PEALS results indicate that bus 9 has the highest voltage participation factor (sensitivity to susceptance change), while buses 10, 4, 8, and 11 also have large factors. Considering that the power transfer might be in either direction between the two areas, bus 8 would obviously be the best choice.

An SVC comprising a fixed capacitor and a thyristor-controlled reactor (TCR) is considered for enhancement of system stability. The rating of the SVC is assumed to be 0 to 200 MVAr (capacitive). Figure 17.17 shows the block diagram of the SVC, including the

<table>
<tr><td colspan="2">Table 17.4 Bus participation
computed by VSTAB</td></tr>
<tr><td>Bus</td><td>Participation</td></tr>
<tr><td>8</td><td>0.5100</td></tr>
<tr><td>9</td><td>0.1711</td></tr>
<tr><td>7</td><td>0.1613</td></tr>
<tr><td>10</td><td>0.0668</td></tr>
<tr><td>6</td><td>0.0617</td></tr>
<tr><td>11</td><td>0.0153</td></tr>
<tr><td>5</td><td>0.0138</td></tr>
</table>

<table>
<tr><td colspan="2">Table 17.5 Bus voltage eigenvector magnitude
for interarea mode computed by PEALS</td></tr>
<tr><td>Bus</td><td>Voltage Vector
Magnitude Factor</td></tr>
<tr><td>9</td><td>1.0000</td></tr>
<tr><td>10</td><td>0.8852</td></tr>
<tr><td>4</td><td>0.7294</td></tr>
<tr><td>8</td><td>0.6547</td></tr>
<tr><td>11</td><td>0.6147</td></tr>
<tr><td>3</td><td>0.3645</td></tr>
</table>

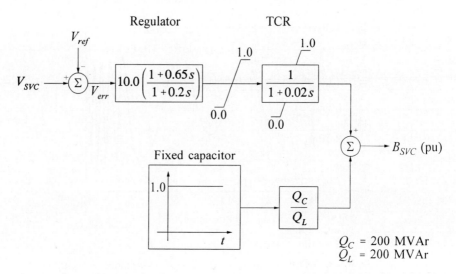

Figure 17.17 Block diagram of SVC and voltage regulator

voltage regulator. The voltage regulator gain is set at 10 to provide a 10% slope in the control range. The parameters of the lead-lag block have been selected so that the voltage regulator enhances system transient stability (i.e., high initial response).

Small-signal stability performance:

Table 17.6 summarizes the frequencies and damping ratios of the interarea mode with and without the SVC, for prefault and postfault system conditions. While the SVC has stabilized the interarea mode, the damping is still very low.

Table 17.6 Effect of SVC on interarea mode (frequency and damping ratio)

System Condition	No SVC		With SVC	
	Frequency	ζ	Frequency	ζ
Prefault	0.540 Hz	0.0064	0.547 Hz	0.0096
Postfault	0.417 Hz	−0.0228	0.476 Hz	0.0154

Supplementary control to improve damping:

The input signal used for supplementary control of the SVC should be responsive to the modes of oscillation to be damped. This can be determined by residues and observability using the MASS program described in Chapter 12. Table 17.7 gives the residues and observability factors for various input signals for both prefault and postfault conditions. Clearly, a good choice for the input signal is the magnitude of current in the line between buses 9 and 10. Figure 17.18 shows the frequency response characteristics of the transfer function between the SVC input and this current. It has a high gain at the frequency of interarea mode – an indication of good selectivity of this signal.

Table 17.7 Residues and observability factors

Signal	Prefault		Postfault	
	Residue	Observability	Residue	Observability
$\Delta\omega$ of G1	$-0.2680-j0.1156$	0.8738	$-0.9522-j0.3651$	0.5471
$\Delta\omega$ of G2	$-0.2121-j0.1004$	0.7025	$-0.7733-j0.3347$	0.4510
$\Delta\omega$ of G3	$0.4588+j0.1121$	1.4140	$2.6200+j0.2945$	1.4140
$\Delta\omega$ of G4	$0.4064+j0.0947$	1.2510	$2.4380+j0.2469$	1.3140
ΔP, line 6-7	$-0.2286+j0.4914$	1.6230	$-0.6551+j1.5160$	0.8861
ΔP, line 10-9	$0.2122-j0.8560$	2.6400	$0.5305-j4.3570$	2.3550
ΔQ, line 6-7	$-0.0310+j0.2107$	0.6375	$0.1289+j1.2020$	0.6847
ΔQ, line 10-9	$-0.0400+j0.2346$	0.7126	$-0.0924+j1.6290$	0.8752
ΔI, line 6-7	$-0.3157+j0.8618$	2.7980	$-0.8919+j3.4640$	1.9120
ΔI, line 10-9	$0.2615-j0.8732$	2.7290	$0.6484-j3.8810$	2.1110

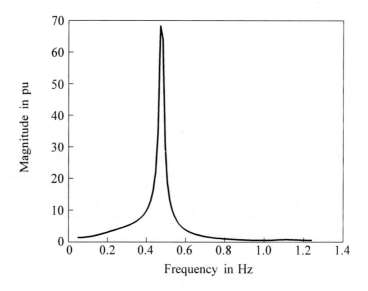

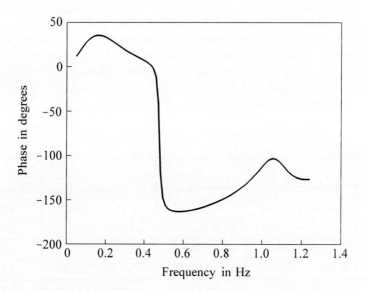

Figure 17.18 Frequency response of the transfer function between the SVC input and the current in line between buses 9 and 10

Control design:

Several alternative control design techniques may be used for determining the control parameters for SVCs [55]:

- Pole placement technique

- Phase and gain margin technique

- H-infinity technique

Here, we will illustrate the use of the pole placement technique. First, we will provide a brief description of the method.

The pole placement technique uses root locus rules to shift a pair of dominant poles to a newly assigned location in the *s*-plane. After the initial design the controller is tested for robustness and, if necessary, appropriate modifications are made. At this stage, the nonlinearities may be included and the compensator limits may be set.

Let the transfer function of the open-loop plant and the controller (compensator) be $G(s)$ and $H(s)$, respectively, as shown in Figure 17.19.

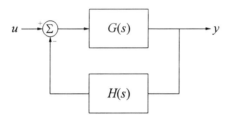

Figure 17.19 Transfer function of a plant with feedback

The closed-loop transfer function of the compensated plant is

$$G_c(s) = \frac{G(s)}{1 + G(s)H(s)}$$

The zeros of $1+G(s)H(s)$ are the poles of the closed-loop system.

Assume that the system eigenvalue λ is to be shifted to a new location in the *s*-plane denoted by λ_0. Since λ_0 must satisfy the characteristic equation of the closed-loop system,

$$H(\lambda_0) = -\frac{1}{G(\lambda_0)}$$

This can be expressed in terms of magnitude and phase, as follows:

$$|H(\lambda_0)| = \frac{1}{|G(\lambda_0)|}$$

$$arg(H(\lambda_0)) = 180° - arg(G(\lambda_0))$$

Thus, the magnitude and phase of the compensator at λ_0 can be calculated from the magnitude and phase of the plant at λ_0; that is, from the complex frequency response of the plant at the new pole location. The new pole location is chosen to satisfy the specified damping ratio. Its imaginary part is usually chosen slightly larger than that of λ.

The compensator generally consists of a washout and a series of lead and/or lag functions to satisfy the argument equation, as shown below.

$$K\frac{sT_W}{1+sT_W}\frac{1+sT_1}{1+sT_2}\cdots\frac{1+sT_{2n-1}}{1+sT_{2n}}$$

The washout is intended to eliminate the dc component or reduce the close-to-dc component of the measured signal and has a large time constant. Each lead or lag block compensation is limited to a maximum of 60° for practical reasons. The maximum angle, θ_m, that the ith block can provide, is given by

$$\sin\theta_m = \frac{1-a_i}{1+a_i}$$

where $a_i = T_{2i}/T_{2i-1}$.

The frequency at which this maximum occurs is

$$\omega_m = \frac{1}{\sqrt{a_i}\,T_{2i-1}}$$

which is usually chosen to be near the frequency of λ_0. Finally, the gain K is chosen to satisfy the magnitude equation.

For the system of Figure 17.15, the magnitude and phase of the transfer function between the SVC input and the current in the line from bus 10 to bus 9 are computed for a number of values of λ_0. This is done for the prefault as well as the postfault system conditions. The objective is to find satisfactory values of λ_0 that result in approximately the same magnitude and phase for both operating conditions. These values of λ_0 are given in Table 17.8.

Table 17.8 The desired location of the eigenvalues and
the value of the system transfer function

Prefault System		Postfault System	
New Eigenvalue Location	Value of Transfer Function	New Eigenvalue Location	Value of Transfer Function
$-0.127 \pm j3.544$	$4.262 \angle 137.8°$	$-0.576 \pm j3.498$	$4.177 \angle 132.3°$

Figure 17.20 shows the block diagram of the supplementary control. The output limits of the controller are set to provide the desired large-signal response.

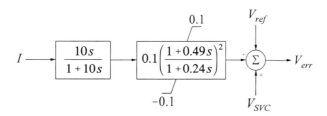

I = magnitude of current in line between buses 9 and 10

Figure 17.20 Supplementary control block diagram

The frequency and damping ratios of the interarea mode with supplementary control of the SVC are as follows:

Prefault: f=0.564 Hz ζ=0.036
Postfault: f=0.557 Hz ζ=0.163

The transient stability performance of the system with and without supplementary control of the SVC is depicted in Figure 17.21. The disturbance considered is, as before, a 74 ms three-phase fault on one of the lines between buses 8 and 9. The SVC with voltage regulation stabilizes the system, but the damping is poor. The oscillations are well damped with the supplementary control of the SVC.

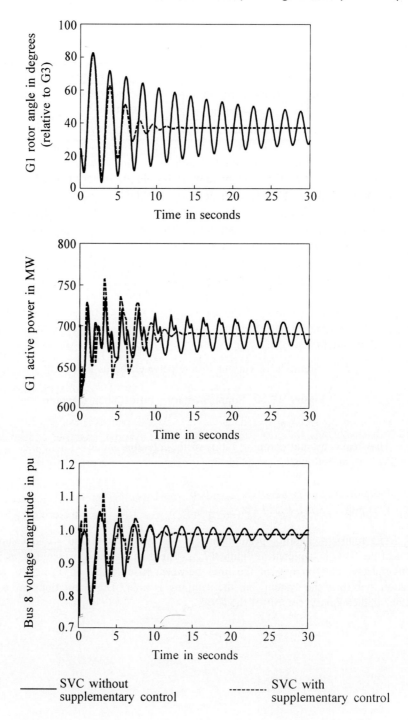

_____ SVC without
           supplementary control

---------- SVC with
           supplementary control

Figure 17.21 System response to a severe disturbance with and
without supplementary control of an SVC at bus 8

17.2.3 Supplementary Control of HVDC Transmission Links

As described in Chapter 10, the basic controlled quantity in an HVDC transmission link is the direct current. Normally, the direct current is controlled by the rectifier, and the dc line voltage is maintained near the rated value by inverter control. Damping of the ac system electromechanical oscillations can be increased by modulating the current order at the rectifier. Alternatively, both the current order at the rectifier and the voltage order at the inverter can be modulated.

The HVDC link may either be embedded in an ac system or form an asynchronous link between two ac systems. Supplementary control with either of these types of HVDC links is effective in damping ac system oscillations. References 42 and 43 describe dc modulation schemes used in several HVDC transmission links.

The following example illustrates the use of supplementary control of HVDC links for the enhancement of system stability.

Example of HVDC link supplementary control

The system used to illustrate the application of HVDC supplementary control is shown in Figure 17.22. It is based on the simple two-area system considered in the previous section (Figure 17.15) and in Example 12.6. A 200 MW bipolar dc link has been added between buses 7 and 9 in parallel with two ac ties. The dc link is represented as a monopolar link with a voltage rating of 56 kV and a current rating of 3,600 A. The dc line resistance is 1.5 Ω, and inductance is 100 mH. The commutating reactance (X_c) associated with each converter is 0.57 Ω. A smoothing reactor of 50 mH is used at each end of the line.

The ac portion of the system is very similar to that of Example 12.6. The principal difference is the reactive support of 125 MVAr provided at the rectifier and inverter buses.

Modelling of converter controls:

Figures 17.23 to 17.28 show models used to represent the dc link controls. The master control, shown in Figure 17.23, determines the current order for the rectifier and inverter. If

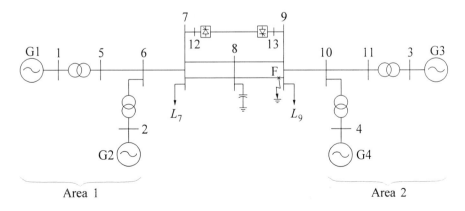

Figure 17.22 A two-area system with parallel dc and ac ties

the rectifier end dc voltage falls below 90% of rated value, the control switches to constant current mode. The hysteresis characteristic represented by block MC1 prevents hunting between constant power and constant current control modes. The current order is limited by the voltage-dependent current order limit (VDCOL) shown in Figure 17.24.

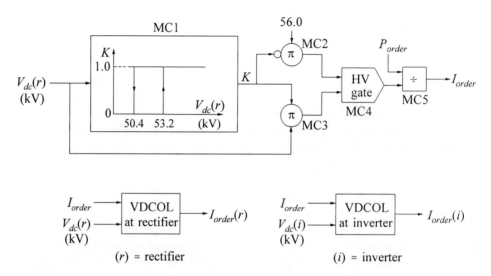

Figure 17.23   Master control

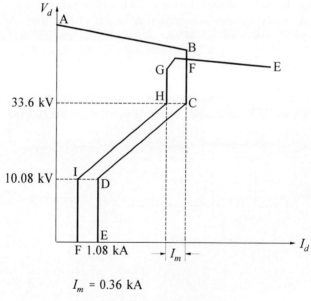

I_m = 0.36 kA

The downtime constant is 5 ms.
The uptime constant is 100 ms.

Figure 17.24 Voltage-dependent current order limits

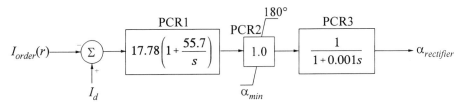

Figure 17.25 Rectifier pole control

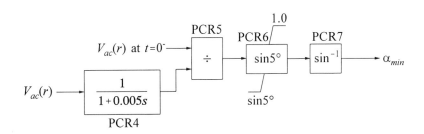

Figure 17.26 α_{min} calculation

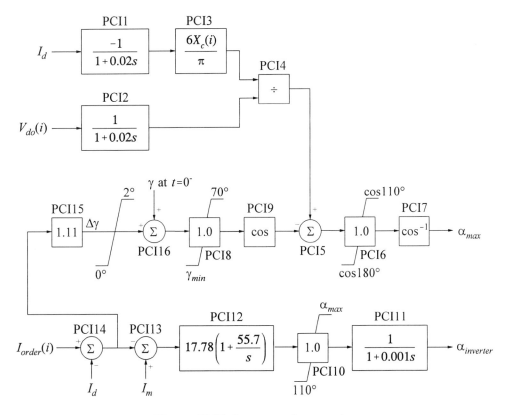

Figure 17.27 Inverter pole control

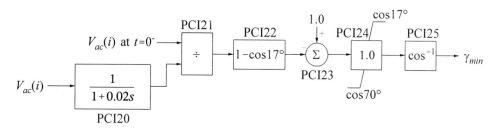

Figure 17.28 γ_{min} calculation

The rectifier pole control representation is shown in Figure 17.25. The normal mode of rectifier operation is current control. A proportional plus integral controller (block PCR1) is used to control the direct line current. The ignition delay angle α is limited between α_{min} and 180°. The lower limit α_{min} ensures that there is sufficient forward voltage across the thyristor for a successful turn-on. This limit, as shown in Figure 17.26, varies with the ac line voltage at the rectifier (blocks PCR4 to PCR7) so that there is always a minimum amount of forward voltage across the thyristor.

Figure 17.27 shows the inverter pole control representation. Under normal conditions, the inverter is on constant extinction angle (CEA) control represented by

$$\cos\alpha = \frac{6X_c I_d}{\pi V_{d0}} - \cos\gamma$$

The control ensures that $\cos\alpha$ corresponds to a condition with γ constant (see Chapter 10, Section 10.9.3).

The inverter is also equipped with a current controller that is active only when the firing angle derived from current control is smaller than the firing angle derived from the CEA controller. The inverter switches to the current control mode when the rectifier hits α_{min} and is unable to maintain the desired direct current.

The lower limit of the extinction angle (γ_{min}) is varied with the inverter ac bus voltage so that a minimum commutation margin area is maintained. Figure 17.28 shows the logic for the calculation of γ_{min}.

If the inverter ac bus voltage drops below 0.5 pu, it is assumed that commutation failure occurs. The rectifier and inverter are blocked if the inverter experiences more than two commutation failures. The inverter recovers following a commutation failure when the ac bus voltage subsequently goes above 0.7 pu.

System performance without supplementary control:

Figures 17.29 and 17.30 show the system response to a three-phase fault near bus 9 on one of the circuits between buses 8 and 9. The fault is cleared by isolating the faulted circuit in 83 ms. We see from the results that the system is transiently stable. However, oscillations between area 1 and area 2 generators have zero or slightly negative damping.

Table 17.9 summarizes the results of small-signal stability analysis. The frequency and damping ratio of the interarea mode are listed for different system conditions. The interarea mode is poorly or negatively damped in all the cases considered.

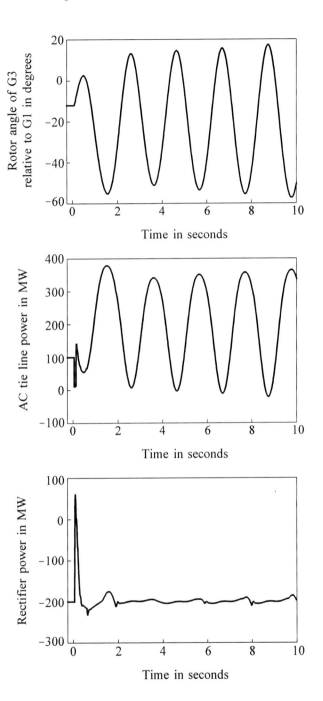

Figure 17.29 System response to an ac system fault near
the inverter, with no supplementary control

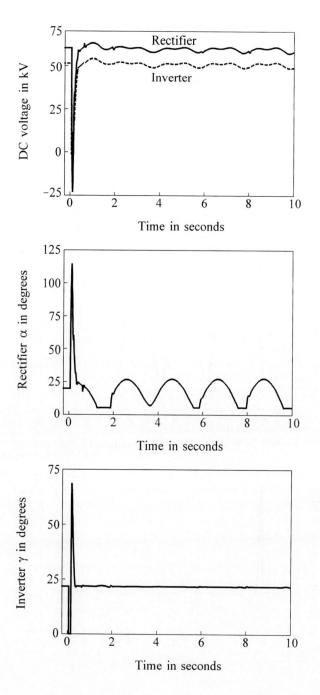

Figure 17.30 HVDC link response to an ac system fault near the inverter, with no supplementary control

Table 17.9 Frequency and damping ratio of interarea
mode without supplementary control

Case No.	MW Flow from Area 1 to Area 2		Circuits out of Service	Interarea Mode	
	DC Tie	AC Tie		Freq. (Hz)	ζ
1(a)	200	200	None	0.575	0.0076
1(b)	200	200	8 to 9 (1 cct)	0.495	−0.0054
1(c)	200	200	7 to 8 (1 cct) 8 to 9 (1 cct)	0.440	−0.0167
2(a)	50	352	None	0.560	0.0052
2(b)	50	352	8 to 9 (1 cct)	0.466	−0.0110
2(c)	50	352	7 to 8 (1 cct) 8 to 9 (1 cct)	0.397	−0.0254

Supplementary control to improve damping:

The general procedure for designing the supplementary control is the same as that for an SVC using the pole placement technique as described in Section 17.2.2.

Based on observability considerations, active power through the line between buses 7 and 8 is selected as the feedback signal. The initial target for the placement of the eigenvalue associated with the interarea mode is chosen to be $-0.64\pm j3.1$. This corresponds to a frequency of 0.5 Hz and a damping ratio of 0.2. The magnitude and phase of the open-loop transfer function between the power order signal of HVDC master control and the active power flow in the line between buses 7 and 8 at a complex frequency $s = -0.64\pm j3.1$ for different operating conditions are listed in Table 17.10. The results show that the phase compensation required is between 85° and 150°. As a compromise, the phase compensation is chosen to be 100°. Figure 17.31 shows the block diagram of the supplementary control.

Table 17.10 Magnitude and phase of the open-loop
transfer function at $s = -0.64\pm j3.1$

Case No.	Open-Loop Transfer Function
1(a)	0.815∠−146.92°
1(c)	0.773∠−86.84°
2(a)	0.716∠−142.20°
2(c)	0.237∠−112.90°

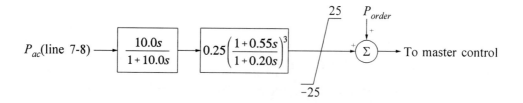

Figure 17.31 Supplementary control block diagram

    The washout time constant is 10 s. The parameters of the phase-lead block are chosen to provide 100° at the complex frequency $-0.64\pm j3.1$. The gain is limited to 0.25 so as to ensure adequate damping of all system modes. Relatively large output limits (±25 MW) are used to allow a high level of contribution from the supplementary controller during large swings.

    The frequency and damping ratio of the interarea mode with the supplementary control are summarized in Table 17.11 for the different operating conditions considered in Table 17.9. The interarea mode is well damped in all cases with the supplementary control.

Table 17.11   Frequency and damping ratio of interarea mode with supplementary control

Case No.	Interarea Mode	
	Freq. (Hz)	ζ
1(a)	0.607	0.1730
1(b)	0.506	0.1572
1(c)	0.467	0.2512
2(a)	0.588	0.1474
2(b)	0.475	0.1007
2(c)	0.422	0.1031

    Figures 17.32 and 17.33 show the transient response of the system with supplementary control for a three-phase fault near bus 9 on a circuit between buses 8 and 9. The oscillations are now well damped.

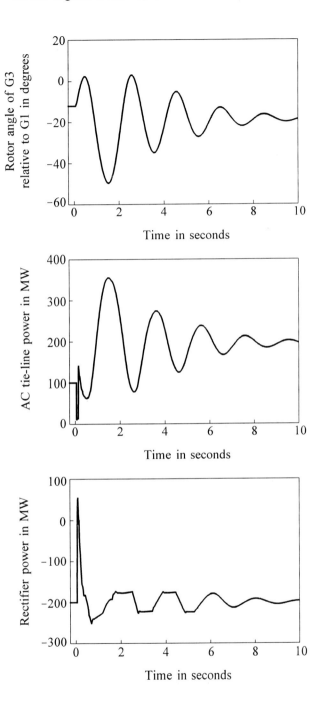

Figure 17.32 System response to an ac system fault near
the inverter, with supplementary control

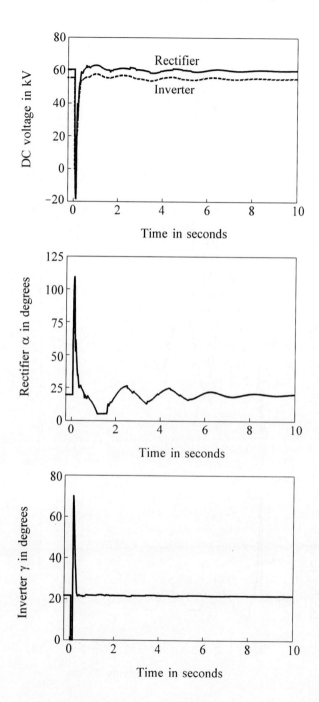

Figure 17.33 HVDC link response to an ac system fault near the inverter, with supplementary control

REFERENCES

[1]   R.O. Berglund, W.A. Mittelstadt, M.L. Shelton, P. Barkan, C.G. Dewey, and
      K.M. Skreiner, "One-Cycle Fault Interruption at 500 kV: System Benefits and
      Breaker Design," *IEEE Trans.* Vol. PAS-93, pp. 1240-1251, September/
      October, 1974.

[2]   J. Esztergalyos, M.T. Yee, M. Chamia, and S. Liberman, "The Development
      and Operation of an Ultra High Speed Relaying System for EHV Lines,"
      CIGRE 34-04, 1978.

[3]   C. Lindsay and V. Shenoy, "Reliability and Selection of Characteristics for
      EHV Transformers," CIGRE 12-08, 1978.

[4]   G.D. Breuer, H.M. Rustebakke, R.A. Gibley, and H.O. Simmons, "The Use of
      Series Capacitors to Obtain Maximum EHV Transmission Capability," *IEEE
      Trans.,* Vol. PAS-83, pp. 1090-1101, November 1964.

[5]   E.W. Kimbark, "Improvement of System Stability by Switched Series
      Capacitors," *IEEE Trans.,* Vol. PAS-85, pp. 180-188, February 1966.

[6]   IEEE Task Force Report, "A Description of Discrete Supplementary Controls
      for Stability," *IEEE Trans.,* Vol. PAS-97, pp. 149-165, January/February 1978.

[7]   H.M. Ellis, J.E. Hardy, A.L. Blythe, and J.W. Skooglund, "Dynamic Stability
      of the Peace River Transmission System," *IEEE Trans.*, Vol. PAS-85, pp. 586-
      600, June 1966.

[8]   M.L. Shelton, R.F. Winkleman, W.A. Mittelstadt, and W.L. Bellerby,
      "Bonneville Power Administration 1400 MW Braking Resistor," *IEEE Trans.,*
      Vol. PAS-94, pp. 602-611, March/April 1975.

[9]   CIGRE SC38-WG02 Report, "State of the Art in Non-Classical Means to
      Improve Power System Stability," *Electra*, No. 118, pp. 88-113, May 1988.

[10] EPRI Report EL-5859, "Technical Limits to Transmission System Operation,"
      Final Report of Project 5005-2, prepared by Power Technologies Inc., June
      1988.

[11] IEEE Working Group, "Single Pole Switching for Stability and Reliability,"
      Report of a panel discussion held at the 1984 PES Summer Meeting, *IEEE
      Trans.,* Vol. PWRS-1, No. 2, pp. 25-36, May 1986.

[12]   R.D. Dunlop, R.M. Maliszewski, and B.M. Pasternack, "Application of Single Phase Switching on the AEP 765-kV System," *Proceedings of the American Power Conference*, Vol. 42, pp. 461-462, 1980.

[13]   A.J. Fakheri et al., "The Use of Reactor Switches in Single Phase Switching," CIGRE 13-06, 1980.

[14]   N. Knudsen, "Single-Phase Switching of Transmission Lines Using Reactors for Extinction of the Secondary Arc," CIGRE 310, 1962.

[15]   E.W. Kimbark, "Suppression of Ground-Fault Arcs on Single-Pole Switched Lines by Shunt Reactors," *IEEE Trans.*, Vol. PAS-83, pp. 285-290, March 1964.

[16]   R.M. Hasibar, A.C. Legate, J.H. Brunke, and W.G. Peterson, "The Application of High-Speed Grounding Switches for Single-Pole Reclosing on 500-kV Power Systems," *IEEE Trans.*, Vol. PAS-100, pp. 1512-1515, April 1981.

[17]   A.J. Gonzales, G.C. Kung, C. Raczkowski, C.W. Taylor, and D. Thonn, "Effects of Single-and-Three Pole Switching and High-Speed Reclosing on Turbine-Generator Shafts and Blades," *IEEE Trans.*, Vol. PAS-103, pp. 3218-3228, November 1984.

[18]   ANSI Standard C50.13-1989.

[19]   R.C. Buell, R.J. Caughey, E.M. Hunter, and V.M. Marquis, "Governor Performance during System Disturbances," *AIEE Trans.*, Vol. 50, pp. 354-369, March 1931.

[20]   F.P. deMello, D.N. Ewart, M. Temoshok, and M.A. Eggenberger, "Turbine Energy Controls Aid in Power System Performance," *Proceedings of the American Power Conference*, Vol. 28, pp. 438-445, 1966.

[21]   R.H. Park, "Improved Reliability of Bulk Power Supply by Fast Load Control," *Proceedings of the American Power Conference*, Vol. 30, pp. 1128-1141, 1968.

[22]   R.H. Park, "Fast Turbine Valving," *IEEE Trans.*, Vol. PAS-92, pp. 1065-1073, May/June 1973.

[23]   E.W. Cushing, G.E. Drechsler, W.P. Killgoar, H.G. Marshall and H.R. Stewart, "Fast Valving as an Aid to Power System Transient Stability and Prompt Resynchronization and Rapid Reload after Full Load Rejection," *IEEE Trans.*, Vol. PAS-91, pp. 1624-1636, July/August 1972.

[24]   H.F. Martin, D.N. Tapper and T.M. Alston, "Sustained Fast Valving Applied
       to TVA's Watts Bar Nuclear Units," Paper 76-JPGC-PWR-5, presented at the
       1976 Joint Power Generation Conference, Buffalo, N.Y., September 1976,
       ASME Trans., Series A, Vol. 99, No. 1, 1977.

[25]   P.L. McGaha and T.L. Dresner, "A Nuclear Steam Turbine Intercept Valve
       and Control System for Fast Valving," Paper presented at the 1977 Joint
       Power Generation Conference, Long Beach, Calif., September 1977.

[26]   L. Edwards, R.J. Thomas, D.C. Hogue, P. Hughes, W. Novak, G. Weiss and
       J.E. Welsh, "Sustained Fast Valving at TVA's Cumberland Steam Plant:
       Background and Test Results," *Proceeding of the American Power Conference*,
       Vol. 43, pp. 142-152, April 1981.

[27]   IEEE Working Group Report of panel discussion, "Turbine Fast Valving to
       Aid System Stability: Benefits and Other Considerations," *IEEE Trans.*, Vol.
       PWRS-1, pp. 143-153, February 1986.

[28]   W.A. Morgan, H.B. Peck, D.R. Holland, F.A. Cullen, and J.B. Ruzek,
       "Modern Stability Aids for Calvert Cliffs Units," *IEEE Trans.*, Vol. PAS-90,
       pp. 1-10, January/February 1971.

[29]   P. Kundur and J.P. Bayne, "A Study of Early Valve Actuation Using Detailed
       Prime Mover and Power System Simulation," *IEEE Trans.*, Vol. PAS-94, pp.
       1275-1287, July/August 1975.

[30]   D.L. Osborn, "Fast Valving and Neutral Resistor Application on a 600 MW
       Fossil Unit," Paper A76608-0, presented at the 1976 Joint Power Generation
       Conference, Buffalo, N.Y., September 1976.

[31]   N.J. Balu, "Fast Turbine Valving and Independent Pole Tripping Breaker
       Applications for Plant Stability," *IEEE Trans.*, Vol. PAS-99, pp. 1330-1342,
       July/August 1980.

[32]   R.M. Maliszewski, B.M. Pasternack, and R.D. Rana, "Temporary Fast Turbine
       Valving on the AEP System," *Proceedings of the American Power Conference*,
       Vol. 44, pp. 118-125, 1982.

[33]   T.D. Younkins, J.H. Chow, A.S. Brower, J. Kure-Jensen, and J.B. Wagner,
       "Fast Valving with Reheat and Straight Condensing Steam Turbines," *IEEE
       Trans.*, Vol. PWRS-2, pp. 397-405, May 1987.

[34]   P. Kundur, R.E. Beaulieu, C. Munro, and P.A. Starbuck, "Steam Turbine Fast
       Valving: Benefits and Technical Considerations," ST267 position paper,

presented at the Canadian Electrical Association, spring meeting, March 1986.

[35]   R.H. Hillery and E.D. Holdup, "Load Rejection Testing of Large Thermal Electric Generating Units," *IEEE Trans.*, Vol. PAS-87, pp. 1440-1453, June 1968.

[36]   V.F. Carvalho, "Use of Generator and Customer Load Rejection to Increase Power Transfer Limits," CIGRE Paper 32-79-00-56, presented at the Study Committee 32 Meeting, Minneapolis, Minn., 1979.

[37]   P. Kundur and W.G.T. Hogg, "Use of Generation Rejection in Ontario Hydro to Increase Power Transfer Capability," Paper presented at panel session on generator tripping, IEEE PES winter meeting, New York, January/February 1982.

[38]   R.G. Farmer, "Independent Detection Scheme to Initiate Western System Islanding for Pacific AC Intertie Outages," Paper presented at the panel session on controlled separation and load shedding, IEEE PES summer meeting, July 1985.

[39]   C.W. Taylor, F.R. Nassief, and R.L. Cresap, "Northwest Power Pool Transient Stability and Load Shedding Controls for Generation-Load Imbalances," *IEEE Trans.*, Vol. PAS-100, pp. 3486-3495, July 1981.

[40]   D.C. Lee and P. Kundur, "Advanced Excitation Control for Power System Stability Enhancement," CIGRE 38-01, 1986.

[41]   C.W. Taylor, J.R. Mechenbier, and C.E. Matthews, "Transient Excitation Boosting at Grand Coulee Third Power Plant: Power System Application and Field Test," Paper 92 SM 533-0 PWRS, presented at the IEEE PES summer meeting, Seattle, July 12-16, 1992.

[42]   IEEE Committee Report, "Dynamic Performance Characteristics of North American HVDC Systems for Transient and Dynamic Stability Evaluations," *IEEE Trans.*, Vol. PAS-100, pp. 3356-3364, July 1981.

[43]   IEEE Committee Report, "HVDC Controls for System Dynamic Performance," *IEEE Trans.*, Vol. PWRS-6, No. 2, pp. 743-752, May 1991.

[44]   C.E. Grund, G.D. Breuer, and R.P. Peterson, "AC/DC System Dynamic Performance - Transient Stability Augmentation with Dynamic Reactive Power Compensation," *IEEE Trans.*, Vol. PAS-99, pp. 1493-1502, July/August 1980.

[45]   P.L. Dandeno, A.N. Karas, K.R. McClymont, and W. Watson, "Effect of High-

Speed Rectifier Excitation Systems on Generator Stability Limits," *IEEE Trans.*, Vol. PAS-87, pp. 190-201, January 1968.

[46]   W. Watson and G. Manchur, "Experience with Supplementary Damping Signals for Generator Static Excitation Systems," *IEEE Trans.*, Vol. PAS-92, pp. 199-203, January/February 1973.

[47]   W. Watson and M.E. Coultes, "Static Exciter Stabilizing Signals on Large Generators - Mechanical Problems," *IEEE Trans.*, Vol. PAS-92, pp. 204-211, January/February 1973.

[48]   P. Kundur, D.C. Lee, and H.M. Zein El-Din, "Power System Stabilizers for Thermal Units: Analytical Techniques and On-Site Validation," *IEEE Trans.*, Vol. PAS-100, pp. 81-95, January 1981.

[49]   F.P. deMello, L.N. Hannett, and J.M. Undrill, "Practical Approaches to Supplementary Stabilizing from Accelerating Power," *IEEE Trans.*, Vol. PAS-97, pp.1515-1522, September/October 1978.

[50]   D.C. Lee, R.E. Beaulieu, and J.R.R. Service, "A Power System Stabilizer Using Speed and Electrical Power Inputs - Design and Field Experience," *IEEE Trans.*, Vol. PAS-100, pp. 4151-4167, September 1981.

[51]   E.V. Larsen and D.A. Swan, "Applying Power System Stabilizers, Parts I, II, and III," *IEEE Trans.*, Vol. PAS-100, pp. 3017-3046, June 1981.

[52]   F.P. deMello, J.S. Czuba, P.A. Ruche, and J. Willis, "Developments in Application of Stabilizing Measures through Excitation Control," CIGRE Paper 38-05, 1986.

[53]   P. Kundur, M. Klein, G.J. Rogers, and M.S. Zywno, "Application of Power System Stabilizers for Enhancement of Overall System Stability," *IEEE Trans.*, Vol. PWRS-4, pp. 614-626, May 1989.

[54]   M. Klein, G.J. Rogers, S. Moorty, and P. Kundur, "Analytical Investigation of Factors Influencing Power System Stabilizer Performance," *IEEE Trans.*, Vol. EC-7, pp. 382-388, September 1992.

[55]   Canadian Electrical Association Report, "Investigation of Low Frequency Inter-Area Oscillation Problems in Large Interconnected Power Systems," Report of Project 294 T 622, prepared by Ontario Hydro, 1993.

[56]   E.V. Larsen and J.H. Chow, "SVC Control Design for System Dynamic Performance," IEEE Special Symposium on Application of SVS for System

Dynamic Performance, Publication 87TH0187-5-PWR, 1987.

[57]   N. Martins and L.T.G. Lima, "Determination of Suitable Locations of PSS and
        SVC for Damping Electromechanical Oscillations in Large Power System,"
        Proceedings of the 1989 Power Industry Computer Application Conference,
        pp. 74-82, May 1989.

Index